Indentation Fracture

Indentation Fracture

Strength and Toughness for Brittle Materials Design

Robert F. Cook

The American Ceramic Society

WILEY

Published by John Wiley & Sons, Inc., Hoboken, New Jersey.
Published simultaneously in Canada.

Library of Congress Cataloging-in-Publication Data applied for:

Hardback ISBN: 9781394207206

Cover Design: Wiley
Cover Image: © Robert F. Cook

Set in 9.5/12.5pt STIXTwoText by Straive, Chennai, India

To Brian and Shelley,
who gave me every opportunity to learn about fracture

Contents

Preface *xiii*
Abbreviations and Symbols *xv*

1 Brittle Fracture Fundamentals *1*
1.1 Brittle Components and Materials *1*
1.1.1 Components *2*
1.1.2 Materials *3*
1.2 Transformations on Brittle Fracture *4*
1.2.1 Observations *4*
1.2.2 Energy Balance *6*
1.3 Equilibrium Fracture on Uniform Loading *7*
1.3.1 Work and Elastic Energy *7*
1.3.2 Mechanical Energy and Surface Energy *8*
1.3.3 The Griffith Equation *9*
1.3.4 Configurational Forces: $\mathcal{G}$ and R *12*
1.4 Equilibrium Fracture on Localized Loading *13*
1.4.1 Mechanical Energy *13*
1.4.2 The Roesler Equation *13*
1.5 Stress-Intensity Factor and Toughness *17*
1.6 Nonequilibrium Fracture *20*
1.6.1 Unstable Adiabatic Fracture *20*
1.6.2 Stable Isothermal Fracture *22*
1.7 Fracture by Spatially Varying Loading *26*
1.7.1 Stress-Intensity Factor for Fracture at an Inclusion *26*
1.7.2 Crack Initiation *28*
1.8 Summary *31*
References *33*

2 Indentation Cracking Behavior *37*
2.1 Sharp Indentation of Brittle Materials *37*
2.1.1 Observations: Morphology *37*
2.1.2 Observations: Materials *41*
2.1.3 Residual Stress Field Model *44*
2.2 Indenter Shape Effects *47*
2.3 Material Effects *53*
2.4 Initiation *57*
2.5 Summary and Discussion *61*
References *63*

3 Indentation Strengths: Invariant Toughness Materials *67*
3.1 Indentation Strength of Brittle Materials *67*
3.1.1 Historical Context *67*
3.1.2 Experimental Methodology *68*
3.2 Fracture Mechanics of Indentation Strength *70*
3.2.1 Ideal Indentations *70*
3.2.2 Precursor Crack Extension *74*
3.2.3 Generalized Analysis *77*
3.3 Indentation Strength Variations *79*
3.3.1 Contact Angle *79*
3.3.2 Spherical, Rounded, and Flat Punch Contacts *80*
3.3.3 Lateral Cracked Contacts *87*
3.3.4 Impacts *93*
3.3.5 Contacts in Stressed and Layered Components *96*
3.3.6 Uncracked Contacts *104*
3.4 Indentation Strength Observations *108*
3.4.1 Contact Angle Effects *108*
3.4.2 Contact Radius Effects *110*
3.4.3 Lateral Crack Effects *112*
3.4.4 Impact Energy Effects *114*
3.4.5 Surface Stress Effects *119*
3.4.6 Subthreshold Effects *126*
3.5 Summary and Discussion *131*
References *132*

4 Indentation Strengths: Toughened Materials *141*
4.1 Introduction *141*
4.2 Microstructural Toughening in Brittle Materials *142*
4.2.1 Thermodynamics and the Energy Framework *142*
4.2.2 Mechanics and the Stress Framework *143*
4.2.3 Crack Wake Toughening: Overview *144*
4.2.4 Crack Wake Toughening: Analysis *149*
4.3 Toughened Materials Strength Analyses *150*
4.4 Toughened Materials Strength Observations *153*
4.4.1 Alumina *153*
4.4.2 Silicon Nitride *161*
4.4.3 Glass-Ceramics *163*
4.4.4 Ferroelastics *165*
4.4.5 Alumina Composites *168*
4.4.6 Zirconia *172*
4.4.7 Alumina Impact *177*
4.5 Summary and Discussion *180*
References *182*

5 Indentation Strengths: Toughness Estimation *187*
5.1 Introduction *187*
5.2 Methods *188*
5.2.1 Analysis *188*
5.2.2 Experiment *190*
5.3 Results *192*
5.3.1 Relation to Conventional Method *192*
5.3.2 Refinement of Modified Technique *193*

5.3.3 Application of Modified Technique *196*
5.4 Discussion *198*
5.5 Conclusions *200*
References *200*

6 Indentation Strengths: Reactive Environments *203*
6.1 Introduction *203*
6.1.1 Background Concepts *203*
6.1.2 Background Observations *205*
6.2 Reactive Strength and Lifetime Analyses *210*
6.2.1 Empirical Crack Velocity: Griffith Flaws *210*
6.2.2 Empirical Crack Velocity: Indentation Flaws *212*
6.2.3 Effect of Lateral Cracks and Microstructure *214*
6.2.4 Thermally Activated Crack Velocity *216*
6.3 Reactive Strengths Observations *220*
6.3.1 Single Contact Loads *220*
6.3.2 Multiple Contact Loads *221*
6.3.3 Predictions *223*
6.3.4 Surface Stress Effects *227*
6.3.5 Probabilistic Methods *231*
6.4 Summary and Discussion *233*
References *236*

7 Dynamic Fragmentation Patterns in Flexed Ceramic Disks *241*
7.1 Introduction *241*
7.2 Number of Fragments *243*
7.3 Distribution of Fragment Sizes *248*
7.4 Crack Branching Distances *251*
7.5 Application Examples *254*
7.5.1 Si Die Failure *255*
7.5.2 Ceramic Microstructure Development *256*
7.5.3 Branching Radius Variation *256*
7.6 Conclusions *257*
References *258*

8 Edge Chipping at Small Scales and Strengths of Diced Components *263*
8.1 Introduction *263*
8.2 Sharp Contact Cracking at Edges *265*
8.2.1 Contact Morphology *265*
8.2.2 Crack Lengths *266*
8.2.3 Chipping Threshold *269*
8.3 Edge Chip Size Variation *270*
8.4 Edge Chip Strength Variation *273*
8.4.1 Strength Measurements *273*
8.4.2 Strength Analysis and Prediction *274*
8.5 Summary *277*
References *278*

9 Scratches and Lateral Cracking at Linear Sharp Contacts *279*
9.1 Introduction *279*
9.2 Analysis *282*
9.2.1 Contact Deformation *283*

9.2.2 Lateral Cracking *285*
9.3 Results *288*
9.3.1 Point Contacts *288*
9.3.2 Dragged and Rolled Contacts *290*
9.4 Mohs Scale Minerals *293*
9.5 Summary and Discussion *298*
References *300*

10 Strengths of Silicon Devices Determined by Wafer Backgrinding *303*
10.1 Introduction *303*
10.1.1 Semiconductor Device Fabrication *303*
10.1.2 Wafer Backgrinding and Die Strength *304*
10.2 Linear Flaws in Si *306*
10.2.1 Single Scratches *306*
10.2.2 Multiple Striations *307*
10.3 Strength of Si *309*
10.3.1 Indentation Fracture Mechanics—Linear Contact Flaws *309*
10.3.2 Indentation Fracture Mechanics—General Contact Flaws *313*
10.3.3 Observations—Controlled Flaws *316*
10.3.4 Observations—Grinding Flaws *318*
10.4 Grinding Controlled Strength Model *322*
10.4.1 Analysis *322*
10.4.2 Results *323*
10.5 Summary *327*
References *328*

11 Strength and Toughness of Cordierite Glass-Ceramic Composites *331*
11.1 Introduction *331*
11.2 Experimental Methods *332*
11.2.1 Materials *332*
11.2.2 Contact Properties *334*
11.2.3 Toughness *338*
11.2.4 Strength *338*
11.3 Analysis Methods *339*
11.3.1 Toughness Variation *339*
11.3.2 Indentation Strength Variation *340*
11.3.3 Intrinsic Strength Variation *340*
11.4 Results *341*
11.4.1 Toughness Measurements *341*
11.4.2 Strength Measurements *343*
11.5 Conclusions *345*
References *345*
Appendix *347*

12 Crack Propagation in Toughened Materials *351*
12.1 Introduction *351*
12.2 Indentation Crack Extension Analyses *353*
12.2.1 Residual Field Variation *353*
12.2.2 Similar Driving and Resistance Fields *354*
12.2.3 Dissimilar Driving and Resistance Fields *355*
12.2.4 Piecewise Continuous Resistance Field *356*

12.3 Crack Propagation Observations *359*
12.3.1 Polycrystalline Alumina *359*
12.3.2 Phase-Transforming Zirconia *361*
12.3.3 Ion-Exchanged Glass *364*
12.4 Summary *366*
References *367*

13 Summary and Future *369*
13.1 Introduction *369*
13.2 Materials Comparisons *370*
13.2.1 Fracture in Inert Environments *370*
13.2.2 Fracture in Reactive Environments: Master Maps *373*
13.3 Future Topics *375*
13.3.1 Thermal Shock *375*
13.3.2 Concurrent Flaw Distributions *377*
13.3.3 Indentation Measurements *378*
References *379*

Index *381*

Preface

This is a book for experimentalists. In particular, it is for all those, including me, who obtain experimental results using the techniques of indentation fracture and ask themselves the same questions I have often asked myself: How do my results compare with those of other researchers? Are the values the same? Are the uncertainties the same? How can I use the results? How can I analyze the results? The goal of this book is to provide answers to these questions and to advance the field of fracture within the broader fields of mechanics and physics.

Indentation fracture techniques are one of the most effective means of determining the mechanical properties of brittle materials. Brittle materials—glasses, ceramics, semiconductors, rocks, minerals, bones, teeth, and seashells—typically provide many advantageous engineering properties: lightweight, resistant to elastic and plastic deformation, electrically and thermally insulating, often optically transparent, usually corrosion resistant, and formable from raw materials commonly found on earth. The uses of brittle materials thus range over domestic, architectural, industrial, cyber, and biological applications. The major limitation, however, is that they are brittle. On tensile loading, components of brittle materials fail by fracture after limited elastic deformation—they are usually not very tough. Consequently, the fracture strengths of such materials are typically small, often variable, and often time dependent, significantly limiting structural applications, despite the other, superior, properties. This book describes and illustrates advanced applications of indentation fracture in evaluating strength, toughness, and related properties of brittle materials.

In the 1921 fracture study of Griffith, the physics of brittle fracture was shown to be a manifestation of the first law of thermodynamics: Work performed on a component by applied loading is converted into internal elastic strain energy within the component that is then converted by fracture into surface energy on newly formed surfaces of the component. The mechanism by which the final conversion to surface energy takes place is crack propagation, the expansion of a plane of ruptured bonds in a component—a crack—leading to component failure. Griffith showed that the critical stress required to propagate a crack was an inverse function of crack size. Hence, variation of fracture strengths within a group of components is explained by variation in the size of the strength controlling cracks or flaws.

Alternatively, deliberate introduction of controlled flaws into components can circumvent strength variability. This obvious method to enable assessment of material strengths was implemented from the earliest studies. Griffith used annealed scratches. In the 1970s, Marshall and Lawn developed experimental techniques and fracture mechanics analyses to describe the cracks formed by sharp diamond indentation contacts on brittle surfaces. Subsequently, thousands of studies have implemented indentation techniques to evaluate fracture properties of brittle materials. The vast majority of these studies make use of the simplicity and economy of indentations to generate measurable stable surface cracks from which material toughness is estimated. Very few make use of the controlled flaw properties to investigate material strengths and the underlying flaw and toughness variations or make use of the myriad additional applications. Design, manufacturing, and application of brittle materials is impeded by this limited use of indentation fracture.

This book demonstrates a diversity of indentation fracture techniques with an emphasis on indentation strength tests. In these tests indentations are the controlled flaws in strength measurements of components. The indenter used in the experiments described throughout the book is primarily the four-sided Vickers diamond pyramid, although other indenters such as the cone and three-sided Berkovich and cube corner are considered. The book begins with an overview of the fundamentals of brittle fracture, clearly defining strength and toughness and making connections with broader physical concepts. This background is followed by a study of the lengths of cracks generated by indentations prior to strength testing. This in turn is followed by detailed and extensive considerations of indentation strengths. The considerations are initially for materials that exhibit invariant toughness (primarily glass) and then for materials that exhibit increases in toughness with crack length (primarily alumina). The two chapters concerned with these topics are the longest in the book and are the foundation for much that follows. Measurement of toughness of brittle materials at strength relevant length scales is then described. The measurement is based on the simultaneous determination of crack length and failure stress at fracture instability in indentation strength tests. This is followed by consideration of static and dynamic fatigue—the time-dependent failure of brittle materials in reactive environments. For many brittle materials this environment is simply water or moisture. Indentation techniques are demonstrated for determination of fatigue kinetics and the critical threshold stress for fatigue failure. The next three chapters demonstrate the power of the indentation technique to control the size, location, and geometry of flaws in studies of the common phenomena of fragmentation, chipping, and scratching, including shedding light on the well-used mineralogical Mohs scale. The three following chapters have a materials focus, demonstrating application of indentation techniques to characterize the fracture properties of semiconductors, ceramic composites, and toughened materials such as the chemically treated glasses used in mobile phone screens. The final chapter brings all the indentation strength measurements together to illustrate the power of indentation fracture techniques in designing with brittle materials.

Some elements of the book may be surprising. Very little attention is given to measurements of indentation crack lengths. This oft-debated subject is discussed in many publications. Much attention is given to the effects of lateral cracks on strength. Lateral cracks markedly influence the indentation strength behavior of nearly all brittle materials over the most commonly used indentation load domains. Failure to account for lateral crack-related relaxation of indentations stress fields can lead to overestimation of toughness and misidentification of indentation fracture data as indicating microstructural effects, crack blunting, crack healing, phase transformations, densification, or surface stresses. As will be noted it is almost impossible to discuss the effects of microstructure on the fracture strength of polycrystals without considering the effects of lateral cracks.

Many of the ideas presented and developed here have formed over the years from discussions and interactions with colleagues, including S. J. Bennison, T. P. Dabbs, E. R. Fuller Jr, B. R. Lawn, D. B. Marshall, M. L. Oyen, G. M. Pharr IV, M. V. Swain, R. Tandon, and S. M. Wiederhorn.

Raw data for many of the results contained in Chapters 4, 6–8, and 10–12 were obtained with the extremely able experimental assistance of K. L. Donsbach, M. P. Hughey, E. G. Liniger, R. L. Mendelson, D. J. Morris, R. Tandon, and A. M. Vodnick. Dr M. V. Swain is thanked for provision of the raw data for Figure 9.10.

The thinking and writing for this book were conducted on the Atlantic Ocean coast and in the Appalachian mountains of North Carolina—in Duck, Manteo, Wilmington, and Asheville—with the patient and much appreciated encouragement of my partner Michelle.

June 2024

Robert F. Cook
Wilmington, NC

Abbreviations and Symbols

ASG	aluminosilicate glass
BSG	borosilicate glass
CTE	coefficient of thermal expansion
FSG	fused silica glass
pdz	plastic deformation zone
SIF	stress-intensity factor
SLG	soda lime silicate glass
TZP	tetragonal zirconia polycrystal
a	contact impression semi-diagonal; radius of inclusion; radius of force application or contact
b	component dimension; characteristic pdz dimension
c	crack length
f	function; microstructural line force; frequency
h	component dimension; distance to edge; Planck's constant
g	function
i,j	indices
k	component stiffness
k_B	Boltzmann's constant
p	pressure; partial pressure
r	radial coordinate
s	scallop dimension
m	mass
t	time
u	displacement
v	crack affected volume; crack velocity; particle velocity
z	spatial coordinate
A	crack area
B	cracking susceptibility
D	strength degradation; scallop dimension
E	Young's modulus
F	force applied to component
$\mathcal{G}$	mechanical energy release rate
H	hardness
K	stress-intensity factor
L	component length
M	elastic modulus
N	Mohs number
P	indentation load
Pr	empirical distribution function
Q	heat
R	component radius; fracture resistance

S	entropy; contact strength
T	toughness
U	energy
$\mathcal{U}$	energy density
V	component volume; pdz volume
W	work
α	linear CTE; numerical coefficient
β	numerical coefficient
γ	surface energy
δ	load-point displacement; indentation depth; microstructural traction zone length
ε	strain
η	thermal energy/area
κ	bulk elastic modulus
λ	component compliance; material grain size; crack branching distance
μ	chemical potential; microstructural geometry term
ν	Poisson's ratio
ξ	residual stress geometry term
θ	temperature; angular coordinate
ϕ	indenter included semi-angle
ω	angular coordinate
ρ	material mass density
σ	stress; strength
χ	residual stress geometry term
ψ	applied stress geometry term
ζ	relative crack length
Δ	finite change in quantity
Γ	adsorption coverage
Λ	entropy production rate
Φ	thermodynamic state function
Ω	crack nucleus geometry term
$\sim$	"varies as"
$\approx$	"approximately equal to"

Acronyms include the plural. Subscripts, superscripts, indices, dummy variables, bold variables, annotated variables, and less frequent uses of these and other symbols are defined in context.

1

Brittle Fracture Fundamentals

This chapter defines and describes in clear, simple mathematical terms the physical, thermodynamic, and mechanical basis and associated terminology of brittle fracture. The transformation on fracture of mechanical energy (including work and elastic energy) into surface energy is shown as the fundamental brittle fracture process. From an energy viewpoint, fracture equilibrium is defined by the balance between configurational forces associated with these energies: the mechanical energy release rate and the fracture resistance. From an equivalent stress viewpoint, fracture equilibrium is defined by the balance between the stress-intensity factor and the toughness. The limiting cases of uniform and localized loading of cracks are shown to correspond to mechanical energy release rates that increase and decrease with crack area, respectively, leading to the well-established Griffith and Roesler equations of fracture. The limiting cases are similarly described by stress-intensity factors that increase and decrease with crack length. Nonequilibrium fracture is described by the addition of a kinetic energy term into analysis of adiabatic fast fracture and the addition of an entropy term into analysis of isothermal slow fracture. Fracture in a spatially varying loading field is considered through the example of crack initiation at a misfitting inclusion. The example is based on stress-intensity factor behavior and highlights considerations of fracture stability.

1.1 Brittle Components and Materials

Many of the objects encountered in everyday life are composed of brittle materials. On mechanical loading such materials are characterized by several distinctive responses—of particular distinction is the nature of failure on tensile loading. On exposure to excess tension brittle materials exhibit limited deformation followed by instability and failure in the form of abrupt fracture and separation into two or more fragments. After failure the fragments recover their initial unloaded and undeformed shapes. The fragments include newly formed surfaces on which the fracture and separation took place. Prior to failure, deformation is elastic and reversible. In a related phenomenon, on excess compressive loading at surface contacts brittle materials exhibit a mixture of localized reversible and irreversible deformation followed by adjacent cracking and chipping. After contact the surface recovers its unloaded shape, usually incompletely, and newly formed fracture surfaces remain in the form of stable cracks and chips. These two characteristic responses are often related: components formed of brittle materials often fail in applied tension from cracks generated by prior compressive contacts.

This book combines the two related responses just discussed in a deliberate way, showing how indentation fracture (intentionally introduced contact cracks) can be used to both evaluate parameters for brittle materials design (and avoid component failure) and understand everyday brittle fracture phenomena (such as fragmentation and chipping). This chapter develops and summarizes the fundamental mechanics and thermodynamics of brittle fracture as a basis for Chapters 2–12 to follow. The chapter begins with phenomenological descriptions of brittle fracture and failure and the transformations of energy that characterize such behavior. This is followed by development from an energy viewpoint of analyses describing equilibrium fracture in limiting stable and unstable configurations, and how such analyses are implemented in practice from a stress viewpoint. The limiting stability configurations are shown to correspond to the two characteristic brittle fracture responses discussed earlier. Experimental data from the last 100 years that exemplify stable and unstable equilibrium fracture are given. Nonequilibrium fracture is then considered, also in two limits, as adiabatic and isothermal behaviors that pertain to these responses. Finally, a practical example of fracture mechanics is given, that of cracking in a brittle material at a misfitting inclusion. The example highlights the stress-based analysis method used throughout the book and demonstrates unstable, stable, adiabatic, and isothermal fracture behavior in a single system. The inclusion system exhibits

Indentation Fracture: Strength and Toughness for Brittle Materials Design, First Edition. Robert F. Cook.

the phenomenon of crack initiation and provides a basis for description of many other fracture systems. Although intended as a basis for the analysis and application of indentation fracture, the material in this chapter is general and applies to all fracture.

1.1.1 Components

Figure 1.1 illustrates a brittle bar loaded in tension. The bar is gripped at each end and loading consists of forces F applied at each grip. The directions of the forces are outward, tensile, and parallel to the bar axis. The bar is rectangular in cross section with dimensions b and h and has length L between the grips. Under the actions of the forces, the bar extends by a load-point displacement δ. F and δ are taken as positive in tension and directed as drawn. The force–displacement F–δ response of the bar is shown in Figure 1.2a. At a maximum force F_{max}, the bar fractures and separates into fragments, and the force supported by the bar decreases to zero with negligible further increase in displacement. The fragments recover to their unloaded, undeformed configurations. Schematic diagrams of the unloaded bar (dashed outline), loaded bar (shaded), and unloaded fragments (solid outline) are also shown in Figure 1.2a. This process is *brittle failure.*

The force–displacement behavior of most brittle components is linear elastic, most noticeably prior to component failure. The behavior is linear, as the F–δ relationship is given by $F = k\delta$, where k is a displacement-independent *stiffness.* In Figure 1.2a, k is the stiffness of the bar. The behavior is elastic, as k is also time-invariant, such that the $F = k\delta$ relation represents reversible equilibrium bar and fragment configurations. Prior to failure these are the $F \leq F_{max}$ loaded configurations of the bar. After failure, these are the $F = 0$ unloaded configurations of the fragments. Macroscopically, the prefailure equilibrium states are balances between the applied forces at the grips. Microscopically, these equilibrium states are balances between the applied forces and the restoring forces exerted by atomic interactions within the bar. The parameters F, δ, k, and F_{max} are indicated in Figure 1.2a. Linear elastic behavior for most components usually requires δ/L to be very small. After failure, the bar ceases to function as a *structural component* capable of supporting a mechanical force. Equivalently, linear elastic behavior and failure may also be caused by an imposed displacement δ, in which the applied loading generates an associated force F such that $\delta = \lambda F$, where λ is the *compliance* of the bar. For linear elastic behavior $\lambda = 1/k$. A structural component is thus also capable of supporting an imposed deformation. In Figure 1.1 the applied forces are shown as dark shaded arrows and loading fixtures indicated by hatching—unless otherwise stated, this scheme is used throughout the book. Terms in italics are mostly *definitions* to be used throughout—some are specific to this book.

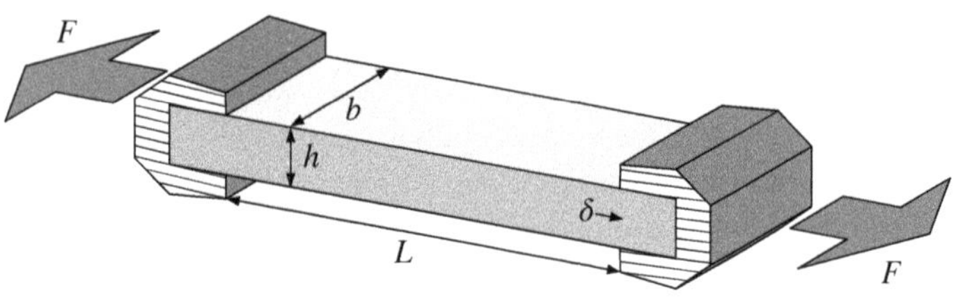

Figure 1.1 Schematic diagram of a bar under uniform tensile loading. On application of forces F the bar ends between the grips undergo relative displacement δ.

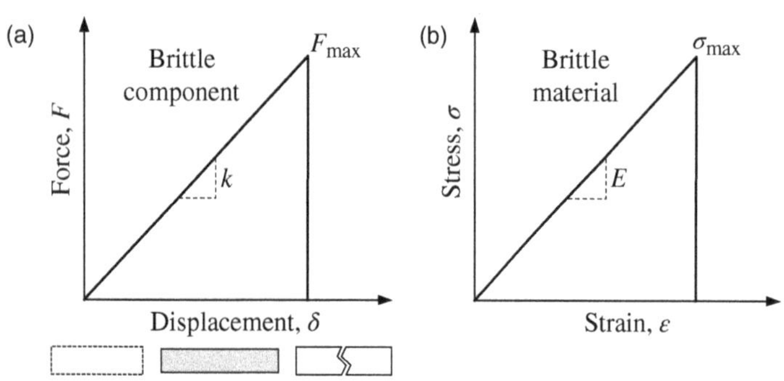

Figure 1.2 Brittle failure behavior of the bar and beam shown in Figures 1.1 and 1.3: (a) as a component, shown in a force–displacement plot and (b) as the underlying material, shown in a stress–strain plot. On application of loading, the force or stress increases linearly with displacement or strain. At a maximum force or stress, the supported load abruptly decreases to zero. The resulting fragments are undeformed. Schematic diagrams of the bar responses are shown in (a).

To perform as a structural component the condition $F_{max} > F$ must pertain for all forces and displacements imposed in application. To include an engineering "safety factor," this operational condition is usually extended to $F_{max} \gg F$. However, it is often not obvious that a functional brittle component is also acting as a structural component and that comparisons of F_{max} and F are thus necessary to guarantee component performance. Atomic interactions within brittle materials often impart useful material properties that lead to more obvious material and component functions. For example, the thermal insulating nature and small manufacturing costs of ceramics lead to their common use as house bricks and tableware; the controllable electrical conductivity of silicon leads to its extensive use as a semiconductor in myriad microelectronic devices; the optical transparency of glasses leads to their wide use as building windows, camera lenses, and mobile device screens; the electrical insulating nature of glasses and ceramics is also critical in microelectronic devices; the chemical and wear resistance of ceramics and glasses lead to their use as coatings in corrosive and abrasive environments; and, finally, most brittle materials are also resistant to elastic deformation and have small densities, rendering them ideal for applications in which dimensional stability is required, such as telescope mirrors.

The brittle materials used in microelectronic devices provide a pervasive example of the structural requirements "hidden" in many functional applications. To span a range of electrical properties—conducting, semiconducting, and insulating—a wide range of materials is required in the fabrication of microelectronic devices, e.g., copper, silicon, silica, and polyimide. This range leads to ubiquitous thermal expansion mismatch effects between the disparate materials. The inevitable consequence is the development of large tensile forces arising during thermal excursions in device fabrication and operation and the equally inevitable consequence of brittle component failure, e.g., of the silicon and silica glass components. In such cases, the device ceases to perform not because of a change in material electrical properties but because of inherent mechanical design factors. (Hence, significant attention is given in this book to brittle failure of materials used in microelectronic devices—Chapters 8–11.)

The force required to cause a structural component to fail in its load-bearing function is the component *failure force* (sometimes referred to as the supportable force, Ashby 1999). The terms *loaded* and *loading* are used throughout as broad descriptors to refer to any configuration that generates applied forces on components. Loading may arise from the direct application of a force, as in an overlying mass giving rise to a gravitational weight (a "load"). Loading may arise from an imposed displacement, in which case the force arises indirectly as a reaction to the displacement, as in thermal expansion mismatch configurations. Loading may also arise from an imposed change in displacement rate, in which case the applied force arises indirectly as a consequence of deceleration of a moving mass, as in impact configurations. A *loading cycle* is thus composed of *load–unload* segments, separated by the *peak force* generated in the cycle.

A key aspect of brittle component design is thus to ensure that the component failure force always exceeds the operational peak force. Design with brittle components thus requires knowledge of F_{max}. In turn, F_{max} depends on the component dimensions and the mechanical properties of the brittle materials forming the component. This dependence and brittle material properties are considered in Section 1.1.2.

1.1.2 Materials

Attention here is focused on the linear elastic behavior $F = k\delta$ in Figure 1.2a, prior to bar failure. To compare bars of different dimensions, it is convenient to define the *engineering stress* σ acting on the bar by $\sigma = F/bh$ and the *engineering strain* ε of the bar by $\varepsilon = \delta/L$. The linear elastic behavior of the bar expressed in terms of these quantities is thus $\sigma = (kL/bh)\varepsilon$ or $\varepsilon = (\lambda bh/L)\sigma$. The usual condition for the observation of linear elasticity can thus be restated as one in which the strain is very small, $\varepsilon \ll 1$. Using an invariant combination of *extensive* (size dependent) parameters, kL/bh, experiments show that the stress–strain expressions describe the deformation of bars formed of the same material but with differing dimensions. The implications are that σ and ε are *intensive* (size independent) quantities describing the loading on the bar and the deformation of the bar, respectively, and that the invariant extensive combination can be expressed as $k = Ebh/L$, where E is an invariant intensive material property. If the bar is long or "slender" (Ashby 1999), $L/(bh)^{1/2} > 10$, the quantity E is termed the *Young's modulus* of the material. E is positive. The expression for the intensive linear elastic stress–strain response of the material forming the bar is thus $\sigma = E\varepsilon$. The brittle *failure stress* or *strength* of the bar material σ_{max} is given by $\sigma_{max} = F_{max}/bh$. Figure 1.2b shows the brittle failure behavior of the bar material in a stress–strain plot. In this example, σ, ε, and σ_{max} are all positive. A major focus of this book is measurement and prediction of the material *brittle fracture strength* σ_{max}. Such measurements and predictions enable determination of component failure forces F_{max} for use in design with brittle materials.

Two additional aspects of elastic deformation are factors in brittle fracture strength considerations. The first is the transverse strain ε_T generated by the axial loading in Figure 1.1. The second is the *work* W performed by the loading. Much more detailed information regarding elasticity can be found elsewhere (Mase 1970; Timoshenko and Goodier 1970; Sadd 2009). Atomic interactions within the bar lead to internal connectivity of the bar material and transverse contractions in the b and h dimensions associated with extension in the L dimension. For the bar here, as $b \approx h$ and both dimensions are unconstrained transversely along L ($\approx$ indicates "approximately equal to"), the transverse strain ε_T along the length of the bar is uniform and given by $\varepsilon_T = -\nu\varepsilon$. ν is an invariant material property termed the *Poisson's ratio*. ν is intensive and usually positive, ≈ 0.3. An elastically isotropic material is one in which the stress–strain response of the material is identical in all directions and for which E and ν completely describe material deformation. Such a material is also elastically homogeneous, in which the deformation response is identical at all points. If the bar of Figure 1.1 is uniformly composed of an elastically isotropic material and is in equilibrium with the applied force, stress and strain are distributed homogeneously throughout the bar. If one of the transverse dimensions of the bar in Figure 1.1 is large relative to the other, say $b \gg h$, the stress–strain relation becomes $\sigma = E'\varepsilon$, where $E' = E/(1-\nu^2)$ is termed the "plane strain modulus." Other geometrical constraints lead to other moduli expressed as functions of (E, ν) and are used in this book, for example, the "biaxial modulus" and the "indentation modulus" (Mase 1970; Timoshenko and Goodier 1970; Sadd 2009). (The symbol M is used for an arbitrary elastic modulus defined in context.)

The work W performed by the applied force in deformation of the bar is expressed in extensive terms as $W = F\delta/2 = \lambda F^2/2 = k\delta^2/2$ (the area under the F–δ response in Figure 1.2a). Substituting the expressions for stress and strain into the extensive work expressions gives $W = (\sigma\varepsilon/2)(bhL) = (\sigma^2/2E)(bhL) = (E\varepsilon^2/2)(bhL)$. These expressions embody the First Law of Thermodynamics and make clear that the work performed by the applied force generates a change in the internal energy of the bar. The change can be expressed as $W = U_E$, where the E subscript indicates *internal elastic energy*, and $U_E = 0$ is set at $\delta = 0$. For example, internal elastic energy can thus be expressed as $U_E = (\sigma^2/2E)(bhL)$; a product of a combination of intensive parameters within the first set of parentheses and a combination of extensive parameters in the second set. The other combinations of intensive parameters are similarly expressed. The combination of extensive parameters is recognized as the volume of the bar, $V = bhL$. The implication is that the combinations of parameters within the first sets of parentheses must represent the mean *elastic energy density* (energy/volume) of the bar. Denoting the elastic energy density as $\mathcal{U}_E$ gives $U_E = \mathcal{U}_E V$, where, for example, $\mathcal{U}_E = \sigma^2/2E$ (the area under the σ–ε response in Figure 1.2b). U_E is extensive and $\mathcal{U}_E$ is intensive, both are positive. As stress and strain are distributed homogeneously throughout the bar, elastic strain energy is also distributed homogeneously and $\mathcal{U}_E$ describes the elastic energy density at all points in the bar material.

1.2 Transformations on Brittle Fracture

1.2.1 Observations

Figure 1.3 illustrates a beam loaded in three-point bending. The applied loading consists of force F imposed through an upper movable roller and directed toward two lower fixed rollers. The direction of the force and the roller axes are perpendicular to the beam axis. Under the action of the force the beam bends and the beam center exhibits relative displacement δ. F and δ are taken as positive when directed as drawn. At a maximum force F_{max} the beam fractures and separates and the force supported by the beam decreases to zero with negligible further increase in displacement, identical to that shown in Figure 1.2a. However, in the bending example of Figure 1.3, the stress, strain, and elastic strain energy are not homogeneously distributed throughout the beam. The maximum longitudinal stress σ_L is modified by the aspect ratio of the beam, $\sigma_L = (6L/h)F/bh$, and occurs on the lower face directly beneath the central roller (using the same parameters as Figure 1.1). Defining $\mathcal{U}_L = \sigma_L^2/2E$ gives $U_E = (2/3)\mathcal{U}_L V$, reflecting the strain energy density variation. Other inhomogeneously loaded systems exhibit similar variations (Mase 1970; Timoshenko and Goodier 1970; Ashby 1999; Sadd 2009).

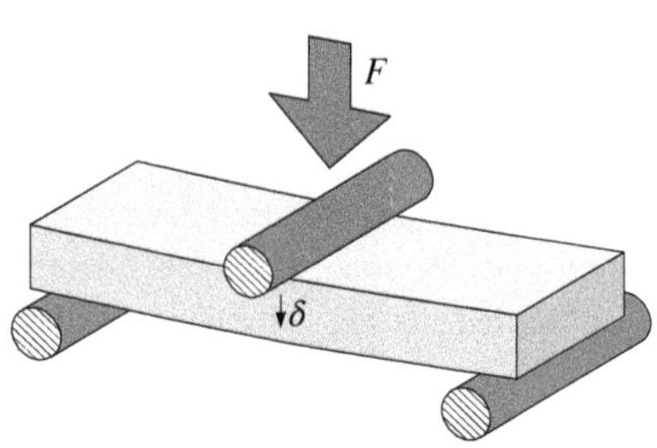

Figure 1.3 Schematic diagram of a beam under loading by three-point bending, leading to uniaxial flexure. On application of force F the center of the beam undergoes displacement δ perpendicular to the beam axis. Loading in bending is common for brittle materials.

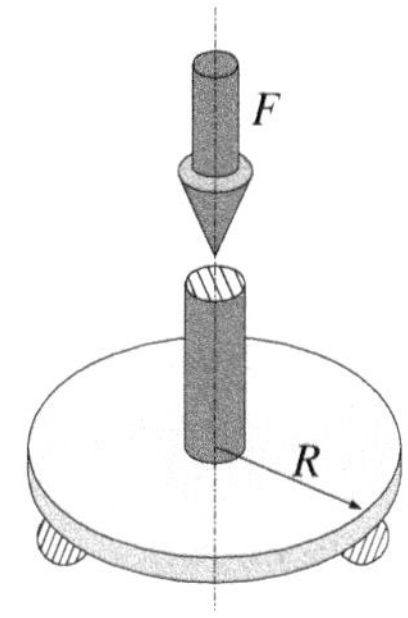

Figure 1.4 Schematic diagram of a circular plate under loading in a flat on three spheres geometry, leading to biaxial flexure. On application of force F the center of the plate undergoes displacement δ (not shown) parallel to the plate axis. Loading in flexure is common for brittle materials.

In particular, Figure 1.4 illustrates a disk, radius R, loaded in biaxial flexure. The applied loading consists of force F imposed through an upper movable punch directed toward lower fixed supports arranged in a circle. The direction of the force is parallel to the disk axis. Under the action of the force the disk flexes and the disk center exhibits relative displacement δ (not shown). F is taken as positive when directed as drawn. At a maximum force F_{max}, the disk fractures and separates and the force supported by the disk decreases to zero with negligible further increase in displacement, identical to that shown in Figure 1.2a. The maximum biaxial stress, σ_R, occurs on the lower face directly beneath the punch and $\sigma_R \sim F/R^2$ (where $\sim$ indicates "varies as"). Stress analyses for bending and flexure are considered in detail elsewhere (Timoshenko and Goodier 1970).

To avoid contact damage associated with gripping (Figure 1.1), brittle materials are usually mechanically tested in bending (Figure 1.3) or flexure (Figure 1.4). In addition, although the tensile bar of Figure 1.1 is a simple loading example, most brittle components also fail in bending or flexure, Figures 1.3 and 1.4. Familiar examples include: a dropped ceramic plate experiences bending on impact with a floor, between the moving upper sections of the plate and the lower stationary sections at the fixed floor; a glass window experiences flexure on loading by the wind, between the center and the fixed frame. Less familiar, more technological, examples are shown in Figures 1.5 and 1.6 and provide insight into the energy transformation that occurs on brittle fracture. Figure 1.5a shows an image of an intact semiconductor device. The metal and glass lines are visible on the underlying silicon substrate. Figure 1.5b shows an image of the two fragments from a device loaded in bending (Figure 1.3) until fracture. The fracture surfaces are visible in the center of the image. Figure 1.5c shows an image of the two fragments from Figure 1.5b reassembled into the original unloaded configuration. A faint separation

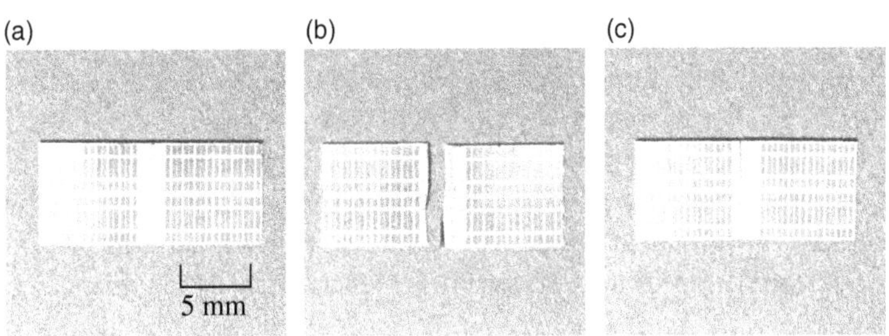

Figure 1.5 Optical images of a brittle silicon semiconductor device. (a) Intact after fabrication. (b) Broken after loading in bending. (c) Outline restored after fragments reassembled.

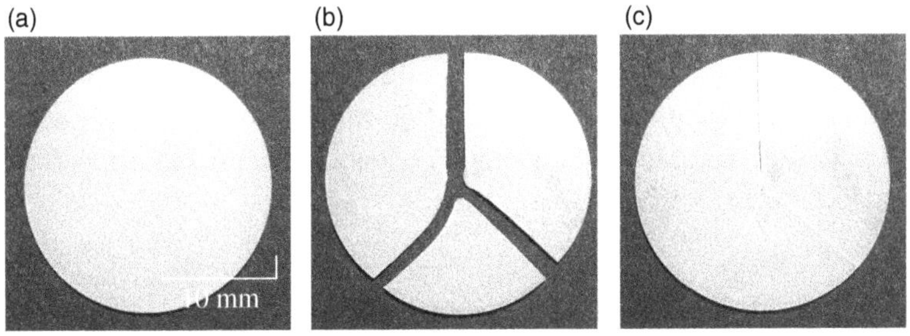

Figure 1.6 Optical images of a brittle alumina material in disk form. (a) Intact after fabrication. (b) Broken after loading in biaxial flexure. (c) Outline restored after fragments reassembled.

is visible in the center of the image but otherwise the configuration is identical to Figure 1.5a. Figure 1.6a shows an image of a disk of translucent polycrystalline alumina (Al_2O_3) similar to that used as the bulb material in Na vapor street lamps. Figure 1.6b shows an image of three fragments from a disk loaded in biaxial flexure until fracture. Figure 1.6c shows an image of the three fragments from Figure 1.6b reassembled into the original unloaded configuration. Clear separations are visible in the image but otherwise the configuration is identical to Figure 1.6a. The implication of Figures 1.5 and 1.6 is that the process of brittle fracture is one of energy transformation, from elastic energy U_E to surface energy U_S.

Fracture in brittle materials—encompassing rocks, minerals, ceramics, glasses, semiconductors, salts, bones, teeth, and seashells—has been well studied and analyzed (Lawn 1993; Maugis 2000). In brief, brittle fracture is a process of sequential interatomic bond rupture. On tensile loading, the vast majority of a component formed from a brittle material is deformed elastically. Fracture is restricted to an expanding plane of ruptured bonds—a *crack*—delimited by a surrounding linear boundary region, localized to a few bond spacings, separating intact from ruptured bonds—the *crack front*, often referred to as the *crack tip*. A *flaw* in a material—typically including a preexisting crack—concentrates stress on loading such that the resulting deformation local to the flaw is accommodated by bond rupture. If the flaw does not include a preexisting crack, this process is known as crack initiation. If the flaw does include a crack, this process is known as crack propagation. If the loaded fracture system is unstable, the crack propagates through the material, separating the component into two or more pieces by rupturing bonds one at a time. The maximum stress at which this process occurs is the strength. The fracture process can be viewed from an energy perspective: The work performed by the applied loading external to the component, W, generates an internal energy change within the component associated with elastic strain energy. On fracture, an internal energy conversion takes place: the elastic strain energy, U_E, distributed throughout the component, is transformed by crack propagation into surface energy, U_S, localized to the newly formed fracture surfaces within the component. The final state is one in which the component pieces, the fragments, are in their initial undeformed configuration but possess new peripheral surfaces formed by broken bonds. As brittle crack propagation is restricted to planar bond rupture, the disruption to atomic order on fracture surfaces is negligible. Consequently, fragments resulting from brittle fracture of a component can often be reassembled into the form of the original component. Examples of this phenomenon are shown in Figures 1.5 and 1.6.

1.2.2 Energy Balance

Fracture systems may be considered from the viewpoints of either the energy variations or the stress variations that occur in loaded cracked components. Although the energy viewpoint usually provides greater insight into physical fracture phenomena, the stress viewpoint is usually more amenable to fracture analyses, especially in systems with multiple sources of applied loading. Fracture analysis from the stress viewpoint, often termed "fracture mechanics," for metals, ceramics, polymers, and biomaterials is considered in detail elsewhere (Broek 1982; Williams 1984; Atkins and Mai 1985; Kanninen and Popelar 1985; Lawn 1993; Wei 2010; Ritchie and Liu 2021) and is the primary analysis method used throughout the book. However, the important fracture concepts of equilibrium and stability are defined more easily in the energy viewpoint and then reexpressed in the stress viewpoint by simple transformations. Hence, the development of fracture analysis begins here from the energy viewpoint, with consideration of the energy changes associated with a crack in a loaded bar.

Figure 1.7 shows a schematic diagram of a cylindrical tensile bar, length L, radius R, loaded at the ends by axial forces F. The bar includes a crack, indicated in light gray, in the center of the bar on a plane perpendicular to the loading axis. The crack has a circular outline, radius c. The First Law of Thermodynamics for an adiabatic change between equilibrium states of the bar (heating is absent) is given by

$$\Delta U = W, \tag{1.1}$$

where ΔU is the change in internal energy of the bar and W is the work performed on the bar (Planck convention). On fracture, the internal energy change has two elements, the change in elastic energy ΔU_E, distributed throughout the bar, and the change in surface energy ΔU_S, localized to the crack surfaces, such that

$$\Delta U = \Delta U_E + \Delta U_S. \tag{1.2}$$

For an adiabatic change of state of the applied loading mechanism giving rise to the forces F, the First Law gives an equation similar to Eq. (1.1),

$$\Delta U_A = W_A, \tag{1.3}$$

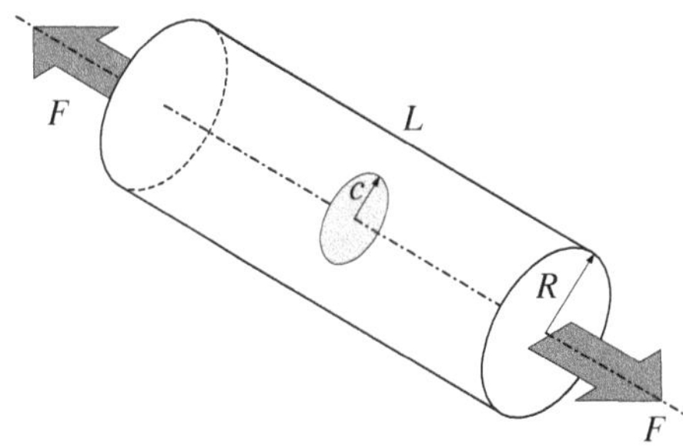

Figure 1.7 Schematic diagram of a loaded cylindrical bar containing a crack with circular perimeter, radius c, under uniform tensile stress.

where ΔU_A is the change in internal energy of the applied loading mechanism and W_A is the work performed on the mechanism. The work performed by the mechanism is thus $-W_A$ and if the applied loading and the bar are coupled to form a composite system such that the applied loading does work on the bar, $W = -W_A$. Hence

$$W = -\Delta U_A. \tag{1.4}$$

Combining Eqs. (1.2) and (1.4) into Eq. (1.1) gives a relation between the energy changes on transformations between equilibrium states of the composite bar–applied loading system:

$$\Delta U_E + \Delta U_A = -\Delta U_S. \tag{1.5}$$

The *mechanical energy* change ΔU_M is defined by

$$\Delta U_M = \Delta U_E + \Delta U_A, \tag{1.6}$$

such that the First Law equilibrium statement of Eq. (1.1) can be rewritten for consideration of fracture as

$$-\Delta U_M = \Delta U_S. \tag{1.7}$$

Specifying these changes and this balance in terms of crack length c, applied force F, and component geometry L and R is one of the major goals of fracture mechanics. Section 1.3 considers Eq. (1.7) for cracks loaded by uniform tension and localized force.

1.3 Equilibrium Fracture on Uniform Loading

1.3.1 Work and Elastic Energy

The bar in Figure 1.7 is loaded homogeneously by forces F applied uniformly over the cross-sectional end areas πR^2. The bar is composed of linear elastic material, Young's modulus E. Figure 1.8 shows a force–displacement diagram of possible loading responses of the bar. In the absence of a crack, $c = 0$, loading from zero force, state O, to a peak force F, state A, leads to a linearly increasing end displacement δ to the peak value δ_0. The linear response is characterized by $\delta = \lambda F$ or $F = k\delta$, where λ is the compliance and $k = 1/\lambda$ is the stiffness. For the uncracked bar, the compliance is $\lambda_0 = L/\pi R^2 E$ and is the reciprocal of the slope of the line OA in Figure 1.8. The work W_0 performed by the applied loading in extension of the

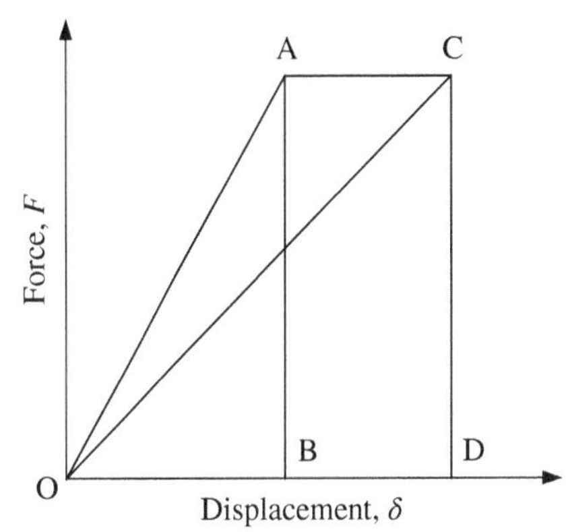

Figure 1.8 Linear elastic force–displacement responses of an uncracked component, path OA, and a cracked component, path OC.

uncracked bar is the area enclosed by OAB in Figure 1.8, $W_0 = F\delta_0/2$. As the system is linear elastic, the work generates identical elastic energy in the bar, $U_{E0} = F\delta_0/2$ (taking the initial condition $\delta = F = U_E = 0$).

In the presence of cracks, many systems also exhibit linear force–displacement behavior, identical qualitatively but differing quantitatively, from uncracked systems. Such behavior is the basis of linear elastic fracture mechanics and is assumed here. In this case, loading the cracked bar, $c > 0$, from zero force to the peak value F, with no change in c, leads to an increasing linear end displacement to the peak value δ_c, state C, as shown in Figure 1.8. The linear response in the cracked configuration is characterized by $\delta = \lambda_c F$ or $F = k_c\delta$, where λ_c and k_c are the compliance and stiffness of the cracked bar. The work W_c performed by the applied loading in extension of the cracked bar is the area enclosed by OCD in Figure 1.8, $W_c = F\delta_c/2$, equal to the total elastic energy in the cracked bar, $U_{EC} = F\delta_c/2$. The difference in elastic energy for the cracked and uncracked bars loaded to identical force is $\Delta U_E = U_{EC} - U_{E0} = F(\delta_c - \delta_0)/2 = F^2(\lambda_c - \lambda_0)/2$, and it is noted that $\Delta U_E > 0$.

The work performed by the applied loading in the cases leading to state A or state C is for static system configurations loaded serially, either uncracked (0) or cracked (c). Work can also be performed by the applied loading during the transformation from state A to state C for a dynamic system configuration in which the crack length changes from 0 to c in parallel with application of a force. Such a transformation is shown for a constant force path, line AC in Figure 1.8. The work performed by the applied loading on the cracking system in this case is the area enclosed by BACD, $W_F = F(\delta_c - \delta_0)$. This value illustrates Clapeyron's theorem, which states that the work performed on a system during a constant force transformation between two linear elastic states is twice the elastic energy difference between the states. Here, $W_F = 2\Delta U_E$. From Eq. (1.3) $W_F = -W_A$, and thus from Eq. (1.4) $\Delta U_A = -2\Delta U_E$. From Eq. (1.7), $\Delta U_M = \Delta U_E - 2\Delta U_E = -\Delta U_E$. The description of the mechanical energy of the system in which the crack extends at constant force thus devolves to specification of the change in elastic energy on crack extension.

1.3.2 Mechanical Energy and Surface Energy

The tensile stress generated homogeneously in the uncracked cylindrical bar is $\sigma = F/\pi R^2$, and thus the elastic energy in the uncracked bar is given in extensive terms by $U_{E0} = \lambda_0 F^2/2$ and in intensive terms by $U_{E0} = (\sigma^2/2E)(\pi R^2 L)$. The first term in the second expression is recognized as the homogeneous elastic energy density of the uncracked bar, $\mathcal{U}_{E0}$, and the second term is the volume of the bar, V, such that $U_{E0} = \mathcal{U}_{E0}V$. For the cracked bar, a simple scaling extension is then

$$U_{EC} = U_{E0}\left[1 + \upsilon(c)/V\right], \tag{1.8}$$

where $\upsilon(c)$ is the volume of material affected by the crack. Physically, elastic strain energy density is increased in material adjacent to the crack perimeter and decreased in material adjacent to the crack faces such that the increase dominates the decrease. For systems in which the cracks are small, $c \ll L, R$, the volume $\upsilon(c)$ scales as the crack dimension alone,

$$\upsilon(c) = (4\pi\psi^2/3)c^3. \tag{1.9}$$

ψ^2 is a dimensionless geometrical parameter of order 1. The elastic energy change is given by

$$\Delta U_E = U_{EC} - U_{E0} = U_{E0}(\upsilon/V) = \mathcal{U}_{E0}\upsilon. \tag{1.10}$$

Equation (1.10) can be interpreted as representing a cracked bar containing a zone of material centered on the crack in which the strain energy density is greater than the surrounding material. In the case here the zone of material is near spherical (ψ can account for ellipsoidal perturbations). The cross-sectional area of the crack A is given by $A = \pi c^2$, such that υ for the circular crack may be expressed in terms of crack area as

$$\upsilon(A) = (4\psi^2/3\pi^{1/2})A^{3/2}. \tag{1.11}$$

Combining the above expressions for ΔU_M, ΔU_E, $\mathcal{U}_{E0}$, and $\upsilon(A)$ gives the mechanical energy change in terms of crack area for cracks extending under constant stress. For the circular crack here

$$\Delta U_M = -(2\psi^2/3E\pi^{1/2})\sigma^2 A^{3/2}. \tag{1.12}$$

Equation (1.12) provides an expression for the term on the left side of the fracture equilibrium expression, Eq. (1.7).

The term on the right side of the fracture equilibrium expression, Eq. (1.7), is the surface energy change. This change can be expressed most simply as a linear proportionality in crack area:

$$\Delta U_S = 2\gamma A. \tag{1.13}$$

The proportionality constant γ is the excess areal density of energy associated with a cracked surface relative to the bulk material. The factor of 2 arises as there are two crack surfaces. Usually, the term "areal density" is omitted and implied, with the factor of 2 used explicitly, such that, in fracture contexts, 2γ is usually referred to as the "surface energy." For most materials, $2\gamma \approx (1\text{–}10)\,\text{J m}^{-2}$.

1.3.3 The Griffith Equation

Setting $U_M = U_S = 0$ as initial conditions at $A = 0$ enables the total energy U_T of the loaded cracked system to be expressed in terms of the changes, Eqs. (1.12) and (1.13), as

$$U_T = 2\gamma A - (2\psi^2/3E\pi^{1/2})\sigma^2 A^{3/2}. \tag{1.14}$$

Figure 1.9 shows a plot of $U_T(c)$ using Eq. (1.14) and $A = \pi c^2$, with material properties $E = 200$ GPa and $2\gamma = 10\,\text{J m}^{-2}$, crack geometry term $\psi = 1.25$, and applied stress $\sigma = 100$ MPa—values typical for brittle materials (Lawn 1993). The crack area A is shown on the upper axis. The dashed lines show the asymptotes of the first and second terms, Eqs. (1.13) and (1.12), respectively, and the solid line shows the full response, Eq. (1.14). The total energy passes through a maximum. Setting $\text{d}U_T/\text{d}A = 0$, equivalent to imposing the equilibrium condition of Eq. (1.7) in the infinitesimal limit, gives the crack area at the maximum of

$$A = \left[\frac{\pi^{1/2}}{\psi^2}\right]^2 \left[\frac{2\gamma E}{\sigma^2}\right]^2, \tag{1.15}$$

approximately 0.052 mm^2, and crack length, c, of approximately 0.128 mm. The variation of U_T has $\text{d}^2U_T/\text{d}A^2 < 0$ at this point and hence the crack area and length at the maximum is a point of unstable equilibrium, indicated by the arrow in Figure 1.9. An interpretation of Figure 1.9 can be made in terms of progression along the path of fixed force, and thus fixed stress, AC in Figure 1.8. As this path is traversed, the crack grows in area until a point of unstable equilibrium is reached. Perturbations in F or δ at this point that cause an increase in crack area lead to monotonic decrease in total energy, runaway crack extension, and component failure. A more intuitive interpretation of Figure 1.9 can be made in

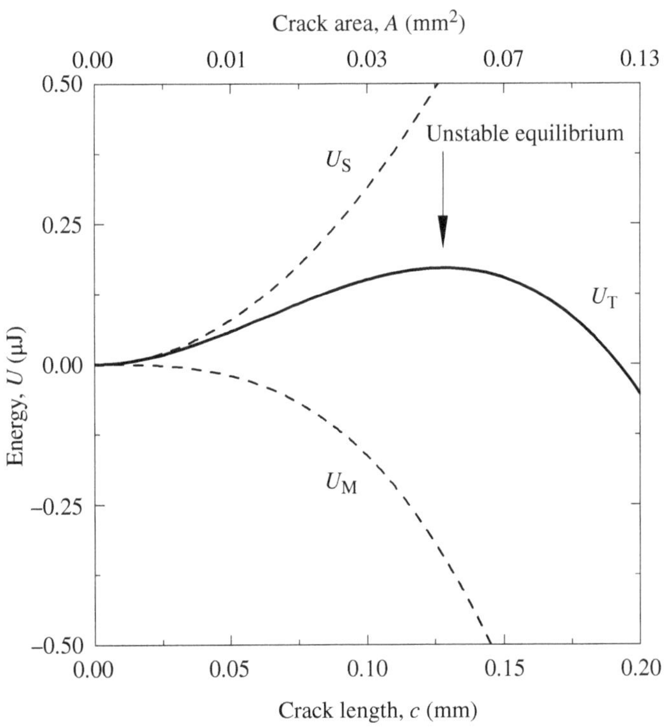

Figure 1.9 Energy versus crack area for a component containing a circular crack in uniform tension. The surface energy U_S and mechanical energy U_M are shown as dashed lines. The total energy U_T is shown as the solid line and passes through a maximum of unstable equilibrium. Such behavior is the basis of the Griffith equation for brittle fracture strength.

terms of progression along the path of fixed crack length, OC in Figure 1.8. As this path is traversed, the crack length is static and the force, and thus stress, increases until a point of unstable equilibrium is reached. Continued increase in F or δ at this point lead to monotonic decrease in total energy, runaway crack extension, and component failure. The stress at the point of unstable equilibrium is the strength of the cracked component. Inverting Eq. (1.15) and replacing A with c gives this stress as

$$\begin{aligned}\sigma &= (2E\gamma)^{1/2}/\psi c^{1/2}\\ &= Dc^{-1/2},\end{aligned} \tag{1.16}$$

which is the Griffith equation (Griffith 1921). An analysis based on a straight crack in a bar of rectangular section yields an identical result (Cook 2023).

Equation (1.16) makes clear that the factor $D = (2\gamma E)^{1/2}/\psi$ linking strength and crack size can be interpreted as the ratio of two terms—both recognized by Griffith. The first is a material properties term involving the product of resistance to elastic deformation and the excess energy associated with surface creation. The second is a dimensionless geometry term connecting crack length to the volume of stress-enhanced material adjacent to the crack (such that D is different for different shaped cracks in the same material). The value of D evaluated from the geometry and material parameters discussed earlier and Eq. (1.16) is comparable to that used to describe many glass, ceramic, rock, and mineral materials, $D = 1.13$ MPa m$^{1/2}$ (Lawn 1993). D characterizes the *strength degradation* of a material; how much the strength decreases as flaw size (here crack length) increases. The inverse of D characterizes *flaw tolerance*. Strength degradation, flaw tolerance, and D variations are discussed in detail in Chapters 3 and 4. Here, D is taken to be invariant. The analysis here differs from that of Griffith in emphasizing the physical scaling of the volume of material perturbed by a crack, $v(c)$, in favor of the evaluation of the numerical value of ψ. Griffith considered straight cracks and used the description developed by Inglis (1913) for the stress field about an elliptical cavity in a plate to estimate ψ. It is recognized that application of the Griffith equation requires that rates of crack retraction in nonequilibrium states are much slower than those of crack extension.

In the century-plus since Griffith (1921) the Griffith equation in the form $\sigma = Dc^{-1/2}$ has been verified in experiments many times. Figure 1.10 shows examples of the effects of flaw size, characterized by measured crack lengths c, on brittle material strengths σ, in a series of logarithmic plots using data derived from the published works cited. The plots are drawn to the same scale (vertical × horizontal = 10 × 100) and include a diversity of materials and flaw types and are arranged in approximate order of increasing strength. Note decreasing crack length scales on right. Some of the earliest work in this area, and the most influential, was that of Griffith (1921), in which the strengths of glass bulb components were measured by pressure loading. Prior to strength testing, scratches and cracks of various lengths were formed in the bulbs and annealed. The results are shown in Figure 1.10b. Symbols represent individual strength measurements at the crack lengths indicated (converted into modern units). The solid line represents the strength behavior predicted by Griffith, based on the energy perspective discussed earlier and the stress analysis of Inglis (1913). Griffith estimated D from measurements of the surface energy of glass as a function of temperature. The agreement between measurements and prediction, both quantitatively and qualitatively is clear.

Figure 1.10d shows more recent strength measurements of soda-lime glass (SLG), uniaxially flexed bars containing scribed flaws from the work of Shand (1965) (filled symbols) and biaxially flexed square plates containing annealed Vickers indentation flaws from the work of Glaesemann et al. (1987) (open symbols). Symbols represent individual strength measurements. The solid line represents a best fit to the indentation data consistent with Eq. (1.16). The fit well describes the data and is consistent with the scribe data; as noted by Glaesemann et al. the indentation data are also consistent with other indented glass measurements. Figure 1.10a shows strength measurements of notched bend beams of titania (TiO_2) agglomerates from the work of Kendall (1988). Symbols represent individual strength measurements. The solid line represents a best fit to the data consistent with Eq. (1.16) and again, the fit well describes the data. Figure 1.10c shows strength measurements of bend beams of SLG and polycrystalline Si_3N_4 ceramic from the works of Kirchner and Isaacson (1982) and Kirchner (1984). The beams were formed by single- and multipoint grinding, leaving scratched test surfaces. Symbols represent individual strength measurements. The solid lines represent best fits to the data consistent with Eq. (1.16) and fit the data well, demonstrating a clear difference in the strength degradation properties of the glass and the ceramic. Figure 1.10e shows strength measurements of biaxially flexed circular cordierite plates containing Vickers indentation flaws. Symbols and bars represent means and standard deviations of multiple crack length and strength measurements at different indentation forces. The solid line represents a best fit to the data consistent with Eq. (1.16) and

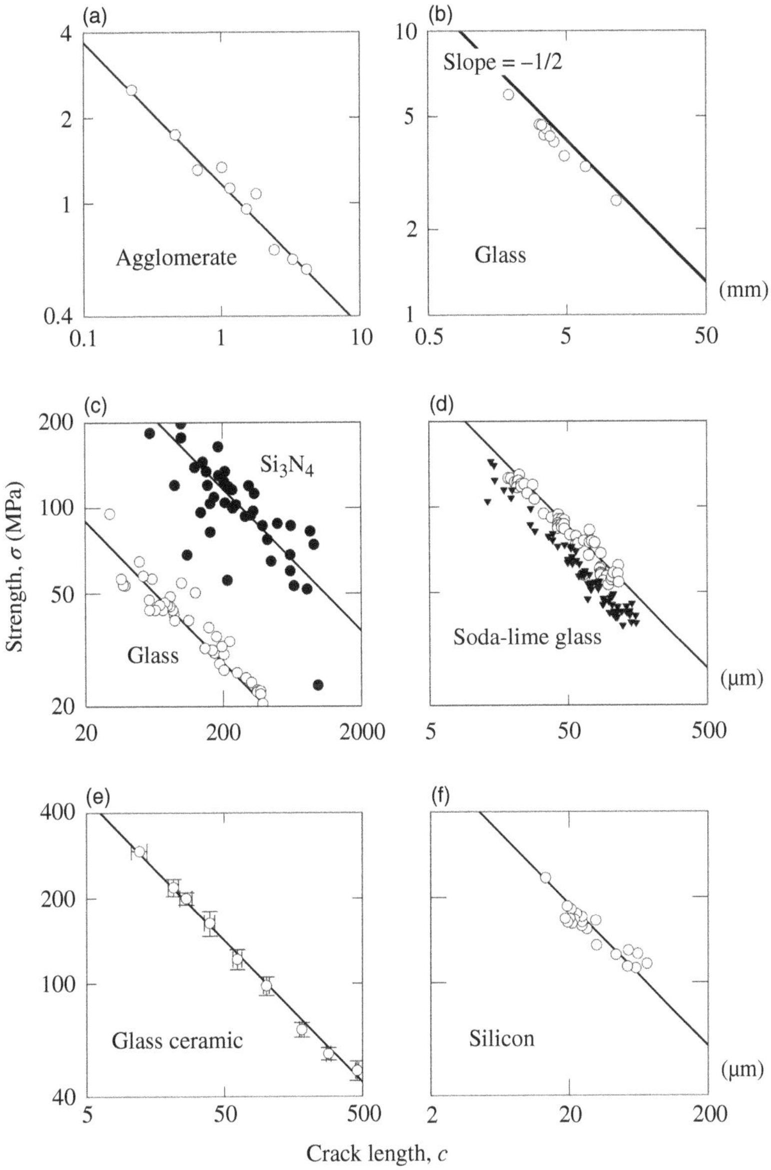

Figure 1.10 A series of strength σ versus crack length c plots in logarithmic coordinates for brittle materials, illustrating the Griffith equation, Eq. (1.16). Lines indicate empirical best fits. (a) Titania (TiO_2) powder compact agglomerates containing notches. Source: Adapted from Kendall (1988). (b) Glass bulbs containing annealed scratches. Source: Adapted from Griffith (1921). (c) Glass and polycrystalline silicon nitride (Si_3N_4) beams containing single- and multipoint grinding scratches. Source: Adapted from Kirchner and Isaacson (1982) and Kirchner (1984). (d) Glass plates containing annealed Vickers indentations. Source: Adapted from Glaesemann et al. (1987). (e) Cordierite ($2MgO{\cdot}2Al_2O_3{\cdot}5SiO_2$) glass ceramic plates containing Vickers indentations. (f) Silicon beams containing annealed Knoop indentations. Source: Adapted from Yasutake et al. (1986). The Griffith equation has been well verified by experiment for over 100 years.

well describes the data. Figure 1.10f shows strength measurements of silicon bend beams containing annealed Knoop indentation flaws from the work of Yasutake et al. (1986). Symbols represent individual strength measurements. The solid line represents a best fit to the data consistent with Eq. (1.16) and well describes the data. Similar behavior up to strengths of 2.5 GPa was observed by Engel and Hübner (1978) in WC–Co.

There are two major conclusions to be drawn from Figure 1.10. The similarity between the Griffith and subsequent strength data, spanning a range of materials and flaw types and factors of 10^3 in crack size and strength, is apparent. Hence, the first conclusion is analytical: The scaling relations of Sections 1.3.1 and 1.3.2 implemented in the First Law of Thermodynamics are extremely effective in describing brittle fracture strengths. The reduced experimental scatter of the indentation strength based data, Figure 1.10d–f, relative to the grinding strength data, Figure 1.10c, is also apparent. Hence, the second conclusion is experimental: Controlled flaw indentation techniques are extremely effective for reducing uncertainty in measuring brittle fracture strengths.

1.3.4 Configurational Forces: $\mathcal{G}$ and *R*

Configurational forces acting on the crack area A may be defined from the mechanical energy U_M and the surface energy U_S. The *mechanical energy release rate* $\mathcal{G}$ is given by

$$\mathcal{G} = -\frac{dU_M}{dA}. \tag{1.17}$$

The *fracture resistance* R is given by

$$R = \frac{dU_S}{dA}. \tag{1.18}$$

The First Law fracture equilibrium statement of Eq. (1.7) can then be written generally as

$$\mathcal{G} = R. \tag{1.19}$$

The equilibrium of a crack system can then be interpreted as the equal and opposite action of two forces on the crack area, one acting to increase the crack area, $\mathcal{G}$, and the other acting to decrease the crack area, R. (R here is not to be confused with radius.) Similar configurational forces include the Peach–Koehler force acting on the positions of dislocations in plastically deforming materials (Peach and Koehler 1950; Weertman and Weertman 1992), the force analyzed by Eshelby, acting on the position of an elastic singularity (e.g. inclusion, void) in a loaded component (Eshelby 1951), and the osmotic forces acting on separated solutions and on dislocations in fields of lattice defects (Adkins 1983; Weertman and Weertman 1992). Extensive mathematical details are given in Maugin (2011).

For the specific case of the uniform applied stress, the mechanical energy release rate for both circular and straight cracks can be written as

$$\mathcal{G} = \psi^2\sigma^2 c/E. \tag{1.20}$$

For different or more complicated applied stress or crack geometry configurations, the crack length dependence is more complicated than linear. Similarly, in general, R is crack length dependent, $R = R(c)$, reflecting microstructural influences on the resistance to fracture. For the simple case here, R was invariant:

$$R = 2\gamma. \tag{1.21}$$

2γ is a lower bound to R. R may be increased in some materials by microstructural effects leading to irreversible deformation and energy loss (plasticity in metals and polymers are the prime examples). R may be decreased by environmental effects leading to rate-dependent chemical reactions and irreversible entropy production (stress corrosion cracking is the prime example). Chapters 4 and 6 consider these phenomena for brittle materials along with the implied variation in $D = (RE)^{1/2}/\psi$.

The stability of a crack system is determined by the relative signs of the derivatives of $\mathcal{G}$ and R. For $d\mathcal{G}/dA > dR/dA$, the system is unstable, and for $d\mathcal{G}/dA < dR/dA$, the system is stable. For the case here of uniform stress acting on a crack in a homogeneous material, $d\mathcal{G}/dA > 0$ and $dR/dA = 0$, and the system is unstable.

Section 1.3.1 showed that the change in elastic energy of a system transformed at constant force F between equilibrium configurations was $\Delta U_E = F^2\Delta\lambda/2$, where $\Delta\lambda$ is the change in compliance between the configurations. Noting that for this condition $\Delta U_M = -\Delta U_E$, using Eq. (1.17) gives

$$\mathcal{G} = \frac{F^2}{2}\frac{d\lambda}{dA}. \tag{1.22}$$

In the analysis of uniform stress discussed earlier, the compliance of the cracked bar can be expressed similarly to the elastic energy as $\lambda_c = \lambda_0(1 + v/V)$. Applying Eq. (1.22) and the relevant equations thus leads to $\mathcal{G}$ as in Eq. (1.20). Simple scaling of the compliance by the projected area of the fractured section leads to an increase that scales as $\Delta\lambda \sim A \sim c^2$ and assumes that the stress field in a bar remains homogeneous in the presence of the crack. The increased crack length dependence determined here, $\Delta\lambda \sim v \sim c^3$, is a consequence of the inhomogeneous stress field in the bar generated by the crack.

1.4 Equilibrium Fracture on Localized Loading

1.4.1 Mechanical Energy

A schematic diagram of a loaded crack configuration at the opposite extreme to Figure 1.7 is shown in Figure 1.11. The crack is similar to that in Figure 1.7 as it has a circular outline, radius c, is embedded in the center of a large block of material, characteristic dimension L, and has tensile loading perpendicular to the crack plane. However, the crack in Figure 1.11 is loaded by a force F that is applied over a very small region at the center of the crack, characteristic dimension a, and the geometry is known as a center-loaded penny crack. Similar to the crack in Figure 1.7, the elastic strain energy in the block of material in Figure 1.11 increases as c increases at fixed F, but with decreasing rather than increasing crack length derivative. Hence, in contrast to the analysis discussed earlier, focused on describing the unstable $F(c)$ configuration that defined component strength, attention here is focused on describing the *stable* $c(F)$ configuration that defines initial flaw size. In developing this description, the analysis uses energy scaling ideas first articulated by Roesler (1956). These ideas are here incorporated into the framework and notation used earlier.

Analysis begins by recognizing that, as $c \ll L$, the characteristic stress σ in the fracture system of Figure 1.11 varies as $\sigma \sim F/c^2$. Similarly, the volume of material υ surrounding the penny crack affected by the force F, and thus stress σ, varies as $\upsilon \sim c^3$. A characteristic strain energy density in the system is thus $\sigma^2/2E \sim (F/c^2)^2/E$, and a characteristic energy is $(F^2/Ec^4)(c^3) = F^2/Ec$. Noting that $A \sim c^2$, scaling of the elastic energy change on crack introduction can then be expressed as

$$U_E \sim U_a - \frac{F^2}{EA^{1/2}}, \tag{1.23}$$

where U_a is the maximum elastic energy for large cracks given by $U_a \approx F^2/Ea$. Note that the scaling of Eq. (1.23) leads to $\Delta U_M \sim F^2/EA^{1/2}$ and that the signs of all quantities remain unchanged from those discussed earlier. Applying Eqs. (1.17)–(1.23) gives

$$\mathscr{G} = -\frac{dU_M}{dA} \sim \frac{F^2}{EA^{3/2}} = \frac{\chi^2 F^2}{Ec^3}, \tag{1.24}$$

where the last term is expressed in terms of crack length c and a proportionality term χ^2 that includes all geometrical factors. Note that the amplitude of Eq. (1.24) has several similarities to that in Eq. (1.20): quadratic dependence on a loading term, inverse dependence on modulus, and a geometrical dependence. The difference is in the crack length dependence.

1.4.2 The Roesler Equation

Setting $U_S = 0$ at $A = 0$ and $U_M = U_{max}$ at a bounding large crack area $A_{max} \approx L^2$ enables the total energy U_T of the loaded cracked system to be expressed using Eqs. (1.13) and (1.23) as

$$U_T = 2\gamma A + 2\pi^{3/2}\chi^2 F^2/EA^{1/2} - U_{max}. \tag{1.25}$$

Figure 1.12 shows a plot of $U_T(c)$ using Eq. (1.25) and $A = \pi c^2$, the material properties $E = 200$ GPa and $2\gamma = 10\ \mathrm{J\,m^{-2}}$, and the χF product adjusted as described in the following. The equivalent crack area A is shown on the upper axis. The dashed

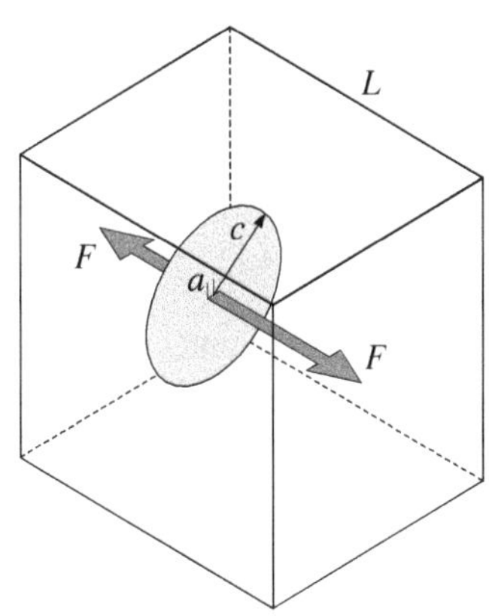

Figure 1.11 Schematic diagram of a block of material containing a crack loaded by a localized force. The crack has circular perimeter, radius c, and the fracture configuration is known as a center-loaded penny crack.

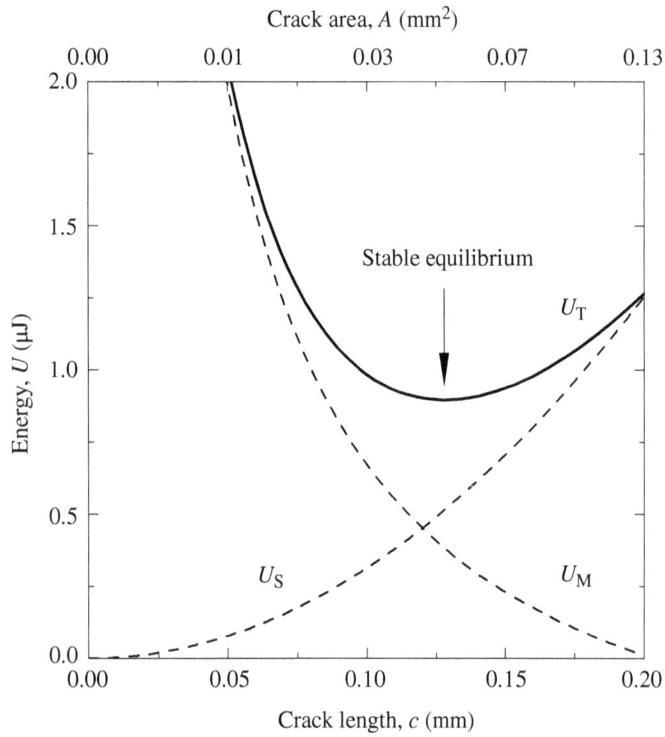

Figure 1.12 Energy versus crack area for a component containing a circular crack loaded by a localized force. The surface energy U_S and mechanical energy U_M are shown as dashed lines. The total energy U_T is shown as the solid line and passes through a minimum of stable equilibrium. Such behavior is the basis of the Roesler equation for brittle crack lengths.

lines show the asymptotes of the first and second terms, Eqs. (1.13) and (1.23), respectively, and the solid line shows the full response, Eq. (1.25). The total energy passes through a minimum. Setting $dU_T/dA = 0$, equivalent to imposing the equilibrium condition of Eq. (1.7) in the infinitesimal limit, gives the crack area at the minimum of

$$A = \left[\frac{\pi^{3/2}(\chi F)^2}{2\gamma E}\right]^{2/3}, \tag{1.26}$$

where $\chi F \approx 2$ was adjusted to give the minimum at $A \approx 0.05\,\text{mm}^2$ and $c \approx 0.13\,\text{mm}$, similar to that of the maximum in Figure 1.9. (For typical values, $\chi \approx 0.1$, $F \approx 20\,\text{N}$. Details are considered in Chapter 2.) The variation of U_T has $d^2U_T/dA^2 > 0$ at this point and hence the crack area and length at the minimum is a point of stable equilibrium, indicated by the arrow in Figure 1.12. An interpretation of Figure 1.12 can be made in terms of progression along the path of fixed force, AC in Figure 1.8. As this path is traversed, the crack grows in area until a point of stable equilibrium is reached. Perturbations in F or δ at this point that cause an increase in crack area lead to increases in total energy and return of the crack to the equilibrium state.

Crack length as a function of applied force at the point of stable equilibrium is a key experimental observation that provides insight into the likely size of strength controlling flaws in a material. Replacing A with c in Eq. (1.26) gives crack length as

$$\begin{aligned} c &= \left[(\chi F)/(2E\gamma)^{1/2}\right]^{2/3} \\ &= BF^{2/3} \end{aligned} \tag{1.27}$$

a result found by Roesler (1956). Equation (1.27) makes clear that the factor B linking crack length and local force is the ratio of geometry and material terms, similar to D discussed earlier. B characterizes the *cracking susceptibility* of a material; how much crack length increases as local force increases (Cook 2020). Cracking susceptibility and B are discussed in detail in Chapter 2. Here, B is taken to be invariant for a given material and crack system. The Roesler equation, Eq. (1.27), and the Griffith equation, Eq. (1.16), stand at opposite ends of fracture scaling. In the case here, the force is applied over an area small relative to the crack, leading to fracture stability. In the Griffith case, the force is applied over an area large relative to the crack, leading to fracture instability. In both cases, the product $(2\gamma E)^{1/2}$ appears as a proportionality constant linking crack length to either force or stress.

The most common method of assessing Eq. (1.27) is through *indentation*, in which the flat surface of a block of material is contacted in a controlled loading cycle by a small probe. Indentations are typically characterized by P, the peak force normal to the surface in the contact cycle. P is usually the experimentally controlled variable in indentation testing and is referred to as the *indentation load*; the terms "peak" and "normal" are usually omitted and implied, and the term "load" is taken to indicate a physical force. The probe may be spherical, leading to "blunt" indentation in which the contact is elastic and described by the Hertzian stress field (Johnson 1985; Lawn 1998), or conical or pyramidal, leading to "sharp" indentation in which the contact is elastic–plastic and described by a combination of the Boussinesq stress field and a blister field (Yoffe 1982; Johnson 1985; Cook and Pharr 1990; Cook 2019). If the indentation load is great enough, cracks form adjacent to the contact at both blunt and sharp indentations on brittle materials. The cracks used in the strength degradation plots of Figure 1.10d–f were generated by sharp indentations. Figure 1.13 shows schematic cross-section diagrams of (a) blunt and (b) sharp indentations. In Figure 1.13a, a cone crack is generated, with truncated apex at the contact site. In Figure 1.13b, a set of perpendicular half-penny cracks is generated, emanating from a near hemispherical zone of compaction and plastic deformation. Both the cracks and the zone are centered at the contact site. In both geometries, the diameter of the cracks is taken as $2c$ as shown, noting that the dimension for the cone crack is subsurface. For both spherical and conical or pyramidal indentations the effective force F normal to the indentation cracks (Figure 1.11) is taken to vary linearly with the experimental indentation load, $F \sim P$. This linearity is not general and other indentation geometries lead to other, nonlinear $F(P)$ relationships. Indentation crack initiation and loading details are discussed in Chapter 2. Flat punches of hardened metal are often used in blunt indentation and Vickers diamond pyramids are often used in sharp indentation.

Similar to the Griffith equation, in the 60+ years since Roesler (1956), the Roesler equation in the form $c = BP^{2/3}$ has been verified in experiments many times. Figure 1.14 shows logarithmic plots of measurements of stable crack length as a function of indentation load, $c(P)$, for a range of contact systems, using data derived from the published works cited. The plots are all at the same relative scale, vertical × horizontal of 500 × 10,000, and are arranged in approximate order of decreasing indentation load. Note that the absolute load scales, right, differ by a factor of 10^6, from kN to mN. The fine solid lines are guides to the eye of slope 2/3, consistent with Eq. (1.27). Figure 1.14a shows in situ measurements of cone cracks in fused silica glass (FSG) indented with flat punches, (0.4–6.4) mm in diameter, from the work of Benbow (1960). Almost identical results over a similarly broad load domain were obtained by Mouginot and Maugis (1985) from in situ measurements of cone cracks in borosilicate glass; no difference in cracking was observed by Mouginot and Maugis for indentation by spheres or flat punches. Similar fracture results were obtained for FSG and SLG over smaller indentation load domains by Roesler (1956), Lawn and Fuller (1975), and Zeng et al. (1992a,b). The data are well described by the 2/3 trend of the Roesler equation. A point to note is that cone crack systems are large: Forces of 0.1 kN to over 20 kN are required to generate millimeter-scale cracks. As the contacts are elastic, the cracks close on force removal, practically limiting extensive measurements to transparent materials under load. Figure 1.14b illustrates this effect, showing

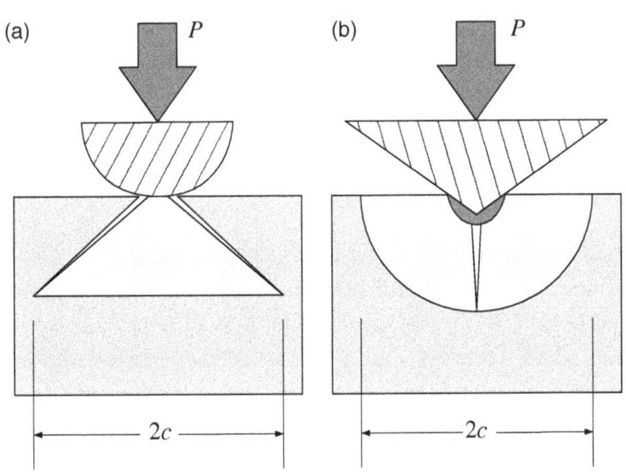

Figure 1.13 Schematic cross-section diagrams of indentation cracks generated by indentation loads P. (a) Blunt, spherical indentation generating cone cracks of diameter $2c$. (b) Sharp, pyramidal or conical indentation generating plastic deformation zone (shaded dark) and half-penny cracks of diameter $2c$.

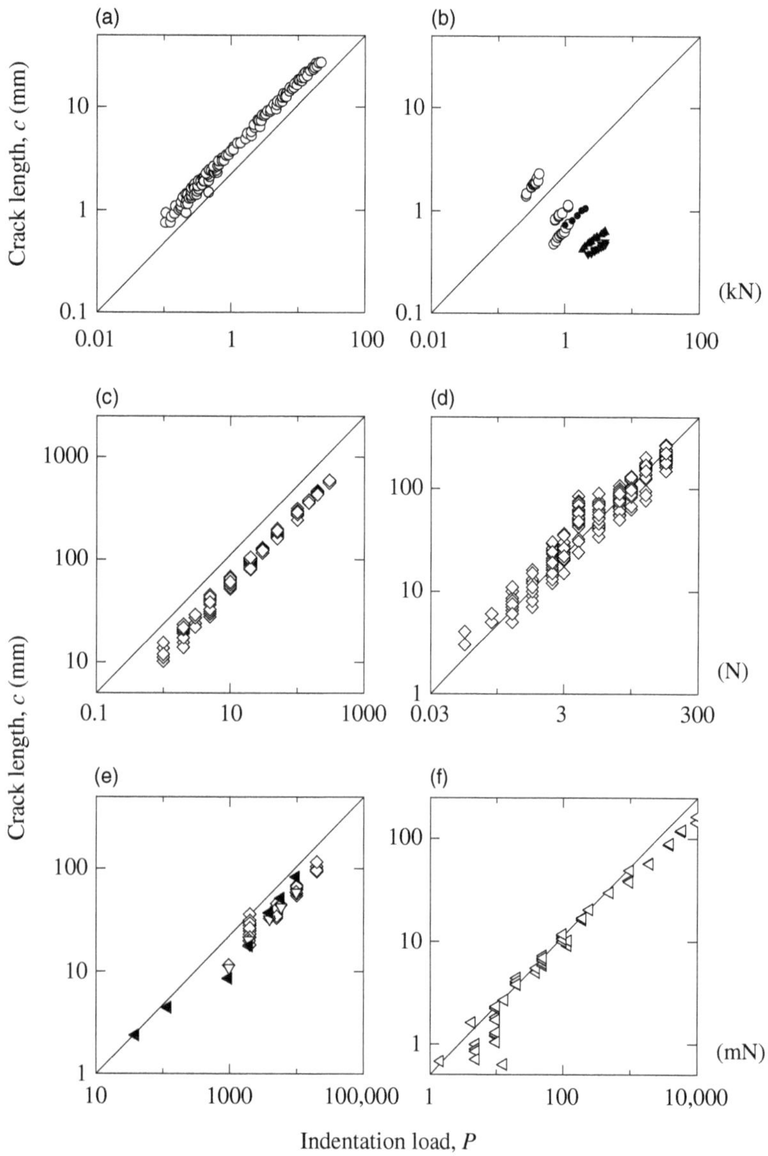

Figure 1.14 A series of crack length c versus indentation load P plots in logarithmic coordinates for brittle materials, illustrating the Roesler equation, Eq. (1.27). Lines are guides to the eye. (a) Cone cracks in glass. Source: Adapted from Benbow (1960). (b) Cone cracks in glass and ceramics. Source: Adapted from Zeng et al. (1992b). (c) Half-penny cracks in cordierite glass ceramic. (d) Half-penny cracks in polycrystalline alumina. (e) Surface traces of indentation cracks in sapphire (corundum, Al_2O_3). Open symbols: Vickers and Berkovich indenters. Filled symbols: cube corner indenter. Source: Berkovich and cube corner data adapted from Pharr et al. (1993). (f) Surface traces of cube corner indentation cracks in soda-lime glass. Source: Includes data adapted from Pharr et al. (1993); and Harding et al. (1995). The Roesler equation has been well verified by experiment for over 60 years.

ex situ post-contact measurements on polished sections or split samples of SLG and ceramics indented with tungsten carbide (WC) spheres from the work of Zeng et al. (1992b), Lee et al. (1997), and Lawn et al. (1998). The open symbols (Zeng) are measurements from glass (upper), a fine-grained Al_2O_3 (center), and a Al_2O_3–SiC whisker composite (lower). The filled symbols (Lee, Lawn) are adjusted for cone crack surface truncation and are from a silica-based glass ceramic composite (upper), a fine-grained Si_3N_4 (center), and a medium-grained Si_3N_4 (lower). Measurements in the opaque ceramics are clearly restricted to ex situ techniques, and the post-contact measurements are clearly restricted in load domain. For similar indentation loads the ranges of cone crack lengths in the ceramics are much smaller than that of glass. The 2/3 trend describes the limited ranges of data (from which Zeng et al. estimated material fracture properties, see Chapter 2).

Figure 1.14c shows half-penny crack measurements in a fine-grained glass ceramic indented with Vickers pyramids (Cook 2020). This is the same material represented in the strength measurements of Figure 1.10e. Symbols represent individual measurements. The width of the force domain (a factor of 300), the relative dispersion of crack lengths at a given force (about a factor ±20%), and the description of the data by the 2/3 trend are all similar to those of Figure 1.14a. The obvious point of difference to note is that half-penny crack systems are small: Forces of 1–300 N generate cracks tens of micrometers in scale. A characteristic of sharp indentations is that the contacts are elastic–plastic, generating localized zones of plastic deformation on loading that continue to force the surface traces of the half-penny cracks open on unloading. As a consequence, measurements are extended to opaque materials after force removal. Figure 1.14d shows half-penny crack measurements in a large-grained polycrystalline alumina indented with Vickers pyramids (Cook 2020). Symbols represent individual measurements. The data are described by the 2/3 trend, although there is a great deal of relative dispersion in the crack lengths at a given indentation force (about a factor ±50%) and scatter about the 2/3 trend is significant. An extensive survey (Cook 2020) shows that the results of Figure 1.14c and d are representative of sharp indentation cracks in a wide range of materials: the 2/3 trend is observed over wide domains of contact force and dispersion increases with microstructural scale.

The lower bounds of the data domains in Figure 1.14c and d indicate the loads required to form cracks on indentation. The lower bound load is termed the *threshold load*, and indentations at loads less than the threshold do not exhibit indentation cracking. Thresholds are observed for both blunt and sharp indentations, and the threshold load depends on both the material and indenter geometry. In particular, the threshold is decreased significantly on sharp indentation for more acute (smaller included tip angle) indenters. This effect is illustrated in Figure 1.14e, which shows the lengths of surface traces of indentation cracks in sapphire. Open symbols show the responses for Vickers (four-sided) and Berkovich (three-sided) pyramidal indenters (Pharr et al. 1993; Cook 2020), using data derived from the published works cited. The crack lengths are independent of indenter shape and exhibit a threshold load of about 1000 mN (1 N, Vickers and Berkovich indenters have similar equivalent included tip angles). The filled symbols show the responses for an acute cube corner pyramidal indenter. The crack lengths exhibit the same 2/3-slope trend as the less acute indenters but the threshold load is reduced significantly to 30 mN. The reduced threshold is illustrated even more strikingly in Figure 1.14f, which shows the lengths of surface traces of indentation cracks in SLG formed by cube corner indentation alone, including the observations of Pharr et al. (1993) and Harding et al. (1995), using data derived from the published works cited. The crack lengths exhibit the 2/3-slope trend of the Roesler equation over a factor of 10^4 in indentation load, in large part due to a significantly reduced threshold of about 1 mN.

There are two major conclusions to be drawn from Figure 1.14; the first is similar to that drawn from Figure 1.10. The similarity between the Roesler, Benbow, and subsequent indentation crack length data, spanning a range of materials, indenters, flaw types, and a factor of 10^3 in crack size (and 10^7 in indentation load), is apparent. Hence, the first conclusion is again analytical: The scaling relations of Section 1.4.1 implemented in the First Law of Thermodynamics are extremely effective in describing brittle fracture crack lengths. The clear differences in indentation crack lengths for different materials and the increased dispersion of crack lengths in a polycrystalline material relative to glass and single crystal materials, Figure 1.14d, are also apparent. Hence, the second conclusion is experimental: Controlled flaw indentation techniques are extremely effective for assessing material differences and microstructural influences in brittle fracture. The predominant analytical method used in experimental application of indentation fracture is discussed in Section 1.5.

1.5 Stress-Intensity Factor and Toughness

The physical basis of brittle fracture as the conservation of energy is made clear in Sections 1.3 and 1.4. However, an energy viewpoint becomes unwieldy analytically when applied to crack loading geometries more complicated than the considered limiting cases of uniform and local loading. In particular, energy-based analyses become very difficult for the often encountered case of a sharp contact flaw in a component with cracks under the superposed influences of contact and applied stresses and material microstructural forces. The chief difficulty arises as superposed loading geometries usually lead to mechanical energy expressions in which the loading terms are not independent. However, most fracture behavior in linear elastic materials is analytically tractable from a stress viewpoint, in which the loading terms are independent. This viewpoint, which is the basis for analysis of indentation fracture throughout the book, is now considered. The definitions and terms used in the stress viewpoint and the relation to the energy viewpoint are discussed here in Section 1.5. Relevant

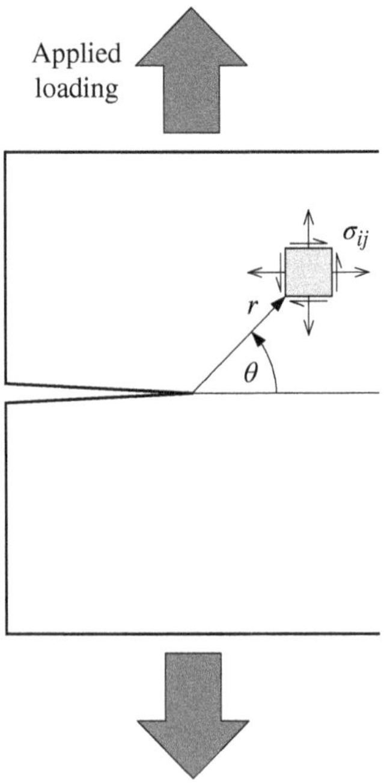

Figure 1.15 Schematic cross-section diagram of a loaded planar crack illustrating the polar crack tip coordinates (r, θ) and stress field components σ_{ij} used to define the stress-intensity factor. The coordinate out of the diagram is linear z for planar systems and angular ω for spherical systems.

example applications are discussed in Section 1.6. [Fracture geometries in which the superposed forces are orthogonal lead to analyses in which mechanical energy loading terms *are* independent, (Lawn 1993; Wei 2010). Such mixed-mode fracture is not considered here.]

Analyses of brittle fracture from a stress viewpoint are significantly aided by the fact that stress distributions surrounding cracks in materials that behave as linear elastic continua are universal functions of relative crack tip coordinates. Figure 1.15 shows a schematic cross section of a planar crack in such a material, along with a crack tip polar (r, θ) coordinate system and a representative material element in that system. The element is located at a distance r from the crack tip and at an angle θ relative to an axis in the crack plane. The crack is loaded in the far field, remote from the crack tip, indicated by the representative dark arrows. The components of the stress tensor, σ_{ij}, for the element are given by

$$\sigma_{ij} = \frac{K}{(2\pi r)^{1/2}} f_{ij}(\theta), \tag{1.28}$$

where $f_{ij}(\theta)$ depends on ij. The derivation of Eq. (1.28), based on complex potential analysis, is given elsewhere (e.g., Broek 1982; Kanninen and Popelar 1985; Wei 2010). An important aspect of Eq. (1.28) is that the amplitude, radial, and angular functional components are separable. The radial component decreases with increasing distance from the crack tip (and formally diverges at the tip, a consequence of the continuum assumption), reinforcing the energy-based ideas discussed earlier that strain energy changes associated with fracture are localized to the crack. The tensor angular functions are listed in many works (e.g. Broek 1982; Williams 1984; Atkins and Mai 1985; Kanninen and Popelar 1985; Lawn 1993; Wei 2010; Ritchie and Liu 2021) and in many cases $f_{ij}(0) = 1$.

The scalar amplitude component K in Eq. (1.28) is known as the *stress-intensity factor* (SIF). K includes all information regarding the boundary conditions of loading—the magnitude, orientation, and location of all applied forces and displacements, and, importantly, the crack length c. As might be anticipated, there is a close connection between K and $\mathcal{G}$. The connection between the two is known as the Irwin relation, derivable in a number of ways (see, e.g., Lawn 1993), and is given by

$$\mathcal{G} = \frac{K^2}{E}. \tag{1.29}$$

Equation (1.29) has the familiar structure of an energy term ($\mathcal{G}$) related to the ratio of a stress term squared (K^2) and a modulus (E). As $K(c)$ describes the crack tip stress field in a linear elastic material and is a linear stress amplitude term, $K(c)$ fields arising from different sources of applied loading on a crack are linearly additive to give the total SIF. For example, three separate sources of applied loading p, q, and r in a fracture system each give rise to a SIF determined independently from the separate p, q, and r boundary conditions, $K_p(c), K_q(c)$, and $K_r(c)$. The total SIF $K(c)$ is given by the sum

$$K(c) = K_p(c) + K_q(c) + K_r(c). \tag{1.30}$$

$K_p(c)$, $K_q(c)$, and $K_r(c)$ can all be different functions of c and all need not be positive, provided the sum $K(c)$ is positive (so that a crack exists). The total mechanical energy release rate $\mathscr{G}$ in this example is given by

$$\mathscr{G} = \left[K_p^2 + K_q^2 + K_r^2 + 2K_pK_q + 2K_qK_r + 2K_rK_p\right]/E, \tag{1.31}$$

illustrating the lack of independence of loading terms in the energy viewpoint, Eq. (1.31), relative to the stress viewpoint, Eq. (1.30).

$K(c)$ functions for a wide variety of loading configurations are listed in several compendia (e.g., Sih 1973; Tada et al. 1973; Murakami 1987) and in many fracture texts (e.g., Lawn 1993; Ritchie and Liu 2021). Fracture mechanics analysis for a particular application typically involves selecting and adding the $K(c)$ SIF function or functions appropriate to the considered fracture configuration, imposing equilibrium and stability conditions, and solving for the quantity of interest (e.g., a force or a crack length). The equilibrium condition in the SIF framework is defined by direct correspondence with that in the energy framework, Eq. (1.19):

$$\mathscr{G} = R.$$

Implementing the Irwin relation on both sides of this equation gives

$$\frac{K^2}{E} = \frac{T^2}{E} \tag{1.32}$$

or

$$K = T, \tag{1.33}$$

where T is the *toughness* of the material, given by $T = (RE)^{1/2}$, and defines the value of the SIF in fracture equilibrium, noting that K is the net SIF. T may depend on crack length, $T = T(c)$. In microstructurally inhomogeneous materials that exhibit toughening $\mathrm{d}T(c)/\mathrm{d}c > 0$. Toughening effects are considered in Chapters 4 and 5. T may exhibit environmental dependence in materials that react with chemical species. In an inert environment, T takes a maximum value T_0, although the 0 is often omitted and implied. In a reactive environment, often ambient moisture, $T < T_0$. The surface energy is similarly notated, such that for homogeneous materials in inert environments $T_0 = (2\gamma_0 E)^{1/2}$ (and again the 0 of inert γ_0 is often omitted). Environmental and reaction effects, including reaction rates, are discussed briefly in Section 1.6.2 and Chapter 2 and more thoroughly in Chapter 6.

Using the Irwin relation, the SIF K_A for uniform applied stress σ from Eq. (1.20) is

$$K_A = \psi\sigma c^{1/2}. \tag{1.34}$$

The SIF K_F for a penny crack center loaded by a localized force F from Eq. (1.24) is

$$K_F = \chi F/c^{3/2}. \tag{1.35}$$

Exact stress analyses give $\psi = 2/\pi^{1/2}$ and $\chi = \pi^{-3/2}$ for circular cracks and $\psi = \pi^{1/2}$ for a straight crack (Sih 1973; Tada et al. 1973; Murakami 1987). Imposing the equilibrium condition $K_A = T$ in Eq. (1.34) returns the Griffith equation

$$\sigma = (T/\psi)c^{-1/2} \tag{1.36}$$

and the strength degradation parameter D previously mentioned in Eq. (1.16) is recognized as T/ψ. Similarly, imposing the equilibrium condition $K_F = T$ in Eq. (1.35) returns the Roesler equation

$$c = (\chi F/T)^{2/3} \tag{1.37}$$

and the cracking susceptibility parameter B previously mentioned in Eq. (1.27) is recognized as $(\chi/T)^{2/3}$.

Inversion of Eq. (1.37) gives $T = \chi(F/c^{3/2})$, such that measuring $F/c^{3/2}$ from indentation crack lengths can provide an estimate of material toughness T (if χ is known). This was the intent of Zeng et al. (1992b) using cone cracks and of many other studies using sharp indentation half-penny cracks in ceramics and glasses (see Cook 2020 for a review). In these cases K_F is used alone. Variations in B with material and indenter geometry are discussed in detail in Chapter 2. Similarly, inversion of Eq. (1.36) gives $T = \psi(\sigma c^{1/2})$, such that measuring $\sigma c^{1/2}$ from strength tests can also provide an estimate of toughness (if ψ is known). To avoid the difficulties of estimating strength controlling flaw sizes on fracture surfaces, many studies of brittle material strengths use controlled indentation flaws and replace c with the indentation load P in toughness analyses. In these cases, the sum $K_F + K_A$ is used. Such analyses, and variations in D with material and flaw size, are discussed in Chapters 3–6. [The definition of the term B here is unchanged from Cook (2020). The term D here was notated B in Cook (2023).]

Stability of a fracture system is determined by the relative magnitudes of the crack length derivatives of K and T, recognizing that these derivatives are related to the crack area derivatives of the mechanical and surface energies. For $\mathrm{d}K/\mathrm{d}c > \mathrm{d}T/\mathrm{d}c$, a fracture system is destabilized on loading, and for $\mathrm{d}K/\mathrm{d}c < \mathrm{d}T/\mathrm{d}c$, the system is stabilized. For $K = T$, if $\mathrm{d}K/\mathrm{d}c > \mathrm{d}T/\mathrm{d}c$, the fracture system is in a state of *unstable equilibrium*, and if $\mathrm{d}K/\mathrm{d}c < \mathrm{d}T/\mathrm{d}c$, the system is in a state of *stable equilibrium*. Assuming that T is single valued and thus $\mathrm{d}T/\mathrm{d}c = 0$, equilibrium in the uniform stress system is unstable (Section 1.3) and in the localized force system is stable (Section 1.4). For $K \neq T$ (or equivalently $\mathcal{G} \neq R$), the fracture system is not in equilibrium and the system exhibits *nonequilibrium fracture* and crack motion. The two most common and important nonequilibrium fracture phenomena in brittle materials are rapid fracture associated with component failure and slow fracture associated with flaw generation. Section 1.6 develops analyses describing these phenomena.

1.6 Nonequilibrium Fracture

Crack motion in nonequilibrium fracture is characterized by a *crack velocity* $v = \mathrm{d}c/\mathrm{d}t$ describing crack behavior within a loaded component (t is time). A goal of nonequilibrium fracture analyses is thus to arrive at expressions for v subject to the mechanical and thermodynamic constraints on the fracture system. The first constraint considered here is mechanical, in which the *kinetic energy* U_K of material elements within a cracked component is significant and controls crack propagation. Analysis of fracture behavior in such systems is developed under adiabatic conditions as in Sections 1.2–1.4. The analysis applies to fast-moving cracks departing from unstable equilibrium configurations leading to component failure. The second constraint here is thermodynamic, in which the *irreversible entropy production* S_{irr} of material elements within a cracked component is significant and controls crack propagation. Analysis of fracture behavior in such systems is developed under isothermal conditions. The analysis applies to slow-moving cracks. Here, attention is focused on cracks tending to stable equilibrium configurations leading to flaw generation. (Cracks departing from unstable equilibrium configurations leading to delayed failure of components in reactive environments are considered in Chapter 6.)

1.6.1 Unstable Adiabatic Fracture

Analysis of adiabatic fracture begins with extension of the expression for internal energy change, Eq. (1.2), to include a kinetic energy term ΔU_K,

$$\Delta U = \Delta U_E + \Delta U_S + \Delta U_K. \tag{1.38}$$

Kinetic energy U_K is distributed throughout the component as coherent lattice vibrations in the form of propagating phonons (Ashcroft and Mermin 1976). The phonons are generated by the rapidly altering displacement field associated with a fast-moving crack. The scattering of such phonons from fracturing surfaces in breaking brittle components leads to the common experiences of sound wave production in air and rapid emission of broken fragments. The forces (or, more correctly, the stresses) on materials leading to significant kinetic energy associated with fast-moving cracks, and the resulting sound production and fragment emission, are often large and rapidly applied, and often by impact. Hence, fracture in which kinetic energy is a significant factor is often known as "dynamic fracture" (see, e.g., Freund 1990; Lawn 1993).

The total energy including kinetic energy is given by extension of Eq. (1.14) for uniform loading of a component containing a circular crack

$$U_T = 2\gamma(\pi c^2) - (2\psi^2/3E\pi^{1/2})\sigma^2(\pi c^2)^{3/2} + U_K(c) \tag{1.39}$$

recognizing that for an adiabatic system U_T is invariant. Here, the energies are specified in terms of crack length c. The first two terms, the surface and mechanical energies, vanish for $c = 0$ as applied earlier in Eq. (1.14). The third, kinetic energy term vanishes for the crack at rest. This condition pertains to the equilibrium instability state $c = c_0$ given by the Griffith condition Eq. (1.16), such that $U_K(c_0) = 0$. Taking advantage of these conditions enables U_T to be evaluated as

$$U_T = \psi^2\sigma^2\pi c_0^3/3E = 2\gamma\pi c_0^2/3. \tag{1.40}$$

Combining Eqs. (1.39) and (1.40) gives an expression for the kinetic energy in terms of crack extension:

$$U_K(c) = (\psi^2\sigma^2\pi c^3/3E)(1 - c_0/c)^2(2 + c_0/c). \tag{1.41}$$

The kinetic energy in the component increases in an unbounded manner with crack extension. An almost identical expression describes the kinetic energy associated with propagation of a straight crack, as first given by Mott (1948) and more recently by Lawn (1993).

Development of an expression for kinetic energy associated with fracture that includes crack velocity begins with the contribution dU_K from a material element of volume dV:

$$dU_K = \rho\left(\frac{\partial u_i}{\partial t}\right)^2\frac{dV}{2}, \tag{1.42}$$

where u_i is the vectorial displacement of the element and the Einstein summation convention is assumed. Application of the chain rule gives

$$dU_K = \rho\left(\frac{\partial c}{\partial t}\right)^2\left(\frac{\partial u_i}{\partial r}\right)^2\left(\frac{\partial r}{\partial c}\right)^2\frac{dV}{2}, \tag{1.43}$$

where r is the radial coordinate in the crack tip coordinate system of Figure 1.15. The first term in parentheses is recognized as v^2. The second term can be evaluated by assuming that the form of the displacement field is given by the SIF description and is

$$u_i = \frac{K}{2E}f_i(\theta)\left(\frac{r}{2\pi}\right)^{1/2}, \tag{1.44}$$

where $f_i(\theta)$ depends on i. As with Eq. (1.28), an important aspect of Eq. (1.44) is that the amplitude, radial, and angular functional components are separable. The radial component increases with increasing distance from the crack tip. The tensor angular functions are listed in many works (e.g., Broek 1982; Williams 1984; Atkins and Mai 1985; Kanninen and Popelar 1985; Lawn 1993; Wei 2010; Ritchie and Liu 2021) and in many cases $f_i(0) = 1$. Hence, the second term becomes (omitting all numerical factors)

$$\left(\frac{\partial u_i}{\partial r}\right)^2 \sim \frac{K^2}{E^2}f_i^2(\theta)\frac{1}{r}. \tag{1.45}$$

Combining all the above equations gives for the contribution to the kinetic energy from the material element (again, omitting numerical factors)

$$dU_K \sim v^2\frac{\rho}{E}\frac{K^2}{E}f_i^2(\theta)\left(\frac{\partial r}{\partial c}\right)^2\frac{dV}{r}. \tag{1.46}$$

The second term is recognized as the reciprocal of an acoustic velocity $1/v_L^2$, and the third term is recognized as $\mathcal{G}$. Hence, the total kinetic energy from all material elements in the component is given by the integral

$$U_K \sim \frac{v^2}{v_L^2}\mathcal{G}\int f_i^2(\theta)\left(\frac{\partial r}{\partial c}\right)^2\frac{r^2 dr \sin\theta d\theta d\omega}{r}, \tag{1.47}$$

where spherical polar (r, θ, ω) coordinates are used. For the circular crack, all sections as Figure 1.15 are equivalent on variation in ω, such that the integral over ω devolves to a multiplicative factor of 2π. Similarly, consistent with the assumption leading to Eq. (1.44), the crack tip field is taken to be geometrically invariant with respect to crack length c. Hence, the integral over θ also devolves to a multiplicative factor. As the field is geometrically invariant, all dimensions scale with crack length, $r \sim c$, such that the overall integral becomes

$$U_K \sim \frac{v^2}{v_L^2}\mathcal{G}\int c dc \sim \frac{v^2}{v_L^2}\mathcal{G}c^2. \tag{1.48}$$

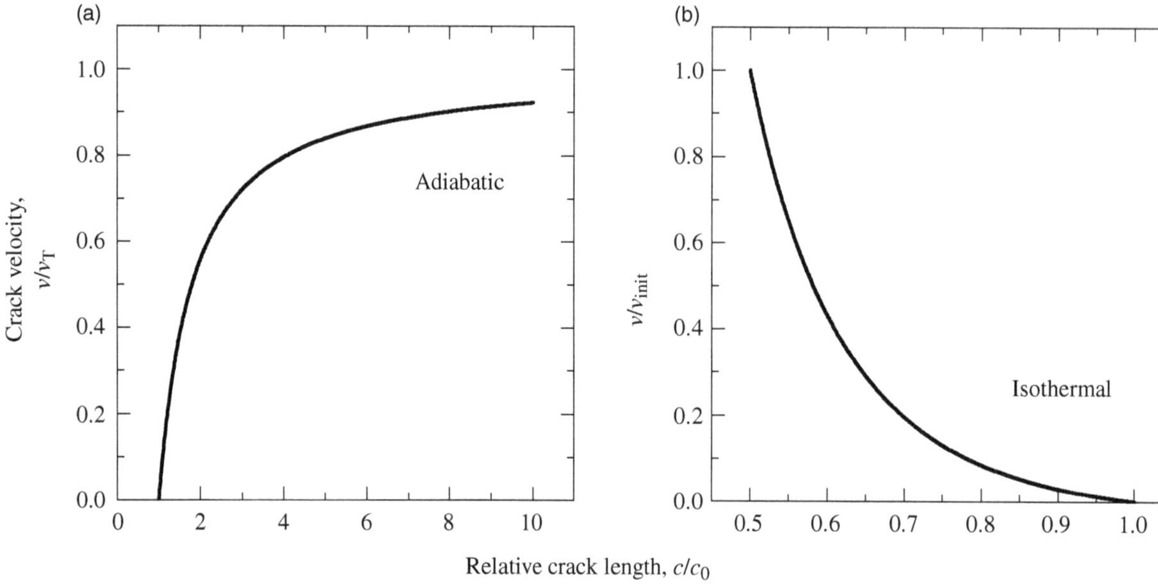

Figure 1.16 Plots of crack velocity v versus crack length c for (a) unstable adiabatic fracture and (b) stable isothermal fracture. Note in (a) that the velocity approaches a dynamic upper bound at large crack lengths $v \rightarrow v_T$ and in (b) the crack length approaches a static upper bound at zero velocity $c \rightarrow c_0$.

For a uniformly applied stress leading to the Griffith condition, $\mathscr{G} \sim \sigma^2 c/E$. Combining Eq. (1.48) with Eq. (1.41) gives an expression for the variation of crack velocity with crack length in a uniformly stressed component containing a propagating circular crack:

$$v \sim v_L(1 - c_0/c)[(2 + c_0/c)/2]^{1/2}. \tag{1.49}$$

As might be anticipated, an almost identical expression for a straight crack was first given by Mott (1948) and more recently by Lawn (1993), using similar, but less formally stated, geometrical similarity principles. Figure 1.16a shows a plot of crack velocity versus crack length using Eq. (1.49). Velocity increases rapidly at first before approaching a limiting plateau velocity v_L for extended cracks. Representative values are $v_L \approx 1\ \mathrm{km\ s^{-1}}$ and $c_0 \approx 100\ \mu\mathrm{m}$, such that when the crack has reached about 1 mm in length (10 c_0), the crack velocity is about 900 m s^{-1}. This is in accordance with experience—brittle components break extremely rapidly, almost instantaneously in human perception.

Comparison shows that Eq. (1.48) has the same form as Eq. (1.12) modified by a v^2 term, consistent with the scaling laws used earlier. The advantage of the approach taken here is that the development of Eq. (1.48) and its straight crack analog make direct connection to linear elastic fracture mechanics. They are general in the sense of not assuming particular forms for $\mathscr{G}(c)$. Hence, different forms of $\mathscr{G}(c)$ and thus different forms of U_T and U_K, Eqs. (1.39)–(1.41), can be assumed, leading to crack velocity behavior different from Eq. (1.49). This procedure will be followed in Chapter 2 in considering indentation crack initiation.

1.6.2 Stable Isothermal Fracture

Analysis of isothermal fracture begins with a restatement of Eq. (1.1) as the full First Law of Thermodynamics. The restatement allows for isothermal paths, including heating, between equilibrium states of the cracked bar in Figure 1.7,

$$\Delta U = Q + W. \tag{1.50}$$

ΔU is the change in internal energy of the bar and W is the work performed on the bar, as stated previously. An added term Q is the heat transferred to the bar. Also as stated previously, the internal energy change on fracture consists of two elements, Eq. (1.2): the change in elastic energy ΔU_E and the change in surface energy ΔU_S. The Second Law of Thermodynamics for an isothermal change in state of the bar, in thermal contact with an ideal heat reservoir, is

$$\Delta S = \frac{Q}{\theta} + \Delta S_{irr}, \tag{1.51}$$

where θ is the absolute temperature of the reservoir and ΔS is the change in the total entropy of the bar. This entropy change is composed of two parts: a reversible component Q/θ associated with heating of the bar by the reservoir (the change in entropy of the reservoir is $-Q/\theta$) and an irreversible component $\Delta S_{\text{irr}} \geq 0$, where the equality applies for reversible changes (Mase 1970). The combined statement of the First and Second Laws is then

$$\Delta U_{\text{E}} + \Delta U_{\text{S}} = \theta \Delta S - \theta \Delta S_{\text{irr}} + W.$$

For an infinitesimal change in state of the bar, this combined statement becomes

$$\text{d}U_{\text{E}} + \text{d}U_{\text{S}} = \theta \text{d}S - \theta \text{d}S_{\text{irr}} + F\text{d}L, \tag{1.52}$$

in which the infinitesimal work increment performed on the bar by the applied loading mechanism is generalized from $-\text{d}U_{\text{A}}$, Eq. (1.4), to $F\text{d}L$. For a reversible infinitesimal change of state $\text{d}S_{\text{irr}} = 0$ and Eq. (1.52) can be written

$$\text{d}U_{\text{E}} - F\text{d}L = \theta \text{d}S - \text{d}U_{\text{S}}, \tag{1.53}$$

as a generalization of Eq. (1.5). The state variables of applied force and length, F and L, are *external variables* or work variables of the bar system as they are *measurable* and *controllable* and are identified explicitly with a work term in the statement of the First Law. The state variable of the crack area A is an *internal variable* of the bar system that is *observable* and *measurable*, but *not* controllable and not associated explicitly with a work term. All the variables in Eq. (1.52) (except invariant θ) are implicit functions of A, and the irreversible evolution of A under isothermal conditions is the focus here. The theory and application of internal variables in irreversible and relaxation processes are discussed in detail in Maugin (1999) and briefly elsewhere (Haase 1969; de Groot and Mazur 1984).

The energy terms discussed earlier can be used to extend the configurational forces defined in Section 1.3.4 to isothermal conditions. A generalized infinitesimal change in mechanical energy is given by

$$\text{d}U_{\text{M}} = \text{d}U_{\text{E}} - F\text{d}L, \tag{1.54}$$

and thus a generalized mechanical energy release rate is given by

$$\mathcal{G} \equiv -\frac{\text{d}U_{\text{M}}}{\text{d}A} = -\left(\frac{\partial U_{\text{E}}}{\partial A}\right)_L. \tag{1.55}$$

The *complementary elastic energy* U_{E}^* is defined by

$$U_{\text{E}}^* + U_{\text{E}} = FL \tag{1.56}$$

and thus the infinitesimal change in mechanical energy is also given by

$$\text{d}U_{\text{M}} = L\text{d}F - \text{d}U_{\text{E}}^* \tag{1.57}$$

and hence

$$\mathcal{G} = \left(\frac{\partial U_{\text{E}}^*}{\partial A}\right)_F. \tag{1.58}$$

[For linear elastic systems $U_{\text{E}} = U_{\text{E}}^*$ and thus Eqs. (1.55) and (1.58) are often confused.] A generalized equilibrium fracture resistance is given by

$$R \equiv \frac{\text{d}U_{\text{S}}}{\text{d}A} - \theta\frac{\text{d}S}{\text{d}A}, \tag{1.59}$$

making clear that both surface and entropic effects resist fracture. The fracture equilibrium statement $\mathcal{G} = R$, Eq. (1.19), thus remains, but now based on a broader statement of thermodynamic equilibrium encompassing the First and Second Laws, Eq. (1.53), and including temperature. [A common and intuitive form for U_{S} and S on fracture is simple lumped proportionality to A: $U_{\text{S}} - \theta S = 2\gamma A$, where $\gamma = \gamma(\theta)$ is temperature dependent.]

Constraints on nonequilibrium, $\mathcal{G} \neq R$, crack velocities under isothermal conditions can be developed using the parameters discussed earlier following the analysis of Rice (1978) in consideration of the loaded bar. Development begins with the definition of the appropriate Gibbs state function Φ_{G} (Adkins 1983),

$$\Phi_{\text{G}} = U - FL - \theta S. \tag{1.60}$$

Forming the differential $\text{d}\Phi_{\text{G}}$ from this equation and using Eq. (1.52) gives

$$\text{d}\Phi_{\text{G}} = -L\text{d}F - S\text{d}\theta - \theta \text{d}S_{\text{irr}}. \tag{1.61}$$

For $dF = d\theta = 0$, Eq. (1.61) gives $d\Phi_G = -\theta dS_{irr}$, implying that the Gibbs function for the bar is minimized at equilibrium under isotonic (fixed force) isothermal conditions, similar to that for other fluid and solid systems (Adkins 1983). In this case, minimization of Φ_G takes place through nonequilibrium irreversible crack motion quantified by changes in the internal variable A. For $dA = d\theta = 0$, Eq. (1.61) gives

$$L = -\left(\frac{\partial \Phi_G}{\partial F}\right)_{\theta,A} \tag{1.62}$$

as the nonequilibrium variation in the length of the bar under fixed fracture and isothermal conditions. In this case, the crack is metastably trapped at a fixed nonequilibrium area and there is no crack motion or change in the internal variable A as the length of the bar L changes.

The time derivative of Eq. (1.60) gives

$$\dot{\Phi}_G = \dot{U} - \dot{F}L - F\dot{L} - \theta\dot{S}, \tag{1.63}$$

(where $\dot{x}$ indicates time derivative of x). Time derivatives of Eqs. (1.50) and (1.51) give

$$\dot{U} = \dot{Q} + F\dot{L}, \tag{1.64}$$

$$\dot{S} = \frac{\dot{Q}}{\theta} + \Lambda, \tag{1.65}$$

where

$$\Lambda = \dot{S}_{irr} \tag{1.66}$$

is the *entropy production rate* associated with irreversible crack motion. Incorporating the time dependence of Φ_G through the dependence on U, L, and S, by combining Eqs. (1.63)–(1.66), gives

$$\Lambda\theta = -\dot{\Phi}_G - \dot{F}L, \tag{1.67}$$

leaving only the time dependence on A and F (recognizing that θ is invariant). Thus, rewriting Eq. (1.67) using the chain rule gives

$$\Lambda\theta = -\left(\frac{\partial \Phi_G}{\partial A}\right)_{\theta,F}\dot{A} - \left(\frac{\partial \Phi_G}{\partial F}\right)_{\theta,A}\dot{F} - \dot{F}L \tag{1.68}$$

recognizing that the last two terms on the right side cancel. Using Eqs. (1.2) and (1.56) in Eq. (1.60) gives $\Phi_G = -U_E^* + U_S - \theta S$. The reduced Eq. (1.68) is thus expressed entirely in terms of crack area dependencies:

$$\Lambda\theta = \left[\left(\frac{\partial U_E^*}{\partial A}\right)_{\theta,F} - \left(\frac{\partial U_S}{\partial A}\right)_{\theta,F} + \theta\left(\frac{\partial S}{\partial A}\right)_{\theta,F}\right]\dot{A}. \tag{1.69}$$

Using the definitions of $\mathcal{G}$ and R from Eqs. (1.58) and (1.59) thus gives the isothermal equation of motion of the crack:

$$\Lambda\theta = (\mathcal{G} - R)\dot{A}. \tag{1.70}$$

An identical result was originally derived in a similar fashion by Rice (1978) using a Helmholtz state function $\Phi_H = U - \theta S$ appropriate to a bar under fixed-length isothermal constraints. Helmholtz function approaches were also used by Rice and Chuang (1981) to arrive at a similar constraint for isothermal cavity growth in a stressed grain boundary system and by Khonsari and Amiri (2013) to describe damage and crack growth in cyclically stressed fatigue systems.

As both Λ and θ are ≥ 0, the product $(\mathcal{G} - R)\dot{A}$ also must be ≥ 0 and thus the quantities $(\mathcal{G} - R)$ and $\dot{A}$ must both have the same sign or must both be identically zero. When $\mathcal{G} = R$, the reversible condition discussed earlier is obtained and the fracture system is in equilibrium with $\dot{A} = 0$ and $\Lambda = 0$. That is, the crack is at rest with zero velocity and irreversible entropy production in the body is zero. When $\mathcal{G} > R$, the crack driving force exceeds the fracture resistance, the fracture system is not in equilibrium and the crack propagates with $\dot{A} \geq 0$ and $\Lambda \geq 0$. That is, the crack propagates forward with positive velocity and irreversible entropy production is positive. In this context, the equilibrium value R is known as the *crack velocity threshold*. (Kinetic limitations on crack velocity behavior for crack driving forces greater than the threshold are considered in Chapter 6.) When $\mathcal{G} < R$, the crack driving force is less than the fracture resistance (and may even be negative), the fracture system is not in equilibrium and the crack closes with $\dot{A} \leq 0$ and $\Lambda \geq 0$. That is, the crack "heals" with negative velocity, although irreversible entropy production is still positive. (In these cases, kinetic limitations on the crack velocity are quite common; it is very difficult to get cracks to heal.) The explicit appearance of θ in Eq. (1.70) strongly suggests that if the entropy production rate is dominated by configurational entropy changes (e.g., by significant fracture surface atomic rearrangements) rather than thermal entropy changes (e.g., by increased fracture surface atomic vibrations), then the crack velocity will have a strong temperature dependence. Hence, the equation discussed earlier contains three important elements concerning crack motion: the equilibrium result is recovered as the crack velocity threshold, the crack velocity is determined by the signed difference between the configurational forces on the crack—the separation from equilibrium, and temperature is expected to be a dominant factor in determining crack velocity.

The crack velocity threshold in Eq. (1.70) is one of the most important factors determining isothermal brittle fracture behavior. Fracture behavior for systems close to the threshold is now considered. The pervasive application of Eq. (1.70) is for systems in which newly formed fracture surfaces react chemically with species in the environment. This is very common for brittle materials and is typified by oxide surfaces that react with water molecules in moist air gaseous environments or with liquid water. In simple brittle materials not influenced by material toughening (e.g., silicate glasses), R and Λ in Eq. (1.70) are determined solely by the energetics and kinetics of such surface chemical reactions. In these cases $R = 2\gamma$ is invariant and the surface energy γ reflects the thermodynamics of the reaction or *adsorption* of reactive species from the environment onto the surface. Briefly, the chemical potential μ of reactive species in a gaseous environment is given by (Adkins 1983)

$$\mu = \mu_0 + k_B\theta \ln p, \tag{1.71}$$

where p is the partial pressure of the reactive species in the environment, k_B is Boltzmann's constant, and μ_0 is the chemical potential for a pure phase of the reactive species. μ is the energy required to add another reactive molecule to a system and has units of energy/molecule. For environments consisting almost entirely of reactive species (e.g., very humid air), $p \to 1$ and $\mu \to \mu_0$. For inert environments in which reactive species are very sparse (e.g., dry N_2 gas), $p \to 0$ and $\mu \to -\infty$. A freshly formed fracture surface (e.g., a surface formed by rapid fracture as in Section 1.6.1) effectively has $p \approx 0$ and thus $\mu \ll \mu_0$. Chemical equilibrium is defined by spatially uniform μ, implying that a surface must react with or adsorb environmental species to establish the equilibrium surface state. Adsorption is quantified by Γ and has units of molecules/area. The surface energy γ is thus given by (Rice 1978)

$$\gamma = \gamma_0 - \int_{-\infty}^{\mu} \Gamma(\mu')\mathrm{d}\mu', \tag{1.72}$$

where μ' is a dummy variable. γ_0 is the upper bound surface energy in an inert environment ($\Gamma \to 0$), and it is recognized that γ is an equilibrium quantity. Hence, if insufficient time is given for adsorption or reaction such that reactive species coverage is depressed from Γ, the integral in Eq. (1.72) is depressed and the surface energy is elevated to γ', where $\gamma < \gamma' < \gamma_0$.

The simplest expression yielding $\Lambda > 0$ for all mechanical perturbations of an isothermal fracture system from equilibrium is quadratic:

$$\Lambda \sim (\mathcal{G} - 2\gamma)^2. \tag{1.73}$$

Equation (1.70) thus implies that in this case the areal crack expansion rate $\dot{A}$ is linear in the perturbation:

$$\dot{A} \sim (\mathcal{G} - 2\gamma), \tag{1.74}$$

noting that Eq. (1.73) could have started with $\Lambda \sim \dot{A}^2$. A commonly encountered perturbed isothermal fracture system for which these conditions apply is one in which a crack slowly extends from an initial length to a fixed equilibrium length. In such systems $\mathcal{G}$ is stabilizing. An example includes contact-induced penny-like cracks that are subsequently perturbed by greater contact forces. For the center-loaded penny crack system of Section 1.4, the equilibrium crack length c_0 (at the greater force F) is given by the Roesler equation, Eq. (1.27). The initial condition is crack length $c_{init} < c_0$, such that the initial perturbation is $\mathcal{G} > 2\gamma$. The initial crack velocity is $v_{init} > 0$, noting that $\dot{A} \sim vc$. Hence, for the initial perturbation

$$v_{init}c_{init} \sim \left(\frac{c_0^3}{c_{init}^3} - 1\right) \tag{1.75}$$

and for an arbitrary perturbation

$$vc \sim \left(\frac{c_0^3}{c^3} - 1\right). \tag{1.76}$$

The crack velocity v as a function of crack length c relative to the equilibrium crack length c_0 is thus given by

$$v \sim v' \frac{c_0}{c}\left(\frac{c_0^3}{c^3} - 1\right), \tag{1.77}$$

where v' depends on the selected calibration conditions. Figure 1.16b shows a plot of crack velocity versus crack length using Eq. (1.77), setting calibration at $c_{init}/c_0 = 0.5$. Velocity decreases rapidly at first before approaching zero velocity more slowly at a doubled crack length. Representative values are $v_{init} \approx 1000\,\mu m\,s^{-1}$ and $c_0 \approx 50\,\mu m$, such that when the crack has reached about 90 μm in length (1.8 c_{init}), the crack velocity is about $3\,\mu m\,s^{-1}$. This is in accordance with experience—crack *generation* in brittle components is extremely slow, almost imperceptible to humans.

Comparison of the energy-based equations of motion for fast adiabatic motion in Section 1.6.1, Eq. (1.48), essentially $U_K \sim \mathscr{G}\dot{A}^2$, and slow isothermal fracture here, Eq. (1.70), $\Lambda\theta = (\mathscr{G} - R)\dot{A}$, highlights the differences in the two phenomena. In fast fracture, work performed by the applied loading characterized by $\mathscr{G}$ is partially transformed into kinetic energy within the component. Temperature is of little relevance. For unstable loading, the fraction of kinetic energy in the component increases as the crack extends until the crack approaches a terminal velocity. In slow fracture, work performed by the applied loading is partially transformed into entropy within the component. Temperature appears directly in the equation of motion. For stable loading, the fraction of entropy in the component decreases until the crack approaches a terminal length. The differences in behavior are clear in Eqs. (1.49) and (1.77) and Figure 1.16.

1.7 Fracture by Spatially Varying Loading

Crack motion in Section 1.6 was largely analyzed from an energy viewpoint of fracture. The analyses are simply translated into a SIF-based stress viewpoint through the use of the Irwin relation, Eq. (1.29), such that $\mathscr{G} \to K^2/E$ and $2\gamma \to T^2/E$. Some care must be taken in the translation that K represents the total SIF acting on a crack, summing all sources of loading, positive and negative, and that T is restricted to surface energy effects, bond breaking and chemical reactions alone, in a crack tip region. If these conditions are met, both adiabatic and isothermal fracture are simply described by consideration of $K(c)$ variations relative to T. That procedure will be followed in Chapter 6 in considering indentation crack propagation.

This section considers an example of the SIF framework that is relevant to indentation crack *initiation*. The example illustrates the additive nature of the SIF and equilibrium, nonequilibrium, stable, and unstable fracture. In particular, the section considers the fracture mechanics of some common flaw characteristics in brittle materials, those of a crack associated with a misfitting expanding inclusion and fracture behavior in a spatially varying loading field. In this example, the SIF of the inclusion stress field is constructed through the common technique of integration using a weighting function derived from a simpler field. The resulting inclusion SIF field consists of a destabilizing-to-stabilizing transition that describes crack initiation at inclusions and indentations. The transition demarcates a boundary between unstable adiabatic fracture and stable isothermal fracture.

1.7.1 Stress-Intensity Factor for Fracture at an Inclusion

Analysis of fracture at an expanding inclusion begins with consideration of a simple geometry that is an extension of Figure 1.11. A schematic diagram is shown in Figure 1.17a. The fracture geometry consists of a circular crack, radius c, in an infinite body. Tensile forces F are applied on the opposing crack faces along circles of radius a centered on the crack. The forces are applied as uniform line forces, F_L (forces per line length), such that here $F_L = F/2\pi a$. The SIF for this crack is (Sih 1973; Tada et al. 1973; Murakami 1987)

$$K = \frac{2F_L}{(\pi c)^{1/2}} \frac{a}{(c^2 - a^2)^{1/2}}. \tag{1.78}$$

Cracks very large relative to the radius of the loaded circles are given by $c \gg a$. Taking this limit and replacing F_L using the expression in F gives Eq. (1.35) with the ideal value of χ, $K = F/(\pi c)^{3/2}$. For large cracks, the details of the loading configuration of F are unimportant and the fracture system behaves as a center-loaded penny crack.

A fracture configuration related to Figure 1.17a is that of a crack loaded by a uniform tensile stress σ_a over disks of radius a, as shown in the schematic diagram of Figure 1.17b. The SIF for this configuration is obtained by recognizing

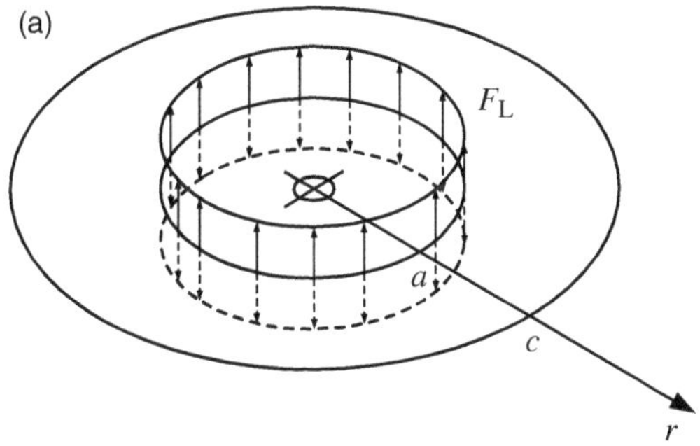

Figure 1.17 Schematic diagrams of loaded circular cracks, radius c. (a) Crack loaded by a line force F_L on a circle radius a. (b) Crack loaded by uniform tensile stress σ_a over a disk radius a.

that the uniform stress can be recast as a series of line forces $dF_a = \sigma_a dr$ over the radial domain $0 \leq r \leq a$, where dr is an infinitesimal line width. An infinitesimal contribution dK to the disk loaded crack is then, from Eq. (1.78),

$$dK = \frac{2\sigma_a dr}{(\pi c)^{1/2}} \frac{r}{(c^2 - r^2)^{1/2}}. \tag{1.79}$$

The total SIF arising from the uniformly loaded disks is then given by a standard integral

$$\begin{aligned} K = \int_0^a dK &= \frac{2\sigma_a}{(\pi c)^{1/2}} \int_0^a \frac{r dr}{(c^2 - r^2)^{1/2}} \\ &= \frac{2\sigma_a}{(\pi c)^{1/2}} \left[c - (c^2 - a^2)^{1/2}\right], \end{aligned} \tag{1.80}$$

in agreement with known SIF solutions (Sih 1973; Tada et al. 1973; Murakami 1987). For the condition $a = c$, the crack is completely uniformly stressed and Eq. (1.80) devolves to Eq. (1.34) with the ideal value of ψ for a circular crack, $K = (2/\pi^{1/2})\sigma_a c^{1/2}$. For cracks very large relative to the loaded disk, $c \gg a$. Taking this limit and replacing σ_a with $F = \sigma_a \pi a^2$ again gives Eq. (1.35) characterizing a center-loaded penny crack with the ideal value of χ, $K = F/(\pi c)^{3/2}$.

Figure 1.18a shows a schematic diagram of a spherical inclusion, radius a in an infinite body, surrounded by an annular crack of radial width c. The combined inclusion + crack system constitutes a flaw with an extent given by a radius $e = a + c$. In this case, the crack dimension c is not measured from the center of the flaw but from the edge of the inclusion, as in experimental observations. The material within the inclusion has undergone a volume increase relative to the matrix material, from the effects of either phase transformation or coefficient of thermal expansion (CTE) difference. Such an increase is constrained by the elastic matrix, leading to a positive hydrostatic pressure, p_a, compressive stress, within the

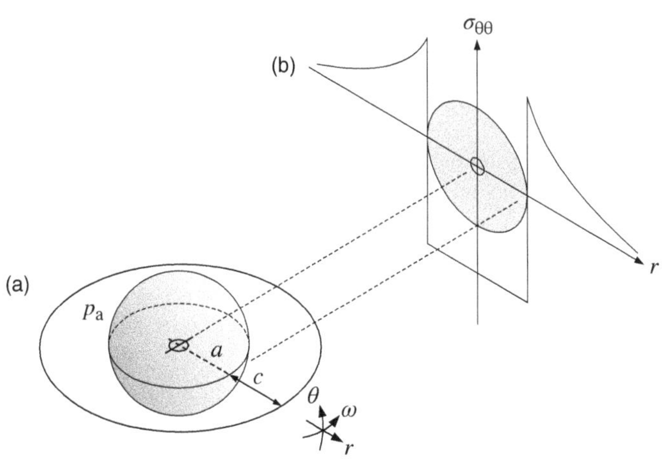

Figure 1.18 (a) Schematic diagram of an annular crack, radial width c, loaded by a misfitting spherical inclusion, radius a, with internal hydrostatic pressure p_a. Note the spherical polar coordinate system (r, θ, ω). (b) The variation of the circumferential stress perpendicular to the crack plane, $\sigma_{\theta\theta}$, with radial coordinate, r, through the center of the inclusion and crack.

inclusion. The expression for the stress developed in the matrix, exterior to the inclusion, is well known (e.g., Selsing 1961), and given in terms of radial coordinate r as

$$\sigma_{rr} = -2\sigma_{\theta\theta} = -2\sigma_{\omega\omega} = -p_a\left(\frac{a}{r}\right)^3, \quad r > a, \tag{1.81}$$

using the spherical polar coordinate system of Figure 1.18a. The radial stress component $\sigma_{rr} < 0$ is compressive and the circumferential stress components $\sigma_{\theta\theta} = \sigma_{\omega\omega} > 0$ are half the magnitude and tensile. $\sigma_{\theta\theta}$ acts perpendicular to the crack plane and is the stress component of interest here, as illustrated in Figure 1.18b. The SIF acting on the crack at the misfitting inclusion is given here by a generalization of Eq. (1.80) for a radially varying stress field. Combining Eq. (1.81) into the integral of Eq. (1.80) and adjusting the limits gives

$$\begin{aligned} K &= \frac{2}{(\pi e)^{1/2}} \int_a^e \frac{p_a}{2}\frac{a^3}{r^3}\frac{r\mathrm{d}r}{(e^2 - r^2)^{1/2}} \\ &= \frac{p_a a^3}{(\pi e)^{1/2}} \int_a^e \frac{\mathrm{d}r}{r^2(e^2 - r^2)^{1/2}}. \end{aligned} \tag{1.82}$$

It is noted that although the expression encompasses the entire flaw dimension e, the integral extends over only the stressed crack dimension c. A similar approximation has been used for cracks at indentation deformation zones (Lawn and Evans 1977), pores (Green 1980), and transforming particles (Green 1981; Swain 1981). Equation (1.82) is also a standard integral and the SIF $K(c)$ for a crack at the inclusion is given by

$$K(c) = \frac{p_a a^{1/2}}{\pi^{1/2}}\left[\frac{(1+\zeta)^2 - 1}{(1+\zeta)^5}\right]^{1/2}, \quad \zeta = c/a, \tag{1.83}$$

noting the definition of c for this geometry. A similar equation was obtained by similar means by Green (1981). The symmetry assumptions and forms of related equations are discussed in Cook (2023). In the limit of a crack large relative to the inclusion, $c \gg a$ and thus $\zeta \gg 1$. In this case, Eq. (1.83) approaches the response of the SIF for a center-loaded penny crack, $K(c) = F/(\pi c)^{3/2}$ with $F = p_a \pi a^2$. The stress field exterior to the inclusion, although distributed, is concentrated by the cubic radial dependence such that the distribution is not perceived by large cracks. In the limit of a crack small relative to the inclusion, $c \ll a$ and thus $\zeta \ll 1$. In this case, Eq. (1.83) approaches the response of the SIF for a uniformly stressed crack, $K(c) = (2/\pi)^{1/2} p_a c^{1/2}$. The amplitude term reflects the geometry of the system and is not identical to that for either a penny crack or a straight crack uniformly stressed by p_a. (The maximum tensile stress experienced by the annular crack is $p_a/2$, and the crack does not extend to the center of the flaw.)

Figure 1.19 is a logarithmic plot of the $K(c)$ response of Eq. (1.83), shown as the bold solid line. The asymptotic behaviors for short $K \sim c^{1/2}$ (slope 1/2) and long $K \sim c^{-3/2}$ (slope −3/2) cracks discussed earlier are shown as dashed lines. It is straightforward to show that the SIF passes through a maximum between these two asymptotes at $c^*/a = (5/3)^{1/2} - 1 \approx 0.29$, shown as the central vertical line. It is clear from Figure 1.19 that this line separates destabilizing, $\mathrm{d}K/\mathrm{d}c > 0$, behavior at short crack lengths from stabilizing, $\mathrm{d}K/\mathrm{d}c < 0$, behavior at long crack lengths, and that the maximum represents a neutral transition point. The stability behavior in terms of crack initiation at inclusions is discussed in Section 1.7.2.

1.7.2 Crack Initiation

The equilibrium of a fracture system is given by $K = T$. The matrix toughness T for the inclusion + crack system of Figure 1.18 was set as $T = 0.7\ \mathrm{MPa\ m^{1/2}}$ (typical of silicate glass) indicated by the horizontal line in Figure 1.19. The $K(c)$ variation was set so as to generate two equilibrium crack lengths, c_1 (left, $c_1 < c^*$) and c_0 (right, $c_0 > c^*$) indicated by the vertical lines in Figure 1.19. The existence and magnitudes of the equilibrium crack lengths were set by the composite amplitude term combining inclusion pressure and size, $p_a(a/\pi)^{1/2}$, from Eq. (1.83). The size of the inclusion was set as $a = 100\ \mu\mathrm{m}$, and the crack length $c = c_1 = 1\ \mu\mathrm{m}$ was taken to represent a small crack nucleus on the inclusion periphery. The inclusion was taken to undergo a volume increase relative to the material matrix, characterized by pressure p_a and generating the SIF field as described earlier. The pressure $p_a \approx 0.9\ \mathrm{GPa}$ was set such that the c_1 crack nucleus was in a state of unstable equilibrium. Note that this pressure in a material of elastic modulus $M \approx 70\ \mathrm{GPa}$ (typical of silicate glass) corresponds to an inclusion volume strain of $p_a/M \approx 1.3\%$. This value is typical of phase transforming particle systems, e.g., NiS in glass (on cooling), 4% (Swain 1981), ZrO_2 in ceramics (on cooling), 5% (Muddle and Hannink 1986), and SiO_2 quartz in rocks (on heating), 5% (Deer et al. 1966). Similarly, such a strain in a system with CTE mismatch between the (less expansive) inclusion and matrix of $\Delta\alpha \approx 9 \times 10^{-6}\ \mathrm{K^{-1}}$ corresponds to a temperature decrease of $p_a/\Delta\alpha M \approx 1400\ \mathrm{K}$, both values typical

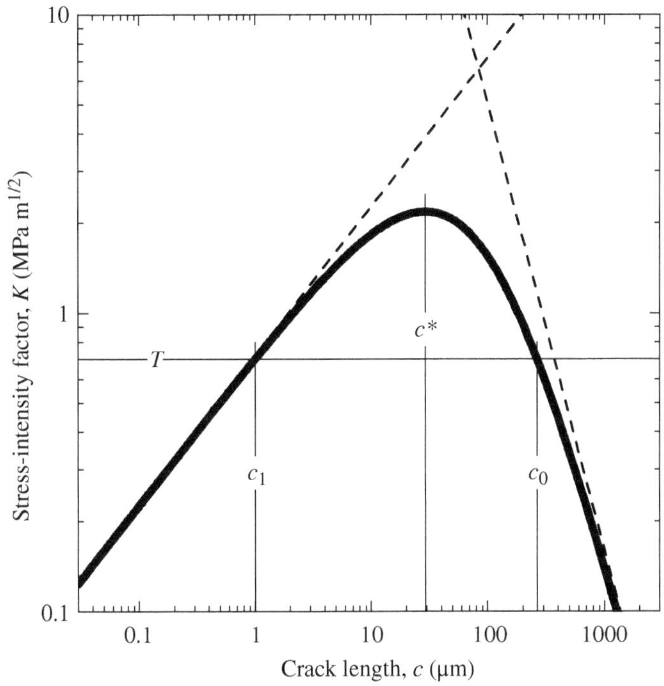

Figure 1.19 Plot of the stress-intensity factor K as a function of crack length c for the inclusion and crack system of Figure 1.18 (solid line). The asymptotic behaviors at short and long crack lengths are shown as dashed lines. The full solution passes through a maximum.

of ceramic and rock systems (Deer et al. 1966; Kingery et al. 1976). A consequence of the choice of amplitude term is that the system exhibits a state of stable equilibrium at a crack length $c = c_0 = 265\ \mu m$. The two equilibrium configurations are separated by the maximum in the SIF at $c = c^* = 29\ \mu m$.

The fracture behavior in Figure 1.19 represents *crack initiation*—unstable "Griffith"-like fracture at small crack lengths followed by a transition to stable "Roesler"-like fracture at large crack lengths. Initiation is also characterized by a transition from adiabatic behavior in an inert environment as in Section 1.6.1 to isothermal behavior in a reactive environment as in Section 1.6.2. The process is as follows. The crack nucleus c_1 is in an unstable configuration for the amplitude term selected—any perturbation that increases c also increases K and thus leads to a nonequilibrium state with positive fracture force that further increases the crack length. For such a perturbation to $K > T$ (or $\mathcal{G} > 2\gamma$), the system follows the dynamic crack velocity behavior described in Section 1.6.1. However, in this case as the crack propagates to crack lengths $c > c_1$ three additional circumstances develop. The first circumstance is that K approaches a maximum (at c^*) and the proportion of elastic energy in the system available for conversion into kinetic energy is thus decreased and the crack velocity decreases (see Lawn 1993 for an analysis of this phenomenon). The second circumstance is that as the crack length evolves in time the fracture system transforms from an initial thermally isolated adiabatic state (U conserved) toward a final state in thermal contact and isothermal equilibrium with the environment (θ maintained). The third circumstance is that as the crack expansion rate decreases sufficient time is allowed for reaction between the newly created surface and species in the environment (such that there is a transition from γ_0 to γ). Hence, after some time, the system is better viewed as one perturbed from the stable equilibrium configuration of crack length c_0. In this case, as the crack extends K decreases and the system follows the kinetic crack velocity behavior described in Section 1.6.2.

In the crack initiation process described a distinction in size is apparent between nuclei, $c < c^*$, and cracks, $c > c^*$. In the example discussed earlier, all nuclei and cracks in the domain $c_1 \leq c \leq c_0$ initiate or propagate to crack length c_0 on increase of inclusion pressure to 0.9 GPa. If the pressure is less than this value, cracks in the domain $c^* \leq c \leq c_0$ can propagate through a sequence of stable equilibria as the pressure increases. Similarly, increasing numbers of nuclei in the domain $c_1 \leq c \leq c^*$ can initiate from a sequence of unstable equilibria as the pressure increases. In both cases c^* marks the minimum stable equilibrium crack length for the inclusion. For $K(c^*) < T$, no equilibria are possible and thus the pressure required to generate $K(c^*) = T$ is a critical threshold value for crack initiation. If the increase in pressure is rapid such that the condition $K(c^*) > T$ is applied quickly, all nuclei and cracks in the domain $c_1 \leq c \leq c_0$ are placed in nonequilibrium states and a sequence of crack "pop-in" events occurs that is dependent on the perturbation kinetics. It is apparent from Eq. (1.83) that the condition for unstable equilibrium of a crack nucleus depends only on the product $p_a a^{1/2}$. Hence, the discussion above could have been framed equally well in terms of fixed pressure and varying inclusion size, or fixed pressure and inclusion size and variations in nucleus size. Similarly, the condition for stable equilibrium of a large crack ($c \gg a$) depends only on the product $p_a a^2$ and is thus not affected by the details of initiation or the separate contributions of p_a or a^2. If p_a is invariant, then $c_0 \sim a^{4/3}$ in the limit of large cracks.

Crack initiation in brittle materials at inclusions or at large grains in polycrystals has predominantly focused on observations of the critical inclusion or grain size for cracking. Swain (1981) observed crack initiation behavior in SLG from NiS inclusions. On cooling from the glass processing temperature, NiS undergoes an expansive phase transformation leading to a thermoelastic estimate of $p_a \approx 600$ MPa. Swain observed a cracking threshold of inclusion radius $a > 30$ μm. Inclusions smaller than this exhibited no associated cracks. Swain also noted that the stable crack lengths were short relative to the initiating inclusion size and that a description as center-loaded penny cracks was not appropriate and a clear $c \sim a^{4/3}$ trend was not obvious. Todd and Derby (2004) observed crack initiation in ceramic composites of Al_2O_3 containing 20% by volume of graded sizes a of SiC inclusions. SiC exhibits a smaller CTE than Al_2O_3 and hence on cooling from the hot pressing temperature of 1700 °C, a compressive stress developed in the SiC inclusions and compensating tensile stress developed in the polycrystalline Al_2O_3 matrix. A threshold for cracking at the SiC inclusions after cooling of $a = (10 \pm 3)$ μm was observed. Direct measurements of stress in the hot-pressed composites were performed using neutron diffraction showing approximately −1.3 GPa in the inclusions and +0.4 GPa in the matrix for inclusion sizes smaller than the cracking threshold and −0.8 and +0.3 GPa in the inclusions and matrix, respectively, for larger inclusions. Cracking thresholds as a function of inclusion size were also observed by Padture and colleagues (Padture et al. 1993a,b; Lawn et al. 1993) and Watts (2010) in ceramic composites of Al_2TiO_5–Al_2O_3 and SiC–ZrB_2, respectively. In the first system, Al_2TiO_5 has a smaller CTE than Al_2O_3, and in the second, SiC has a smaller CTE than ZrB_2. In both systems, the first named constituents were incorporated into the composites in the form of duplex microstructural grains. In the first system, the Al_2TiO_5 inclusions developed a compressive stress estimated from thermoelastic constants of $p_a \approx 7.7$ GPa on cooling from the sintering temperature of approximately 1600 °C. In the second system, the SiC inclusions developed a compressive stress of approximately $p_a \approx 880$ MPa, determined by X-ray diffraction and Raman spectroscopy, on cooling from the hot pressing temperature of 1800 °C. In both studies, the threshold for matrix cracking associated with a critical inclusion size was identified by direct observation and by the clear onset of precipitous decreases in strengths of the composites. The Al_2TiO_5–Al_2O_3 materials exhibited cracking thresholds of 7–12 μm to 2.5–4 μm, decreasing with increasing Al_2TiO_5 fraction. The SiC–ZrB_2 materials exhibited a SiC inclusion size cracking threshold of approximately 11.5 μm. Detailed analyses of inclusion stress, including the effects of inclusion shape, have been made by Li and Bradt (1989a,b) consistent with these estimates.

In many polycrystalline materials, a characteristic stress σ_λ is developed at grain boundaries on cooling from the processing temperature. This stress is given by $\sigma_\lambda \sim \Delta\alpha\Delta\theta M$. Here $\Delta\alpha$ is the maximum difference in CTE values as a function of orientation within the constituent phase. $\Delta\theta$ is a characteristic change in temperature from the processing conditions, and M is a characteristic elastic modulus. Figure 1.20a shows a schematic diagram of a polyhedral grain in a polycrystal with the direction of least CTE of the grain in the z direction of the (r, θ, z) cylindrical polar coordinate system shown. A characteristic grain dimension λ is in the perpendicular r direction. Crack propagation c in the r direction is in the form of a segmented ring on grain boundary facets primarily in the $r\theta$ plane. The component of thermal expansion mismatch stress σ_λ controlling fracture is $\sigma_{zz}(r)$, shown in Figure 1.20b. The physical similarities of Figure 1.18 and Figure 1.20 are clear. Typically, the $\Delta\alpha$ values in single phases generated by anisotropy are less than those generated in multi-phase composites by inhomogeneity, 2–3 × 10^{-6} K^{-1} rather than 5–7 × 10^{-6} K^{-1}, and as a consequence the stress values σ_λ are less, ≈100s of MPa rather than ≈1 GPa, as discussed earlier. Nevertheless, the amplitude term in the SIF describing crack initiation, Eq. (1.83), can be significant and in this case varies as $\sigma_\lambda \lambda^{1/2}$. The amplitude term thus increases with increasing grain size λ, and λ is often interpreted as the size of the largest grains in a polycrystal. For a polycrystalline material with grain boundary toughness T_λ, the crack initiation condition can thus be written fully as $\Delta\alpha\Delta\theta M\lambda^{1/2} \sim T_\lambda$, a scaling relation shown by Bradt and colleagues in early consideration of crack initiation in ceramic polycrystals (Kuszyk and Bradt 1973; Cleveland and Bradt 1978) and later reiterated by Lawn (1991).

The above relation can be interpreted in several ways with regard to crack initiation in a polycrystalline material on cooling from a stress-free maximum processing temperature. The simplest interpretation is if $\Delta\alpha = 0$, as in polycrystals of cubic materials, and no cracking is possible. The next simplest is if λ is small, as in fine-grained polycrystals, and the equilibrium fracture condition is not met for any grain–grain boundary combination and no cracking is possible. However, if λ

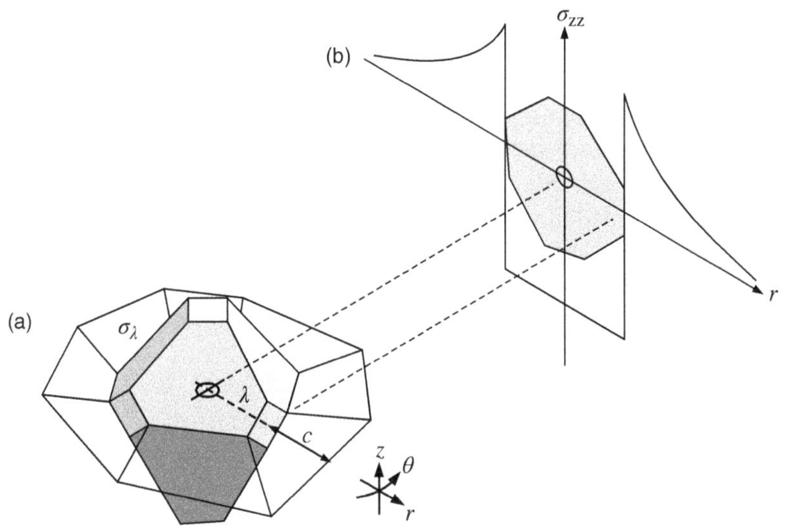

Figure 1.20 (a) Schematic diagram of a faceted intergranular crack, characteristic width c, loaded by a thermal expansion mismatched grain, characteristic size λ, with internal stress σ_λ. Note the cylindrical polar coordinate system (r, θ, z). (b) The variation of the circumferential stress perpendicular to the crack plane, σ_{zz}, with radial coordinate, r, through the center of the grain and crack.

is large enough, the equilibrium condition for crack initiation can be met and cracking is expected. Experiments on Al_2O_3 in which λ was varied were performed by Blendell and Coble (1982), in which extensive cracking was observed to develop as grain size was increased from an apparent crack-free $\lambda \approx 50\,\mu m$, to the point that the samples had no integrity and could not be handled for $\lambda > 150\,\mu m$. More detailed interpretations are possible. The grain size required to observe thermal expansion anisotropy induced cracking effects, such as decrease in elastic modulus or decrease in strength, scales as $\lambda \sim (\Delta\alpha)^{-2}$ for different materials and was demonstrated by Cleveland and Bradt (1978). The temperature difference required to observe such effects scales as $\Delta\theta \sim \lambda^{-1/2}$ for a single material and was demonstrated by Case et al. (1980) and Ohya et al. (1987) (the latter verified cracking through in situ acoustic emission). The equilibrium fracture condition can be written as $\Delta\alpha\Delta\theta M\lambda^2/c^{3/2} \sim T_\lambda$, noting that this equation applies to cracks long relative to the large-grained precursor, $c \gg \lambda$ and thus provides a lower bound for cracks in the surrounding small-grained matrix. The scaling relation for grain boundary crack length as a function of grain size is thus $c \sim \lambda^{4/3}$. The nonlinearity explains why "spontaneous microcracking" in polycrystalline anisotropic materials appears so grain size dependent.

1.8 Summary

This chapter has established the broad fundamentals of brittle fracture and developed physical and analytical foundations for Chapters 2–13. The underlying physical principle of brittle fracture was shown to be the transformation of energy: from work on a component into elastic energy within a component into surface energy associated with cracks and fragments generated by fracture of a component. The mathematical description of brittle fracture from an energy viewpoint was shown to be based on the comparison of mechanical energy, encompassing work and elastic energy, and surface energy. Configurational forces associated with the mechanical and surface energies were defined by derivatives with respect to crack area, leading to the concepts of mechanical energy release rate and fracture resistance. The equilibrium of a fracture system was then stated compactly as the equality of these configurational forces, similar to the equilibrium force balance in other physical systems.

Limiting forms of the mechanical energy release rate were shown to be related to the extremes of uniform or localized loading on a crack. The limiting forms described unstable and stable fracture, respectively, and described the common brittle fracture phenomena of component failure and material flaw generation. The two phenomena are encapsulated in equations due to Griffith (1921) and Roesler (1956). The mathematical description of brittle fracture from a stress viewpoint was shown to be based on the parameters of SIF and toughness. The equilibrium of a fracture system could then be stated compactly, as in the energy viewpoint, by the equality of these two parameters. Although the energy and stress viewpoints are equivalent, SIF and toughness are much easier to manipulate analytically in descriptions of fracture phenomena (and hence are used throughout the book).

More complete descriptions of component failure and material flaws were provided by considering crack motion and nonequilibrium fracture. The analytical framework describing fracture from the energy viewpoint was extended by a kinetic energy term to describe unstable fast fracture on component failure and by an entropy term to describe stable slow fracture on flaw generation. Flaw generation was considered in greater detail through an example of the SIF formulation that described crack initiation at a misfitting inclusion. The formulation was characterized by a unified SIF expression that combined all the fracture phenomena described in this chapter: equilibrium and nonequilibrium, stable and unstable.

The focus of this book is indentation fracture—cracking in brittle materials that is associated with indentations. Indentations are deliberate and controlled surface contacts that have long been used to assess the plastic deformation properties of metals (Tabor 1951). In brittle materials, indentations can lead to stable and unstable fracture, providing a powerful method to study and rank brittle material behavior and predict the performance of brittle components. Each of Chapters 2–13 examines a specific experimental aspect of indentation fracture, along with the required extensions to the analytical frameworks so far developed. The experimental methods take advantage of controlled indenter shape and, particularly, controlled indentation load. Examples of indentation fracture techniques are shown in this chapter in the strength degradation and cracking susceptibility plots of Figures 1.10 and 1.14, illustrating indentation shape and load control. The indentation fracture phenomena considered in each chapter are now briefly summarized in the context of analyses in Sections 1.1–1.7.

Chapters 2–6 demonstrate indentation fracture techniques in measurements of a wide range of fundamental behaviors of brittle materials.

Chapter 2 considers the simplest form of indentation fracture, that of stable indentation crack lengths. Greatest attention is given to the most commonly applied experimental condition of Vickers indentation in air and comparisons between materials of the resulting cracking susceptibility parameter B defined in Eq. (1.27) with indentation load P replacing F. The variations of the indentation geometry term χ and the crack initiation dimension c^* (Figure 1.19) with materials properties and indenter shape are considered, along with observed isothermal crack velocities v of indentation cracks. This chapter builds on Sections 1.4–1.7.

Chapter 3 considers the simplest perturbation of indentation cracks, that of measurement of inert strengths by superposition of a uniform applied stress. The resulting system is described by the addition of stabilizing and destabilizing SIF, as given in Eqs. (1.34) and (1.35), leading to a strength prediction. Greatest attention is given to the most commonly applied experimental condition of Vickers indentation in air followed by strength measurement in inert conditions, yielding the *inert strength*. Comparisons between materials are made here with variation of the contact strength parameter S, extending the meaning of the strength degradation parameter D defined in Eq. (1.16) with P replacing c. Attention is restricted to materials with invariant toughness, although variations of χ with indentation load are considered. This chapter builds on Sections 1.3 and 1.5.

Chapter 4 extends the perturbation considered in Chapter 3 to include microstructural interactions with a propagating crack and material toughening. The now three-term SIF leads to a strength expression that includes flaw tolerance and a prediction of the intrinsic strength of a material. Greatest attention is once again given to inert strengths as a function of indentation load. Comparisons between materials are made here through variations in S, the intrinsic strengths, and the implicit variations of toughness with crack length. The concept of frames of reference for fracture is introduced. This chapter builds on Sections 1.3 and 1.5.

Chapter 5 extends experimental techniques to include explicit measurement of a strength–crack length combination similar to that in Eq. (1.16). The extension enables a clear estimate of toughness through the analysis of Chapter 3 but treats indentation load as an implicit parameter rather than a key variable. The measurement philosophy is similar to that of traditional toughness measurements but has the advantage of providing fracture information at strength-relevant small crack lengths. This chapter builds on Section 1.3.

Chapter 6 extends experimental techniques to include measurements of strengths in a reactive environment, typically liquid water such that $\mu = \mu_0$ in Eq. (1.71). Strengths are measured as a function of applied stressing rate, such that the kinetics of reaction are assessed along with estimates of inert and reactive equilibrium surface energies, γ_0 and γ, Eq. (1.72). The measurements extend the study of isothermal fracture from stable, Chapter 2, to unstable. The concept of reactive strength is introduced. This chapter builds on Section 1.6.2.

Chapters 7–9 are case studies of indentation fracture application in the interpretation of three commonly observed brittle fracture phenomena: component fragmentation, edge chipping, and surface scratching.

Chapter 7 addresses fragmentation on brittle component failure, as illustrated in Figure 1.6, through detailed experimental measurements of fragment characteristics. Fragmentation is a clear implication that the invariant angular dependence $f_i(\theta)$ of the moving crack tip stress field assumed in Eq. (1.44) does not apply to all adiabatic fracture and that crack path instability is a factor. The measurements here provide experimental insights into such instability. This chapter builds on Sections 1.2 and 1.6.1. Chapter 8 addresses chipping associated with contacts close to the edges of brittle components. Detailed experimental measurements of indentation chip characteristics, especially the dependence on edge proximity, are the basis for strength analyses of chipped components. Such analyses rely on knowledge of edge variation of χ and ψ in Eqs. (1.34) and (1.35), and knowledge of chipped edge strengths are critical in the design of sawn components. This chapter builds on Section 1.4. Chapter 9 addresses scratching associated with moving sharp contacts on the surfaces of brittle components. A fracture mechanics analysis is developed that extends Section 1.4 and Chapter 2 in two ways: the contacts considered are linear not point-like and the cracks are parallel to the surface not perpendicular (and thus somewhat similar to those considered in Chapter 8). The analysis illustrates a method of SIF construction and provides insight into experimental observations of the Mohs scale used in mineralogy.

Chapters 10–12 are case studies of indentation fracture application in the development of brittle materials and processes. The materials are single crystal silicon, cordierite glass ceramic whisker-reinforced composite, polycrystalline alumina, tetragonal zirconia polycrystal (TZP), and surface stressed ion-exchanged glass.

Chapter 10 addresses the CTE issue inherent in semiconductor fabrication (Section 1.1) by applying indentation strength techniques to assess the effects of wafer backgrinding on the strength of silicon devices. The strengths are investigated experimentally and analytically. This chapter builds on Chapters 3 and 9. Chapter 11 illustrates a materials development application for a conventionally toughened material—a matrix containing whiskers that constrain fracture by exerting localized tractions across crack faces. Indentation fracture techniques are applied to assess the effects of whisker loading on strength and toughness. This chapter builds on Chapters 2–5. Chapter 12 illustrates a materials development application for toughened materials. In situ crack propagation measurements, applied in Chapters 5 and 11, are used to assess the effects of microstructural variables for crack wake, crack front, and component level toughening mechanisms. Chapter 13 summarizes the many results presented in Chapters 1–12 and illustrates the power of indentation techniques in comparisons of materials fracture properties in both inert and reactive environments. Chapter 13 also outlines some areas of future fracture research that build on the analytical and experimental methods developed in this book.

References

Adkins, C.J. (1983). *Equilibrium Thermodynamics, 3rd ed.* Cambridge.

Ashby, M.F. (1999). *Materials Selection in Mechanical Design, 2nd ed.* Butterworth-Heinemann.

Ashcroft, N.W. and Mermin, N.D. (1976). *Solid State Physics*. Holt, Rinehart and Winston.

Atkins, A.G. and Mai, Y.-W. (1985) *Elastic and Plastic Fracture*. Ellis Horwood.

Benbow, J.J. (1960). Cone cracks in fused silica. *Proceedings of the Physical Society* 75: 697–699.

Blendell, J.E. and Coble, R.L. (1982). Measurement of stress due to thermal expansion anisotropy in Al_2O_3. *Journal of the American Ceramic Society* 65: 174–178.

Broek, D. (1982) *Elementary Engineering Fracture Mechanics, 3rd ed.* Martinus Nijhoff.

Case, E.D., Smyth, J.R., and Hunter, O. (1980). Grain-size dependence of microcrack initiation in brittle materials. *Journal of Materials Science* 15: 149–153.

Cleveland, J.J. and Bradt, R.C. (1978). Grain size/microcracking relations for pseudobrookite oxides. *Journal of the American Ceramic Society* 61: 478–481.

Cook, R.F. (2019). Fracture sequences during elastic–plastic indentation of brittle materials. *Journal of Materials Research* 34: 1633–1644.

Cook, R.F. (2020). A critical evaluation of indentation crack lengths in air. *Journal of the American Ceramic Society* 103: 2278–2295.

Cook, R.F. (2023). *Particle Strengths*. Wiley.

Cook, R.F. and Pharr, G.M. (1990). Direct observation and analysis of indentation cracking in glasses and ceramics. *Journal of the American Ceramic Society* 73: 787–817.

Deer, W.A., Howie, R.A., and Zussman, J. (1966) *An Introduction to the Rock Forming Minerals*. Longman.

de Groot, S.R. and Mazur, P. (1984) *Non-Equilibrium Thermodynamics*. General.

Engel, U. and Hübner, H. (1978). Strength improvement of cemented carbides by hot isostatic pressing (HIP). *Journal of Materials Science* 13: 2003–2012.
Eshelby, J.D. (1951). The force on an elastic singularity. *Philosophical Transactions of the Royal Society of London. Series A: Mathematical and Physical Sciences* 244: 87–112.
Freund, L.B. (1990). *Dynamic Fracture Mechanics*. Cambridge.
Glaesemann, G.S., Jakus, K., and Ritter Jr, J.E. (1987). Strength variability of indented soda-lime glass. *Journal of the American Ceramic Society* 70: 441–444.
Green, D.J. (1980). Stress intensity factor estimates for annular cracks at spherical voids. *Journal of the American Ceramic Society* 63: 342–344.
Green, D.J. (1981). Stress-induced microcracking at second-phase inclusions. *Journal of the American Ceramic Society* 64: 138–141.
Griffith, A.A. (1921). The phenomena of rupture and flow in solids. *Philosophical Transactions of the Royal Society London A* 221: 163–198.
Haase, R. (1969) *Thermodynamics of Irreversible Processes*. Addison-Wesley.
Harding, D.S., Oliver, W.C., and Pharr, G.M. (1995). Cracking during nanoindentation and its use in the measurement of fracture toughness. In *MRS Proceedings Volume 356: Thin Films: Stresses and Mechanical Properties V* (Eds. S.P. Baker, P. Børgesen, and C.A. Ross) pp. 663–668. Pittsburgh, PA: Materials Research Society.
Inglis, C.E. (1913). Stress in a plate due to the presence of cracks and sharp corners. *Transactions of the Institution of Naval Architects* 55: 219–241.
Johnson, K.L. (1985). *Contact Mechanics*. Cambridge.
Kanninen, M.F. and Popelar, C.H. (1985) *Advanced Fracture Mechanics*. Oxford.
Kendall, K. (1988). Agglomerate strength. *Powder Metallurgy* 31: 28–31.
Khonsari, M.M. and Amiri, A. (2013). *Introduction to Thermodynamics of Mechanical Fatigue*. CRC.
Kingery, W.D., Bowen, H.K., and Uhlmann, D.R. (1976). *Introduction to Ceramics*. Wiley.
Kirchner, H.P. (1984). Comparison of single-point and multipoint grinding damage in glass. *Journal of the American Ceramic Society* 67: 347–353.
Kirchner, H.P. and Isaacson, E.D. (1982). Residual stresses in hot-pressed Si_3N_4 grooved by single-point grinding. *Journal of the American Ceramic Society* 65: 55–60.
Kuszyk, J.A. and Bradt, R.C. (1973). Influence of grain size on effects of thermal expansion anisotropy in $MgTi_2O_5$. *Journal of the American Ceramic Society* 56:420–423.
Lawn, B.R. (1991). Fundamental condition for the existence of microcrack clouds in monophase ceramics. *Journal of the European Ceramic Society* 7: 17–20.
Lawn, B.R. (1993). *Fracture of Brittle Solids, 2nd ed.* Cambridge.
Lawn, B.R. (1998). Indentation of ceramics with spheres: a century after Hertz. *Journal of the American Ceramic Society* 81: 1977–1994.
Lawn, B.R. and Evans, A.G. (1977). A model for crack initiation in elastic/plastic indentation fields. *Journal of Materials Science* 12: 2195–2199.
Lawn, B.R. and Fuller Jr, E.R. (1975). Equilibrium penny-like cracks in indentation fracture. *Journal of Materials Science* 10: 2016–2024.
Lawn, B.R., Lee, S.K., Peterson, I.M., and Wuttiphan, S. (1998). Model of strength degradation from Hertzian contact damage in tough ceramics. *Journal of the American Ceramic Society* 81: 1509–1520.
Lawn, B.R., Padture, N.P., Braun, L.M., and Bennison, S.J. (1993). Model for toughness curves in two-phase ceramics: I, Basic fracture mechanics. *Journal of the American Ceramic Society* 76: 2235–2240.
Lee, S.K., Wuttiphan, S., and Lawn, B.R. (1997). Role of microstructure in Hertzian contact damage in silicon nitride: I, Mechanical characterization. *Journal of the American Ceramic Society* 80: 2367–2381.
Li, Z. and Bradt, R.C. (1989a). Micromechanical stresses in SiC-reinforced Al_2O_3 composites. *Journal of the American Ceramic Society* 72: 70–77.
Li, Z. and Bradt, R.C. (1989b). Thermoelastic micromechanical stresses associated with a large α-silicon carbide grain in a polycrystalline β-silicon carbide matrix. *Journal of the American Ceramic Society* 72: 459–466.
Mase, G.E. (1970). *Continuum Mechanics*. McGraw-Hill.
Maugin, G.A. (1999). *The Thermodynamics of Nonlinear Irreversible Processes*. World Scientific.
Maugin, G.A. (2011). *Configurational Forces*. CRC.

Maugis, D. (2000). *Contact, Adhesion and Rupture of Elastic Solids*. Springer-Verlag.

Mott, N.F. (1948). Brittle fracture in mild steel plates—II. *Engineering* 165: 16–18.

Mouginot, R. and Maugis, D. (1985). Fracture indentation beneath flat and spherical punches. *Journal of Materials Science* 20: 4354–4376.

Muddle, B.C. and Hannink, R.H. (1986). Crystallography of the tetragonal to monoclinic transformation in MgO-partially-stabilized zirconia. *Journal of the American Ceramic Society* 69: 547–555.

Murakami, Y. (Ed.) (1987) *Stress Intensity Factors Handbook*. Pergamon.

Ohya, Y., Nakagawa, Z.E., and Hamano, K. (1987). Grain-boundary microcracking due to thermal expansion anisotropy in aluminum titanate ceramics. *Journal of the American Ceramic Society* 70: C-184–C-186.

Padture, N.P., Bennison, S.J., and Chan, H.M. (1993a). Flaw-tolerance and crack-resistance properties of alumina-aluminum titanate composites with tailored microstructures. *Journal of the American Ceramic Society* 76: 2312–2320.

Padture, N.P., Runyan, J.L., Bennison, S.J., Braun, L.M., and Lawn, B.R. (1993b). Model for toughness curves in two-phase ceramics: II, Microstructural variables. *Journal of the American Ceramic Society* 76: 2241–2247.

Peach, M. and Koehler, J. (1950). The forces exerted on dislocations and the stress fields produced by them. *Physical Review* 80: 436–439.

Pharr, G.M., Harding, D.S., and Oliver, W.C. (1993). Measurement of fracture toughness in thin films and small volumes using nanoindentation methods. In *Mechanical Properties and Deformation Behavior of Materials having Ultra-Fine Microstructures* (Eds. M. Nastasi, D.M. Parkin, and H. Gleiter) pp. 449–461. Netherlands: Kluwer.

Rice, J.R. (1978). Thermodynamics of the quasi-static growth of Griffith cracks. *Journal of the Mechanics and Physics of Solids* 26: 61–78.

Rice, J.R. and Chuang, T.-J. (1981). Energy variations in diffusive cavity growth. *Journal of the American Ceramic Society* 64: 46–53.

Ritchie, R.O. and Liu, D. (2021). *Introduction to Fracture Mechanics*. Elsevier.

Roesler, F.C. (1956). Brittle fractures near equilibrium. *Proceedings of the Physical Society Section B* 69: 981–992.

Sadd, M.H. (2009) *Elasticity, 2nd ed.* Elsevier.

Selsing, J. (1961). Internal stresses in ceramics. *Journal of the American Ceramic Society* 44: 419.

Shand, E.B. (1965). Strength of glass–the Griffith method revised. *Journal of the American Ceramic Society* 48: 43–49.

Sih, G.C. (1973) *Handbook of Stress-Intensity Factors*. Lehigh University.

Swain, M.V. (1981). Nickel sulphide inclusions in glass: an example of microcracking induced by a volumetric expanding phase change. *Journal of Materials Science* 16: 151–158.

Tabor, D. (1951). *The Hardness of Metals*. Oxford.

Tada, H., Paris, P.C., and Irwin, G.C. (1973). *The Stress Analysis of Cracks Handbook*. Del Research.

Timoshenko, S.P. and Goodier, J.N. (1970). *Theory of Elasticity, 3rd ed.* McGraw-Hill.

Todd, R.I. and Derby, B. (2004). Thermal stress induced microcracking in alumina–20% SiC_p composites. *Acta materialia* 52: 1621–1629.

Watts, J.L. (2011). Stress measurement and development of zirconium diboride-silicon carbide ceramics. Doctoral Dissertation. 2010. https://scholarsmine.mst.edu/doctoral dissertations/2010.

Weertman, J. and Weertman, J.R. (1992) *Elementary Dislocation Theory*. Oxford.

Wei, R.P. (2010). *Fracture Mechanics*. Cambridge.

Williams J.G. (1984) *Fracture Mechanics of Polymers*. Ellis Horwood.

Yasutake, K., Iwata, M., Yoshii, K., Umeno, M., and Kawabe, H. (1986). Crack healing and fracture strength of silicon crystals. *Journal of Materials Science* 21: 2185–2192.

Yoffe, E.H. (1982). Elastic stress fields caused by indenting brittle materials. *Philosophical Magazine A* 46: 617–628.

Zeng, K., Breder, K., and Rowcliffe, D.J. (1992a). The Hertzian stress field and formation of cone cracks—I. Theoretical approach. *Acta metallurgica et materialia* 40: 2595–2600.

Zeng, K., Breder, K., and Rowcliffe, D.J. (1992b). The Hertzian stress field and formation of cone cracks—II. Determination of fracture toughness. *Acta metallurgica et materialia* 40: 2601–2605.

2
Indentation Cracking Behavior

This chapter uses the analyses of Chapter 1 in a survey of the contact fracture behavior of brittle materials in air, primarily by Vickers indenters, but also including conical, cube corner, and Berkovich indenters. Measurements of crack lengths and deformation at Vickers indentations are one of the most common and powerful assessments of the mechanical properties of brittle materials and provide a basis for consideration of cracking at other indentation geometries and of indentation strengths. The survey provides a critical evaluation of observed behavior, including crack morphology, as functions of material type, indentation load, and indenter acuity. Materials include single crystals, polycrystals, transforming materials, glasses, multiphase materials, including cermets, glass-ceramics, and tooth enamel, and structural and electronic ceramics, including oxides, carbides, nitrides, and titanates. Analyses describing the indentation stress field, stable equilibrium crack lengths, indentation crack initiation, nonequilibrium fracture, and indenter shape effects are developed and applied in interpretations of extensive experimental data sets. Crack length data are presented in a single format for ease of interpretation in terms of idealized indentation fracture and for inter-material comparisons. A simple quantitative evaluation and ranking scheme based on indentation cracking susceptibility is demonstrated.

2.1 Sharp Indentation of Brittle Materials

Sharp indentation contacts are a pervasive and effective tool for assessing the deformation and fracture properties of brittle materials (Lawn and Wilshaw 1975; Lawn and Cook 2012; Marshall et al. 2015). The most common implementation of such contacts is the application of four-sided Vickers diamond pyramid indenters onto material surfaces in controlled loading cycles. The ensuing residual contact impressions formed in material surfaces by Vickers indentations provide a measure of material resistance to plastic deformation and yield (Tabor 1951). Cracks are also generated at Vickers indentations and, as shown throughout this book, enable many methods for measuring the fracture properties of materials. This chapter considers the dimensions of surface traces of indentation cracks and the susceptibility of brittle materials to contact fracture. Chapters 3–12 consider the effects of stress applied to indentation cracks and the large number of additional fracture properties thus measurable. This section summarizes qualitative phenomena and descriptive aspects of indentation deformation and fracture, providing background and context for the quantitative assessments of indentation cracking behavior in Sections 2.2–2.4 and for discussions of indentation strengths in Chapters 3–6.

2.1.1 Observations: Morphology

Figure 2.1 is a schematic diagram of a typical Vickers indentation sequence on a brittle material surface, showing Figure 2.1a preload, Figure 2.1b peak load, and Figure 2.1c unload. Prior to loading, a flat material surface is presented to the indenter, usually perpendicular to the indenter axis (in this case, the pyramid axis). At peak load the diamond pyramid indents or penetrates the surface and short surface traces of cracks in the material are often visible emanating from intersections of the pyramid edges with the surface. On unload a square pyramidal contact impression remains and the surface traces of cracks are visible, usually longer and more open than at peak load, emanating from the impression corners parallel to the impression diagonals. The residual impression and cracks are together regarded as forming a sharp *indentation flaw*. The peak indentation loads P are typically a few tens of newtons ($10\,\mathrm{N} \approx 2.2\,\mathrm{lb}$) leading to flaws typically a few hundred μm in extent.

Indentation Fracture: Strength and Toughness for Brittle Materials Design, First Edition. Robert F. Cook.

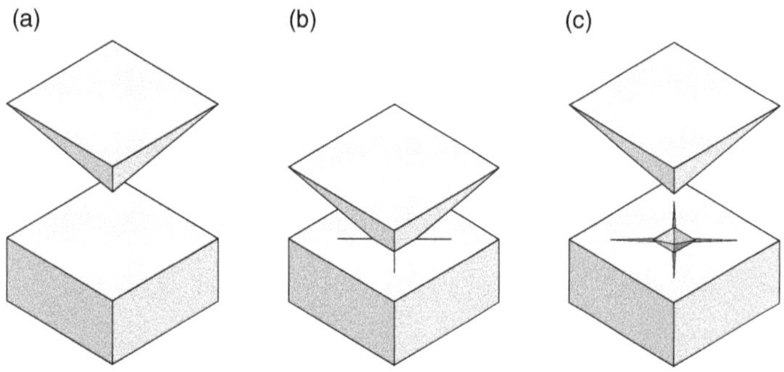

Figure 2.1 Schematic diagram of a contact sequence for a Vickers diamond pyramid indentation. (a) Preload; (b) peak load; (c) unload. A square contact impression is formed irreversibly at peak load. Small surface traces of cracks appear at peak load and extend and open on unload.

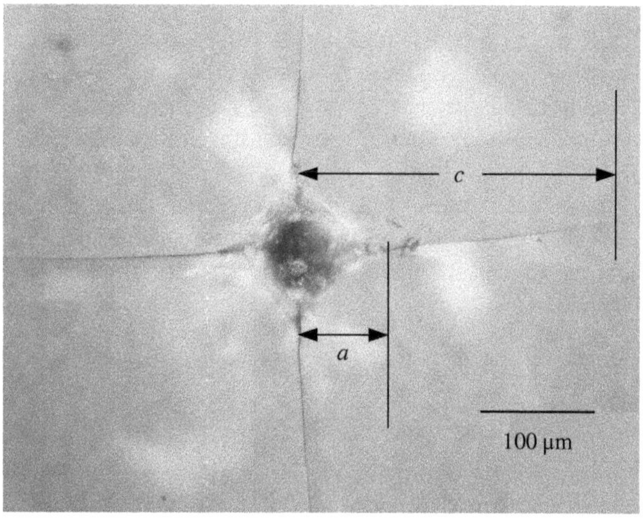

Figure 2.2 Reflected light optical image of a Vickers indentation flaw in aluminosilicate glass. Indentation load $P = 30$ N. The residual square contact impression has semidiagonal a. The surface traces of half-penny cracks emanating from the impression corners have length c (measured from the impression center). Reflections from the subsurface faces of lateral cracks emanating from the impression depth are visible as bright regions.

Figure 2.2 shows a reflected light optical image of a $P = 30$ N Vickers indentation in aluminosilicate glass (ASG, often used as the base glass in mobile device screens). A square contact impression and surface traces of cracks are clearly visible. The impression semidiagonal is dimension a. Many direct observations (Tabor 1951; Cook and Pharr 1990 and references therein) show that a is set by the peak load P and that the impression diagonal does not alter or recover on unloading. Hence, the contact area at peak load is given by $2a^2$, and the mean supported contact stress, termed the *hardness* H, is given by

$$H = P/2a^2. \tag{2.1}$$

Over broad load domains H is invariant and is thus a convenient measure of resistance of a material to plastic deformation on contact. H is determined by measurements of $a(P)$. Hardness values vary from $H \approx 0.1$ GPa for annealed copper to $H \approx 25$ GPa for sapphire. The hardness of most glasses is $H \approx 6$ GPa. The length of the surface traces of the cracks in Figure 2.2 is c, measured postindentation from the center of the impression. The surface traces represent a set of two cracks centered at the contact impression on planes perpendicular to the surface and each other. Following the Roesler scaling arguments of Chapter 1, the tendency to such indentation fracture is termed the cracking susceptibility B (Cook 2020) and is given by

$$B = cP^{-2/3}. \tag{2.2}$$

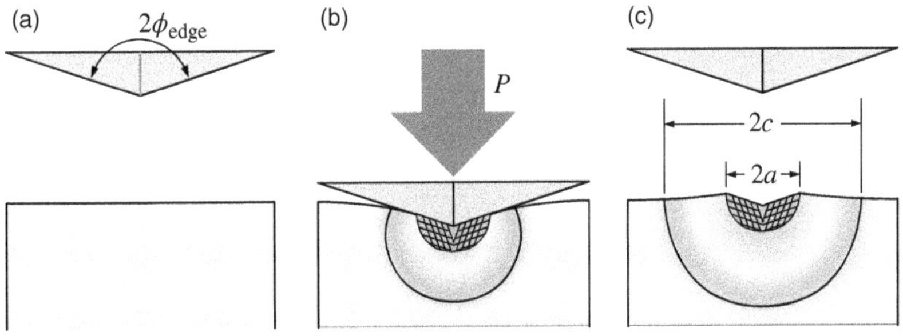

Figure 2.3 Schematic side view diagram of a contact sequence for a Vickers indenter. (a) Preload, showing the included angle between the edges of the indenter, $2\phi_{edge}$; (b) peak load P, showing the formation of a localized plastic deformation zone (hatched) and largely subsurface median crack (shaded); (c) unload, showing elastic uplift and the formation of a half-penny crack and associated surface traces. The residual contact impression diagonal is $2a$. The half-penny crack surface trace diameter is $2c$. Compare with Figure 2.1.

Over broad load domains B is invariant and is thus a convenient measure of the tendency of a material to contact fracture. B is determined by measurements of $c(P)$. Susceptibility values vary from $B \approx 3\,\mu m\,N^{-2/3}$ for tungsten carbide (WC)-based hard metals to $B \approx 24\,\mu m\,N^{-2/3}$ for lead zirconate titanate. The susceptibility of most glasses is $B \approx 20\,\mu m\,N^{-2/3}$. A physical interpretation is that B (in these units) represents the crack length formed by a $P = 1$ N indentation.

Figure 2.3 is a schematic side view diagram of the indentation process of Figure 2.1 showing deformation and fracture in more detail. The *included angle* for the indenter is specified as $2\phi_{edge}$ (Figure 2.3a). For the Vickers indenter this angle (between the indenter axis and an inclined pyramidal edge) is $\phi_{edge} = 74°$. Other angles such as the face angle ϕ_{face} and the equivalent cone angle ϕ can be specified and are discussed in detail in Section 2.2. The cone angle characterizes indentation deformation and is given by $\tan^2\phi = (\pi/2)\tan^2\phi_{edge}$ for a Vickers indenter such that $\phi \approx 70°$. During loading, a *plastic deformation zone* (pdz) of shear and compaction forms irreversibly in the material beneath the indenter. The zone is localized beneath the contact impression and is approximately hemispherical in shape. In brittle materials the pdz consists primarily of shear faults—multiple, near-parallel, shear cracks separating small blocks of material that are simultaneously compressed and displaced so as to accommodate the imposed indentation strain (of order $\cot\phi \approx 0.36$ for Vickers indenters). Extensive shear fault and pdz measurements and observations in ceramics and glasses are given in the works of Hagan, Lawn, and Dabbs (Arora et al. 1979; Hagan 1979a,b, 1980; van der Zwaag et al. 1980; Lawn et al. 1983; Hagan and van der Zwaag 1984; Lathabai et al. 1991) and also in the early indentation works on metals and rocks of Mulhearn, Samuels, and Brace (Samuels and Mulhearn 1957; Mulhearn 1959; Brace 1960). The faults are indicated schematically by hatching in Figure 2.3b and, as shown in Figure 2.3c, the pdz remains on unloading.

During an indentation cycle, material interior and exterior to the pdz also deforms elastically and reversibly. The elastic deformation is noted in direct observation experiments and, particularly, by in situ measurements of the axial load–displacement response of the indenter (Cook and Pharr 1990, 1994; Pharr and Cook 1990; Cook et al. 1994b; Thurn and Cook 2002; Thurn et al. 2002). The cyclic load–displacement responses during Vickers indentation are thus hysteretic. Such responses are typical of sharp contacts and reflect elastic + plastic deformation on loading and elastic recovery on unloading (Oliver and Pharr 1992). At peak load the surface exterior to the contact is elastically depressed (Figure 2.3b). On unload this surface rebounds and more than recovers, leading to small uplifts (Figure 2.3c) observable in topographic and strain measurements (Vaudin et al. 2011; Friedman et al. 2016). For Vickers and similar indenters the form of the load–displacement responses are not significantly influenced by the peak load or by crack formation.

Figure 2.4 highlights indentation fracture in a sequence of side view optical images of Vickers indentation in soda-lime silicate glass (SLG, the predominant material for domestic containers and windows, and microscope slides). The loading–unloading sequence is a–h. The peak load, $P = 200$ N, is shown in Figure 2.4e. Unload is shown in Figure 2.4h. Fracture is made visible by interference fringes, also shown in schematic form in Figure 2.3. At $P \approx 100$ N, a circular penny-like crack forms beneath the contact and on continued loading expands downward and outward, truncated slightly by intersection with the surface. The perpendicular crack system and other indentation damage are visible as dark contrast. On unloading, the surface traces extend outward with little associated downward crack extension and the interference fringes broaden indicating partial crack closure. Complete unload leads to the formation of a *half-penny crack*. A similar sequence was shown in SLG by Lawn and Swain (1975) (see also Lawn and Evans 1977). The postindentation geometry

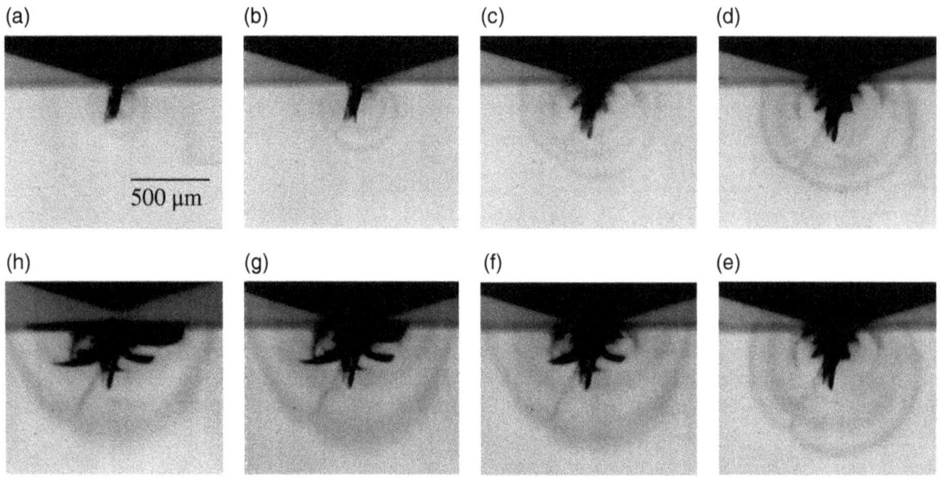

Figure 2.4 Transmitted light side view optical images of a Vickers indentation contact sequence in soda-lime silicate glass. (a), (b), (c) Loading; (d) peak load, $P = 200$ N; (e), (f), (g), (h) unloading. Median crack formation on loading and transformation to half-penny crack geometry on unloading appear as shaded fringes. Lateral crack formation on unloading appears as dark contrast. Compare with Figure 2.3.

of a half-penny crack centered on a surface localized pdz, Figure 2.4h, is the basis for consideration of most indentation fracture and is used throughout the book.

Figure 2.5 shows schematic diagrams of the half-penny crack system and other indentation cracks. The diagrams are cross sections through the diagonals of Vickers indentations and illustrate limiting crack forms. The most surface localized are the *radial cracks* (Figure 2.5a). These exhibit surface traces that extend (or "radiate") from the impression corners and typically extend downward to depths comparable to that of the pdz. Radial cracks are sometimes termed "Palmqvist" cracks after the original work on such cracks formed at Vickers indentations in WC–Co hard metal composites

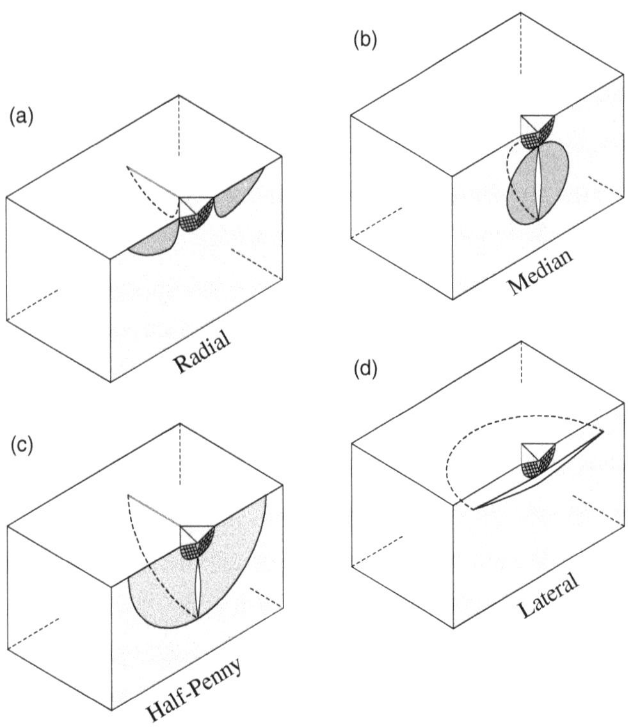

Figure 2.5 Schematic cross-section diagrams of crack morphologies generated at Vickers indentations. (a) Radial; (b) median; (c) half-penny; (d) lateral.

(Palmqvist 1957). Radial cracks are predominantly perpendicular to the surface. The least surface localized are the *median cracks* (Figure 2.5b). These exhibit circular outlines that are completely subsurface and form on the central or "median" planes beneath the pdz. Median cracks at indentations are analogous to the median vents that are formed beneath the tracks of glass-cutting rollers (Cook 1994). Median cracks are predominantly perpendicular to the surface. Hybridized forms of the radial and median cracks that exhibit both surface traces and extensions into the depth of the material are the *half-penny cracks* (Figure 2.5c). These exhibit semielliptical outlines that are centered on the pdz and approach a semicircular true "half-penny" form in many cases (e.g., Figure 2.4h). Half-penny cracks are predominantly perpendicular to the surface. It is recognized that it is nearly impossible to distinguish the radial and half-penny geometries from observations of crack surface traces in opaque materials. Distinct from the previous three crack systems are the *lateral cracks*, as these propagate predominantly parallel to the surface (Figure 2.5d). In short form, these exhibit no surface trace and extend laterally away from the indentation axis, typically at depths comparable to the pdz. In long form, lateral crack fronts turn upward and exit the material surface, exposing or removing chips of material. The terms lateral cracking and chipping are often synonymous. Short lateral cracks are visible in Figure 2.4f–h. Lateral cracks that have not exited the material surface are visible as the bright reflections in Figure 2.2 between the half-penny crack traces.

2.1.2 Observations: Materials

Vickers indentation fracture and deformation patterns on different materials provide a rich set of variations on the theme established in glass, Figure 2.2. The indentation patterns most similar to those of glass are those formed in fine-grained polycrystals. Figure 2.6 shows reflected light optical images of Vickers indentations in such materials, all with grain sizes in the range of 3–5 μm: Figure 2.6a, cordierite glass-ceramic $P = 10$ N; Figure 2.6b, Al_2O_3-TiC composite $P = 100$ N; Figure 2.6c, Y-doped tetragonal zirconia polycrystal $P = 100$ N. In all cases, the indentation edges and cracks are straight, suggesting very little microstructural perturbation of the deformation and fracture fields. The major distinction between the indentation responses of the fine polycrystals and glass is the aspect ratio of the indentation pattern. The c/a ratio of the polycrystals here is 2–3, whereas that for the ASG of Figure 2.2 is approximately 4. Combining Eqs. (2.1) and (2.2) gives $c/a \sim BH^{2/3}a^{1/3}$, showing that the aspect ratio has both material and scale dependence with the dominant influence that of material cracking susceptibility. Comparison here suggests that the fine-grained polycrystals have less cracking susceptibility than glass.

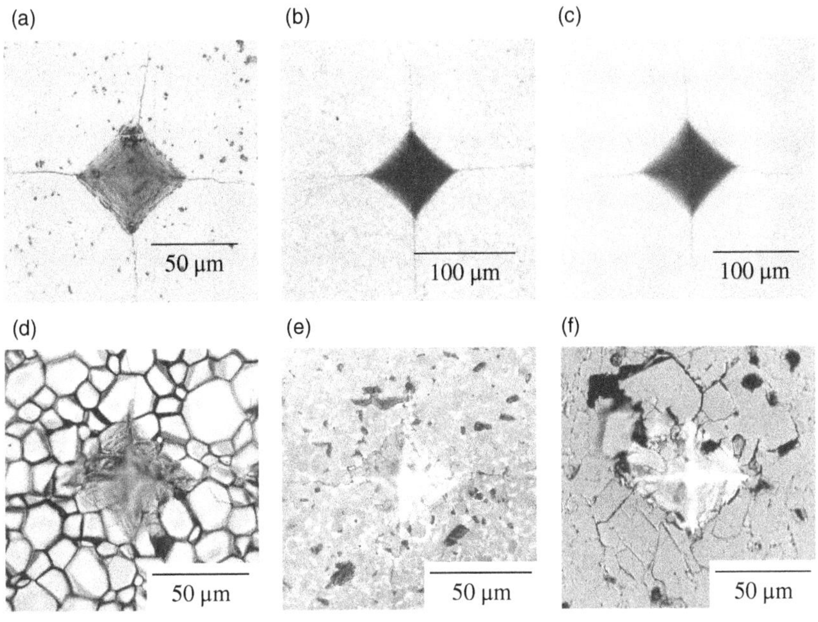

Figure 2.6 Reflected light optical images of Vickers indentation flaws in polycrystalline materials. (a) Cordierite glass-ceramic; (b) Al_2O_3–TiC composite; (c) tetragonal zirconia polycrystal; (d) Al_2O_3 polycrystal; (e) glass-bonded Al_2O_3 polycrystal; (f) MgO polycrystal. (a)–(c) are "fine" grained. (d)–(f) are "coarse" grained.

The indentation patterns most distinct from those of glass are those formed in large-grained polycrystals. Figure 2.6 shows reflected light optical images of Vickers indentations in such materials, all with grain sizes in the range of 10–20 μm: Figure 2.6d, Al_2O_3–0.1% MgO $P = 50$ N; Figure 2.6e, Al_2O_3–4% glass $P = 20$ N; Figure 2.6f, MgO $P = 10$ N. In all cases, the indentation edges are predominantly straight, suggesting only weak microstructural perturbation of the plastic deformation behavior. However, in all cases, the indentation crack traces are extremely difficult to detect, suggesting significant microstructural perturbation of fracture. Transgranular fracture segments are visible, suggesting linking intergranular fracture segments that are not clearly distinct from grain boundary contrast, particularly in Figure 2.6d. In addition, disruption of the indentation pattern by grain-localized lateral cracking is clear in Figure 2.6f. The small c/a ratios suggest that the large-grained polycrystals also have less cracking susceptibility than glass.

Indentation patterns in single crystals are somewhat similar to those in glass. Figure 2.7 shows reflected light optical images of Vickers indentations in some brittle single crystals: Figure 2.7a, MgO (cubic) $P = 40$ N; Figure 2.7b, Si (cubic) $P = 40$ N; Figure 2.7c, Al_2O_3 (sapphire, hexagonal) $P = 5$ N; Figure 2.7d, $LaGaO_3$ (orthorhombic) $P = 5$ N. In Figure 2.7a and b, the indentations were aligned so as to obtain surface crack traces parallel with the indentation diagonals; in Figure 2.7c and d, such alignment was not possible. In all cases, crystallographic cleavage effects were apparent. However, the indentation edges are predominantly straight, suggesting only weak crystallographic perturbation of the plastic deformation behavior. Lateral cracks were apparent in all cases, either as subsurface reflections or as chips. The c/a ratios were comparable to those of glass, suggesting comparable cracking susceptibility.

Figure 2.8 shows transmitted light optical images of $P = 40$ N Vickers indentations in three glasses: Figure 2.8a, SLG; Figure 2.8b, borosilicate glass (BSG, a heat resistant glass used in chemical and domestic applications); and Figure 2.8c, fused silica glass (FSG, amorphous SiO_2, used in optical and microelectronic components). The indentation patterns of the glasses display a gradation of behavior: SLG exhibits well-developed radial and lateral cracks and FSG exhibits extremely short radial cracks and considerable damage in the form of short circumferential cracks surrounding the contact. BSG is intermediate between these extremes. Not visible in these images is that cone cracks formed in BSG and FSG on loading and closed on unloading. All the glasses demonstrated radial cracking on unloading. The differences in glass behavior are attributed to differences in densification within the pdz on indentation. Glasses with open structures with few network modifiers (such as FSG and BSG) are regarded as "anomalous," and those with less open structures and many network modifiers (such as SLG) are regarded as "normal." Anomalous glasses tend to densify on indentation more than normal

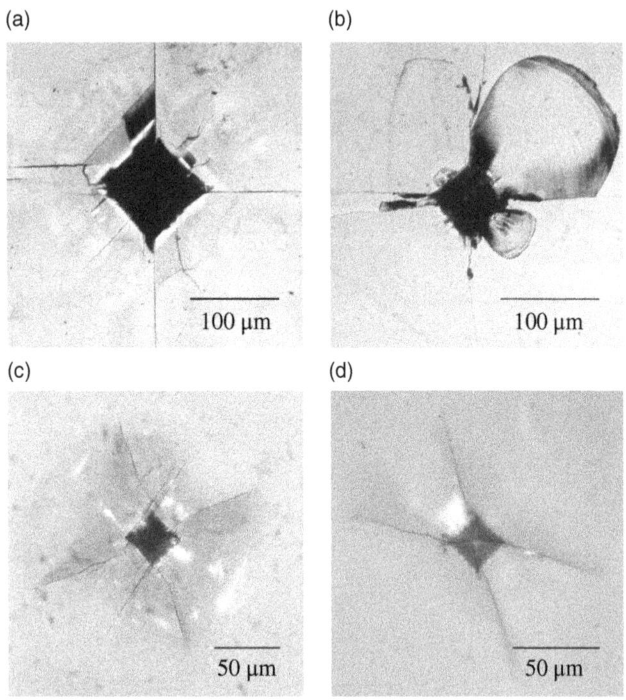

Figure 2.7 Reflected light optical images of Vickers indentation flaws in single-crystal materials. (a) MgO; (b) Si; (c) Al_2O_3; (d) $LaGaO_3$. (a) and (b) have cubic symmetry.

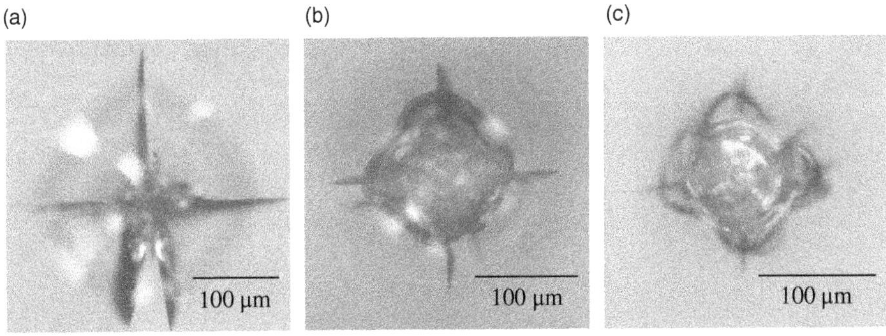

Figure 2.8 Transmitted light optical images of $P = 40$ N Vickers indentation flaws in glass materials. (a) Soda-lime silicate; (b) borosilicate; (c) fused silica. (a) is the prototypical "normal" glass and (c) is the prototypical "anomalous" glass.

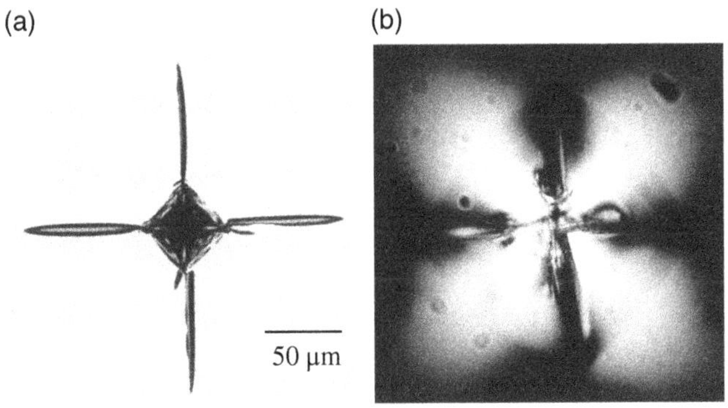

Figure 2.9 Optical images of a Vickers indentation in soda-lime silicate glass. (a) Reflected light; (b) transmitted cross-polarized light showing birefringent contrast indicative of a residual stress field.

glasses, which exhibit greater shear deformation. The degree of densification alters the indentation stress field and thus indentation cracking—behavior noted for some time (Arora et al. 1979; Hagan 1979a; Tandon and Cook 1992).

Indentation stress fields are most easily observed by stress birefringence measurements in glass. Figure 2.9 shows optical images of $P = 10$ N indentations in SLG. Figure 2.9a shows a reflected light optical image. The indentation impression and surface traces of radial cracks are clearly visible. Figure 2.9b shows a transmitted cross-polarized light image. Lobes of bright contrast appear between the indentation crack traces, indicating significant birefringence within the material. The birefringence arises as a consequence of a *residual stress field* (strictly the difference between the principal stresses, Cloud 1998). The residual stress field is localized to the indentation and extends on a scale comparable to the radial crack lengths. Extensive sets of similar images, providing direct evidence of the residual stress field, and variations with the type of glass and the indentation load, are shown in Arora et al. (1979) and Marshall and Lawn (1979). Direct evidence of the residual field is also provided by spectroscopic and diffraction measurements in indented Si (Vaudin et al. 2008, 2011; Gerbig et al. 2009, 2010; Friedman et al. 2016; Gayle et al. 2017; Cook and Friedman 2018).

Indentation shear faults on the contact surfaces of indentation impressions are observed easily by optical microscopy in reflected light, or at greater magnification in scanning electron microscope images. The faults appear as rows of crack openings parallel to the impression edges. Figure 2.10a shows a scanning electron microscope image of the residual contact impression of $P = 2$ N Vickers indentation in Si. The surface has been etched, highlighting the parallel shear faults. On the right of the image, a radial crack can be observed emanating from the indentation. Close inspection shows that the crack originates from a shear fault intersection within the contact impression adjacent to the impression vertex. Figure 2.10b shows a reflected light optical image of the residual contact impression of $P = 30$ N Vickers indentation in SLG. The parallel shear faults within the contact impression are clear (such faults are also visible in Figures 2.2 and 2.6). On the right of the image, a radial crack can be observed emanating from the indentation. Inspection shows that the crack originates from a shear fault intersection within the contact impression adjacent to the impression vertex and that there are several other short cracks originating in such a manner. Similar behavior is observed at the top and bottom of the image. Similar observations have been shown in previous work on SLG and have formed the basis of crack initiation models (Lawn et al. 1983; Lathabai et al. 1991).

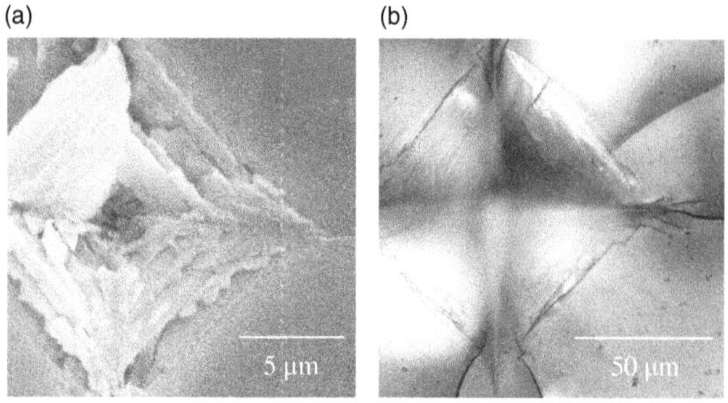

Figure 2.10 Images of shear faults within Vickers indentation contact impressions. (a) Scanning electron microscope image of etched indentation in silicon; (b) transmitted light optical image of indentation in soda-lime silicate glass.

2.1.3 Residual Stress Field Model

Indentation stress fields and indentation fracture mechanics have been considered in detail in many prior studies. Especially notable are the works of Lawn, Marshall, and Evans regarding half-penny and median cracks (Lawn and Evans 1977; Marshall and Lawn 1979; Marshall et al. 1979; Lawn et al. 1980). Radial cracks have also been considered by these authors (Chiang et al. 1982a,b; Lathabai et al. 1991) and others, including Lankford and Davidson (1979a), Lankford (1982), and Laugier (1985, 1987), as have lateral cracks (Marshall et al. 1982; Chen et al. 2005). The major feature common to all these studies is a crack driving force generated on indentation by a localized pdz within a restraining elastic matrix. In reaction to the permanent formation of the pdz, the matrix deforms elastically and generates a residual stress field. The stress field is similar to that generated by an expanding inclusion (Figure 1.18) and thus acts in the far field similar to a point force F (Figure 1.11). A schematic cross-section diagram of an indentation half-penny crack system is shown in Figure 2.11a, in direct correspondence to Figure 1.18a. The indentation contact impression a and half-penny crack surface trace c are shown, along with the pressure p_a exerted by the pdz at the center of the crack. The spherical polar coordinate system indicated is similar to that used in indentation crack formation studies (Cook and Pharr 1990; Cook 2019). The stress field driving the half-penny cracks is $\sigma_{\omega\omega} = (p_a/2)(a/r)^3$; the variation for the (r, θ) crack is shown in Figure 2.11b, in direct

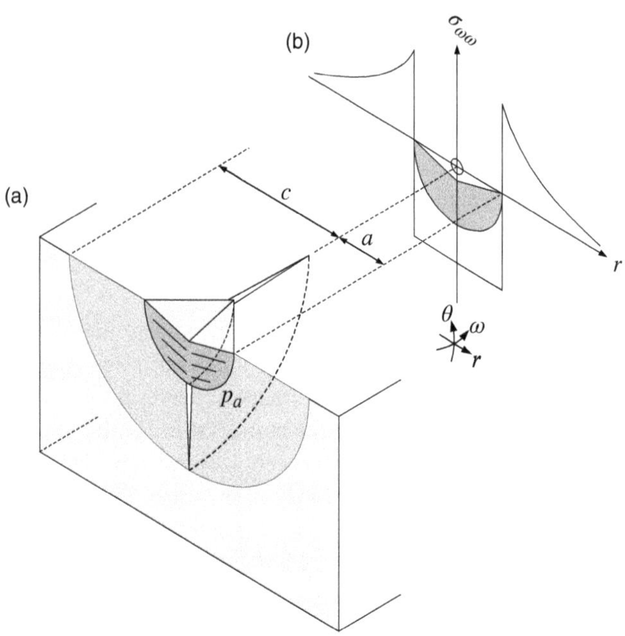

Figure 2.11 (a) Schematic diagram of a half-penny crack, surface trace length c, loaded by a plastic deformation zone, radius a, with internal hydrostatic pressure p_a. Note the spherical polar coordinate system (r, θ, ω). (b) The variation of the circumferential stress perpendicular to the crack plane, $\sigma_{\omega\omega}$, with radial coordinate, r, through the center of the inclusion and crack.

correspondence to Figure 1.18b. Scaling analysis, in combination with a result from plasticity, is used in the following to express the effective force F in terms of indentation and material parameters. The analysis enables the indentation crack driving force to be expressed as a stress-intensity factor (SIF) K_r characterizing the residual field.

Residual SIF analysis begins by recognizing that the characteristic residual pressure p_a exerted by the pdz on the surrounding matrix and crack center scales as

$$p_a \sim -\kappa\left(\frac{\Delta V}{V}\right), \tag{2.3}$$

where κ is the material bulk modulus, V is the initial volume of material forming the pdz, and ΔV is the imposed change in volume as a consequence of the formation of the indentation impression. Observations of Vickers indentation pdz morphology show that

$$V \sim a^3 \tag{2.4}$$

and

$$\Delta V \sim -a^2\delta \sim -a^3 \cot\phi, \tag{2.5}$$

scaling the indentation impression depth with the effective cone angle as $\delta \sim a\cot\phi$. Combining Eqs. (2.4) and (2.5) into Eq. (2.3) and omitting terms in material Poisson's ratio gives

$$p_a \sim E\cot\phi, \tag{2.6}$$

where E is the material Young's modulus. The force exerted by the pdz on the crack thus scales as

$$F \sim p_a a^2 \sim E\cot\phi\frac{P}{H}, \tag{2.7}$$

using Eq. (2.1) to eliminate the impression size a in favor of the indentation load P and the material hardness H. Analysis of the isotropic volume strain generated by an internally pressurized spherical cavity in an elastic–plastic material gives (Hill 1986)

$$\left(\frac{\Delta V}{V}\right) \sim \left(\frac{H}{E}\right)^{3m}, \tag{2.8}$$

where m is a best-fit exponent with $m \approx 1/2$. The indentation system of Figure 2.11 is *not* isotropic, however, as the impression area and the characteristic pdz cross section are constrained to vary identically as a^2. There is thus only one spatial degree of freedom, the effective impression depth, that scales as

$$\left(\frac{\Delta V}{V}\right)^{1/3} \sim \cot^{1/3}\phi \sim \left(\frac{H}{E}\right)^{1/2}. \tag{2.9}$$

Analysis by Chiang et al. (1982a) of pdz observations from their own work and from that of Hagan and others confirmed the included angle and materials properties scaling of Eq. (2.9). Combining Eq. (2.9) into Eq. (2.7) gives

$$F \sim \cot^{2/3}\phi\left(\frac{E}{H}\right)^{1/2} P. \tag{2.10}$$

An identical result was obtained originally by Lawn et al. (1980) using an unconstrained pdz dimension b and a constrained cavity dimension a such that $\Delta V/V \sim a^3/b^3$, as in the isotropic elastic–plastic spherical cavity analysis (Hill 1986). In indentation contexts, b is a *defined* quantity, given by the indentation dimension a and the volume strain imposed in the pdz. The convenience of b is that the ratio a/b simplifies to $a/b \sim (H/E)^{1/2}$ and that $b > a$ can be regarded as a characteristic indentation pdz dimension. [Although applied in indentation fracture analyses (Chiang 1982a,b), $b > a$ does not describe the surface extent of the compressed pdz. In fact, as shown in Figure 2.11 and used in the initiation analysis of Section 2.4, the surface extent of compression is $a_1 < a$.]

Following Chapter 1, the residual field SIF for half-penny cracks can then be written as

$$K_r = \chi P/c^{3/2}, \tag{2.11}$$

where $c \gg a$ has been assumed and all details of indenter geometry, material elastic and plastic properties, pdz zone shape, and crack shape have been absorbed into the dimensionless prefactor χ. Equation (2.10) implies $\chi \sim \cot^{2/3}\phi(E/H)^{1/2}$. However, a single indenter, usually Vickers, is most commonly used in testing and hence ϕ and thus $\cot\phi$ are invariant and χ is written as

$$\chi = \xi(E/H)^{1/2}, \tag{2.12}$$

(a)

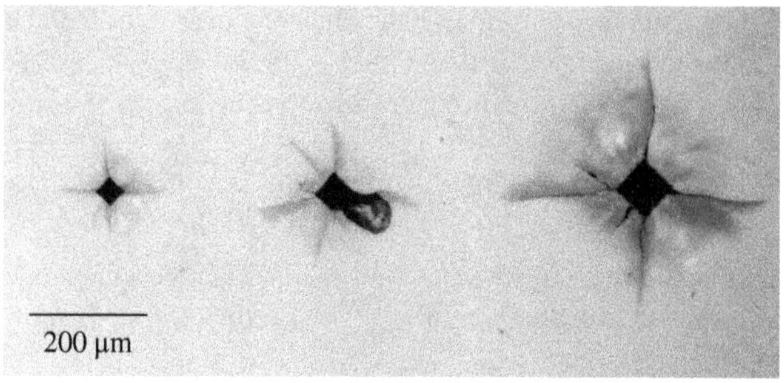

(b)

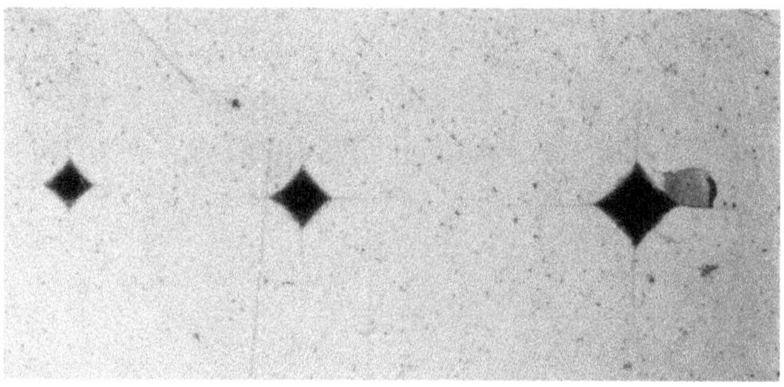

Figure 2.12 Reflected light optical images of Vickers indentation sequences demonstrating effects of increasing contact load left to right. (a) $P = 5$, 10, and 20 N indentations in polycrystalline Y_2O_3; (b) $P = 20$, 30, and 50 N indentations in cordierite glass-ceramic.

where ξ includes all geometrical factors. Applying the equilibrium condition $K_r = T_{air}$ in Eq. (2.11), where T_{air} is the toughness of the material in air, returns the most commonly used form of the Roesler equation, Eq. (1.37),

$$c_{air} = (\chi_{air} P / T_{air})^{2/3} = B_{air} P^{2/3}, \tag{2.13}$$

where χ_{air} and $B_{air} = (\chi_{air}/T_{air})^{2/3}$ are the fracture parameters appropriate to indentations in a moisture-containing reactive air environment. Equations (2.11)–(2.13) are the bases for much of the analysis in this book.

Implementation of the indentation technique is illustrated in Figure 2.12, which shows two sets of indentation sequences: (a) $P = 5$, 10, and 20 N Vickers indentations in polycrystalline Y_2O_3 and (b) $P = 20$, 30, and 50 N Vickers indentations in cordierite glass-ceramic. Both sequences were formed in air. The location, orientation, and size of the flaws is clearly controllable. The differences in impression dimensions and crack lengths enable the different material tendencies to plastic deformation and fracture to be determined quantitatively through estimates of H and B_{air}. The contacts demonstrate the progression from radial cracking to lateral cracking as contact load increases and the difference in this progression between materials. The power of the indentation technique to generate controlled flaws for strength testing is also clear.

It is possible, with care (Anstis et al. 1981), to perform indentation fracture experiments excluding ambient moisture. Equation (2.13) can then be written as

$$c_0 = (\chi_0 P / T_0)^{2/3} = B_0 P^{2/3}, \tag{2.14}$$

where the 0 subscript indicates inert conditions and, using Eq. (2.12),

$$\chi_0 = \xi_0 (E/H)^{1/2}. \tag{2.15}$$

If inert indentation crack length measurements are performed on a material an estimate of B_0 is obtained. If the material properties E, H, and T_0 are also measured, Eqs. (2.14) and (2.15) can be combined and inverted to provide an estimate of ξ_0:

$$\xi_0 \approx B_0^{3/2} T_0 (H/E)^{1/2}. \tag{2.16}$$

In principle, ξ_0 is material invariant and determinable from the cavity analysis and SIF calculations such as Eq. (1.82). In practice, estimates of ξ_0 are based on the scaling analyses discussed earlier and determined from Eq. (2.16) and experimental measurements. Such estimates show that the relative uncertainty in ξ_0 is ±25% if the materials are restricted (Anstis et al. 1981). Restrictions exclude the large-grained polycrystals of Figure 2.6, the anomalous glasses of Figure 2.8, and any other materials that do not exhibit "ideal" indentation patterns, e.g., the crystals of Figure 2.7. This approach to indentation fracture is not pursued here. Attention here, as earlier (Cook 2020), will focus on determination and application, without materials restrictions, of the common experimental parameter characterizing indentation cracks in air, B_{air}. Building on the analysis here, Section 2.2 considers the effects of indenter shape on indentation cracking susceptibility, providing context for Section 2.3 that considers material effects.

2.2 Indenter Shape Effects

A direct test of the indentation scaling and fracture mechanics leading to Eq. (2.13) is the measurement of cracking susceptibility B for indentations on a single material for different contact geometries. In particular, Eq. (2.10) implies that the susceptibility should vary with the included angle ϕ of the indenter as $B \sim \chi^{2/3} \sim \cot^{4/9}\phi$, as noted earlier (Lawn and Fuller 1975; Chiang et al. 1982a; Morris and Cook 2004; Morris et al. 2004a; Jang and Pharr 2008). That is, materials should be more susceptible to indentation fracture for more acute, smaller ϕ, indenters. This section examines this expectation. (Hereafter, unless stated otherwise, the "air" subscripts for c and B are omitted and implied.)

Figure 2.13 shows schematic diagrams of the shapes of commonly used and modeled indenters. The indenters are shown tip upward, and the diagrams are Figure 2.13a, the Vickers four-sided pyramid, Figure 2.13b, the Berkovich three-sided pyramid, Figure 2.13d, the Cube Corner indenter, and Figure 2.12c and e, two conical indenters. Cross sections perpendicular to the indenter axes are shown shaded, and all cross sections have the same projected area. The distances of the cross sections from the indenter tips in Figure 2.12a–c and in Figure 2.12d and e are δ_1 and δ_2, respectively. Hence, if the indenters were inverted and imbedded in perfectly plastic materials, the indenters in Figures 2.12a–c would have identical indentation depth/area ratios, as would those in Figure 2.12d and e. The indenters are characterized by face angle ϕ_{face} (between an indenter axis and an inclined face). For ease of visualization, the indenter heights are exaggerated by a factor of approximately 2, and thus the face angles are reduced by approximately 2. Table 2.1 list the values of ϕ_{face} for the indenters in Figure 2.13 and some other indenters. Simple geometry shows that the edge angle ϕ_{edge}, noted previously, is given by

$$\cot\phi_{edge} = \cot\phi_{face} \cdot \cos(\pi/n_p), \tag{2.17}$$

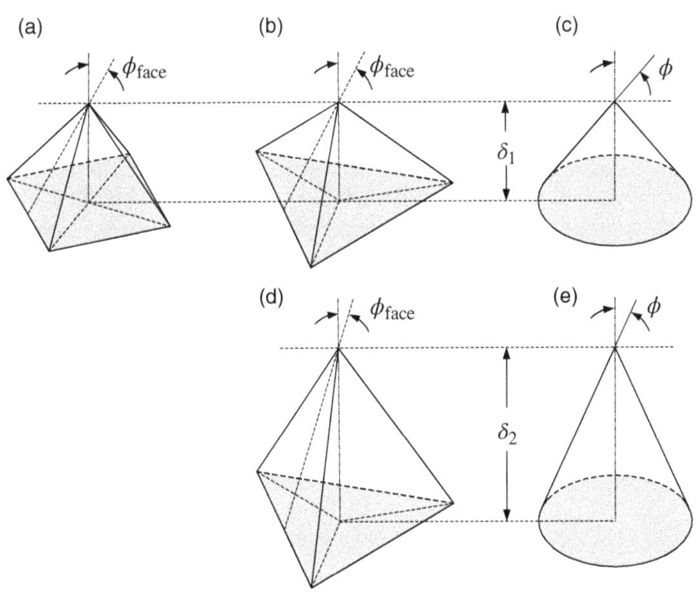

Figure 2.13 Schematic diagrams illustrating indenter geometries. Projected areas identical (shaded). Face angles ϕ_{face}, cone angles ϕ, and depths δ_1 and δ_2 indicated. Vertical dimensions exaggerated by a factor of 2. (a) Vickers; (b) Berkovich; (c) equivalent cone to (a) and (b); (d) cube corner; (e) equivalent cone to (d).

Table 2.1 Indenter geometry parameters.

Indenter	ϕ_{face} (°)	ϕ_{edge} (°)	ϕ (°)	cot ϕ	α_δ
3 Pyramid[a]	85.0	87.5	86.1	0.07	679
3 Pyramid[a]	75.0	82.4	78.2	0.21	72.4
Vickers	68.0	74.1	70.3	0.36	24.5
Berkovich	65.3	77.0	70.3	0.36	24.5
Equivalent Cone 1[b]	—	—	70.3	0.36	24.5
3 Pyramid[c]	52.2	68.8	58.9	0.60	8.64
3 Pyramid[c]	42.4	61.3	49.6	0.85	4.33
Cube Corner	35.3	54.7	42.3	1.10	2.60
Equivalent Cone 2[b]	—	—	42.3	1.10	2.60
4 Pyramid[d]	30.0	39.2	33.1	1.53	1.33
Cone[e]	—	—	30	1.73	1.04

a) Jang and Pharr (2008).
b) Figure 2.13.
c) Morris et al. (2004a).
d) Chiang et al. (1982a).
e) Lawn and Fuller (1975).

where n_p is the number of pyramidal faces and edges (3 for a Berkovich, 4 for a Vickers). An equivalent conical indenter is defined by a cone with identical depth/area ratio as a pyramidal indenter (as in Figure 2.13). The equivalent cone angle ϕ is given by

$$\cot^2\phi = (\pi/n_p)\cot(\pi/n_p)\cdot\cot^2\phi_{face}. \tag{2.18}$$

The indenter cross-sectional area A_δ expressed in terms of indentation depth is given by

$$A_\delta = \pi\tan^2\phi\ \delta^2 = \alpha_\delta\ \delta^2, \tag{2.19}$$

where $\alpha_\delta = \pi\tan^2\phi$ is often termed the "area coefficient." Table 2.1 also lists the values of ϕ_{edge}, ϕ, cot ϕ, and α_δ. It can be observed from Table 2.1 that the conical indenter in Figure 2.13c is equivalent to those in Figure 2.12a and b (the Berkovich and Vickers are equivalent) and the conical indenter in Figure 2.13e is equivalent to that in Figure 2.13d. A measure of indenter acuity is cot ϕ, and the indenters of Table 2.1 are listed in order of increasing acuity. The importance of cot ϕ is that the relative indentation depth varies as $\cot\phi = \delta/a$. For geometrically similar indenters (cones, pyramids), cot ϕ is thus a measure of volume of material displaced on indentation at fixed contact area and thus driving force for indentation fracture.

Figure 2.14 shows an example of the effects of indenter acuity on indentation deformation and fracture. Figure 2.14a shows a $P = 0.1$ N three-sided Berkovich diamond pyramid indentation in SLG. The residual contact impression is

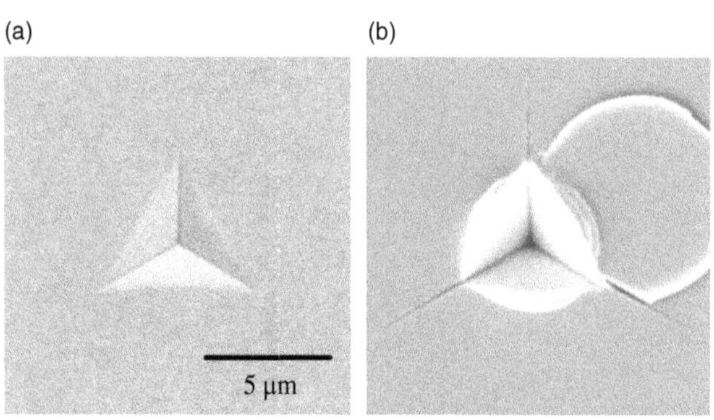

Figure 2.14 Scanning electron microscope images of (a) Berkovich and (b) cube corner $P = 0.1$ N indentations in soda-lime silicate glass. Note that the contact impression area for both indentation is identical but that the Berkovich indentation is crack free and the cube corner indentation exhibits surface traces of both radial and lateral cracks.

triangular rather than square, as for Vickers indentations, cf Figure 2.2 and following, although, as noted previously, the Berkovich and Vickers indenters have identical acuity. Distinct from the previous indentations, the indentation in Figure 2.14a exhibits no cracking and is referred to as *subthreshold*, as the threshold for crack initiation has not been met at the indentation load used (the indentation impression in Figure 2.14a is about 20 times smaller than that in Figure 2.2). Threshold and initiation effects are considered in Section 2.4 and Chapter 10. Figure 2.14b shows a $P = 0.1$ N three-sided cube corner diamond pyramid indentation in SLG. The residual contact impression is triangular and of the same size as the Berkovich indentation of Figure 2.14a, although, as noted previously, the cube corner indenter has markedly greater acuity than the Berkovich. In particular, the cube corner indenter displaces about three times the volume as the Berkovich (and thus, Vickers) at the same projected contact area. As a consequence, the threshold load for indentation cracking is reduced considerably for cube corner indentation, and cube corner cracks are longer at comparable *postthreshold* indentation loads. The cube corner indentation in Figure 2.14b is clearly postthreshold and exhibits half-penny and lateral cracks. In addition, material displaced by the indenter from beneath the contact has been extruded in the form of sheared layers along the less constrained indenter edges (thereby reducing somewhat the volume strain generated by the indentation impression in the pdz). Overall, the cube corner geometry is much more potent than Vickers and Berkovich indenters for generating controlled indentation flaws. Although this is an advantage for studying small load half-penny cracking, the concomitant lateral cracking and extrusion effects, along with greater susceptibility to tip breakage, increase experimental and analytical difficulties in application.

Indenters with varying acuity were used in prior works, often with equally spaced angles. Lawn and Fuller (1975) used WC cones with $\phi = (30, 40, 50, 60, 70, 80)$ °, Chiang et al. (1982a) used four-sided sapphire pyramids with $\phi_{\mathrm{face}} = (30, 45, 60, 68, 70, 80)$ °, Morris et al. (2004a) used three-sided diamond pyramids with approximately equally spaced values of $\cot \phi = (0.36, 0.60, 0.85, 1.10)$, and Jang and Pharr (2008) used three-sided diamond pyramids with $\phi_{\mathrm{face}} = (35, 45, 55, 65, 75, 85)$ °. Note that the indenters of Morris et al. span the domain of acuity from Berkovich (or Vickers) to cube corner

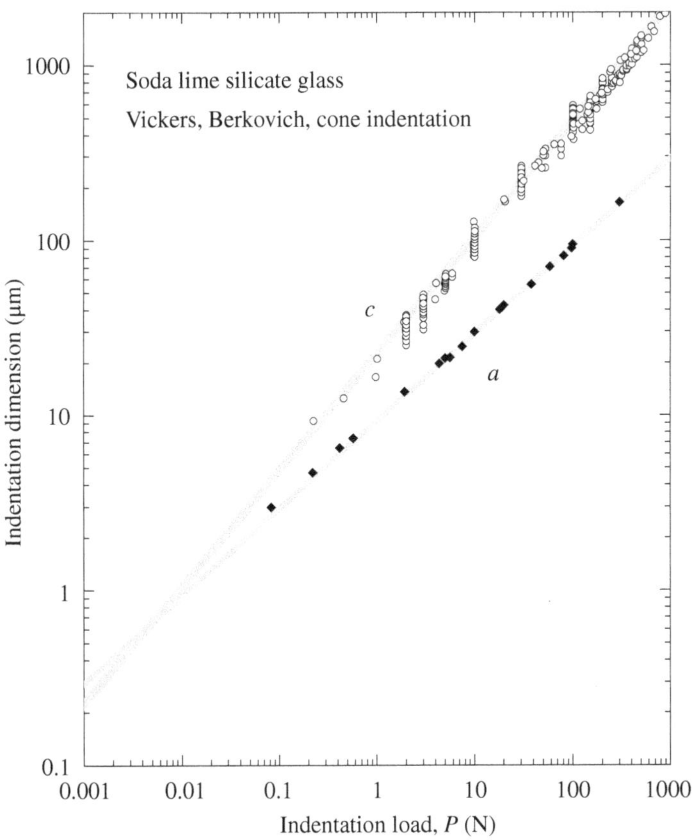

Figure 2.15 Plot in logarithmic coordinates of indentation dimensions, crack length *c* (open symbols) and contact impression semidiagonal *a* (filled symbols), as a function of indentation load *P* for Vickers, Berkovich, and conical indentation of soda-lime silicate glass. The shaded bands indicate mean and standard deviation limits of fits to the data consistent with invariant cracking susceptibility, Eq. (2.2), and invariant hardness, Eq. (2.1). Source: Includes data adapted from Arora et al. (1979), Charles et al. (2006), Cook (1989, 2020), Dériano et al. (2004), Kedir et al. (2018), Lawn and Swain (1975), Lawn and Wilshaw (1975), Pharr et al. (1993).

and appear in the center of Table 2.1; the extremes of indenters of greater or less acuity are also listed. Direct imaging measurements on a wide variety of materials in these acuity studies showed that the projected contact areas of residual indentation impressions did not change with indenter acuity on indentations to fixed peak load (Chiang et al. 1982a; Morris et al. 2004a; Jang and Pharr 2008), as in Figure 2.14. The hardness H given by Eq. (2.1) is thus invariant with respect to ϕ; an assumption implicit in the fracture analysis of Section 2.1.3.

Indentation crack length measurements for indenters of various shapes are included in the latest version of an open online repository for indentation fracture and deformation data (Cook 2024, 45 materials, more than 4400 indentation measurements). The repository, initially established as a catalog to assess measurements of indentation cracks in air (Cook 2020), also includes a broad range of material behavior, considered in Section 2.3. Examples of Vickers (or equivalent) and cube corner indentation data included in the repository are shown in Figures 2.15–2.18 for SLG and single-crystal silicon (Si). The data are shown in logarithmic plots of impression dimension (closed symbols) and crack length (open symbols) as functions of indentation load, $a(P)$ and $c(P)$, using data derived from the published works cited. The plots are approximately the same relative scale, vertical × horizontal of $10^4 \times 10^6$. The shaded bands and solid lines are best fits to the data of slope 1/2 (impressions, a) or slope 2/3 (cracks, c), consistent with Eqs. (2.1) or (2.2), respectively. The Vickers, Berkovich, and cone indentation measurements for SLG, Figure 2.15, include extensive data from a historically broad range of works: Lawn and Fuller (1975), Lawn and Swain (1975), Arora et al. (1979), Cook (1989, 2020), Pharr et al. (1993), Dériano et al. (2004), Charles et al. (2006), and Kedir et al. (2018), supplemented with data from the present work. The Vickers and Berkovich indentation measurements for Si, Figure 2.17, also include data from a broad range of works: Lawn et al. (1981), Tsai and Mecholsky (1991, 1992), Dukino and Swain (1992), Pharr et al. (1993), Swadener and Nastasi (2002), Cook (2006), and Jang and Pharr (2008). The cube corner indentation measurements for SLG, Figure 2.16, include data from Pharr et al. (1993) and Harding et al. (1995), supplemented with data from the

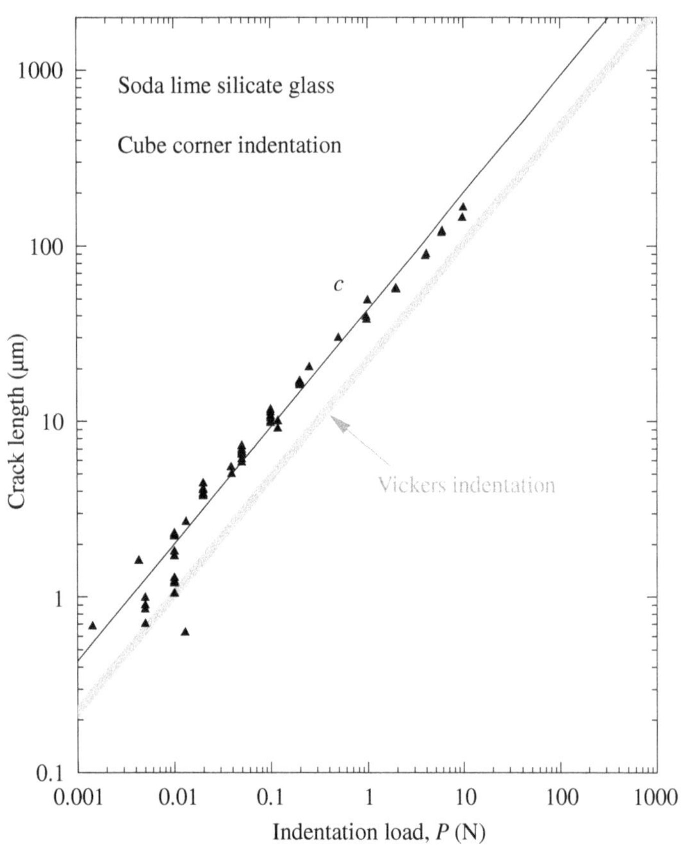

Figure 2.16 Plot in logarithmic coordinates of indentation crack length c (symbols) as a function of indentation load P for cube corner indentation of soda-lime silicate glass. The solid line indicates a fit to the data consistent with invariant cracking susceptibility, Eq. (2.2). The shaded band indicates the mean and standard deviation limits of the Vickers indentation response. Source: Includes data adapted from Harding et al. (1995) and Pharr et al. (1993).

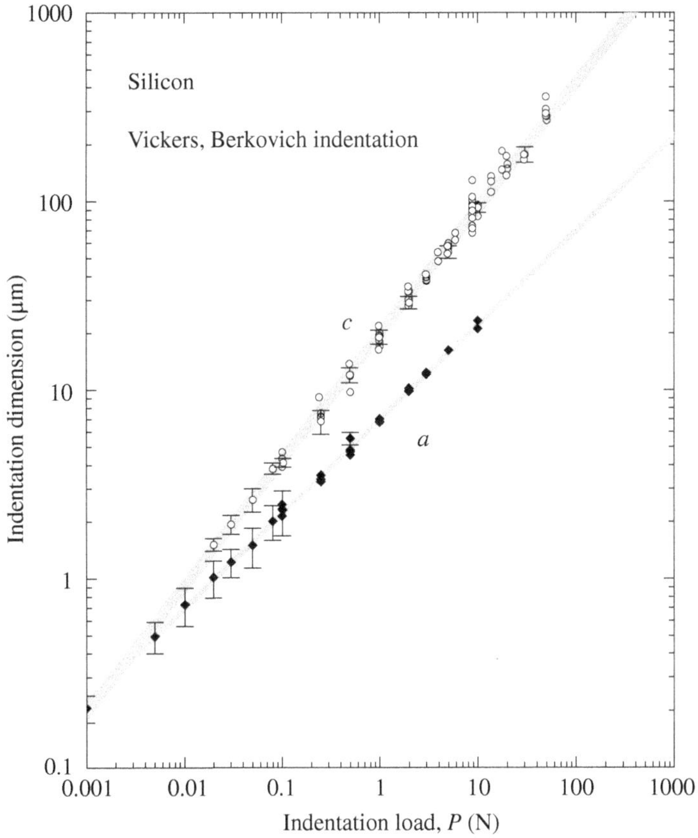

Figure 2.17 Plot in logarithmic coordinates of indentation dimensions, crack length *c* (open symbols) and contact impression semidiagonal *a* (filled symbols), as a function of indentation load *P* for Vickers and Berkovich indentation of single-crystal silicon. The shaded bands indicate mean and standard deviation limits of fits to the data consistent with invariant cracking susceptibility, Eq. (2.2), and invariant hardness, Eq. (2.1). Source: Includes data adapted from Cook (2006), Dukino and Swain (1992), Jang and Pharr (2008), Lawn et al. (1981), Pharr et al. (1993), Swadener and Nastasi (2002), Tsai and Mecholsky (1991, 1992).

present work. The cube corner indentation measurements for Si, Figure 2.18, include data from Pharr et al. (1993) and Jang and Pharr (2008).

For both SLG and Si, in measurements using both Vickers and cube corner indenters and indentation loads extending over factors of 10^4, there is agreement between the data from the various works and with indentation load variations reflecting invariant hardness and cracking susceptibility. The Vickers indentation cracking responses from Figures 2.15 and 2.17 are reproduced in Figures 2.16 and 2.18. For both SLG and Si, indentation cracks generated by cube corner indentation are longer at a given indentation load, indicating greater cracking susceptibility, and are observed at much smaller indentation loads, indicating less resistance to crack initiation. Combining Eqs. (2.10), (2.11), (2.12), and (2.13) gives the variation of cracking susceptibility as a function of indenter included angle

$$\frac{B(\phi)}{B_{\text{Vickers}}} = \left(\frac{\cot\phi}{\cot\phi_{\text{Vickers}}}\right)^{4/9}, \tag{2.20}$$

in this case expressed relative to the Vickers indenter. The solid line in Figure 2.19 is a plot of Eq. (2.20) showing the predicted increase in cracking susceptibility with indenter acuity (decreasing ϕ); the Vickers and cube corner indenter positions are indicated. The symbols in Figure 2.19 show the relative cracking susceptibility ratios determined from analysis of experimental observations such as those in Figures 2.15–2.18. Observations include conical indentation on SLG (Lawn and Fuller, 1975) and three- and four-face pyramidal indentations on As_2S_3 glass, SLG, BSG, FSG, sapphire, Si, germanium, and silicon nitride (Chiang et al. 1982a; Pharr et al. 1993; Harding et al. 1995; Tandon and Buchheit 2007; Jang and Pharr 2008; present work). Notwithstanding some scatter, there is a clear trend, extending across material and indenter geometries, of increased indentation cracking susceptibility with increased indenter acuity. Experimental uncertainty precludes definitive agreement with Eq. (2.20) but the physical basis of increased displaced volume leading to increased

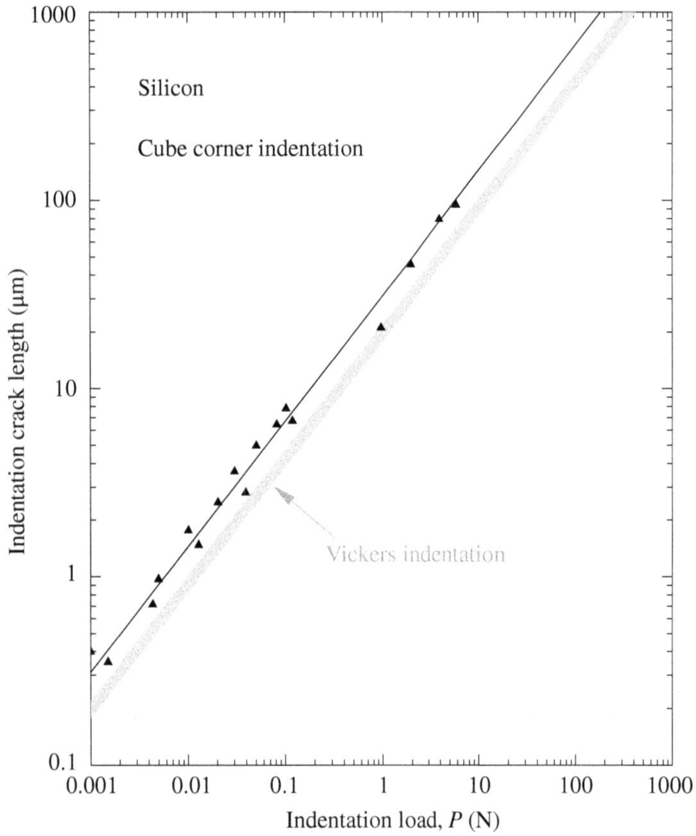

Figure 2.18 Plot in logarithmic coordinates of indentation crack length *c* (symbols) as a function of indentation load *P* for cube corner indentation of single-crystal silicon. The solid line indicates a fit to the data consistent with invariant cracking susceptibility, Eq. (2.2). The shaded band indicates the mean and standard deviation limits of the Vickers indentation response. Source: Includes data adapted from Jang and Pharr (2008) and Pharr et al. (1993).

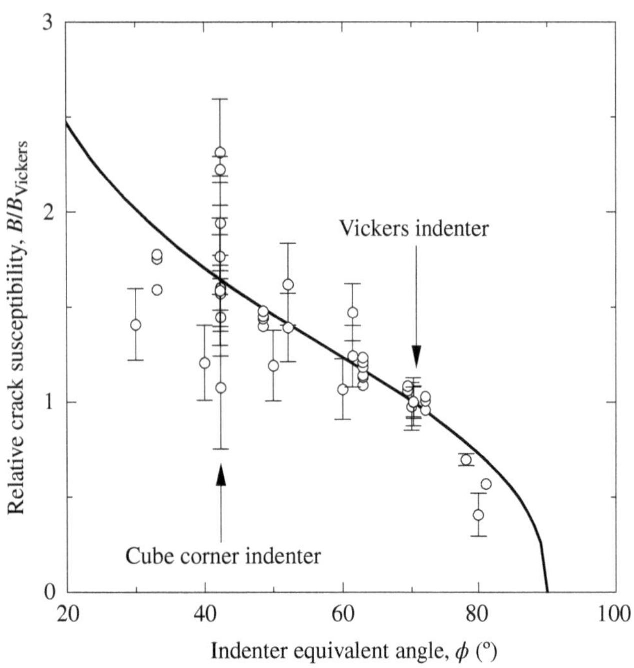

Figure 2.19 Plot of cracking susceptibility B relative to that of Vickers indentation $B_{Vickers}$ as a function of indenter included angle ϕ. Solid line is a prediction from the compressed plastic deformation zone model of indentation fracture. Symbols represent observations for a range of materials and indenters.

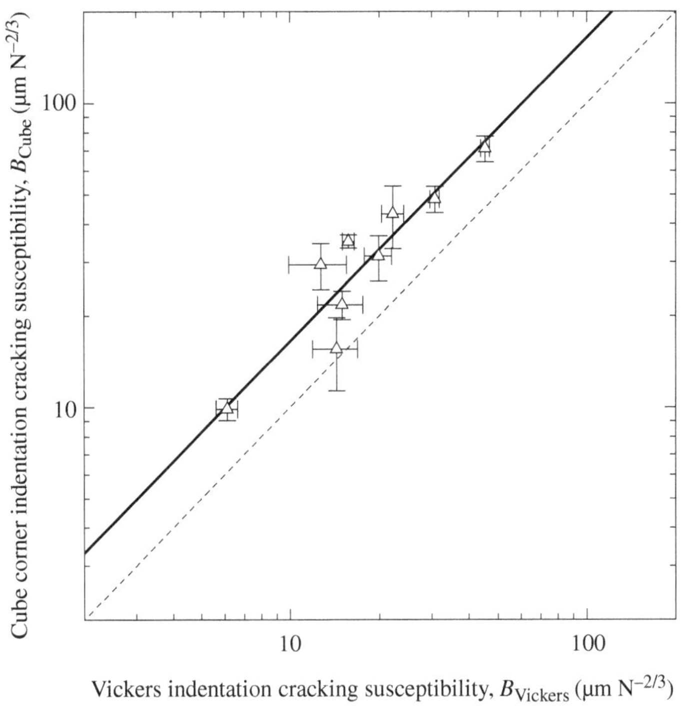

Figure 2.20 Plot in logarithmic coordinates of cracking susceptibility on cube corner indentation, B_{Cube}, as a function of cracking susceptibility on Vickers indentation, $B_{Vickers}$. The solid line indicates an increase by a factor of 1.65 based on considerations of contact impression volume. Symbols represent the mean and standard deviation limits of behavior for several materials.

cracking susceptibility appears sound. In more practical detail, Figure 2.20 shows a logarithmic plot of cracking susceptibilities for a range of materials determined by "acute" indentation with cube corner indenters and by more conventional "obtuse" indentation with Vickers indenters. The dashed line indicates no effect of acuity and the solid line indicates the enhancement predicted by Eq. (2.20), a factor of approximately 1.65. Although there is some scatter, cracking susceptibility on cube corner indentation is largely in agreement with that for Vickers indentation.

Section 2.3 shifts attention from the effects of different indenter shapes on indentation crack lengths to the effects of different indented materials.

2.3 Material Effects

The predominant application of indentation testing as embodied in Eq. (2.13) is the measurement of cracking susceptibility B for different materials using the single contact condition of Vickers indentation in air. Assessment of B is the focus of the online repository of indentation fracture and deformation data (Cook 2024). An example of Vickers (and Berkovich equivalent) indentation behavior from the repository is shown in Figure 2.21 for polycrystalline silicon carbide. The format is as Figures 2.15 and 2.17, using data derived from the published works cited: Lankford and Davidson (1979a,b), Li et al. (1989), Dukino and Swain (1992), and Padture (1994). The material in Figure 2.21, a polycrystalline ceramic, complements those in Figure 2.15, a silicate glass, and Figure 2.17, a single crystal. All three materials exhibit the invariant hardness and cracking susceptibility behaviors given by Eqs. (2.1) and (2.2) and exhibit similar behaviors for Vickers and Berkovich indentations. Consideration of Figures 2.15, 2.17, and 2.21 shows that SLG and Si exhibit similar crack lengths that are longer than those of silicon carbide, and that contact impression dimensions decrease in size from SLG to Si to silicon carbide. These observations indicate the lesser cracking susceptibility B and greater hardness H of silicon carbide relative to SLG and Si, consistent with the intended use of silicon carbide as a structural ceramic.

Quantitative details of the trends discussed are given in Table 2.2, which lists the microstructural form and B and H values for the materials included in the online repository of indentation data (Cook 2024). The full range of microstructural variants are represented in Table 2.2—single crystals, polycrystals, glasses, glass ceramics, and composites. More details and

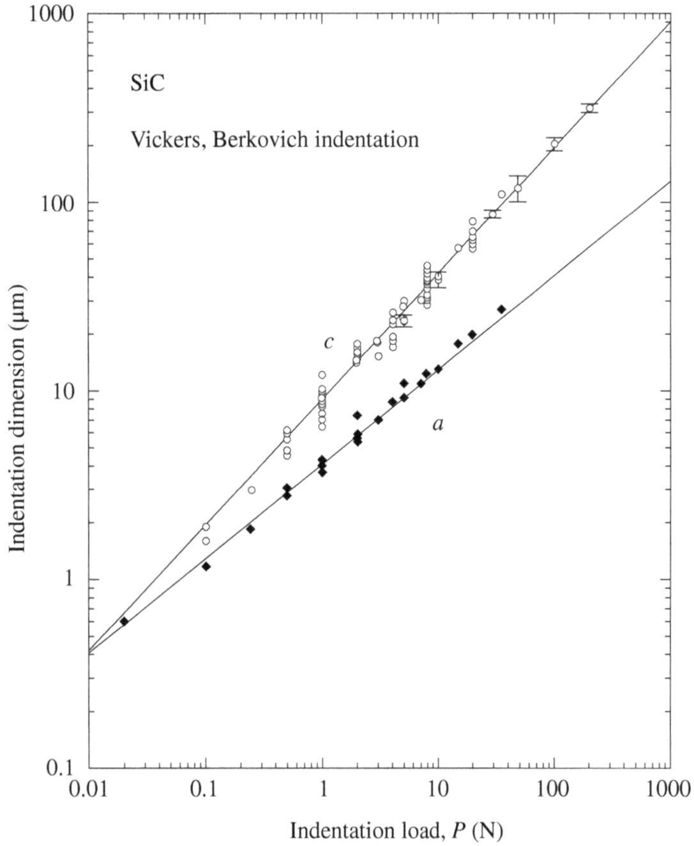

Figure 2.21 Plot in logarithmic coordinates of indentation dimensions, crack length c (open symbols) and contact impression semidiagonal a (filled symbols), as a function of indentation load P for Vickers and Berkovich indentation of polycrystalline SiC. The solid lines indicate fits to the data consistent with invariant cracking susceptibility, Eq. (2.2), and invariant hardness, Eq. (2.1). Source: Includes data adapted from Dukino and Swain (1992), Lankford and Davidson (1979a,b), Li et al. (1989), Padture (1994).

data sources can be found online. Some broad trends emerge on inspection of Table 2.2: Glasses and single crystals tend to exhibit greater susceptibility to indentation cracking than do composites and polycrystals, and there is a weak inverse correlation between hardness and cracking susceptibility, suggesting that materials that deform easily on contact usually also crack easily on contact.

A more application-oriented insight for the variation of values given in Table 2.2 is provided in Figure 2.22, which is a semilogarithmic chart of the B values in a ranked format and identifies representative materials and groups of materials. The groups of brittle materials, left to right, extend from those least susceptible to contact fracture to those most susceptible. (Figure 2.22 is an update of an earlier similar chart, Cook 2020.) The groups of materials are shown as shaded gray regions and extend over factors of approximately 2 in susceptibility and are identified as cermets, structural ceramics, multifunctional ceramics, thermal, electrical, and optical ceramics, and semiconductors. The cermets are metal matrix composites (and extremely wear resistant). The structural ceramics are typified by zirconia and silicon carbide (very strong and tough). The multifunctional ceramics are typified by alumina (strong and electrically insulating). The thermal, electrical, and optical ceramics (often termed "functional" ceramics) are typified by barium titanate (effective dielectric) and SLG (transparent and resists chemical attack). The semiconductors are mostly nonoxides (and have tunable electrical conductivity).

Figure 2.22 has two principal engineering uses: (i) Contact cracking behavior of all brittle materials can be assessed on a single chart. Families of materials, and outliers, can be identified quickly and easily. (ii) Quantitative comparisons of materials behavior, for new materials or between existing materials, are made readily. The chart also serves to place bounds on interpretation of measurements. A factor of 2 variation in B within a family of materials suggests a relative range about the family median of approximately ±35%. This value compares with the average relative uncertainty in estimation of B of approximately ±25% (even for the extensive measurements in the online repository). Hence, the

Table 2.2 Indentation fracture and deformation parameters.

Material	Microstructure	Susceptibility, B (μm $N^{-2/3}$)	Hardness, H (GPa)
WC–Co	Composite	2.81 ± 0.26	14.4 ± 1.4
WC–Co	Composite	4.41 ± 0.47	10.4 ± 0.3
WC–Co	Composite	4.90 ± 0.40	18.2 ± 0.5
ZrO_2	Polycrystal	5.76 ± 0.24	12.9 ± 0.4
Si_3N_4	Polycrystal	6.14 ± 0.52	16 ± 2
Al_2O_3	Polycrystal	6.22 ± 0.42	6.3 ± 0.3
Al_2O_3–TiC	Composite	6.33 ± 0.65	15 ± 2
Al_2O_3	Polycrystal	7.11 ± 0.49	12.7 ± 0.8
SiC	Polycrystal	7.31 ± 0.70	26.1 ± 2.2
ZrO_2	Polycrystal	7.42 ± 0.45	13.2 ± 0.5
Al_2O_3–TiC	Composite	7.57 ± 0.57	20.0 ± 1.1
$2MgO \cdot 2Al_2O_3 \cdot 5SiO_2$[a)]	Glass ceramic	8.54 ± 0.73	8.4 ± 0.5
Al_2O_3	Polycrystal	8.81 ± 0.67	18 ± 2
SiC	Polycrystal	9.06 ± 1.28	30.0 ± 4.4
Al_2O_3	Polycrystal	9.26 ± 1.28	13.0 ± 0.6
Al_2O_3	Polycrystal	9.28 ± 0.55	16.5 ± 0.8
Al_2O_3	Polycrystal	10.2 ± 1.1	14.4 ± 1.2
$2MgO \cdot 2Al_2O_3 \cdot 5SiO_2$[a)]	Glass ceramic	10.8 ± 1.0	7.5 ± 0.5
Al_2O_3	Polycrystal	12.0 ± 3.1	14 ± 2
Al_2O_3	Polycrystal	12.1 ± 3.8	18.3 ± 3.0
SiO_2	Glass	12.7 ± 2.8	7.6 ± 0.4
$2MgO \cdot 2Al_2O_3 \cdot 5SiO_2$[a)]	Glass ceramic	12.9 ± 1.1	7.5 ± 0.2
Li_2SiO_3–SiO_2	Glass ceramic	13.0 ± 1.4	4.8 ± 0.2
$2MgO \cdot 2Al_2O_3 \cdot 5SiO_2$[a)]	Glass	14.2 ± 1.8	9.5 ± 1.5
Al_2O_3	Single crystal	14.4 ± 2.5	19.6 ± 2.6
$K_2O \cdot Al_2O_3 \cdot 4SiO_2$[b)]	Glass ceramic	14.4 ± 0.7	6.0 ± 0.3
MgO	Polycrystal	14.8 ± 3.1	5.7 ± 0.4
$MgAl_2O_4$	Polycrystal	15.0 ± 2.6	14.1 ± 0.5
$CaTiO_3$	Polycrystal	15.2 ± 2.0	9.1 ± 1.0
MgO	Polycrystal	16.0 ± 5.3	6.2 ± 1.2
$3Ca_3(PO_4,CO_3)_2 \cdot Ca(OH)_2$[c)]	Polycrystal	18.2 ± 1.0	3.53 ± 0.12
$BaTiO_3$	Polycrystal	18.5 ± 3.0	6.0 ± 0.7
$BaTiO_3$	Polycrystal	18.7 ± 2.8	7.9 ± 0.4
SiO_2–Al_2O_3	Glass (ASG)	18.7 ± 2.6	6.6 ± 0.3
Y_2O_3	Polycrystal	19.7 ± 3.1	7.7 ± 0.4
Si	Single crystal	19.9 ± 2.1	10.3 ± 0.9
MgO	Single crystal	21.6 ± 2.7	5.9 ± 2.2
SiO_2–Na_2O–CaO	Glass (SLG)	22.3 ± 1.9	5.9 ± 0.5
$(Pb,Zr)TiO_3$	Polycrystal	24.4 ± 2.2	3.4 ± 0.2
NaCl	Single crystal	28.8 ± 3.0	0.24
$LiNbO_3$	Single crystal	30.0 ± 5.0	5.5
GaAs	Single crystal	30.4 ± 1.5	6.9 ± 0.3
Ge	Single crystal	30.8 ± 1.1	7.5 ± 0.6
ZnS	Polycrystal	31.6 ± 1.4	1.9 ± 0.1
As_2S_3	Glass	45.5 ± 1.6	1.5 ± 0.1

Compounds:

a) Cordierite.

b) Leucite.

c) Carbo-hydroxyapatite (tooth enamel).

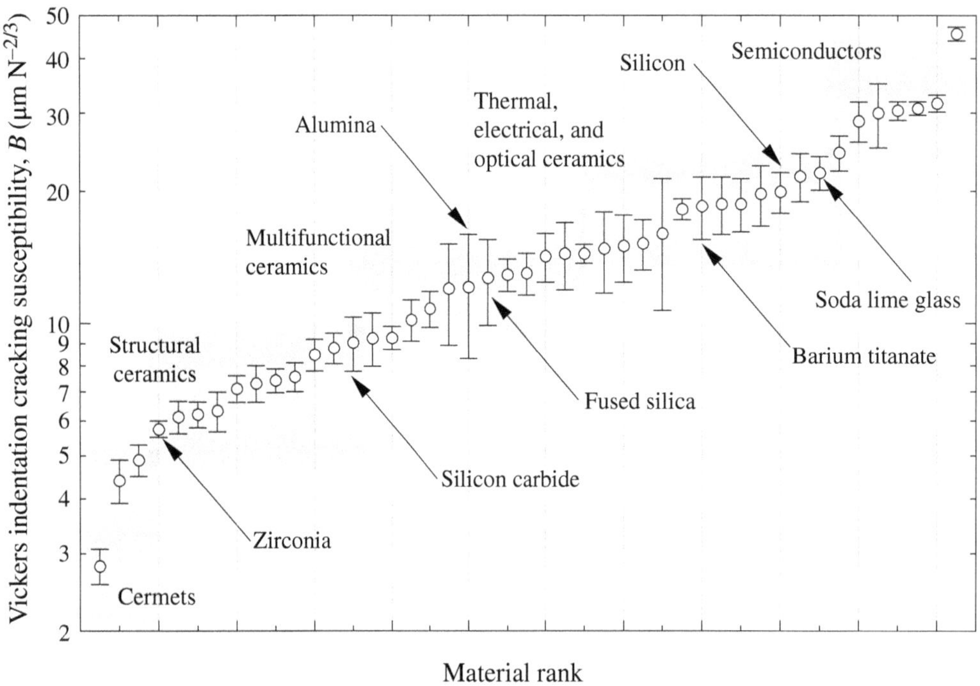

Figure 2.22 Semilogarithmic plot of Vickers indentation cracking susceptibility values B ranked for a large set of brittle materials. Ranking increases left to right such that materials that exhibit the greatest susceptibility are on the extreme right. Families of similar materials are indicated by shaded regions. Individual materials of note are indicated by arrows.

contact fracture susceptibilities of midfamily materials such as barium titanate and alumina can probably be distinguished in indentation experiments, as can the extremes of single family materials such as zirconia and silicon carbide. However, more refined distinctions are likely to be much more difficult and deconvolution of B values into toughness T values is made even more difficult by uncertainties in ξ. Measurement uncertainties are discussed elsewhere (Cook 2020).

Two materials stand out in Figure 2.22 as ranked separately from their groups. First, FSG exhibits significantly less susceptibility to contact fracture than SLG, a phenomenon known for some time (Arora et al. 1979). As all silicate glasses exhibit approximately identical values of elastic modulus E, hardness H, and toughness T, Eqs. (2.12) and (2.13) suggest that the cracking susceptibility B of the two materials should be similar. However, this assessment assumes that the proportion of non-volume conserving deformation (compaction or densification) on indentation is also identical for the two materials, and thus ξ is material invariant. FSG and the related "anomalous glasses," including BSG, exhibit considerable densification on indentation. Thus, these glasses exhibit significantly reduced residual stress fields relative to the "normal" glasses such as SLG and ASG and are characterized by significantly decreased values of ξ (Arora et al. 1979) and thus B. The second material that stands out is Si, which exhibits significantly less susceptibility to contact fracture than other semiconductors. Interpretation in terms of material differences is less straightforward here as the modulus/hardness E/H ratios and toughness T values of semiconductors are very similar, although Si is slightly tougher. In addition, semiconductor indentation is complicated by crystallographic anisotropy and variable tendencies to phase transformation, extrusion, and chipping (Jang et al. 2005; Oliver et al. 2007, 2008; Gerbig et al. 2009; and references therein), such that ξ is very poorly known. Whatever the underlying reason, an implication of the reduced cracking susceptibility is that Si is the pervasive material used in semiconductor and microelectromechanical systems fabrication as it is more tolerant of mechanical loading during device manufacture and operation (Cook 2006; DelRio et al. 2015; also see Chapter 10).

The chart of Figure 2.22 provides a conservative approach to engineering considerations of contact fracture in that the load dependence of cracks assumed to have formed at contacts on brittle materials is quantified. Another important aspect of design with brittle materials is the minimum or "threshold" contact load at which a crack can be initiated. Crack initiation, a less conservative design approach, is considered in Section 2.4.

2.4 Initiation

Crack initiation consists of three sequential fracture states: (i) an unstable equilibrium, followed by (ii) a nonequilibrium path passing through a crack driving force maximum, followed by termination in (iii) a stable equilibrium. Attention is commonly focused on state (i), often referred to as *the* initiation event. In Chapter 1 (Section 1.7), the fracture concepts and analyses of crack initiation at expanding inclusions and anisotropic grains were considered. Here, these concepts and analyses are applied in analysis of crack initiation at sharp indentation contacts. Emphasis is placed on the conditions required to attain the unstable equilibrium state (i) and in particular the specification of the minimum, *threshold*, indentation load required to initiate a radial or half-penny crack. Consideration is then given to the final stages of state (ii) as the crack approaches stable equilibrium state (iii) via a nonequilibrium path.

In developing a description of indentation crack initiation, attention shifts from the lengths of well-developed stable cracks considered previously, $c > a$ (Sections 2.1–2.3) to the behavior of small unstable crack nuclei, $c < a$. Figure 2.23 shows a schematic diagram of an indention impression and crack initiation location similar to those in Figure 2.10. The square contact impression of the pdz surface perimeter and the contact diagonals are indicated by dashed lines and the contact impression semidiagonal a is indicated. A compressively stressed zone of pressure p_a within the pdz is indicated by the shaded region. The zone does not extend to the contact impression vertices, consistent with crack initiation observations such as Figure 2.10 and prior analyses (Lawn and Evans 1977; Chiang et al. 1982a,b; Lathabai et al. 1991; Cook et al. 1994a; Cook and Braun 1994). The zone has dimensions along the contact diagonals of $a_1 < a$. Attention here is focused on the small *crack nucleus* of dimension $c_1 = a - a_1$ exterior to the compressive zone. The nucleus is formed by the intersection of shear faults within the pdz adjacent to the impression vertices. The nucleus is in a state of circumferential tension $\approx p_a/2$ (see Chapter 1). (For the long cracks considered previously the difference between a and a_1 is not significant.)

Following Section 1.7, the SIF K_n for the crack nucleus in the tensile indentation stress field is

$$K_n = \Omega_1 p_a c_1^{1/2}, \tag{2.21}$$

where Ω_1 is a dimensionless geometrical coupling constant that characterizes the mixed-mode natures of the crack nucleus and the indentation stress field (Cook 2019). The SIF origin is taken at the compressive zone periphery. In Section 1.7 crack initiation was considered in terms of invariant c_1 and increasing p_a. Here the opposite is considered: the magnitude of the indentation stress field remains invariant, as the indentation deformation field maintains geometrical similarity, and the nucleus size increases with impression size, also as a consequence of geometrical similarity. Hence, the invariant stress is given by Eq. (2.6),

$$p_a \sim E \cot \phi,$$

and the varying nucleus size is given by

$$c_1 = \Omega_2 a, \tag{2.22}$$

where Ω_2 is a dimensionless proportionality constant. Combining Eqs. (2.6) and (2.22) into Eq. (2.21) and imposing $K_n = T$ equilibrium, as in the inset of Figure 2.23, gives

$$K_n = \Omega E \cot \phi a^{1/2} = T, \tag{2.23}$$

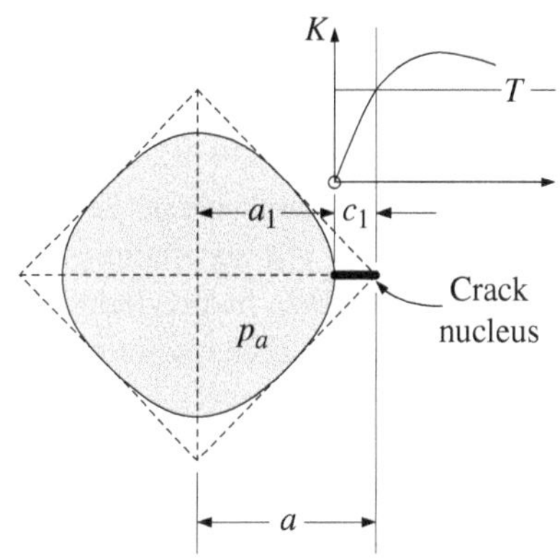

Figure 2.23 Schematic plan view diagram of a Vickers indentation contact impression at crack initiation. The impression exterior boundary is indicated by a dashed line. An interior zone of compression p_a is indicated by a shaded region. A crack nucleus on the zone periphery is indicated by a bold line. The inset graph shows the variation of the crack stress-intensity factor at initiation.

where $\Omega = \Omega_1 \Omega_2^{1/2}$ is a combined geometry constant (Cook 2019), similar to ψ used previously. (As the environment of the unstable segment of indentation crack initiation is usually constrained and the process rapid T here is usually assumed to approximate the inert toughness, $T \approx T_0$. The stable segment, in which $T \approx T_{air}$, is considered at the end of this section.)

Eliminating a in favor of the contact load by $P \sim Ha^2$ and inverting gives

$$P_\Omega \sim \frac{T^4}{H^3} \left(\frac{H}{E\Omega \cot \phi} \right)^4, \tag{2.24}$$

as the scaling of the crack initiation load P_Ω with material and indenter characteristics. The first term, T^4/H^3, is a well-established crack initiation material scaling combination (Lawn and Evans 1977). The crack initiation contact dimension a_Ω is given by

$$a_\Omega \sim \frac{T^2}{H^2} \left(\frac{H}{E\Omega \cot \phi} \right)^2, \tag{2.25}$$

which is similarly well established. Equations (2.24) and (2.25) make clear that the initiation conditions in terms of the threshold load P_Ω and impression size a_Ω are very sensitive functions of indenter angle ϕ, much more than the cracking susceptibility B. Figure 2.20 shows an increase in indentation cracking susceptibility B of factor 1.65 on use of a cube corner indenter ($\phi \approx 42°$) rather than a Vickers indenter ($\phi \approx 70°$). Figure 1.14 shows a decrease in cracking threshold load P_Ω of approximately 10^3 for the same indenter comparison!

A lower bound to the threshold load, P^*, can be obtained by setting $\Omega_2 = 1$, such that the crack nucleus fills the semidiagonal of the contact impression, $c_1 = a$ and $\Omega = \Omega_1$ in Eqs. (2.22) and (2.24). This bound is similar to the P^* value obtained by Lawn and Evans (1977) calculated by setting equilibrium for the complete SIF field maximum, $K(c^*) = T$ (as $c^* \approx a$, Section 1.7), and identical to that used later by Jung et al. (2004), calculated by setting $c = a$ explicitly. In both cases, P^* is well approximated by extrapolation of contact impression and crack length variations from the well-developed crack domain to small indentation loads. The intersection point in the small indentation load domain is given by $c(P^*) = a(P^*)$, described by B and H,

$$B(P^*)^{2/3} = (P^*/2H)^{1/2}. \tag{2.26}$$

This equation results in $P^* \sim (B^2H)^{-3} \sim T^4/H^3$ as in Eq. (2.24). As implied earlier, P_Ω can be estimated from the smallest experimental indentation load at which cracking is observed. Such observations provide very close estimates of the threshold in experiments in which contact impression measurements are extended to even smaller, subthreshold, loads. Experimental estimates of P_Ω can be compared with the lower bound prediction of P^* determined from Eq. (2.26). The preceding analysis enables physical interpretation of such a comparison through the scaling of Eq. (2.24): $P_\Omega/P^* = (1/\Omega_2)^4$.

Figure 2.24 shows a schematic logarithmic plot of the Vickers indentation cracking and deformation behavior of SLG, using information from Figure 2.15. The domains of the experimental observations are shown as gray bands and the best-fit responses of Eqs. (2.1) and (2.2) are shown as dashed lines. The domain of indentation impression data (a) extends to smaller indentation loads than the domain of observed indentation crack lengths (c). The lower bound indentation load of the observed crack domain provides an experimental estimate of the indentation cracking threshold P_Ω. Indentations formed at loads $P < P_\Omega$ are termed subthreshold. Experimental indentation cracking thresholds are clear in Figures 2.15, 2.17, and 2.21. The best-fit responses of Eqs. (2.1) and (2.2) in Figure 2.24 intersect at indentation load P^*, which provides a lower bound to the indentation cracking threshold (cracks measured from the contact impression center cannot be shorter than the impression semidiagonal). Lower bound indentation cracking thresholds marked by intersections are also clear in Figures 2.15, 2.17, and 2.21.

Figure 2.25 shows a logarithmic plot of experimentally observed threshold indentation loads P_Ω as a function of lower bound threshold loads P^* determined from indentation dimension measurements. The data encompass the many materials listed in Table 2.2 and were obtained from plots similar to Figures 2.15, 2.17, and 2.21 contained in the online repository of indentation data. The uncertainties in the calculated P^* values were obtained by propagation of variance from uncertainties in B and H. The hatched region indicates the subthreshold domain in which no cracks should form. Although there is significant scatter (perhaps anticipated in considerations of initiation), it is clear that observed threshold loads P_Ω scale linearly with lower bound loads P^* over a factor of 10^5 in load. The solid line in Figure 2.25 is a guide to the eye of this trend obtained by setting $\Omega_2 = 0.35$ in the scaling relation described earlier. This value is consistent with *surface* observations of initiated cracks at indentations, for example, in Figure 2.10 and in Lawn et al. (1983), Lathabai et al. (1991), Cook et al. (1994a), and Cook and Braun (1994) which suggest that $\Omega_2 > 0.1$–0.2.

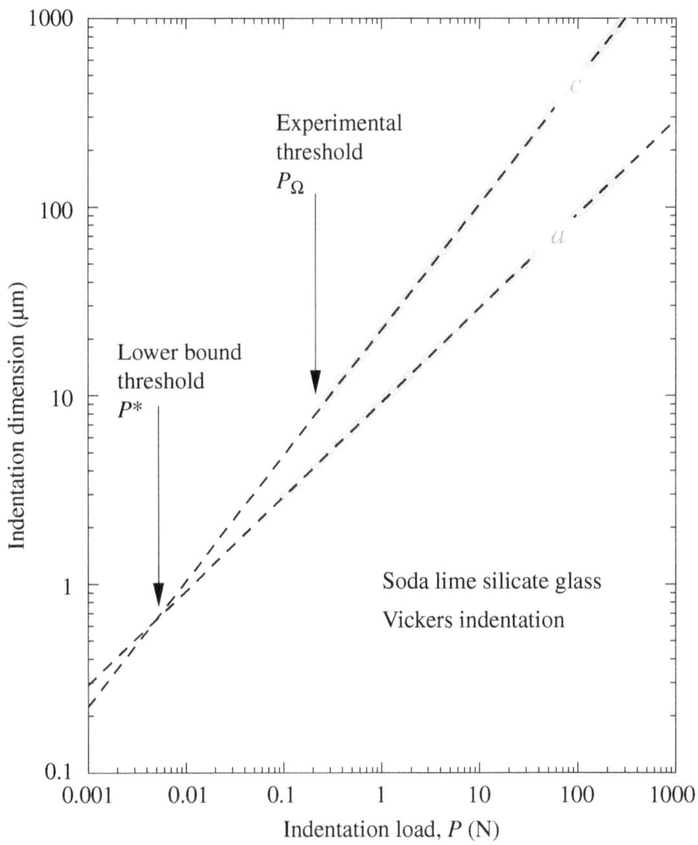

Figure 2.24 Plot in logarithmic coordinates of indentation dimensions, crack length c and contact impression semidiagonal a, as functions of indentation load P for Vickers indentation of soda-lime silicate glass. The shaded lines indicate the domains of experimental observations from Figure 2.15. The dashed lines indicate best fits to the observations. The experimental cracking threshold P_Ω is determined from the least load of the cracking observations. A lower bound prediction of the threshold P^* is determined from the load at intersection of the cracking and contact impression trends.

In completing a description of indentation crack initiation, attention shifts to the variation in length of well-developed cracks approaching a stable equilibrium configuration, $c \rightarrow c_{air}$. Such variation is described by motion along a nonequilibrium fracture trajectory, Figure 1.19, and decreasing crack velocity. (c_0 was used in Chapter 1 to emphasize an equilibrium configuration, independent of environment.) Figure 2.26 shows the variation of radial crack surface trace lengths c with time after indentation t in environments of water, air, and silicone oil, from the work of Cook and Liniger (1993). The crack lengths clearly increase with both time and decreasing velocities toward equilibrium values that are slightly different in the three environments. Similar kinetic behaviors, over smaller time domains, were reported by Anstis et al. (1981), Gupta and Jubb (1981), Lawn et al. (1985), and Cook (1989) for crack length measurements in SLG and Anusavice et al. (1991) and Ali (1993) for crack length measurements in leucite ("dental porcelain"). Similar equilibrium separation is notable from in situ cone crack measurements in soda-lime glass under vacuum (Lawn and Fuller 1975) that exhibited shorter cracks than measurements in (moist) air. The shortening reflects the environmental influence and decreased surface energy discussed in Chapter 1.

Differentiation of crack length data as shown in Figure 2.26 enables direct assessment of decreasing crack velocity v as separation from equilibrium decreases in indentation fracture systems. Following Eq. (1.77), taking the behavior at $c/c_{air} = 0.9$ as a common calibration point, Figure 2.27 shows a semilogarithmic plot of relative crack velocity as a function of relative crack length using the data of Figure 2.26 and those of Lawn et al. (1985), Anusavice et al. (1991), and Ali (1993). The data are shown as symbols; the solid line represents a similarly calibrated variation of Eq. (1.77). The significant decrease in crack velocity as the equilibrium state is approached is clear in both the experimental observations and the prediction of Eq. (1.77). The behavior of Eq. (1.77) was based on the assumption that the fracture system exhibited only a small perturbation from equilibrium and that the crack velocity was thus linear in the perturbation. As a consequence, the solid line in Figure 2.27 is a good description of crack velocity behavior close to equilibrium, $c/c_{air} \geq 0.8$. For larger separations from equilibrium,

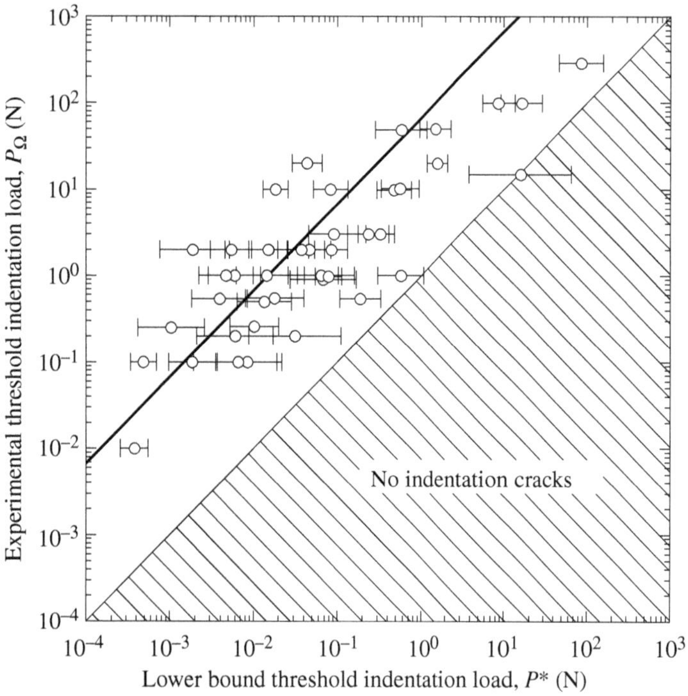

Figure 2.25 Plot in logarithmic coordinates of observed Vickers indentation cracking threshold load P_Ω as a function of lower bound load P^*. Values predicted from well-developed crack and contact dimension measurements for many brittle materials. Shaded region indicates no cracking for indentation loads less than the lower bound prediction. Symbols and bars represent mean and standard deviation limits determined from experimental observations of the threshold and propagation of variance. Solid line indicates the threshold trend from initiation analysis assuming effective crack nuclei approximately one third of the contact impression size.

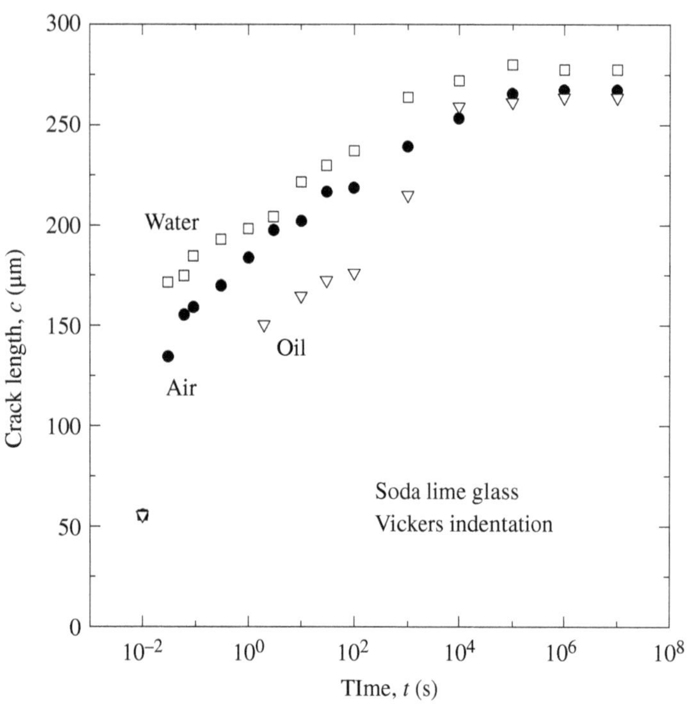

Figure 2.26 Semilogarithmic plot of indentation crack length c as a function of time after indentation t for soda-lime glass in various environments. The cracks extend significantly immediately after indentation before reaching plateau values that depend on environment long after indentation. Source: Adapted from Cook and Liniger (1993).

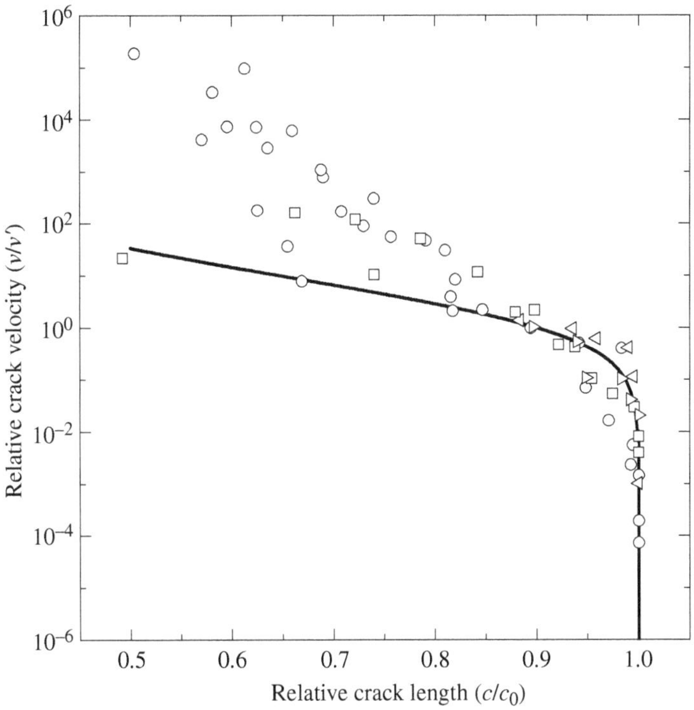

Figure 2.27 Semilogarithmic plot of indentation crack velocity v as a function of crack length c for soda-lime glass in various environments. Symbols represent data calibrated to values at $c = 0.9c_0$, where c_0 is the terminal equilibrium crack length, in this case c_{air}. The solid line is the predicted variation in velocity for isothermal crack propagation in stable fracture systems weakly perturbed from equilibrium.

$c/c_{air} < 0.8$, the crack velocity is underestimated. In these cases, crack velocity exhibits much greater values and exhibits an exponential dependence on the separation from equilibrium. The full crack velocity response is considered in detail in Chapter 6, but Figure 2.27 shows that isothermal nonequilibrium crack motion well describes the termination of the crack initiation process.

2.5 Summary and Discussion

This chapter has surveyed indentation cracking at sharp contacts in brittle materials. Experimental observations have focused on cracking at the most common indentation flaw, Vickers indentations. The primary source of quantitative information has been an extensive online repository of Vickers and related indentation cracking measurements (Cook 2024). The residual square contact impression characteristic of plastic deformation at Vickers indentations has been identified, along with the major cracking systems—median, radial, half-penny, and lateral. The variations in these features between indentation flaws in materials of different microstructures—crystalline, polycrystalline, and amorphous—has been noted. Modeling descriptions have focused on the effects of the subsurface plastic deformation zones developed beneath the contact impressions at sharp indentations. The zones are in states of compression leading to residual stress fields in the surrounding elastic matrices. Tensile elements of the residual fields are responsible for the initiation and stabilization of indentation cracks. Attention here was focused on the radial and half-penny crack geometries, as these cracks form perpendicular to an indented material surface and exert the greatest influence on the strengths of indented components. Median cracks were shown to be formed during indentation loading as precursors to half-penny cracks developed on complete unload. Lateral cracks, forming parallel to an indented surface, also on unloading, are associated mostly with chipping and material removal at indentations.

The surface traces of radial and half-penny cracks are a clear feature of Vickers indentations. The variation of crack length c, the surface trace, as a function of indentation load P, the peak load, is the most common indentation fracture measurement, usually performed on indentations and cracks in air. Experimental crack length variation, $c(P)$, is characterized by cracking susceptibility B. The minimum indentation load required to generate cracks is characterized by the experimental threshold load P_Ω. Modeling here was directed toward establishing the dependence of B and P_Ω on the indenter included angle, or acuity, ϕ, and the material properties (in air) of elastic modulus, E, hardness, H, and toughness, T. In summary, crack length was shown to vary with indentation load as $c = BP^{2/3}$, where $B \sim \cot^{4/9}\phi(E/H)^{1/3}/T^{2/3}$, with threshold load $P_\Omega \sim (T^4/H^3)(H/E\cot\phi)^4$. Consideration of these expressions shows that indentation cracking susceptibility varies moderately with material resistance to fracture as characterized by toughness and weakly with indenter angle and material deformation characteristics, whereas indentation cracking threshold varies strongly with indenter angle and material toughness and moderately with material deformation characteristics. Initiation criteria for other indentation cracks are similar, e.g., the lateral and cone systems that have secondary effects on component strength (Cook 2019). The onset of initiation in many physical systems is typified, as here, by enhanced sensitivity to system parameters: compare the Griffith and Roesler equations (Chapter 1) and the Gibbs–Thomson and Avrami equations in considerations of phase nucleation and growth (Cantor 2020).

Fracture and plastic deformation at indentations were shown to be described by invariant cracking susceptibility B and hardness H, respectively, for variations in indentation load of over factors of 10^3, see Figures 2.15, 2.17, 2.21, and the online repository (Cook 2024). The variation in B with indenter angle ϕ, notwithstanding some scatter, was shown to be consistent with the expected dependence mentioned earlier, giving rise to an increase in B by a factor of approximately 2 for cube corner indenters relative to Vickers indenters, Figures 2.19 and 2.20. The variation in B observed over an extensive domain of brittle materials (drawn from the data repository of 45 materials and more than 4400 crack measurements) is a factor of approximately 15, Figure 2.22. The variation in P_Ω observed over the same domain is a factor of more than 10^4, Figure 2.25. The process of indentation crack initiation was shown to be a nonequilibrium transition from an unstable fracture equilibrium determined by P_Ω to a stable equilibrium determined by B. The decreasing crack velocities observed as crack surface traces approached stable equilibrium conditions were described by considerations of isothermal behavior for fracture systems weakly perturbed from equilibrium, Figure 2.27.

The extensive observations and modeling of indentation cracking behavior described here provide a firm foundation for many aspects of design with brittle materials. First, plots such as Figure 2.22 provide a direct means of assessing the precision and accuracy of an indentation crack length measurement technique. $c(P)$ measurements on control materials easily obtained in polished form, such as SLG, Si, sapphire, or Al_2O_3–TiC, can be used to assure that the behavior and B values obtained are consistent with those in Figure 2.22. Second, plots such as Figure 2.22 provide a means of assessing the contact fracture properties of developmental materials in comparison with the families of materials identified in Figure 2.22. For example, a value of crack susceptibility $B = 12\,\text{MPa m}^{1/2}$ for a new material might be regarded as acceptable for a prototype alumina, unacceptable for use as a structural ceramic, and exceptional for a barium titanate. Third, variability in materials manufacturing can be assessed through monitoring the batch to batch variability in B. Fourth, knowledge of B and the spectrum of sharp contact loads likely to be encountered by a material in operation enables prediction of the range of crack lengths likely to generated in the material. Such a range can be used to provide guidance for material use in operation or selection of nondestructive evaluation techniques. Alternatively, if the range of crack lengths generated in operation is known (by inspection or by a nondestructive evaluation technique) the spectrum of sharp contact loads encountered in operation can be inferred. Applications of P_Ω in brittle materials design are similar.

The observations and modeling of indentation cracking described here in Chapter 2 also provide a foundation for consideration of indentation fracture topics in the remainder of this book. In particular, the majority of additional topics involve the superposition of an externally applied stress field on a preexisting indentation residual stress field. The introduction of an applied stress enables consideration of the important engineering design factor of *strengths of components* containing sharp contact flaws. The introduction of applied stress also highlights the significant influence of two equilibrium fracture phenomena that are minor factors in determining indentation crack lengths but major factors in determining indentation strengths. The first phenomenon is the relaxation of the residual stress field by lateral cracking and chipping at large indentation loads. The second phenomenon is the increase in toughness by microstructural influences at long crack lengths. The first of these phenomena is considered in Chapter 3, and simultaneous action of both phenomena are considered in Chapters 4 and 5. [It is possible to discern lateral crack and microstructural effects through extensive indentation crack length measurements (Cook et al. 1990, 1994b; Cook and Liniger 1992), but the effects are small.] Finally, the introduction of applied stress makes clear the significant influence of nonequilibrium, activated isothermal fracture in determining observed strengths. The effects of nonequilibrium fracture, also commonly termed "slow crack growth" or "stress corrosion cracking," are considered in Chapter 6, building on the analyses in Chapters 3–5. Chipping, microstructural effects, and thermal activation act at three different length scales in fracture such that controlled indentation flaws are ideal for separating multiscale effects on strength (Cook 2015). [Controlled indentation flaws are also ideal for identifying length scale effects in fracture of surface stressed components, e.g., machined components (Cook et al. 1981), laminated composites (Cook 2005), and ion-exchanged glasses (Morris et al. 2004b). Such effects are not considered here.]

Chapter 3 makes perhaps the simplest extension to the fracture mechanics of indentation flaws by considering the superposed action of a uniform applied stress. Analysis of such action leads to the concepts of stable equilibrium crack extension and indentation strength. The indentation strengths of materials of invariant toughness are surveyed, including the destabilizing effects of lateral cracks.

References

Ali, S.A. (1993). Prediction of the service life of dental porcelains by the measurements of post-indentation slow crack growth. *Dental Materials Journal* 12: 45–53.

Anstis, G.R., Chantikul, P., Lawn, B.R., and Marshall, D.B. (1981). A critical evaluation of indentation techniques for measuring fracture toughness: I, Direct crack measurements. *Journal of the American Ceramic Society* 64: 533–538.

Anusavice, K.J., Gray, A., and Shen, C. (1991). Influence of initial flaw size on crack growth in air-tempered porcelain. *Journal of Dental Research* 70: 131–136.

Arora, A., Marshall, D.B., Lawn, B.R., and Swain, M.V. (1979). Indentation deformation/fracture of normal and anomalous glasses. *Journal of Non-Crystalline Solids* 31: 415–428.

Brace, W.F. (1960). Behavior of rock salt, limestone, and anhydrite during indentation. *Journal of Geophysical Research* 65: 1773–1788.

Cantor, B. (2020) *The Equations of Materials.* Oxford.

Charles, Y., Hild, F., Roux, S., and Vandembroucq, D. (2006). Material-independent crack arrest statistics: application to indentation experiments. *International Journal of Fracture* 142: 51–67.

Chen, X., Hutchinson, J.W., and Evans A.G. (2005). The mechanics of indentation induced lateral cracking. *Journal of the American Ceramic Society* 88: 1233–1238.

Chiang, S.S., Marshall, D.B., and Evans, A.G. (1982a). The response of solids to elastic/plastic indentation. I. Stresses and residual stresses. *Journal of Applied Physics* 53: 298–311.

Chiang, S.S., Marshall, D.B., and Evans, A.G. (1982b).The response of solids to elastic/plastic indentation. II. Fracture initiation. *Journal of Applied Physics* 53: 312–317.

Cloud, G. (1998). *Optical Methods of Engineering Analysis.* Cambridge.

Cook, R.F. (1989). Influence of crack velocity thresholds on stabilized nonequilibrium fracture. *Journal of Applied Physics* 65: 1902–1910.

Cook, R.F. (1994). Deformation and fracture by sharp rolling contacts. *Journal of the American Ceramic Society* 77: 1263–1273.

Cook, R.F., 2005. Toughening of a cordierite glass–ceramic by compressive surface layers. *Journal of the American Ceramic Society* 88: 2798–2808.

Cook, R.F. (2006). Strength and sharp contact fracture of silicon. *Journal of Materials Science* 41: 841–872.

Cook, R.F. (2015). Multi-scale effects in the strength of ceramics. *Journal of the American Ceramic Society* 98: 2933–2947.

Cook, R.F. (2019). Fracture sequences during elastic–plastic indentation of brittle materials. *Journal of Materials Research* 34: 1633–1644.

Cook, R.F. (2020). A critical evaluation of indentation cracks in air. *Journal of the American Ceramic Society* 103: 2278–2295.

Cook, R.F. (2024). Indentation fracture and deformation v.2. *Zenodo* DOI: 10.5281/zenodo.1149515.

Cook, R.F. and Braun, L.M. (1994). Trapped cracks at indentations: Part II Fracture mechanics model. *Journal of Materials Science* 29: 2192–2204.

Cook, R.F., Braun, L.M., and Cannon, W.R. (1994a). Trapped cracks at indentations: Part I Experiments on yttria-tetragonal zirconia polycrystals. *Journal of Materials Science* 29: 2133–2142.

Cook, R.F., Liniger E.G., and Pascucci, M.R. (1994b) Indentation fracture of polycrystalline cubic ceramics. *Journal of Hard Materials* 5: 191–212.

Cook, R.F. and Friedman, L.H. (2018). Weakly anisotropic residual contact stress in silicon demonstrated by electron backscatter diffraction and expanding cavity models. *Applied Physics Letters* 113: 231903.

Cook, R.F., Lawn, B.R., Dabbs, T.P., and Chantikul, P. (1981). Effect of machining damage on the strength of a glass-ceramic. *Journal of the American Ceramic Society* 64: C-121–C-122.

Cook, R.F. and Liniger, E.G. (1992). Grain-size effects in the indentation fracture of MgO. *Journal of Materials Science* 27: 4751–4761.

Cook, R.F. and Liniger, E.G. (1993). Kinetics of indentation cracking in glass. *Journal of the American Ceramic Society* 76: 1096–1105.

Cook, R.F., Pascucci, M.R., and Rhodes, W.H. (1990). Lateral cracks and microstructural effects in the indentation fracture of yttria. *Journal of the American Ceramic Society* 73: 1873–1878.

Cook, R.F. and Pharr, G.M. (1990). Direct observation and analysis of indentation cracking in glasses and ceramics. *Journal of the American Ceramic Society* 73: 787–817.

Cook, R.F. and Pharr, G.M. (1994). Indentation load–displacement behaviour during conventional hardness testing. *Journal of Hard Materials* 5: 179–190.

DelRio, F.W., Cook, R.F., and Boyce, B.L. (2015). Fracture strength of micro-and nano-scale silicon components. *Applied Physics Reviews*, 2: 021303.

Dériano, S., Jarry, A., Rouxel, T., Sangleboeuf, J.C., and Hampshire, S. (2004). The indentation fracture toughness (KC) and its parameters: the case of silica-rich glasses. *Journal of Non-Crystalline Solids* 344: 44–50.

Dukino, R.D. and Swain, M.V. (1992). Comparative measurement of indentation fracture toughness with Berkovich and Vickers indenters. *Journal of the American Ceramic Society* 75: 3299–3304.

Friedman, L.H., Vaudin, M.D., Stranick, S.J., Stan, G., Gerbig, Y.B., Osborn, W., and Cook, R.F. (2016). Assessing strain mapping by electron backscatter diffraction and confocal Raman microscopy using wedge-indented Si. *Ultramicroscopy* 163: 75–86.

Gayle, A.J., Friedman, L.H., Beams, R., Bush, B.G., Gerbig, Y.B., Michaels, C.A., Vaudin, M.D., and Cook, R.F. (2017). Two-dimensional strain-mapping by electron backscatter diffraction and confocal Raman spectroscopy. *Journal of Applied Physics* 122: 205101.

Gerbig, Y.B., Stranick, S.J., Morris, D.J., Vaudin, M.D., and Cook, R.F. (2009). Effect of crystallographic orientation on phase transformations during indentation of silicon. *Journal of Materials Research* 24: 1172–1183.

Gerbig, Y.B., Stranick, S.J., and Cook, R.F. (2010). Measurement of residual stress field anisotropy at indentations in silicon. *Scripta Materialia* 63: 512–515.

Gupta, P.K. and Jubb, N.J. (1981). Post-indentation slow growth of radial cracks in glasses. *Journal of the American Ceramic Society* 64: C-112–C-114.

Hagan, J.T. (1979a). Cone cracks around Vickers indentations in fused silica glass. *Journal of Materials Science* 14: 462–466.

Hagan, J.T. (1979b). Micromechanics of crack nucleation during indentations. *Journal of Materials Science* 14: 2975–2980.

Hagan, J.T. (1980). Shear deformation under pyramidal indentations in soda-lime glass. *Journal of Materials Science* 15: 1417–1424.

Hagan, J.T. and van der Zwaag, S. (1984). Plastic processes in a range of soda-lime-silica glasses. *Journal of Non-Crystalline Solids* 64: 249–268.

Harding, D.S., Oliver, W.C., and Pharr, G.M. (1995) Cracking during nanoindentation and its use in the measurement of fracture toughness. *Materials Research Society Symposium Proceedings* 365: 663–668 (MRS, Pittsburgh).

Hill, R. (1986). *The Mathematical Theory of Plasticity*. Oxford.

Jang, J.I., Lance, M.J., Wen, S., and Pharr, G.M. (2005). Evidence for nanoindentation-induced phase transformations in germanium. *Applied Physics Letters*, 86: 131907.

Jang, J.I. and Pharr, G.M. (2008). Influence of indenter angle on cracking in Si and Ge during nanoindentation. *Acta Materialia* 56: 4458–4469.

Jung, Y.G., Pajares, A., Banerjee, R., and Lawn, B.R. (2004). Strength of silicon, sapphire and glass in the subthreshold flaw region. *Acta Materialia* 52: 3459–3466.

Kedir, N., Faucett, D.C., Choi, S.R., and Bansal, N.P. (2018). Slow-crack-growth and indentation damage in calcium magnesium aluminosilicate (CMAS) glass from desert sand. *Ceramics International* 44: 2676–2682.

Lankford, J. (1982). Indentation microfracture in the Palmqvist crack regime: implications for fracture toughness evaluation by the indentation method. *Journal of Materials Science Letters* 1: 493–495.

Lankford, J. and Davidson, D.L. (1979a). The crack-initiation threshold in ceramic materials subject to elastic/plastic indentation. *Journal of Materials Science* 14: 1662–1668.

Lankford, J. and Davidson, D.L. (1979b). Indentation plasticity and microfracture in silicon carbide. *Journal of Materials Science* 14: 1669–1675.

Lathabai, S., Rödel, J., Dabbs, T., and Lawn, B.R. (1991). Fracture mechanics model for subthreshold indentation flaws: Part I equilibrium fracture. *Journal of Materials Science* 26: 2157–2168.

Laugier, M.T. (1985). Palmqvist crack extension and the center-loaded penny crack analogy. *Journal of the American Ceramic Society* 68: C-51–C-52.

Laugier, M.T. (1987). New formula for indentation toughness in ceramics. *Journal of Materials Science Letters* 6: 355–356.

Lawn, B.R. and Cook, R.F. (2012). Probing material properties with sharp indenters: a retrospective. *Journal of Materials Science* 47: 1–22.

Lawn, B.R., Dabbs, T.P., and Fairbanks, C.J. (1983). Kinetics of shear-activated indentation crack initiation in soda-lime glass. *Journal of Materials Science* 18: 2785–2797.

Lawn, B.R. and Evans, A.G. (1977). A model for crack initiation in elastic/plastic indentation fields. *Journal of Materials Science* 12: 2195–2199.

Lawn, B.R., Evans, A.G., and Marshall, D.B. (1980). Elastic/plastic indentation damage in ceramics: the median/radial crack system. *Journal of the American Ceramic Society* 63: 574–581.

Lawn, B.R. and Fuller, E.R. (1975). Equilibrium penny-like cracks in indentation fracture. *Journal of Materials Science* 10: 2016–2024.

Lawn, B.R., Jakus, K., and Gonzalez, A.C. (1985). Sharp vs blunt crack hypotheses in the strength of glass: a critical study using indentation flaws. *Journal of the American Ceramic Society* 68: 25–34.

Lawn, B.R., Marshall, D.B., and Chantikul, P. (1981). Mechanics of strength-degrading contact flaws in silicon. *Journal of Materials Science* 16: 1769–1775.

Lawn, B.R. and Swain, M.V (1975). Microfracture beneath point indentations in brittle solids. *Journal of Materials Science* 10: 113–122.

Lawn, B. and Wilshaw, R. (1975). Indentation fracture: principles and applications. *Journal of Materials Science* 10: 1049–1081.

Li, Z., Ghosh, A., Kobayashi, A.S., and Bradt, R.C. (1989). Indentation fracture toughness of sintered silicon carbide in the Palmqvist crack regime. *Journal of the American Ceramic Society* 72: 904–911.

Marshall, D.B., Cook, R.F., Padture, N.P., Oyen, M.L., Pajares, A., Bradby, J.E., Reimanis, I.E., Tandon, R., Page, T.F., Pharr, G.M., and Lawn, B.R. (2015). The compelling case for indentation as a functional exploratory and characterization tool. *Journal of the American Ceramic Society* 98: 2671–2680.

Marshall, D.B. and Lawn, B.R. (1979). Residual stress effects in sharp contact cracking: Part 1 Indentation fracture mechanics. *Journal of Materials Science* 14: 2001–2012.

Marshall, D.B., Lawn, B.R., and Chantikul, P. (1979). Residual stress effects in sharp contact cracking: Part 2 strength degradation. *Journal of Materials Science* 14: 2225–2235.

Marshall, D.B., Lawn, B.R., and Evans, A.G. (1982). Elastic/plastic indentation damage in ceramics: the lateral crack system. *Journal of the American Ceramic Society* 65: 561–566.

Morris, D.J. and Cook, R.F. (2004). In situ cube-corner indentation of soda–lime glass and fused silica. *Journal of the American Ceramic Society* 87: 1494–1501.

Morris, D.J., Myers, S.B., and Cook, R.F. (2004a). Sharp probes of varying acuity: instrumented indentation and fracture behavior. *Journal of Materials Research* 19: 165–175.

Morris, D.J., Myers, S.B., and Cook, R.F. (2004b).Indentation crack initiation in ion-exchanged aluminosilicate glass. *Journal of Materials Science* 39: 2399–2410.

Mulhearn, T.O. (1959). The deformation of metals by Vickers-type pyramidal indenters. *Journal of the Mechanics and Physics of Solids* 7: 85–88.

Oliver, D.J., Bradby, J.E., Williams, J.S., Swain, M.V., and Munroe, P. (2007). Giant pop-ins and amorphization in germanium during indentation. *Journal of Applied Physics* 101: 043524.

Oliver, D.J., Lawn, B.R., Cook, R.F., Reitsma, M.G., Bradby, J.E., Williams, J.S., and Munroe, P. (2008). Giant pop-ins in nanoindented silicon and germanium caused by lateral cracking. *Journal of Materials Research* 23: 297–301.

Oliver, W.C. and Pharr, G.M. (1992). An improved technique for determining hardness and elastic modulus using load and displacement sensing indentation experiments. *Journal of Materials Research* 7: 1564–1583.

Padture, N.P. (1994). *In situ*-toughened silicon carbide. *Journal of the American Ceramic Society* 77: 519–523.

Palmqvist, S. (1957). A method to determine the toughness of brittle materials, especially hard metals. *Jernkontorets Annaler* 141: 300–307.

Pharr, G.M. and Cook, R.F. (1990). Instrumentation of a conventional hardness tester for load–displacement measurement during indentation. *Journal of Materials Research* 5: 847–851.

Pharr, G.M., Harding, D.S., and Oliver, W.C. (1993). Measurement of fracture toughness in thin films and small volumes using nanoindentation methods. In *Mechanical Properties and Deformation Behavior of Materials having Ultra-Fine Microstructures* (Eds. Nastasi, M., Parkin, D.M., and Gleiter, H.) pp. 449–461. Dordrecht: Springer.

Samuels, L.E. and Mulhearn, T.O. (1957). An experimental investigation of the deformed zone associated with indentation hardness impressions. *Journal of the Mechanics and Physics of Solids* 5: 125–134.

Swadener, J.G. and Nastasi, M. (2002). Effect of dopants on the fracture toughness of silicon. *Journal of Materials Science Letters* 21: 1363–1365.

Tabor, D. (1951). *The Hardness of Metals*. Oxford.

Tandon, R. and Buchheit, T.E. (2007). Use of cube-corner nano-indentation crack length measurements to estimate residual stresses over small spatial dimensions. *Journal of the American Ceramic Society* 90: 502–508.

Tandon, R. and Cook, R.F (1992). Cone crack nucleation at sharp contacts. *Journal of the American Ceramic Society* 75: 2877–2880.

Thurn, J. and Cook, R.F. (2002). Simplified area function for sharp indenter tips in depth-sensing indentation. *Journal of Materials Research* 17: 1143–1146.

Thurn, J., Morris, D.J., and Cook, R.F. (2002). Depth-sensing indentation at macroscopic dimensions. *Journal of Materials Research* 17: 2679–2690.

Tsai, Y.L. and Mecholsky, J.J. (1991a). Fractal fracture of single crystal silicon. *Journal of Materials Research* 6: 1248–1263.

Tsai, Y.L. and Mecholsky, J.J. (1992b). Fracture mechanics description of fracture mirror formation in single crystals. *International Journal of Fracture* 57: 167–182.

van der Zwaag, S., Hagan, J.T., and Field, J.E. (1980). Studies of contact damage in polycrystalline zinc sulphide. *Journal of Materials Science* 15: 2965–2972.

Vaudin, M.D., Gerbig, Y.B., Stranick, S.J., and Cook, R.F. (2008). Comparison of nanoscale measurements of strain and stress using electron back scattered diffraction and confocal Raman microscopy. *Applied Physics Letters* 93: 193116.

Vaudin, M.D., Stan, G., Gerbig, Y.B., and Cook, R.F. (2011). High resolution surface morphology measurements using EBSD cross-correlation techniques and AFM. *Ultramicroscopy* 111: 1206–1213.

3

Indentation Strengths: Invariant Toughness Materials

This chapter uses the analysis of Chapter 1 in a survey of Vickers (and conical and spherical) indentation strengths of brittle materials. The materials are glasses, single crystals, and fine-grained polycrystals and composites that exhibit little effect of microstructure on fracture such that the materials exhibit invariant toughness. The strength tests are largely conducted in inert environments. At small indentation loads, the indentation-strength behavior approaches the ideal response. At large indentation loads most of the materials exhibit strengths well above the ideal response, reflecting indentation stress relaxation associated with multiple (especially lateral) crack patterns and chipping, such that in many cases the strengths approach a lower-bound chipping controlled invariant strength. The Vickers indentation strength behavior is compared with strength degradation resulting from sharp particle impact and erosive wear conditions. At very small indentation loads, indentations are crack free, leading to sub threshold strengths that are often considerably greater than the larger load post threshold strengths. Fracture mechanics analyses are developed to describe indentation strengths and the effects of indenter geometry, lateral cracking, impact, surface stresses, and sub threshold flaws. Strength data are presented in a single format for ease of interpretation in terms of idealized indentation fracture and for inter-material comparisons.

3.1 Indentation Strength of Brittle Materials

3.1.1 Historical Context

In the early study of Griffith (1921), the physics of brittle fracture was shown clearly as a manifestation of the First Law of Thermodynamics regarding the interchangeability of energy. Chapter 1 considers such energy changes in detail and can be briefly summarized as follows: Work performed on a component by applied loading is converted into internal elastic strain energy within the component that is converted on fracture into surface energy on newly formed surfaces of the component. The mechanism by which the final conversion to surface energy takes place is crack propagation; the expansion in area of a plane of ruptured bonds in a component, a crack, leading to component failure. Griffith showed that the critical stress required to propagate a crack was an inverse function of crack size. Hence, variation of fracture strengths within a group of components is explained by variation in the size of the underlying strength-controlling crack or flaw in each component, as initially considered by Weibull (1939) and Epstein (1948).

Alternatively, deliberate introduction of controlled flaws into components can circumvent strength variability. This obvious method to enable assessment of material strength properties was implemented from the earliest studies. For example, Griffith (1921) used annealed scratches, Levengood (1958) used roller tracks, Mould and Southwick (1959a,b) and Nordberg et al. (1964) used abrasions, Gruver and Kirchner (1973) used lathe generated scratches, and Petrovic et al. (1975, 1976) and Petrovic and Jacobson (1976) used Vickers and Knoop indentations. In somewhat less controlled manners, Baker and Preston (1946) used manufacturing damage, Shand (1959) used uniformly ground surfaces, and in the 1930s and 1950s, Hirata and Terao used thermal gradients and thermal shock to initiate cracks (for a brief review see Terao 1978). It is now known that the scratches, tracks, abrasions, and indentations were all manifestations of the First Law and interchangeability of energy during brittle fracture by contact. Such energy changes were first analyzed in this way by Roesler (1956) and are considered in detail in Chapter 2. Roesler showed that the crack length generated on contact was an increasing function of contact load. Hence, variation of fracture strengths within a group of contacted components is explained by variation in the contact load that generated the strength-controlling crack or flaw in each component, as was implicit in the works of Griffith, Levengood, Mould and Southwick, and Petrovic and colleagues.

Indentation Fracture: Strength and Toughness for Brittle Materials Design, First Edition. Robert F. Cook.

In an overview considering application of controlled indentation flaws in strength measurements, Petrovic and Mendiratta (1979) noted the advantages and disadvantages of indentations and highlighted the importance of indentation contact deformation and the attendant residual stress field in influencing the strength of indented components. At about the same time, an indentation fracture mechanics formulation incorporating the residual field was established by Marshall and Lawn (1979) and Marshall et al. (1979), in which crack propagation at sharp indentations and the mechanics of indented component strengths were described. Subsequently, thousands of studies have implemented indentation techniques to evaluate fracture properties of brittle materials. The vast majority of these studies make use of the apparent simplicity and economy of indentations to generate measurable stable cracks, considered in detail in Chapter 2, from which material fracture resistance or toughness can be estimated (Anstis et al. 1981; see Cook 2020a,b for a recent review and commentary). Very few studies make use of the controlled flaw properties to investigate material strengths and the underlying flaw and toughness variations or make use of the myriad additional applications (Lawn and Cook 2012; Marshall et al. 2015).

This chapter considers the indentation strength behavior of materials with invariant toughness. The major focus is the decrease in strength with increased indentation load—*strength degradation*—of indented components, and how material properties and indentation geometry affect such degradation. The fracture mechanics of indentation strength is framed in terms of the propagation of indentation half-penny cracks under the combined influences of the localized residual indentation contact field (see Chapter 2) and a globally applied stress field. The importance and pervasive influence of lateral cracks in mediating the amplitude of the indentation field and thus indentation strengths are demonstrated and emphasized. The commonly encountered, domestically and industrially, indentation process of strength degradation by sharp particle impact is considered (Sections 3.2 and 3.3), and the analyses and observations considered here provide background and context for other commonly encountered and subsequently discussed indentation processes of scratching (Chapter 9) and wear (Chapter 10). Chapters 4 and 5 consider the indentation strength behavior of materials with varying toughness and an indentation technique for toughness estimation, respectively, and build on the analyses in this chapter. The observations here (Sections 3.1–3.4) considerably extend two earlier indentation strength surveys (Lawn et al. 1980; Chantikul et al. 1981b), and the similarities and differences between indentation strength behavior and thermal shock measurements are considered in discussion (Section 3.5). The remainder of this section describes the experimental basis of the indentation strength methodology and presents some "ideal" indentation strength measurements.

3.1.2 Experimental Methodology

An indentation strength test consists of two steps, illustrated in Figure 3.1: Step 1, controlled indentation of a component surface, followed by Step 2, strength measurement of the component. The indenter is usually a Vickers four-sided diamond pyramid, and the peak indentation load normal to the surface P is the controlled or independent test variable. Indentations are usually performed in ambient (moist) air. The strength measurement is usually performed under inert conditions, using rapid applied stressing rates, GPa s^{-1} or greater, and in dry environments, N_2 gas or under a silicone oil film. Both rate and environment are implemented so as to exclude or impede atmospheric moisture from reacting with the strength-controlling flaw during strength testing. The strength thus measured is the *inert strength* σ_m and is the dependent variable. For sharp indenters such as the Vickers pyramid, the introduced flaw usually consists of a square plan surface contact impression, a near-hemispherical subsurface plastic deformation zone (pdz), and near-semicircular outline subsurface half-penny cracks exhibiting surface traces collinear with the impression diagonals. The strength

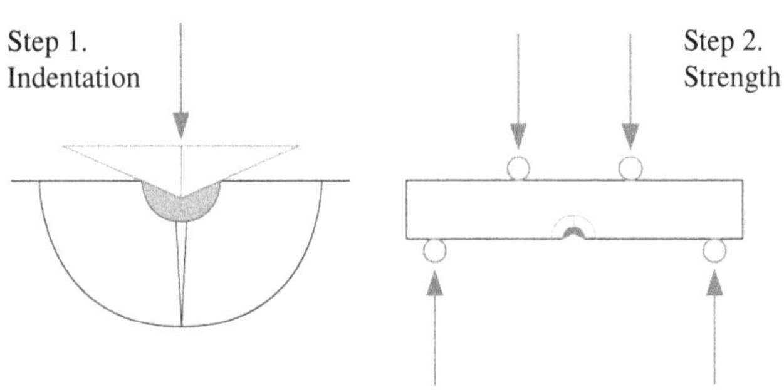

Figure 3.1 A schematic diagram of the two-step experimental indentation strength procedure for measuring fracture properties of brittle materials.

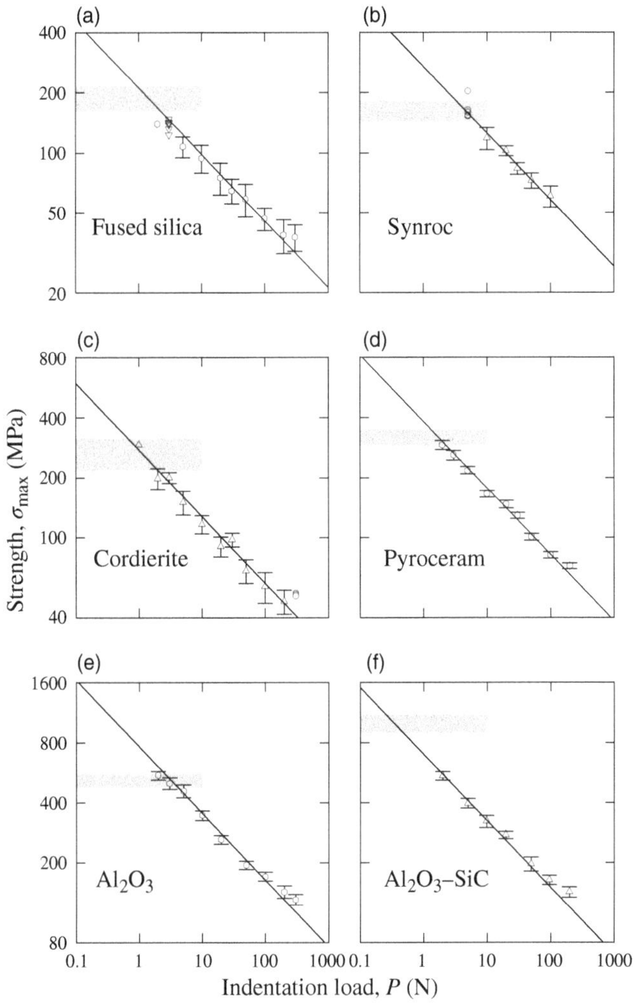

Figure 3.2 Logarithmic plots of indentation strength measurements for materials exhibiting near-ideal Vickers indentation strength behavior, $\sigma_m \sim P^{-1/3}$. (a) Fused Silica. (b) Synroc. (c) Cordierite. (d) Pyroceram. (e) Alumina. (f) Alumina-SiC composite. Source: Includes data adapted from Cook (2015), Cook et al. (1981, 1982a,b, 1985, 1987), and Zhao et al. (1993).

testing method usually employs one of two geometries: either uniaxial bending of rectangular section beams or circular section rods or biaxial flexure of square plates or disks. In both uniaxial and biaxial geometries the schematic elevation of Figure 3.1 pertains.

Figure 3.2 shows examples of $\sigma_{max}(P)$ observations for material systems that exhibited near-ideal Vickers indentation strength behavior. Some of the materials were introduced previously (Cook et al. 1981, 1982a,b, 1985, 1987; Zhao et al. 1993; Cook 2015), but nearly all data sets of strength are significantly enhanced from those in the prior works. The general $\sigma_{max}(P)$ data are shown in logarithmic plots with fixed domain:range aspect ratios of 500:1; note that the absolute value of the strength scale increases from top to bottom. The symbols with bars in each plot represent the means and standard deviation limits of typically 8–20 individual strength measurements at each designated indentation load. Isolated symbols represent individual strength measurements. Only test specimens ascertained to have failed from the introduced indentation flaws are included in the symbol sets. The shaded bar in each plot represents the mean and standard deviation limits of strength measurements of specimens that did not fail from indentation flaws—either as no indentation was introduced or the specimen did not fail from an introduced indentation but from an intrinsic or preexisting flaw. These upper strength levels represent the *intrinsic strength* σ_i of the components. The diagonal line in each plot is a best fit to the $\sigma_m(P)$ data of the form $\sigma_m = \hat{S}P^{-1/3}$, shown in Section 3.2 to describe the strength behavior of idealized indentations, where $\hat{S}$ is the material-dependent *mean contact strength*. The $\hat{S}$ values for the materials in Figure 3.2 are given in Table 3.1 as means and standard deviations of the experimental data sets over the load domains indicated.

Table 3.1 Mean contact strengths.

Material	Microstructure	Load Domain (N)	Contact Strength, $\hat{S}$ (MPa $N^{1/3}$)
Fused silica	Glass	2–300	211.9 ± 37.2
Synroc	Composite	5–100	271.4 ± 29.6
Cordierite	Glass-ceramic	1–300	276.9 ± 32.0
Pyroceram[a)]	Glass-ceramic	2–200	383.4 ± 20.9
Al_2O_3[b)]	Polycrystal	2–300	765 ± 64
Al_2O_3–SiC	Composite	2–200	810 ± 37
Y–ZrO_3	Polycrystal	4–300	1000 ± 150

a) Corning C9606.
b) Friedrichsfeld F99.

There is no doubt that ideal indentation strength responses describe the behavior of the materials in Figure 3.2. The values of $\hat{S}$ are thus a useful tool for designers of brittle materials: Strength degradation of components by sharp contacts of known load can be determined and use conditions and component size adjusted accordingly. $\hat{S}$ is thus a natural engineering successor to the strength degradation parameter D introduced in Chapter 1 to describe crack length limited Griffith strengths. In addition, an effective contact load P_i characterizing the intrinsic strength can be estimated by $P_i = (\hat{S}/\sigma_i)^3$ and provides an upper bound to allowed sharp contact loads in use if the intrinsic strength is required. An example of intrinsic strength and contact load evaluation is given for cordierite composites in Chapter 11. A feature of materials exhibiting near-ideal indentation strength behavior is that the transition from indentation to intrinsic flaw-controlled strength is abrupt. The "success" rate of indentation failures abruptly decreases to zero as the indentation load is decreased and approaches P_i. The isolated symbols in Figure 3.2a–c attest to this lack of success. The strength degradation behavior is easily placed in domestic context: The mass of a 200 mm diameter pyroceram dinner plate is about 100 g, leading to about 1 N weight. Hence, simply placing a plate on a sharp particle under its own weight is enough to generate a flaw greater in size than those responsible for the intrinsic strength, as shown in Figure 3.2d. At the other extreme, strength measurements using large indentations is often problematic as many indenters lack large load capability and large load indentation is often difficult and fear inducing. The fear is well founded: a 300 N indentation (about 66 lb) is like a fully grown Labrador retriever balanced on a diamond ring. The isolated symbols at $P = 300$ N in Figure 3.2c attest to this difficulty.

Nevertheless, although there are small and large indentation caveats, an ideal indentation response characterized by $\hat{S}$ describes the leading indentation strength behavior for many materials. Section 3.2 develops fracture mechanics analyses to describe this behavior.

3.2 Fracture Mechanics of Indentation Strength

3.2.1 Ideal Indentations

The analytical basis of the indentation strength methodology is a fracture mechanics description by two stress-intensity factors (SIF) of the half-penny indentation crack in the strength test of Figure 3.1. The SIF describe the applied stress field during strength testing, K_a, given by Eq. (1.34),

$$K_a = \psi \sigma_a c^{1/2} \tag{3.1}$$

and the postindentation residual contact field, K_r, given by Eq. (2.11),

$$K_r = \chi P / c^{3/2}. \tag{3.2}$$

σ_a is the instantaneous applied stress and P is the peak indentation load. In the two-step sequence of Figure 3.1, K_a and K_r are superposed. ψ and χ are dimensionless geometry terms, discussed in detail in Chapters 1 and 2. Taken separately K_a

leads to the Griffith equation, and K_r leads to the Roesler equation. During an indentation strength test, the total SIF K is given by the sum of K_a and K_r

$$K = K_a + K_r = \psi\sigma_a c^{1/2} + \chi P/c^{3/2}. \tag{3.3}$$

The fracture system is idealized in that: K_a and K_r are both taken to represent crack opening modes and are therefore simply additive with no cross terms (Lawn 1993); free surface effects are included in the dimensionless numerical factors, taken as invariant ψ and χ; each term is linear in its amplitude factor σ_a and P; and each term has a simple functional factor $c^{a/2}$ and $c^{r/2}$, where $a = 1$ and $r = 3$ (Fuller et al. 1983; Cook 1989). A further idealization is that the toughness of the material of which the component is formed is an invariant value T, pertaining in the test environment during both indentation and stressing steps of Figure 3.1.

The equilibrium condition for the above system is $K = T$ such that, using Eq. (3.3),

$$\psi\sigma_a c^{1/2} + \chi P/c^{3/2} = T. \tag{3.4}$$

The functional forms of K_a and K_r have opposing stability influences: K_r is stabilizing, $dK_r/dc < 0$, and dominates at small crack lengths; K_a is destabilizing, $dK_a/dc > 0$, and dominates at large crack lengths. An intermediate crack length of neutral stability, marking incipient instability on increasing crack length, is thus given by $dK/dc = dT/dc$, which for invariant T devolves to $dK/dc = 0$. The instability condition for the above system is thus, using Eq. (3.4) and $c \neq 0$,

$$\psi\sigma_a c^{1/2} - 3\chi P/c^{3/2} = 0. \tag{3.5}$$

Consideration of the forms of K_a and K_r shows that the total SIF K of Eq. (3.3) passes through a minimum of incipient instability given by the solution of Eq. (3.5). Simultaneous imposition of Eqs. (3.4) and (3.5) constrains this instability to be an equilibrium configuration. The joint constraints are written compactly in matrix form as

$$\begin{bmatrix} 1 & 1 \\ 1 & -3 \end{bmatrix} \begin{bmatrix} \mathbf{K_a} \\ \mathbf{K_r} \end{bmatrix} = \begin{bmatrix} 1 \\ 0 \end{bmatrix}, \tag{3.6}$$

where $\mathbf{K_a} = K_a/T$ and $\mathbf{K_r} = K_r/T$ are scaled SIF variables. Solution of the simultaneous equations represented by Eq. (3.6) yields the crack length–stress combination (c_m, σ_m), where

$$c_m = (4\chi P/T)^{2/3} \tag{3.7}$$

and

$$\sigma_m = 3T/4\psi c_m^{1/2} = 3T^{4/3}/4^{4/3}\psi(\chi P)^{1/3}. \tag{3.8}$$

The second expression for σ_m in Eq. (3.8) uses Eq. (3.7) and can be written simply as

$$\sigma_m = \hat{S}P^{-1/3} \tag{3.9}$$

in analogy with Eqs. (1.16) and (1.27). The mean contact strength $\hat{S}$ is given by a combination of material and flaw geometry parameters,

$$\hat{S} = 3T^{4/3}/4^{4/3}\psi\chi^{1/3}, \tag{3.10}$$

where the caret indicates the value of S for an ideal indentation system. Equation (3.9) describes the best-fit lines in Figure 3.2. The invariance of T throughout the entire indentation strength measurement implies that postindentation, prior to the application of applied stress, at $\sigma_a = 0$, an initial zero-stress crack length c_0 is given by an expression similar to Eq. (2.14),

$$c_0 = (\chi P/T)^{2/3} = c_m/4^{2/3} \approx c_m/2.52. \tag{3.11}$$

If, in addition, a step is inserted between indentation and strength testing such that the residual stress is removed, $\chi \to 0$, [by annealing (Marshall et al. 1979; Yasutake et al. 1986; Anderson and Braun 1990), surface grinding (Petrovic and Mendiratta 1979; Cook et al. 1981), or corrosion (Cook 2018b)] without altering the crack length, the strength of the indented component exhibits the Griffith strength σ_0,

$$\sigma_0 = T/\psi c_0^{1/2} = (4^{4/3}/3)\sigma_m \approx 2.12\sigma_m, \tag{3.12}$$

or $\sigma_m = (3/4^{4/3})\sigma_0 \approx 0.47\sigma_0$. Implicit in these considerations is that the toughness T is that for an inert environment and thus the "m" subscript indicates the configuration of *maximum* supportable stress, or strength, of a component containing indentation P. This inert condition need not be the case and, in fact, this constraint is relaxed in most indentation systems, such that the initial crack length is more typically given by Eq. (2.13) and $c_{air} > c_0$. This issue is addressed in Section 3.2.2 and Chapters 5 and 6.

Fracture behavior during an indentation strength test is conveniently examined using scaled variables. In addition to T used in Eq. (3.6), previous choices have used crack length and strength sets that deliberately excluded an explicit dependence on the indentation load: (c_m, σ_m) (Marshall et al. 1979; Fuller et al. 1983) and (c_0, σ_0) (Marshall and Lawn 1980; Chantikul et al. 1981a). As the variation of indentation strength with indentation load is a focus of this book, indentation load is introduced explicitly here through scaling using the experimental threshold indentation load P_Ω as defined in Chapter 2. The crack length and strength set is thus $(c_\Omega, \sigma_\Omega)$, where

$$c_\Omega = (\chi P_\Omega / T)^{2/3} \tag{3.13}$$

and

$$\sigma_\Omega = 3T^{4/3}/4^{4/3}\psi(\chi P_\Omega)^{1/3}, \tag{3.14}$$

where χ is the amplitude factor for the residual field that pertains at crack initiation (Cook 2019) and during a strength test. Setting the ensuing scaling relations

$$\begin{aligned} \mathbf{K} &= K/T \\ \mathbf{C} &= c/c_\Omega \\ \mathbf{S_a} &= \sigma_a/\sigma_\Omega \\ \mathbf{P} &= P/P_\Omega \end{aligned} \tag{3.15}$$

enables the SIF expression of Eq. (3.3) to be written in simple scaled form as

$$\mathbf{K} = (3/4^{4/3})\mathbf{S_a}\mathbf{C}^{1/2} + \mathbf{P}\mathbf{C}^{-3/2}, \tag{3.16}$$

recognizing that the terms $\mathbf{S_a}$ and $\mathbf{P}$ are now simply numerical weighting coefficients and that for ideal indentations χ and T are invariant and implicitly take maximal values.

Figure 3.3 shows logarithmic plots of the $\mathbf{K}(\mathbf{C})$ variation from Eq. (3.16) for (a) $\mathbf{P} = 1$ and (b) $\mathbf{P} = 10$ for various indicated values of $\mathbf{S_a}$. In both plots, the $\mathbf{S_a} = 0$ responses are shown as dashed straight lines. For $\mathbf{P} = 1$, the postindentation configuration is indicated by the intersection of the $\mathbf{S_a} = 0$ line with the horizontal $\mathbf{K} = 1$ equilibrium line at crack length $\mathbf{C} = 1$. As $\mathbf{P} = 1$ represents the threshold indentation load, this configuration represents the smallest indentation crack in this system. For a small increase in applied stress to $\mathbf{S_a} = 0.25$, a $\mathbf{K}(\mathbf{C})$ response with a minimum results, including

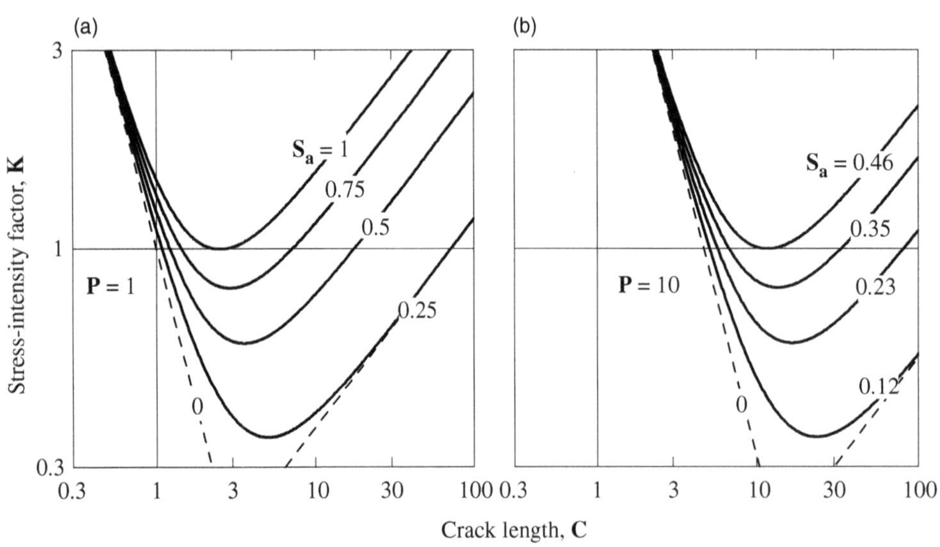

Figure 3.3 Logarithmic plots of scaled stress-intensity factor versus crack length for ideal indentation behavior at two scaled indentation loads. (a) $\mathbf{P} = 1$. (b) $\mathbf{P} = 10$. The stress-intensity factors exhibit minima between the small crack limit, dominated by the indentation residual field, and the large crack limit, dominated by the applied stress field.

an additional large crack asymptotic response indicated by a dashed line. The $\mathbf{K}(\mathbf{C})$ function now exhibits two equilibria. At small crack lengths, the initial stable equilibrium configuration has now shifted to a slightly greater crack length, $\mathbf{C} \gtrsim 1$. At large crack lengths, a new unstable equilibrium configuration at $\mathbf{C} \gg 1$ now exists. The interpretation is that the new stable configuration reflects an extension of the initial crack through a series of near-equilibrium configurations. The unstable configuration is inaccessible as the system must pass through a series of subequilibrium $\mathbf{K} < 1$ states between the two equilibrium configurations. As the applied stress is increased to $\mathbf{S_a} = 0.5$ and $\mathbf{S_a} = 0.75$, the stable equilibrium crack length increases, reflecting crack extension, and the unstable equilibrium crack length decreases, although still inaccessible. For $\mathbf{S_a} = 1$ the stable and unstable branches merge at $\mathbf{C} = 2.52$. Under the action of applied stress, the stable crack has extended to a point of incipient instability and any positive perturbation of the system (in crack length or stress) leads to a $\mathbf{K} > 1$ nonequilibrium unstable configuration, rapidly increasing crack length and crack velocity (see Section 1.6.1), and dynamic component failure. The stress $\mathbf{S_a} = 1$ is thus the indentation strength and represents the greatest strength in this system for an ideal indentation. It is noted that the stable–unstable transitions of Figure 3.3 are the reverse of the unstable–stable transition of Figure 1.19: Crack propagation in a destabilizing field is the reciprocal of crack initiation in a stabilizing field.

The physical behavior of the system at a greater indentation load $\mathbf{P} = 10$ is identical to that at $\mathbf{P} = 1$. There are, however, different numerical values of the crack length and strength parameters. Using Eq. (3.15), it is straightforward to show that the scaled initial and final crack lengths expressions are $\mathbf{C_0} = \mathbf{P}^{2/3}$ and $\mathbf{C_m} = 4^{2/3}\mathbf{C_0}$ and that the scaled indentation strength expression is $\mathbf{S_m} = \mathbf{P}^{-1/3}$. The previous considerations for $\mathbf{P} = 1$ are seen to be simple examples of these expressions. For $\mathbf{P} = 10$, the postindentation configuration, indicated as before by the intersection of the $\mathbf{S_a} = 0$ line with $\mathbf{K} = 1$ line, is thus at crack length $\mathbf{C_0} \approx 4.6$. On increase of applied stress, the stable crack length increases and an inaccessible unstable crack length decreases, also as before, although both crack lengths are greater and the required applied stress is smaller. At $\mathbf{S_m} \approx 0.46$ the stable and unstable branches merge at $\mathbf{C_m} \approx 11.6$ and the system fails.

A hallmark of the indentation strength test is the subcritical equilibrium extension of indentation cracks under the influence of applied stress. All the indentation cracks represented in the strength degradation plots of Figure 3.2 must have exhibited such precursor extension prior to component failure. The SIF plots of Figure 3.3 demonstrate the physical behavior of such stressed indentation cracks but restrict visualization of the precursor extension to a single horizontal line along $\mathbf{K} = 1$ and require separate lines and plots for different values of $\mathbf{S_a}$ and $\mathbf{P}$. A constrained alternative plotting scheme enables greater focus on distinguishing the behavior of the experimental variables $\mathbf{S_a}$, $\mathbf{C}$, and $\mathbf{P}$. The scheme sets $\mathbf{K} = 1$ in Eq. (3.16) and inverts the equation to gain

$$\mathbf{S_a} = \frac{4^{4/3}}{3\mathbf{C}^{1/2}}\left[1 - \mathbf{P}\mathbf{C}^{-3/2}\right]. \tag{3.17}$$

Figure 3.4 is a logarithmic plot of $\mathbf{S_a}(\mathbf{C})$ variations from Eq. (3.17) for $\mathbf{P} = 1, 3, 10$, and 30. The fine lines for each indentation load show the complete responses. The overlaid bold lines show the experimentally accessible regions of $\mathbf{C} = \mathbf{C_0}$ to $\mathbf{C_m}$ and $\mathbf{S_a} = 0$ to $\mathbf{S_m}$ for each case. The dashed line is the $\mathbf{S_0}$ ($\approx 2.12\,\mathbf{S_m} = 2.12\,\mathbf{C}^{-1/2}$) behavior to which all responses are asymptotic at $\mathbf{C} \gg 1$. The precursor crack length-applied stress trajectories during failure of ideal indentations and the effects of indentation load are clear in Figure 3.4. On loading by an applied stress, indentation cracks extend through a sequence of equilibrium configurations that become progressively less stable as the crack extends until stability vanishes at maximum extension and maximum supported applied stress. Halting the loading at an intermediate stress halts the crack extension at an intermediate stable equilibrium crack length. Such trajectories and effects are thus directly amenable to experiment and allow assessment of the assumption of ideal indentation behavior. Weaker assessments of ideal indentation behavior are provided in other ways. Combining Eqs. (3.8) and (2.13) implies

$$\sigma_m = Dc_{air}^{-1/2}, \tag{3.18}$$

which is a version of the Griffith equation, Eq. (1.16). Confirmation of Eq. (3.18) is provided in Figure 1.10e, which suggests (only) that the ratio $(\chi_{air}T_0/\chi_0T_{air})$ is indentation load invariant for the material. Similarly, combining Eqs. (3.9) and (2.13) implies

$$\hat{S} \sim B^{-2}. \tag{3.19}$$

This relation will be examined after further consideration of indentation strength data, but even if shown to be consistent with experimental observations suggests (only) that the ratio $c_m/c_{air} \sim (T_{air}/T_0)$ is material invariant. In indentation strength experiments the ratios mentioned above figure prominently in application and validation

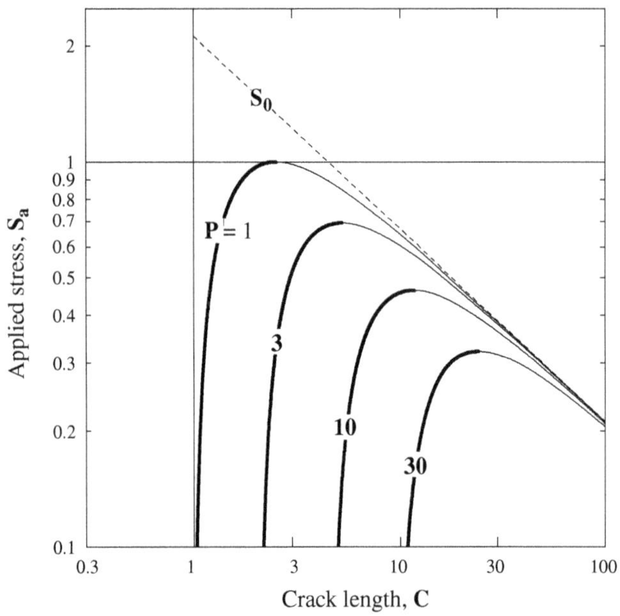

Figure 3.4 Logarithmic plots of scaled applied stress versus crack length for ideal equilibrium indentation fracture behavior at several scaled indentation loads. The applied stress responses exhibit maxima indicating the indentation strengths. The bold lines indicate experimentally accessible behavior.

of ideal indentation fracture mechanics. In particular, the ratios appear directly in consideration of experimental measurements of precursor crack extension during indentation strength tests. Such an extension is the subject of Section 3.2.2.

3.2.2 Precursor Crack Extension

A major point to be recognized in crack extension during indentation strength tests is that the cracks have often passed through a sequence of environments before and during loading by an applied stress. The environments are characterized by the inert toughness T_0 and the reactive toughness, typically that for moist air, $T_{air} < T_0$. The sequence of environments experienced by the cracks is usually inert–reactive–inert. The sequence of mechanical processes is crack initiation–arrest–loading, characterized by a sequence of fracture fields, destabilizing–stabilizing–destabilizing. The environmental sequence requires refinement of the crack length expressions applicable during an indentation strength test.

In the first step of the sequence, half-penny crack initiation takes place dynamically in near-inert conditions, from a crack nucleus c_1 to an initial crack length c_0 (see Chapter 2). Equation (2.14) thus applies:

$$c_0 = (\chi_0 P/T_0)^{2/3}. \tag{3.20}$$

However, most indentations are performed in air, which leads to the second step of the sequence. There is both a decrease in material toughness and an increase in indentation lateral cracking. These effects lead (simultaneously) to nonequilibrium extension of the half-penny crack (as described in Chapter 1) and to partial relaxation of the residual field (Lawn et al. 1985; Cook and Roach 1986; Cook and Liniger 1993). As a consequence, the half-penny crack extension takes place kinetically in reactive conditions to an arrested length $c_{air} > c_0$ given by Eq. (2.13):

$$c_{air} = (\chi_{air} P/T_{air})^{2/3}. \tag{3.21}$$

Unless special precautions are taken (Anstis et al. 1981; Chantikul et al. 1981b), c_{air} is the postindentation crack length applicable, and measurable, in most indentation strength tests (Cook 2020a). The third step of the sequence, loading of an indented component in an inert environment, leads to precursor extension of the crack from c_{air} to c_m such that Eq. (3.7) applies,

$$c_m = (4\chi P/T_0)^{2/3}, \tag{3.22}$$

and the inert strength is attained, Eq. (3.8),

$$\sigma_m = 3T_0^{4/3}/4^{4/3}\psi(\chi P)^{1/3}. \tag{3.23}$$

It is recognized that c_{air} is trapped in a metastable state on imposition of inert conditions from reactive conditions. However, provided $c_{air} \leq c_m$, Eqs. (3.22) and (3.23) describe the inert failure configuration, and σ_m is the measurable strength. It is also recognized that just as lateral crack formation and extension reduces the residual field amplitude from χ_0 to χ_{air}, the imposition of applied stress reduces the amplitude to χ during a strength test. Three of the idealizations of Section 3.2.1 are thus removed here: T is not invariant, χ is not invariant, and the SIF expression, Eq. (3.3), includes a cross term as $\chi = \chi(\sigma_a)$.

Comparison of precursor crack extension for a range of test systems is best performed using scaled variables that exclude indentation load dependence. Hence, (in contrast to Section 3.2.1) the set used here is the directly measurable experimental combination (c_{air}, σ_m), and was that used by Lawn et al. (1985). The revised scaling relations are

$$\begin{aligned} \mathbf{T} &= T_{air}/T_0 \\ \mathbf{X} &= \chi/\chi_{air} \\ \mathbf{C} &= c/c_{air} \\ \mathbf{S_a} &= \sigma_a/\sigma_m. \end{aligned} \tag{3.24}$$

The SIF expression of Eq. (3.3) is now written and inverted as

$$\mathbf{S_a} = \frac{4^{4/3}(\mathbf{XT})^{1/3}}{3\mathbf{C}^{1/2}}\left[1 - (\mathbf{XT})\mathbf{C}^{-3/2}\right]. \tag{3.25}$$

If an adjusted crack length quantity $\mathbf{c} = (\mathbf{XT})^{-2/3}\mathbf{C}$ is defined, it is clear that Eq. (3.25) is simply the minimal idealized form, Eq. (3.17) with $\mathbf{P} = 1$, expressed in the adjusted crack length,

$$\mathbf{S_a} = \frac{4^{4/3}}{3\mathbf{c}^{1/2}}\left[1 - \mathbf{c}^{-3/2}\right]. \tag{3.26}$$

The limiting applied stress values $\mathbf{S_a} = 0$ and $\mathbf{S_a} = 1$ are obtained for $\mathbf{c} = 1$ and $\mathbf{c} = 4^{2/3}$ or $\mathbf{C} = (\mathbf{XT})^{2/3}$ and $\mathbf{C} = (4\mathbf{XT})^{2/3}$, respectively. At the threshold indentation load and in the absence of relaxation and environmental effects, $\mathbf{P} = \mathbf{X} = \mathbf{T} = 1$, all expressions converge to the idealized form of Eq. (3.26).

In materials and indentation systems in which environmental and relaxation effects are unknown (often the case), it is not possible to separate these effects through application of Eq. (3.25) alone. However, Eq. (3.25) is general and may be applied effectively to limiting variations of the environmental sequence, as shown in examples here in this chapter, and in Chapters 4, 5, 11, and 12. In particular, the sequence inert–inert–inert may be implemented for components indented and strength tested under moisture-excluding films of oil (Anstis et al. 1981; Chantikul et al. 1981b; Ramachandran and Shetty 1991; Kaur et al. 2009) or for components of materials that do not react with atmospheric moisture, such as Si_3N_4 (Marshall et al. 1983; Li et al. 1992) and Si (Cook 2006). In these cases, T_0 replaces T_{air} in Eqs. (3.21) and (3.24) such that $\mathbf{T} = 1$, and experimental data are interpreted through Eq. (3.25) in terms of variations in $\mathbf{X}$ and residual field relaxation. The sequence inert–reactive–reactive may also be implemented for components indented and strength tested, slowly, in air. In these cases, T_{air} replaces T_0 in Eq. (3.24) such that $\mathbf{T} = 1$, and experimental data are again interpreted through Eq. (3.25) in terms of variations in $\mathbf{X}$ and residual field relaxation.

Figure 3.5 shows an example of stable crack extension under applied tensile stress at an indentation flaw in cordierite glass-ceramic in air. The stress direction is horizontal in the images: Figure 3.5a shows the as-indented zero stress state; Figure 3.5c shows the near-maximum stress state, just prior to failure; and Figure 3.5b shows an intermediate state. In Figure 3.5a, all surface traces of the half-penny cracks are of similar length. On application of the applied stress, as in Figure 3.5b and c, the vertical cracks extend and the horizontal cracks remain at the as-indented lengths. Figure 3.5b and c depicts stable equilibrium configurations: Unless the applied stress was increased, the vertical cracks remained at the stable extended lengths. Consideration of Figure 3.5 shows that the configuration and lengths of the lateral cracks also changed under the action of the applied stress, consistent with a change in the value of χ on remote loading. The effect of lateral cracks on χ is considered in detail in Section 3.3.3. Nevertheless, Figure 3.5 shows stable indentation crack extension by a factor of approximately 2.5, consistent with the ideal behavior of Eq. (3.11).

Measurements of near-ideal indentation crack extension behavior are shown in Figure 3.6 for (a) polycrystalline Si_3N_4 from the works of Marshall et al. (1983) (open symbols) and Li et al. (1992) (filled symbols) and for (b) the cordierite

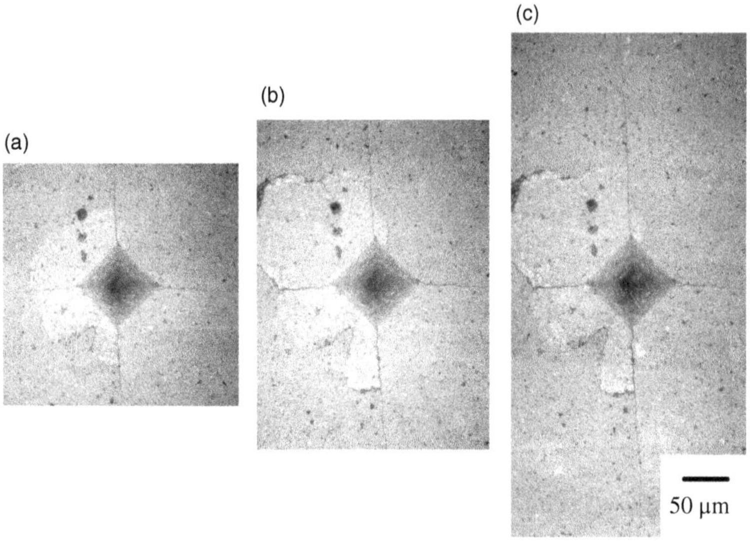

Figure 3.5 Sequence of images of indentation crack extension in a cordierite glass-ceramic in air. Applied stress direction is horizontal. (a) As indented. (b) Partially stressed. (c) Immediately prior to failure.

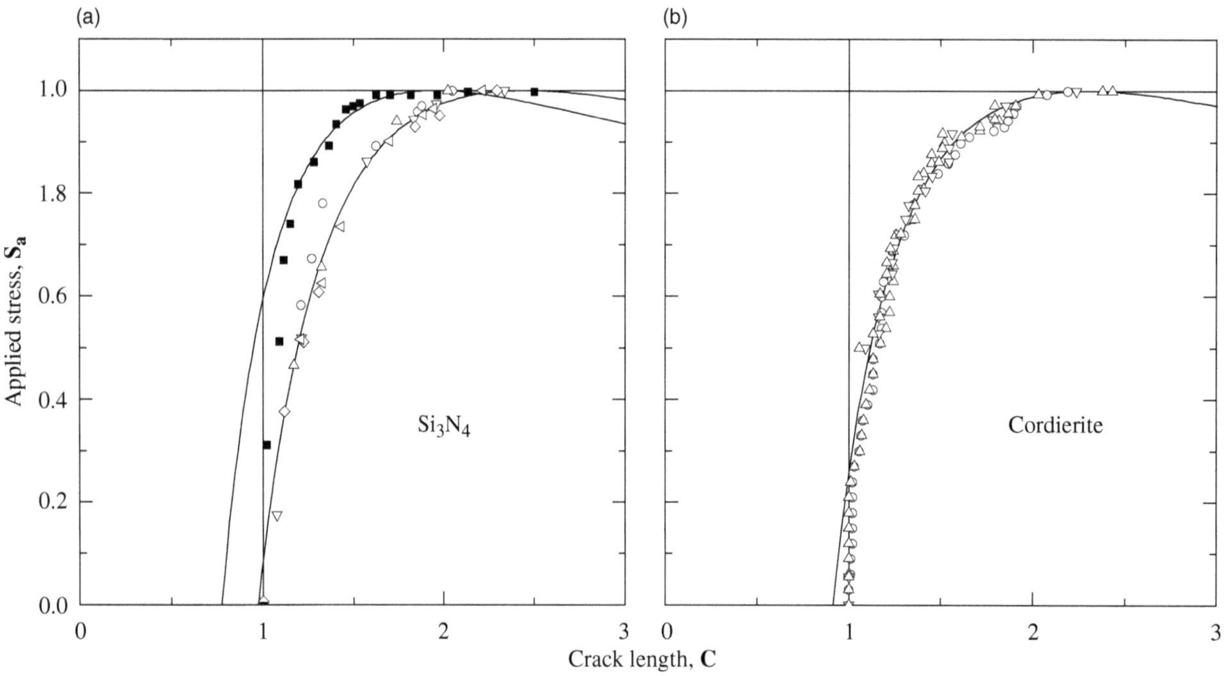

Figure 3.6 Plots of indentation crack extension behavior as scaled applied stress versus scaled crack length. (a) Polycrystalline Si_3N_4 and (b) cordierite glass-ceramic (see Figure 3.5). Source: (a) Data adapted from Marshall et al. (1983) and Li et al. (1992).

glass-ceramic of Figure 3.5. The data are scaled using Eq. (3.24), and different symbols represent results from separate indentations. The vertical lines indicate the initial crack lengths, $\mathbf{C} = 1$, and the horizontal lines indicate the maximum applied stresses $\mathbf{S_a} = 1$, noting that the environmental sequence in (a) is inert–inert–inert and in (b) inert–reactive–reactive. The curved lines are the response of Eq. (3.25) fit to the maximum crack length and applied stress observations for each material, treating $\mathbf{X}$ as a fitting variable. The responses clearly describe the experimental data, and values of $\mathbf{X} = 0.98$ and $\mathbf{X} = 0.84$ were used for Si_3N_4 and $\mathbf{X} = 0.94$ for cordierite, respectively, indicating near-ideal $\mathbf{X} = 1$ behavior, as do the relative maximum crack extensions of approximately 2.5.

In contrast, measurements of near-invariant crack behavior under the influence of applied stress are shown in Figure 3.7 for glass, from the work of Marshall and Lawn (1979), polycrystalline alumina (Al_2O_3) from the work of Kovar et al. (2000),

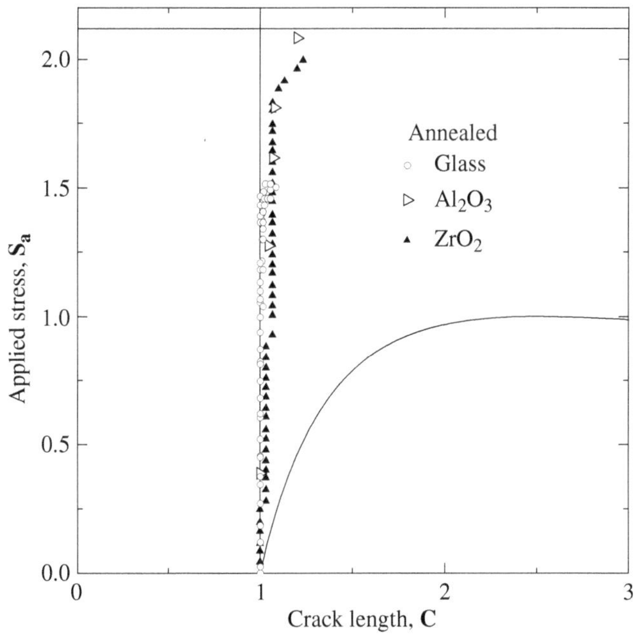

Figure 3.7 Plots of indentation crack extension behavior as scaled applied stress versus scaled crack length for annealed indentations. Source: Includes data adapted from Marshall et al. (1979) and Kovar et al. (2000).

and polycrystalline zirconia (ZrO_2), considered in detail in Chapter 12. For each material, components were annealed after indentation to remove the residual stress field prior to strength testing and crack observations. The data are scaled using Eq. (3.24), using the appropriate σ_m values from strength tests on unannealed indented components. Different symbols represent results from individual materials. The vertical line indicates the initial crack length, $\mathbf{C} = 1$. The horizontal line indicates the ideal residual stress-free maximum applied stresses $\mathbf{S}_a = 2.12$, noting that the environmental sequence for glass and alumina was inert–inert–inert and for zirconia was inert–reactive–reactive. The responses clearly indicate almost no crack extension during loading prior to failure for all three materials. Although the peak stress significantly exceeded $\mathbf{S}_a = 1$ in all cases, it did not approach the ideal value of 2.12 for the glass, and the alumina and zirconia did exhibit some small precursor extension and $\mathbf{C} > 1$. These discrepancies from an ideal residual stress-free invariant crack length response will be considered in Section 3.3.3 and Chapters 4 and 12.

Taken together, Figures 3.5–3.7 show the clear effects of the pdz-based residual stress field in controlling indentation crack extension and indentation strength. The stabilizing nature of the residual field modifies the influence of a destabilizing applied stress field, leading to stable precursor subcritical crack extension and reduced component strengths. If the residual field is removed after indentation, precursor crack extension is removed and strengths return to values given by the Griffith equation for stress-free flaws.

Precursor crack extension during strength tests is considered in its own right in Chapter 12 and used as an indicator of microstructural effects in the fracture of zirconia and other materials. Here, in this chapter and again in Chapters 4 and 11, precursor crack extension is considered as an aid to explaining and interpreting indentation strength behavior. It will be shown that lateral crack effects tend to destabilize half-penny cracks and diminish precursor extension and microstructural effects tend to stabilize half-penny cracks and enhance precursor extension. A generalized indentation fracture mechanics framework that allows for lateral crack and other effects in invariant toughness materials is now developed. The framework is further generalized to include microstructural effects in toughened materials and is developed in Chapters 5 and 6.

3.2.3 Generalized Analysis

Most indentation strength observations vary in some way from a description based on the SIF of Eq. (3.3). To examine these variations, Eq. (3.3) is rewritten in a generalized form as

$$K = \psi(c)\sigma_a c^{a/2} + \chi(P) f(P)/c^{r/2}. \tag{3.27}$$

Nonlinear effects are incorporated through the geometry terms $\psi(c)$ and $\chi(P)$, the related crack length exponents a and r, and an indentation load function $f(P)$, usually expressed as a power law $f(P) = P^x$. The geometry and indentation load functions may include dependencies on component, indenter, and material parameters. Simple implementations of Eq. (3.27) are to set $f(P) = 0$, $a = 1$, and invariant $\psi = \pi^{1/2}$ or $\psi = 2/\pi^{1/2}$ for straight or circular cracks, respectively, to arrive at the Griffith equation (see Chapter 1). Variable $\psi = \psi(c)$ generates variations on the Griffith equation, with functional dependencies on the ratios of crack length to crack width, to imbedded pore or inclusion size, or to component size (Murakami 1987; Lawn 1993; Cook 2023). Other simple implementations are to set $\sigma_a = 0$ as in common indentation crack length measurements with $x = 1$, $r = 3$, and material- and indenter-dependent $\chi = \chi(E, H, \phi)$, to arrive at variations on the Roesler equation (see Chapter 2). E and H are the material elastic modulus and hardness, respectively, and ϕ is the included angle of the indenter (Anstis et al. 1981; Chiang et al. 1982a,b; Cook 2020a,b). Related variations are to set $r = 1$ and $\chi = \chi(E, H, \phi, R)$ to arrive at the length of median cracks (see Chapter 10) at static long-wedge sharp contacts ($x = 1$), at dragged short-wedge sharp contacts ($x = 1/2$), at rolled conical contacts ($x = 2/3$), and at rolled spherical contacts ($x = 3/4$), where R is the radius of the rolling object. Or, similarly, for the length of lateral cracks (see Chapter 9), $r = 4$, at long-wedge sharp contacts ($x = 5/2$), at indentations or dragged short-wedge sharp contacts ($x = 5/4$), at rolled conical contacts ($x = 5/3$), and at rolled spherical contacts ($x = 15/8 \approx 2$).

Many of the above variations and indentation geometries have been examined in indentation strength tests. Analyses describing these tests are considered in Section 3.3, using Eq. (3.27) as a basis with a focus on point indentations, $r = 3$. Schematic cross-section diagrams of the considered indentation variations are shown in Figures 3.8 (indentation shape variation), 3.9 (indentation size variation), 3.10 (indentation impact), and 3.11 (indentations in stressed layered structures). The analyses of Section 3.3 develop indentation strength expressions that are assessed against experimental observations in Section 3.4. Although not a focus of the work here, the analyses of Section 3.3 are supported in many cases by indentation cracking observations.

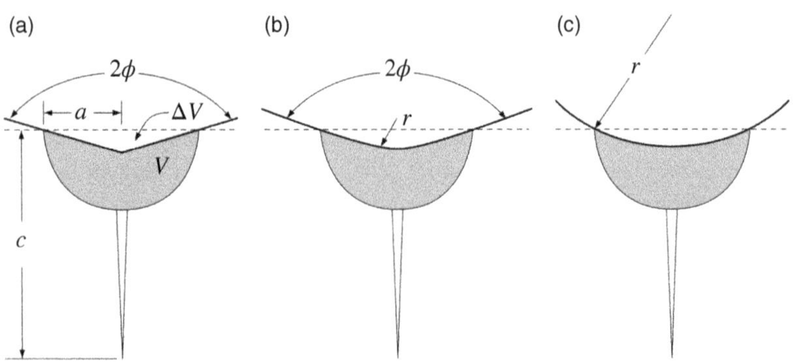

Figure 3.8 Schematic cross-section diagrams of elastic–plastic indentation by indenters of various shapes. The initial surface is shown dashed. Common features include the plastic deformation zone shown shaded, the volume displaced by indenter between the dashed line and shaded region, the contact radius a, and half-penny crack radius c. (a) Conical contact, included cone angle 2ϕ. (b) Conospherical contact, cone angle 2ϕ, tip radius r. (c) Spherical contact, radius r.

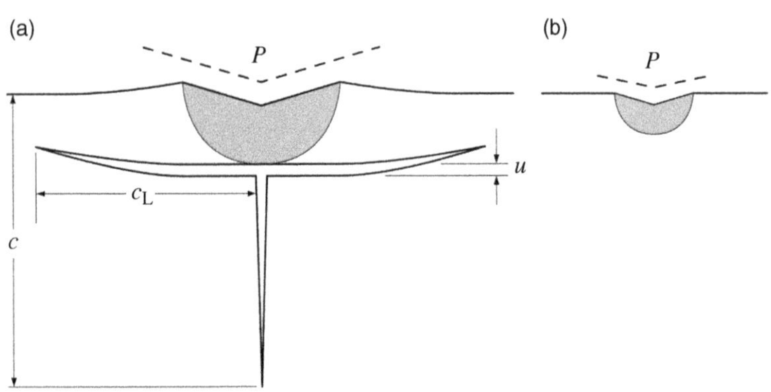

Figure 3.9 Schematic cross-section diagrams of elastic–plastic indentation by a sharp indenter at two contact loads P. The plastic deformation zones are shown shaded. (a) Large, postthreshold load generating half-penny cracks c and lateral cracks c_L and u. (b) Small, subthreshold load generating no cracks.

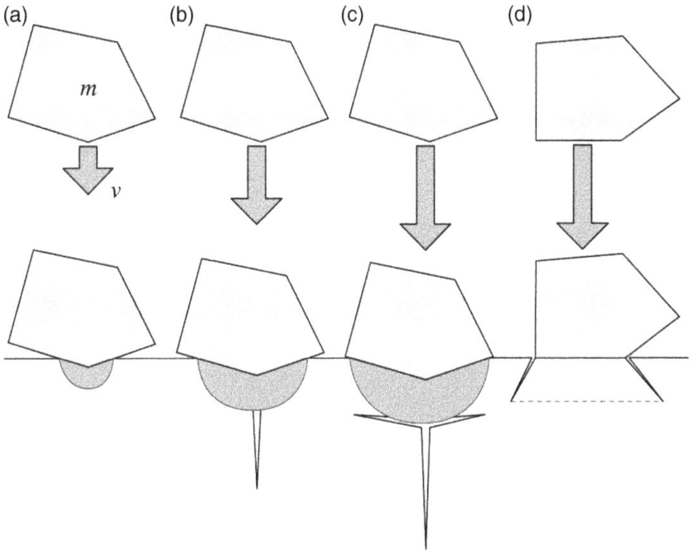

Figure 3.10 Schematic cross-section diagrams of elastic–plastic impact by a particle of mass m at various contact velocities v. The plastic deformation zones are shown shaded. (a) Small velocity sharp contact, subthreshold flaw. (b) Moderate velocity sharp contact, postthreshold flaw with half-penny crack. (c) Large velocity sharp contact, postthreshold flaw with half-penny crack and lateral crack. (d) Large velocity blunt contact with no plastic deformation zone, postthreshold flaw with cone crack.

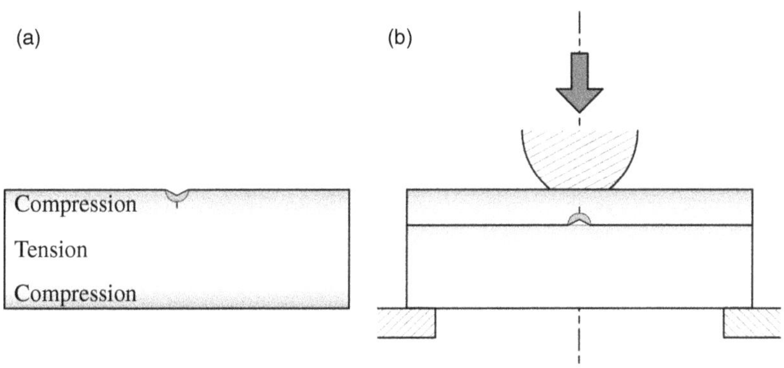

Figure 3.11 Schematic cross-section diagrams of indentation flaws in components with stress gradients. (a) A symmetrically stressed component with exterior compression and interior tension. (b) An asymmetrically stressed component in biaxial flexure with upper compression and lower tension.

3.3 Indentation Strength Variations

3.3.1 Contact Angle

One of the simplest variations in indentation strength measurements is to test various materials with the same indenter such that $\chi = \chi(E, H)$. This was the approach of Chantikul et al. (1981b) in the evaluation of indentation strength measurements for toughness estimation and is here not considered directly (see Chapters 2 and 5 for details). A related simple variation is to test the same material with various geometrically similar indenters such that $\chi = \chi(\phi)$. This was the approach of Lawn et al. (1976) and is illustrated in Figure 3.8a. The analysis of postindentation, pdz-driven half-penny cracks of Chapter 2 shows

$$\chi(\phi) \sim \cot^{2/3}\phi, \tag{3.28}$$

using Eq. (2.10). Thus, using Eq. (3.10), the mean contact strength varies as

$$\hat{S} \sim \cot^{-2/9}\phi. \tag{3.29}$$

Early sharp indentation studies used very large contact loads such that half-penny cracks invariably initiated during indentation loading and attained near-full size at peak load, as illustrated in Figures 2.3 and 2.4. These cracks are driven by the reversible elastic contact field and are akin to the annealed or surface ground cracks discussed in connection with Eq. (3.11). Analysis of such peak load, elastic contact-driven half-penny cracks gives (Lawn and Fuller 1975; Lawn et al. 1976)

$$\chi(\phi) \sim \cot \phi'. \tag{3.30}$$

ϕ' is an apparent contact angle that takes friction at the inclined indenter–surface interface into account,

$$\phi' = \phi + \tan^{-1}\mu, \tag{3.31}$$

where μ is the coefficient of friction at the interface. [Equation (3.30) is simply derived by resolving the indentation load into contact reaction and frictional forces at the interface.] Combining Eq. (3.30) into Eqs. (3.9)–(3.12) gives

$$\hat{S} \sim \cot^{-1/3}\phi'. \tag{3.32}$$

It is noted that both Eqs. (3.29) and (3.32) describe ideal indentations, although not restricted to the Vickers geometry ($\phi \approx 70°$). The angular dependencies for pdz and elastic indentation cracks in these equations are very similar, suggesting that experimental differentiation might be difficult for large indentations, and is considered in Section 3.4.1.

3.3.2 Spherical, Rounded, and Flat Punch Contacts

A distinction is often made between "blunt" and "sharp" indentations in considerations of contact fracture: Blunt indentations typically occur at Hertzian spherical or flat punch contacts, exhibit near-reversible or near-elastic load–displacement responses, and, for contact loads greater than threshold values, generate ring and cone cracks. Sharp indentations typically occur at pyramidal or conical contacts, exhibit hysteretic irreversible or elastic–plastic load–displacement responses, and for contact loads greater than a different set of threshold values generate radial, median, half-penny, and lateral cracks. However, the distinction, although extremely useful for broad descriptive purposes (Lawn and Wilshaw 1975; Lawn and Marshall 1978, 1979; Lawn 1992, 1993), really only pertains to limiting conditions: large spheres at small contact loads are blunt and acute cones at large contact loads are sharp, for example, where the terms large and small are particular to a specific indenter–material system. In these limits, the equilibrium crack configuration on blunt contact cycles is determined by the peak indentation load, and the equilibrium crack configuration on sharp contact cycles is determined by the postindentation residual field associated with elastic–plastic deformation.

In the context of the indentation strength analyses of Section 3.2.1, blunt contact flaws exhibit the ultimate limit for an absence of residual stress field effects during component failure. Strengths of components containing dominant sharp indentation flaws in which the residual contact stress was removed *after* indentation by annealing, grinding, or corrosion were considered in Eqs. (3.11) and (3.12). Strictly, χ in Eq. (3.11) should be written as χ_r, indicating the amplitude of the irreversible component of the residual sharp contact field, and residual field removal after indentation represented by $\chi_r \to 0$. During both sharp and blunt contacts there is also a reversible contact field that, as transitory, is usually neglected in sharp contact postindentation fracture analyses. The transient field is important in indentation cracking *during* sharp contact (Marshall and Lawn 1979) and has angular and radial symmetry different from that of the residual pdz field; it is often modeled by the Boussinesq contact field (Johnson 1985; Cook and Pharr 1990; Cook 2019). For blunt contacts, Eq. (3.11) is rewritten in a version of the Roesler equation as

$$c_0 = (\chi_e P/T)^{2/3}, \tag{3.33}$$

where χ_e indicates the amplitude of the reversible elastic component of the contact field. c_0 is determined by the peak contact load P. The strengths of components containing dominant blunt indentation flaws are then determined by a contact stress that is removed during the indentation cycle. On indentation unloading contact stress removal is represented by $\chi_e \to 0$ but the crack remains metastably trapped at c_0. The strength σ_0 is thus a version of the Griffith equation, Eq. (3.12),

$$\sigma_0 = T/\psi c_0^{1/2} = T^{4/3}/\psi(\chi_e P)^{-1/3}. \tag{3.34}$$

[On sharp indentation at large contact loads, the elastic field dominates the pdz field at peak load, $\chi_e \gg \chi_r$, and half-penny cracks are also described by Eqs. (3.33) and (3.34), with contact angle effects described by Eqs. (3.30)–(3.32). Such cracks at sharp indentations are not considered further here.]

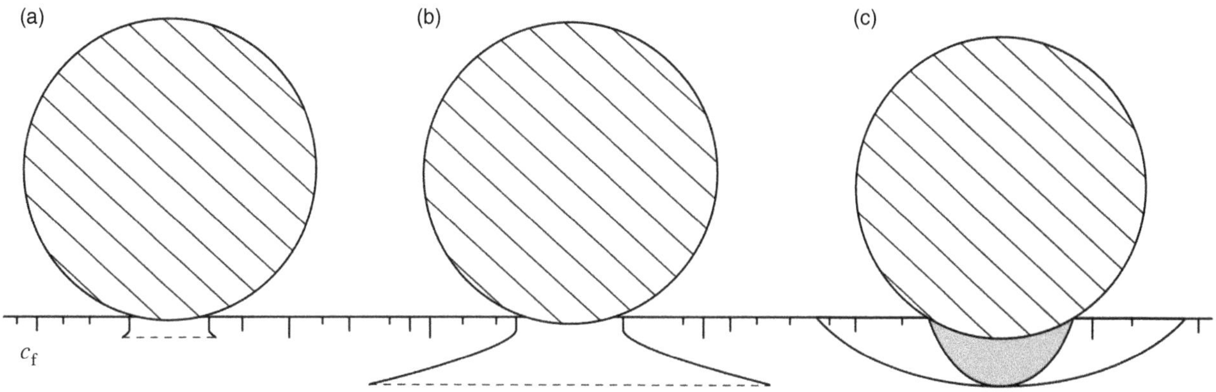

Figure 3.12 Schematic cross-section diagrams of flaws formed at spherical contacts. Surface flaws c_f are distributed on the surface prior to contact. (a) Small load contact, leading to ring crack formation, initiated from a surface flaw. (b) Moderate load contact, leading to cone crack formation, initiated from a ring crack. (c) Large load contact, leading to plastic deformation zone formation (shown shaded) and radial crack formation, initiated from the deformation zone.

Examples of blunt, spherical contact flaws on the surface of a component are shown in the schematic cross-section diagrams of Figure 3.12. The surface is shown with a population of small flaws, characteristic size c_f, preexisting in the component prior to indentation. In the absence of an indentation contact of sufficiently large scale, the most potent of these preexisting flaws determines the intrinsic strength of the component σ_i (Cook 2023). The contacting sphere is shown hatched. Figure 3.12a shows a *ring crack* formed on the component surface by spherical contact of small load ("small" is a relative term—such loads are often ≈ 0.5–1 kN for typical 5 mm diameter spheres). The ring crack initiates in the tensile field of the indentation from a preexisting flaw on the component surface, exterior to the contact perimeter, and propagates to form a cylindrical or near-cylindrical crack with a ring-like surface trace. Beneath the surface the crack may flare out from the contact axis to form an embryonic truncated cone. The surface traces of ring cracks are often visible on surface inspection. Figure 3.12b shows a *cone crack* formed on the component surface by spherical contact of moderate load (≈ 1–2 kN). The cone crack initiates in the mixed-mode tensile + shear field of the indentation from the ring crack and propagates to form a subsurface cone, typically making an angle of ≈20° to the component surface. Often a separate ring stage is not apparent in section observations. As cone cracks are subsurface and close (although not "heal") on indentation unloading, cone crack formation is usually not visible by surface inspection, especially in opaque materials such as ceramics and semiconductors. Details of ring crack and cone crack development at spherical elastic contacts are given in Wilshaw (1971) and at cylindrical flat punch contacts by Mouginot and Maugis (1985). Figure 3.12c shows radial cracks formed on the component surface by spherical contact of large load (>2 kN), such that the contact stress exceeds the material hardness and the contact is elastic–plastic. The subsurface pdz is shown shaded. The radial cracks initiate in the tensile surface field exterior to the contact from shear faults at the periphery of the contact generated pdz, as described in Chapter 2. The radial cracks are perpendicular to the component surface and usually exhibit clear surface traces. Details of radial crack development at spherical elastic–plastic contacts are given in Evans and Wilshaw (1976).

Figure 3.13 shows optical images of a ring + cone crack generated by a 1.9 kN indentation on the surface of a borosilicate glass (BSG) component. The load–displacement response during indentation was completely reversible. The image was acquired using a split specimen section technique and oblique lighting was used to highlight the crack. Figure 3.13a shows a plan view perpendicular to the loading axis. The section edge is at the bottom of the image and the near-semicircular outline of the ring crack trace is visible on the edge. Some scallop cracks are also visible on the component edge (see Chapter 8). The dark contrast surrounding the ring crack trace indicates reflected illumination by the subsurface cone crack surfaces. Figure 3.13b shows a section view parallel to the loading axis. The section edge is at the top of the image, and edge damage is clearly visible. The traces of the cone cracks are clear, although no subsurface ring crack stage is obvious. Images similar to Figure 3.13 have been presented since the earliest indentation works (Frank and Lawn 1967; Lawn 1968; Williams et al. 1970; Mikosza and Lawn 1971; Wilshaw 1971; Hartley and Wilshaw 1973; see also Nadeau 1973 for digitized traces), as well as more recently (Zeng et al. 1992; Lee et al. 1997). Images of radial cracks at spherical elastic–plastic contacts were also presented in early works (Lawn and Wilshaw 1975; Evans and Wilshaw 1976).

Ring cracks and cone cracks generated by large enough blunt elastic contacts may control the strength of indented components. For example, the cone cracks in Figure 3.13 are clearly larger than the edge damage. Radial cracks generated

(a)

(b)

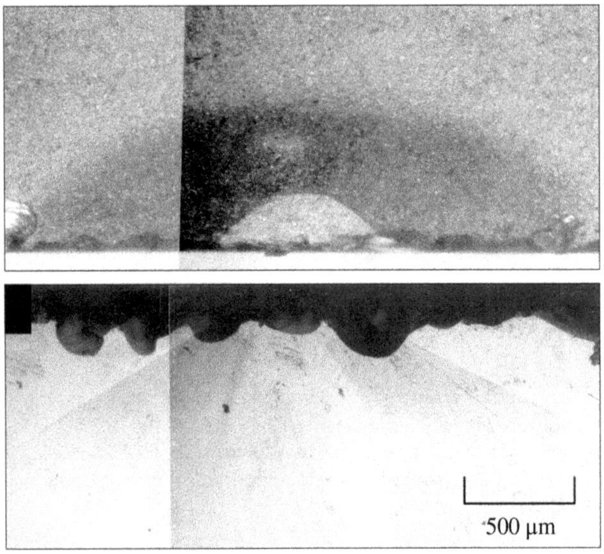

Figure 3.13 Optical images of cone crack formation obtained using a split specimen of borosilicate glass. (a) Plan view; split edge is at the bottom of the image. (b) Section view; split edge is at the top of the image.

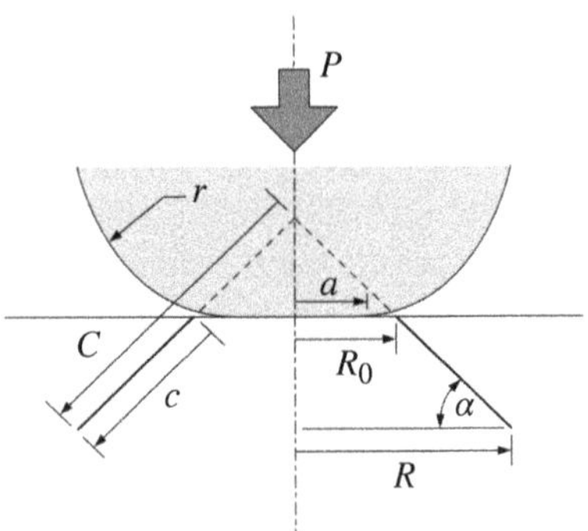

Figure 3.14 Schematic cross-section diagram of ring and cone crack formation on spherical contact (sphere, radius r, shown shaded). Contact radius is a. Ring crack radius is R_0 and cone crack radius is R.

by the very large loads required for elastic–plastic spherical contacts are nearly always strength controlling. The fracture mechanics of indentation strengths for blunt contacts is now considered, beginning with ring and cone cracks. Figure 3.14 is a schematic cross-section diagram with dimensions indicated of a ring and cone crack system at a spherical contact. (The nomenclature and vertical exaggeration of the diagram are common in such descriptions; surface deformation is not shown.) A sphere of radius r contacts a surface with load P such that the contact radius is a. A ring crack of radius R_0 is initiated in the surface exterior to the contact and propagates as a truncated cone into the material at angle α. The equilibrium termination of the cone is at radius R. The length of the crack between R_0 and R is c. A virtual cone, indicated by the extension of the cracks along the dashed lines, can be constructed by setting $C = R/\cos\alpha$. Similarly, the crack length c is given by

$$c = (R - R_0)/\cos\alpha. \tag{3.35}$$

The two radius terms in Eq. (3.35) can be expressed in terms of the experimentally more accessible variable of the contact load P by

$$\begin{aligned} R &= (\chi P/T)^{2/3} \\ R_0 &= \beta(rP/M)^{1/3}, \end{aligned} \tag{3.36}$$

where the first equation is the Roesler relation for well-developed cone cracks, and the second represents the observation that $R_0 = \beta a$ where $\beta \approx 1.1$ is a constant for a given contact system and $a = (rP/M)^{1/3}$ is the Hertzian contact relation where M is the contact modulus. Combining Eqs. (3.35) and (3.36) gives an expression for the crack length in terms of indentation load,

$$c = c_R \left(\frac{P}{P_R}\right)^{2/3} \left[1 - \left(\frac{P}{P_R}\right)^{-1/3}\right], \quad P \geq P_R. \tag{3.37}$$

P_R and c_R are characteristic indentation load and crack length parameters for the contact system, given by

$$P_R = \frac{\beta^3 T^2 r}{\chi^2 M} \tag{3.38}$$

and

$$c_R = \left(\frac{\beta^3 T r}{\chi M}\right)^{2/3}. \tag{3.39}$$

Inspection of Eq. (3.37) shows that for $P \gg P_R$ the Roesler equation expressed in terms of crack length is recovered, $c \sim P^{2/3}$, and that for $P \to P_R$, the crack lengths are shorter than those extrapolated from large loads, consistent with the early observations of Roesler (1956) and Mikosza and Lawn (1971). Equation (3.37) describes the equilibrium shortening of ring cracks relative to cone cracks as the indentation load is decreased. [A similar analysis can be performed for the virtual cone dimension using $C = (\chi P/T)^{2/3}$, and these were the adjusted data (Lee et al. 1997) used in Figure 1.14.] Combining Eq. (3.37) into Eq. (3.12) gives a version of the Griffith equation for the indentation strength σ_0 of ring + cone cracked components:

$$\sigma_0 = T/\psi c^{1/2} = \sigma_R \left(\frac{P}{P_R}\right)^{-1/3} \left[1 - \left(\frac{P}{P_R}\right)^{-1/3}\right]^{-1/2}, \quad P \geq P_R, \tag{3.40}$$

where σ_R is a characteristic strength for the contact system, given by

$$\sigma_R = T/\psi c_R^{1/2}. \tag{3.41}$$

For $P \gg P_R$, Eq. (3.34) is recovered as $\sigma_0 \sim P^{-1/3}$. For $P \to P_R$, the strength diverges, $\sigma_0 \gg \sigma_R$. Note that for a fixed material system only r is a contact variable and that $P_R \sim r$ and $\sigma_R \sim r^{-1/3}$, thus providing a means of adjusting strength behavior with the radius of a contacting sphere.

Insight into the overall indentation strength variation can be gained by setting scaling relations similar to Eqs. (3.15) and (3.24):

$$\begin{aligned} \mathbf{S_0} &= \sigma_0/\sigma_R \\ \mathbf{P} &= P/P_R, \end{aligned} \tag{3.42}$$

and thus Eq. 3.40 becomes

$$\mathbf{S_0} = \mathbf{P}^{-1/3}\left(1 - \mathbf{P}^{-1/3}\right)^{-1/2}, \quad \mathbf{P} \geq 1. \tag{3.43}$$

Ideal indentation strength behavior is given by $\mathbf{S_0} = \mathbf{P}^{-1/3}$. As noted in the discussion of Figure 3.12, preexisting flaws of size c_f determine component intrinsic strength σ_i in the absence of a large enough indentation flaw, e.g., indentations smaller than depicted in Figure 3.12a. The intrinsic strength and preexisting flaw size can be regarded as linked by a Griffith equation, $\sigma_i = T/\psi c_f^{1/2}$. The scaled intrinsic strength and a related effective indentation load for intrinsic flaws are given by

$$\begin{aligned} \mathbf{S_i} &= \sigma_i/\sigma_R \\ &= \mathbf{P_i}^{-1/3}\left(1 - \mathbf{P_i}^{-1/3}\right)^{-1/2}, \end{aligned} \tag{3.44}$$

where Eq. (3.44) defines $\mathbf{P_i}$. For small indentation loads $\mathbf{P} < \mathbf{P_i}$, components exhibit the intrinsic strength Eq. (3.44), and for large indentation loads $\mathbf{P} \geq \mathbf{P_i}$, components exhibit the indentation strength Eq. (3.43). Figure 3.15 shows a logarithmic indentation strength plot for the ring + cone crack system, using the scaled indentation load and strength variables **P** and **S**. The intrinsic strength level is set to $\mathbf{S_i} = 1.75 \pm 0.25$ and is indicated by the hatched band. The ideal $\mathbf{P}^{-1/3}$ indentation strength response is indicated by the dashed line. The ring + cone crack strength response Eq. (3.43) is indicated by the bold gray line. Note that this response approaches ideal behavior asymptotically at large loads $\mathbf{P} \gg 1$ and diverges as $\mathbf{P} \to 1$. The full strength response is indicated by the bold black line. For this case Eq. (3.44) gives $\mathbf{P_i} \approx 2$. Hence, for small indentation loads $\mathbf{P} < 2$, components exhibit the invariant intrinsic strength $\mathbf{S} = \mathbf{S_i} = 1.75$. For larger indentation loads

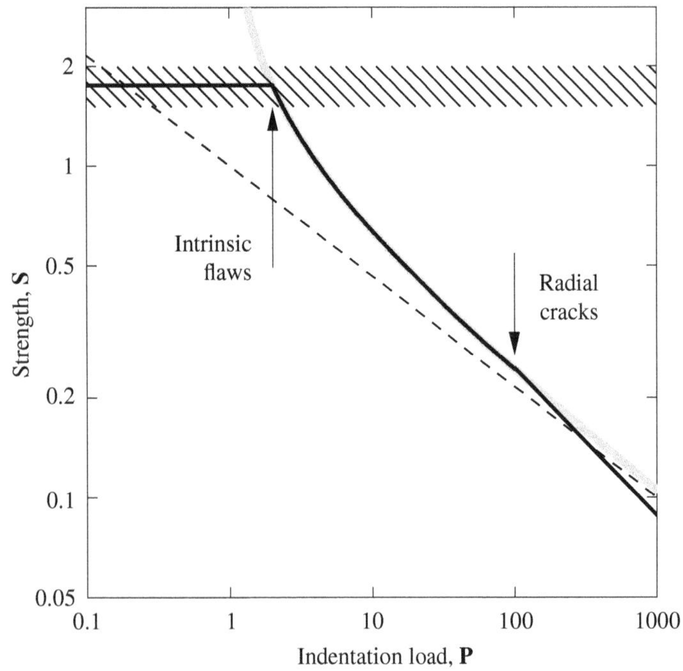

Figure 3.15 Logarithmic indentation strength plot in scaled coordinates for spherical contacts. Full behavior is shown as the bold solid line. At small loads, intrinsic flaws and strengths dominate. At large loads, radial cracks dominate. At intermediate loads, cone and ring cone cracks dominate.

$\mathbf{P} \geq 2$, components exhibit the decreasing indentation strength $\mathbf{S} = \mathbf{S}_0(\mathbf{P})$. Note, as might be inferred from Figure 3.2, that there is no discontinuity in the strength response in Figure 3.15 at $\mathbf{P} = \mathbf{P}_\mathrm{i} = 2$, although a discontinuity in the derivative as strength control passes from intrinsic flaws to indentation flaws. Note that there is also a discontinuity in the derivative of the strength response in Figure 3.15 at $\mathbf{P} = 100$. This discontinuity is associated with the formation of radial cracks at very large, $\mathbf{P} \geq 100$, spherical indentation loads. Hence, in this case, the blunt indentation ring + cone crack indentation strength response, set by Figure 3.12a and b, is restricted to the domain $2 \leq \mathbf{P} \leq 100$ (between the arrows of Figure 3.15). The blunt indentation *radial* crack indentation strength response, set by Figure 3.12c, is now considered.

Exceptions to the sharp-blunt distinction outside the limiting conditions are common. The exception encountered most frequently is a subthreshold flaw, known since early indentation studies to exhibit plastic deformation and densification at pyramidal indentation or scratch contacts, but no cracks (Diehl and Schulze 1957; Palmqvist 1957; Evers 1967; Ernsberger 1968). Threshold conditions for half-penny, lateral, and median cracks are considered in Chapters 2, 9, and 10, and subthreshold indentation and scratch strengths are considered in Section 3.3.4 and Chapter 10, respectively. Another exception to the distinction, also frequently encountered, is the generation of ring and cone cracks at elastic–plastic pyramidal indentation contacts, particularly in glasses (Arora et al. 1979; Cook and Pharr 1990; Tandon and Cook 1992; Cook 2019). A final exception is the generation of radial, half-penny, and lateral cracks at elastic–plastic spherical indentations, typically formed by large load indentation by small radius indenters, and also known since early studies (Peter 1964; Swain and Hagan 1976; Kirchner and Ragosta 1983; Marshall 1984). This phenomenon is the next subject and will be shown to fit naturally within the pdz-based contact fracture framework developed for cracks at indentations exhibiting geometric similarity, e.g., conical, and the subsequent indentation strengths.

The starting point for fracture mechanics analysis of rounded and spherical elastic–plastic contacts is to recognize that such contacts are simply members of a set, illustrated in Figure 3.8, that includes conical and spherical indentations as end members. The residual subsurface deformation and fracture phenomena for contacts by a cone, Figure 3.8a, cone with rounded tip, Figure 3.8b, and sphere, Figure 3.8c, are identical: a near-hemispherical pdz, volume V, subtended at the surface by contact radius a and exerting an outward force F on a half-penny crack of radius c. The relation of the peak contact load P to the residual surface impression dimension a is also identical and given by the material hardness $H = P/\pi a^2$. The difference between the contacts is the relationship between the volume of material displaced by the indenter, ΔV, and the contact radius. This relationship depends on the shape of the indenter. However, the characteristic volume strain generated in the pdz, $-\Delta V/V$, does not depend on this shape (at least for near-hemispherical pdz) and is given by the material hardness/elastic modulus ratio, $\Delta V/V \sim (H/E)^{3/2}$, where E is the material Young's modulus. Based on these similarities, a general expression for the residual SIF K_r acting on half-penny cracks at elastic–plastic contacts can be developed. The analyses of half-penny (Section 2.1.3) and median (Section 10.3.1) cracks at geometrically similar point and line contacts are specific examples of this development, which was originally articulated, for spherical contacts, by Marshall (1984) (and later outlined for rolled or dragged spherical contacts by Cook 2018a).

A general expression for K_r is developed as follows. From Eq. (2.3), the pressure p_a exerted by the pdz on the surrounding matrix is

$$p_a \sim E(\Delta V/V). \tag{3.45}$$

Generalizing Eq. (2.7), the force F exerted by the pdz at the center of a half-penny crack in the matrix is

$$F \sim p_a V^{2/3} \sim E(\Delta V/V)^{1/3}(\Delta V)^{2/3}. \tag{3.46}$$

Using the expression $\Delta V/V \sim (H/E)^{3/2}$ for the characteristic pdz volume strain and that for the SIF of a center loaded crack, $K \sim F/c^{3/2}$, Eq. (1.35), gives

$$K_r \sim (EH)^{1/2}(\Delta V)^{2/3}/c^{3/2}. \tag{3.47}$$

Equation (3.47) expresses the essential physics of the indentation field: Dependency on material modulus and hardness are included, representing the elastic–plastic nature of the contact; dependency of indentation volume is included, representing indentation shape and scale; and inverse dependency on crack length is included, representing a stabilizing fracture field. The volume displaced by the indenter is conveniently expressed as

$$(\Delta V)^{2/3} \sim a^2 g(a/r, \phi), \tag{3.48}$$

where $g(a/r, \phi)$ is a dimensionless function including, if applicable, information regarding the size (r) and shape (ϕ) of the indenter. Note that in the progression of indenters in Figure 3.8, the geometric similarity of the conical indenter of Figure 3.8a exhibits no intrinsic length scale and is defined by ϕ alone and that the spherical indenter of Figure 3.8c has no intrinsic angular scale and is defined by r alone. Only the rounded cone, Figure 3.8b, requires specification of both r and ϕ to define the shape. Combining Eqs. (3.47) and (3.48) gives a general expression for the residual SIF K_r as

$$K_r \sim (EH)^{1/2}a^2 g/c^{3/2}, \tag{3.49}$$

in which terms in the material properties, indentation size, indentation shape, and crack size are clearly separated. For the conical indenter, $\Delta V \sim a^3 \cot\phi$ and thus $g \sim \cot^{2/3}\phi$, which, using $a^2 \sim P/H$, returns Eqs. (2.11) and (2.12) as noted previously. For the spherical indenter, $g \sim (a/r)^{2/3}$ (Marshall 1984) and thus, using $a^2 \sim P/H$ again,

$$K_r \sim (P/r^{1/2})^{4/3}/c^{3/2}. \tag{3.50}$$

For the rounded cone indenter,

$$g \sim \left(\cot\phi - \frac{h(\phi)r^3}{a^3}\right)^{2/3}, \tag{3.51}$$

where $h(\phi)$ is a function of the cone angle (D.B. Marshall, private communication 1983). [The second term in Eq. (3.51) is sometimes referred to as the "tip deficiency," referring to the lack of displaced volume associated with a rounded indenter tip.] In the limit of very large indentations, $a \gg r$, tip rounding goes unnoticed, $g \to \cot^{2/3}\phi$ and Eqs. (2.11) and (2.12) pertain. For $a < r\cos\phi$, the spherical analysis of Eq. (3.50) applies.

In terms of the generalized SIF of Eq. (3.27), elastic–plastic spherical indentations are described by $\chi \sim 1/r^{2/3}$ and $f(P) = P^{4/3}$. Setting $\sigma_a = 0$ and $K = T$, or using Eq. (3.50) directly, gives an expression for the stable equilibrium crack lengths at such indentations

$$c \sim \frac{P^{8/9}}{r^{4/9}T^{2/3}}. \tag{3.52}$$

Similarly, replacing χP with $P^{4/3}/r^{2/3}$ in Eq. (3.8) gives an expression for the indentation strength of elastic–plastic spherical indentations:

$$\sigma_m \sim \frac{T^{4/3}r^{2/9}}{P^{4/9}}. \tag{3.53}$$

or, similar to Eq. (3.9),

$$\sigma_m = SP^{-4/9}, \tag{3.54}$$

where S here is a spherical indenter- and material-dependent contact strength parameter. As noted by Marshall (1984), the near-linear $c \sim P^{8/9}$ variation of crack length with indentation load for elastic–plastic spherical indentations is very different from the near-ubiquitously observed $P^{2/3}$ dependence exhibited by pyramidal and conical indentations (Chapter 2; see Cook 2020a,b, 2024 for extensive reviews). However, the $\sigma_m \sim P^{-4/9}$ dependence for indentation strength is barely different from the ideal $P^{-1/3}$ dependence discussed previously. Using the scaling of Eq. (3.42) enables Eq. (3.54) to be written as $\mathbf{S_m} \sim \mathbf{P}^{-4/9}$, where the constant of proportionality depends on the strength continuity conditions at the radial crack initiation load. This dependence for the radial crack-controlled strength response is shown as the black line in Figure 3.15 for $\mathbf{P} \geq 100$, calculated assuming strength continuity. There is a very small discontinuity in the strength derivative that would be extremely difficult to detect from strength measurements alone. Indentation strength measurements using blunt indenters such that there is a well-developed cone crack load domain are thus unlikely to be much affected by the large load onset of radial cracking. Much wider indentation load domains for radial cracking can be observed using small radius probes that effectively eliminate cone cracking, and such observations are now considered.

Figure 3.16 shows logarithmic plots of indention dimensions, contact impression radius a and crack surface trace length c, as functions of indentation load P for elastic–plastic spherical indentations, using data derived from the published works cited. Observations from Swain and Hagan (1976) on a calcium aluminosilicate glass (ASG) are shown in Figure 3.16a, and observations from Kirchner and Ragosta (1983) (filled symbols) and Marshall (1984) (open symbols) on polycrystalline ZnS are shown in Figure 3.16b. The data of Swain and Hagan and Kirchner and Ragosta represent individual measurements at the designated indentation loads; those of Marshall represent the means and standard deviations of multiple measurements. The lower symbols in Figure 3.16a indicate contact impression radius dimensions a formed by spherical indenters of diameter 0.4 and 1 mm. The lower solid line represents the predicted $a(P)$ response from Vickers indentation hardness tests. There is agreement in both trend, $P^{1/2}$, and numerical value, set by H. The upper symbols in Figure 3.16a indicate crack lengths c (measured from impression center) for a spherical indenter of diameter 0.4 mm. The upper solid line represents a visual fit to the data of the predicted $c(P) \sim P^{8/9}$ trend from Eq. (3.52). There is clear agreement. As a guide to the eye, the dashed line in Figure 3.16a represents the $P^{2/3}$ trend commonly observed for crack lengths at Vickers indentations and does not describe the data. The lower symbols in Figure 3.16b indicate contact impression radius dimensions a for spherical indenters of diameter 0.36–3.18 mm. As before, the lower solid line represents the predicted $a(P)$ response from Vickers indentation hardness tests and there is agreement in trend, numerical value, and invariance with probe diameter. The contact impressions for ZnS are larger than those of glass, reflecting the smaller hardness, 2 versus 6.8 GPa. The upper symbols in Figure 3.16b indicate crack lengths c. The upper solid line represents the *predicted* $c(P)$ behavior

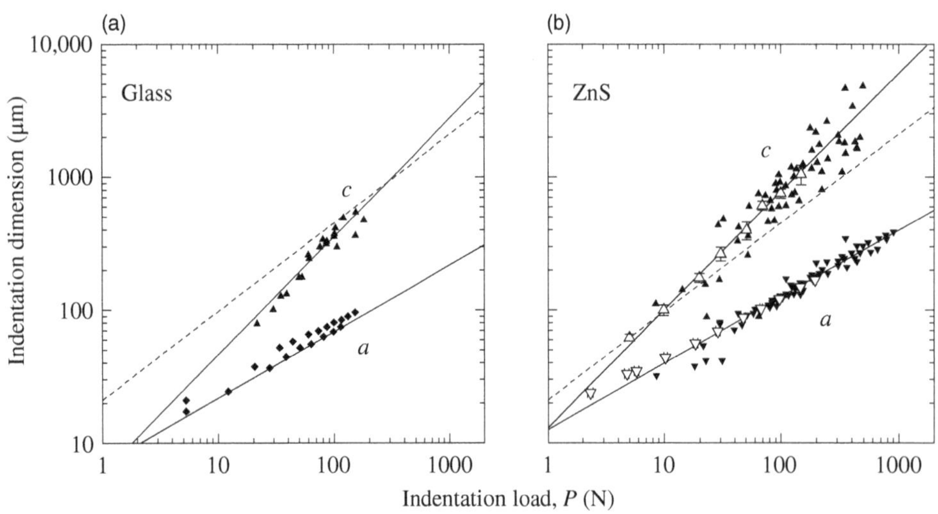

Figure 3.16 Logarithmic plots of indentation dimensions, crack length c and impression size a, for elastic–plastic spherical contacts, load P. Solid lines indicate predicted trends. (a) Aluminosilicate glass. Source: Data adapted from Swain and Hagan (1976). (b) Polycrystalline ZnS. Source: Data adapted from Kirchner and Ragosta (1983) and Marshall (1984).

for an 0.4 mm spherical indenter, using Eq. (3.52) and crack length measurements from Vickers indentation tests (Marshall 1984). There is clear agreement. The data of Kirchner and Ragosta exhibit a range of a factor of approximately 3, consistent with the factor of 2.6 anticipated from Eq. (3.52) (larger diameter indenters exhibited smaller cracks). The dashed line guide to the eye from Figure 3.16a is repeated in Figure 3.16b, emphasizing the differences between indentation cracking at spherical and pyramidal elastic–plastic indentations and between ZnS and glass (glass is slightly tougher, 0.75 MPa $m^{1/2}$ vs 0.6 MPa $m^{1/2}$).

Overall, elastic–plastic spherical indentation dimension measurements are in agreement with pdz-based fracture mechanics analyses and emphasize the lack of clear boundary between "sharp" and "blunt" indentations. The analyses above of elastic and elastic–plastic spherical indentation strength responses suggest that the sharp–blunt boundary may be even less clear in spherical indentation strength measurements. Spherical indentation strength measurements are discussed in Section 3.4.2.

3.3.3 Lateral Cracked Contacts

The most visible features of sharp contact flaws are usually lateral cracks and the associated chips of removed surface material. Such cracks are shown in the schematic cross section of Figure 3.9a and described in Chapter 2. Relative to the size of the contact impression, such cracking and chipping increases with contact load. An example of the increase with load is shown in Figure 3.17, which shows (a) $P = 30$ N and (b) $P = 100$ N Vickers indentations in cordierite glass-ceramic. A single lateral crack is visible in Figure 3.17a, indicated by a bright region resulting from subsurface fracture and surface tilt adjacent to the contact impression. Two lateral cracks are visible in Figure 3.17b, indicated by scallop shaped dark regions resulting from chipping and material ejection adjacent to the contact impression. A threshold indentation load P_{0L} is observed for lateral cracking, such that for small loads $P < P_{0L}$ no lateral cracks are formed. The threshold is material- and indenter-dependent, similar to P_{Ω} observed for half-penny cracks described in Chapter 2 and usually $P_{0L} > P_{\Omega}$. (A small load, uncracked, subthreshold $P < P_{\Omega}$ indentation is shown in the schematic diagram of Figure 3.9b. Strengths of uncracked indentations are considered in Section 3.3.6.)

For $P \geq P_{0L}$, the increase in observed lateral cracking with load is quantified empirically by *lateral crack development* L_D given by

$$L_D = (2V + R)/12 \qquad (3.55)$$

Here, V is the number of lateral cracks visible at an indentation, and R is the number of lateral crack chips removed from the surface (Cook et al. 1990). The weighting of V:R as 2:1 reflects the observations that most of the effects of lateral cracking on disruption and relaxation of indentation stress fields are caused by the occurrence of the lateral cracks, rather than by any subsequent material removal (although the latter is the most obvious). L_D takes limiting values of 0 for no lateral cracking (for $P < P_{0L}$) and 1 for complete removal of lateral crack chips in all four quadrants (usually for $P \gg P_{0L}$). For example, the 30 N indentation in Figure 3.17a has $L_D = 0.17$, and the 100 N indentation in Figure 3.17b has $L_D = 0.5$. Figure 3.18 shows a semilogarithmic plot of L_D as a function of indentation load for Vickers indentations of single crystal, glass-ceramic, and polycrystalline materials. The load coordinate is plotted relative to the lateral crack threshold P/P_{0L} for each material, and

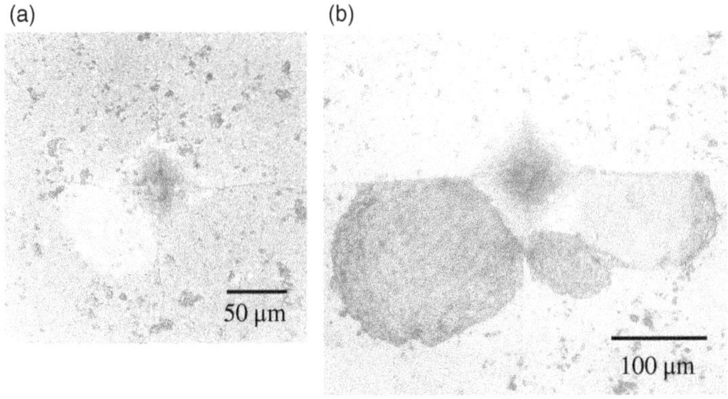

Figure 3.17 Optical images of (a) $P = 30$ N and (b) $P = 100$ N Vickers indentations in cordierite glass-ceramic illustrating lateral crack development with contact load. Multiple large lateral cracks and removed chips of material are visible in (b).

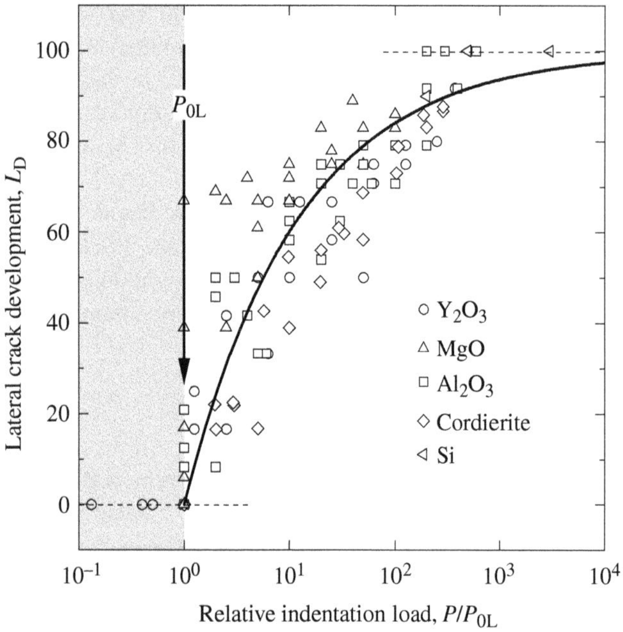

Figure 3.18 Plot of lateral crack development as a function of scaled indentation load for a range of materials. Source: Includes data adapted from Cook et al. (1990, 1994) and Cook and Liniger (1992).

data are included from Cook et al. (1990, 1994) and Cook and Liniger (1992). The shaded band indicates the lateral crack subthreshold domain $P/P_{0L} < 1$. The solid line is a guide to the eye and represents a logarithmic increase in lateral crack development L_D between the bounds of 0 and 1 as P/P_{0L} increased from 1 to approximately 10^3. L_D provides a direct measure of surface wear and optical properties degradation by sharp contacts and a semiquantitative guide to stress relaxation effects active during indentation cracking and strength: cracks are shorter and strengths are greater for large values of L_D.

Figure 3.19 places the effects of lateral cracking on indentation strength on a more quantitative basis in a plot of indentation contact strength as a function of relative lateral crack length. The plot describes Vickers indentation behavior on

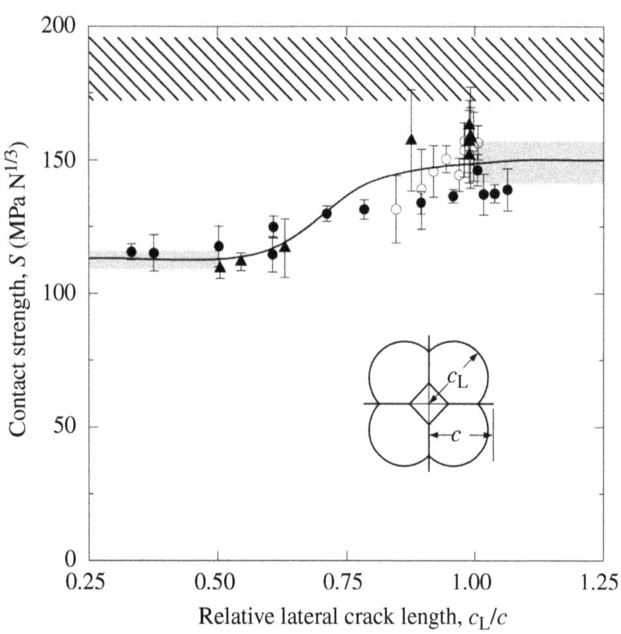

Figure 3.19 Plot of contact strength versus relative lateral crack length for soda-lime silicate glass. Source: Data adapted from Chantikul et al. (1981) and Lawn et al. (1985).

soda-lime silicate glass (SLG) from the works of Chantikul et al. (1981b) and Lawn et al. (1985), using data derived from the published works cited. Relative lateral crack length c_L/c is given as the ratio of surface trace lengths of lateral and radial cracks, c_L and c respectively, as shown in the inset. Contact strength S is given as an indentation load adjusted strength parameter by inversion of Eq. (3.9) to $S = \sigma_m P^{1/3}$, where σ_m is the measured strength of a component containing an indentation of load P. Note that if strength is affected by load-dependent lateral cracking the experimental value of S deviates from the ideal invariant value $\hat{S}$. Symbols represent the means and standard deviations of load-adjusted strengths for aged indentations over the load domain $P = 5$–100 N and aging periods between indentation and strength measurement of 10–10^6 s. Aging was carried out in silicone oil (filled symbols) and water (open symbols), leading to increases in c_L/c and S with aging time; in the scaled coordinates used there was no significant effect of P or environment on strength behavior. The shaded band on the left indicates the initial contact strength prior to lateral crack extension. The hatched band indicates the strength σ_0, Eq. (3.12), of annealed specimens; aging time had no significant effect on annealed strengths. Consistent with Figure 3.7, the annealed contact strength was a factor of approximately 1.65 greater than the initial contact strength. On aging, c_L/c increased, typically slowing to a plateau value of $c_L/c \approx 1$ after approximately 10^4 s (several hours). The contact strength S increased with increasing c_L/c, also to a plateau value. The solid line is a guide to the eye indicating the increase. The final plateau value is shown as the shaded band on the right at a value approximately 1.35 times greater than the initial value. The interpretation of the increase toward the annealed value, as noted in Chantikul et al. (1981b), is that the residual field relaxes and the amplitude term χ decreases, Eq. (3.10), on lateral crack extension. Almost identical strength results and interpretation in terms of lateral crack extension were reported by Shand (1965) in terms of post-contact aging time for as-scribed and annealed linear contacts. (Ironically, lateral crack effects on indentation cracking and strength were rediscovered over 50 years later by Liu et al. 2020 and "explained" in atomistic terms without recourse or reference to any indentation works.)

Various fracture mechanics aspects of these responses have been analyzed in prior works: lateral crack initiation during indentation contact cycles and the material dependencies of P_{0L} were considered by Cook (2019); the dimensions of lateral cracks at point contacts, in particular lateral crack length c_L and associated surface uplift u, see Figure 3.9a, as a function of indentation load at Vickers indentations, were developed by Marshall et al. (1982); similar fracture mechanics of lateral cracks at line contacts and scratches are developed in Chapter 9; the effects of lateral crack development in diminishing the residual field driving half-penny and radial cracks at large indentation loads were considered by Cook et al. (1990) and Cook and Liniger (1992); and the effects of such diminished residual fields to increase indentation strengths at large contact loads were considered by Cook and Roach (1986). The fracture mechanics analysis of this last aspect is considered here. The analysis follows the pdz-based volume strain formalism of Section 3.3.2 and the strength analysis of Cook and Roach. An indentation load dependent residual field amplitude term $\chi(P)$ as defined in Section 3.2.3 is developed and applied in crack extension and strength analyses.

The physical principle of lateral crack influence on indentation residual stress fields is that lateral cracking decreases the mechanical constraint on a contact generated pdz. During strength testing, the decreased constraint reduces the driving force on half-penny cracks and thus leads to increased strengths. Mathematically, this influence is modeled as an increase in the volume into which the material from the contact impression is distributed. Thus, Eq. (3.45) is modified as

$$p_a \sim E(\Delta V/V'), \tag{3.56}$$

where V' is the increased constraining volume. V' is expressed as the sum of the pdz volume V and the volume V_L associated with lateral cracking and decohesion of the pdz from the matrix of surrounding material:

$$V' = V + V_L. \tag{3.57}$$

The opening force F exerted at the center of a half-penny crack in the matrix is now given by modification of Eq. (3.46) as

$$F \sim E(\Delta V/V)^{1/3}(\Delta V)^{2/3}/(1 + V_L/V). \tag{3.58}$$

All but one of the terms on the right side of this expression are those used previously:

$$\begin{aligned} (\Delta V/V)^{1/3} &\sim (H/E)^{1/2} \\ (\Delta V)^{2/3} &\sim a^2 \sim P \\ V &\sim P^{3/2}(E/H^2)^{3/2}. \end{aligned} \tag{3.59}$$

Referring to Figure 3.9a, the term V_L scales as $V_L \sim uc_L^2$, where u is the crack opening displacement of the lateral crack and decohesion of the pdz from the matrix and c_L is the lateral crack radius. From the analysis of Marshall et al. (1982)

for lateral cracks at point contacts, these dimensions scale with indentation load as $u \sim P^{1/2}$ and $c_L \sim P^{5/8}$ (see Chapter 9 for analogous line contact scaling). Hence, the increase in the constraining volume associated with near-planar subsurface lateral cracking in the postindentation equilibrium state is $V_L \sim P^{7/4}$. This is a lower-bound dependence on indentation load. As noted by Marshall et al. this variation does not account for nonequilibrium lateral crack extension and increased compliance as lateral cracks turn and approach the surface at large indentation loads. Similarly, as noted by Swain and Lawn (1976), although geometrical similarity requires $c_L \sim a \sim P^{1/2}$ at very large indentation loads at which the indentation flaw is completely chipped ($L_D = 1$), u/c_L diverges at such loads, such that the entire flaw is not geometrically similar and V_L/V also diverges. Consistent with these statements, a simple analytical approximation is $V_L \sim P^{5/2}$ such that $V_L/V \sim P$. Combining this approximation into Eq. (3.58) and the previous relations gives

$$F \sim P/(1 + P/P_L), \tag{3.60}$$

where P_L is an empirical scaling parameter, such that $\chi(P)$ leading to nonlinearity in Eq. (3.27) is

$$\chi(P) = \chi_{air}/(1 + P/P_L). \tag{3.61}$$

[It is recognized that there is a semiempirical element here, such that Eq. (3.61) is expected to apply predominantly during indentation strength tests in which chipping is exacerbated. Prior to testing, a weaker load dependence is likely, especially in inert environments for which χ_{air} is replaced by χ_0. This issue is addressed in the discussion.]

Replacing χ in Eq. (3.8) with $\chi(P)$ from Eq. (3.61) gives an expression for the maximum sustainable stress or strength of components containing lateral cracked dominant flaws:

$$\begin{aligned}\sigma_m &= (3T^{4/3}/4^{4/3}\psi\chi_{air}^{1/3}P^{1/3})(1 + P/P_L)^{1/3} \\ &= \sigma_L[(1 + P/P_L)/(P/P_L)]^{1/3},\end{aligned} \tag{3.62}$$

where

$$\sigma_L = 3T^{4/3}/4^{4/3}\psi\chi_{air}^{1/3}P_L^{1/3} \tag{3.63}$$

is an invariant strength, characteristic of components containing strength-controlling flaws with well-developed lateral cracks. The indentation load limits of Eq. (3.62) are then expressed as

$$\begin{aligned}\sigma_m &\rightarrow \sigma_L(P/P_L)^{-1/3} \quad (P \ll P_L) \\ &\rightarrow \sigma_L \quad (P \gg P_L).\end{aligned} \tag{3.64}$$

The indentation load P_L and strength σ_L are constants for a material and contact system and characterize lateral crack effects in indentation strength responses. Interpretation of Eqs. (3.62) and (3.64) is aided by setting scaled load and strength parameters of

$$\begin{aligned}\mathbf{P} &= P/P_L \\ \mathbf{S_m} &= \sigma_m/\sigma_L.\end{aligned} \tag{3.65}$$

The scaled residual stress parameter of Eq. (3.24) is now expressed with explicit load dependence

$$\mathbf{X} = 1/(1 + \mathbf{P}) \tag{3.66}$$

and thus

$$\mathbf{S_m} = (\mathbf{P}/\mathbf{X})^{-1/3} = [(1 + \mathbf{P})/\mathbf{P}]^{1/3}. \tag{3.67}$$

These equations are now used to assess experimental indentation crack extension and strength behavior.

Using Eq. (3.25), variable forms of $\mathbf{S_a}(\mathbf{C})$ describing "nonideal" indentation crack extension behavior can be predicted using Eq. (3.66), noting that ideal behavior is represented by $P \ll P_L$ or $\mathbf{P} \ll 1$. Figure 3.20 shows plots of crack extension under applied stress using (a) $\mathbf{P} = 0.2$ and (b) $\mathbf{P} = 0.8$. The plots are in the same format as Figure 3.6. The fine lines represent Eq. (3.25) in each case; that in Figure 3.20a approximates the responses in Figure 3.6, noting that here $\mathbf{X} \approx 0.83$. The bold lines represent the full experimental behavior: On initial application of stress, the cracks remain metastably trapped at fixed length $\mathbf{C} = 1$ until the applied stress reaches a value at which the fracture system comes into a stable equilibrium configuration (marked by the intersection of the vertical line and the curved line). Further increases of the applied stress lead to quasiequilibrium stable crack extension (along the curved line) until the applied stress reaches a value at which

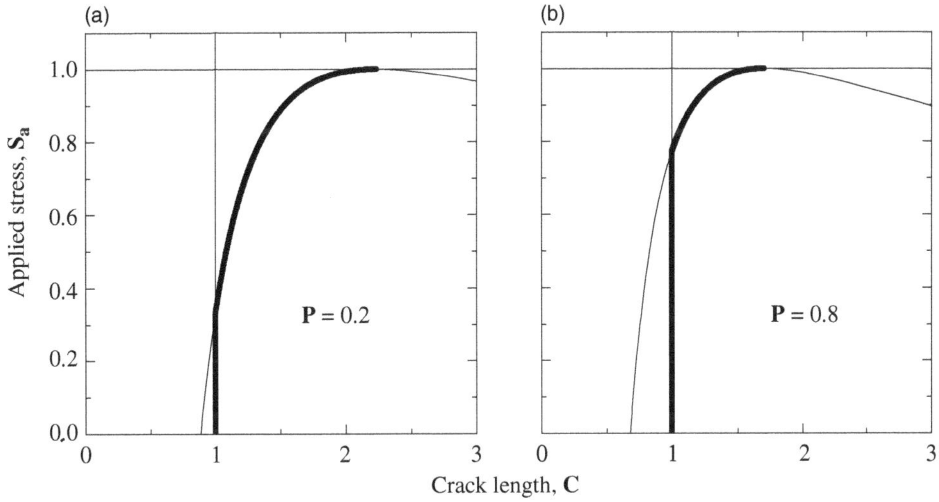

Figure 3.20 Plots of indentation crack extension behavior as scaled applied stress versus scaled crack length, illustrating the effects of lateral cracking and indentation stress relaxation. (a) Small indentation load and relaxation, corresponding to Figure 3.6. (b) Large indentation load and relaxation, corresponding to Figure 3.22.

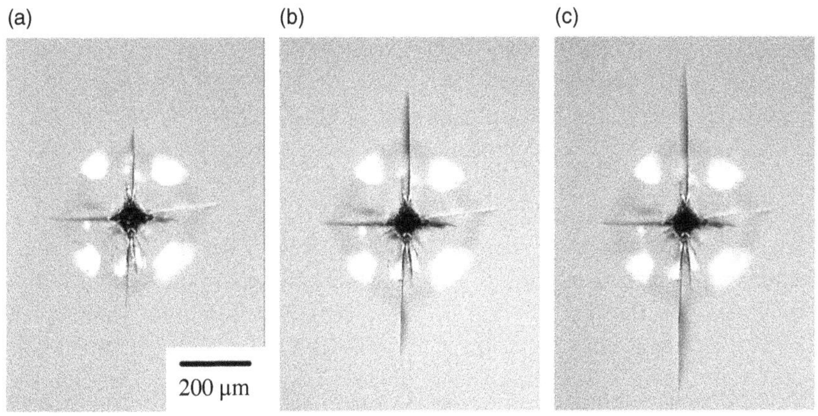

Figure 3.21 Optical images of stable indentation crack extension in aluminosilicate glass. Applied stress is horizontal in images. Surface traces of half-penny cracks are dark. Subsurface facets of lateral cracks are bright. (a) Initial configuration. (b) Intermediate configuration. (c) Configuration immediately prior to failure.

the fracture system comes into an unstable equilibrium configuration (marked by the intersection of curved line and the horizontal line). At this point the component fails at $\mathbf{S_a} = 1$. Note that the nonzero value of **P** leads to truncation of the stable precursor crack extension.

Figure 3.21 shows an example of reduced stable crack extension under applied tensile stress at a lateral cracked indentation flaw in ASG in air. The stress direction is horizontal in the images: Figure 3.21a shows the as-indented zero stress state, Figure 3.21c shows the near-maximum stress state, just prior to failure, and Figure 3.21b shows an intermediate state. The surface traces of the half-penny cracks are visible as the dark lines and the subsurface openings of the lateral cracks are visible as the intervening bright regions. In Figure 3.21a, all surface traces of the half-penny cracks are of similar length. On application of the applied stress, as in Figure 3.21b and c, the vertical cracks extend and the horizontal cracks remain at the as-indented lengths. Figure 3.21b and c depicts stable equilibrium configurations: Unless the applied stress was increased, the vertical cracks remained at the stable extended lengths. Unlike Figure 3.5, here the configuration and lengths of the lateral cracks did not change under the action of the applied stress. The effect of lateral cracks to reduce χ is illustrated by the stable indentation crack extension of a factor of approximately 1.7, considerably less than the value of 2.52 predicted for ideal indentation fracture behavior.

Measurements of such a reduced, typical, nonideal indentation crack extension behavior are shown in Figure 3.22 for (a) SLG, from the work of Marshall et al. (1979) (open symbols), and ASG (filled symbols) and for (b) silicon (Si), from

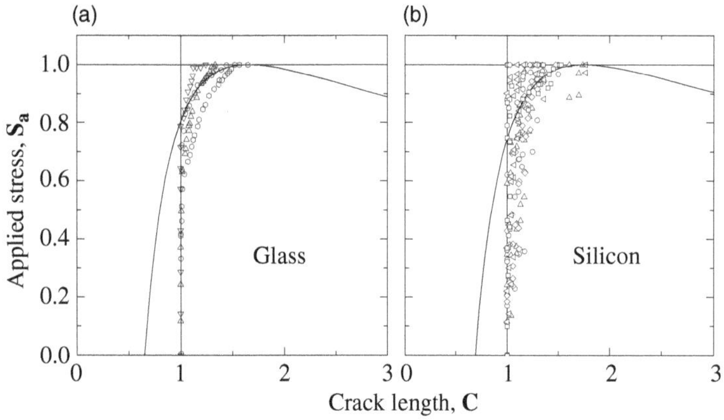

Figure 3.22 Plots of indentation crack extension behavior as scaled applied stress versus scaled crack length. (a) Glass. Source: Includes data adapted from Marshall et al. (1979). (b) Silicon. Source: Data adapted from Lawn et al. (1981) and Cook (2006).

the works of Lawn et al. (1981) and Cook (2006). The plots use the same scaling and format as Figure 3.6 and different symbols represent different indentation loads, noting that the environmental sequence is inert–inert–inert for the open symbol data and reactive–reactive–reactive for the filled symbol data. The curved lines are the response of Eq. (3.25) fit to the maximum crack length and applied stress observations for each material, treating $\mathbf{X}$ as a fitting variable. The responses clearly describe the experimental data, and values of $\mathbf{X} = 0.53$ and $\mathbf{X} = 0.58$ were used for glass and Si, corresponding to $\mathbf{P} = 0.89$ and $\mathbf{P} = 0.72$, respectively. These values clearly indicate nonideal behavior, as do the relative maximum crack extensions of approximately 1.5. Comprehensive crack extension measurements such as shown in Figures 3.6 and 3.22 are rarely available to complement indentation strength measurements. However, the ratio of final/initial crack lengths c_m/c_{air} is often measured. Experiments that measured this ratio are considered in detail in Chapter 5 and show a median value of 1.34, consistent with those shown in Figure 3.22 and much less than the ideal value of 2.52.

The fine line in Figure 3.20b represents Eq. (3.25) and approximates the responses in Figure 3.22, noting that here $\mathbf{X} \approx 0.55$. As before, the bold line represents the full experimental behavior: On initial application of stress, the crack remains trapped at fixed length $\mathbf{C} = 1$ until the applied stress reaches a large value $\mathbf{S_a} \approx 0.8$ at which the fracture system comes into a stable equilibrium configuration. Further increases of the applied stress lead to a small quasiequilibrium stable crack extension until the fracture system comes into an unstable equilibrium configuration. At this point, the component fails at $\mathbf{S_a} = 1$. Note that the nonzero value of $\mathbf{P} \approx 0.8$ leads to significant truncation of the stable precursor crack extension, as observed in the experimental measurements of Figure 3.22.

Using Eq. (3.67), lateral crack affected indentation strength behavior can be predicted, noting that ideal strength behavior is represented by $\mathbf{P} \ll 1$ such that $\mathbf{X} \to 1$. The physical interpretation of Eq. (3.67) is that indentation strength is controlled by the ratio of the indentation load and the residual stress field amplitude. For lateral crack affected strengths, the amplitude decreases with indentation load, thereby increasing strengths. Figure 3.23 shows a logarithmic plot of $\mathbf{S_m}(\mathbf{P})$ using Eq. (3.67). The bold line indicates the full convex response. The fine lines represent the scaling parameters $\mathbf{P} = 1$ and $\mathbf{S_m} = 1$, and the dashed line represents ideal $\mathbf{P}^{-1/3}$ behavior. The scaled indentation load $\mathbf{P} = 1$ provides a demarcation between indentation load-controlled behavior and lateral crack-controlled behavior, indicated by the arrows. At large indentation loads, the increase in strength to values greater than the ideal extrapolation and tendency to an invariant strength are clear. At $\mathbf{P} = 1$, the strength is approximately 25% greater than the ideal value, and at $\mathbf{P} = 10$, the strength is approximately twice the ideal value. These increases should be considered in terms of the typical dispersion of approximately 5% for an indentation strength measurement—lateral crack effects are easily distinguished in indentation strength measurements (and may in fact be discerned in some of the near-ideal indentation strength responses in Figure 3.2).

Lateral crack effects are pervasive in indentation experiments. Here, quantification of lateral cracking and chipping at indentations was quantified by lateral crack development and illustrated in Figures 3.17 and 3.18. A model of the effect of lateral cracking on the pdz-based indentation residual stress field was developed. The model was shown to account for truncation of half-penny crack extension during application of applied stress in Figures 3.20 and 3.22. The model was also applied in Figure 3.23 to predict the tendency to strength invariance for all materials at large indentation loads. A feature of the model is a material-dependent indentation load P_L at which lateral crack effects become dominant in both cracking and strength. Lateral crack effects in indentation strength responses of materials with invariant toughness are discussed in Section 3.4.3.

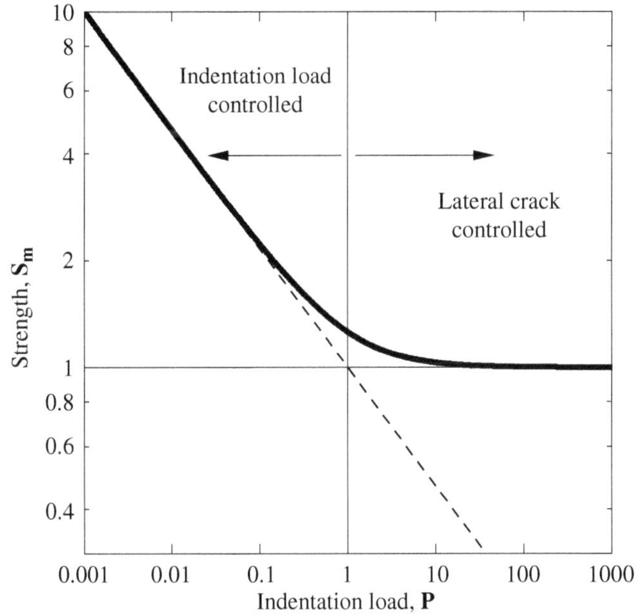

Figure 3.23 Logarithmic indentation strength plot in scaled coordinates for indentation contacts exhibiting lateral crack effects. Full behavior is shown as bold solid line. At small loads, strengths approach the ideal indentation response (shown as dashed line). At large loads, strengths approach an invariant plateau (shown as horizontal fine line).

3.3.4 Impacts

Impact flaws, in which a free particle moving with kinetic energy U_K contacts a surface, have many of the same characteristics as quasi-static flaws formed by indentation contacts of load P. These characteristics are noted in the overview diagram of Figure 3.10. The diagram shows schematic cross sections of an angular particle of mass m approaching a surface with various velocities v indicated by the arrows. The particle is assumed to be harder, stiffer, and tougher than the surface such that contact damage is restricted to the surface (see Cook 2023 for a detailed discussion of the opposite assumption, in which damage is restricted to the particle). Elastic–plastic impacts, in which a vertex or edge of the particle contacts the surface, Figure 3.10a–c, are considered first. For small velocities, Figure 3.10a, contact damage is restricted to a crack-free localized pdz, similar to the subthreshold indentation flaw shown in Figure 3.9b. For moderate velocities, Figure 3.10b, half-penny cracks are initiated and stabilized at the contact, emanating from the pdz periphery, similar to the indentation cracks shown in Figure 2.11. For large velocities, Figure 3.10c, lateral cracks are also initiated and stabilized at the contact, emanating from the pdz base, similar to the indentation cracks shown in Figure 3.9a. Elastic impacts occur when a face of the particle contacts the surface, Figure 3.10d, initiating and stabilizing cone cracks, similar to Figure 1.13. Determination of component strengths controlled by dominant impact flaws thus requires fracture mechanics analysis-similar to that developed previously for indentation flaws in Section 3.2.1. In particular, much of the analysis can be applied directly if a correspondence is formed between particle kinetic energy U_K and indentation load P. Such a correspondence is now developed.

Figure 3.24 shows detailed schematic cross sections of contact and deformation on impact for (a) a spherical particle and (b) an angular particle. Both particles are of material density ρ, with radius r for the spherical particle and characteristic radius r for the angular particle. In both cases, the mass m of the particle is thus given by $m \sim \rho r^3$. The angular particle presents a vertex of conical symmetry with included angle 2ϕ to the surface. Initially, dashed outlines, both particles are moving with speed v toward and normal to the stationary surface such that the kinetic energy U_K of both particles is $U_K \sim mv^2 \sim \rho r^3 v^2$. On impact, shaded areas, the surface decelerates the particles such that the speeds are decreased to zero, $v = 0$, at maximum penetrations or *total displacements* h of the particles into the surface. In both cases, the total displacement h consists of two components: a *contact displacement* h_c over which the particle is in contact with surface and an *elastic displacement* $(h - h_c)$ in which the surface displaces and deflects elastically but does not contact the particle. The ratio of total/contact displacements is $\beta = h/h_c$. In both cases, the contact radius over which the particle contacts the surface is a such that the contact area is πa^2. The surface profiles within the contact radii are defined

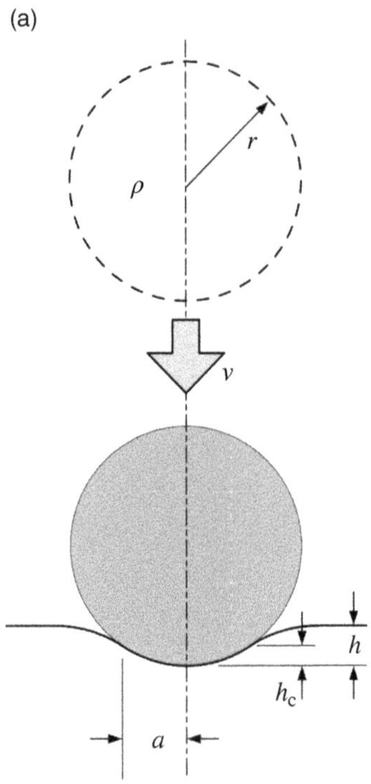

Figure 3.24 Schematic cross-section diagrams of particle impacts on surfaces. Particle material density ρ, characteristic radius r, and velocity v. Incoming particle positions are dashed. Contacting particles are shaded. (a) Spherical particle, elastic contact. (b) Angular particle, elastic–plastic contact, plastic deformation zone hatched.

by the shapes of the particles in each case. For the spherical particle, $a = h_c(2r/h_c - 1)^{1/2}$, and for the angular particle, $a = h_c \tan \phi$.

The nature of the deformation within the contact areas is taken to be (a) elastic, for the spherical contact, and (b) elastic–plastic, for the conical contact, such that a hemispherical pdz, shown hatched, is generated beneath the contact of the angular particle. The work W performed by the particles in deforming the surface is

$$W = \int_0^h P(h')\mathrm{d}h', \tag{3.68}$$

where $P(h)$ is the force–displacement relation specific to the particle-material impact combination and h' is a dummy displacement variable. Typical particle impact velocities, $v \approx 1\text{–}100\ \mathrm{m\ s^{-1}}$, are much slower than the velocities of acoustic waves in solids (Marshall et al. 1983; Cook 2023), and thus impact force–displacement relations are well approximated by quasi-static force–displacement relations. For the spherical elastic impact, the Hertzian contact relation is applicable

$$P(h) = Mr^{1/2}h^{3/2}, \tag{3.69}$$

where M is the elastic contact modulus and β takes an upper bound of $\beta = 2$ for a rigid particle (Johnson 1985). For the elastic–plastic conical impact, the geometrical similarity contact relation is applicable

$$P(h) = H\pi \tan^2 \phi h^2/\beta^2, \tag{3.70}$$

where β can be approximated as $\beta \approx 1 + 2H \tan \phi/\pi M$ for acute contacts (Marshall et al. 1983) and $\beta \approx 1 + (2H \tan \phi/M)^{1/2}$ for obtuse contacts (Cook and Oyen 2007). For typical material and particle parameters of $H/M \approx 0.07$ and $\phi \approx 70°$, these give β in the range of 1.1–1.5, noting that the exact value for elastic contact by a cone is $\beta = \pi/2 \approx 1.57$. As might be anticipated, relative to elastic spherical contacts elastic–plastic conical contacts exhibit much reduced elastic displacements, as indicated in Figure 3.24. For conical indentation of ideal plastic materials, $\beta = 1$, and Eq. (3.70) reverts to $P = H\pi a^2$.

Evaluating Eq. (3.68) using Eqs. (3.69) and (3.70) gives

$$W \sim M^{-2/3}r^{-1/3}P^{5/3} \tag{3.71}$$

and

$$W \sim H^{-1} \tan^{-1}\phi P^{3/2}, \tag{3.72}$$

respectively, omitting all constants, where P in these equations is taken as the maximum contact load. If the contacts are rapid enough to be adiabatic, the First Law of Thermodynamics applies as $W = U_K$, such that all particle kinetic energies are converted into works of deformation. The sought relations between the incoming particle kinetic energies and the equivalent contact loads are then

$$P \sim M^{2/5} r^{1/5} U_K^{3/5} \tag{3.73}$$

and

$$P \sim H^{1/3} \tan^{2/3}\phi U_K^{2/3}. \tag{3.74}$$

For both spherical elastic contacts and conical elastic–plastic contacts, the cone and half-penny cracks generated scale in size c according to Roesler's law $c \sim P^{2/3}$ (see Chapter 2). The strengths of components containing such dominant flaws scale as the Griffith equation (Chapter 1 and Section 3.2.1) $\sigma \sim c^{-1/2}$, such that for both forms of contact component strengths scale with contact load as $\sigma \sim P^{-1/3}$. Using Eqs. (3.73) and (3.74) in this relation gives expressions for the strength of impacted components in terms of the kinetic energy of the contacting particle:

$$\sigma \sim M^{-2/15} r^{-1/15} U_K^{-1/5} \quad \text{(Spherical elastic)} \tag{3.75}$$

and

$$\sigma \sim H^{-1/9} \tan^{-2/9}\phi U_K^{-2/9} \quad \text{(Conical elastic–plastic).} \tag{3.76}$$

Expressing the particle kinetic energy as $U_K \sim \rho r^3 v^2$ gives

$$\sigma \sim M^{-2/15} \rho^{-1/5} r^{-2/3} v^{-2/5} \quad \text{(Spherical elastic)} \tag{3.77}$$

and

$$\sigma \sim H^{-1/9} \tan^{-2/9}\phi \rho^{-2/9} r^{-2/3} v^{-4/9} \quad \text{(Conical elastic–plastic).} \tag{3.78}$$

The fracture mechanics development beginning with Eq. (3.68), and the resulting expressions for strength in terms of particle kinetic energy or velocity, Eqs. (3.75)–(3.78), are identical to those in earlier works (Wiederhorn and Lawn 1977, 1979), although somewhat more succinctly stated.

There are several points to note regarding Eqs. (3.75)–(3.78). First, for both types of impact, the strength expressions consist of a sequence of power laws in material properties and particle attributes. For elastic contacts, the relevant material property is the elastic contact modulus M, and for elastic–plastic contacts, the relevant material property is the mean supported contact stress or hardness H. Second, although the pairs of strength expressions Eqs. (3.75)–(3.78) describe the responses of very different impact flaws, the power law dependencies on impacting particle kinetic energy and velocity are very similar, differing only by a factor of 10/9. This small ratio is to be compared with the ratio 4/3 relating the contact load dependencies of elastic–plastic and elastic spherical contact flaws (Section 3.3.2) and the factor of 2 relating the contact load dependencies of static and dragged sharp wedge flaws (Chapter 9). The implication here is that without direct observation it might be very difficult to distinguish strength degradation by "sharp" and "blunt" impacts from strength tests alone. Third, elastic–plastic conical impact flaws are susceptible to lateral cracking, chipping, and relaxation of the residual field. The effects on impact flaws are identical to those discussed in Section 3.3.3 regarding indentation flaws. It is straightforward to use Eq. (3.74) to define a characteristic impact kinetic energy U_{KL} analogous to the characteristic indentation load P_L of Eq. (3.61). Impacts with kinetic energy $U_K > U_{KL}$ exhibit lateral crack effects in strength responses. Defining

$$\mathbf{U} = U_K/U_{KL} \tag{3.79}$$

enables the full lateral crack affected impact strength response to be written as a modification of Eq. (3.67),

$$\mathbf{S_m} = \left[(1 + \mathbf{U}^{2/3})/\mathbf{U}^{2/3}\right]^{1/3}. \tag{3.80}$$

The strength response is similar in form to that of Figure 3.23: at small impact energies $\mathbf{S_m} \to \mathbf{U}^{-2/9}$, consistent with Eq. (3.76); at large impact energies, the strength becomes invariant, $\mathbf{S_m} \to 1$.

Particle impacts are a pervasive cause of strength-controlling flaws on brittle component surfaces. Here, the major particle impact geometries, spherical and conical, and contact severity amplitude, particle kinetic energy or velocity, were incorporated through the First Law of Thermodynamics into an indentation strength framework. Indentation strength responses of components containing dominant impact flaws in materials of invariant toughness are discussed in Section 3.4.4.

3.3.5 Contacts in Stressed and Layered Components

As noted throughout this book, brittle materials such as ceramics and glasses fail by fracture and the propagation of cracks, primarily under the action of tensile stresses. The tensile stresses can arise from many sources (Kingery et al. 1976; Lawn 1993; Cook and Pharr 1994): in all fracture systems by direct mechanical loading; in inhomogeneous or anisotropic systems by differences in densification behavior, thermal expansion behavior, or elastic modulus behavior; and in systems with limited thermal conductivity by thermal shock. In ceramics and glasses, the overwhelming majority of cracks occur on component surfaces (Lawn 1993; Cook and Pharr 1994). Cracks can be initiated at surfaces by accidental or deliberate sharp particle contact or by the attachment of surface structures. An obvious conclusion therefore is that a method for failure prevention in ceramic and glass components is the incorporation of compressive surface stress (Kingery et al. 1976; Cooper 1982). Surface stresses can be introduced by many physical methods: equilibrium differences in thermal expansion within a component, known as cladding or coating (Oel and Fréchette 1967; Krohn and Cooper 1969; Cooper 1982; du Merac et al. 2018); transient differences in thermal expansion, known as thermal tempering (Kingery et al. 1976; Cooper 1982; Tandon and Cook 1993); differences in densification behavior (Kingery et al. 1976; Cai et al. 1997); machining (Cook et al. 1981; Lange et al. 1983); ion-exchange, known as chemical tempering (Cooper and Krohn 1969; Kingery et al. 1976; Cooper 1982; Green et al. 1999; Morris et al. 2004a); chemically or mechanically induced phase transformation (Green et al. 1983; Kishino et al. 1996) or irradiation (Lawn and Fuller 1984).

Whatever the method of stress formation, mechanical equilibrium requires balanced forces and moments across all planes in a component. A consequence of this requirement is that the introduction of a surface stress into a component also generates a reaction stress of opposite sign in the component bulk. Several limiting component and internal stress state geometries are of interest. The first limiting geometry is if the thickness of the stressed surface layer is very small relative to the dimensions of the component. In this case, the bulk stress, although still of opposite sign to the surface stress, is negligible and the component as a whole is undeformed. The second limiting geometry is if the stiffness values of the surface layer and bulk are comparable, usually because the layer and bulk materials and thickness dimensions are similar. In this case, the magnitude of the bulk stress is comparable to that of the surface stress and an internal bending moment is generated. The moment causes component deformation in the form of external curvature that balances the internal moment. (The stereotypical example is the "bimetallic strip," in which a bilayer of materials of different coefficients of thermal expansion deforms in response to temperature changes.) The third limiting geometry is if surface layers are symmetrically distributed about a line or plane in the component. In this case, bulk and surface stresses are of opposite sign but no bending moment is generated and the component, although stressed internally, remains undeformed externally, see Figure 3.11a. In most instances of these three geometries, the spatial distribution of stress within the surface layer is not uniform but decreases in magnitude with increasing distance from the component surface.

By implementing one of the fabrication methods listed above, many commercial components of brittle materials incorporate surface compressive stresses in one or a combination of these three limiting geometries. Films and coatings applied in semiconductor manufacturing exhibit the very thin layers of the first geometry and rarely deform the Si wafer substrate. However, machining of the back faces of wafers (see Chapter 10) leads to stressed layers that are not negligible in size relative to the remaining wafer thickness. Back face ground wafers thus often exhibit the second geometry and wafer "bow." Thermally tempered glasses, used pervasively in thick sheets for windows in automobiles and commercial construction, exhibit the symmetric stress distribution of the third geometry (about the sheet center plane) on length scales comparable to the sheet thicknesses. Chemically tempered glasses, used pervasively as thin sheets in the display screens of hand held telephones and tablets, also exhibit symmetric distributions of surface compression, but in this case as very thin surface layers, thereby combining the first and third geometries. Irradiated or phase transformed components typically lead to a state between the first and second geometries as surface material is usually altered on only a single face of the component. Component curvature thus increases as alteration proceeds and stress in the altered layer increases. Conversely, if initially stress-free components are deliberately deformed into curved deformation states by applied loading, asymmetric stress distributions are developed that exhibit maxima in magnitude at surface layers, see Figure 3.11b. This bending or flexing geometry is used pervasively in the strength testing of brittle components.

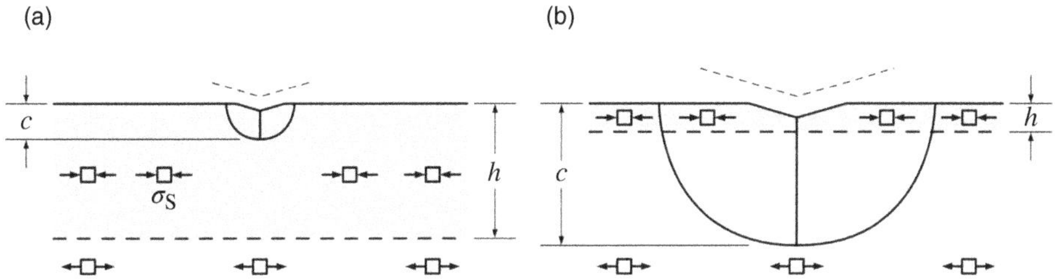

Figure 3.25 Schematic cross-section diagrams of compressive surface stressed structures containing indentation flaws. (a) Flaw contained within exterior surface stressed layer. (b) Flaw with crack extended into tensile interior bulk.

This section will develop fracture mechanics analyses required to describe indentation cracking and indentation strength in surface stressed and layered components. The focus of the analysis is on the effects of compressive surface stress but the fracture formulations developed will include the effects of compensating bulk tensile stress. Fracture in inhomogenous stress fields is considered and in particular the frequently occurring geometry of a circular crack propagating into a rectilinear stress field. This geometry pertains to indentation cracks in stress fields generated interior to components by surface stresses, as in Figure 3.11a, and exterior to components by bending loading, as in Figure 3.11b. More extensive analyses of fracture in surface-related stress fields can be found elsewhere (Lawn and Marshall 1977; Lawn and Fuller 1984; Cook 2005).

Figure 3.25 shows detailed schematic diagrams of indentation and cracking in surface stressed structures similar to the exterior of the layered component of Figure 3.11a. In both diagrams of Figure 3.25, a compressive stressed surface layer is shown shaded. The tensile stressed bulk is shown unshaded. The stress orientations are indicated by squares and arrows. In both diagrams, the *magnitude* of the compressive stress is σ_S; the formal sign of the stress is negative. Figure 3.25a shows the effects of a small indentation load that generated half-penny cracks, length c, that are completely contained within the surface stressed layer, depth h, such that $c \leq h$. Figure 3.25b shows the effects of a larger indentation load that generated half-penny cracks that extended beyond the surface stressed layer into the bulk, such that $c > h$. Fracture mechanics analysis begins with consideration of the contained crack of Figure 3.25a. The SIF for the half-penny crack in this case consists of two terms, similar to Eq. (3.3):

$$K = K_r + K_S = \chi P/c^{3/2} + \psi\sigma_S c^{1/2}. \tag{3.81}$$

K_r characterizes the effects of the indentation residual stress field, and K_S characterizes the effects of the surface stress. As the indentation crack is completely contained within the surface stressed layer, the surface stress σ_S is uniform over the crack and thus

$$K_S = \psi\sigma_S c^{1/2}, \tag{3.82}$$

where ψ is a dimensionless geometry factor and it is noted that K_S is a signed quantity. The equilibrium condition (assumed in air) $K = T$ is thus

$$\chi P/c^{3/2} + \psi\sigma_S c^{1/2} = T, \tag{3.83}$$

noting that the emphasis here is slightly different from that of Eq. (3.4). Here, σ_S is fixed and $c(P)$ is sought, rather than $\sigma(P)$ sought as previously. As $\sigma_S < 0$, both SIF terms in Eq. (3.83) are stabilizing ($\mathrm{d}K/\mathrm{d}c < 0$) and a stable equilibrium crack length, c_S, for $K_S \neq 0$ can be found by simple inversion of Eq. (3.83). In the absence of a surface stressed layer, $K_S = 0$, the standard indentation fracture mechanics of Chapter 2 prevails and the stable equilibrium crack length is c_r. These crack lengths are given by

$$\chi P/c_S^{3/2} = T - \psi\sigma_S c_S^{1/2} \tag{3.84}$$

and

$$\chi P/c_r^{3/2} = T. \tag{3.85}$$

Combining these equations gives

$$P/c_S^{3/2} = P/c_r^{3/2}\left[1 - (\psi\sigma_S/T)c_S^{1/2}\right], \tag{3.86}$$

which provides a linearized plotting scheme between $P/c_S^{3/2}$ and $c_S^{1/2}$ that allows the estimation of σ_S (Gruninger et al. 1987; Cook 2005; Tandon and Buchheit 2007). A somewhat more intuitive analysis scheme is to rearrange Eqs. (3.85) and (3.86) to express c_r in two ways, first as the Roesler equation,

$$c_r = BP^{2/3}, \tag{3.87}$$

and then in terms of the surface stressed crack length,

$$c_r = c_S\left[1 - (\psi\sigma_S/T)c_S^{1/2}\right]^{2/3}. \tag{3.88}$$

As $c_r/c_S > 1$ for $\sigma_S < 0$, Eq. (3.88) makes clear that indentation cracks generated in surfaces with compressive stress are shorter than cracks in unstressed surfaces.

Figure 3.26 shows logarithmic plots of indentation crack length as a function of indentation load for four surface stressed systems fabricated by the methods listed above (Gruninger et al. 1987; Morris et al. 2004a; Cook 2005; Tandon and Buchheit 2007); the plots are shown in order of increasing material homogeneity. The plots in the left column represent the contained crack stress states of Figure 3.25a, and the plots in the right column represent the extended crack stress states of Figure 3.25b. In each plot, the filled symbols represent crack length measurements in unstressed "control" surfaces and open symbols represent crack length measurements in stressed surfaces. Bars indicate published uncertainties in experimental measurements or uncertainties determined from analysis of published data. The diagonal solid lines indicate best fits of Eq. (3.87) to measurements in unstressed surfaces. The vertical bars in each plot are guides to the eye and represent

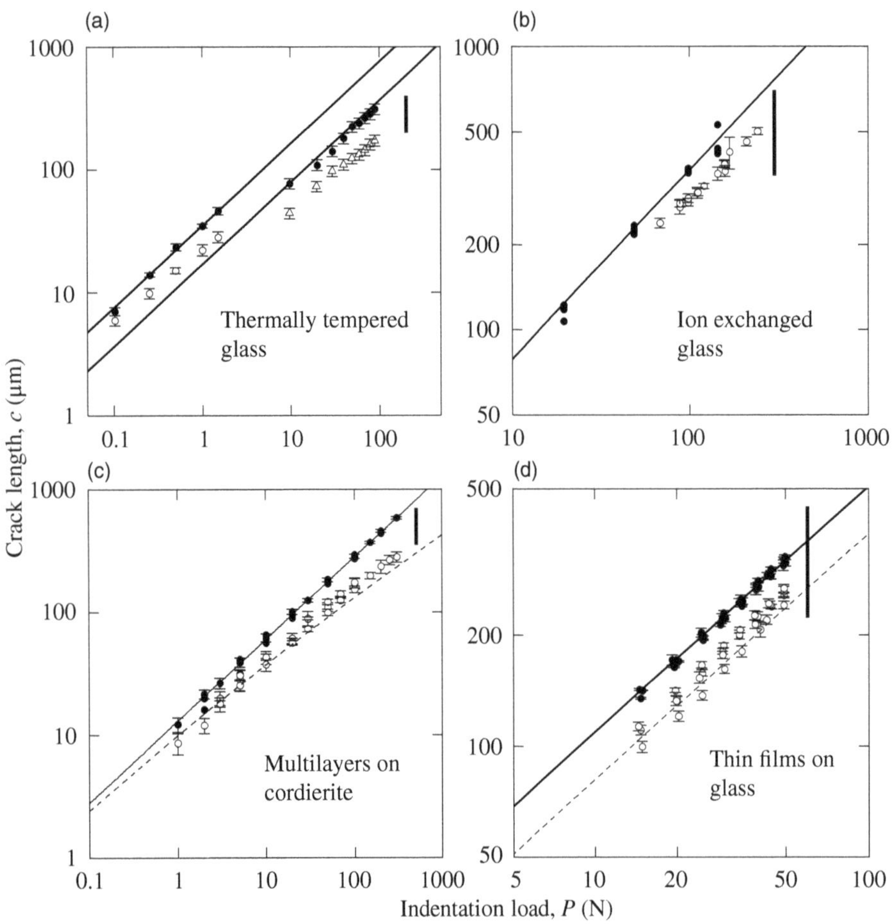

Figure 3.26 Logarithmic plots of indentation crack length as a function of indentation load for four compressive surface stressed systems. Filled symbols represent unstressed control surface observations. (a) Thermally tempered soda-lime silicate glass. Source: Data adapted from Tandon and Buchheit (2007). (b) Chemically tempered aluminosilicate glass. Source: Data adapted from Morris et al. (2004b). (c) Clad cordierite glass-ceramic. Source: Data adapted from Cook (2006). (d) Thin film coated soda-lime silicate glass. Source: Data adapted from Gruninger et al. (1987).

factors of 2 in the coordinates used. Figure 3.26a shows the effects of thermal tempering of SLG from the work of Tandon and Buchheit (2007). The upper filled symbols and line represent cube corner indentation cracks and the lower filled symbols and line represent Vickers indentation cracks. The longer cube corner cracks and depressed threshold are consistent with the angular analyses of Chapter 2. Figure 3.26b shows the effects of chemical tempering (often termed ion-exchange) in ASG from the work of Morris et al. (2004a). The open symbols in this case indicate the lengths of cracks at the indentation threshold condition for different levels of tempering. Figure 3.26c shows the effects of multilayer cladding of cordierite glass-ceramic from the work of Cook (2005). The bulk substrate material had greater coefficient of thermal expansion (CTE) than the surface material, such that surface compression was generated on cooling. Figure 3.26d shows the effects of thin film coatings on SLG from the work of Gruninger et al. (1987). Different open symbols represent silica, alumina, silicon nitride, and zirconia films. In each subplot of Figure 3.26, the surface stress was compressive and the observed indentation cracks in the surface stressed components were shorter. Equations (3.87) and (3.88) form a parametric set for the indentation crack length response $c_S(P)$ of cracks contained in a surface stressed layer. The dashed line in Figure 3.26c is a best fit of such a response to the shortest measured cracks. The fit describes the observations and in particular the increase in the ratio c_r/c_S as P, and thus c_S, increased, consistent with Eq. (3.88). The physical explanation is that larger cracks experience a larger area of compressive stress and the lengths are more effectively diminished relative to unstressed cracks. A similar trend with similar explanation is observed in Figure 3.26a.

The trend of increasing c_r/c_S with P in Figure 3.26b and the lack of trend and near-invariant c_r/c_S response in Figure 3.26d have different explanations from those above. The explanation in these cases relies on a difference in crack geometry. The fracture system for the indentation cracks of Figure 3.26b and d is shown in the schematic diagram of Figure 3.25b. Distinct from Figure 3.25a, the half-penny crack extends beyond the surface compression layer into the bulk. Detailed fracture mechanics description of this geometry can be quite complicated as the relative sizes of the crack and layer, c/h, and layer thickness and component dimension, h/L, must be taken into account (Cook 2005), and become more complicated if the stress in the layer is nonuniform (Lawn and Marshall 1977; Morris et al. 2004a). However, considerable insight can be gained for cracks assumed large relative to the layer thickness, $c/h \gg 1$, and layers assumed small relative to component dimensions, $h/L \ll 1$. In these cases, the SIF associated with the surface stress is given by (Lawn and Fuller 1984)

$$K_{S\infty} = 2\psi\sigma_S h^{1/2}, \tag{3.89}$$

where the subscript ∞ indicates that $c \gg h$. In this case, the total SIF for the indentation crack is $K = K_r + K_{S\infty}$ and the equilibrium condition is given by

$$\chi P/c^{2/3} + 2\psi\sigma_S h^{1/2} = T. \tag{3.90}$$

This equation is different in character from Eq. (3.4) as the $K_{S\infty}$ term is entirely associated with the component and includes no crack length dependence. Separation of the indentation loading term from the loading-independent material and component terms is accomplished by defining a surface stress *toughening* term T_S as

$$T_S = -K_S. \tag{3.91}$$

In this case, $T_S = T_{S\infty} = -K_{S\infty}$, and a long crack toughness T_∞ is defined by

$$T_\infty = T_{env} + T_{S\infty} = T_{env} - 2\psi\sigma_S h^{1/2}, \tag{3.92}$$

where the subscript "env" indicates that the base toughness takes the value appropriate to the environment, air or inert. It is emphasized that all terms in Eq. (3.92) are invariant, the first determined by the material and environment and the second by the component. Equation (3.90) is then rewritten as

$$\chi P/c^{2/3} = T_\infty, \tag{3.93}$$

and simply inverted as

$$c_S = (\chi P/T_\infty)^{2/3}, \tag{3.94}$$

which is the Roesler equation. As $\sigma_S < 0$ here, $T_\infty > T_{air}$ and thus $c_S < c_r$ for a given indentation load: the component and, by implication, the material, appears more resistive to indentation fracture.

The physical interpretation of the algebraic manipulation of Eqs. (3.91)–(3.94) is that the frame of reference has been shifted from the crack tip frame in Eq. (3.90) to the indentation frame in Eq. (3.93). In the reference frame of the indentation, the material appears tougher as the indentation cracks are shorter. In general, toughening effects are perceived in a frame of reference related to the applied loading. Such effects are characterized algebraically by setting material- and component-related SIF terms as toughness terms, as in Eq. (3.91), and then separating applied loading terms from toughness terms, as in Eq. (3.90). Also, in general, toughening effects are crack length dependent, $T = T(c)$, although this was not the case here. Extensive use is made of changes in reference frames and the concept of toughening in consideration of microstructural effects on fracture and is a key element of the analyses in Chapter 4.

The indentation cracking behaviors of the thin layer surface stressed systems of Figure 3.26b and d are simply interpreted using Eq. (3.94). In Figure 3.26d, indentations of increasing load were performed on the same film on glass system and σ_S was invariant. Thus, in this case, T_S and T_∞ were invariant with P and Roesler behavior, $c_S \sim P^{2/3}$, and invariant c_r/c_S predicted. The dashed line in Figure 3.26d shows a fit of Eq. (3.94) to the thin film systems data consistent with this prediction. In Figure 3.26b, indentations of increasing load were required to initiate cracks in increasingly ion-exchanged glass systems. In these measurements, σ_S thus increased with indentation load and hence T_S and T_∞ increased with P. As a consequence, the ratio c_r/c_S is predicted to increase and is observed in Figure 3.26b.

An obvious consideration for indentation cracking behavior in surface stressed components is if the constraint of a very small layer relative to the component dimensions $h \ll L$ is relaxed. In this case the total SIF for the system is

$$K = K_r + K_{S\infty} + K_B = \chi P/c^{2/3} + 2\psi\sigma_S h^{1/2} - \beta\psi\sigma_S c^{1/2}, \tag{3.95}$$

where $\beta \approx h/L$ (e.g., Gruninger et al. 1987) and K_B is a SIF characterizing the mechanical equilibrating reaction stress in the bulk of the component. K_B is given by

$$K_B = -\beta\psi\sigma_S c^{1/2} \tag{3.96}$$

such that if $\sigma_S < 0$, $K_B > 0$ and is destabilizing as $\mathrm{d}K_B/\mathrm{d}c > 0$. Using the definition of T_∞ gives the equilibrium condition for the system of Eq. (3.95) as

$$-\beta\psi\sigma_S c^{1/2} + \chi P/c^{3/2} = T_\infty. \tag{3.97}$$

Equation (3.97) is identical in mathematical structure to Eqs. (3.4) and (3.83) and describes a stabilizing to destabilizing transition as in Section 2.1. Here, as in Eq. (3.83), the emphasis is different from Eq. (3.4) as an instability in the indentation load occurs. Replacing σ_a in Eq. (3.8) with $-\sigma_S$ from Eq. (3.97) and rearranging gives an expression for the maximum sustainable contact load P_m of compressively surface stressed and bulk tensile stressed components:

$$P_m = -(27/256)T_\infty^4/\psi^3\chi(\beta\sigma_S)^3. \tag{3.98}$$

For contact loads $P < P_m$, the generated cracks are longer than those in the base material, $c_S > c_r$. At $P = P_m$, the crack length c_S diverges and the component fails, as the effects of the bulk tensile stress overcome those of the surface compressive stress. Measurement of this critical failure load was a key element of the fragmentation studies on ion-exchanged glasses of Tandon and Glass (2015). Other studies of failure and fragmentation under contact loading are considered in Chapter 7. It is noted that the expression for the critical indentation load at fracture instability, Eq. (3.98) scales as the familiar ratio of toughness^4/stress^3 as in indentation crack initiation (the stress is the hardness, Section 2.4), and other ductile–brittle transitions (the stress is the yield stress, Cook 2023). The sensitivity to variation in material and component characteristics, T_∞^4 and σ_S^3, accounts for the extreme variability often encountered in contact-induced fragmentation of brittle, stressed, objects.

Indentation strength analysis of surface stressed components begins with consideration of the effects of an applied stress σ_a on the contained crack of Figure 3.25a. Using Eq. (3.1) the total SIF for this system is

$$K = K_a + K_r + K_S = \psi\sigma_a c^{1/2} + \chi P/c^{3/2} + \psi\sigma_S c^{1/2}. \tag{3.99}$$

Replacing σ_a in Eq. (3.8) with $(\sigma_a + \sigma_S)$ from Eq. (3.99) gives an expression for the maximum sustainable stress or strength of components with surface stressed, contained, dominant flaws:

$$\begin{aligned}\sigma_m &= (3T^{4/3}/4^{4/3}\psi\chi^{1/3}P^{1/3}) - \sigma_S \\ &= \sigma_\Omega(P_\Omega/P)^{1/3} - \sigma_S,\end{aligned} \tag{3.100}$$

The second line of Eq. (3.100) has been written using the experimental threshold indentation load P_Ω as defined in Chapter 2 and used previously and the maximum indentation strength in unstressed components σ_Ω as defined in Eq. (3.14). Setting scaling relations of

$$\begin{aligned} \mathbf{S_m} &= \sigma_m/\sigma_\Omega \\ \mathbf{S_S} &= \sigma_S/\sigma_\Omega \\ \mathbf{P} &= P/P_\Omega \end{aligned} \tag{3.101}$$

enables a scaled inert strength relation to be written for stressed components:

$$\mathbf{S_m} = \mathbf{P}^{-1/3} - \mathbf{S_S}. \tag{3.102}$$

Ideal indentation strength behavior is given by setting $\mathbf{S_S} = 0$ and thus

$$\mathbf{S_m} = \mathbf{P}^{-1/3}. \tag{3.103}$$

Figure 3.27 shows a logarithmic plot of $\mathbf{S_m}(\mathbf{P})$ using Eq. (3.102). The diagonal dashed line represents ideal $\mathbf{P}^{-1/3}$ behavior, Eq. (3.103). The bold line indicates the full response for $\mathbf{S_S} = -0.5$. The horizontal dashed line represents $|\mathbf{S_S}| = 0.5$. The full response is asymptotic to the dashed responses at small and large values of $\mathbf{P}$; only the indentation domain greater than the threshold load, $\mathbf{P} \geq 1$, is physically meaningful. Although superficially similar, the full response of Figure 3.27 is very different from that of Figure 3.23. The most noticeable differences are that at small indentation loads the compressive surface stressed response is shifted to greater strengths and does not overlap the ideal response and that overall the compressive surface stressed response exhibits less convexity. In addition, together with material and indenter characteristics the compressive surface stressed response also depends on the component. This last point is shown in the increased strength of components with greater surface compression, indicated by the fine line in Figure 3.27. The effectiveness of compressive surface stresses to increase the strength of components with dominant contact flaws is clear. In the example of $\mathbf{S_S} = -0.5$, typical for stressed components, at a moderate contact load of $\mathbf{P} = 10$ the strength is approximately twice the ideal value and at $\mathbf{P} = 100$ the strength is approximately five times the ideal value.

The point made previously regarding impact flaws and lateral cracking also applies here. That point lead to Eqs. (3.79) and (3.80) and a similar set of equations can be developed for elastic–plastic conical impact flaws and surface stress

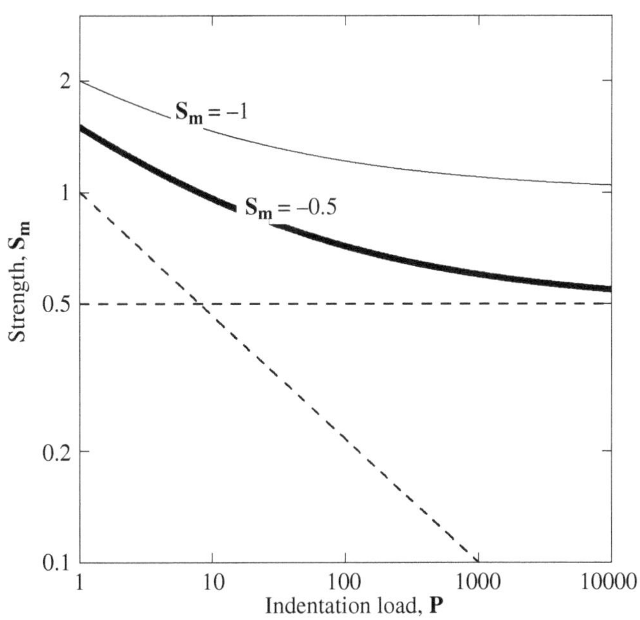

Figure 3.27 Logarithmic indentation strength plot in scaled coordinates for indentation contacts with cracks contained in surface compressive regions. Full behavior is shown as solid lines. At small loads, strengths approach the ideal indentation response (shown as diagonal dashed line). At large loads, strengths approach an invariant plateau (shown as horizontal dashed line).

effects. The effects on impact flaws are identical to those on indentation flaws. As before, it is straightforward to use Eq. (3.74) to define a threshold impact kinetic energy $U_{K\Omega}$ analogous to the threshold indentation load P_{Ω} of Chapter 2. Defining

$$\mathbf{U} = U_K/U_{K\Omega} \tag{3.104}$$

enables a surface stress affected impact strength response to be written as a modification of Eq. (3.102),

$$\mathbf{S_m} = \mathbf{U}^{-2/9} - \mathbf{S_S}. \tag{3.105}$$

The strength response is similar in form to that of Figure 3.27: at small impact energies $\mathbf{S_m}$ is shifted from the response related to ideal behavior, Eq. (3.76), $\mathbf{U}^{-2/9}$; at large impact energies, the strength becomes invariant, $\mathbf{S_m} \to -\mathbf{S_S}$.

Indentation strength analysis of surface stressed components with the extended crack of Figure 3.25b is based on the total SIF for this system of

$$K = K_a + K_r + K_{S\infty} + K_B = \psi\sigma_a c^{1/2} + \chi P/c^{3/2} + 2\psi\sigma_S h^{1/2} - \beta\psi\sigma_S c^{1/2}. \tag{3.106}$$

Using the definition of T_∞ gives the equilibrium condition in this case as

$$\psi\sigma_a c^{1/2} - \beta\psi\sigma_S c^{1/2} + \chi P/c^{3/2} = T_\infty. \tag{3.107}$$

Equation (3.107) is identical in mathematical structure to Eqs. (3.4), (3.83), and (3.97) and describes a stabilizing to destabilizing transition as in Section 2.1. Replacing σ_a in Eq. (3.8) with $(\sigma_a - \beta\sigma_S)$ from Eq. (3.107) gives an expression for the strength of components with surface stressed, extended, dominant flaws:

$$\begin{aligned}\sigma_m &= (3T_\infty^{4/3}/4^{4/3}\psi\chi^{1/3}P^{1/3}) + \beta\sigma_S \\ &= \sigma_\Omega(T_\infty/T_0)^{4/3}(P_\Omega/P)^{1/3} + \beta\sigma_S\end{aligned} \tag{3.108}$$

using σ_Ω and P_Ω as before and $T_{env} = T_0$ here. Using an additional scaling relation of $\mathbf{T}_\infty = T_\infty/T_0$ enables the scaled inert strength in this case to be written as

$$\mathbf{S_m} = \mathbf{T}_\infty^{4/3}\mathbf{P}^{-1/3} + \beta\mathbf{S_S}. \tag{3.109}$$

Two cases of Eq. (3.109) are important. In the first, applicable to components that are large relative to the stressed layer, $\beta \to 0$ and Eq. (3.109) devolves to

$$\mathbf{S_m} = \mathbf{T}_\infty^{4/3}\mathbf{P}^{-1/3}, \tag{3.110}$$

which is simply a shifted version of Eq. (3.103). For $\sigma_S < 0$, $\mathbf{T}_\infty > 1$ and strengths at a given indentation load are increased. In the second case, distinct from Eq. (3.102), it is noted that as $\mathbf{S_S} < 0$ it is possible to find a nonzero value of $\mathbf{P} = \mathbf{P_m}$ at which $\mathbf{S_m}$ vanishes. Hence, setting $\mathbf{S_m} = 0$ in Eq. (3.109) gives

$$\mathbf{P_m} = -\mathbf{T}_\infty^4/(\beta\mathbf{S_S})^3, \tag{3.111}$$

which, in scaled form, is identical to Eq. (3.98). $\mathbf{P_m}$ represents the indentation load at which the component spontaneously fails on indentation with no applied stress.

Figure 3.28 shows a logarithmic plot of $\mathbf{S_m}(\mathbf{P})$ using Eq. (3.109). As in Figure 3.27, the diagonal dashed line represents ideal $\mathbf{P}^{-1/3}$ behavior and the upper bold line indicates the contained crack response for $\mathbf{S_S} = -0.5$. The lower bold line represents the extended crack response for $\mathbf{S_S} = -0.5$, $\mathbf{T}_\infty = 1.2$, and $\beta = 0.1$, which gives through Eq. (3.111), $\mathbf{P_m} = 1.66 \times 10^4$. The vertical dashed line represents $\mathbf{P_m}$. The full response is parallel to the ideal failure dashed response at small values of $\mathbf{P}$ and asymptotic to the spontaneous failure dashed response at large values of $\mathbf{P}$. The extended crack indentation strength response is completely different from that of the contained crack response. The extended response is concave, closely follows the ideal failure response at small and moderate indentation loads, and vanishes as the load approaches the upper-bound spontaneous failure load.

The contained and extended cracks considered here are limiting configurations that bound full surface stress fracture mechanics analyses. Such analyses consider crack extension and change in configuration in spatially varying stress fields (Lawn and Marshall 1977; Lawn and Fuller 1984; Cook 2005). In particular, an indentation crack may be regarded as nearly fully contained at small indentation loads and nearly fully extended at large indentation loads, and may pass between these bounds on extension under the action of applied stress. Hence, an estimate of the intermediate indentation strength behavior is required. Such an estimate for strength may be made by interpolating

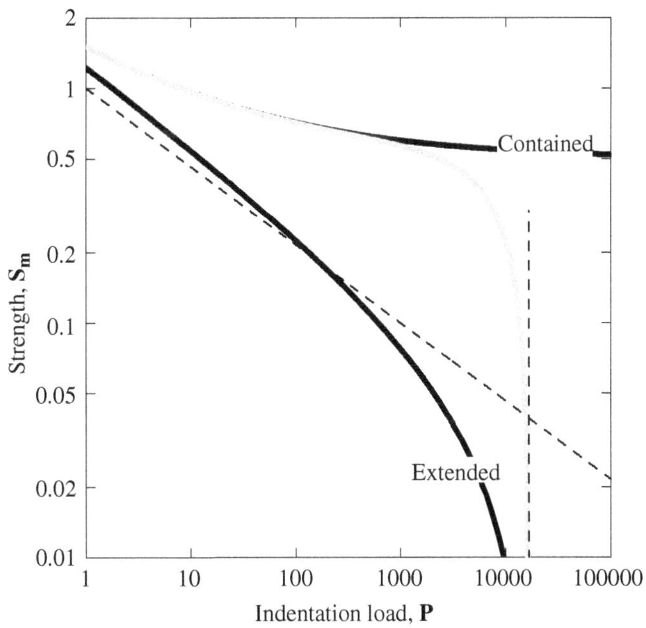

Figure 3.28 Logarithmic indentation strength plot in scaled coordinates for indentation contacts with cracks extending from surface compressive regions into bulk tensile regions. Limiting contained and extended behavior are shown as bold lines. Ideal indentation response is shown as diagonal dashed line. Interpolated full behavior is shown as grayed line.

between the contained and extended strength limits using indentation load as a parameter. A linear interpolation is given by

$$\mathbf{S_m} = (1 - \mathbf{P}/\mathbf{P_m})\mathbf{S_m}(\text{contained}) + (\mathbf{P}/\mathbf{P_m})\mathbf{S_m}(\text{extended}), \tag{3.112}$$

where $\mathbf{S_m}$(contained) and $\mathbf{S_m}$(extended) are the scaled strength solutions from Eqs. (3.102) and (3.109), respectively. Equation (3.112) is shown as the gray line in Figure 3.28. A feature of this estimate, and the exact solution (Cook 2005), is the abrupt decrease in strength as the indentation load approaches the upper-bound load. A prediction from this estimate is that compressive surface stress effects will enhance indentation strengths with little effect of compensating bulk tension unless the indentation load is very close to the spontaneous failure load, and then strength will tend to vanish with little further increase in load. In an engineering context, surface compression toughening alone is thus not consistent with conservative design against contact failure.

Finally, a different approach to analysis of indentation cracking and strength in layered structures is application of Eq. (3.27) directly. This approach is different from that pursued in Eq. (3.81) and following in which replacements and additions were applied to Eq. (3.3). The direct approach is somewhat more suited to incorporation of stress field gradients. In particular, fracture of components in many bending configurations involves crack propagation in a continuously decreasing tensile field that is not well suited to modeling by the discrete layers depicted in Figure 3.25. Such crack propagation typically occurs if the crack length c becomes comparable to the thickness dimension of the component h. The example of interest here is illustrated in Figure 3.11b and is common in many coated or clad glass and ceramic systems (Rhee et al. 2001): A thin, stiff, brittle layer h is adhered to a thick, compliant, elastic substrate and the entire assembly placed in biaxial flexure. In the diagram, the thin layer is uppermost. The lower surface of the thin layer contains a contact flaw. On loading of the assembly, a stress gradient develops in the thin layer as a consequence of tension on the lower surface of the layer and compression on the upper surface. For large contact loads, the related cracks c thus experience a stress gradient and the average tensile stress over the crack decreases with crack size.

The most common fracture mechanics approach to specifying the SIF K_a for a fracture configuration including an applied stress gradient similar to Figure 3.11b is to expand ψ in Eq. (3.27) as a power-law series in (c/h): $\psi = \psi_0[1 - a_1(c/h) + a_2(c/h)^2 + \cdots]$ (e.g., Murakami 1987). Such expansions are empirical fits to other, numerical or analytical, calculations. These power-law forms for K_a are intended principally for use as the sole applied loading in a fracture system. As a consequence, even to leading order in (c/h), this approach does not allow indentation strengths to be expressed in closed form and is thus not suitable for interpreting experimental observations. A simple approximation is to choose a single value of c/h, representative of those used in strength tests, and thence evaluate a single, representative, value of ψ (e.g., Petrovic and Roof 1984; Kaur et al. 2007) to obtain closed form indentation strength expressions. Here, a different approximation is demonstrated that also enables closed form indentation strength expressions to be developed. The approximation uses a simple functional form of ψ that is applicable over a limited c/h domain representative of that used in strength tests. The approximation is empirical and provides insight into the effects of stress gradients on indentation strength responses.

Analysis begins with expressing $\psi(c)$ of Eq. (3.27) as a decreasing power-law function

$$\psi = \psi_0 (c/h)^{(a-1)/2}, \tag{3.113}$$

where ψ_0 and $a \leq 1$ are empirical amplitude and exponent terms. Equation (3.27) then becomes

$$K = \psi_h \sigma_\mathrm{a} c^{a/2} + \chi P/c^{r/2}, \tag{3.114}$$

where ψ_h, χ, and P are invariant and the applied stress geometry term ψ_h is given by

$$\psi_h = \psi_0 h^{(a-1)/2}. \tag{3.115}$$

The equilibrium and stability conditions remain as before such that the joint condition is written in matrix form similar to Eq. (3.6) as

$$\begin{bmatrix} 1 & 1 \\ 1 & -r/a \end{bmatrix} \begin{bmatrix} \mathbf{K_a} \\ \mathbf{K_r} \end{bmatrix} = \begin{bmatrix} 1 \\ 0 \end{bmatrix}, \tag{3.116}$$

where $\mathbf{K_a} = K_\mathrm{a}/T$ and $\mathbf{K_r} = K_\mathrm{r}/T$ are now scaled, generalized, SIF variables from Eq. (3.114). Solution of the simultaneous equations represented by Eq. (3.116) yields the indentation strength σ_m:

$$\sigma_\mathrm{m} = \frac{(r/a)T}{(1 + r/a)\psi_h} \left[\frac{T}{(1 + r/a)\chi P} \right]^{a/r}. \tag{3.117}$$

The expression for σ_m can be written simply as

$$\sigma_\mathrm{m} = S P^{-a/r}, \tag{3.118}$$

which reverts to the ideal solution, Eq. (3.9), for $a/r = 1/3$ characterizing a spatially invariant applied stress field. Here, the effects of a decreasing stress field were approximated by a power-law SIF of the form $K_\mathrm{a} \sim c^{-a/2}$, where $a < 1$. As a consequence, the inert strength σ_m does not decrease with indentation load P as rapidly as the ideal response. In experiments, it is expected that Eq. (3.9) would pertain at small indentation loads and Eq. (3.118) would pertain at large indentation loads, similar to the lateral crack affected strength response limits of Eq. (3.62). In practice it might be difficult to distinguish these two effects from strength measurements alone.

3.3.6 Uncracked Contacts

At small contact loads, flaws formed by conical and pyramidal indenters can be crack free. The flaws consist solely of surface contact impressions and associated subsurface pdz, as shown in the schematic cross section of Figure 3.9b. Such flaws are referred to as subthreshold, as the indentation load P_Ω or impression size a_Ω for crack initiation are not exceeded by the contacts. A subthreshold Berkovich indentation in SLG is shown in Figure 2.14a. Subthreshold Vickers indentations in polycrystalline ZrO_2 and Al_2O_3 are shown in Figure 3.29a and b, respectively. Fracture mechanics of indentation thresholds and crack initiation are analyzed in detail in Section 2.4.

Subthreshold Vickers indentations and scratches have been observed, primarily in glass, since the earliest indentation studies (e.g., Diehl and Schulze 1957; Marsh 1964a,b; Evers 1967; Ernsberger 1968; Hagan 1979b; see Kurkjian et al. 1995 for a review), along with observations of near-subthreshold indentations with very small surface crack traces (Palmqvist 1957). More recent indentation studies have confirmed such observations, in glass (Lathabai et al. 1991a; Kurkjian et al. 1995; Jung et al. 2004a; Morris et al. 2004b; Yoshida et al. 2015; Salem 2023), BaF_2 and NaCl (Morris et al. 2004b), single-crystal Si and Si with inorganic films (Jung et al. 2004a,b; Pajares et al. 2004), amorphous Al_2O_3 and hybrid organic–inorganic thin films (Kim et al. 2004; Thurn and Cook 2004), single-crystal Al_2O_3 (sapphire, Jung et al. 2004b), and polycrystalline ZrO_2 (Cook et al. 1994; Readey and McCallen 1995; Zhang and Lawn 2005; Alves et al. 2022). Related contact-scale observations are those of impacts on glass by sharp abrasive particles, which exhibit plastic deformation and localized material removal by "gouging" and chipping but no real evidence of radial or half-penny crack surface traces (Wiederhorn and Lawn 1979; Lawn 1985).

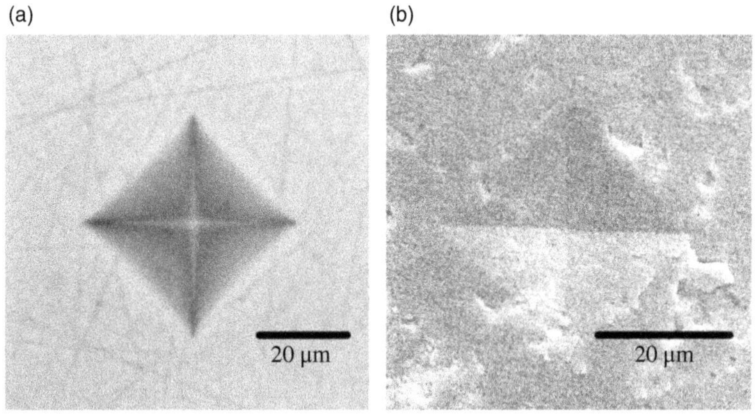

Figure 3.29 Images of subthreshold (crack free) Vickers indentations in polycrystalline materials. (a) Optical image, ZrO_2. (b) Scanning electron microscope image, Al_2O_3.

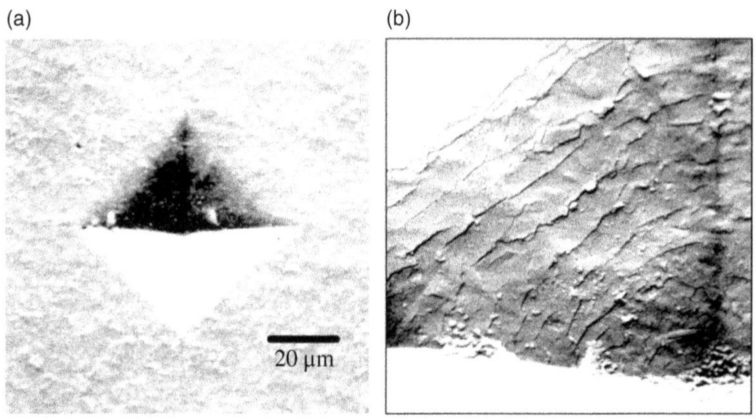

Figure 3.30 Scanning electron microscope images of near-subthreshold Vickers indentation in cordierite glass-ceramic. (a) Overall indentation. (b) Expanded view of contact impression face showing parallel shear faults exiting the surface.

Of relevance to this section and perhaps the most important observations for analyzing crack initiation at indentations and subthreshold indentation strengths are the pdz-scale cross-sectional observations of Hagan and colleagues (Hagan and Swain 1978; Hagan 1979a,b, 1980; Hagan and Van Der Zwaag 1984) followed by those of Dabbs, Lawn, and colleagues (Lawn et al. 1983; Lathabai et al. 1991a). These observations show that irreversible deformation within the pdz is caused by the action of shear faults—the separation of small, discrete blocks of material translated relative to each other by mixed-mode crack propagation—the faults—in shear and compression (fracture modes II and III, see Lawn 1993). The faults are often visible on indentation impression faces as cracks parallel to the impression edges. Examples are shown in Figures 2.6, 2.10, and 3.30. Figure 3.30a shows a near-subthreshold indentation in a glass-ceramic. Figure 3.30b shows an enlarged detail of an impression face and the exiting shear faults parallel to the impression edge. Extensive observations show that intersections of the shear faults adjacent to the impression vertices give rise to crack nuclei that lead to well-developed indentation cracks. The observations also show that separation of the faults is regular and material- and indenter-dependent, but not indentation load dependent, such that the fault intersections and crack nuclei scale with the contact impression size, $c_1 \sim a$, consistent with the geometrical similarity of deformation fields at pyramidal indentations.

The primary motivation of Dabbs, Lawn, and colleagues in studying subthreshold flaws was to explain the strength behavior of optical fibers. In the late 1970s, investigations of strength distributions and reliability of silica optical fibers often lead to confusing and conflicting results. The results were often rationalized by the idea that extended (km) lengths of fiber contained two subpopulations of flaws (Maurer 1975, 1977; Kurkjian et al. 1976; Kalish et al. 1977; Scott and Gaddipati 1978; Snowden 1978; Tariyal and Kalish 1978) that were variously active in individual measurements. The subpopulation of larger flaws lead to "low strength" fibers and strength and reliability behavior that were seemingly consistent with the known fracture behavior of glass. The subpopulation of smaller flaws lead to potentially "high strength" fibers that were

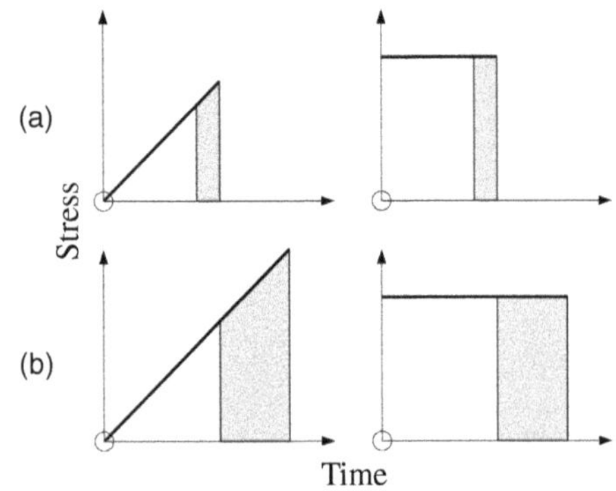

Figure 3.31 Schematic diagrams of the strength and lifetime behavior of the two subpopulations of flaws in optical fibers. (a) Large consistent flaws, leading to reduced strengths (left) and lifetimes (right) with little dispersion. (b) Small variable flaws, leading to increased strengths and lifetimes with significant dispersion.

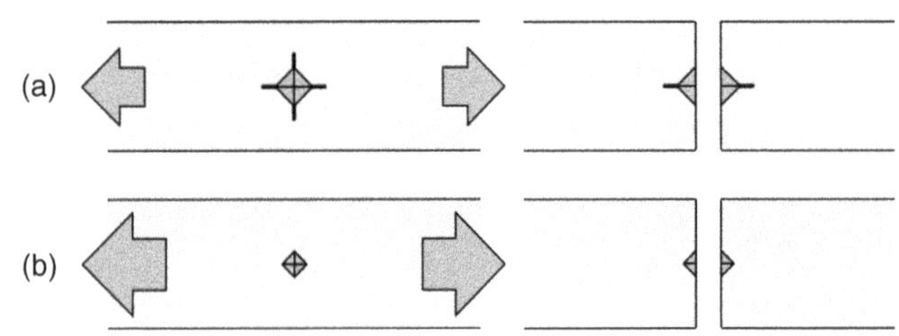

Figure 3.32 Schematic diagrams of the contact strength failure modes corresponding to the subpopulations of flaws inferred from the strength and lifetime tests of Figure 3.31. (a) Postthreshold, cracked, flaws. (b) Subthreshold, uncracked, flaws.

useful for application but which exhibited great variability and for which the strength and reliability were not consistent with predictions using conventional glass fracture properties. Kurkjian et al. (1989) give a historical summary of the observations and ideas leading to this assessment. A schematic diagram of the stress–time responses of the two failure modes is shown in Figure 3.31. Dabbs et al. (1980) provided a key piece of evidence explaining the observations by showing that fibers containing dominant postthreshold and subthreshold indentation flaws exhibited very different strengths. The postthreshold flaws were weaker and well described by known crack propagation parameters. The subthreshold flaws were stronger and, by implication, described by (unknown) crack initiation parameters and (likely) greater variability. A schematic diagram of interpretation of the two failure modes in terms of postthreshold and subthreshold contact flaws is shown in Figure 3.32. Dabbs, Lawn, and colleagues subsequently built on this evidence in a series of experimental and analytical works that addressed the equilibrium and kinetic behavior of subthreshold indentation crack initiation and indentation strength (Dabbs and Lawn 1982a,b, 1985; Dabbs et al. 1982 Lawn et al. 1983; Lathabai et al. 1991a,b), as did others, particularly Ritter, Jakus, and colleagues (e.g., Choi et al. 1990 and references therein).

The fundamental physics of subthreshold flaw indentation strengths is that crack nuclei generated on indentation are initiated to form cracks by the subsequent action of an applied stress that then also propagates the cracks so as to cause component failure, Figure 3.32b. Analysis of the strength of components controlled by subthreshold contact flaws is then a matter of assessing the effects of applied stress on crack initiation. As discussed in Chapters 1 and 2, such an analysis begins with a statement of the SIF driving fracture of a crack nucleus. In this case, the total SIF K is composed of two terms: the short crack limit of the indentation field, K_n, Eq. (2.21), and the applied stress field, K_a, Eq. (3.1). The equilibrium condition $K = T$ for a crack nucleus c_1 is then

$$K = K_n + K_a = \Omega_1 E \cot\phi c_1^{1/2} + \psi\sigma_a c_1^{1/2} = T \tag{3.119}$$

Equation (3.119) is a departure from Sections 3.3.1–3.3.5 in that the SIF K_n characterizing the indentation field is altered to a destabilizing form identical to K_a. In the absence of applied stress, $\sigma_a = 0$, Eq. (3.119) reverts to the initiation analysis of Section 2.4. In general form, Eq. (3.119) describes a combination of Figures 2.22 and 3.32b. For geometrical similar contacts and indentations, the crack nucleus size c_1 is given by

$$c_1 \sim a \sim (P/H)^{1/2} \tag{3.120}$$

such that Eq. (3.119) can be written as

$$\Omega E \cot\phi (P/H)^{1/4} + \psi\sigma_a (P/H)^{1/4} = T, \tag{3.121}$$

where Ω is a dimensionless geometrical factor that takes Eqs. (3.119) and (3.120) into account. For $\sigma_a = 0$, the spontaneous initiation load $P = P_\Omega$ is recovered. As both K_n and K_a are destabilizing ($dK/dc > 0$), the general solution for the maximum sustainable applied stress or strength σ_m for a subthreshold indentation flaw is given by simple inversion of Eq. (3.121) as

$$\sigma_m = \left[\frac{\Omega E \cot \phi}{\psi}\right]\left[\left(\frac{P_\Omega}{P}\right)^{1/4} - 1\right], \; P \le P_\Omega. \tag{3.122}$$

Interpretation of Eq. (3.122) is aided by eliminating ψ using Eqs. (3.13) and (3.14) such that the first bracket is reexpressed as $\alpha\sigma_\Omega$, where $\alpha \sim (\Omega E c_\Omega^{1/2}/T)$ is a dimensionless proportionality factor. Using the relations of Eq. (3.101) thus leads to a scaled strength relation

$$\mathbf{S_m} = \alpha\left(\mathbf{P}^{-1/4} - 1\right), \; \mathbf{P} \le 1, \tag{3.123}$$

which has similar mathematical structure to Eq. (3.102).

There are several points to note regarding Eq. (3.123): First, the value of α, which depends on both material and indenter factors, is large. Using typical values of the above parameters gives $\alpha \approx 3$–100. Subthreshold indentation strengths can be much greater than conventional postthreshold indentation strengths. Second, for large subthreshold indentation loads, $\mathbf{P} \rightarrow 1$, the subthreshold strength vanishes, similar to Eq. (3.109). Third, for small subthreshold indentation loads, $\mathbf{P} \ll 1$, the subthreshold strength approaches the asymptotic behavior of $\mathbf{S_m} \rightarrow \mathbf{P}^{-1/4}$. This is similar, although not identical, to the ideal postthreshold indentation strength response of $\mathbf{S_m} = \mathbf{P}^{-1/3}$, Eq. (3.103). Experimentally, it might be difficult to distinguish the two flaw regions from trends in indentation strength measurements alone, especially if there is significant variability in the subthreshold geometry factor Ω.

Figure 3.33 shows a logarithmic plot of postthreshold and subthreshold indentation strength behavior and Eq. (3.123). The hatched region indicates postthreshold behavior $\mathbf{P} \ge 1$. The straight line in this region indicates the ideal postthreshold indentation strength response using Eq. (3.103) terminating with a maximum of $\mathbf{S_m} = 1$ at $\mathbf{P} = 1$. The open region indicates subthreshold behavior. The curved lines in this region indicate subthreshold indentation strength responses using Eq. (3.123) and $\alpha = 3$, 10, and 30, shown bottom to top, respectively. These responses vanish at minima of $\mathbf{S_m} = 0$ at $\mathbf{P} = 1$. The dashed line is the continuation of the ideal postthreshold response into the subthreshold region, indicating nonequilibrium spontaneous crack "pop-in" events that generate postthreshold indentation configurations for $\mathbf{P} < 1$ (Lawn et al. 1983). A full indentation strength response thus consists of ideal postthreshold $\mathbf{P}^{-1/3}$ behavior at large indentation loads and subthreshold $\mathbf{P}^{-1/4}$ behavior at small loads, separated by a region of mixed behavior in which the subthreshold strengths converge on the postthreshold response.

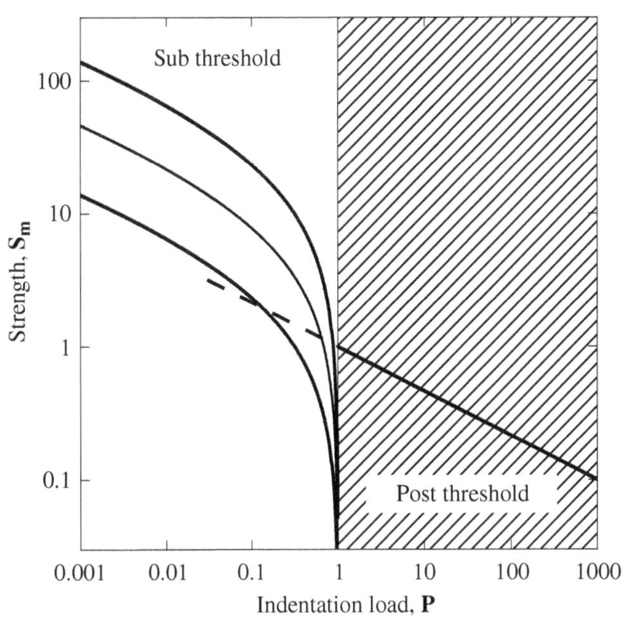

Figure 3.33 Logarithmic indentation strength plot in scaled coordinates for subthreshold (uncracked) and postthreshold (cracked) indentation contacts. The behavior of metastable postthreshold flaws in the subthreshold domain is indicated by dashed line.

The analyses of this section can be viewed in two contexts considering indentation crack initiation behavior. In the context of this book, the analyses here are intermediate between the zero applied stress crack initiation considerations of Section 2.4 and the extension to considerations of subthreshold strength phenomena exhibited by linear flaws in Section 10.3. In the context of prior work, the analyses here lie at the simpler end of the range of subthreshold considerations. In particular, previous analyses of much greater complication have used equations similar to Eqs. (1.31) and (1.80) to determine the detailed crack length variation of the initiating SIF $K_n(c)$ from the mixed-mode stress field variation associated with the post-contact pdz (Lawn and Evans 1977; Chiang et al. 1982a,b; Lathabai et al. 1991a,b), including the effects of phase transformation (Cook et al. 1994c) and ion-exchange (Morris et al. 2004a). Other previous analyses have used full tensorial descriptions of the pdz stress field during contact to predict initiation of the various indentation cracks (see Figure 2.5), initially by a "strength of materials" approach (Cook and Pharr 1990) and subsequently by a fracture mechanics approach (Cook 2019). An analysis approximating a single scalar component of the postcontact pdz stress tensor was used to analyze half-penny crack initiation and component reliability (Choi et al. 1990) and is similar to the initiation analysis in Chapter 1 and earlier (Cook 2023). By far, the simplest approach is that demonstrated in the scalar analysis here and earlier (Jung et al. 2004a,b), in which a single component of the pdz stress tensor is approximated by an invariant uniform indentation stress and a variable superposed applied stress.

3.4 Indentation Strength Observations

Section 3.3 developed analyses to describe the major physical influences on indentation strength behavior, including the effects of indenter shape, indentation lateral cracking, indentation impact, component surface stresses, and indentation crack initiation. This section considers experimental indentation strength measurements affected by each of these influences and compares the measurements with the appropriate analysis. The background and context for Section 3.4 are the observations and analysis of ideal indentation strength behavior considered in Sections 3.1 and 3.2. The structure of Section 3.4 is identical to that of Section 3.3.

3.4.1 Contact Angle Effects

Analysis of contact angle effects on conical and pyramidal indentation strength for indenters exhibiting geometric similarity is considered in Section 3.3.1. The included contact angle 2ϕ for such indenters is illustrated in Figure 3.8a.

Figure 3.34 is a logarithmic plot of the indentation strength behavior of SLG beams from the work of Lawn et al. (1976), using data derived from the published work. The beams were indented with tungsten carbide cones of various included angles. Symbols represent individual strength σ measurements at the indentation loads P and cone angles ϕ indicated, noting that the effective cone angle for the Vickers indenter is $\phi_{\text{Vickers}} \approx 70°$. The solid line is a slope $-1/3$ (in logarithmic coordinates) best fit to the 70° data consistent with an ideal indentation strength response, Eq. (3.9), of $\sigma = \hat{S}P^{-1/3}$. The mean contact strength for the SLG was $\hat{S} = (157.0 \pm 19.4)$ MPa N$^{1/3}$, less than that for the Vickers indentation contact strength of fused silica glass (FSG) (212 MPa N$^{1/3}$) given in Table 3.1.

Figure 3.34 shows that increased or decreased indenter cone angles ϕ at a given indentation load lead to increased or decreased strengths, respectively. The mean contact strength as a function of cone angle from these measurements, expressed relative to the Vickers equivalent indenter, is shown in Figure 3.35 as $\hat{S}(\phi)/\hat{S}_{\text{Vickers}}$. The symbols and bars represent means and standard deviations of best fits to the strength data; the Vickers indenter position is indicated. The plot is analogous to the relative cracking susceptibility angular plot of Figure 2.18. The variation of mean contact strength as a function of indenter included angle can be predicted from the pdz-based fracture mechanics analysis of Section 3.3.1 by writing Eq. (3.29) as

$$\frac{\hat{S}(\phi)}{\hat{S}_{\text{Vickers}}} = \left(\frac{\cot\phi}{\cot\phi_{\text{Vickers}}}\right)^{-2/9}, \tag{3.124}$$

analogous to Eq. (2.20). The solid line in Figure 3.35 is a plot of Eq. (3.124) showing the predicted increase in mean contact strength with decreasing indenter acuity (increasing ϕ). Both the experimental and predicted variation of contact strength with indenter angle exhibit the same qualitative trend: very little variation for angles less than the Vickers equivalent of $\phi \approx 70°$ and significant increase for greater angles. However, the quantitative agreement between the observations and the

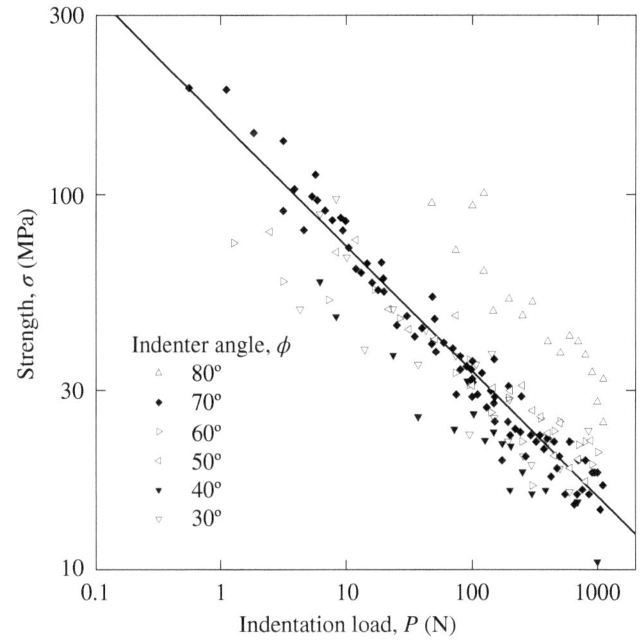

Figure 3.34 Logarithmic plot of conical indentation strength of soda-lime silicate glass as a function of cone angle. Solid line is a best fit to the 70° behavior (Vickers equivalent). Source: Data adapted from Lawn et al. (1976).

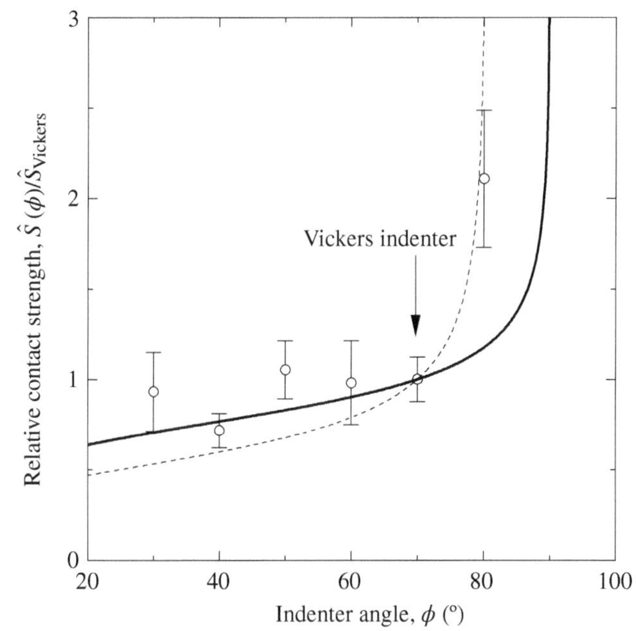

Figure 3.35 Plot of indentation contact strength relative to Vickers equivalent indentation behavior as a function of indenter angle. The solid and dashed lines represent the trends expected from elastic–plastic and elastic + friction contact behavior, respectively.

prediction is barely within the standard deviation bounds of the experimental uncertainties; a shift of cone angle in the pdz prediction from Eq. (3.124) of $\approx -10°$ is required to achieve agreement.

In addition, the variation of mean contact strength as a function of indenter included angle can be predicted from the reversible elastic contact field analysis by writing Eq. (3.32) as

$$\frac{\hat{S}(\phi')}{\hat{S}_{\text{Vickers}}} = \left(\frac{\cot\phi'}{\cot\phi'_{\text{Vickers}}}\right)^{-1/3}. \tag{3.125}$$

ϕ' is an effective contact angle that is made less acute than the physical angle by the effects of contact friction. The dashed line in Figure 3.35 is a plot of Eq. (3.125) showing the predicted increase in mean contact strength with increasing ϕ, using a best-fit friction coefficient for the indenter–surface interface from Eq. (3.30) of $\mu = 0.17$. As anticipated, the predicted angular dependencies for pdz- and elastic + friction-based indentation strengths are very similar; the elastic + friction analysis provides a better prediction of the increase in contact strength at large ϕ but is no better otherwise. Detailed considerations of the effects of indenter acuity on indentation cracking (Morris and Cook 2004, 2005; Morris et al. 2004b, 2005) show that "conventional" pdz-based indentation fracture mechanics alone are not sufficient to describe indentation fracture across the full domain of included angles. In particular, during contact with acute probes reversible "wedging" effects are significant such that angular divergences from both Eqs. (3.124) and (3.125) are likely. Figure 3.35 suggests that this is the case and that although the major trends are correct some refinement of angular indentation strength analysis is required.

3.4.2 Contact Radius Effects

Analysis of contact radius effects on spherical indentation strength is considered in Section 3.3.2. The contact radius r for such indenters is illustrated in Figure 3.8c.

The results from two studies of spherical indentation strength are considered here. In the first study (Lawn et al. 1975), the strength of a single material, SLG, was examined using variations in preindentation surface damage and variations in indenter radii. Surface damage was generated by SiC grit blasting of specimen surfaces prior to indentation and strength testing. The grit size was controlled, 100–320 grade, such that flaws on the specimen surfaces were varied in size over the domain 10–23 μm. Indentations were performed with tungsten carbide (WC) spheres of radius r from 0.40 to 6.38 mm using peak contact loads P extending to 4.5 kN. Strength testing was by uniaxial bending of beams. In the second study (Widjaja et al. 1996), the indentation strengths of three materials, SLG and two grades of commercial alumina containing different volume fractions of amorphous phases, were examined using approximately single-sized spherical indenters. Indentation prior to strength testing of SLG was by an $r = 2.38$ mm spherical steel probe to loads of 1.5 kN. Indentation of the aluminas was by an $r = 2.0$ mm WC probe to loads of 2.5 kN. Strength testing was by biaxial flexure of plates and disks.

Figure 3.36 shows logarithmic plots of the spherical indentation strength behavior of SLG beams from the work of Lawn et al. (1975), using data derived from the published work. Symbols with bars on the left of Figure 3.36a indicate the means and standard deviation limits of intrinsic strengths σ_i; these strengths increase with increasing grade of grit treatment prior to testing and thus with decreasing preexisting flaw size c_f. Symbols without bars in both Figure 3.36a and b represent individual strength measurements σ at the indentation loads P indicated. Figure 3.37 shows logarithmic plots of the spherical indentation strength behavior of (a) SLG disks and (b) and (c) alumina plates from the work of Widjaja et al. (1996), using data derived from the published work. The hatched bands in each plot indicate the mean and standard deviation limits of intrinsic strengths for each material; polycrystalline alumina is clearly stronger than glass. Symbols with bars

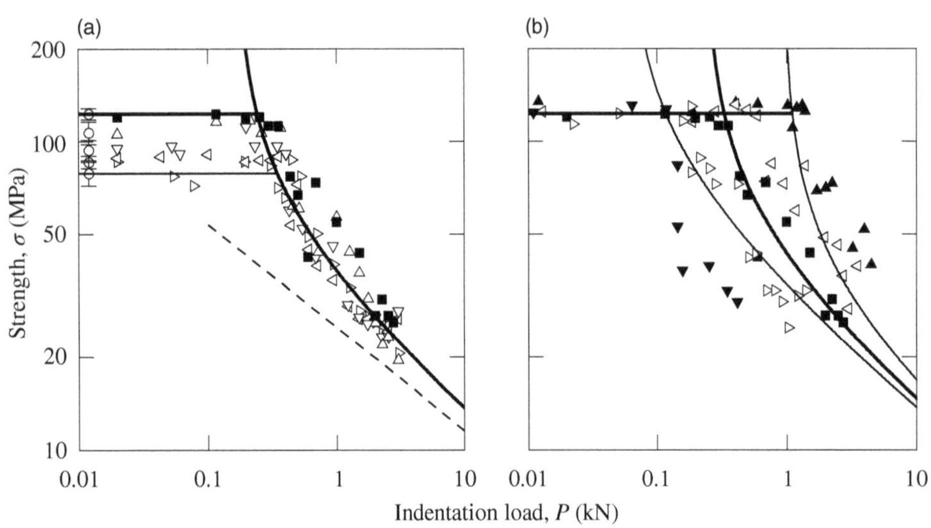

Figure 3.36 Logarithmic plots of spherical indentation strength of soda-lime silicate glass. (a) Variation in surface treatment. (b) Variation in indentation radius. Solid lines are fits and predictions assuming elastic contact. Source: Data adapted from Lawn et al. (1975).

represent means and standard deviations of strength measurements at the indentation loads indicated. The left vertical line in each plot indicates the median indentation load at which ring cracks were observed postindentation for each material. The right vertical lines indicate the median indentation load at which radial cracks were observed (no radial cracks were observed for SLG).

In more detail, Figure 3.36a represents the indentation strength degradation behavior of specimens indented with a single radius indenter, $r = 1.58$ mm. The different symbols indicate different surface treatments over the full range of grit sizes; treatment with the finest grit is indicated by the filled symbols. As discussed in Section 3.3.2, the strengths exhibit two domains of behavior. For indentation loads $P > 0.2$ kN, strengths decrease with increasing indentation load and there is no significant effect of surface finish. For indentation loads $P < 0.2$ kN, the strengths are invariant with load but decrease with increasing grit size. The two domains are clearly the indentation strengths and intrinsic strengths, respectively, and $P_i \approx 0.2$ kN. The curved solid line in Figure 3.36a is a visual best fit to the indentation strength data using Eq. (3.40). The dashed line indicates the ideal indentation $P^{-1/3}$ asymptotic trend. The solid horizontal lines indicate the bounds of strengths in the intrinsic domains. The resemblance to Figure 3.15 is clear. The significant underestimate of the observed indentation strength variation by an ideal response was noted in the earlier work (Lawn et al. 1975).

Figure 3.36b represents the indentation strength degradation behavior of specimens surface treated with a single grit, the finest, 320 grade SiC. The different symbols indicate different indenters over the full range of radii; the data for the mid-radius indenter, $r = 1.58$ mm, are identical to those in Figure 3.36a and are indicated by the same filled symbols. As in Figure 3.36a there are two clear domains of indentation and intrinsic strength behavior. However, now the intrinsic strength is invariant and the indentation strength is dispersed, a consequence of a single surface treatment and varying indenter radius and reversal of the constraints in Figure 3.36a. The bold solid line is a best fit to the $r = 1.58$ mm data using Eq. (3.40). The fine lines either side are predictions for the bounds of behavior for the different-sized indenters using the radius scaling relations of Eqs. (3.38) and (3.39) (noting that the radii vary by a factor of 16). The bounds are of the correct scale, although there is considerable experimental scatter. Nevertheless, it appears that the predicted width of the bounds is somewhat less than that observed, probably as a consequence of less than ideal sphericity of the probes.

The solid line in each plot of Figure 3.37 represents a visual best fit to the indentation strength data using Eq. (3.40). The dashed lines indicate the ideal indentation $P^{-1/3}$ asymptotic trends for each fit. Again, the resemblance to Figure 3.15 in each case is clear. The indentation strength behavior of the SLG in Figure 3.37a is similar to that in Figure 3.36a as might be anticipated for strengths set by similar radii indenters. However, the intrinsic strength levels appear to persist to greater indentation loads in Figure 3.37a, suggesting greater difficulty in initiating strength-controlling indentation cracks on the glass surface in the latter. The indentation strengths of the aluminas are greater than that of glass, more so in Figure 3.37c, which has a smaller amorphous content than that of Figure 3.37b. Less extensive indentation strength data (Lawn 1998; Lawn et al. 1998) for polycrystalline Al_2O_3, polycrystalline Si_3N_4, glass-ceramics, and porcelain can be fitted in a similar way and enable microstructural interpretation. Although there are clear strength transitions in Figure 3.37 from intrinsic flaws to indentation flaws, there appear to be no clear transitions in strength from indentation ring → cone → radial cracks, despite much speculation (Widjaja et al. 1996). The postindentation inferred crack geometry thresholds appear to have little correlation with strength behavior, again despite earlier interpretation.

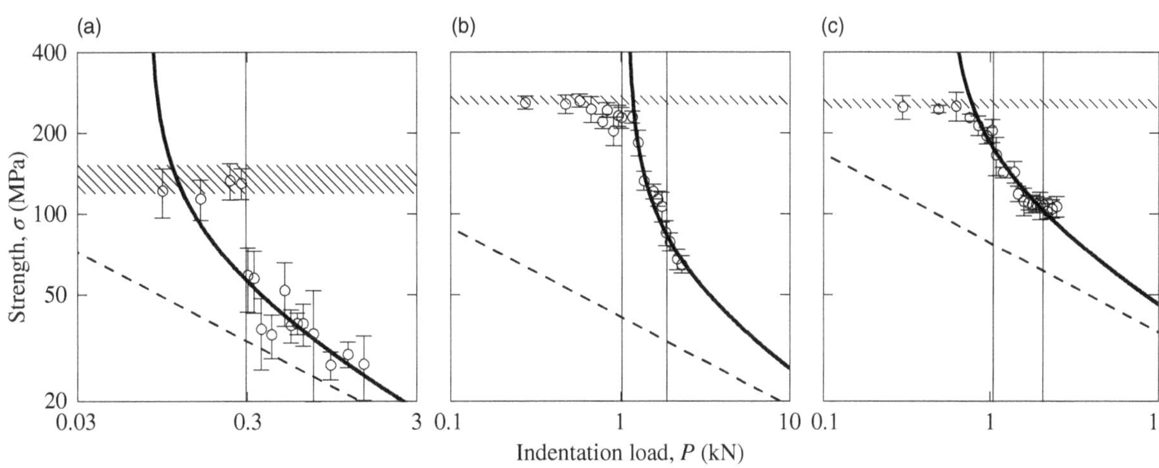

Figure 3.37 Logarithmic plots of spherical indentation strengths. (a) Soda-lime silicate glass. (b) Polycrystalline alumina (AD90). (c) Polycrystalline alumina (AD995). Solid lines are fits assuming elastic contact. Source: Data adapted from Widjaja et al. (1996).

Overall, the effects of component surface finish and indenter radius in spherical indentation strength measurements are well accounted for by the analyses of Section 3.3.2. In particular, a strength analysis extended from a unified, fracture mechanics based, formulation of ring + cone cracking at elastic spherical indentations was developed. In addition, a strength analysis based on fracture mechanics of radial cracking at elastic–plastic spherical indentations was also developed. The elastic analysis described the significantly greater variation in strength with spherical indentation load observed in experiments relative to that predicted by ideal indentation behavior. The analysis accounts for the strength deviation, originally noted by Lawn et al. (1975), without recourse to explanations in terms of changes in crack morphology forwarded by Widjaja et al. (1996). The elastic–plastic analysis showed that a transition from cone to radial cracking would probably not be discernible in strength measurements, also consistent with experimental strength observations.

3.4.3 Lateral Crack Effects

Analysis of lateral crack effects on indentation strength is considered in Section 3.3.3. The geometry of such indentations is illustrated in Figure 3.9a.

Experimental results from eight studies of Vickers indentation strengths are considered here. The studies exhibit clear lateral crack effects on strength. The pervasive influence of lateral cracking and chipping on strength behavior is illustrated in the variety of brittle material microstructures included—glass, glass-ceramic, single crystal, and polycrystal. The primary applications of the materials are both structural and nonstructural, with a range of contact strengths greater than a factor of 10. Figure 3.38 shows logarithmic plots of the indentation strength behavior, divided into two parts: (i) Figure 3.38a–d, primarily nonstructural materials, and (ii) Figure 3.38e–h, primarily structural materials. The majority of the observations

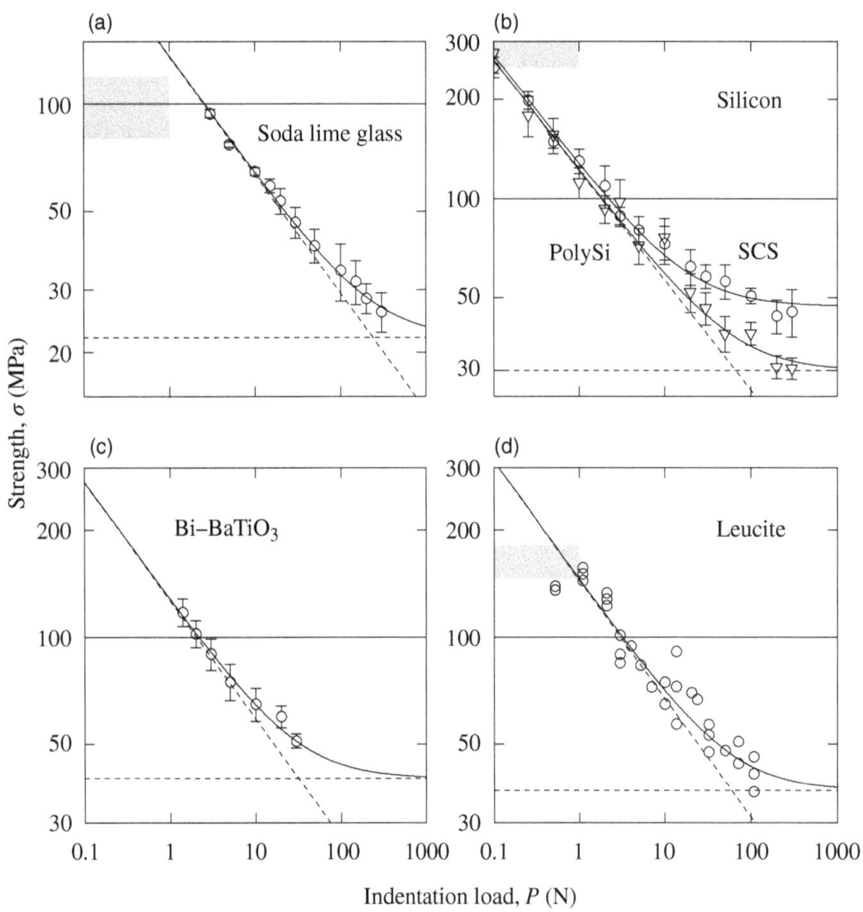

Figure 3.38 Logarithmic plots of Vickers indentation strengths demonstrating the effects of lateral cracking. (a) Soda-lime silicate glass. (b) Silicon. (c) Polycrystalline Bi–$BaTiO_3$. (d) Leucite glass-ceramic. (e) Polycrystalline Y_2O_3. (f) Polycrystalline AlN. (g) Polycrystalline SiC. (h) Sapphire (single-crystal Al_2O_3). Source: Includes data adapted from Cook and Roach (1986), Cook et al. (1994a), Cook (2006), and Alves et al. (2022).

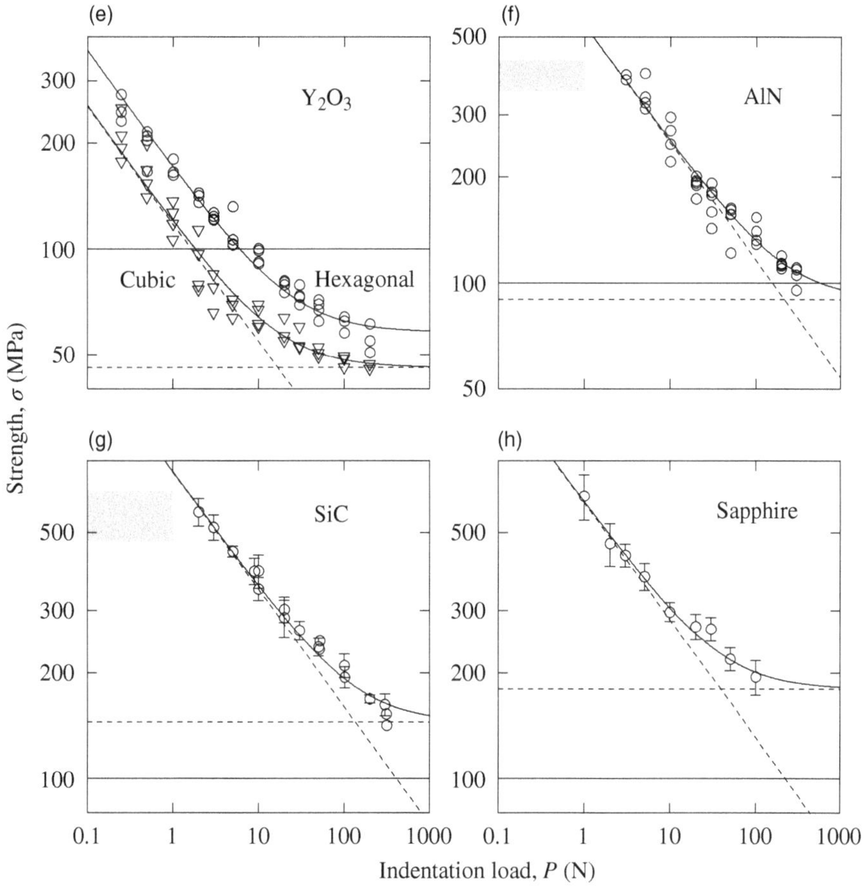

Figure 3.38 *(Continued)*

for each material were published previously, including contributions for SLG, polycrystalline Bi–BaTiO$_3$, polycrystalline SiC, sapphire (single crystalline Al$_2$O$_3$) in Figure 3.38a, c, g, and h (Cook and Roach 1986), Si in Figure 3.38b (Cook 2006), leucite glass-ceramic in Figure 3.38d (Alves et al. 2022), and polycrystalline Y$_2$O$_3$ in Figure 3.38e (Cook et al. 1994a).

Figure 3.38 was constructed using data derived from the published works cited. The plots in Figure 3.38 are all the same aspect ratio, 10×10^4. Symbols with bars indicate the means and standard deviation limits of strengths σ measured at the indentation loads P indicated; isolated symbols indicated individual strength measurements. Different symbols in Figure 3.38b indicate the responses of p-type single-crystal Si, SCS, and polycrystalline Si, polySi, respectively. Different symbols in Figure 3.38e indicate the responses single phase, cubic, and mixed phase, (cubic+)hexagonal, polycrystalline Y$_2$O$_3$. The means and standard deviation limits of intrinsic strength measurements, where available, are indicated by the shaded bands on the left of the plots. The often measured intrinsic strength of SLG, 100 MPa, is indicated by the horizontal straight solid line and is repeated in each plot as a guide to the eye indicating relative strengths. The indentation strength responses in all of Figure 3.38a–h clearly resemble that described by Eq. (3.62) in Figure 3.23. The curved solid lines in each plot are visual best fits to the strength data using Eq. (3.62) and the material specific lateral crack indentation load and strength parameters, P_L and σ_L. The asymptotes of the fits are indicated by the diagonal and horizontal straight dashed lines and are given by $\sigma = SP^{-1/3}$ and $\sigma = \sigma_L$, respectively, as shown in Eqs. (3.63) and (3.64). Here, the contact strength S incorporates lateral crack effects and is given by $S = \sigma_L P_L^{1/3}$. The parameters S, P_L, and σ_L are given in Table 3.2; the uncertainties reflect the uncertainties in the strength measurements and were determined by visual best fits to the experimental strength bounds.

There are several points to note from Figure 3.38. First, lateral crack effects are a very clear factor in indentation strength degradation measurements. At the upper bounds of the experimental indentation load domains (typically $P \approx 200$ N), the measured indentation strengths are factors of 1.2–2 greater than the values extrapolated from lower-bound measurements assuming ideal behavior. As relative strength uncertainties are typically ±0.05, lateral crack effects are clearly significant.

Table 3.2 Lateral crack affected contact strength parameters.

Material	Contact Strength, S (MPa $N^{1/3}$)	Load, P_L (N)	Strength, σ_L (MPa)
Soda-lime glass	137 ± 3	241 ± 165	22 ± 5
Borosilicate glass	190 ± 25	200 ± 100	33 ± 14
Silicon p-type	126 ± 6	19 ± 9	47 ± 7
Silcon poly	122 ± 10	67 ± 37	30 ± 5
Bi–$BaTiO_3$	123 ± 9	32 ± 8	40 ± 2
Leucite glass-ceramic	145 ± 12	60 ± 29	37 ± 5
Y_2O_3 cubic	118 ± 11	17 ± 5	46 ± 2
Y_2O_3 hexagonal	169 ± 11	25 ± 8	58 ± 5
AlN	540 ± 40	216 ± 75	90 ± 8
Sapphire	610 ± 60	39 ± 20	180 ± 25
SiC	745 ± 55	136 ± 41	145 ± 10

Second, the importance of lateral crack effects is only observed if indentation strengths are measured over a sufficiently broad domain of indentation loads. In these cases, the large load data can be placed in the context of the small load data and deviations from ideal behavior thus noted. Corollaries to this point are that if a narrow load domain is used (e.g., between the "easy" indentation loads of $20\,N \leq P \leq 50\,N$), misleading interpretations of both magnitude and trend in indentation strength can result. Examples of this effect on indentation strength trends are considered with regard to varying stress fields in Eq. (3.118) and, more prevalent in reported strength measurements, with regard to microstructural toughening effects in Chapter 4. Note that the misinterpretations are not consistent with conservative design: strengths are overestimated and degradation is underestimated if lateral crack effects are neglected. Third, the experimental indentation strength measurements of Figure 3.38 are well described by the fracture mechanics analysis of Section 3.3.3. The relative uncertainties (about ±7%) in the lateral crack affected contact strength parameters S of Table 3.2 are slightly less than those describing ideal indentation strength responses $\hat{S}$ of Table 3.1 (about ±10%). Similarly, the relative uncertainties (about ±15%) in the lateral crack-controlled strength minimum σ_L are slightly less than those typically observed for the intrinsic strength maximum σ_i (about ±25%). These decreased uncertainties are consistent with conservative design that incorporates lateral crack effects into brittle fracture strength assessments.

3.4.4 Impact Energy Effects

Analysis of impact effects on indentation strength is considered in Section 3.3.4. The overall geometry of impact contacts is illustrated in Figure 3.10. Details of contact deformation on impact are shown in Figure 3.24 and details of contact fracture on impact are shown in Figures 3.9, 3.12, and 3.14.

Experimental results from several studies of postimpact strength measurements are considered here, divided into four groups. The first group considers spherical elastic impacts on SLG surfaces using contacting particles of different radii and materials. Particle velocity is the independent test variable. The second group considers angular elastic–plastic impacts on SLG surfaces containing varying levels of surface stress, using contacting particles of different sizes. Particle energy is the test variable. The third group considers angular elastic–plastic impacts on polycrystalline alumina surfaces using contacting particles of different sizes. Particle energy is the test variable. The fourth group considers liquid water jet impacts on glass and zinc sulfide surfaces. Impact velocity is the test variable. For the second and third groups, comparison is made with Vickers indentation strength measurements on the same material surfaces. Data for the first group are from the works of Wiederhorn and Lawn (1977, 1979), for the second group from Marshall and Lawn (1978), Wiederhorn and Lawn (1979), Lawn et al. (1979, 1980), for the third group from Ritter et al. (1984), Ritter (1985), and Breder et al. (1988), and for the fourth group from Van der Zwaag and Field (1982, 1983). Data from the first group are analyzed using Eq. (3.77), from the second group using Eqs. (3.76) and (3.105), from the third group using Eqs. (3.67), (3.76), and (3.80), and from the fourth group using Eq. (3.78).

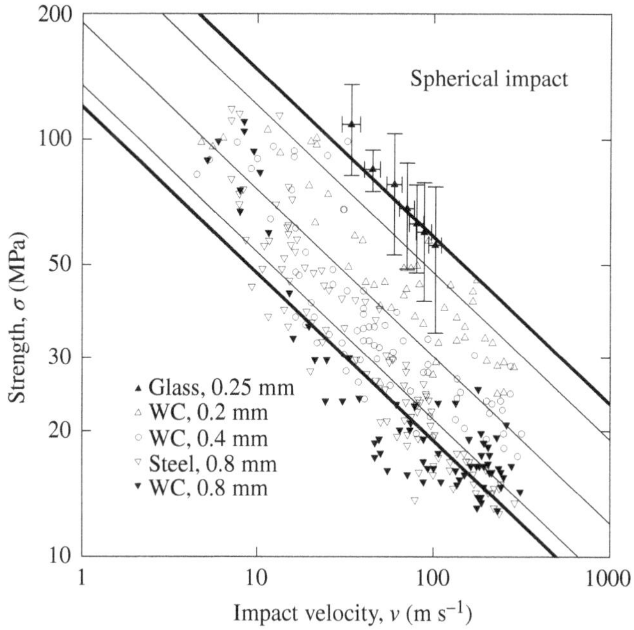

Figure 3.39 Logarithmic plot of spherical impact strengths of soda-lime silicate glass as a function of impact velocity. Impacting materials and sphere radii indicated. Source: Data adapted from Wiederhorn and Lawn (1977, 1979).

Figure 3.39 is a logarithmic plot of the strength of SLG specimens containing spherical contact flaws generated at the impact velocities indicated (Wiederhorn and Lawn 1977, 1979), using data derived from the published works cited. Symbols represent individual strength measurements following impacts with WC spheres of radius 0.8, 0.4, and 0.2 mm, steel spheres of radius 0.8 mm, or glass spheres of radius 0.25 mm, noting that the mass densities of the three materials are WC, 14.1 $\mathrm{Mg\,m^{-3}}$, steel, 7.86 $\mathrm{Mg\,m^{-3}}$, and glass 2.5 $\mathrm{Mg\,m^{-3}}$. Following Eq. (3.77) for flaws generated by near-elastic impacts, the subsequent specimen strength should decrease with impact velocity υ as $\sigma \sim \upsilon^{-2/5}$ and with sphere density ρ and radius r as $\sigma \sim \rho^{-1/5} r^{-2/3}$. For a given velocity, the ratios of the strengths for the different particle impacts in the order WC(0.8):steel:WC(0.4):WC(0.2):glass are thus 1:1.12:1.59:2.52:3.07, noting that contact by the largest, densest spheres leads to the least strength. Different symbols indicate different spheres in Figure 3.39 and the extremes of size and density are shown as filled symbols. The bold solid lines indicate best fits of Eq. (3.77) to the strength data using the given scaling relations. The fine lines indicate the strength behavior for the intermediate spheres. The strength data are described well by Eq. (3.77). Wiederhorn and Lawn (1977) noted changes in contact fracture morphology with impact velocity, particularly the formation of median, half-penny, and lateral cracks and increased curvature of cone cracks as velocity increased. Such changes were not apparent in the strength degradation of Figure 3.39, perhaps for two reasons: Wiederhorn and Lawn noted that the changes were gradual and, as noted in Section 3.3.4, a transition from elastic contact to elastic–plastic contact and from Eqs. (3.77) to (3.78) does not cause a great change in the trend of strength with velocity. In addition, the scatter in impact strength experiments is greater than that in the more controlled indentation strength experiments. Similar assessments were made in Section 3.4.2 regarding strength degradation by quasi-static spherical contacts.

Figure 3.40 is a logarithmic plot of the strength of SLG specimens containing angular contact flaws generated at the impact energies indicated (Wiederhorn and Lawn 1979; Lawn et al. 1979, 1980), using data derived from the published works cited. Contacts were generated by gas stream impelled impacts with SiC grit of characteristic radii 290–37 μm (30–200 mesh). Impact energies and uncertainties were determined from particle size estimates and impact velocity measurements. Symbols and bars represent means and standard deviation limits for strength measurements following impacts. The lower filled symbols represent the strengths of annealed glass specimens. The upper filled symbols represent the strengths of thermally tempered specimens with surface compression $\sigma_S = -128$ MPa ± 15 MPa. The central open symbols represent the strengths of as-received acid washed specimens. There was no significant effect of particle size on strength as a function of impact energy. The shaded bands on the left represent the means and standard deviation limits of the intrinsic strengths σ_i of the annealed and tempered specimens. The bold solid line represents a best fit of Eq. (3.76), $\sigma \sim U_K^{-2/9}$, to the strength measurements of the annealed specimens and well describes the data. The upper fine line represents a prediction for the strengths

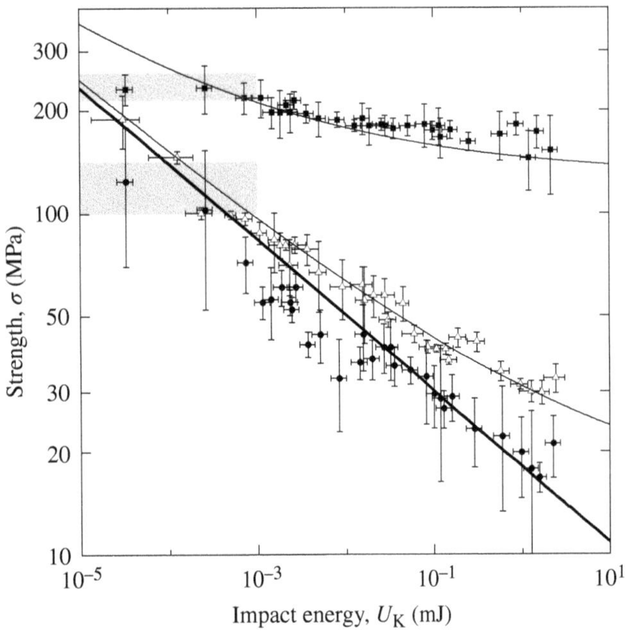

Figure 3.40 Logarithmic plot of angular sharp impact strengths of soda-lime silicate glass as a function of impact energy. Impacts were by SiC grit. Lower symbols represent annealed specimens. Upper symbols represent thermally tempered specimens. Source: Data adapted from Wiederhorn and Lawn (1979) and Lawn et al. (1979, 1980).

of the tempered specimens using Eq. (3.105) and the given value of σ_S (setting $U_{K\Omega} = 3 \times 10^{-4}$ mJ consistent with σ_i). The prediction also well describes the data. The central fine line represents a fit to the strengths of the as-received specimens using Eq. (3.105) and $\sigma_S = -13$ MPa. The clear fit suggests that the glass fabrication process retained a small amount of surface compression. In both the as-fabricated and tempered states, the indentation strength responses are well described by a compressively stressed layer that is uniform over the flaw, as shown in Figure 3.25a.

It is instructive to compare the impact strength responses of Figure 3.40 with those of indentation strength responses. Figure 3.41 is a logarithmic plot of the Vickers indentation strength behavior of SLG materials nearly identical to those used in Figure 3.40 (Marshall and Lawn 1978; Lawn et al. 1980), using data derived from the published works cited. Symbols with bars represent means and standard deviation limits for strength measurements following indentation; isolated symbols represent individual strength measurements. The lower filled symbols represent the strengths of the annealed glass specimens. The upper filled symbols represent the strengths of the thermally tempered specimens. Here, the central open symbols represent the strengths of partially tempered specimens with surface compression $\sigma_S = -48$ MPa ± 5 MPa. As before, the shaded bands represent the means and standard deviation limits of the intrinsic strengths of annealed and tempered specimens. The bold solid line represents a best fit of the ideal indentation strength response, Eq. (3.9), $\sigma = \hat{S}P^{-1/3}$, to the strength measurements of the annealed specimens and well describes the data using $\hat{S} = 122 \pm 16$ MPa N$^{-1/3}$. The upper fine line represents a prediction for the strengths of the tempered specimens using Eq. (3.102) and the given value of σ_S (setting $P_\Omega = 1$ N consistent with σ_i). The prediction also well describes the data. The central fine line represents a fit to the strengths of the partially tempered specimens using Eq. (3.102) and $\sigma_S = -48$ MPa. Again, the fit well describes the data. The resemblance of Figures 3.40 and 3.41 is striking. The resemblance provides further support for the assertions that controlled indentation flaws are good models for "naturally" occurring contact flaws and that fracture mechanics analyses based on the residual field effects arising from contact generated pdz are effective descriptions and predictions of behavior (Lawn and Cook 2012; Marshall et al. 2015).

Another set of strength experiments in alumina systems also illustrates the similarity of impact and indentation behavior. Figure 3.42 is a logarithmic plot of the strength of SLG and polycrystalline alumina specimens containing angular contact flaws generated at the impact energies indicated (Ritter et al. 1984; Breder et al. 1988), using data derived from the published works cited. Contacts were generated primarily by a slinger apparatus that impelled SiC grit of characteristic radii 765–45 μm (16–150 mesh). Some contacts were gas stream impelled. Impact energies were determined from particle size estimates and impact velocity measurements. Symbols and bars represent means and standard deviation limits for strength measurements following impacts. The upper filled and open symbols represent the strengths of alumina specimens of two

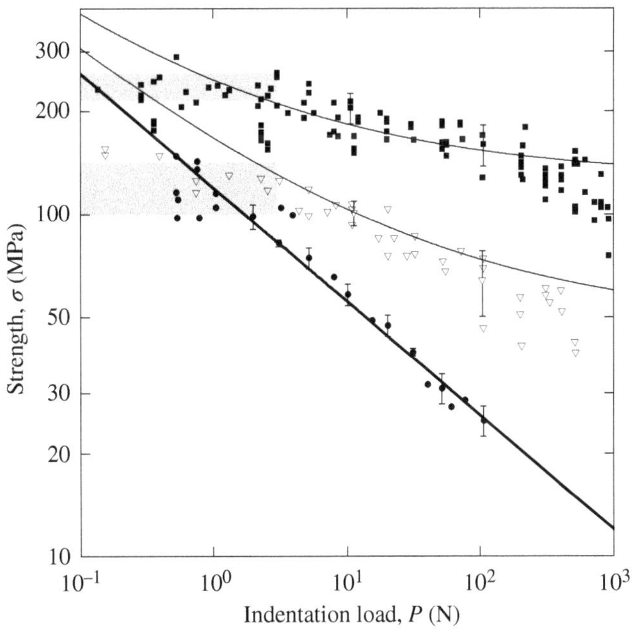

Figure 3.41 Logarithmic plot of Vickers indentation strength of soda-lime silicate glass. Lower symbols represent annealed specimens. Upper symbols represent thermally tempered specimens. Compare with Figure 3.40. Source: Data adapted from Marshall et al. (1978) and Lawn et al. (1980).

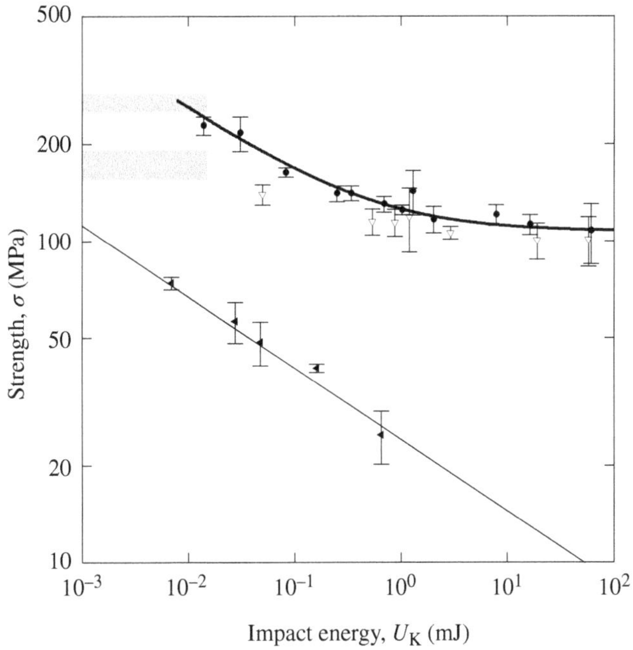

Figure 3.42 Logarithmic plot of angular sharp impact strengths of soda-lime silicate glass and alumina as a function of impact energy. Impacts were by SiC grit. Lower symbols represent soda-lime silicate glass specimens. Upper symbols represent alumina specimens. Source: Data adapted from Ritter et al. (1984) and Breder et al. (1988).

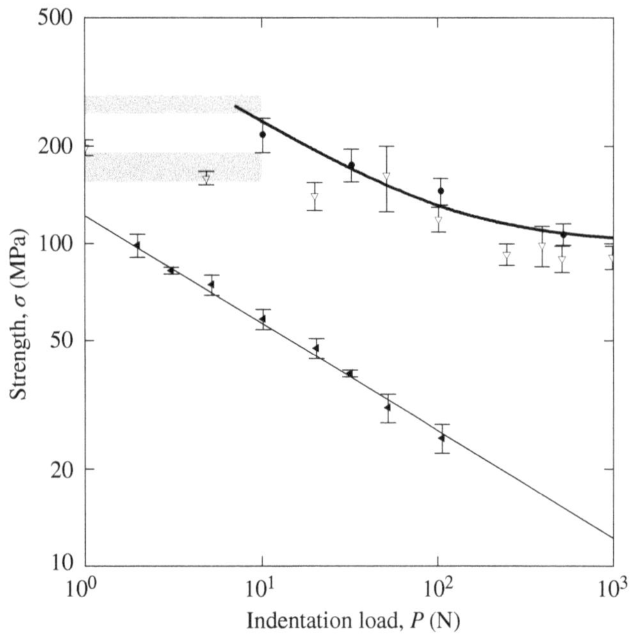

Figure 3.43 Logarithmic plot of Vickers indentation strength of soda-lime silicate glass and alumina. Lower symbols represent soda-lime silicate glass specimens. Upper symbols represent alumina specimens. Compare with Figure 3.42. Source: Data adapted from Marshall et al. (1978) and Breder et al. (1988).

different materials. The lower filled symbols represent the strengths of annealed glass specimens. The shaded bands on the left represent the means and standard deviation limits of the intrinsic strengths σ_i of the alumina specimens. Observations of impact damage on SLG showed the expected pdz-related residual contact impressions and surface traces of half-penny and lateral cracks, along with material removal by lateral crack chips. Hence, the fine solid line represents a best fit of Eq. (3.76), $\sigma \sim U_K^{-2/9}$, to the strength measurements of the glass specimens and well describes the data. Observations of impact damage on the alumina specimens showed extreme versions of pdz-related contacts. Extensive lateral cracking and chipping typically removed a substantial portion of the pdz, the associated surface impression, and much surrounding material; the contact flaws appeared in the form of "pits" formed by intergranular fracture surfaces. The upper bold line thus represents a fit of the lateral crack affected impact strength response, Eq. (3.80), to the data from the alumina specimens indicated by the filled symbols. The fit well describes the data, particularly the lack of variation in strength at large impact energies. The open symbol data behaved similarly at slightly lesser strengths.

Figure 3.43 is a logarithmic plot of the Vickers indentation strength behavior of the same alumina materials as in Figure 3.42 (Breder et al. 1988), using data derived from the published work. Also shown, as a convenient reference, are the annealed SLG indentation strength data from Figure 3.41. The shaded bands represent the means and standard deviation limits of the intrinsic strengths of the alumina specimens. The lower fine solid line represents ideal indentation strength $P^{-1/3}$ behavior. The upper bold solid line represents a fit of the lateral crack affected indentation strength response, Eq. (3.67), to the data from the alumina specimens indicated by the filled symbols. Again, the fit well describes the data, particularly the lack of variation in strength at large impact energies, and the fit is entirely analogous to that in Figure 3.42. Here as well the open symbol data behaved similarly at slightly lesser strengths. Once again, the resemblance of Figures 3.42 and 3.43 is striking. The resemblance provides further evidence that indentation flaws are good models for contact flaws by extending the ensuing fracture mechanics analyses to include polycrystalline materials and lateral crack effects. Ritter advanced a different, pit-based, model for the impact strength of polycrystalline materials based on weak intergranular fracture (Ritter et al. 1984; Ritter 1985). The model leads to strength invariance at large impact energies but lacks correspondence with strength behavior at small impact energies and with indentation contacts. The microstructure focused pit-based model is considered in Chapter 4.

A final set of experiments measuring strength after impact considers liquid water jets as simulations of raindrop impacts on aircraft. A typical raindrop is about 2 mm in diameter and approaches a terminal falling velocity of about 10 m s^{-1} (Fletcher 1969). However, drops occur in size distributions containing larger drops and supersonic aircraft travel at speeds greater than 760 miles per hour, about 340 m s^{-1}. Hence, raindrop impact collision is a major design factor for brittle

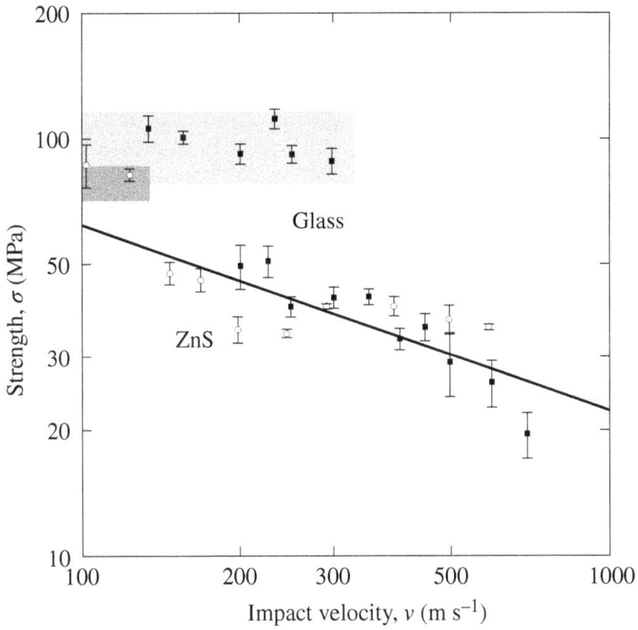

Figure 3.44 Logarithmic plot of water jet impact strengths of soda-lime silicate glass and polycrystalline ZnS as a function of jet velocity. Compare with Figure 3.16. Source: Data adapted from Van der Zwaag and Field (1982, 1983).

materials, e.g., glass, Y_2O_3, ZnS, used in visible and infrared light transmitting aircraft windows. Figure 3.44 is a logarithmic plot of the strength of SLG and polycrystalline ZnS specimens containing water jet flaws generated at the impact velocities indicated (Van der Zwaag and Field 1982, 1883), using data derived from the published works cited. The jets simulated impact of a moving window with stationary 4 mm diameter raindrops and the domain of impact velocities covered subsonic–supersonic flight. Symbols with bars represent the means and standard deviation limits of strength measurements performed using a biaxial burst geometry. Filled symbols represent SLG and open symbols represent ZnS. The shaded bands at left represent the means and standard deviations of the intrinsic strengths; glass upper, ZnS lower. The line is a guide to the eye of the form $\sigma \sim \upsilon^{-4/9}$, characterizing elastic–plastic angular contact, Eq. (3.78). Consistent with the elastic–plastic spherical contact dimension observations of Figure 3.16, there is very little difference in the liquid water jet impact strength of the two materials. In addition, observations of the water jet impacts in both materials show surface traces of multiple incomplete ring cracks. These observations suggest that the impacts could have behaved as spherical elastic contacts with strength given by $\sigma \sim \upsilon^{-2/5}$, Eq. (3.77). The scatter in the data precludes definitive distinction, but there is no doubt that high-velocity water impacts cause brittle component strength degradation. Comparison with Figure 3.41 suggest that a 4 mm water droplet impact at barely supersonic speeds is equivalent to a static 30 N Vickers indentation contact.

3.4.5 Surface Stress Effects

Analysis of the effects of surface stresses and layered structures on indentation strengths is considered in Section 3.3.5. The geometries of the stressed and layered components considered here are illustrated in Figures 3.11 and 3.25.

Surface stress effects on fracture were demonstrated in exemplary fashion in Section 3.4.4 by indentation strength measurements of thermally tempered glass. As shown in Figure 3.41, tempered glass components containing compressive surface stress exhibited greater strengths and less strength degradation than components of the base glass. Both effects were described by Eq. (3.102). An effect similar to tempering in glass was demonstrated by Insley and Barczak (1964) in experiments on "thermal conditioning" of polycrystalline alumina, in which both mechanical and dielectric strength were increased. (The experiments were performed in the context of spark plug manufacture. Thermal conditioning effects had been known much earlier—see Insley and Barczak for a brief review.) Similar work was later performed by Kirchner and colleagues in experiments on quenching polycrystalline alumina components from high temperatures. Kirchner and colleagues demonstrated strengthening by quenching as a function of grain size and for scratched and indented alumina components (Gruver and Kirchner 1973; Kirchner et al. 1973; Marshall et al. 1978) that lead to a patented fabrication process (Kirchner and Walker 1975). Figure 3.45 is a logarithmic plot of indentation strength measurements from sets of base and

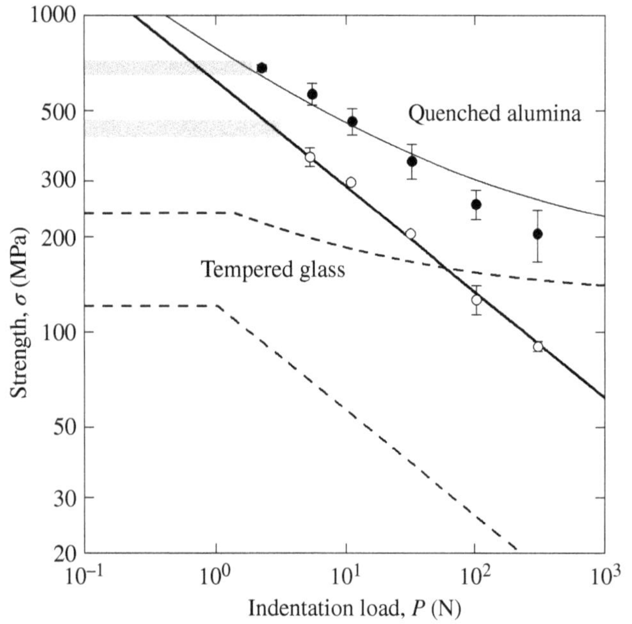

Figure 3.45 Logarithmic plot of Vickers indentation strength of polycrystalline alumina. Lower symbols represent base material specimens. Upper symbols represent quenched specimens. Compare with Figure 3.41. Source: Data adapted from Marshall et al. (1978).

quenched polycrystalline alumina rods (Marshall et al. 1978), using data derived from the published work. The open and filled symbols indicate the means and standard deviation limits of strength measurements of base and quenched alumina rods respectively at the indentation loads indicated. The lower and upper shaded bands indicate the means and standard deviation limits of intrinsic strengths of the base and quenched rods. The bold solid line represents a fit to the base component data of the ideal indentation strength response Eq. (3.9) using $\hat{S} = 620\,\text{MPa} \pm 40\,\text{MPa}$, shown in scaled form in Eq. (3.103). The fine solid line represents a fit to the quenched component data of the indentation strength response with superposed compression, Eq. (3.102), using a surface stress $\sigma_S = -170\,\text{MPa}$. Both fits well describe the data. The lower and upper dashed lines in Figure 3.45 represent similar best fits to the tempered glass data of Figure 3.41. The similarity of the glass and polycrystalline alumina responses is clear. The implication is that on initial quenching the inner "core" of the alumina components relaxes by crystalline mechanisms or deforms viscously and provides little mechanical constraint to the solidifying and contracting outer "shell." On further quenching, the core solidifies and contracts, placing the shell in compression (and the core in tension). This is the process by which glass is thermally tempered for automobile and construction applications. It is important to recognize that the surface stressed systems of tempered glass, Figures 3.40 and 3.41, and quenched alumina, Figure 3.45, are all well described by contact fracture mechanics in which the surface stress is well approximated as uniform over the strength-controlling flaw, Figure 3.25a. Systems are now considered in which this is not the case.

Many polycrystalline ceramic and glass-ceramic materials are machined in manufacturing processes to obtain the final sizes and shapes of components. Machining applies abrasive sharp hard particles in a myriad of linear contacts to component surfaces and removes material by ductile plowing and lateral cracking (Lawn et al. 2021; Huang et al. 2021). A consequence of the multiple contacts during machining is the formation of overlapping linear pdz leading to a layer of surface compression residual to the machining process (Green et al. 1983; Lange et al. 1983). A further consequence is that the cracking and strength behavior of machined components are thus affected by the layer of surface compression. Figure 3.46 shows logarithmic plots of the indentation strength behavior of machined components. The materials here in Figure 3.46a–c are identical to those in Figure 3.2d–f, Pyroceram glass-ceramic, polycrystalline Al_2O_3, and polycrystalline Al_2O_3–SiC composite. Symbols with bars represent means and standard deviation limits of strength measurements at the indentation loads indicated. Isolated symbols indicate individual strength tests. The open symbols and dashed lines are repeated from Figure 3.2 and represent ideal indentation strength responses. The filled symbols indicate components indented and strength tested as bend beams in the machined state—this was, as is common, the "as-received" state. The surfaces of specimens sampled from the as-received beams in Figure 3.46a and b were polished with increasingly fine diamond paste, from 30 to 1 μm, such that >100 μm was removed from the surfaces, as described in Cook et al. (1981). A sample of

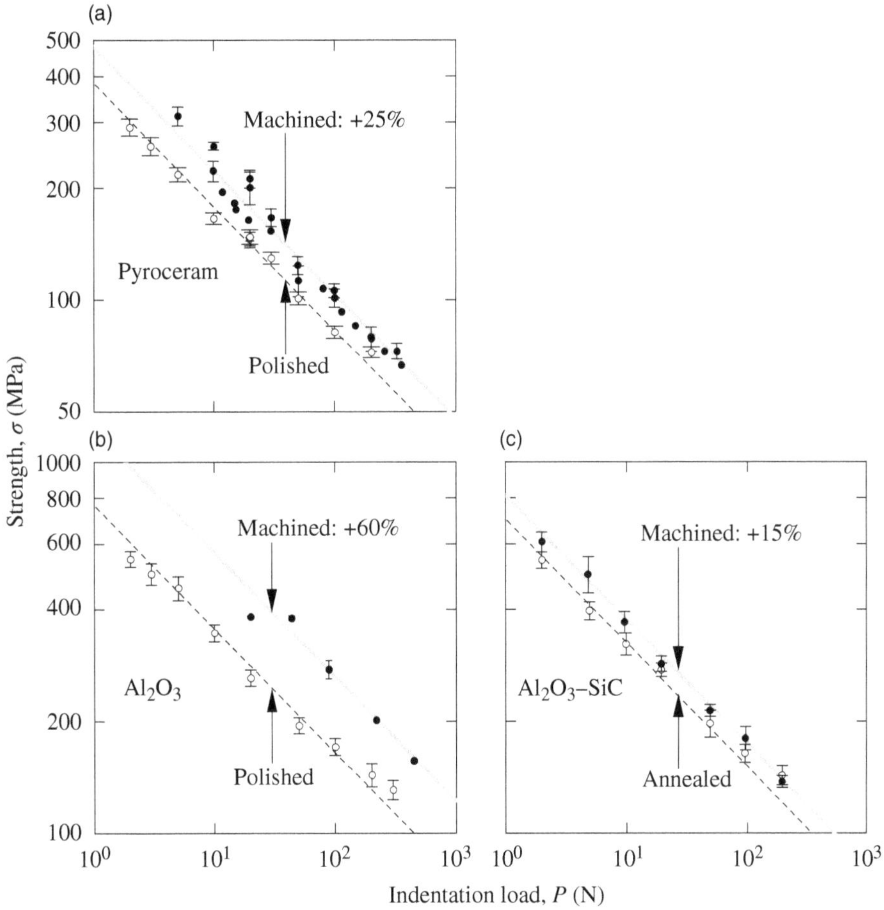

Figure 3.46 Logarithmic plots of Vickers indentation strengths of polycrystalline materials, illustrating the effects of machined surfaces. Lower symbols represent base material specimens (see Figure 3.2). Upper symbols represent machined specimens. (a) Pyroceramc. (b) Alumina. (c) Alumina-SiC composite. Source: Includes data adapted from Lawn et al. (1980) and Zhao et al. (1993).

the as-received beams in Figure 3.46c were annealed, as described in Zhao et al. (1993). The polishing and annealing steps generated beams with stress-free surfaces for subsequent indentation strength testing. The remaining as-received beams were tested with surfaces in the machined states. The polished and annealed beams exhibited ideal indentation strength behavior as shown in Figure 3.2 and the open symbols in Figure 3.46. The machined beams exhibited the indentation strength behavior as shown by the filled symbols in Figure 3.46, including from the works of Lawn et al. (1980) and Zhao et al. (1993) using data derived from the published works cited. In all cases, the machined strengths at a given indentation load were greater. The bold gray lines represent $P^{-1/3}$ best fits to these data consistent with Eq. (3.110). Such fits reflect flaw geometries described by Figure 3.25b in which the cracks are large relative to the surface stressed layer such that $dK_S/dc \approx 0$ and the components are large such that $K_B \approx 0$. The increases in the strengths are indicated. Although the increases were substantial, the indentation strengths of the machined components exhibited $P^{-1/3}$ strength degradation behavior. Hence, from strength tests alone, without separate polishing or annealing experiments it is often difficult to assess the effects of machining on surface stress states and component strengths or to assess whether strength tests are representative of the material or (more commonly) the machining used in specimen fabrication.

Machining can be placed in context by noting that a term used similarly in manufacturing is “grinding” and that “sawing” is simply machining or grinding limited to a (saw-)blade width. The less intentional phenomenon of material removal by multiple sharp linear contacts is often termed “wear.” The related material removal by multiple point-like contacts is often termed “erosion” and was a focus of some of the sharp impact studies cited previously (Ritter et al. 1984; Ritter 1985; Breder et al. 1988). The damage generated on surfaces by multiple sharp particle contacts, point-like or line-like, is often termed “abrasion” and such surfaces are often termed abraded, as in works also cited previously (Lawn et al. 1975, 1976). Material removal is not considered here. However, the remnant damage from machining and sawing figures prominently throughout the book in the form of preexisting flaw populations and thus intrinsic strengths in indentation strength tests (early examples are shown in Lawn et al.). Chipping damage, "scallops," generated by single point contacts adjacent to edges, and strengths associated with sawing of brittle materials are considered in Chapter 8. Single linear contacts, termed "scratches," are considered in detail in Chapter 9. Multiple linear contacts and strengths associated with grinding to remove material are considered in Chapter 10. Specimen polishing and annealing effects can be placed in context by noting the early indentation strength experiments on SiC and Si_3N_4 by Petrovic and colleagues (Petrovic et al. 1975, 1976; Petrovic and Jacobson 1976; Petrovic and Mendiratta 1979). In these experiments, indented specimens were polished or annealed prior to strength testing and the subsequently measured strengths observed to increase. In these cases, the tensile fields of the isolated single contacts were removed, leading to strength increases. In Figure 3.46, the compressive fields of the adjacent multiple contacts were removed, leading to strength decreases. In more recent work, Odén (1992) demonstrated the potential for application of surface stresses by superposing quenching effects, Figure 3.45, and machining effects, Figure 3.46, on polycrystalline SiC and Si_3N_4. Surface stresses of approximately −400 MPa were generated, arising from quenching (−300 MPa) ground (−100 MPa) specimens. Surface stresses of approximately −150 MPa were generated by quenching alone in Al_2O_3–SiC composites.

Tempering, quenching, and machining generate surface stresses in components by processes applied to single materials. Encapsulation, coating, or "cladding" can also lead to surface stresses in components by surrounding a first material, the core, by a second material, the cladding, with different properties from the first. The clad components are now composite structures. Of particular interest is cladding of the core at an elevated temperature by cladding with a lesser CTE material. On cooling, the equilibrium contraction of the cladding is thus less than that of the core and compressive stress develops in the clad surface and tension in the core (Boley and Wiener 1960). Other behaviors of the cladding that lead to reduced contraction relative to the core also lead to surface compression. Such behaviors include those due to differences in coefficient of chemical expansion (Adler 2001; Ram et al. 2009; Bishop et al. 2014) and differences in densification on sintering (Warshaw 1957). In all these cases, providing the behavior of surface flaws and cracks in the component is dominated by the compressive effects of the surface layer, the component is strengthened.

Strengthening of components by cladding with deliberately CTE mismatched materials has been implemented since ancient times in the form of glazes on pottery and ceramic ware (Rado 1969). Glazes are glassy coatings on ceramic bodies that provide hermetic, decorative, and mechanical functions. On cooling from the firing temperature most glazes contract less than the ceramic body leading to surface compression and component strengthening. Modern day glazes can be quite sophisticated as shown, for example, by the glazes developed for a glass-ceramic that produced an increase of >4 factor in strength, from approximately 85–350 MPa (Duke et al. 1968). Another, related, modern day example is the cladding of a 180 μm diameter BSG optical fiber with FSG of lesser CTE, leading to an increase of >2 factor in strength, from approximately 140–295 MPa (Krohn and Cooper 1969). A variation on the glaze concept is to chemically alter the component by forming a solid solution in a surface layer that has a lesser CTE than the bulk. This technique was demonstrated on polycrystalline Al_2O_3, TiO_2, and $MgAl_2O_4$ by forming solid solutions through infiltration of Cr_2O_3 and SnO_2 to produce increases of ≈1.3 factors in strength, from approximately 190–250 MPa (Kirchner and Gruver 1966; Kirchner et al. 1967). An extension of this idea is to chemically alter the component by forming a solid solution in a surface layer that has an atomic-scale unit cell volume greater than the bulk in a chemical expansion effect independent of temperature and CTE. This technique was demonstrated on a glass-ceramic by forming an altered solid solution through infiltration of K^+ ions to replace Na^+ ions (transforming nepheline to kalsilite) and produce very large increases of up to ≈15 factor in strength, from approximately 90 MPa–1.4 GPa (Duke et al. 1967). The end point of this approach is to physically alter the component by cladding a ceramic core in the unfired state with a surface layer of a different material that exhibits less shrinkage on sintering. This technique was demonstrated on an alumina porcelain in which the core and cladding were formed by slip casting and subsequent cofiring to produce an increase of ≈1.3 factor in strength, from approximately 55–75 MPa (Warshaw 1957). More recent combinations of materials with disparate CTE and sintering characteristics are slip-cast multilayered (not clad) components of Al_2O_3 and $3Al_2O_3{\cdot}2SiO_2$ (mullite) that exhibited an increase of >3 factor in strength, from approximately 150–480 MPa (Rao et al. 1999). The fracture of such microstructurally manipulated components are considered in Chapter 4.

Figure 3.47 shows a logarithmic plot of the indentation strength behavior of clad components from Cook (2005). The material here is cordierite glass-ceramic, identical to that in Figure 3.2c, and the components were in the form of 3 mm thick disks. Symbols with bars represent means and standard deviation limits of biaxial strength measurements at the indentation loads indicated. Isolated symbols indicate individual strength tests. The open symbols and dashed diagonal line are repeated from Figure 3.2 and represent ideal indentation strength responses from tests on homogeneous

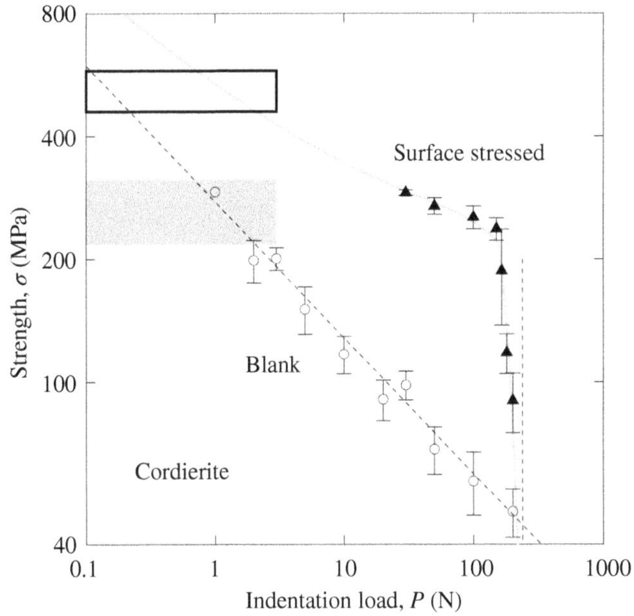

Figure 3.47 Logarithmic plot of Vickers indentation strengths of cordierite glass-ceramic materials, illustrating the effects of compressive stress surface cladding. Lower symbols represent base material specimens (see Figure 3.2). Upper symbols represent clad specimens. Source: Includes data adapted from Cook (2005).

"blank" components. The shaded band represents the intrinsic strength of blanks. The filled symbols indicate the strengths of components clad in a 155 μm thick layer of cordierite with slightly different composition and lesser CTE than the blank core. The cladding and core were otherwise identical mechanically. The cladding generated surface stress of $\sigma_S = -178$ MPa in the cladding layer and compensating bulk core stress of $\sigma_B = +20$ MPa. In all cases, the strengths of the clad components at a given indentation load were greater. The bold gray line represents a best fit to the strengths of the clad components using a full fracture mechanics analysis of the indented components (Cook 2005). The analysis takes into account details of the transition from an indentation crack fully contained in the compressive surface layer, Figure 3.25a, to one controlled by extension into the tensile bulk core, Figure 3.25b. The resemblance of the fit line and the predicted interpolated response of Figure 3.28 is clear. The fit and interpolation are based on the same physical principles, and two strength regions are apparent: At small indentation loads, the strengths of the clad components reflect the effects of surface compression and exhibit elevated and less variable strengths than the ideal response. Approaching a large, limiting, indentation load, the strengths of the clad components reflect the effects of core tension and exhibit a precipitous decrease in strength. The limiting indentation load represents spontaneous indentation failure and in this case $P_m \approx 240$ N, indicated by the vertical dashed line. The maximum experimental effect of cladding can be observed in Figure 3.47 at an indentation load of $P = 150$ N. At this load, the blank strength is ≈50 MPa and the clad strength is ≈230 MPa, an increase of >4 factor in strength and consistent with the surface compression of ≈180 MPa. The maximum predicted effect of cladding can be observed at an indentation load of $P_i \approx 1$ N; the effective intrinsic flaw scale observed in the blank strength measurements. The open bar indicates the extrapolation of the clad strength response to P_i and an intrinsic clad strength of about 500 MPa.

Strengthening of glassy components by chemical alteration of the surface in a process known as "ion-exchange" is a pervasive commercial technique: The display screens for nearly all mobile telephones and other hand-held electronic devices are strengthened by ion-exchange. The technique is related to the solid solution effect described previously in that glassy components are chemically altered by forming an expanded surface layer with an atomic-scale volume greater than the bulk. Expansion in the layer is achieved by counterdiffusion and surface exchange of small monovalent cations (Na^+, Li^+) in the glass with large monovalent cations (usually K^+) in an effect independent of CTE. (The effect is not strictly one of chemical expansion as the glass and the chemical gradient formed are not equilibrium structures and independent of temperature.) The ion-exchange process is accomplished by immersing the glass in a molten salt (e.g., KNO_3) for some time at elevated temperature. Common exchange parameters are depths of 10s of μm and compressive surface stresses of order −500 to −1000 MPa (and distinct from thermal tempering in glass with compression depths of several mm and surface

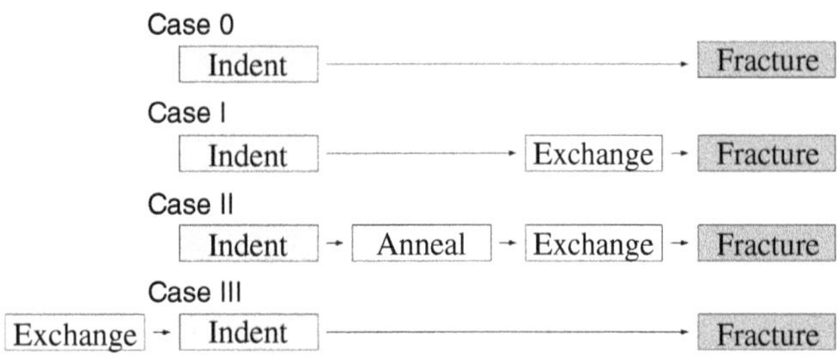

Figure 3.48 Ion-exchange framework illustrating and notating different possible sequences of indentation, exchange, annealing, and fracture as Cases 0–III. Source: Scheme adapted from Tandon and Green (1990).

stresses of −100 to −200 MPa). An early brief review of glass strengthening by surface processes is given by Nordberg et al. (1964), including a 1924 reference to the cladding work of Otto Schott.

The effects of ion-exchange on strengthening of glass are discussed simply in the sequence framework established by Tandon and Green (1990) and shown in Figure 3.48. The framework separates exchange effects into several related cases. Case 0 refers to the conventional indentation strength sequence of Figure 3.1 applied to unexchanged base glass. The term "Indent" here refers to flaw generating contacts broadly, including indentations, sharp contacts from handling and machining damage, impacts as in Figure 3.40, and linear scratches as discussed in Chapters 9 and 10. The term "Fracture" here refers to failure of components but could be used to include crack lengths as in Figure 3.26. Case I refers to the amelioration of flawed glass component strengths by application of an ion-exchange process after flaw generation and prior to component failure. Case II refers to removal of residual contact stresses at flaws by application of an annealing step after flaw generation and prior to exchange. Cases I and II bound the domain of relaxation for contact flaws exposed to exchange in manufacturing processes, with full and zero residual stress, respectively. Case III refers to protection of glass components from strength degradation by contact flaws in service through implementation of an ion-exchange process prior to both flaw generation and exposure to applied stress. By far the most studied element of this sequence framework is Case I, although quantifying the size of the strength-controlling flaws is often left to inference.

Figure 3.49 shows the strengths of Cases 0, I, and II for ion-exchanged SLG. The strengths are shown in the form of a logarithmic indentation strength plot from over 50 years of studies (Kistler 1962; Cooper and Krohn 1969; Tandon and

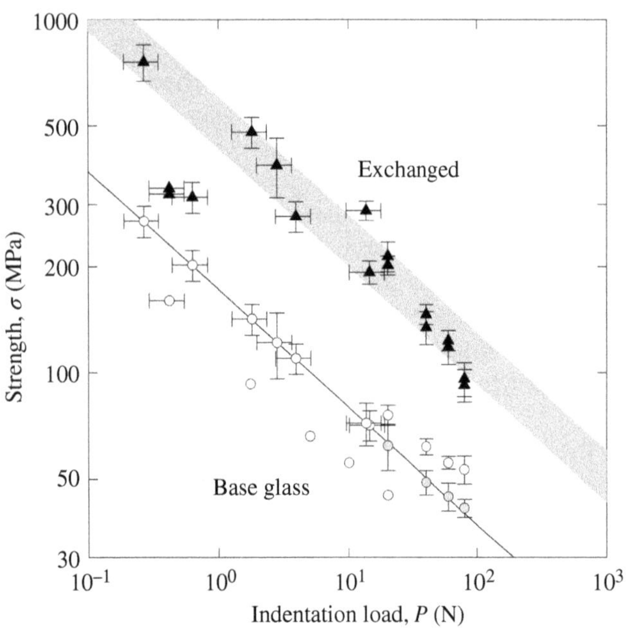

Figure 3.49 Logarithmic plot of Vickers indentation strengths of ion-exchanged soda-lime silicate glass treated as in Cases 0, I, and II (see Figure 3.48). Lower symbols represent base glass specimens. Upper symbols represent ion-exchanged specimens. Source: Includes data adapted from Kistler (1962), Cooper and Krohn (1969), Tandon and Green (1990), Fillery and Lange (2007), Gy (2008), Talimian et al. (2017), and Macrelli (2018).

Green 1990; Fillery and Lange 2007; Gy, 2008; Macrelli 2018; Talimian et al. 2017), using data derived from the published works cited. In some cases (Tandon and Green 1990; Gy 2008), strengths σ were reported as a function of Vickers indentation load P. In most cases, strengths represented the effects of strength-controlling handling flaws. The Case 0 strength data of Tandon and Green are taken as a reference and described by the ideal indentation response $\sigma_0 = \hat{S}_0 P^{-1/3}$ with $\hat{S}_0 = 172 \pm 16$ MPa. Effective indentation loads for the remaining strength data were determined by inversion of the ideal response as described in Section 3.1.2 for intrinsic strengths, $P_0 = (\hat{S}_0/\sigma_0)^3$. The P_0 values provide a convenient empirical ranking scheme. The symbols in Figure 3.49 represent mean indentation strength responses. Vertical bars represent the standard deviation limits of reported strengths. Horizontal bars represent the uncertainties in effective indentation load determined from the uncertainty in $\hat{S}_0$. Case 0 data are shown as open symbols, except for the reference data shown shaded. Case I and II data (within the experimental scatter there was no significant difference) are shown as filled symbols. The fine line shows the best-fit ideal indentation response to the reference data. The shaded band shows a shifted version of this response, $\sigma_I = S_I P^{-1/3}$ with $S_I = (3 \pm 0.5)\hat{S}_0$. For both controlled and uncontrolled flaws, an ion-exchange process implemented post flaw generation appears to increase the strength of SLG by a factor of approximately 3 over a wide domain of flaw sizes. Similar behavior, with often significantly increased strengthening, has been observed for various ASG that are the preferred choice for display panels (Tandon, private communication 1989; Aaldenberg et al. 2016; Beall et al. 2016; Kuzmin et al. 2016; Jiang et al. 2017).

The strength behavior of Case III ion-exchanged glasses is much more varied than those of Cases I and II. Consideration of the progression Case 0→Case I→Case III from the work of Gy (2008) explains the variation. Figure 3.50a shows a logarithmic indentation strength plot of the 0–I–III progression using data derived from the published work. The measurements in this system used Vickers indentations as controlled flaws. The dashed line indicates the ideal indentation response exhibited by Case 0 specimens. Case 0 intrinsic strength of the base glass is indicated by the lower solid line on the left of the plot, intersecting the ideal response at an effective indentation load of approximately 0.3 N. Case I intrinsic strengths of the exchanged glass are indicated by the upper solid lines on the left of the plot. Case III indentation strengths of the exchanged glass are indicated by the filled symbols. For indentation loads $P < 30$ N, the strengths decrease only slightly from the intrinsic level. For $P \approx 50$ N, the strengths decrease precipitously from $\sigma \approx 250$ MPa to $\approx$70 MPa or less, indicated by the arrow. The physical explanation of this behavior is as described previously: for small indentation loads and cracks, components are strengthened by the action of surface compression; for large indentation loads and cracks, components are significantly weakened by the action of bulk tension. An analytical description of this behavior is given by the interpolation formula of Eq. (3.112) shown as the gray line in Figure 3.28. The gray band in Figure 3.50a shows a visual fit as bounds to the observed behavior using Eq. (3.112). The fit describes well the data and the resemblance to the indentation strength behavior of the clad system of Figure 3.47 is clear.

Case III strengths are then expected to be quite variable, depending on a complicated interaction between three independent variables: exchange (determining surface compression and depth), component size (determining bulk tension), and

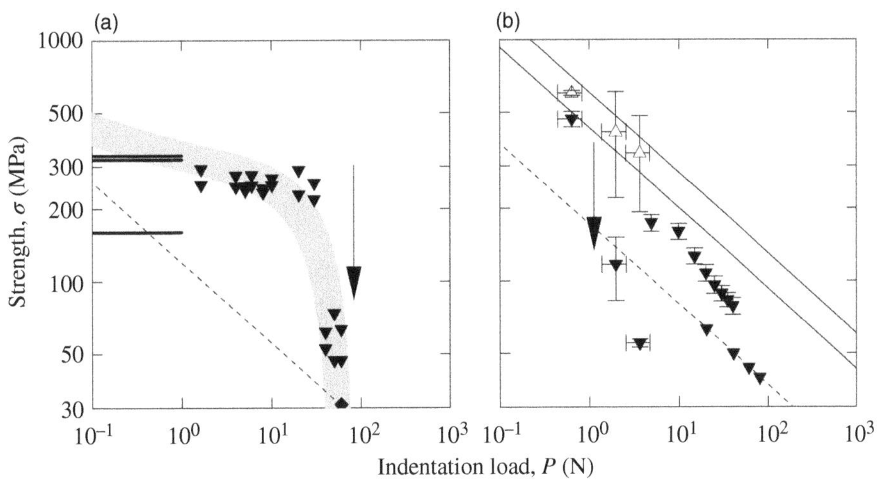

Figure 3.50 Logarithmic plots of Vickers indentation strengths of ion-exchanged glasses treated as in Case III (see Figure 3.48). (a) Single glass. Source: Includes data adapted from Gy (2008). (b) Multiple glasses. Source: Includes data adapted from Nordberg et al. (1964), Tandon and Green (1990), and Fillery and Lange (2007).

contact (determining flaw size). Figure 3.50b shows a logarithmic indentation strength plot of 0–I–III progressions from several exchange studies (Nordberg et al. 1964; Tandon and Green 1990; Fillery and Lange 2007), using data derived from the published works cited. From Figure 3.49, the dashed line represents the Case 0 response and the solid lines represent the bounds of the Case I response. The open symbols represent Case I strengths determined from Nordberg et al. using $P = (S_I/\sigma)^3$. Filled symbols represent Case III strengths and uncertainties at the indentation loads indicated. In all cases, the Case III strengths are decreased from the Case I level; an example similar to that observed in Figure 3.50a is indicated by the arrow. However, the variation in strength degradation is substantial, from almost none to approximately a factor of 5. Case III studies are important. As noted in the early work of Nordberg et al. (1964), "For any strengthening process to be of practical importance, it is necessary that the resulting strength be retained during service of the article." Hence, determining the retained strength as a function of the scale of contacts encountered in service is critical to design with exchanged glasses, and probably much more important from an engineering perspective than the ultimate and oft reported Case I strengths. However, the reduced Case III strengths are less eye catching and require the specification of additional parameters such as flaw size and component size.

A final consideration in the indentation fracture of layered systems are the effects of stress gradients arising in bending or flexing of thin components. In these cases, a stress gradient acting over a flaw arises from the external action of applied loading rather than the actions of an internal layered structure (e.g., cladding). Also distinct from internally layered structures stress gradients arising from externally applied bending are usually characterized by decreasing tension rather than the decreasing compression and increasing tension associated with engineered surface layers. As a consequence, thin components fractured in bending are expected to exhibit weaker strength degradation with flaw size than ideal geometries. Figure 3.51 shows logarithmic plots of indentation strength behavior in thin component systems that support this expectation. The systems are (a) SLG, (b) (111) single-crystal Si, (c) polycrystalline Ba(Bi)TiO_3 and Pb(La)TiO_3 and (d) single-crystal Al_2O_3, sapphire. Symbols with bars represent the means and standard deviation limits of strengths measured at the indentation loads indicated; isolated symbols represent individual strength measurements. The open symbols represent base responses from large specimen tests: those in Figure 3.51a, c, and d are repeated from Figure 3.38a, c, and h, those in Figure 3.51b are from the (111) tests of Lawn et al. (1981) using data derived from the published work. (See discussion in DelRio et al. 2015 regarding the indirect crystallographic orientation effects on Si indentation and fracture.) The dashed lines are guides to the eye representing ideal indentation strength responses for the materials. The responses were determined from the contact strengths S given in Table 3.2 characterizing asymptotic small indentation load behavior. The filled symbols represent strength measurements from thin specimen tests: those in Figure 3.51a, b, and d are from the work of Jung et al. (2004b) on 1 mm thick plates, using data derived from the published work; those in Figure 3.51c are from tests on small, 1.5 mm thick, bars of Pb(La)TiO_3. The solid lines in Figure 3.51 are empirical fits of Eq. (3.118) to the thin specimen data. The fits highlight the significant deviations from an ideal indentation response of strength measurements on thin specimens containing large flaws. The values of $-a/r$ used in Eq. (3.118) characterizing the empirical fits were in the domain −0.03 to −0.15, significantly less than the ideal value of −0.33. Note that ideal behavior is recovered at small loads in Figure 3.51a. The thin specimen flexure geometry used by Jung et al. was intended primarily for assessing the strengths of small load subthreshold indentation flaws and that is the subject of Section 3.4.6.

3.4.6 Subthreshold Effects

Analysis of the indentation strengths of subthreshold, crack free, sharp contact flaws is considered in Section 3.3.6. The geometry of subthreshold flaws is illustrated in Figure 3.9b and images of such flaws are shown in Figures 2.14a, 3.29, and 3.30.

A feature of subthreshold strength variation is that it is derived from consideration of the effects of applied stress on crack initiation at contacts. Hence, in comparison with postthreshold strengths controlled by crack propagation, and in common with many other initiation phenomena, subthreshold strengths are expected to exhibit enhanced sensitivity to material, environmental, and contact parameters. In particular, the characteristic magnitude of subthreshold strengths relative to postthreshold strengths was shown in Eq. (3.123) to depend on the dimensionless parameter α given by $\alpha \sim (\Omega E c_\Omega^{1/2}/T)$. Within this collection of terms, E is the material Young's modulus, T is the material toughness, and Ω and c_Ω are amplitude and length scale parameters of the contact deformation field respectively. Of these, T, Ω, and c_Ω all depend on environment and Ω, and c_Ω depend on contact geometry. As a consequence, subthreshold strengths are expected to vary significantly relative to postthreshold strengths, exhibiting a range between the bounds of very small deviations from the postthreshold level to significant increases, as illustrated in Figure 3.31. Hence, in judging the significance of subthreshold strength

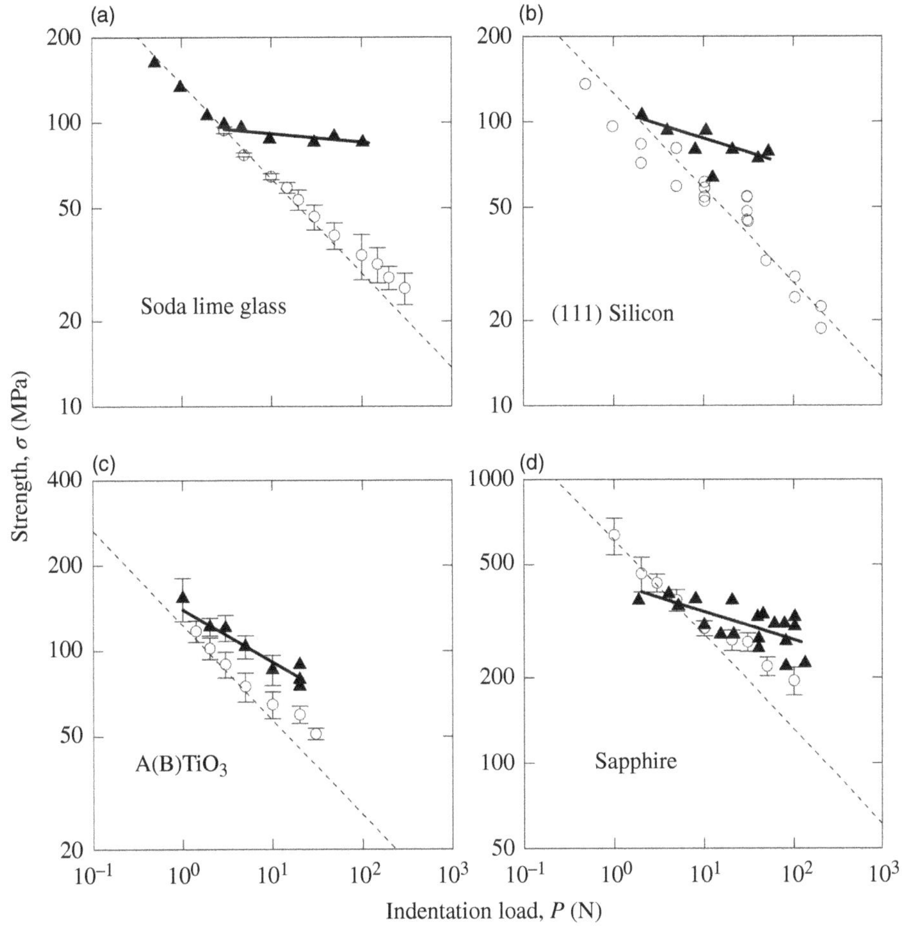

Figure 3.51 Logarithmic plots of Vickers indentation strengths of various materials, illustrating the effects of applied stress gradients during tests of thin components (shown as filled symbols). Source: Includes data adapted from Jung et al. (2004b). (a) Soda Lime Glass. (b) Single crystal Silicon. (c) Polycrystalline titanates. (d) Single crystal Sapphire. Open symbols represent tests of base thick components (see Figure 3.38). Source: Includes data adapted from Lawn et al. (1981).

observations for a material, it is important to consider the variability in the baseline postthreshold strengths of similar materials. Such a variability can be assessed through comparison of postthreshold strengths measured in different laboratories. Postthreshold Vickers indentation strengths measured in various laboratories provide the comparative baselines here in presentation of the subthreshold strengths in Figures 3.52–3.57.

Two key elements of this book are provided in addition by the multilaboratory postthreshold strength data gathered in Figures 3.52–3.57. First, comparison of multilaboratory strength data provides an estimate of the lower-bound variability in controlled flaw strength testing of new materials, new test systems, and new component designs. Figures 3.2, 3.38, 3.45, and 3.46 make clear that such testing can distinguish material behavior. Multilaboratory comparisons enable an assessment of the significance of such distinctions. Second, the materials considered in Figures 3.52–3.57 all exhibit fracture behavior in which microstructural effects are completely absent or negligible. The materials are amorphous, single crystalline, and a very fine-grained polycrystal and as a consequence variations in toughness are absent. The postthreshold Vickers indentation strengths presented here thus provide the baselines for consideration in Chapter 4 of indentation strengths of toughened materials. (Note that subthreshold effects are really only a concern for untoughened materials. For toughened materials, microstructural flaws dominate at small length scales.)

Figures 3.52–3.57 show logarithmic plots in identical coordinates of postthreshold and subthreshold indentation strength measurements for six materials. Symbols with bars represent means and standard deviation limits of reported strengths at the indentation loads indicated; isolated symbols represent individual strength measurements. Postthreshold strengths are represented by open and gray symbols; the gray symbols indicate reference data for fitting. Subthreshold strengths are represented by filled symbols. Means and standard deviation limits of intrinsic strengths are

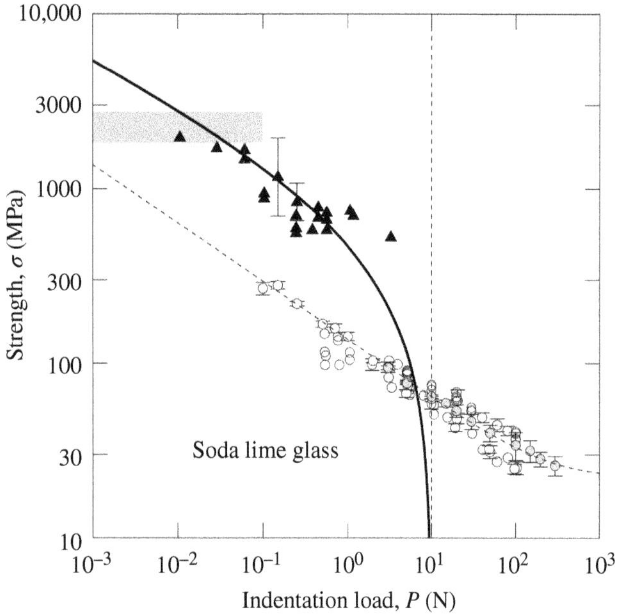

Figure 3.52 Logarithmic plot of Vickers indentation strengths of soda-lime silicate glass showing failure from postthreshold (open symbols) and subthreshold (filled symbols) flaws. Source: Includes postthreshold data adapted from Lawn et al. (1980), Cook and Roach (1986), Tandon and Green (1990), Pajares et al. (2004), Dabbs (1984), and Chantikul (1981); and subthreshold data adapted from Pajares et al. (2004) and Dabbs (1984).

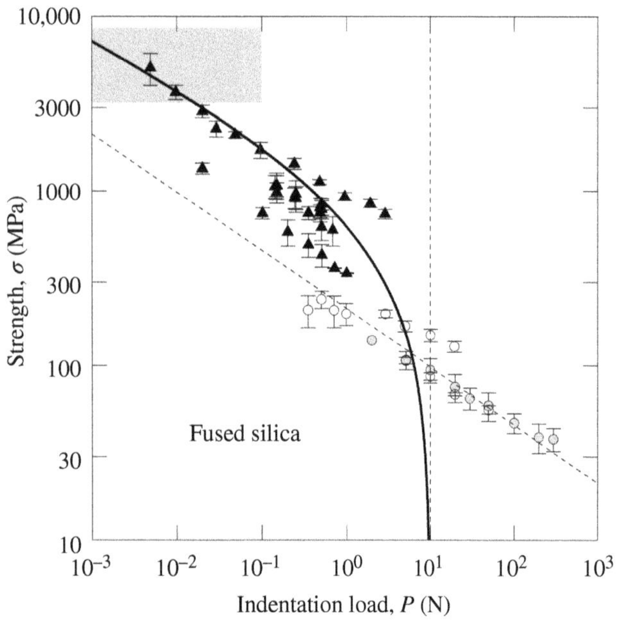

Figure 3.53 Logarithmic plot of Vickers indentation strengths of fused silica glass showing failure from postthreshold (open symbols) and subthreshold (filled symbols) flaws. Source: Includes postthreshold data adapted from Dabbs et al. (1980), Dabbs (1984), Lin et al. (1991), Lin and Matthewson (1996); and subthreshold data adapted from Dabbs (1984), Dabbs et al. (1980), Dabbs and Lawn (1985), Lin et al. (1991), Lin and Matthewson (1996), and Choi et al. (1990).

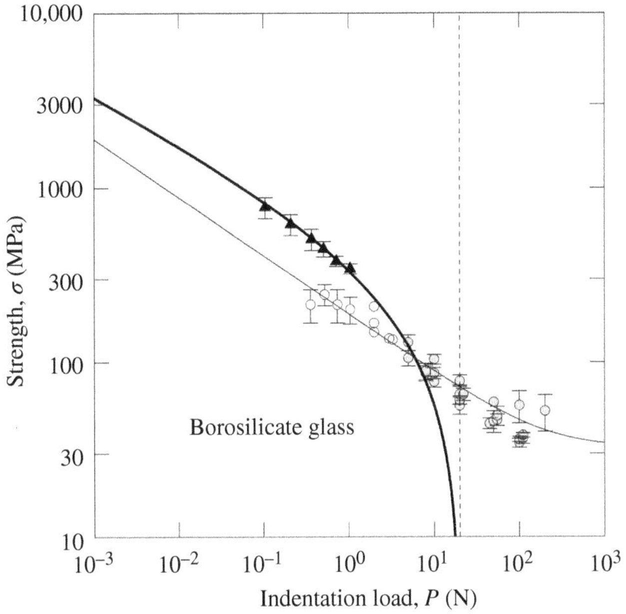

Figure 3.54 Logarithmic plot of Vickers indentation strengths of borosilicate glass showing failure from postthreshold (open symbols) and subthreshold (filled symbols) flaws. Source: Includes postthreshold data adapted from Liniger and Cook (1993), Dabbs (1984), Dabbs et al. (1982), Choi, S.R et al. private communication as cited in Lathabai et al. (1991a); and subthreshold data adapted from Lathabai et al. (1991a).

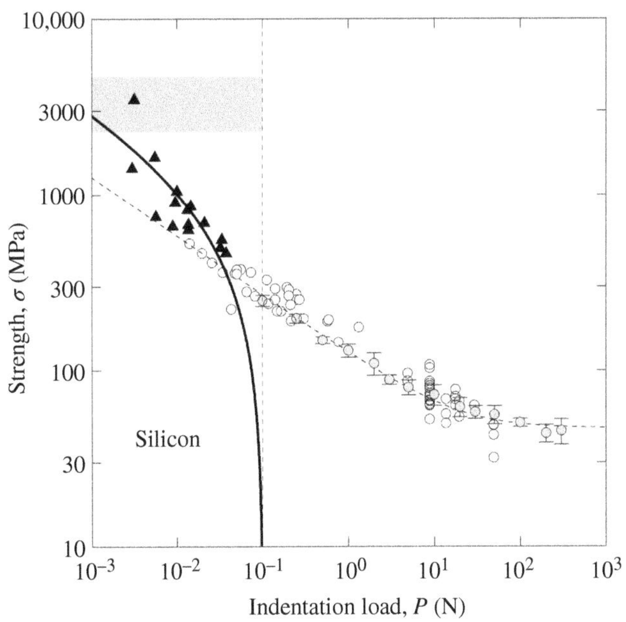

Figure 3.55 Logarithmic plot of Vickers indentation strengths of silicon showing failure from postthreshold (open symbols) and subthreshold (filled symbols) flaws. Source: Includes postthreshold data adapted from Cook (2006), Tsai and Mecholsky (1991), Jung et al. (2004a,b), and Pajares et al. (2004); and subthreshold data adapted from Jung et al. (2004a,b) and Pajares et al. (2004).

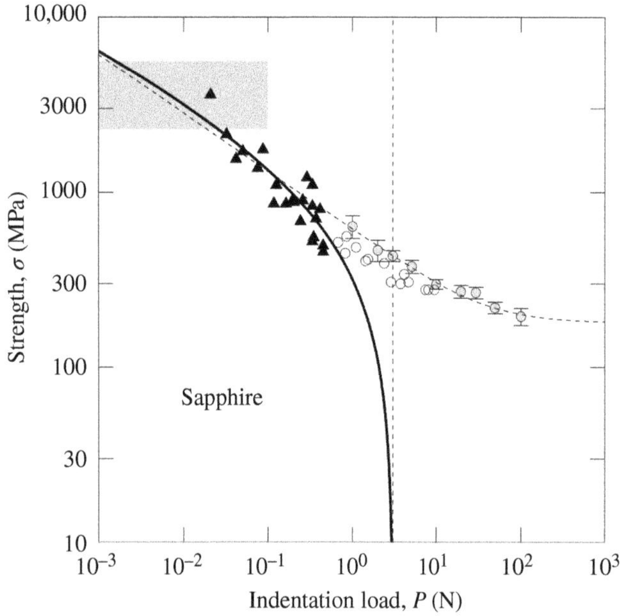

Figure 3.56 Logarithmic plot of Vickers indentation strengths of sapphire showing failure from postthreshold (open symbols) and subthreshold (filled symbols) flaws. Source: Includes postthreshold data adapted from Cook and Roach (1986), Jung et al. (2004b); and subthreshold data adapted from Jung et al. (2004b).

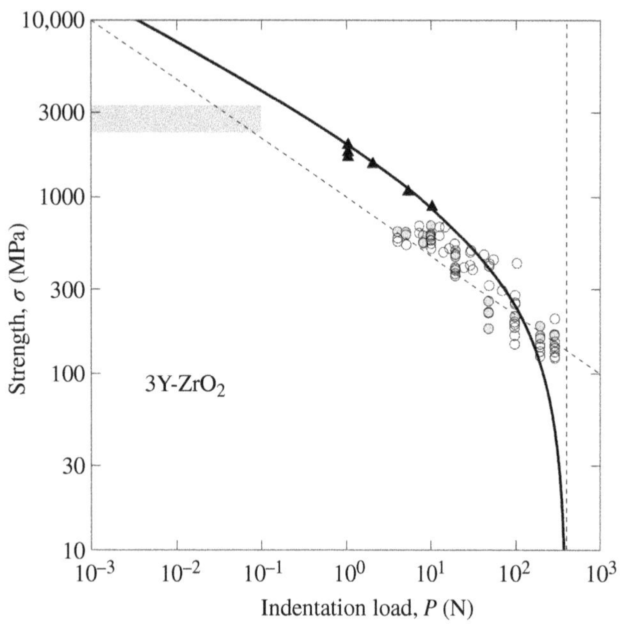

Figure 3.57 Logarithmic plot of Vickers indentation strengths of zirconia showing failure from postthreshold (open symbols) and subthreshold (filled symbols) flaws. Source: Includes postthreshold data adapted from Zhang and Lawn (2005), Alves et al. (2022), Readey and McCallen (1995); and subthreshold data adapted from Zhang and Lawn (2005).

represented by gray bands on the left of the plots. Fits to postthreshold reference strengths describing ideal indentation behavior, Eq. (3.103), and lateral crack affected behavior, Eq. (3.67), are indicated by the dashed diagonal lines (additional $\hat{S}$ and S fit values are given in Tables 3.1 and 3.2). Fits to the subthreshold strengths of Eq. (3.123) are indicated by the bold solid lines; the asymptotic spontaneous initiation loads are indicated by the vertical dashed lines. In all cases, plots were created using data derived from the published works cited: For SLG, Figure 3.52, postthreshold (Lawn et al. 1980, Figure 3.41; Cook and Roach 1986, Figure 3.38a; Tandon and Green 1990, Figure 3.49; Chantikul 1981; Dabbs 1984; Pajares et al. 2004) and subthreshold (Dabbs 1984; Pajares et al. 2004). For FSG, Figure 3.53, postthreshold (Figure 3.2a; Dabbs et al. 1980; Dabbs 1984; Lin et al. 1991; Lin and Matthewson 1996) and subthreshold (Dabbs et al. 1980; Dabbs 1984; Dabbs and Lawn 1985; Lin et al. 1991; Lin and Matthewson 1996; Choi et al. 1990). For BSG, Figure 3.54, postthreshold (Liniger and Cook 1993; Dabbs 1984; Dabbs et al. 1982; Choi et al. private communication as cited in Lathabai et al. 1991a) and subthreshold (Lathabai et al. 1991a,b). For single-crystal Si, Figure 3.55, postthreshold (Cook 2006, Figure 3.38b; Tsai and Mecholsky 1991; Jung et al. 2004a,b Pajares et al. 2004) and subthreshold (Jung et al. 2004a,b; Pajares et al. 2004). For single-crystal Al_2O_3, sapphire, Figure 3.56, postthreshold (Cook and Roach 1986, Figure 3.2e; Jung et al. 2004b) and subthreshold (Jung et al. 2004b). For polycrystalline ZrO_2, Figure 3.57, postthreshold (Readey and McCallen 1995; Zhang and Lawn 2005; Alves et al. 2022) and subthreshold (Zhang and Lawn 2005).

The similarities of Figures 3.52–3.57 to Figure 3.33 are clear, including the anticipated diversity of subthreshold strengths relative to postthreshold strengths and a diversity of indentation load domain overlaps of these strengths. The experimental observations can be classified into two major groups associated with amorphous and crystalline materials. In amorphous materials, the separations of subthreshold and postthreshold strengths are clear, up to a factor of 3, and the domains of indentation load overlap are large, up to more than a factor of 10. The extreme example is SLG, Figure 3.52. In crystalline materials, the separations of subthreshold and postthreshold strengths are barely discernible, and the domains of indentation load overlap are at most a factor of 2. The extreme example is sapphire, Figure 3.56. In all cases, the analytical fits to the subthreshold strengths (bold solid lines) imply threshold indentation loads much greater than those inferred from indentation observations, and much greater than the upper bounds of the overlap domains. The extreme example is ZrO_2, Figure 3.57. Experimental scatter in subthreshold strengths is only noticeably greater for FSG, Figure 3.53; a reflection of measurements from a number of studies.

3.5 Summary and Discussion

This chapter has provided an extensive and detailed assessment of the indentation strength behavior of brittle materials that exhibit invariant toughness. Such materials include many encountered in domestic settings: glass (windows, drinking vessels), glass-ceramics (cookware), single crystals (electronic devices), and fine-grained polycrystals (tooth and bone replacement, drill bits, and saw blades). Indentation strength measurements mimic the most common failure mechanism of components formed from these materials: flaw formation on a component surface by a localized contact—an indentation—followed by loading of the component until the maximum sustainable stress for the flaw is reached—the strength—and component failure occurs. The most common experimental variations within this sequence have been examined: included angle effects at sharp conical and pyramidal contacts; radius effects at blunt spherical contacts; ancillary chipping and lateral crack effects at large contacts; velocity and energy effects at impact contacts; surface stress effects in layered systems; and crack initiation effects at very small contacts. A unified analytical fracture mechanics framework has been developed and expressions derived to describe these indentation strength variations. A focus of the analysis has been description of indentation strength phenomena in experimentally measurable terms with a minimum of notation. Extensive use has been made of scaled variables in expressing indentation strength behavior so as to focus attention on the *form* of the behavior rather than the values of some parameters (e.g., the exact values of the indentation and applied stress geometrical coupling terms, or the material Poisson's ratio; such values are important in engineering applications and these are a focus of Chapters 7–13.)

The usefulness of the indentation strength technique and the ability of indentation fracture mechanics analyses to describe the above diversity of phenomena derive from several refinements implemented in many of the studies discussed in this chapter. As noted in Section 3.1, there were early brittle fracture works that used "controlled" flaws, including that of Griffith (1921). However, historically, the indentation strength technique was developed and refined as a "technique" in the late 1970s to early 1980s at the University of New South Wales, Australia and the US National Bureau of Standards (NBS, now National Institute of Standards and Technology, NIST). Early works focused on the phenomenology of indentation strength with blunt and sharp contacts (e.g., Lawn et al. 1975, 1976), followed by detailed consideration of residual stress effects at sharp contacts (e.g., Marshall and Lawn 1979; Marshall et al. 1979). Two instrumental refinements implemented about that time were (i) gravity-loaded Vickers indenters with broad load ranges (2–300 N) that enabled rapid throughput testing of large numbers of indented specimens and (ii) piezoelectric load cells that enabled strength testing at fast applied stressing rates approaching inert conditions. As a consequence, by the early 1980s, indentation strength measurements were used to compare material fracture responses, assess intrinsic strengths, evaluate effects of tempering, impacts, and subthreshold flaws, and estimate toughness (e.g., Dabbs et al. 1980; Lawn et al. 1980; Chantikul et al. 1981b). Soon after, indentation strength measurements were extended to nonequilibrium fracture and environmental effects (e.g., Dabbs et al. 1982). Two specimen refinements implemented with greater frequency about that time were (i) polished specimens that removed significant compressive surface stresses associated with specimen machining and (ii) disk or plate strength test specimens that were loaded biaxially and removed the effects of intrinsic edge flaws, greatly extending strength measurements to small indentation loads. Hence, by the mid-1980s, lateral crack effects at large indentation loads and microstructural effects at small indentation loads were identified as pervasive indentation strength effects (e.g., Cook et al. 1985; Cook and Roach 1986). Since that time, the emphasis in indentation strength testing has been in application as a tool for elucidation of material properties, particularly in terms of fundamental microstructural parameters (e.g., Cook et al. 1987; Bennison and Lawn 1989; Cook 1990). The analytical framework has been advanced by the integration of microstructural effects, lateral crack effects, and nonequilibrium effects into single strength formulations (Cook et al. 1994b; Cook 2015).

The restriction in this chapter to indentation strength behavior of materials with invariant toughness, e.g., glasses and single crystals, has thus focused on indentation strength phenomena that were well known and explained by the mid-1980s. Many experimental aspects of indentation strength testing were refined by that time, including: the need for polishing of machined test specimens; the need for broad indentation load domains; the need for inert conditions; and the advantages of biaxial strength testing. Many analytical aspects were also refined, including: the need to incorporate lateral crack effects into strength descriptions; the usefulness of scaled coordinates in implementing numerical solutions; the existence of critical indentation loads for spontaneous fracture—first introduced for stressed components and then applied to subthreshold flaws; the ability to change frames of reference in fracture and incorporate toughening into strength analyses—also first introduced for stressed components and then applied to microstructural effects. Chapters 4 and 6 build on these experimental and analytical refinements in consideration of microstructural and nonequilibrium effects in indentation strengths. In particular, the restriction of this chapter is lifted, and Chapter 4 considers the indentation strength behavior of materials with varying toughness, e.g., coarse-grained ceramic polycrystals and composites.

References

Aaldenberg, E.M., Lezzi, P.J., Seaman, J.H., Blanchet, T.A., and Tomozawa, M. (2016). Ion-exchanged lithium aluminosilicate glass: strength and dynamic fatigue. *Journal of the American Ceramic Society* 99: 2645–2654.

Adler, S.B. (2001). Chemical expansivity of electrochemical ceramics. *Journal of the American Ceramic Society* 84: 2117–2119.

Alves, L.M.M., Rodrigues, C.S., Vardhaman, S., Saunders, C., Schneider, J.M., Lawn, B.R., and Zhang, Y. (2022). Exploring ductility in dental ceramics. *Journal of Dental Research* 101: 1467–1473.

Anderson, R.M. and Braun, L.M. (1990). Technique for the R-curve determination of Y-TZP using indentation-produced flaws. *Journal of the American Ceramic Society* 73: 3059–3062.

Anstis, G.R., Chantikul, P., Lawn, B.R., and Marshall, D.B. (1981). A critical evaluation of indentation techniques for measuring fracture toughness: I, Direct crack measurements. *Journal of the American Ceramic Society* 64: 533–538.

Arora, A., Marshall, D.B., Lawn, B.R., and Swain, M.V. (1979). Indentation deformation/fracture of normal and anomalous glasses. *Journal of Non-Crystalline Solids* 31: 415–428.

Baker, T.C. and Preston, F.W. (1946). Fatigue of glass under static loads. *Journal of Applied Physics* 17: 170–178.

Beall, G.H., Comte, M., Dejneka, M.J., Marques, P., Pradeau, P., and Smith, C. (2016). Ion-exchange in glass-ceramics. *Frontiers in Materials* 3: 41.

Bennison, S.J. and Lawn, B.R. (1989). Role of interfacial grain-bridging sliding friction in the crack-resistance and strength properties of nontransforming ceramics. *Acta Metallurgica* 37: 2659–2671.

Bishop, S.R., Marrocchelli, D., Chatzichristodoulou, C., Perry, N.H., Mogensen, M.B., Tuller, H.L., and Wachsman, E.D. (2014). Chemical expansion: implications for electrochemical energy storage and conversion devices. *Annual Review of Materials Research* 44: 205–239.

Boley, B.A. and Wiener, J.H. (1960). *Theory of Thermal Stresses*. Wiley.
Breder, K., Ritter, J.E., and Jakus, K. (1988). Strength degradation in polycrystalline alumina due to sharp-particle impact damage. *Journal of the American Ceramic Society* 71: 1154–1158.
Cai, P.Z., Green, D.J., and Messing, G.L. (1997). Constrained densification of alumina/zirconia hybrid laminates, II: viscoelastic stress computation. *Journal of the American Ceramic Society* 80: 1940–1948.
Chantikul, P. (1981). *Strength Evaluation of Ceramics using Indentation Fracture Techniques*. Ph.D. thesis. University of New South Wales.
Chantikul, P., Lawn, B.R., and Marshall, D.B. (1981a). Micromechanics of flaw growth in static fatigue: influence of residual contact stresses. *Journal of the American Ceramic Society* 64: 322–325.
Chantikul, P., Anstis, G.R., Lawn, B.R., and Marshall, D.B. (1981b). A critical evaluation of indentation techniques for measuring fracture toughness: II, strength method. *Journal of the American Ceramic Society* 64: 539–543.
Chiang, S.S., Marshall, D.B., and Evans A.G. (1982a). The response of solids to elastic/plastic indentation. I. Stresses and residual stresses. *Journal of Applied Physics* 53: 298–311.
Chiang, S.S., Marshall, D.B., and Evans, A.G. (1982b). The response of solids to elastic/plastic indentation. II. Fracture initiation. *Journal of Applied Physics* 53: 312–317.
Choi, S.R., Ritter, J.E., and Jakus, K. (1990). Failure of glass with subthreshold flaws. *Journal of the American Ceramic Society* 73: 268–274.
Cook, R.F. (1989). Effective-medium theory for the fracture of fractal porous media. *Physical Review B*, 39: 2811–2814.
Cook, R.F. (1990). Segregation effects in the fracture of brittle materials: Ca–Al_2O_3. *Acta Metallurgica et Materialia* 38: 1083–1100.
Cook, R.F. (2005). Toughening of a cordierite glass–ceramic by compressive surface layers. *Journal of the American Ceramic Society* 88: 2798–2808.
Cook, R.F. (2006). Strength and sharp contact fracture of silicon. *Journal of Materials Science* 41: 841–872.
Cook, R.F. (2015). Multi-scale effects in the strength of ceramics. *Journal of the American Ceramic Society* 98: 2933–2947.
Cook, R.F. (2018a). Blunt scratch strength of polycrystalline alumina. *Journal of the American Ceramic Society* 101: 16–19.
Cook, R.F. (2018b). Strength of brittle materials in moderately corrosive environments. *Journal of the American Ceramic Society* 101: 1684–1695.
Cook, R.F. (2019). Fracture sequences during elastic–plastic indentation of brittle materials. *Journal of Materials Research* 34: 1633–1644.
Cook, R.F. (2020a). A critical evaluation of indentation crack lengths in air. *Journal of the American Ceramic Society* 103: 2278–2295.
Cook, R.F. (2020b). 1000 at 1000: an indentation toughness method. *Journal of Materials Science* 55: 15069–15073.
Cook, R.F. (2023). *Particle Strengths*. Wiley.
Cook, R.F. (2024). Indentation fracture and deformation v.2. *Zenodo*. doi: 10.5281/zenodo.1149515.
Cook, R.F., Braun, L.M., and Cannon, W.R. (1994c). Trapped cracks at indentations: Part I Experiments on yttria-tetragonal zirconia polycrystals. *Journal of Materials Science* 29: 2133–2142.
Cook, R.F., Fairbanks, C.J., Lawn, B.R., and Mai, Y.W. (1987). Crack resistance by interfacial bridging: its role in determining strength characteristics. *Journal of Materials Research*, 2: 345–356.
Cook, R.F., Lawn, B.R., Dabbs, T.P., and Chantikul, P. (1981). Effect of machining damage on the strength of a glass-ceramic. *Journal of the American Ceramic Society* 64: C-121–C-122.
Cook, R.F., Lawn, B.R., and Anstis, G.R. (1982a). Fatigue analysis of brittle materials using indentation flaws: Part 2 Case study on a glass-ceramic. *Journal of Materials Science* 17: 1108–1116.
Cook, R.F., Lawn, B.R., Dabbs, T.P., Reeve, K.D., Ramm, E.J., and Woolfrey, J.L. (1982b). A comparative study of the fatigue properties of Synroc and borosilicate glass. *Journal of the American Ceramic Society* 65: C172–C173.
Cook, R.F., Lawn, B.R., and Fairbanks, C.J. (1985). Microstructure-strength properties in ceramics: I, effect of crack size on toughness. *Journal of the American Ceramic Society* 68: 604–615.
Cook, R.F. and Liniger, E.G. (1992). Grain-size effects in the indentation fracture of MgO. *Journal of Materials Science* 27: 4751–4761.
Cook, R.F. and Liniger, E.G. (1993). Kinetics of indentation cracking in glass. *Journal of the American Ceramic Society* 76: 1096–1105.
Cook, R.F., Liniger, E.G., and Pascucci, M.R. (1994a). Indentation fracture of polycrystalline cubic ceramics. *Journal of Hard Materials* 5: 191–212.
Cook, R.F., Liniger, E.G., Steinbrech, R.W., and Deuerler, F. (1994b). Sigmoidal indentation–strength characteristics of polycrystalline alumina. *Journal of the American Ceramic Society* 77: 303–314.

Cook, R.F. and Oyen, M.L. (2007). Nanoindentation behavior and mechanical properties measurement of polymeric materials. *International Journal of Materials Research* 98: 370–378.

Cook, R.F., Pascucci, M.R., and Rhodes, W.H. (1990). Lateral cracks and microstructural effects in the indentation fracture of yttria. *Journal of the American Ceramic Society* 73: 1873–1878.

Cook, R.F. and Pharr, G.M. (1990). Direct observation and analysis of indentation cracking in glasses and ceramics. *Journal of the American Ceramic Society* 73: 787–817.

Cook, R.F. and Pharr, G.M. (1994). Mechanical properties of ceramics. In *Materials Science and Technology* (Eds. Cahn, R.W., Haasen, P., and Kramer, E.J.) pp. 339–407. VCH.

Cook, R.F. and Roach, D.H. (1986). The effect of lateral crack growth on the strength of contact flaws in brittle materials. *Journal of Materials Research* 1: 589–600.

Cooper, A.R. (1982) Residual stress development in ceramics. In *Residual Stress and Stress Relaxation*. Proceedings of the Twenty-Eighth Sagamore Army Materials Research Conference (Eds. Kula, E. and Weiss, V.) pp. 439–465. Plenum.

Cooper, A.R. and Krohn, D.A. (1969). Strengthening of class fibers: II, Ion exchange. *Journal of the American Ceramic Society* 52: 665–669.

Dabbs, T.P. (1984). *Fracture Properties of High Strength Glass with Particular Reference to Optical Fibres*. Ph.D. thesis. University of New South Wales.

Dabbs, T.P. and Lawn, B.R. (1982a). Acid-enhanced crack initiation in glass. *Journal of the American Ceramic Society* 65: C-37–C-38.

Dabbs, T.P. and Lawn, B.R. (1982b).Fatigue of high strength soda-lime glass: a constant stressing rate study using substhreshold indentation flaws. *Physics and Chemistry of Glasses* 23: 93–97.

Dabbs, T.P. and Lawn, B.R. (1985). Strength and fatigue properties of optical glass fibers containing microindentation flaws. *Journal of the American Ceramic Society* 68: 563–569.

Dabbs, T.P., Lawn, B.R., and Kelly, P.L. (1982). A dynamic fatigue study of soda-lime silicate and borosilicate glasses using small scale indentation flaws. *Physics and Chemistry of Glasses* 23: 58–66.

Dabbs, T.P., Marshall, D.B., and Lawn, B.R. (1980). Flaw generation by indentation in glass fibers. *Journal of the American Ceramic Society* 63: 224–225.

DelRio, F.W., Cook, R.F., and Boyce, B.L. (2015). Fracture strength of micro-and nano-scale silicon components. *Applied Physics Reviews* 2: 021303.

Diehl, W. and Schulze, R. (1957). Thermische, plastiche Nachwirkungen an Vickersindrücken in Glas. *Zeitschrift für Angewandte Physik* 11: 251–253 (in German).

du Merac, M.R., Bram, M., Malzbender, J., Ziegner, M., Rasinski, M., and Guillon, O. (2018). Increasing fracture toughness and transmittance of transparent ceramics using functional low-thermal expansion coatings. *Scientific Reports* 8: 15644.

Duke, D.A., MacDowell, J.F., and Karstetter, B.R. (1967). Crystallization and chemical strengthening of nepheline glass-ceramics. *Journal of the American Ceramic Society* 50: 67–74.

Duke, D.A., Megles Jr, J.E., MacDowell, J.F., and Bopp, H.F. (1968). Strengthening glass-ceramics by application of compressive glazes. *Journal of the American Ceramic Society* 51: 98–101.

Epstein, B. (1948). Application of the theory of extreme values in fracture problems. *Journal of the American Statistical Association* 43: 403–412.

Ernsberger, F.M. (1968). Role of densification in deformation of glasses under point loading. *Journal of the American Ceramic Society* 51: 545–547.

Evans, A.G. and Wilshaw, T.R. (1976). Quasi-static solid particle damage in brittle solids—I. Observations analysis and implications. *Acta Metallurgica* 24: 939–956.

Evers, M. (1967). Plastische Verformung von Glas mit Diamanteindrücken. *Glastechische Berichte* 40: 41–43 (in German).

Fillery, S.P. and Lange, F.F. (2007). Ion-exchanged glass laminates that exhibit a threshold strength. *Journal of the American Ceramic Society* 90: 2502–2509.

Fletcher, N.H. (1969). *The Physics of Rainclouds*. Cambridge.

Frank, F.C. and Lawn, B.R. (1967). On the theory of Hertzian fracture. *Proceedings of the Royal Society of London, Series A. Mathematical and Physical Sciences* 299: 291–306.

Fuller, E.R., Lawn, B.R., and Cook, R.F. (1983). Theory of fatigue for brittle flaws originating from residual stress concentrations. *Journal of the American Ceramic Society* 66: 314–321.

Green, D.J., Lange, F.F., and James, M.R. (1983). Factors influencing residual surface stresses due to a stress-induced phase transformation. *Journal of the American Ceramic Society* 66: 623–629.

Green, D.J., Tandon, R. and Sglavo, V.M. (1999). Crack arrest and multiple cracking in glass through the use of designed residual stress profiles. *Science* 283: 1295–1297.

Griffith, A.A. (1921). The phenomena of rupture and flow in solids. *Philosophical Transactions of the Royal Society London A* 221: 163–198.

Gruninger, M.F., Lawn, B.R., Farabaugh, E.N., and Wachtman Jr, J.B. (1987). Measurement of residual stresses in coatings on brittle substrates by indentation fracture. *Journal of the American Ceramic Society* 70: 344–348.

Gruver, R.M. and Kirchner, H.P. (1973). Effect of surface damage on the strength of Al_2O_3 ceramics with compressive surface stresses. *Journal of the American Ceramic Society* 56: 21–24.

Gy, R. (2008). Ion exchange for glass strengthening. *Materials Science and Engineering: B* 149: 159–165.

Hagan, J.T. (1979a). Cone cracks around Vickers indentations in fused silica glass. *Journal of Materials Science* 14: 462–466.

Hagan, J.T. (1979b). Micromechanics of crack nucleation during indentations. *Journal of Materials Science* 14: 2975–2980.

Hagan, J.T. (1980). Shear deformation under pyramidal indentations in soda-lime glass. *Journal of Materials Science* 15: 1417–1424.

Hagan, J.T. and Swain, M.V. (1978). The origin of median and lateral cracks around plastic indents in brittle materials. *Journal of Physics D: Applied Physics* 11: 2091–2102.

Hagan, J.T. and Van Der Zwaag, S. (1984). Plastic processes in a range of soda-lime-silica glasses. *Journal of Non-crystalline Solids* 64: 249–268.

Hartley, N.E.W. and Wilshaw, T.R. (1973). Deformation and fracture of synthetic α-quartz. *Journal of Materials Science* 8: 265–278.

Huang, H., Li, X., Mu, D., and Lawn, B.R. (2021). Science and art of ductile grinding of brittle solids. *International Journal of Machine Tools and Manufacture* 161:103675.

Insley, R.H. and Barczak, V.J. (1964). Thermal conditioning of polycrystalline alumina ceramics. *Journal of the American Ceramic Society* 47: 1–4.

Jiang, L., Wang, Y., Mohagheghian, I., Li, X., Guo, X., Li, L., Dear, J.P., and Yan, Y. (2017). Effect of residual stress on the fracture of chemically strengthened thin aluminosilicate glass. *Journal of Materials Science* 52: 1405–1415.

Johnson, K.L. (1985). *Contact Mechanics.* Cambridge.

Jung, Y.G., Pajares, A., and Lawn, B.R. (2004a). Effect of oxide and nitride films on strength of silicon: a study using controlled small-scale flaws. *Journal of Materials Research* 19: 3569–3575.

Jung, Y.G., Pajares, A., Banerjee, R., and Lawn, B.R. (2004b). Strength of silicon, sapphire and glass in the subthreshold flaw region. *Acta Materialia* 52: 3459–3466.

Kalish, D., Tariyal, B.K., and Pickwick, R.O. (1977) Strength and gage length extrapolations in optical fibers. *American Ceramic Society Bulletin* 56: 491–503.

Kaur, S., Cutler, R.A. and Shetty, D.K., (2009). Short-crack fracture toughness of silicon carbide. *Journal of the American Ceramic Society* 92: 179–185.

Kaur, S., Shetty, D.K. and Cutler, R.A., 2007. R curves and crack-stability map: application to Ce-TZP/Al_2O_3. *Journal of the American Ceramic Society* 90: 3554–3558.

Kim, S., Toivola, Y., Cook, R.F., Char, K., Chu, S.H., Lee, J.K., Yoon, D.Y., and Rhee, H.W. (2004). Organosilicate spin-on glasses: I. Effect of chemical modification on mechanical properties. *Journal of the Electrochemical Society* 151: F37–F44.

Kingery, W.D., Bowen, H.K., and Uhlmann, D.R. (1976). *Introduction to Ceramics, 2nd ed.* Wiley.

Kirchner, H.P. and Gruver, R.M. (1966). Chemical strengthening of polycrystalline ceramics. *Journal of the American Ceramic Society* 49: 330–333.

Kirchner, H.P., Gruver, R.M., and Walker, R.E. (1967). Chemically strengthened, leached alumina and spinel. *Journal of the American Ceramic Society* 50: 169–173.

Kirchner, H.P., Gruver, R.M., and Walker, R.E. (1973). Strengthening of hot-pressed Al_2O_3 by quenching. *Journal of the American Ceramic Society* 56: 17–21.

Kirchner, H.P. and Ragosta, J.A. (1983). Relation of load to radial crack length for spherical indentations in hot-pressed ZnS. *Journal of the American Ceramic Society* 66: 293–296.

Kirchner, H.P. and Walker, R.E. (1975). Method of strengthening alumina ceramic by quenching with liquid medium. US Patent 3,887,524.

Kishino, J., Nishiyama, A., and Sakuma, T. (1996). Mechanical properties of sinter-forged Al_2O_3-ZrO_2 ceramics. *Journal of Materials Science* 31: 4991–4995.

Kistler, S.S. (1962). Stresses in glass produced by nonuniform exchange of monovalent ions. *Journal of the American Ceramic Society* 45: 59–68.

Kovar, D., Bennison, S.J., and Readey, M.J. (2000). Crack stability and strength variability in alumina ceramics with rising toughness-curve behavior. *Acta Materialia* 48: 565–578.

Krohn, D.A. and Cooper, A.R. (1969). Strengthening of glass fibers: I, Cladding. *Journal of the American Ceramic Society* 52: 661–664.

Kurkjian, C.R., Albarino, R.V., Krause, J.T., Vazirani, H.N., DiMarcello, F.V., Torza, S., and Schonhorn, H. (1976). Strength of 0.04–50-m lengths of coated fused silica fibers. *Applied Physics Letters* 28: 588–590.

Kurkjian, C.R., Kammlott, G.W., and Chaudhri, M.M. (1995). Indentation behavior of soda-lime silica glass, fused silica, and single-crystal quartz at liquid nitrogen temperature. *Journal of the American Ceramic Society* 78: 737–744.

Kurkjian, C.R., Krause, J.T., and Matthewson, M.J. (1989). Strength and fatigue of silica optical fibers. *Journal of Lightwave Technology* 7: 1360–1370.

Kuzmin, K.L., Zhukovskaya, E.S., Gutnikov, S.I., Pavlov, Y.V., and Lazoryak, B.I. (2016). Effects of ion exchange on the mechanical properties of basaltic glass fibers. *International Journal of Applied Glass Science* 7: 118–127.

Lange, F.F., James, M.R., and Green, D.J. (1983). Determination of residual surface stresses caused by grinding in polycrystalline Al_2O_3. *Journal of the American Ceramic Society* 66: C-16–C-17.

Lathabai, S., Rödel, J., Dabbs, T., and Lawn, B.R. (1991a). Fracture mechanics model for subthreshold indentation flaws: Part I Equilibrium fracture. *Journal of Materials Science* 26: 2157–2168.

Lathabai, S., Rödel, J., Dabbs, T., and Lawn, B.R. (1991b). Fracture mechanics model for subthreshold indentation flaws: Part II Non-equilibrium fracture. *Journal of Materials Science* 26: 2313–2321.

Lawn, B.R. (1968). Hertzian fracture in single crystals with the diamond structure. *Journal of Applied Physics* 39: 4828–4836.

Lawn, B.R. (1985) Indentation: deformation and fracture processes. In *Strength of Inorganic Glass* (Ed. Kurkjian, C.R.) pp.67–86. Plenum.

Lawn, B. (1992). Friction processes in brittle fracture. In *Fundamentals of Friction: Macroscopic and Microscopic Processes* (Eds. Singer, J.L. and Pollock, H.M.) pp. 137–165. Springer.

Lawn, B.R. (1993). *Fracture of Brittle Solids, 2nd ed.* Cambridge.

Lawn, B.R. (1998). Indentation of ceramics with spheres: a century after Hertz. *Journal of the American Ceramic Society* 81: 1977–1994.

Lawn, B.R., Borrero-Lopez, O., Huang, H., and Zhang, Y. (2021). Micromechanics of machining and wear in hard and brittle materials. *Journal of the American Ceramic Society* 104: 5–22.

Lawn, B.R. and Cook, R.F. (2012). Probing material properties with sharp indenters: a retrospective. *Journal of Materials Science* 47: 1–22.

Lawn, B.R., Dabbs, T.P., and Fairbanks, C.J. (1983). Kinetics of shear-activated indentation crack initiation in soda-lime glass. *Journal of Materials Science* 18: 2785–2797.

Lawn, B.R. and Evans, A.G. (1977). A model for crack initiation in elastic/plastic indentation fields. *Journal of Materials Science* 12: 2195–2199.

Lawn, B.R. and Fuller, E.R. (1975). Equilibrium penny-like cracks in indentation fracture. *Journal of Materials Science* 10: 2016–2024.

Lawn, B.R. and Fuller, E.R. (1984). Measurement of thin-layer surface stresses by indentation fracture. *Journal of Materials Science* 19: 4061–4067.

Lawn, B.R., Fuller, E.R., and Wiederhorn, S.M. (1976). Strength degradation of brittle surfaces: sharp indenters. *Journal of the American Ceramic Society* 59: 193–197.

Lawn, B.R., Jakus, K., and Gonzalez, A.C. (1985). Sharp vs blunt crack hypotheses in the strength of glass: a critical study using indentation flaws. *Journal of the American Ceramic Society* 68: 25–34.

Lawn, B.R., Lee, S.K., Peterson, I.M., and Wuttiphan, S. (1998). Model of strength degradation from Hertzian contact damage in tough ceramics. *Journal of the American Ceramic Society* 81: 1509–1520.

Lawn, B.R. and Marshall, D.B. (1977). Contact fracture resistance of physically and chemically tempered glass plates: a theoretical model. *Physics and Chemistry of Glasses* 18: 7–18.

Lawn, B.R. and Marshall, D.B. (1978). Indentation fracture and strength degradation in ceramics. In *Fracture Mechanics of Ceramics*, Vol 3 (Eds. Bradt, R.C., Hasselman, D.P.H., and Lange, F.F.) pp. 205–229. Plenum.

Lawn, B.R. and Marshall, D.B. (1979). Mechanisms of microcontact fracture in brittle solids. In *Studies in Archaeology: Lithic Use-Wear Analysis* (Ed. Hayden, B.) pp. 63–82. Academic Press.

Lawn, B.R., Marshall, D.B., and Chantikul, P. (1981). Mechanics of strength-degrading contact flaws in silicon. *Journal of Materials Science* 16: 1769–1775.

Lawn, B.R., Marshall, D.B., Chantikul, P., and Anstis, G.R., (1980). Indentation fracture: applications in the assessment of the strength of ceramics. *Journal of the Australian Ceramic Society* 16: 4–9.

Lawn, B.R., Marshall, D.B., and Wiederhorn, S.M. (1979). Strength degradation of glass impacted with sharp particles: II, tempered surfaces. *Journal of the American Ceramic Society* 62: 71–74.

Lawn, B.R., Wiederhorn, S.M., and Johnson, H.H. (1975). Strength degradation of brittle surfaces: Blunt indenters. *Journal of the American Ceramic Society* 58: 428–432.

Lawn, B. and Wilshaw, R. (1975). Indentation fracture: principles and applications. *Journal of Materials Science* 10: 1049–1081.

Levengood, W.C. (1958). Effect of origin flaw characteristics on glass strength. *Journal of Applied Physics* 29: 820–826.

Lee, S.K., Wuttiphan, S., and Lawn, B.R. (1997). Role of microstructure in Hertzian contact damage in silicon nitride: I, Mechanical characterization. *Journal of the American Ceramic Society* 80: 2367–2381.

Li, C.W., Lee, D.J. and Lui, S.C. (1992). R-curve behavior and strength for *in-situ* reinforced silicon nitrides with different microstructures. *Journal of the American Ceramic Society* 75: 1777–1785.

Lin, B. and Matthewson, M.J. (1996). Inert strength of subthreshold and post-threshold Vickers indentations on fused silica optical fibres. *Philosophical Magazine A* 74: 1235–1244.

Lin, B., Matthewson, M.J., and Nelson, G.J. (1991) Indentation experiments on silica optical fibers. In *Fiber Optics Reliability: Benign and Adverse Environments IV*, volume 1366. pp. 157–166. SPIE.

Liniger, E.G. and Cook R.F. (1993). A controlled flaw technique for lifetime characterization. *Journal of the American Ceramic Society* 76: 2123–2126.

Liu, P., Youngman, R.E., Jensen, L.R., Bockowski, M., and Smedskjaer, M.M. (2020). Achieving ultrahigh crack resistance in glass through humid aging. *Physical Review Materials* 4: 063606.

Macrelli, G. (2018). Chemically strengthened glass by ion exchange: Strength evaluation. *International Journal of Applied Glass Science* 9: 156–166.

Marsh, D.M. (1964a). Plastic flow and fracture of glass. *Proceedings of the Royal Society of London, Series A. Mathematical and Physical Sciences* 282: 33–43.

Marsh, D. (1964b). Plastic flow in glass. *Proceedings of the Royal Society of London, Series A. Mathematical and Physical Sciences* 279: 420–435.

Marshall, D.B. (1984). Geometrical effects in elastic/plastic indentation. *Journal of the American Ceramic Society* 67: 57–60.

Marshall, D.B., Evans, A.G., Khuri-Yakub, B.T., Tien, J.W., and King, G.S. (1983). The nature of machining damage in brittle materials. *Proceedings of the Royal Society London A* 385: 461–475.

Marshall, D.B. and Lawn, B.R. (1978). Strength degradation of thermally tempered glass plates. *Journal of the American Ceramic Society* 61: 21–27.

Marshall, D.B. and Lawn, B.R. (1979). Residual stress effects in sharp contact cracking: Part 1 Indentation fracture mechanics. *Journal of Materials Science* 14: 2001–2012.

Marshall, D.B. and Lawn, B.R. (1980). Flaw characteristics in dynamic fatigue: the influence of residual contact stresses. *Journal of the American Ceramic Society* 63: 532–536.

Marshall, D.B., Lawn, B.R., and Chantikul, P. (1979). Residual stress effects in sharp contact cracking: Part 2 Strength degradation. *Journal of Materials Science* 14: 2225–2235.

Marshall, D.B., Lawn, B.R., and Evans, A.G. (1982). Elastic/plastic indentation damage in ceramics: the lateral crack system. *Journal of the American Ceramic Society* 65: 561–566.

Marshall, D.B., Lawn, B.R., Kirchner, H.P., and Gruver, R.M. (1978). Contact-induced strength degradation of thermally treated Al_2O_3. *Journal of the American Ceramic Society* 61: 271–272.

Marshall, D.B., Cook, R.F., Padture, N.P., Oyen, M.L., Pajares, A., Bradby, J.E., Reimanis, I.E., Tandon, R., Page, T.F., Pharr, G.M., and Lawn, B.R. (2015). The compelling case for indentation as a functional exploratory and characterization tool. *Journal of the American Ceramic Society* 98: 2671–2680.

Maurer, R.D. (1975). Strength of fiber optical waveguides. *Applied Physics Letters* 27: 220–221.

Maurer, R.D. (1977). Effect of dust on glass fiber strength. *Applied Physics Letters* 30: 82–84.

Mikosza, A.G. and Lawn, B.R. (1971). Section-and-etch study of Hertzian fracture mechanics. *Journal of Applied Physics* 42: 5540–5545.

Morris, D.J. and Cook, R.F. (2004). *In situ* cube-corner indentation of soda–lime glass and fused silica. *Journal of the American Ceramic Society* 87: 1494–1501.

Morris, D.J. and Cook, R.F. (2005). Radial fracture during indentation by acute probes: I, description by an indentation wedging model. *International Journal of Fracture* 136: 237–264.

Morris, D.J., Myers, S.B., and Cook, R.F. (2004a). Indentation crack initiation in ion-exchanged aluminosilicate glass. *Journal of Materials Science* 39: 2399–2410.

Morris, D.J., Myers, S.B. and Cook, R.F. (2004b). Sharp probes of varying acuity: instrumented indentation and fracture behavior. *Journal of Materials Research* 19: 165–175.

Morris, D.J., Vodnick, A.M., and Cook, R.F. (2005). Radial fracture during indentation by acute probes: II, experimental observations of cube-corner and Vickers indentation. *International Journal of Fracture* 136: 265–284.

Mouginot, R. and Maugis, D. (1985). Fracture indentation beneath flat and spherical punches. *Journal of Materials Science* 20: 4354–4376.

Mould, R.E. and Southwick, R.D. (1959a). Strength and static fatigue of abraded glass under controlled ambient conditions: I, general concepts and apparatus. *Journal of the American Ceramic Society* 42: 542–547.

Mould, R.E. and Southwick, R.D. (1959b). Strength and static fatigue of abraded glass under controlled ambient conditions: II, effect of various abrasions and the universal fatigue curve. *Journal of the American Ceramic Society* 42: 582–592.

Murakami, Y. (Ed.) (1987) *Stress Intensity Factors Handbook*. Pergamon.

Nadeau, J.S. (1973). Hertzian fracture of vitreous carbon. *Journal of the American Ceramic Society* 56: 467–472.

Nordberg, M.E., Mochel, E.L., Garfinkel, H.M., and Olcott, J.S. (1964). Strengthening by ion exchange. *Journal of the American Ceramic Society* 47: 215–219.

Odén, M. (1992). *Residual Stress in Ceramics and Ceramic Composites*. Licentiate Thesis. Linköping University, Linköping, Sweden.

Oel, H.J. and Fréchette, V.D. (1967). Stress distribution in multiphase systems: I, composites with planar interfaces. *Journal of the American Ceramic Society* 50: 542–549.

Pajares, A., Chumakov, M., and Lawn, B.R. (2004). Strength of silicon containing nanoscale flaws. *Journal of Materials Research* 19: 657–660.

Palmqvist, S. (1957). A method to determine the toughness of brittle materials, especially hard metals. *Jernkontorets Annaler* 141: 300–307 (in Swedish).

Peter, K. (1964). Sprödbruch und Mikroplastizitaät von Glas in Eindruckversuchen. *Glastechische Berichte* 37: 333–344 (in German).

Petrovic, J.J., Dirks, R.A., Jacobson, L.A., and Mendiratta, M.G. (1976). Effects of residual stresses on fracture from controlled surface flaws. *Journal of the American Ceramic Society* 59: 177–178.

Petrovic, J.J. and Jacobson, L.A. (1976). Controlled surface flaws in hot-pressed SiC. *Journal of the American Ceramic Society* 59: 34–37.

Petrovic, J.J., Jacobson, L.A., Talty, P.K., and Vasudevan, A.K. (1975). Controlled surface flaws in hot-pressed Si_3N_4. *Journal of the American Ceramic Society* 58: 113–116.

Petrovic, J.J. and Mendiratta, M.G. (1979). Fracture from controlled surface flaws. In *Fracture Mechanics Applied to Brittle Materials*. pp. 83–102. ASTM International.

Petrovic, J.J. and Roof, R.B. (1984). Fracture toughness of a Beta-SiC whisker. *Journal of the American Ceramic Society* 67: C-219–C-220.

Rado, P. (1969). *An Introduction to the Technology of Pottery*. Pergamon.

Ram, M. and Tsur, Y. (2009). Eliminating chemical effects from thermal expansion coefficient measurements. *Journal of Electroceramics* 22: 120–124.

Ramachandran, N. and Shetty, D.K. (1991). Rising crack-growth-resistance (R-curve) behavior of toughened alumina and silicon nitride. *Journal of the American Ceramic Society* 74: 2634–2641.

Rao, M.P., Sánchez-Herencia, A.J., Beltz, G.E., McMeeking, R.M., and Lange, F.F. (1999). Laminar ceramics that exhibit a threshold strength. *Science* 286: 102–105.

Readey, M.J. and McCallen, C.L. (1995). Microstructure, flaw tolerance, and reliability of Ce-TZP and Y-TZP ceramics. *Journal of the American Ceramic Society* 78: 2769–2776.

Rhee, Y.W., Kim, H.W., Deng, Y., and Lawn, B.R. (2001). Contact-induced damage in ceramic coatings on compliant substrates: fracture mechanics and design. *Journal of the American Ceramic Society* 84: 1066–1072.

Ritter, J.E. (1985). Erosion damage in structural ceramics. *Materials Science and Engineering* 71: 195–201.

Ritter, J.E., Strzepa, P., Jakus, K., Rosenfeld, L., and Buckman, K.J. (1984). Erosion damage in glass and alumina. *Journal of the American Ceramic Society* 67: 769–774.

Roesler, F.C. (1956). Brittle fractures near equilibrium. *Proceedings of the Physical Society* B 69: 981–992.

Salem, J. (2023). Fracture mechanics properties of fused silicas used for international space station windowpanes. *International Journal of Applied Ceramic Technology* 20: 938–950.

Scott, W.D. and Gaddipati, A. (1978). Weibull parameters and the strength of long glass fibers. In *Fracture Mechanics of Ceramics: Volume 3, Flaws and Testing* (Eds. Bradt, R.C., Hasselman, D.P.H., and Lange, F.F.) pp. 125–142. Springer.

Shand, E.B. (1959). Breaking stress of glass determined from dimensions of fracture mirrors. *Journal of the American Ceramic Society* 42: 474–477.

Shand, E.B. (1965). Strength of glass–the Griffith method revised. *Journal of the American Ceramic Society* 48: 43–49.

Snowden, W.E. (1978). Surface flaws and the mechanical behavior of glass optical fibers. In *Fracture Mechanics of Ceramics: Volume 3, Flaws and Testing* (Eds. Bradt, R.C., Hasselman, D.P.H., and Lange, F.F.) pp. 143–159. Springer.

Swain, M.V. and Hagan, J.T. (1976). Indentation plasticity and the ensuing fracture of glass. *Journal of Physics D: Applied Physics* 9: 2201–2214.

Swain, M.V. and Lawn, B.R. (1976). Indentation fracture in brittle rocks and glasses. *International Journal of Rock Mechanics and Mining Sciences & Geomechanics Abstracts* 13: 311–319.

Talimian, A., Mariotto, G., and Sglavo, V.M. (2017). Electric field-assisted ion exchange strengthening of borosilicate and soda lime silicate glass. *International Journal of Applied Glass Science* 8: 291–300.

Tandon, R. and Buchheit, T.E. (2007). Use of cube-corner nano-indentation crack length measurements to estimate residual stresses over small spatial dimensions. *Journal of the American Ceramic Society* 90: 502–508.

Tandon, R. and Cook, R.F. (1992). Cone crack nucleation at sharp contacts. *Journal of the American Ceramic Society* 75: 2877–2880.

Tandon, R. and Cook, R.F. (1993). Indentation crack initiation and propagation in tempered glass. *Journal of the American Ceramic Society* 76: 885–889.

Tandon, R. and Glass, S.J. (2015). Fracture initiation and fragmentation in chemically tempered glass. *Journal of the European Ceramic Society* 35: 285–295.

Tandon, R. and Green, D.J. (1990). Indentation behavior of ion-exchanged glasses. *Journal of the American Ceramic Society* 73: 970–977.

Tariyal, B.K. and Kalish, D. (1978). Mechanical behavior of optical fibers. In *Fracture Mechanics of Ceramics: Volume 3, Flaws and Testing* (Eds. Bradt, R.C., Hasselman, D.P.H., and Lange, F.F.) pp. 161–175. Springer.

Terao, N. (1978). Experimental study of crack propagation in brittle amorphous solids: Bifurcation of a crack in glass. *physica status solidi A* 47: 565–575.

Thurn, J. and Cook, R.F. (2004). Mechanical and thermal properties of physical vapour deposited alumina films: Part II Elastic, plastic, fracture, and adhesive behaviour. *Journal of Materials Science* 39: 4809–4819.

Tsai, Y.L. and Mecholsky, J.J. (1991). Fractal fracture of single crystal silicon. *Journal of Materials Research* 6: 1248–1263.

van der Zwaag, S. and Field, J.E. (1982). Liquid jet impact damage on zinc sulphide. *Journal of Materials Science* 17: 2625–2636.

van der Zwaag, S. and Field, J.E. (1983). Rain erosion damage in brittle materials. *Engineering Fracture Mechanics* 17: 367–379.

Warshaw, S.I. (1957). Pre-stressed ceramics. *American Ceramic Society Bulletin* 36: 28–30.

Weibull, W. (1939). A statistical theory of the strength of materials. *Proceedings of the Royal Swedish Institute of Engineering Research* 151: 1–45.

Widjaja, S., Ritter, J.E., and Jakus, K. (1996). Influence of R-curve behaviour on strength degradation due to Hertzian indentation. *Journal of Materials Science* 31: 2379–2384.

Wiederhorn, S.M. and Lawn, B.R. (1977). Strength degradation of glass resulting from impact with spheres. *Journal of the American Ceramic Society* 60: 451–458.

Wiederhorn, S.M. and Lawn, B.R. (1979). Strength degradation of glass impacted with sharp particles: I, annealed surfaces. *Journal of the American Ceramic Society* 62: 66–70.

Williams, J.S., Lawn, B.R., and Swain, M.V. (1970). Cone crack closure in brittle solids. *Physica Status Solidi A* 2: 7–29.

Wilshaw, T.R. (1971). The Hertzian fracture test. *Journal of Physics D: Applied Physics* 4: 1567–1581.

Yasutake, K., Iwata, M., Yoshii, K., Umeno, M., and Kawabe, H. (1986). Crack healing and fracture strength of silicon crystals. *Journal of Materials Science* 21: 2185–2192.

Yoshida, S., Kato, M., Yokota, A., Sasaki, S., Yamada, A., Matsuoka, J., Soga, N., and Kurkjian, C.R. (2015). Direct observation of indentation deformation and cracking of silicate glasses. *Journal of Materials Research* 30: 2291–2299.

Zeng, K., Breder, K., and Rowcliffe, D.J. (1992). The Hertzian stress field and formation of cone cracks—II. Determination of fracture toughness. *Acta Metallurgica et Materialia* 40: 2601–2605.

Zhang, Y. and Lawn, B.R. (2005). Fatigue sensitivity of Y-TZP to microscale sharp-contact flaws. *Journal of Biomedical Materials Research Part B: Applied Biomaterials* 72: 388–392.

Zhao, J., Stearns, L.C., Harmer, M.P., Chan, H.M., Miller, G.A., and Cook, R.F. (1993). Mechanical behavior of alumina–silicon carbide “nanocomposites”. *Journal of the American Ceramic Society* 76: 503–510.

4

Indentation Strengths: Toughened Materials

This chapter extends the analyses of Chapters 1 and 3 in a survey of Vickers indentation strengths of toughened brittle materials. The materials are primarily large-grained polycrystals and composites that exhibit significant effects of microstructure on fracture such that the materials exhibit increasing toughness with crack length. The physical bases of crack wake toughening predominant in polycrystals and crack front toughening in phase transforming materials are described. A fracture mechanics based model of indentation strength in polycrystals is developed. Indentation strength data are presented in the format of Chapter 3. As in Chapter 3, at large indentation loads most of the materials exhibit strengths well above the ideal response reflecting lateral crack induced indentation stress relaxation effects. At small indentation loads most of the materials exhibit strengths well below the ideal response, reflecting diminished toughness, such that in many cases the strengths approach an upper-bound microstructurally controlled invariant strength. The materials considered include alumina, silicon nitride, glass ceramics, barium titanate, and zirconia.

4.1 Introduction

Development and refinement of microstructures in brittle materials to improve material mechanical behavior in application have been known since ancient times. Straw was added to mudbricks by Egyptian and Israelite workers as early as 4000 years before the common era (BCE) (Emery 2009; Littman et al. 2014) and continues worldwide to this day. Fine and coarse rocky and stony aggregates were added to concrete and much attention given to the composition of the mortar matrix by Roman workers from at least 100 BCE, greatly influencing trade in the Mediterranean at that time (Hohlfelder and Oleson 2014; Seymour et al. 2023) and introducing fabrication practices that continue with modern concrete (Mehta and Monteira 2014). Building on a long history of ceramic technology and through careful control of raw ingredients and firing cycles, Chinese workers developed the first glass-ceramic materials (unbeknownst to them) in the form of hard porcelain from about 1000 CE (Rado 1969; Carty and Senapati 1998). The material was not really "improved" until the 1900s by the microstructural substitution of alumina particles for quartz particles, forming high-strength porcelains suitable for use as industrial whitewares (Austin et al. 1946). This chapter applies indentation strength techniques to assess the effects of microstructure on more recently fabricated brittle materials, polycrystalline ceramics in particular. The microstructure-controlled variations in toughness with crack length underlying indentation strength responses are assessed. The materials emphasis is on various forms of polycrystalline alumina, a material that is used widely in refractory, electrical, and optical applications and that is amenable to broad microstructural manipulations of grain size, glassy content, and second phase addition. Other "modern day" ceramics, such as silicon nitride, barium titanate, and zirconia are also considered.

The chapter begins with a description of toughening in brittle materials and connection to the analyses of thermodynamics and mechanics of fracture established in Chapter 1. The analysis here starts with a physically based description of toughening in the energy framework of fracture and moves to a description in the stress framework that is more suitable for analysis of indentation strength behavior. The most pervasive toughening mechanism in brittle materials is then considered: crack wake toughening in polycrystalline ceramics. An indentation strength analysis is then developed for such toughened materials. The analysis follows the development of Chapter 3 in that the effects of applied stress on both crack extension and component failure are considered and use is made of scaled variables to illustrate physical phenomena. The analysis is then used in an extensive survey of the effects of microstructure on the indentation strength behavior of toughened materials. The survey constitutes the bulk of the chapter and is presented predominately in the form of composite plots of indentation strength responses of families of materials. Illustrative examples of particular responses in raw form

Indentation Fracture: Strength and Toughness for Brittle Materials Design, First Edition. Robert F. Cook.

are also included. The survey draws heavily from published sources, although significant unpublished data are included and the majority of data were reanalyzed for this book.

4.2 Microstructural Toughening in Brittle Materials

4.2.1 Thermodynamics and the Energy Framework

The fundamental physical principle underlying toughening in brittle materials is the action of irreversible processes associated with crack extension. In components of such materials, the work performed on a cracked component by applied loading and the elastic energy subsequently stored within the component are at least partially converted on fracture into irreversible entropy. The "downgraded" work and elastic energy are therefore not available to rupture interatomic bonds and expand the crack area. In the frame of reference of the applied loading, the material thus appears tougher. In the frame of reference of the crack tip, the loading appears weaker. For a material to exhibit perceivable toughening, irreversible entropy production must therefore occur in a volume greater than that associated with the crack tip bond breaking process. Hence, material microstructures that lead to irreversible processes on crack extension at scales comparable to the crack length lead to toughening. The effects of such microstructural toughening on the indentation strength of brittle materials are the focus of this chapter.

Analysis of the microstructure-toughening linkage begins with the combined statement of the First and Second Laws of Thermodynamics given in Eq. (1.52). The statement is a full expression of the energy changes that can occur during fracture as it allows for irreversible processes. Hence, a general expression for fracture resistance R taking reversible and irreversible processes into account is given by

$$R = \frac{dU_S}{dA} - \theta\frac{dS_{rev}}{dA} + \theta\frac{dS_{irr}}{dA}. \tag{4.1}$$

Here, U_S is the energy required on fracture to create new surface in the material, A is the fracture surface area, θ is the temperature, and S is the entropy generated on fracture, divided into reversible (rev) and irreversible (irr) components. The expression of Eq. (1.59) for reversible fracture resistance comprises the first two terms on the right of Eq. (4.1). As before, an intuitive form is used for these two terms as a simple lumped proportionality to A: $U_S - \theta S_{rev} = R_0 A$. Here, $R_0 = R_0(\theta)$ is a base or lower-bound reversible fracture resistance that is temperature dependent. The third term on the right of Eq. (4.1) is an example of an expression for dissipation in systems undergoing irreversible deformation and is usually expressed as a time derivative (Mase 1970; Khonsari and Amiri 2013) rather than a spatial derivative as here. Following the usual approach to such dissipation, the irreversible component of Eq. (4.1) is written as an irreversible deformation integral

$$R_{irr} = \frac{d\theta S_{irr}}{dA} = \frac{d}{dA}\left[\iint \sigma_{ij} d\varepsilon_{ij}^{irr} dV_{irr}\right]. \tag{4.2}$$

The term V_{irr} is the volume of material in which irreversible deformation occurs and in which entropy is generated irreversibly. The terms σ_{ij} and ε_{ij}^{irr} are the local stress and *irreversible* strain tensors within this volume, and the material constitutive response, $\sigma_{ij}(\varepsilon_{ij}^{irr})$, is hysteretic. All three terms are temperature and crack area dependent, such that Eq. (4.1) becomes

$$R(A) = R_0 + R_{irr}(A). \tag{4.3}$$

Considerable simplification is possible if the crack area is very large such that a steady-state zone of deformation develops around the crack tip and travels with fixed conformation through the material as the crack extends. In this case, the integral over dV_{irr} devolves to simple proportionality to the crack area, $V_{irr} = 2Aw$. w is a characteristic fixed dimension on either side of the crack plane in which irreversible deformation has occurred. In steady state, the integral over $\sigma_{ij} d\varepsilon_{ij}^{irr}$ reaches an invariant maximum of dissipated energy density and devolves to an invariant value $\mathcal{U}_{irr}$ characterizing the full traversal of the hysteretic irreversible deformation path. With these simplifications, the second term on the right of Eq. (4.3) thus reaches an asymptotic value ΔR characterizing the maximum fracture resistance increase,

$$R_{irr}(A) \rightarrow \Delta R = w\mathcal{U}_{irr}, \tag{4.4}$$

and the total fracture resistance of Eq. (4.3) is maximized at R_∞ and given by

$$R(A) \rightarrow R_\infty = R_0 + \Delta R. \tag{4.5}$$

The physical meaning of ΔR and Eq. (4.4) as based on Eq. (4.2) can be illustrated with a few examples of microstructural toughening, of which there are several pictorial compilations (e.g., Lawn 1993; Ritchie and Liu 2021). In whisker composites, the characteristic peak stress σ is the frictional stress τ between whisker and matrix as the whisker is separated from the matrix by crack opening displacement. The characteristic peak strain in terms of whisker length L is thus $\varepsilon \sim L/L \approx 1$. The characteristic zone width is a whisker length $w \sim L$. The peak fracture resistance increment thus scales as $\Delta R \sim \tau L$. In long fiber composites, L and thus ΔR are large, although reaching a steady-state configuration may be difficult if the fibers are strong and comparable in length to the size of the component. In ductile particle composites, the characteristic peak stress is the ultimate tensile strength of the particle, σ_{UTS}, and the characteristic peak strain is the failure strain of the particle, ε_{f}. The characteristic zone width scales as $w \sim R_0 E/\sigma_{\mathrm{UTS}}^2$ (E is the elastic modulus of the matrix, see Chapter 1). The peak fracture resistance increment thus scales as $\Delta R \sim R_0 E\varepsilon_{\mathrm{f}}/\sigma_{\mathrm{UTS}}$. In phase-transforming materials, the characteristic peak stress is the hydrostatic transformation stress σ_{T} and the characteristic strain is the dilatational transformation volume strain, ε_{T}. The characteristic zone width scales as $w \sim R_0 E/\sigma_{\mathrm{T}}^2$ such that the peak fracture resistance increment scales as $\Delta R \sim R_0 E\varepsilon_{\mathrm{T}}/\sigma_{\mathrm{T}}$.

It is emphasized that while providing physical insight these scaling relations omit the geometrical details required to arrive at quantitative expressions of fracture resistance increases. Such details, even within the limits of the scalar formulations given, include the effects of whisker and particle shape, alignment, and volume fraction as well as the effects of matrix constraints on sliding, deforming, or transforming whiskers or particles (Lawn 1993). In addition, it is exactly these geometrical details and others that enable the area dependence of the fracture resistance, $R(A)$ of Eq. (4.3), to be predicted. That is, the scaling relations provide no information about how the fracture resistance increases with crack area between the invariant bounds of R_0 and R_∞ of Eq. (4.5). In experiments, crack area is usually characterized by a dimension along a single direction of crack extension, the crack length c. Hence, what is lacking is a clear prescription of $R(c)$, the variation of fracture resistance with crack length. $R(c)$ is known as the *R curve* of a material. Materials that exhibit microstructural-related toughening and increases in fracture resistance with crack length are commonly referred to as exhibiting "*R* curve behavior." Such a behavior can be described and analyzed within the energy-based framework of fracture by comparing the mechanical energy release rate with the fracture resistance. For example, the fracture equilibrium statement of Eq. (1.19) is thus extended to $\mathcal{G}(c) = R(c)$, in which both terms have crack length dependence. The equality is met for equilibrium crack lengths. For simple fracture systems in which there is a single source of applied loading characterized by a monotonic $\mathcal{G}(c)$ function, e.g., Griffith or Roesler cracks, fracture behavior is analyzed easily and $R(c)$ can be deconvoluted from experimental measurements in a straightforward manner (as the equality is met by a single crack length).

4.2.2 Mechanics and the Stress Framework

For many fracture systems, there are multiple sources of applied loading and the overall loading is characterized by a nonmonotonic crack length dependence. Indentation strength measurements are an example of such systems. In these systems, a description of fracture within the stress-based framework is preferable. Copious examples are given in Chapters 1–3 that demonstrate the description and analysis of diverse experimental measurements within the stress framework. The quantities within this framework analogous to $\mathcal{G}(c)$ and $R(c)$ are the stress-intensity factor (SIF) $K(c)$ and toughness $T(c)$, respectively. Hence, a crack length dependent toughness $T(c)$ represents the physical effects on fracture of irreversible processes and energy dissipation as entropy. The algebraic correspondence of $T(c)$ with $R(c)$, however, requires some attention as the Irwin relation, Eqs. (1.29) and (1.32), used to connect the two frameworks is quadratic. The steady-state long crack toughness of a material T_∞ may be expressed in terms of the base toughness T_0 and a maximum toughening increment ΔT as

$$T_\infty = T_0 + \Delta T, \tag{4.6}$$

following the ideas of Section 3.3.5. Applying the Irwin relation to both sides of this equation gives

$$\begin{aligned} \frac{T_\infty^2}{E} &= \frac{(T_0 + \Delta T)^2}{E} \\ R_\infty &= R_0 + \Delta T(\Delta T + 2T_0)/E. \end{aligned} \tag{4.7}$$

Comparison with Eq. (4.5) shows that $\Delta R \neq \Delta T^2/E$, reflecting the cross-term effects of Eq. (1.31). Hence, although ΔT of Eq. (4.6) represents the maximum toughening in the fracture system arising from irreversible processes, care should be taken in direct interpretation in terms of energy.

Following Eq. (4.3), a variable toughness of a material $T(c)$ may be expressed in terms of the base toughness T_0 and a crack length- and microstructure-dependent toughness increase $T_\mu(c)$ as

$$T(c) = T_0 + T_\mu(c). \tag{4.8}$$

$T_\mu(c)$ represents energy dissipation and irreversible processes associated with fracture as does $R_{irr}(A)$. The terminology and concepts applied in the energy framework of fracture apply in the stress framework. $T(c)$ is known as the *T curve* of a material, and materials that exhibit microstructural-related toughening with increases in crack length are referred to as exhibiting "*T* curve behavior." Such a behavior is described and analyzed within the stress-based framework of fracture by comparing the SIF with the toughness. The fracture equilibrium statement of Eq. (1.33) is thus extended to $K(c) = T(c)$, in which both terms have crack length dependence. The equality is met for equilibrium crack lengths. For simple fracture systems with a single, monotonic, source of applied loading, fracture behavior is analyzed easily and $T(c)$ can be deconvoluted simply from experimental measurements. [The caveat noted previously applies regarding exact numerical identification of $T_\mu^2(c)/E$ with $R_{irr}(A)$, as shown in Eq. (4.7).]

The advantage of considering toughening effects within the stress framework of fracture becomes apparent by extending crack length dependence to the fracture equilibrium. statement of Eq. (1.33) as

$$K(c) = T_0 + T_\mu(c). \tag{4.9}$$

If the toughness term $T_\mu(c)$ is redefined as a SIF term $K_\mu(c)$ by

$$K_\mu(c) = -T_\mu(c), \tag{4.10}$$

then Eq. (4.9) can be rewritten as

$$K(c) + K_\mu(c) = T_0. \tag{4.11}$$

There are several points to be made regarding this procedure. First, if $T_\mu(c) > 0$ and thus represents processes that toughen the material, then $K_\mu(c) < 0$ and thus represents factors that impede crack extension. Second, Eq. (4.9) represents fracture equilibrium as observed in the frame of reference of the applied loading (the material is tougher), and Eq. (4.11) represents fracture equilibrium as observed in the frame of reference of the crack tip (the loading is weaker). The applied loading reference frame is better suited to considerations of engineering design, and the crack tip reference frame is better suited to considerations of bond rupture. Third, an implication of Eq. (4.11) is that $K_\mu(c)$ is expressible similarly to other SIF as a function of crack length and calculable from forces and dimensions associated with the microstructure. The procedure is similar to that of Eq. (4.2) in that an irreversible dissipation function has been transformed into a mechanics function. Fourth, this transformation imposes a strict constraint on Eqs. (4.9)–(4.11) in that these equations can only be used irreversibly—they are not true reversible equilibrium equations. Hence, only one direction of crack motion can be studied using Eq. (4.11). As strength analyses usually only consider forward crack propagation this is not a great constraint here, but does apply to interpretation of cyclic fatigue measurements (Dauskardt et al. 1987, 1990; Lathabai et al. 1989, 1991; Khonsari and Amiri 2013).

Analysis of indentation strength measurements of materials in which there are microstructural toughening influences thus devolves to specification of a SIF function $K_\mu(c)$. By far the most prevalent of these functions are those based on the ubiquitous crack wake effects exhibited during crack propagation in polycrystalline brittle materials. Such wake effects lead to dissipation of energy *behind* a crack tip by mechanical processes imposed in the microstructure by the advancing crack. The processes are commonly referred to as "ligamentary bridging" as elements of interlocked and unbroken material behind the crack tip act as ligaments to restrain crack opening and crack advance. Crack wake effects are ubiquitous in fracture of polycrystalline materials as ligament formation arises from the inherent granular nature of the materials and the typically weaker fracture resistance of grain boundaries relative to that of grains. The most frequently studied material in this regard is polycrystalline alumina (Al_2O_3). Section 4.2.3 gives a brief overview of the phenomena observed in crack wake toughening. The overview is the basis for the subsequent development of a $K_\mu(c)$ function. The function is implemented simply and easily in indentation strength analyses.

4.2.3 Crack Wake Toughening: Overview

Figure 4.1 shows optical images of the as-fired surfaces of two polycrystalline Al_2O_3 materials. The materials were used in a previous study of microstructural effects on fracture (Cook 1990) and were sintered at 1800 °C for (a) 8 h and

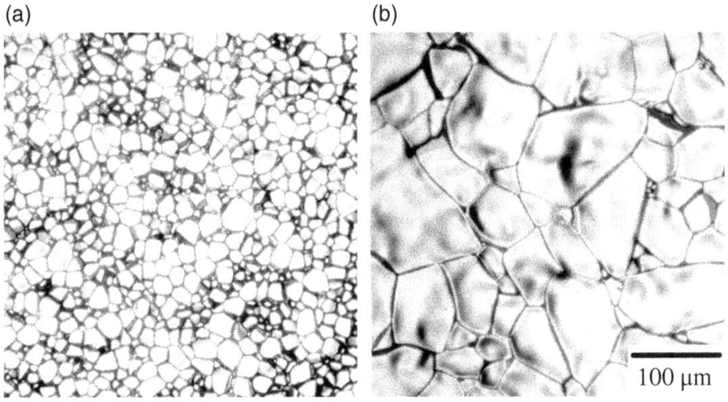

Figure 4.1 Optical images of surfaces of as-fired microstructures of polycrystalline alumina (Al_2O_3). (a) 8 h at 1800 °C. (b) 24 h at 1800 °C.

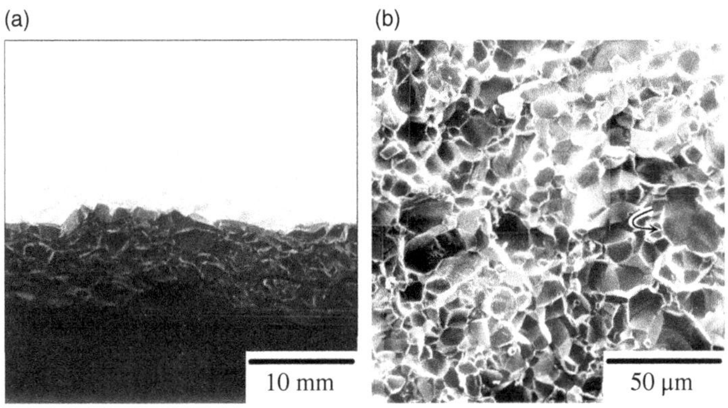

Figure 4.2 Scanning electron microscope images of (a) fracture path and (b) fracture surface of polycrystalline alumina similar to Figure 4.1. Note the microstructural perturbation of fracture from planar and the mix of intergranular and transgranular facets.

(b) 24 h. Grain boundary grooving outlines the grains on the sintered surfaces and the mean grain sizes of the two materials are clearly different; approximately (a) 15 μm and (b) 55 μm, although there is considerable grain size dispersion apparent in both microstructures. The fracture properties that result from such scale changes and dispersion are a major focus of indentation strength studies that examine microstructural effects. The qualitative effects of microstructure on fracture are shown in Figure 4.2 as example scanning electron microscope images of fractured components of Al_2O_3 materials similar to those in Figure 4.1. Figure 4.2a shows an oblique view of a sintered surface and a perpendicular fracture surface. The intersection of the two surfaces in the center of the image is visible along the predominantly intergranular zigzag line and makes clear the microstructural perturbation of the fracture path. Greater detail is shown in a view perpendicular to a fracture surface in Figure 4.2b. The surface is clearly perturbed by the granularity of the microstructure and although predominantly intergranular there are significant large transgranular segments as indicated by the arrow. Images similar to those in Figure 4.2 were presented in some of the early works considering the effects of microstructure on fracture (Kingery et al. 1976; Swain 1986).

Several strands of inquiry lead to quantitative descriptions of the effects of microstructure on fracture. In an early study, Hübner and Jillek (1977) measured the *T* curve of polycrystalline Al_2O_3 using a notched beam geometry. Clear perturbations from straight crack paths by the polycrystalline microstructure were observed, including discontinuous surface traces. Based on the angular nature of the crack propagation, Hübner and Jillek suggested that "interference between the fracture surfaces is expected to lead to additional energy consumption [on fracture] due to friction." In a definitive two-step work, Knehans and Steinbrech (1982) showed that the *T* curve in polycrystalline Al_2O_3 arose from processes behind the crack tip. In a first step, Knehans and Steinbrech (1982) and Steinbrech et al. (1983) showed clear *T* curves in polycrystalline Al_2O_3 using a notched beam geometry. In a second step, Knehans and Steinbrech renotched the wake zones of propagated cracks in the Al_2O_3 specimens and showed that the toughness of the material reverted to the base value. On renewed crack propagation, the *T* curve was redeveloped. Knehans and Steinbrech suggested that the *T* curve might arise from "friction of serrated crack walls." Subsequent similar renotching experiments in a variety of test geometries and polycrystalline materials have confirmed the crack wake toughening inference [Deuerler et al. (1986), double cantilever beam (DCB) Al_2O_3; Sakai et al. (1988), compact tension (CT) polycrystalline graphite; Grimes et al. (1990), notched beam polycrystalline Al_2O_3; Hu and Wittmann (1990), CT cement mortar; White and Kelkar (1991), DCB polycrystalline $MgAl_2O_4$].

Further progress in understanding microstructure-induced crack wake toughening was enabled by strength tests rather than toughness tests per se. In an extensive observational study using indented polycrystalline Al_2O_3 specimens, Swanson et al. (1987) showed that, on crack propagation, load bearing, microstructural scale, crack face tractions formed, developed, and failed in a wake zone adjacent to but behind the crack tip. The tractions from microstructural features were of two forms: (i) frictional interlocks between crack face segments that had components perpendicular to the main crack propagation direction and (ii) unbroken ligaments of material extending across the main crack. A subsequent fracture mechanics model developed by Mai and Lawn (1987) showed that such discrete features, modeled as strain softening traction elements, could describe the *T* curves observed. Soon after Cook et al. (1987) applied the idea of the strain softening traction elements to describe the indentation strength behavior of an extensive range of materials, including many polycrystalline Al_2O_3, $BaTiO_3$, and glass-ceramic microstructures. Earlier works had demonstrated that many, if not most, polycrystalline materials exhibited indentation strength behavior that was described by an underlying *T* curve (Cook et al. 1985a), and that the *T* curve was environment-independent (Cook et al. 1985b), in agreement with the earlier assessment of Knehans and Steinbrech (1982). In another form of two-step experiment, Hay and White (1993, 1995) measured the force–displacement constitutive behavior of traction elements formed in the wake zone. The first step was as before in that cracks were propagated and halted in loaded then unloaded polycrystalline Al_2O_3 specimens in double-notched beam and DCB geometries. In the second step, the specimens were reloaded, perpendicular to the original crack propagation direction. In the case of the notched beam specimens by switching from bending to tensile loading and in the case of DCB specimens by sawing tensile bars from the cracked DCB plates. Hay and White showed that the constitutive response exhibited an elastic response followed by a clear maximum and a longer decreasing tail, consistent with behavior inferred from observations and indentation strength tests. A similar response was later demonstrated for cement mortar (Dasgupta et al. 1998).

The observations above are summarized in Figures 4.3–4.6. The figures provide a platform for development of the $K_\mu(c)$ function required for indentation strength analysis. The summary begins with a system that has very little grain size inhomogeneity. Figure 4.3a shows a schematic cross section of a polycrystalline microstructure similar to that of Figure 4.1a. The grain boundaries are indicated by fine black lines, and all grains have similar shapes and sizes. The microstructure is part of a component that is placed in tension, indicated by the dark arrows, such that a crack propagates through the microstructure along the weak grain boundaries, left to right, indicated by the bold black line. In this material, the

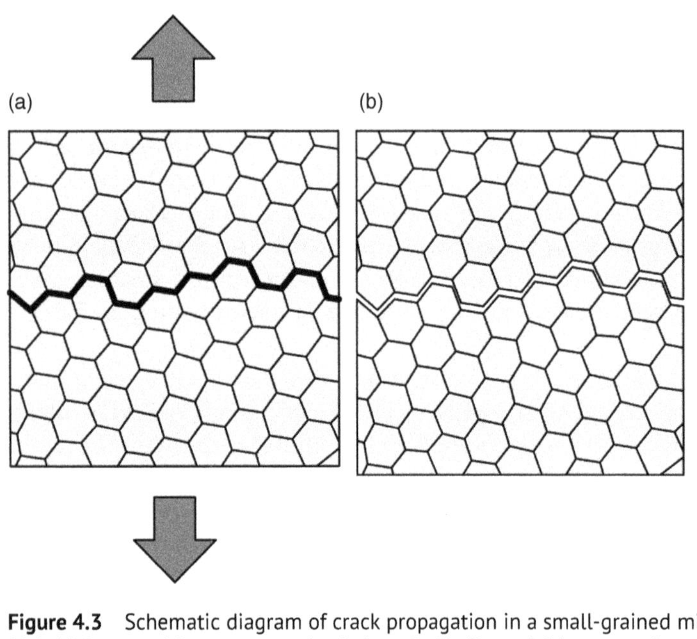

Figure 4.3 Schematic diagram of crack propagation in a small-grained microstructure. Grain boundaries indicated by fine lines. (a) Initial state with component loaded vertically. Potential intergranular crack path outlined in bold. (b) Final state with component fully separated along weakly perturbed fracture surface.

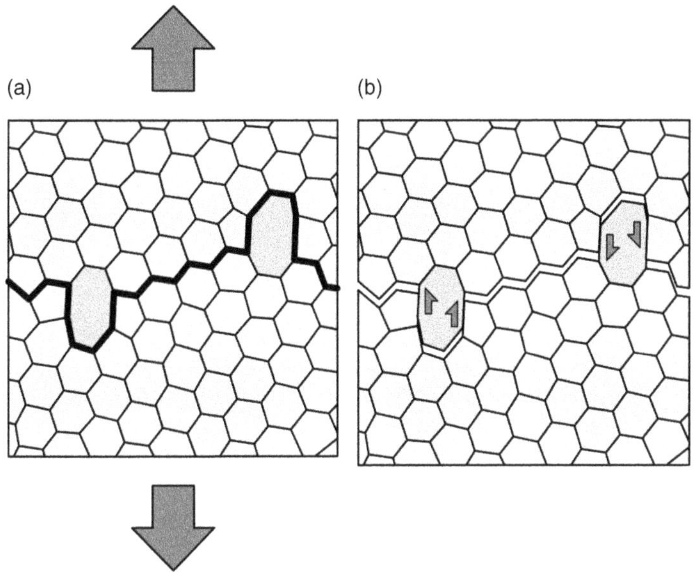

Figure 4.4 Schematic diagram of crack propagation in a microstructure containing a few large grains (shaded). Grain boundaries indicated by fine lines. (a) Initial state with component loaded vertically. Potential intergranular crack path outlined in bold and circumnavigating large grains. (b) Final state with component partially separated along perturbed fracture surface that includes frictional interlocks (arrows) at large grains.

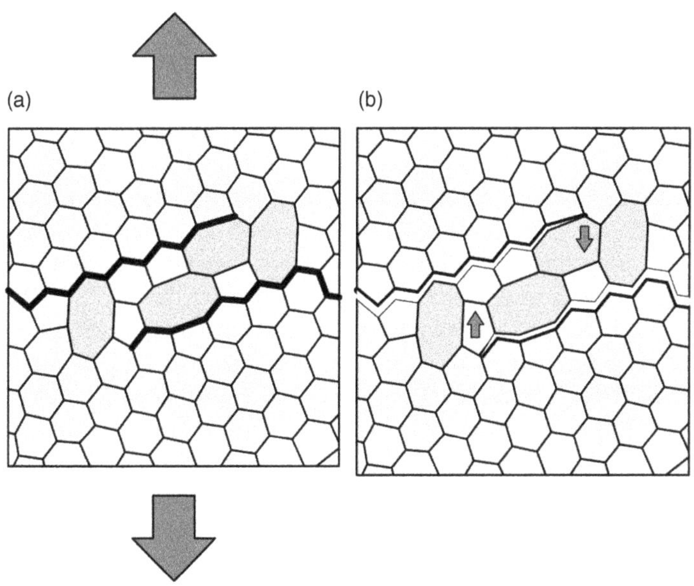

Figure 4.5 Schematic diagram of crack propagation in a microstructure containing a cluster of large grains (shaded). Grain boundaries indicated by fine lines. (a) Initial state with component loaded vertically. Potential intergranular crack path outlined in bold and avoiding large grain cluster. (b) Final state with component partially separated along perturbed fracture surface that includes a ligamentary bridge (arrowed) at large grains. Note that this is a two-dimensional section—the crack path is continuous in three dimensions.

angular and spatial deviations in the crack path from the overall crack propagation direction and extension are small. As a consequence, the material parts easily on the crack surface to generate two fracture surfaces and two fragments, Figure 4.3b. Apart from the slight undulation of the fracture plane and the obvious decrease in fracture resistance of the grain boundaries relative to the crystalline grains, fracture such as this does not exhibit significant microstructural effects.

In contrast, systems that include some grain size inhomogeneity exhibit quite different behavior. Figure 4.4a shows a schematic cross section of a polycrystalline microstructure that includes aspects of Figure 4.1b. In this microstructure,

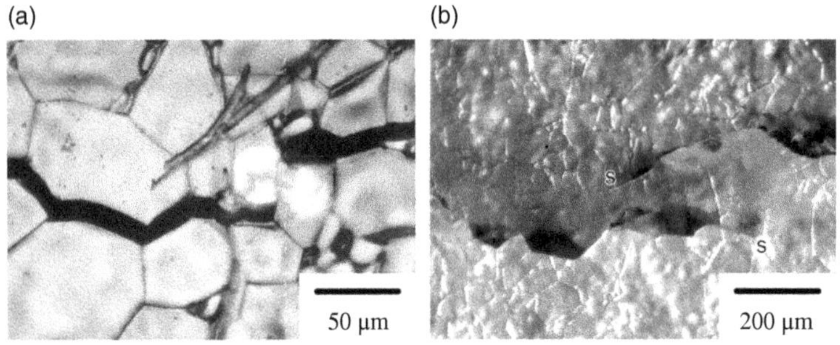

Figure 4.6 Optical images of crack propagation in polycrystalline alumina illustrating the two major forms of crack wake bridging. Crack propagation left to right. (a) Frictional interlock at intergranular fracture path perpendicular to major crack propagation direction (see Figure 4.4). (b) Bridging ligament between intergranular fracture paths separated by intact large grains (see Figure 4.5).

some of the grains are larger and have elongated sections relative to the surrounding matrix of smaller equiaxed grains. As before, the microstructure is part of a component that is placed in tension, shown by the gray arrows, and a crack propagates through the microstructure along the weak grain boundaries, left to right, indicated by the bold black line. In this material, the angular and spatial deviations in the crack path are not small relative to the overall propagation direction and extension. In fact, the large grains in the crack path lead to crack propagation locally perpendicular to the overall propagation direction and deviations in the crack path three to four times the usual grain facet distance. As a consequence, this material does not part easily on the crack surface. The large grains act as frictional interlocks that form localized regions of compressive force across the crack surface and oppose crack opening and propagation. These forces originate from frictional shear tractions acting across contacting fracture surfaces perpendicular to the overall crack propagation direction. Such contacting surfaces and tractions occur between the large grains and the small grain matrix. The frictional tractions exerted by the large grains on the surrounding matrix are indicated by dark arrows in Figure 4.4b.

Figure 4.5a shows a schematic cross section of a polycrystalline microstructure that also includes aspects of Figure 4.1b in the form of larger elongated grains associated in a cluster. As before, the microstructure is part of a component that is placed in tension, shown by the gray arrows. A crack propagates through the microstructure along the weak grain boundaries, left to right, indicated by the bold black lines, such that the crack trace on the section is discontinuous. The crack is envisaged to be continuous in three dimensions such that the discontinuity apparent in the two-dimensional section represents an unbroken ligament of material connecting the two separating elements of the component. The angular deviations in the crack path adjacent to the ligamentary bridge are not large, and the bridge is formed primarily by the large grain cluster. The ligamentary bridge acts as a localized region of compressive force across the crack surface and opposes crack opening and propagation. The force originates from the deformation resistance of the ligament material to the bending moment generated within the ligament by the opposing tractions at the built in ends. The tractions exerted on the surrounding elements are indicated by dark arrows in Figure 4.5b.

Figure 4.6 shows example reflected light images of cracking in a polycrystalline Al_2O_3 similar to Figure 4.1b exhibiting (a) a frictional interlock and (b) a ligamentary bridge. In both cases, the crack path is left to right and the predominant tendency to intergranular fracture is clear. The frictional traction in Figure 4.6a has caused chipping in the adjacent grains that has led to bright contrast. The terminations of the surface traces of the cracks in Figure 4.6b are labeled "S." Note that the frictional interlock involves two grains and is much smaller than the ligamentary bridge that involves many grains. Such a size difference is commonly observed.

The above summary describes microstructural effects in fracture of both polycrystalline and multiphase brittle materials—ceramics in particular. In both microstructural forms, preferred fracture paths exist: In polycrystalline materials, these are weak grain boundaries; in multiphase materials, these are weak interfaces between phases. In addition, in both forms of microstructure, thermal expansion plays a large role: In polycrystalline materials, thermal expansion anisotropy effects often generate large stresses within microstructures on cooling from a processing temperature (particularly in Al_2O_3, Cook and Michaels 2019); similarly, in multiphase materials, thermal expansion inhomogeneity effects often generate large stresses. Most ceramic microstructures contain a much greater diversity of grain sizes and shapes than shown in the schematic diagrams of Figures 4.4 and 4.5. Hence, effects on crack propagation by microstructurally-induced frictional interlocking and ligamentary bridging are pervasive in polycrystalline ceramics.

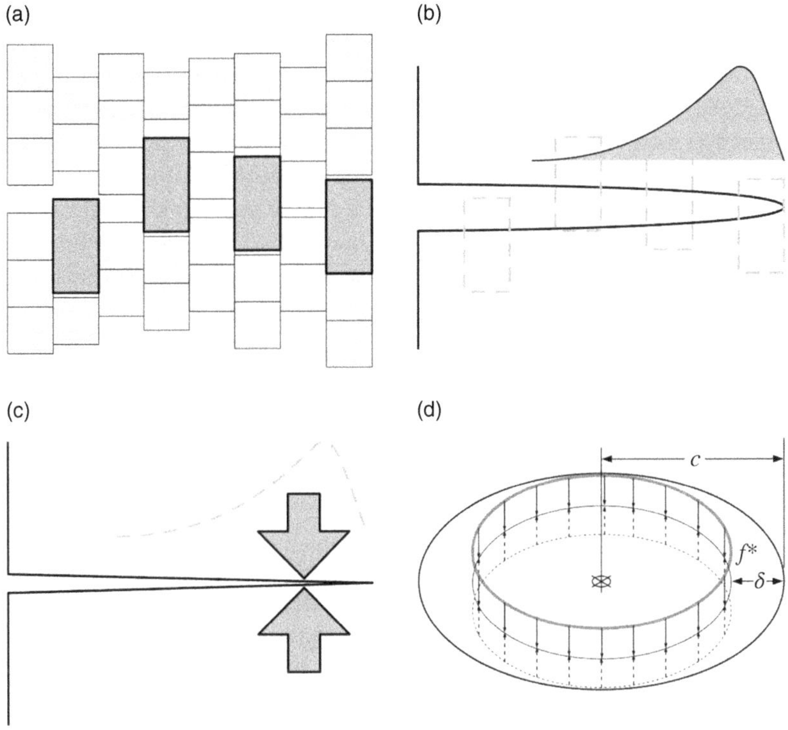

Figure 4.7 Schematic diagrams of crack wake microstructural tractions showing the progression from a discrete representation to a continuum representation. (a) Elongated grains exerting frictional closure tractions as a consequence of imposed crack opening displacement. (b) The magnitude of a traction field exerting a distributed zone of compression over the crack as a consequence of the elongated grains. (c) A single crack closure force characterizing the traction field in magnitude and location. (d) Representation of the closure force in a form applicable to indentation strength analyses.

4.2.4 Crack Wake Toughening: Analysis

Figure 4.7 shows a sequence of schematic diagrams of the effects of microstructural tractions on crack propagation. The sequence, Figure 4.7a–d, proceeds in order from a discrete representation requiring many parameters to a continuum representation requiring two parameters. The discrete representation of Figure 4.7a shows an array of elongated grains acting as frictional interlocks across a crack extending transgranularly in a smaller-grained matrix. Anisotropic coefficient of thermal expansion (CTE) effects place the matrix surrounding the elongated grains in vertical tension and horizontal compression: the tensile stress leads to spontaneous initiation and propagation of short cracks in the matrix (as described in Chapters 1 and 3) and the compressive stress leads to enhanced "clamping" and friction at the interlocks within long cracks. Incorporation of these stresses and the length scales of the microstructure lead to analyses of considerable sophistication for toughening and consequent effects on strength. Such analyses have been developed and applied by Lawn and colleagues for both single-phase materials (Bennison et al. 1989; Chantikul et al. 1990) and multiphase materials (Lawn et al. 1993; Padture et al. 1993a,b).

Figure 4.7b shows a continuum approximation of the discrete crack segment and traction field of Figure 4.7a. The crack is represented as an elliptical solid line, and the elongated grains are indicated as dashed light gray outlines. The effects of the discrete microstructural closure elements are described here by a continuous distribution of compression applied to the crack and indicated by the dark shaded area. The microstructural compressive field increases rapidly adjacent to the crack tip, passes through a maximum, and then decreases to zero on a longer length scale. The exact shape of the distribution is determined by coupling between the effective force–displacement behavior of the frictional and bridging elements and the crack opening displacement function of the material. This was the continuum approach of Mai and Lawn (1987) for toughness estimation, Cook et al. (1987) and Cook (1990) for strength description, and the intended application of the force–displacement constitutive responses measured by Hay and White (1993, 1995).

Figure 4.7c shows an abstracted description of the continuum crack closure field of Figure 4.7b. The crack is represented as a solid line and the closure field is indicated as a dashed light gray line. The effects of the distributed microstructural closure field are described here by a single characteristic compressive force applied to the crack at a characteristic distance from the crack tip and indicated by the dark shaded arrows. The magnitude and relative position of the force are invariant and determined by the distribution of the closure stress; the form of the crack opening field is not considered. This was the basis of the formulation developed by Cook (1986) for application in indentation strength analyses of alumina and later applied, with extensions, to alumina and other materials (Cook et al. 1994c; Cook 2015, 2017). For implementation in Vickers indentation strength analyses, the microstructural force is a line force f^* (force/length) distributed over a circle at an annular distance δ from the crack perimeter, radius c, as shown in Figure 4.7d. Note that f^* contains implicit information regarding the clamping and frictional stresses of the elongated grains and δ contains implicit information regarding the spacing and size of the elongated grains. As will be shown, the great advantage of a microstructural toughening formulation based on Figure 4.7d is that a minimum of additional parameters are required in indentation strength analyses.

Figure 4.7d is similar to Figure 1.17a of Section 1.7.1. The distinction here is that the radius of the circle describing the line force is not fixed (at a in Figure 1.17a) but maintains a fixed separation δ from the crack tip such that here the radius of the circle describing the location of the compressive line force f^* is $(c - \delta)$. Rewriting Eq. (1.78), the SIF in this case is thus

$$K_\mu(c) = -2f^*(c-\delta)/(\pi c)^{1/2}(2c\delta - \delta^2)^{1/2}. \tag{4.12}$$

For use in the strength analyses to follow, a more convenient expression for $K_\mu(c)$ that closely approximates Eq. (4.12) is

$$K_\mu(c) = -\mu f^* \delta^{-1/2}\left[1 - (\delta/c)^{3/2}\right],\ c \geq \delta, \tag{4.13}$$

where the numerical factor $\mu \approx (2/\pi)^{1/2}$. The negative sign indicates that $K_\mu(c)$ acts to oppose crack extension as f^* is the magnitude of a compressive line force. The applicability of Eq. (4.13) is for cracks that have extended into a microstructurally influenced crack length domain, $c \geq \delta$, where δ is envisaged to be about three to four grain diameters in a polycrystalline microstructure.

It is more intuitive to discuss Eq. (4.13) in the frame of reference of the applied loading and use Eq. (4.10) to write the microstructural influence as a toughening factor

$$T_\mu(c) = 0,\ c < \delta, \tag{4.14}$$

$$T_\mu(c) = \mu f^* \delta^{-1/2}\left[1 - (\delta/c)^{3/2}\right],\ c \geq \delta. \tag{4.15}$$

$$T_\mu(c) \to \Delta T = \mu f^* \delta^{-1/2},\ c \gg \delta. \tag{4.16}$$

For $c < \delta$, there is no toughening. For $c \gg \delta$, the microstructural influences apply a maximum toughening increment and the toughness approaches a maximum value

$$T_\infty = T_0 + \mu f^* \delta^{-1/2},\ c \gg \delta. \tag{4.17}$$

The long crack steady-state toughness is increased if the microstructural restraining force is greater and if the restraint is closer to the crack tip. It is important to recognize that Eq. (4.16) characterizes the energy dissipation associated with toughening discussed in Section 4.2.1. This is made clear by taking Eq. (4.7) to leading order and the above expression for ΔT to give Eq. (4.4) as $(f^*)^2/\delta E = w\mathcal{U}_{\text{irr}}$. If the width of the dissipation zone either side of the crack is taken as a grain size $w = \lambda$, the irreversible strain energy dissipated by the microstructural interaction scales as $\mathcal{U}_{\text{irr}} \sim (f^*/\lambda)(f^*/\delta E)$. The first term on the right side is recognized as the characteristic frictional stress and the second term as the characteristic maximum strain at the tail of the traction zone. Section 4.3 applies Eqs. (4.15) and (4.17) in an indentation strength formulation.

4.3 Toughened Materials Strength Analyses

The determination of indentation strength behavior for a material exhibiting microstructural toughening is a straightforward extension of the methods used in Chapters 2 and 3. It is convenient to define two indentation loads, P_δ and P^*, associated with the base toughness and the maximum toughness increment:

$$P_\delta = T_0\delta^{3/2}/\chi, \tag{4.18}$$

and

$$P^* = \Delta T \delta^{3/2}/\chi. \tag{4.19}$$

P_δ is the minimum indentation load required to experience toughening in the material by indentation cracking; it is usually not sensed in indentation strength tests. P^* is a central indentation load that characterizes material toughening in indentation fracture; it is pivotal in indentation strength tests. Expressing the SIF $K(c)$ of Eq. (4.9) as the sum of the SIF arising from the applied stress and the residual indentation fields from Section 3.2.1 and using $T(c)$ from Eq. (4.15) gives

$$\psi \sigma_a c^{1/2} + \chi P/c^{3/2} = T_0 + \mu f^* \delta^{-1/2} \left[1 - (\delta/c)^{3/2}\right]. \quad (4.20)$$

Using the expressions for T_∞ and P^* enables Eq. (4.20) to be written as

$$\psi \sigma_a c^{1/2} + \chi (P + P^*)/c^{3/2} = T_\infty. \quad (4.21)$$

Equation (4.21) is recognized as identical in mathematical form to that for ideal indentations described in Section 3.2.1. [Not uncoincidentally this formulation is also identical to the empirical formulation used by Lawn et al. (1984) and Cook et al. (1985a,b).] The indentation strength for a microstructurally toughened material is then written simply by modification of Eq. (3.8) as

$$\sigma_m = 3T_\infty^{4/3}/4^{4/3}\psi \chi^{1/3}(P + P^*)^{1/3}, \quad (4.22)$$

where the "m" subscript identifies the maximum supported stress or strength of the indentation flaw. Equation (4.22) differs in several important respects from Eq. (3.8). In the large indentation load limit, $P \gg P^*$, σ_m tends to the ideal $P^{-1/3}$ response, but the amplitude is proportional to the long crack steady-state toughness, $T_\infty^{4/3}$ and not base toughness value $T_0^{4/3}$. In the small indentation load limit, $P_\delta < P \ll P^*$, σ_m is not unbounded but tends to an invariant microstructurally related strength

$$\sigma_m^* = 3T_\infty^{4/3}/4^{4/3}\psi(\chi P^*)^{1/3}, \quad (4.23)$$

such that Eq. (4.22) can be written compactly as

$$\sigma_m = \sigma_m^*(1 + P/P^*)^{-1/3}. \quad (4.24)$$

Equation (4.24) shows that microstructural-based T curves introduce significant changes to the form of material indentation strength responses. The changes are characterized by deviations from the ideal power-law response to include strength invariance at small indentation loads. However, Chapter 3 showed that lateral crack related indentation stress relaxation also introduced significant changes to the form of material indentation strength responses. The changes in that case are characterized by deviations from the ideal power-law response to include strength invariance at large indentation loads. See Figure 3.37 for many examples. Figure 4.8 shows images of two similar indentations in two similar polycrystalline alumina materials. The image of Figure 4.8a is focused on the surface of the material and highlights the effect of microstructure on the surface traces of the half-penny cracks, many segments of which are obscured by grain boundary grooves. The image of Figure 4.8b is focused subsurface and highlights the appearance of lateral cracks, visible as regions of bright contrast. Clearly microstructural toughening and lateral cracking occurred simultaneously. Lateral cracking might be expected to be prevalent in microstructurally affected materials, as weak grain boundaries and other weak interfaces would enhance lateral cracking. This is, in fact, what is observed, of which Figure 4.8 is an example. Equation (4.24) is thus incomplete and requires modification to describe lateral crack as well as microstructural effects.

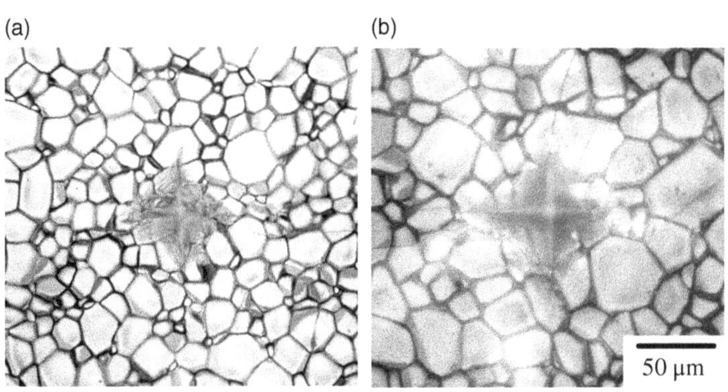

Figure 4.8 Optical images of indentations in polycrystalline alumina illustrating the two indentation crack systems that develop. (a) Surface traces of half-penny cracks exhibiting transgranular fracture adjacent to the indentation vertices and intergranular fracture further from the indentation. (b) Subsurface reflections of lateral cracks surrounding the indentation impression.

An analysis of lateral crack effects on indentation strength was given in Section 3.3.3. Using Eq. (3.62), the strengths of lateral crack affected materials can be rendered into an ideal form as

$$\sigma_m = S\tilde{P}^{-1/3}. \tag{4.25}$$

$\tilde{P}$ is an effective indentation load given by

$$\tilde{P} = \frac{PP_L}{P + P_L}, \tag{4.26}$$

where P_L is an invariant indentation load for a material characterizing lateral crack effects. S is an invariant contact strength given by

$$S = \sigma_L P_L^{1/3}, \tag{4.27}$$

where σ_L is an invariant strength characterizing lateral crack effects. The indentation strength responses of Eqs. (4.25)–(4.27) are discussed in detail in Section 3.3.3. Here, lateral crack and microstructural effects on indentation strengths are combined into a single formulation by replacing P in Eq. (4.24) with $\tilde{P}$,

$$\sigma_m = \sigma_m^*(1 + \tilde{P}/P^*)^{-1/3}, \tag{4.28}$$

and setting S as

$$S = 3T_\infty^{4/3}/4^{4/3}\psi\chi^{1/3} = \sigma^* P^{*1/3} = \sigma_L P_L^{1/3}. \tag{4.29}$$

Equation (4.28) has three asymptotic bounds. At small indentation loads, the strength is microstructure dominated, toughness increases with flaw size to leave the strength invariant at an upper bound,

$$\sigma_m \to \sigma^*, \quad P \ll P^*. \tag{4.30}$$

At large indentation loads, the strength is lateral crack dominated, residual field relaxation increases with flaw size to leave the strength invariant at a lower bound,

$$\sigma_m \to \sigma_L, \quad P \gg P_L. \tag{4.31}$$

At intermediate indentation loads, the strength is flaw size dominated and decreases in an ideal manner,

$$\sigma_m \to SP^{-1/3}, \quad P^* \ll P \ll P_L. \tag{4.32}$$

The scaling ideas of Chapter 3 are easily implemented using

$$\mathbf{S} = \sigma/\sigma_m^*$$
$$\mathbf{P} = P/P^* \tag{4.33}$$

In particular, the scaled strength $\mathbf{S_m}$ and scaled indentation load $\tilde{\mathbf{P}}$ are given by $\mathbf{S_m} = \sigma_m/\sigma_m^*$ and $\tilde{\mathbf{P}} = \tilde{P}/P^*$, such that Eq. (4.28) becomes

$$\mathbf{S_m} = \left(1 + \tilde{\mathbf{P}}\right)^{-1/3}. \tag{4.34}$$

Figure 4.9 is a plot in logarithmic coordinates of Eq. (4.34), noting that $\tilde{\mathbf{P}} = \mathbf{PP_L}/(\mathbf{P} + \mathbf{P_L})$. The microstructural scaling variables $\mathbf{P} = 1$ and $\mathbf{S_m} = 1$ are indicated by the vertical and horizontal fine solid lines. The plot uses $\mathbf{P_L} = 10$, and therefore $\mathbf{S_L} = 0.46$, indicated by the vertical and horizontal fine dashed lines. The ideal $\mathbf{P}^{-1/3}$ response is indicated by the diagonal fine dashed line. The overall strength response is indicated by the fine solid line and a likely experimental domain of 10^3 indicated by the bold gray overline. The overall strength response varies between the two bounds and passes *through* the ideal response at an intermediate strength level.

An expression of perhaps greater usefulness in engineering and design is the application of indentation strength measurements to obtain the indentation load dependence of the effective relative toughness, $T_{eff}(P)/T_\infty$ (Cook et al. 1985a). Generalizing Eq. (4.22) to incorporate effective toughness and lateral crack effects gives

$$\sigma_m \tilde{P}^{1/3} = 3T_{eff}^{4/3}/4^{4/3}\psi\chi^{1/3}, \tag{4.35}$$

where σ_m is the measured strength at indentation load P, and thus at $\tilde{P}$. Combining these equations with Eqs. (4.29) and (4.33) leads to

$$T_{eff}/T_\infty = (\sigma_m \tilde{P}^{1/3}/S)^{3/4} = \mathbf{T_{eff}} = \left[\tilde{\mathbf{P}}/(\tilde{\mathbf{P}} + 1)\right]^{1/4}, \tag{4.36}$$

which provides a means of evaluating the effects of microstructure on toughness by both experimental and analytical means. Note that in distinction to the scaling of Chapter 3, the scaling of toughness here is by the steady-state toughness, T_∞, such that $\mathbf{T_{eff}} \to 1$ as $\tilde{\mathbf{P}} \gg 1$. [For $\tilde{\mathbf{P}} \ll 1$, Eq. (4.36) implies $\mathbf{T_{eff}} \to 0$. In reality, the decrease is halted at a base toughness value T_0 for which only an upper bound can be estimated from indentation strength tests, Cook et al. (1985a, 1994c). This point is considered in the discussion.] Note that Eqs. (4.34) and (4.36) are related transformations of experimental strength measurements. The transformations are based on the same experimental data and fit parameters and convey the same information, expressed in different ways, regarding the effects of microstructure on fracture: Scaled strength plots emphasize the approach to invariant strength at small flaws; effective toughness plots emphasize the approach to invariant toughness at large flaws.

4.4 Toughened Materials Strength Observations

This section presents an extensive survey of the indentation strengths of toughened materials. The survey here differs in emphasis from that of Chapter 3. In Chapter 3, the indentation strengths of a restricted set of materials (particularly glass) were studied, and a range of indentation strength behaviors dependent on a range of fracture phenomena were examined: contact geometry variations, lateral cracking, impact, surface stresses, and subthreshold flaws. Here, in Chapter 4, the opposite approach is taken. Indentation strengths of a broad set of materials are studied, and a single form of indentation strength behavior controlled by two fracture phenomena is examined: sigmoidal strength behavior controlled by microstructural toughening and lateral cracking stress relief. The guiding parts of the analysis of Section 4.3 will be Figure 4.9 and Eqs. (4.28), (4.34), and (4.36).

4.4.1 Alumina

The indentation strength behavior of various forms of Al_2O_3 has been examined in Chapter 3. Polycrystalline Al_2O_3, termed alumina, from commercial sources was used to demonstrate ideal behavior and the effects of machining damage (F99, Figures 3.2 and 3.45b), lateral cracking (AD995, Figure 3.42), and quenching (AlSiMag, Figure 3.44). Single-crystal

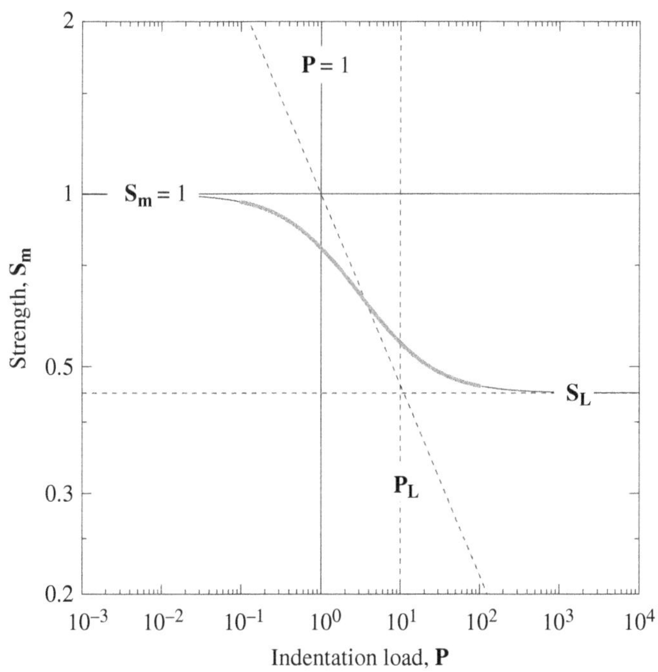

Figure 4.9 Logarithmic indentation strength plot in scaled coordinates for indentation contacts in materials exhibiting microstructural and lateral crack effects. Full behavior shown as the bold solid line. At small loads, strengths approach an upper bound (shown as the fine horizontal line). At large loads, strengths approach a lower bound (shown as horizontal dashed line). At intermediate loads, the strengths pass through the ideal indentation response (shown as the dashed diagonal line).

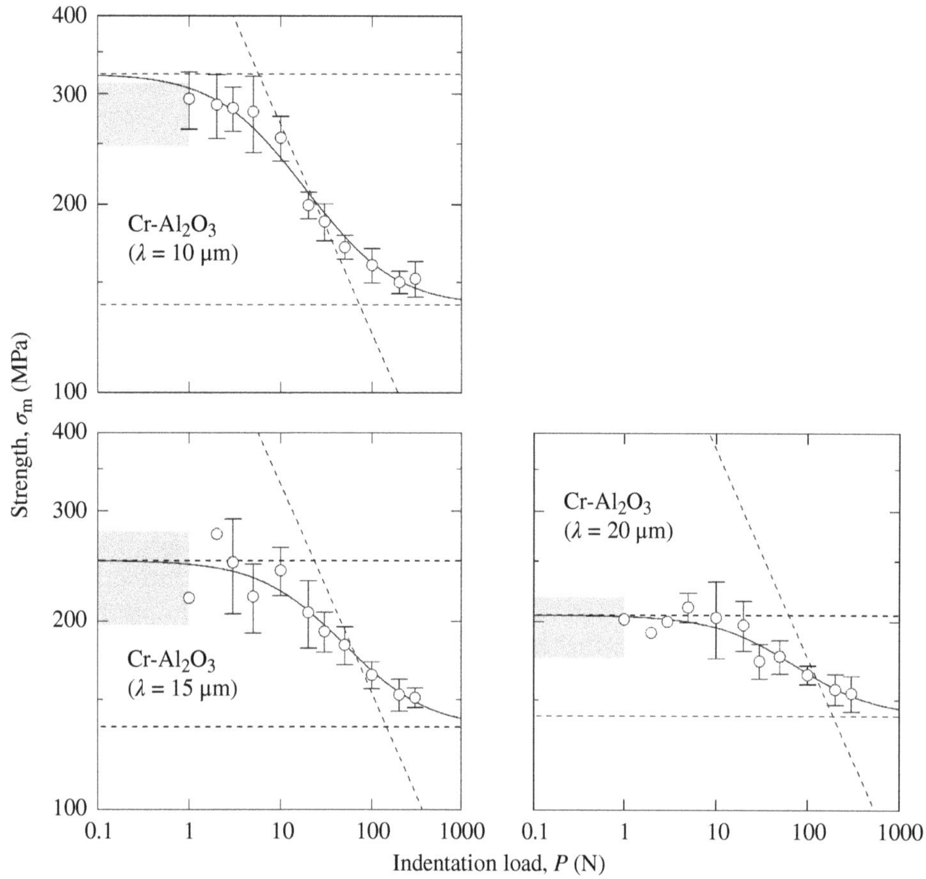

Figure 4.10 Logarithmic plots of Vickers indentation strengths of Cr-doped polycrystalline alumina. Solid lines are fits using a combined microstructural and lateral crack indentation fracture analysis. Dashed horizontal lines indicate strength bounds. Dashed diagonal line indicates ideal indentation response. Source: Data adapted from Cook et al. (1994c).

Al_2O_3, formally corundum, termed sapphire, was used to demonstrate the effects of lateral cracking (Figure 3.37h), stress gradients (Figure 3.50d), and subthreshold flaws (Figure 3.55). (Commercial designations are given for identification purposes only.)

Figure 4.10 shows logarithmic plots of the indentation strength behavior of a customized series of polycrystalline alumina materials of various grain sizes (mode sizes given as λ) from Cook et al. (1994c). The materials were doped with Cr_2O_3 to enable microstructural stress studies (Myers et al. 2016; Michaels and Cook 2016, 2018; Cook and Michaels 2017, 2019), and the grain boundaries were glass free (the basis of the materials was commercial Vistal). (Although not considered here the earlier stress studies provide explicit conformation of the anisotropic CTE-based stresses important in crack wake toughening. See Section 4.2.4). Symbols with bars represent the means and standard deviation limits of experimental strength measurements at the indentation loads indicated. Isolated symbols represent individual strength measurements. The shaded bands represent the means and standard deviation limits of intrinsic strength measurements, σ_i. The fine solid lines represent best fits of Eq. (4.28) to the indentation strength data. The upper and lower horizontal dashed lines represent the asymptotic microstructure controlled and lateral crack controlled strength bounds of σ_m^* and σ_L, respectively. A correlation between the upper indentation strength asymptotes and the intrinsic strengths is apparent. The diagonal dashed lines represent the intermediate asymptotic contact controlled strength behaviors—note that the complete best-fit responses pass clearly through the ideal contact strength asymptotes. The ability of the combined microstructure + lateral crack indentation strength formulation of Section 4.3 to describe the experimental observations of Figure 4.10 is clear, as is the resemblance of the raw data best-fit responses of Figure 4.10 to the universal scaled response of Figure 4.9.

The experimental indentation strength data and best fits of Figure 4.10 can be used in two ways. First, the best fits can be compared in absolute terms thereby enabling differences in strength *values* to be assessed. Second, the experimental indentation strength data can be compared in scaled terms thereby enabling similarities in strength and toughness *behavior*

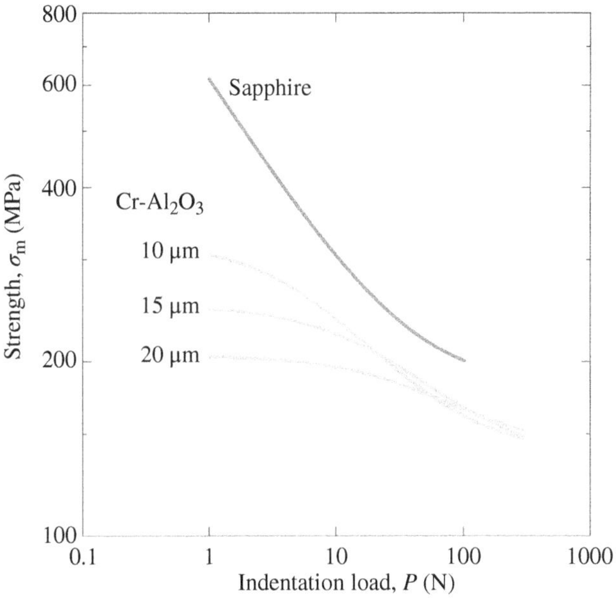

Figure 4.11 Logarithmic composite plot of Vickers indentation strengths of Cr-doped polycrystalline alumina (from Figure 4.10) and single-crystal sapphire (from Figure 3.37). Lines represent fracture mechanics based best-fit responses over the domains of experimental data.

to be assessed. Both comparison methods were used earlier (Cook et al. 1985a) and are used here. Figure 4.11 shows a logarithmic plot of the best fits of the indentation strength behaviors of the polycrystalline Cr–Al_2O_3 from Figure 4.10 and the single-crystal sapphire behavior from Figure 3.37h. In all cases, the plotted fits are restricted to the domains of the experimental measurements. Several points are clear in Figure 4.11. First, the polycrystalline alumina materials are less strong than single-crystal sapphire at all indentation loads, consistent with the idea that grain boundary structures are weaker than the related single crystal. Second, both the polycrystalline materials and sapphire display clear tendencies to strength plateaus at large indentation loads, consistent with residual stress relief from lateral crack effects. Within the domains of the experimental data, the tendency to invariant strength at large flaws sizes appears independent of grain size. Third, the polycrystalline alumina materials display clear tendencies to strength plateaus at small indentation loads, consistent with decreasing toughness associated with microstructural effects. Within the domains of the experimental data, the decreases in toughness at small flaws sizes are greater for the larger grain size materials.

Figure 4.12 shows a logarithmic plot of the scaled experimental indentation strength behaviors of the polycrystalline Cr–Al_2O_3 materials and single-crystal sapphire. The data were scaled using Eq. (4.33) and the best-fit parameters P^*, P_L, and σ_m^* from Figures 4.10 and 3.37h; the value of P^* for microstructurally inactive sapphire was set arbitrarily to a small value $\ll 1$ N consistent with the Cr–Al_2O_3 data; the value does not affect behavior and is simply a convenience for plotting. Symbols with bars represent the means and standard deviation limits of scaled strengths at the scaled loads indicated; uncertainties represent experimental uncertainties and do not include fitting uncertainties. Isolated symbols represent individual scaled values. The fine solid line represents the universal response of Eq. (4.34). The horizontal dashed line represents the asymptotic microstructure controlled scaled strength bound of $\mathbf{S_m} = 1$. The diagonal dashed line represents the asymptotic lateral crack adjusted contact controlled scaled strength bound of $\mathbf{S_m} = \tilde{\mathbf{P}}^{-1/3}$. As might be anticipated from Figures 4.10 and 3.37h, the data now all lie on the universal strength response. The scaled strengths of the polycrystalline Cr–Al_2O_3 materials approach the microstructurally controlled plateau at small scaled loads and the lateral crack adjusted contact response at large scaled loads. Sapphire exhibits an adjusted contact response at all scaled loads. The similarities and differences in behavior are clear.

Figure 4.13 shows a logarithmic plot of the scaled effective toughness behaviors of the polycrystalline Cr–Al_2O_3 materials and single-crystal sapphire. The data were scaled using Eq. (4.36) and the same parameters as in Figure 4.12. Symbols with bars represent the means and standard deviation limits of scaled toughness values at the scaled loads indicated; uncertainties represent experimental uncertainties and do not include fitting uncertainties. Isolated symbols represent individual scaled values. The fine solid line represents the universal response of Eq. (4.36). The horizontal dashed line represents the asymptotic contact controlled scaled bound of $\mathbf{T_{eff}} = 1$. The diagonal dashed line represents the asymptotic microstructure

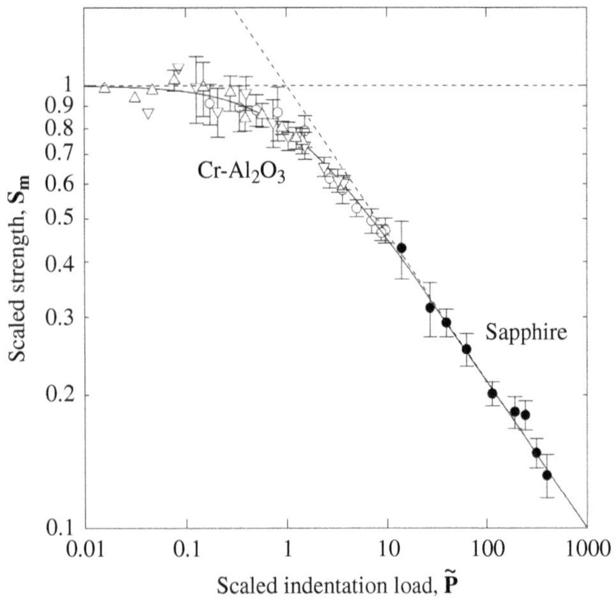

Figure 4.12 Logarithmic plot of scaled Vickers indentation strengths of Cr-doped polycrystalline alumina and single-crystal sapphire (from Figure 4.10). Solid line represents universal response based on indentation fracture mechanics analysis including microstructural and lateral crack effects. Dashed lines represent small and large indentation load asymptotes.

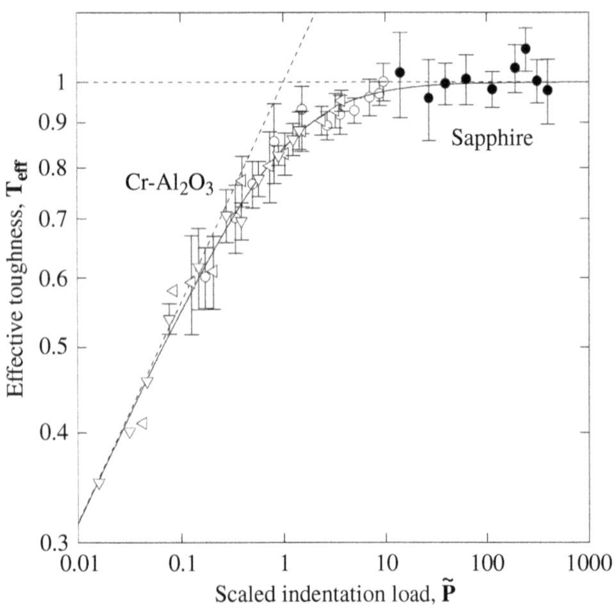

Figure 4.13 Logarithmic plot of scaled effective toughness values of Cr-doped polycrystalline alumina and single-crystal sapphire (from Figure 4.10). Solid line represents universal response based on indentation fracture mechanics analysis including microstructural and lateral crack effects. Dashed lines represent small and large indentation load asymptotes.

controlled scaled bound of $\mathbf{T_{eff}} = \tilde{\mathbf{P}}^{1/4}$. Consistent with Figure 4.12, the data lie on the universal toughness response. Here, however, the increasing toughness of the polycrystalline alumina materials with increasing flaw size at small flaw size and the invariant toughness of sapphire are now both made clear. The inclusion of lateral crack effects into the scaled indentation load has removed the artifact leading to unphysical large load toughness values $\mathbf{T_{eff}} > \mathbf{1}$ incorporated earlier (Cook et al. 1985a) (and similarly for strengths, the artifact of $\mathbf{S_m} > \tilde{\mathbf{P}}^{-1/3}$ has been removed). Consideration of the indentation strength responses of polycrystalline $Cr–Al_2O_3$ materials and single-crystal sapphire has shown that the microstructural toughening analyses of Sections 4.2 and 4.3 are extremely effective in describing experimental observations and that the plotting schemes of Figures 4.10–4.13 are effective data presentation schemes. The analyses and plotting schemes will be used throughout this survey.

It is obvious from the crack wake restraint observations of Section 4.2, the indentation strength analyses of Section 4.3, and the experimental observations of Figure 4.11 that microstructural manipulation can be used to minimize brittle fracture strength degradation of polycrystals. Or, alternatively, that microstructural manipulation can be used to maximize *flaw tolerance*. The observations and analyses make clear that large grains and weak grain boundaries are key steps in this direction. Early works that employed this strategy doped polycrystalline Al_2O_3 with CaO (Cook et al. 1987; Cook 1990; Cook and Schrott 1988). Ca segregates to grain boundaries in alumina, weakening the boundaries by atomic scale substitutional strain and increasing grain size by increasing grain boundary energy. (As above, the base material was commercial Vistal; Cr considered above segregates weakly to grain boundaries in alumina and generates very little strain.) A set of Ca–Al_2O_3 microstructures was obtained through variations in composition and heat treatment. Approximate mean grain sizes of 12–60 μm were generated. Figure 4.14 summarizes the results of subsequent indentation strength tests.

Figure 4.14a shows a logarithmic plot of the scaled experimental indentation strength behaviors of the polycrystalline Ca–Al_2O_3 materials and single-crystal sapphire. The plot uses the same format as Figure 4.12. Symbols represent the means of scaled strengths at the scaled loads indicated; uncertainties are omitted for clarity. A few symbols represent individual scaled values. Lines represent the scaled universal strength response and asymptotes. The data all lie on the universal response. The Ca–Al_2O_3 strengths exhibit clear microstructural effects, and the sapphire strengths exhibit a clear absence of such effects. Of particular note is the observation that the strengths of the polycrystalline Ca–Al_2O_3 materials do not just approach the microstructurally controlled strength plateau at small scaled loads: the strengths exhibit near-ideal flaw tolerance at small flaw sizes extending over approximately a factor of 10^2 in indentation load. Even over a factor of 10^3 in indentation load, the strength degradation is only about a factor of 0.4. Figure 4.14b shows a logarithmic plot of the scaled effective toughness behaviors of the polycrystalline Ca–Al_2O_3 materials and single-crystal sapphire. The plot uses the same format as Figure 4.13. Symbols are as in Figure 4.14a. Lines represent the scaled universal toughness response and asymptotes. The data all lie on the universal response. The explanation for the flaw tolerance exhibited by the Ca–Al_2O_3 materials is apparent in the significant increase in toughness with flaw size (open symbols). The effective toughness increases by approximately a factor of 5 as the flaw size increases by a factor of approximately 10^3. In contrast, the toughness of sapphire remains invariant (filled symbols). There is no doubt that the combined large grain–weak grain boundary strategy was effective in generating flaw tolerance in these materials.

The effects of grain size alone on flaw tolerance in alumina were elucidated in a similarly early study by Lawn and colleagues (Chantikul et al. 1990). In this case, the alumina was doped with trace MgO such that grain growth was inhibited

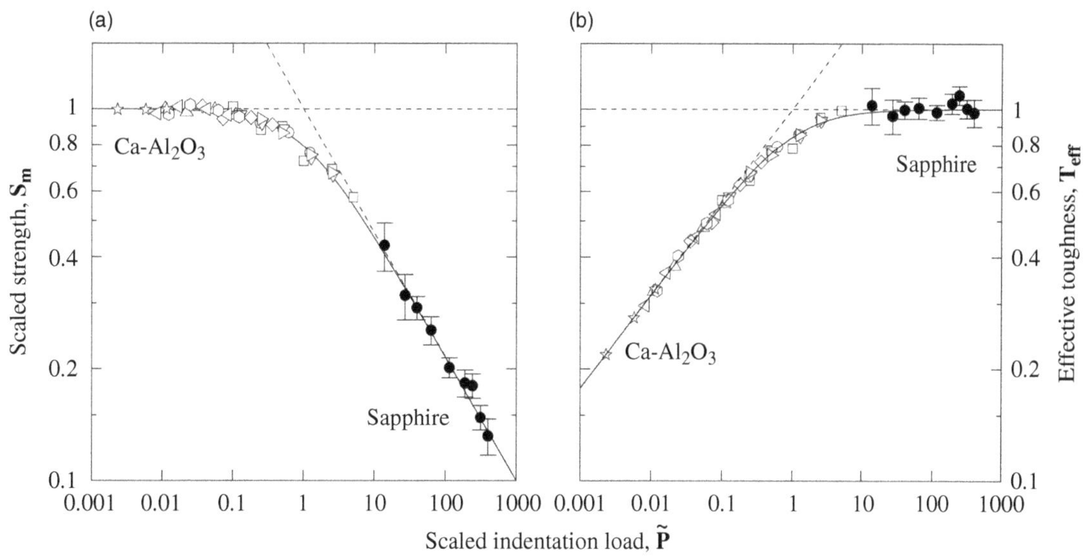

Figure 4.14 Logarithmic plots of (a) scaled Vickers indentation strengths and (b) scaled effective toughness values of Ca-doped polycrystalline alumina and single-crystal sapphire. Solid lines represent universal responses based on indentation fracture mechanics analysis including microstructural and lateral crack effects. Dashed lines represent small and large indentation load asymptotes. Source: Includes data adapted from Cook et al. (1987) and Cook (1990) and Figure 3.37.

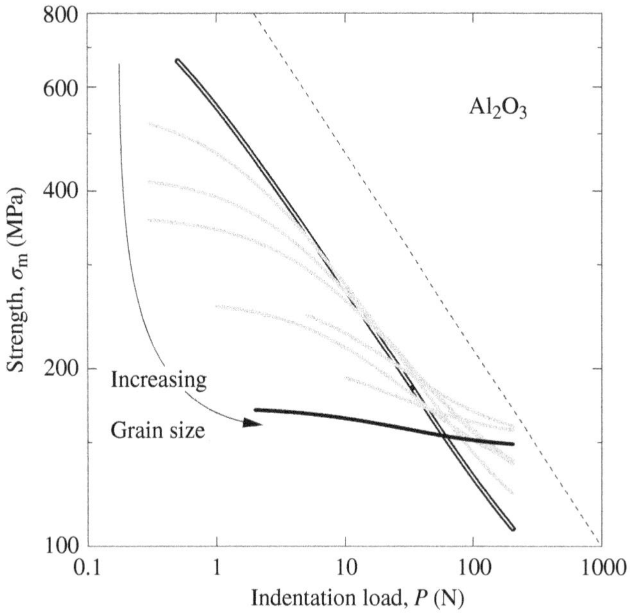

Figure 4.15 Logarithmic composite plot of Vickers indentation strengths of Mg-doped polycrystalline alumina. Lines represent fracture mechanics based best-fit responses over the domains of experimental data. Smallest grain size data shown as open line, largest grain size as bold line, intermediate grain sizes shown as grayed lines. Dashed line indicates ideal indentation behavior. Source: Data adapted from Chantikul et al. (1990).

on an initial firing and grain size was increased in a controlled manner through subsequent heat treatment (the materials were based on commercial Lucalox, similar to Vistal). Approximate mean grain sizes of 2–80 μm were generated. Figure 4.15 summarizes the results of subsequent strength tests of the materials in a logarithmic plot of the indentation strength best-fit behaviors. The format is similar to Figure 4.11 in that the plotted fits are restricted to the domains of the experimental measurements. The behavior of the smallest grain size material is shown as the open line and the behavior of the largest grain size material is shown as the solid line; behaviors of the intermediate grain size materials are shown as the grayed lines. The fine dashed line is a guide to the eye of ideal indentation strength form $\sigma_m \sim P^{-1/3}$. The indentation strength behavior of the smallest grain size material is well approximated by an ideal response. However, the behavior of the largest grain size material exhibits considerable flaw tolerance. The tolerance appears to originate from both microstructural and lateral crack effects as the indentation strengths tend to barely separated plateaus at both small and large flaw sizes. A consequence of these simultaneous effects is that the indentation strength behaviors of the small and large materials intersect and cross at an indentation load, about 50 N, within the experimental domain. Consideration of the behaviors of the intermediate grain size materials shows that the crossing of indentation strength responses as grain size is altered is common and that the indentation load of intersection increases with increasing grain size. Similar crossing behavior is visible in the Cr–Al_2O_3 of Figure 4.11, in the Ca–Al_2O_3 system (Cook 1990), and in the early work in this area (Lawn et al. 1984; Cook et al. 1985a) (although in the last examples the grain boundary chemistry is also altered).

Figure 4.16 plots the results of the indentation strength tests of the Mg–Al_2O_3 materials in a different way, following the format of Figure 4.14. Figure 4.16a shows a logarithmic plot of the scaled experimental indentation strength behaviors of the materials. Symbols represent the means of scaled strengths at the scaled loads indicated; uncertainties are omitted for clarity. Open, grayed, and filled symbols follow the format of Figure 4.15 for small, intermediate, and large grain size materials. Lines represent the scaled universal strength response and asymptotes. The data all lie on the universal response. The strengths exhibit clear microstructural effects, and the small grain size material strengths exhibit a clear absence of such effects. The large grain size material strengths exhibit near-ideal flaw tolerance at small flaw sizes extending over approximately a factor of 10 in indentation load. Figure 4.16b shows a logarithmic plot of the scaled effective toughness behaviors of the materials. Symbols are as in Figure 4.16a. Lines represent the scaled universal toughness response and asymptotes. The data all lie on the universal response. The explanation for the flaw tolerance exhibited by the large grain size materials is apparent in the significant increase in toughness with flaw size (filled symbols). The effective toughness increases by approximately a factor of 3 as the flaw size increases by a factor of approximately 10^3. In contrast, the toughness of the small grain size material remains invariant (open symbols). In this case, it is clear that large grains are effective in generating

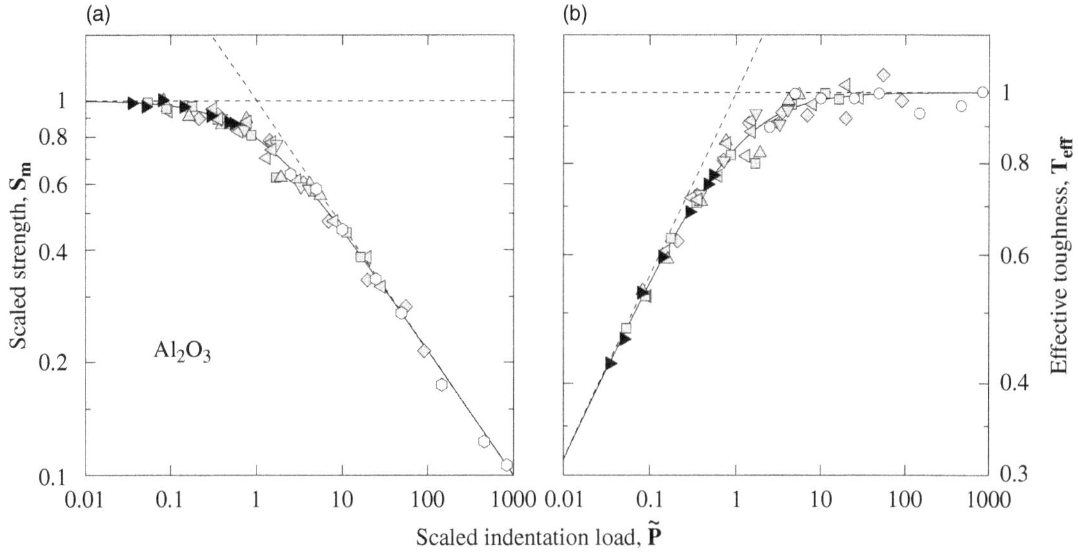

Figure 4.16 Logarithmic plots of (a) scaled Vickers indentation strengths and (b) scaled effective toughness values of Mg-doped polycrystalline alumina. Solid lines represent universal responses based on indentation fracture mechanics analysis including microstructural and lateral crack effects. Dashed lines represent small and large indentation load asymptotes. Source: Data adapted from Chantikul et al. (1990).

flaw tolerance. In a series of extensive studies, Kovar and Readey (1994, 1996) and Kovar et al. (2000)) demonstrated similar results on similarly prepared high-purity polycrystalline alumina materials. The focus of the Kovar works was the effects of grain size distribution and grain shape. The works showed that broader grain size distributions and incorporation of elongated grains lead to greater flaw tolerance although somewhat at the expense of greater strength variability at a given flaw size.

Final attention to the effects of microstructure on indentation strength and toughness behavior of polycrystalline alumina is given here to commercial materials. These materials frequently contain significant glassy fractions (up to 10% by mass) that reduce firing time and temperature requirements and increase manufacturing repeatability, but which render the materials opaque (white or cream). Commercial materials in this category discussed in Chapter 3 include F99, AD995, and AlSiMag. Commercial materials, discussed above, that are "glass free" and translucent include Vistal and Lucalox. The commercial materials now considered, in approximate order of decreasing glassy fraction, are Vistal, AD999, Alsint997, AD995, AD96, Diamonite, MD9211, and AD90. (It is emphasized that commercial designations are for identification purposes only.) The experimental data analyzed are from the published works of Gonzalez et al. (1984), Cook et al. (1985a, 1987), Bennison et al. (1989), and Cook (2017), and from related unpublished work. Figure 4.17 summarizes the results of strength tests of these materials in a logarithmic plot of best-fit indentation strength behaviors. The format is similar to Figure 4.15 in that the plotted fits are restricted to the domains of the experimental measurements. The fine dashed line is repeated from Figure 4.15 and is a guide to the eye of ideal indentation strength form $\sigma_m \sim P^{-1/3}$. It is clear from Figure 4.17 that commercial polycrystalline alumina materials can be divided into two major groups based on indentation strength behavior. The indentation strengths of the first group are only weakly perturbed by microstructural and lateral crack effects and exhibit near-ideal indentation strength behavior. This group is indicated as representative open line and other grayed lines in Figure 4.17. The indentation strengths of the second group are significantly perturbed by microstructural and lateral crack effects and exhibit considerable flaw tolerance and tendencies to strength invariance at small flaw sizes. This group is indicated as the filled lines in Figure 4.17. The indentation strength behaviors of the two groups intersect and cross at an indentation load (about 50 N) within the experimental domains. There is very little crossing within groups. The first group consists mostly of glassy materials in which the corundum grains are bonded by intergranular glass, and the second group consists mostly of glass-free materials.

Figure 4.18 plots the results of indentation strength tests of the commercial alumina materials following the format of Figure 4.14. Figure 4.18a shows a logarithmic plot of the scaled experimental indentation strength behaviors of the materials. Symbols represent the means of scaled strengths at the scaled loads indicated; uncertainties are omitted for clarity. Open, grayed, and filled symbols follow the format of Figure 4.17 for near-ideal and flaw tolerant materials. Lines represent the scaled universal strength response and asymptotes. The data all lie on the universal response. The strengths of the glass-free

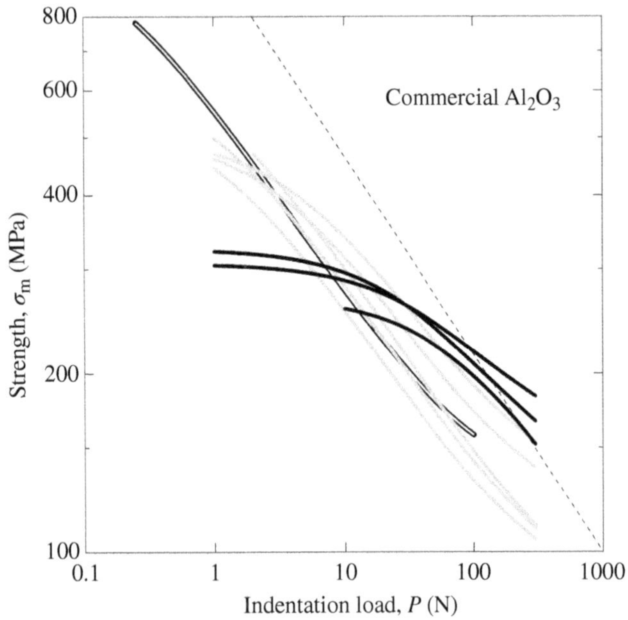

Figure 4.17 Logarithmic composite plot of Vickers indentation strengths of commercial polycrystalline alumina. Bold lines represent fracture mechanics based best-fit responses over the domains of experimental data. Fine dashed line indicates ideal indentation behavior. Near-ideal strength behavior shown as open line, microstructurally perturbed strength data shown as bold lines, intermediate strength behavior shown as grayed lines. Source: Includes data adapted from Gonzalez et al. (1984), Cook et al. (1985a, 1987), Bennison et al. (1989), and Cook (2017).

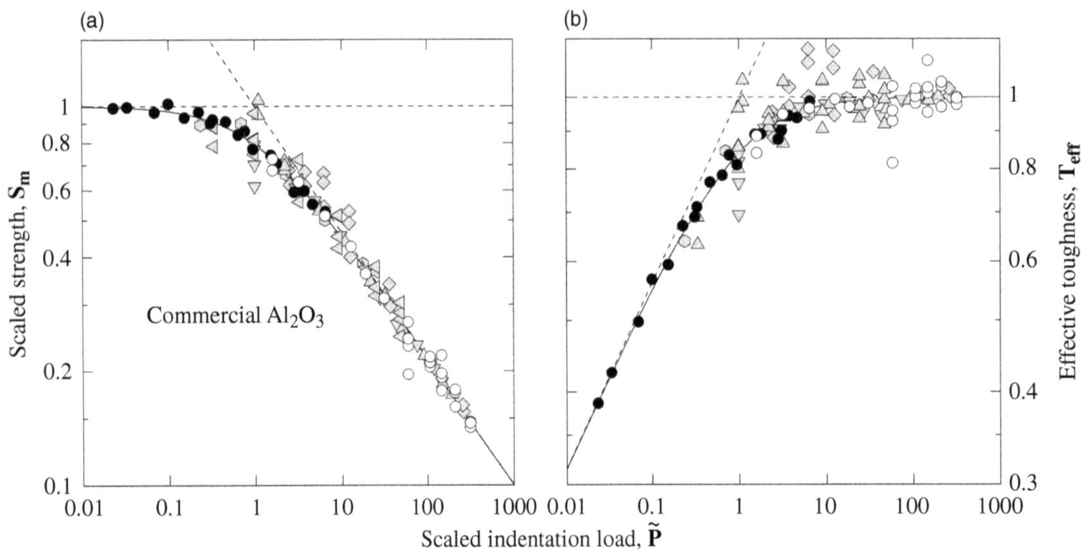

Figure 4.18 Logarithmic plots of (a) scaled Vickers indentation strengths and (b) scaled effective toughness values of commercial polycrystalline alumina. Solid lines represent universal responses based on indentation fracture mechanics analysis including microstructural and lateral crack effects. Dashed lines represent small and large indentation load asymptotes. Source: Includes data adapted from Gonzalez et al. (1984), Cook et al. (1985a, 1987), Bennison et al. (1989), and Cook (2017).

group (filled symbols) exhibit clear microstructural effects and the glass-bonded group exhibit a clear absence of such effects. The glass-free material strengths exhibit flaw tolerance at small flaw sizes extending over approximately a factor of 10 in indentation load, comparable to the microstructurally similar Mg-doped materials of Figure 4.16a. Figure 4.18b shows a logarithmic plot of the scaled effective toughness behaviors of the materials. Symbols are as in Figure 4.18a. Lines represent the scaled universal toughness response and asymptotes. The data all lie on the universal response. The explanation for the flaw tolerance exhibited by the glass-free materials is apparent in the significant increase in toughness with flaw size (filled symbols). The effective toughness increases by approximately a factor of 3 as the flaw size increases by a factor of approximately 10^3. In contrast, the toughness of the glass-bonded materials remains invariant (open and grayed symbols). The explanation in this case is that the majority of commercial alumina materials exhibit small grains and fracture resistant glass-bonded grain boundaries—both of which act against significant crack wake traction and microstructural toughening effects. Bennison et al. (1989) and O'Donnell et al. (1995) demonstrated the ability to impart a degree of flaw tolerance to a glass-bonded alumina materials through heat treatments to coarsen the microstructure.

4.4.2 Silicon Nitride

Polycrystalline Si_3N_4 materials can be divided into three major groups based on the observed indentation strength behavior. The groups correspond approximately to the grain sizes λ of the grouped materials. Figure 4.19 shows logarithmic plots of the indentation strength behavior of polycrystalline Si_3N_4 from Lawn et al. (1980), Choi and Salem (1992), Li et al. (1992), Xu et al. (1995), Choi et al. (1997), and Lee et al. (1997), using data derived from the published works cited. The data are grouped by similar indentation strength behaviors. Symbols with bars represent the means and standard deviation limits of experimental strength measurements at the indentation loads indicated. The shaded bands represent the bounds of intrinsic strength measurements, σ_i, from the various studies included in each group. The fine solid lines represent best fits of Eq. (4.28) to the indentation strength data. The upper and lower horizontal dashed lines represent the asymptotic microstructure controlled and lateral crack controlled strength bounds of σ_m^* and σ_L, respectively. The data and fits reinforce several points noted in consideration of polycrystalline alumina: there is a clear correlation between the upper indentation strength asymptotes and the intrinsic strengths; the diagonal dashed lines represent the intermediate asymptotic contact controlled strength behaviors and the complete best-fit responses pass clearly through these asymptotes; the ability of the

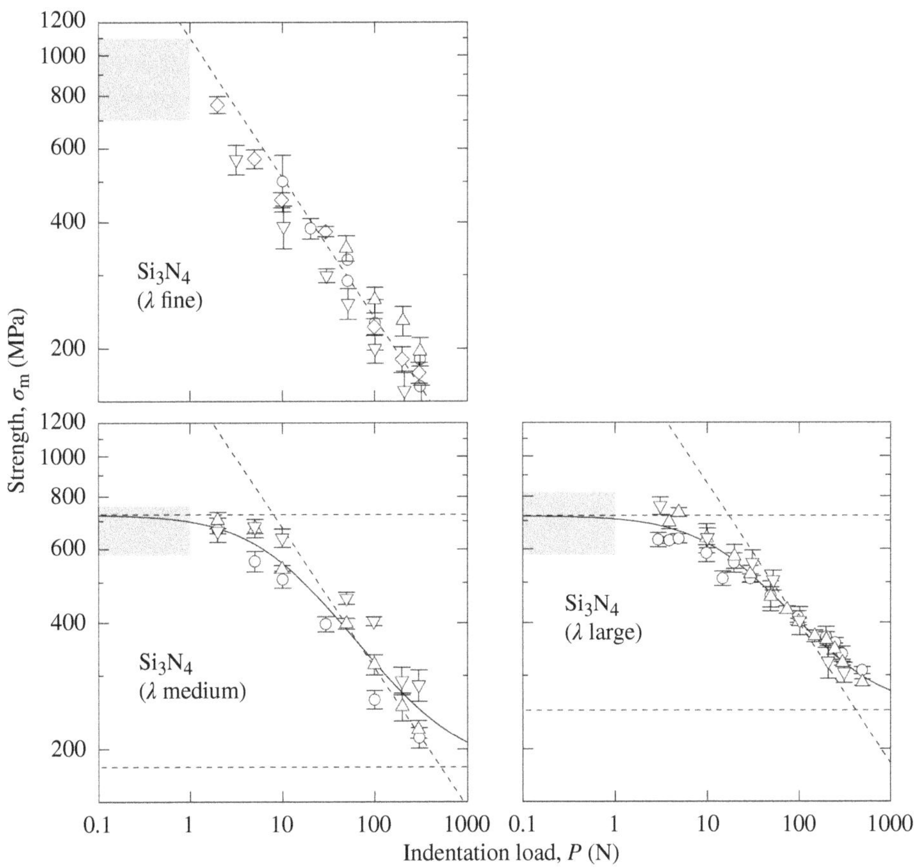

Figure 4.19 Logarithmic plots of Vickers indentation strengths of polycrystalline Si_3N_4. Solid lines are fits using a combined microstructural and lateral crack indentation fracture analysis. Dashed horizontal lines indicate strength bounds. Dashed diagonal line indicates ideal indentation response. Source: Data adapted from Lawn et al. (1980), Choi and Salem (1992), Choi et al. (1997); Li et al. (1992), Xu et al. (1995), and Lee et al. (1997).

combined microstructure + lateral crack indentation strength formulation of Section 4.3 to describe the experimental observations of Figure 4.19 is clear, as is the resemblance of the raw data best-fit responses of Figure 4.19 to the universal scaled response of Figure 4.9.

The indentation strength behaviors of Figure 4.19 correspond to approximate material grain size ranges of $\lambda \approx 0.5$–$2\,\mu m$ (fine), 1–4 μm (medium), and 3–10 μm (coarse), noting that all the Si_3N_4 microstructures presented in the published works exhibited considerable grain size dispersion. Two points of reference are that the group of "fine" materials includes the commercial polycrystalline Si_3N_4 NC132, and that the indentation strengths of Figure 4.19 are about a factor of 5 greater than the grinding controlled Si_3N_4 strengths of Figure 1.10c and reflect much smaller crack sizes (about a factor of $0.04 = 1/5^2$).

As in consideration of alumina in Section 4.4.1, the experimental indentation strength data and best fits of Figure 4.19 can be used in two ways: to compare strength values and to compare strength and toughness behavior. Figure 4.20 shows a logarithmic plot of the best fits of the indentation strength behaviors of the polycrystalline Si_3N_4 materials from Figure 4.19. In all cases, the plotted fits are restricted to the domains of the experimental measurements. The best fit to the designated fine grain size materials exhibits ideal behavior and is indicated by the open line. The best fit to the designated coarse grain size materials exhibits mildly flaw tolerant behavior and is indicated by the solid line. The best fit to the medium grain size materials exhibits intermediate behavior and is indicated by the grayed line. There are two key points to be gained from Figure 4.20. First, from a materials engineering perspective, developing polycrystalline Si_3N_4 materials with increased grain sizes leads to increased strengths and imparts a degree of flaw tolerance (reduced strength degradation) over the frequently encountered contact load range of 10–100 N (both microstructural and lateral crack effects are active). Second, from a materials science perspective, it appears that, similar to the behavior of Al_2O_3, e.g., in Figure 4.17, the indentation strength behaviors of Si_3N_4 materials intersect and cross, in this case, at an indentation load of approximately 5 N, less than the 50 N observed for Al_2O_3 materials. Hence, the same microstructural effects apply to toughness and strength of both these polycrystalline materials with a difference in scale: the grain sizes of the Al_2O_3 materials were approximately 30 μm and the grain sizes of the Si_3N_4 materials were approximately 3 μm. The decreased grain size leads to a decreased indentation load crossing point and thus a reduced domain of experimental strength measurements at small flaw sizes. The correlation of the microstructurally controlled strength asymptotes with the intrinsic strengths in Figure 4.19 provides further support for the implication that the microstructural phenomena in Si_3N_4 and Al_2O_3 are similar and simply changed in scale.

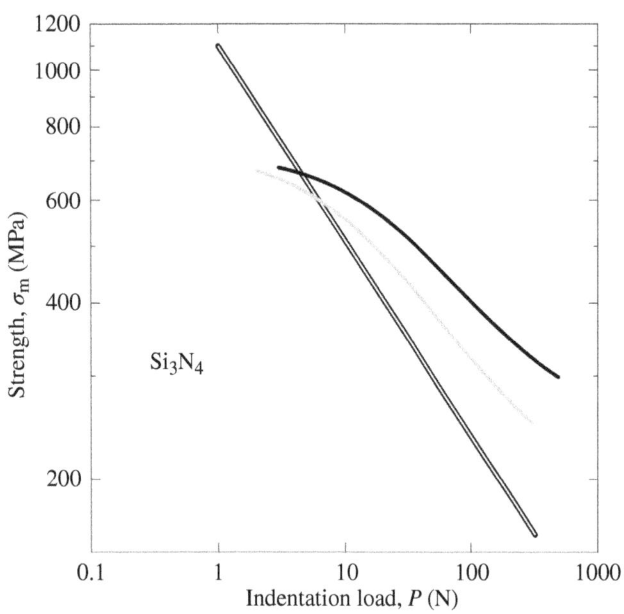

Figure 4.20 Logarithmic composite plot of Vickers indentation strengths of polycrystalline Si_3N_4. Near-ideal strength behavior shown as open line, microstructurally perturbed strength data shown as bold line, intermediate strength behavior shown as grayed line. Source: Data adapted from Lawn et al. (1980), Choi and Salem (1992), Choi et al. (1997), Li et al. (1992), Xu et al. (1995), and Lee et al. (1997).

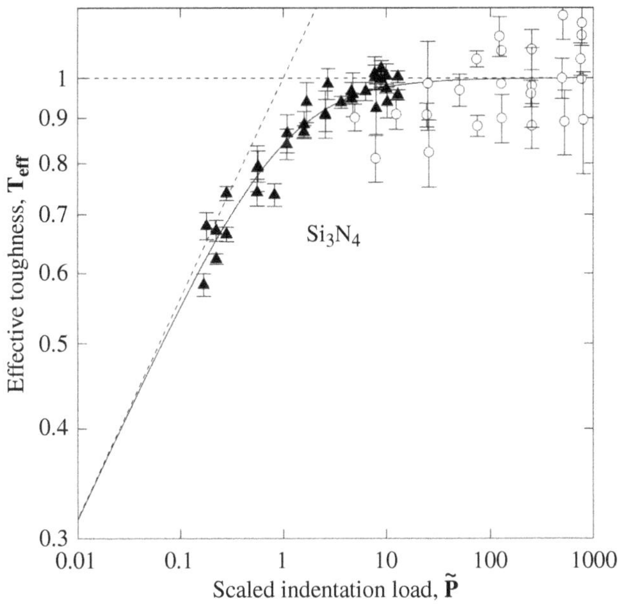

Figure 4.21 Logarithmic plot of scaled effective toughness values of polycrystalline Si_3N_4. Solid line represents universal response based on indentation fracture mechanics analysis including microstructural and lateral crack effects. Dashed lines represent small and large indentation load asymptotes. Source: Data adapted from Lawn et al. (1980), Choi and Salem (1992), Choi et al. (1997), Li et al. (1992), Xu et al. (1995), and Lee et al. (1997).

The similarity in behavior of Si_3N_4 and Al_2O_3 and the role of grain size are emphasized in Figure 4.21, which shows a logarithmic plot of the scaled effective toughness behaviors for the microstructural extremes of the polycrystalline Si_3N_4 materials. The data were scaled using Eq. (4.36) and the same fitting parameters used for the best fits in Figure 4.19. Symbols with bars represent the means and standard deviation limits of scaled toughness values at the scaled loads indicated; uncertainties represent experimental uncertainties and do not include fitting uncertainties. Filled symbols represent the coarse-grained materials and open symbols represent the fine-grained materials. The fine solid line represents the universal response of Eq. (4.36). The horizontal dashed line represents the asymptotic contact controlled scaled bound of $\mathbf{T}_{\mathbf{eff}} = 1$. The diagonal dashed line represents the asymptotic microstructure controlled scaled bound of $\mathbf{T}_{\mathbf{eff}} = \tilde{\mathbf{P}}^{1/4}$. Consistent with similar plots for Al_2O_3 in Section 4.4.1, the data lie on the universal toughness response. Here, however, the increasing toughness of the coarse polycrystalline Si_3N_4 materials with increasing flaw size at small flaw size and the invariant toughness of fine polycrystalline Si_3N_4 materials are now both made clear.

4.4.3 Glass-Ceramics

The indentation strength behavior of various glass-ceramics has been examined in Chapters 1 and 3. Cordierite glass-ceramic was used to demonstrate the Griffith equation and ideal indentation strength behavior in Figures 1.10e and 3.2c, respectively. Pyroceram glass-ceramic was used to demonstrate ideal behavior and the effects of machining damage in Figures 3.2d and 3.45a (see also Shetty et al. 1985) and a leucite glass-ceramic was used to demonstrate the effects of lateral cracking in Figure 3.37d. (Commercial designations are given for identification purposes only.)

Glass-ceramic materials can be divided into three major groups based on the observed indentation strength behavior. The groups correspond approximately to the degree of heterogeneity in the glass-ceramic microstructure, reflecting both the size scale of the crystalline ceramic "particles" imbedded in the (usually) glassy matrix and the chemical difference between the particles and matrix. Figure 4.22 shows a logarithmic plot of observations and best fits of indentation strength behavior of glass-ceramics materials from Morena et al. (1983), Cook et al. (1985a, 1986), Cook (2015), Rosa et al. (2009), and Cesar et al. (2011), using data derived from the published works cited. Symbols with bars represent the means and standard deviation limits of experimental strength measurements for two representative materials at the indentation loads indicated. The shaded bands represent the bounds of intrinsic strength measurements, σ_i, for each material. The fine solid lines represent best fits of Eq. (4.28) to the indentation strength data. The diagonal dashed line represents the intermediate asymptotic contact controlled strength behavior for the stronger material and the complete best-fit response passes clearly

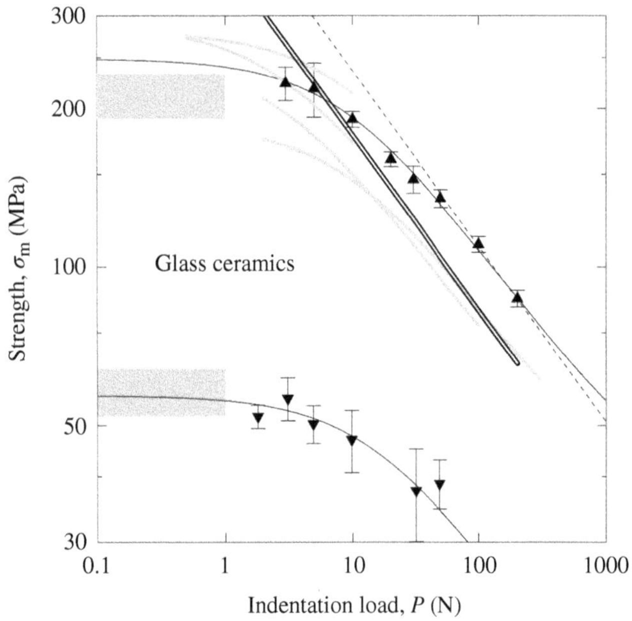

Figure 4.22 Logarithmic composite plot of Vickers indentation strengths of glass-ceramics. Dashed diagonal line indicates ideal indentation response. Upper and lower symbols represent strength responses of lithium silicate- and leucite-based materials, respectively. Fine lines are fits using a combined microstructural and lateral crack indentation fracture analysis. Near-ideal strength behavior shown as open line, microstructurally perturbed intermediate strength behavior shown as grayed lines. Source: Data adapted from Morena et al. (1983), Cook et al. (1985a, 1986), Cook (2015), Rosa et al. (2009), and Cesar et al. (2011).

through this asymptote; the ability of the combined microstructure + lateral crack indentation strength formulation of Section 4.3 to describe the experimental observations is clear. The open and grayed lines represent best fits to indentation strength responses of other glass-ceramic materials. In these cases, the plotted fits are restricted to the domains of the experimental measurements. The best fit to the Pyroceram material exhibits ideal behavior and is indicated by the open line. The best fits to other materials are indicated by grayed lines.

The open line and the two sets of symbols in Figure 4.22 represent the three groups of glass-ceramic indentation strength behavior. No group of glass-ceramics exhibits substantial lateral crack effects on indentation strength. The first group exhibits ideal $\sigma_m \sim P^{-1/3}$ behavior and includes Pyroceram and cordierite. These materials exhibit large values of intrinsic strengths, $\approx$ 300–400 MPa (out of range in Figure 4.22). The second group exhibits substantial microstructural effects in indentation strengths and is typified by the lithium silicate material (Li_2O_3 platelets in SiO_2 cristobalite phase) represented by the upper symbols. This group includes other lithium silicate materials, other cordierite-based materials, and Macor (mica platelets in cristobalite), shown as grayed lines. These materials exhibit moderate values of intrinsic strengths, $\approx$ 200 MPa. The third group also exhibits substantial microstructural effects in indentation strengths and is typified by the leucite-based dental porcelain material represented by the lower symbols. Note that such materials contain very small amounts of crystalline phase and are essentially agglomerated glass. These materials exhibit small (glass-like) values of intrinsic strengths, $\approx$ 50–150 MPa. As noted in consideration of alumina and silicon nitride for the second and third groups, there is a clear correlation between the best-fit upper indentation strength asymptotes and the intrinsic strengths.

Figure 4.23 plots the results of indentation strength tests of the glass-ceramic materials following the format of Figure 4.14. Figure 4.23a shows a logarithmic plot of the scaled experimental indentation strength behaviors of the materials. Symbols represent the means of scaled strengths at the scaled loads indicated; uncertainties are omitted for clarity. Open, grayed, and filled symbols follow the format of Figure 4.22 for the three groups of materials. Lines represent the scaled universal strength response and asymptotes. The data all lie on the universal response. The strengths of the second and third groups (filled and grayed symbols) exhibit clear microstructural effects and the first group exhibit a clear absence of such effects. The strengths of materials from the second and third groups exhibit flaw tolerance at small flaw sizes extending over approximately a factor of 10 in indentation load, comparable to the Mg-doped alumina materials of Figure 4.16a. Figure 4.23b shows a logarithmic plot of the scaled effective toughness behaviors of the materials. Symbols are as in Figure 4.23a. Lines

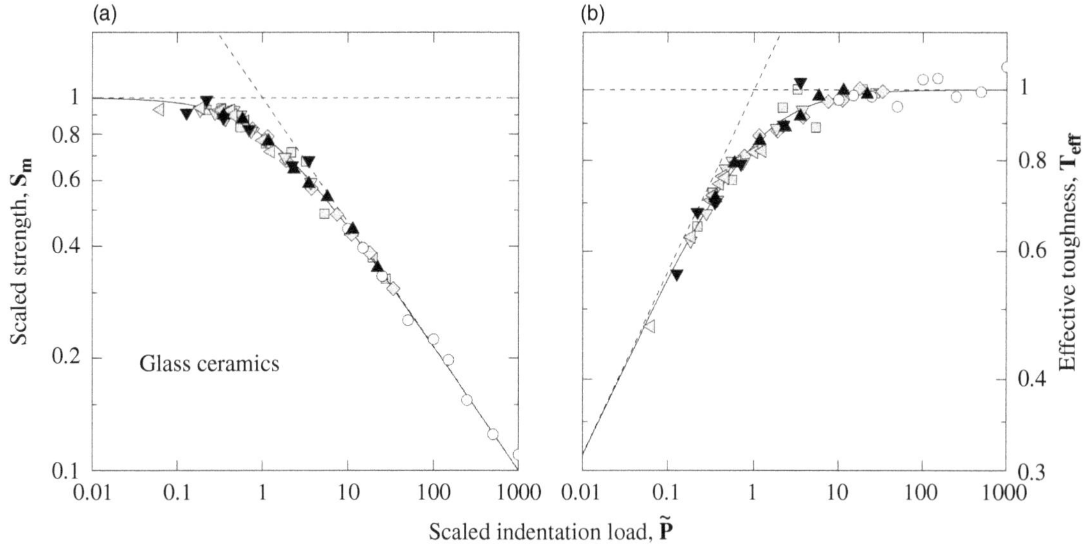

Figure 4.23 Logarithmic plots of (a) scaled Vickers indentation strengths and (b) scaled effective toughness values of glass-ceramics. Solid lines represent universal responses based on indentation fracture mechanics analysis including microstructural and lateral crack effects. Dashed lines represent small and large indentation load asymptotes. Source: Data adapted from Morena et al. (1983), Cook et al. (1985a, 1986), Cook (2015), Rosa et al. (2009), and Cesar et al. (2011).

represent the scaled universal toughness response and asymptotes. The data all lie on the universal response. The explanation for the flaw tolerance exhibited by the second and third groups of glass-ceramic materials is apparent in the increase in toughness with flaw size (filled and some grayed symbols). The effective toughness increases by approximately a factor of 2.5 as the flaw size increases by a factor of approximately 10^2. In contrast, the toughness of the first group of materials remains invariant (open symbols). The explanation in this case is that the majority of commercial glass-ceramic materials (e.g., Pyroceram) exhibit small crystallized grains that are well bonded to fracture resistant glassy grain boundaries—both features act against microstructural toughening effects. The antithesis to these structures is a commercial glass-ceramic (Macor) that incorporates easily cleaved grains (mica) to permit machinability and cracking at small scales. However, the easy cleavage leads to significant microstructural effects and toughening for large cracks.

4.4.4 Ferroelastics

The indentation strength behavior of various polycrystalline titanate materials has been examined in Chapter 3. Multiphase titanate-based SYNROC (a synthetic rock nuclear waste storage material, Ringwood et al. 1981; Ringwood 1982) was used to demonstrate ideal indentation strength behavior in Figure 3.2b. Bi-doped $BaTiO_3$ was used to demonstrate the effects of lateral cracking in Figure 3.37c and then used as a reference material for La-doped $PbTiO_3$ in Figure 3.50c to demonstrate the effects of stress gradients in thin specimen testing. $BaTiO_3$- and $PbTiO_3$-based materials are used extensively as dielectrics in microelectronic capacitor applications. $BaTiO_3$ and $PbTiO_3$ are also *ferroelectrics* that can maintain permanent electrical polarization in the absence of electric fields (Jona and Shirane 1962; Lines and Glass 1977).

The distinguishing feature of ferroelectric materials here is that they are also *ferroelastic*—they exhibit permanent changes in strain in the absence of changes in temperature. The strains are associated with crystallographic phase transitions, from phases of lesser symmetry to phases of greater symmetry as temperature is increased through a critical value. The oft-cited example is $BaTiO_3$, which undergoes a transition from a tetragonal crystal structure to a cubic structure at 120 °C. To accommodate intergrain and interdomain distortions associated with misaligned tetragonal crystal structures, considerable stress and strain distributions are generated throughout polycrystalline and single-crystal $BaTiO_3$ materials at room temperature. Experimental observations of such stresses are discussed in an overview by Howell et al. (2020). The stresses, and their absence, have long been associated with microstructural effects (and their absence) on strengths of polycrystalline $BaTiO_3$ materials measured below and above the phase-transition temperature. The strengths were a focus of the review by Freiman and Pohanka (1989) that also included discussion of the indentation strength experiments and analysis of Cook et al. (1983). Indentation strength measurements of a series of polycrystalline $BaTiO_3$ materials, including

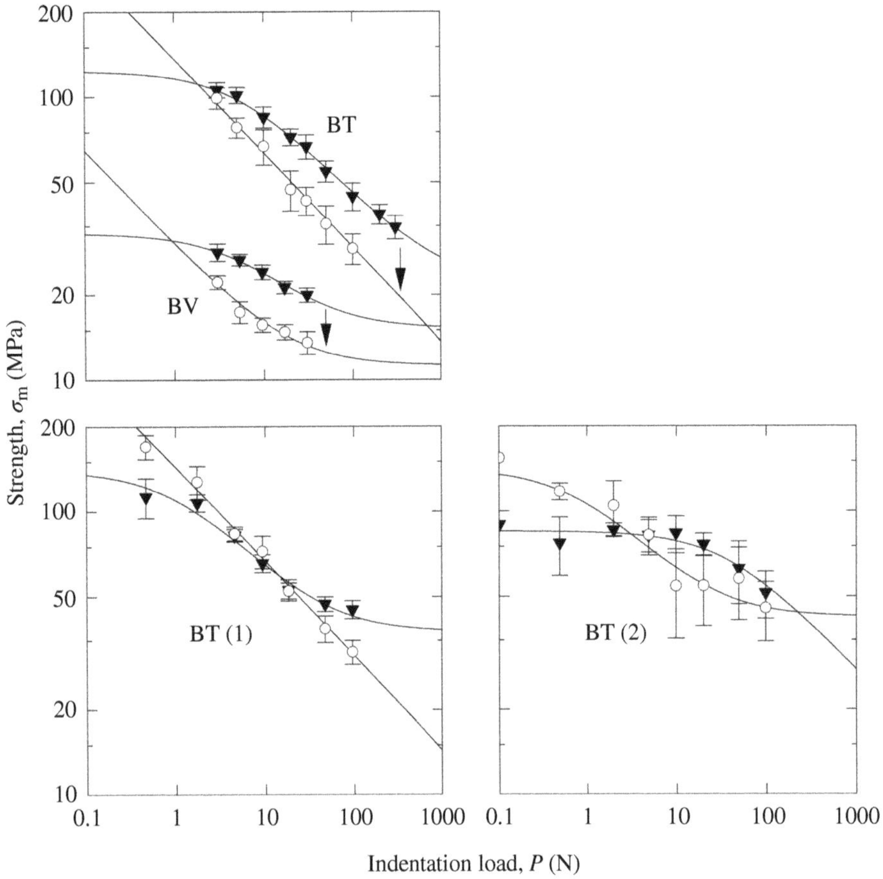

Figure 4.24 Logarithmic plots of Vickers indentation strengths of ferroelastic materials: BT, barium titanate; BV, bismuth vanadate. Filled symbols represent room temperature measurements. Open symbols represent measurements at higher temperature and phases of greater crystalline symmetry. Solid lines are fits using a combined microstructural and lateral crack indentation fracture analysis. Source: Includes data adapted from Cook et al. (1983, 1985a) and Baker et al. (1991).

the earlier work, are considered here, using temperature to control the intensity of the microstructural effect, and placing the measurements in the context of indentation strength measurements of other materials. In addition, the indentation strength properties of a ferroelastic material, polycrystalline $BiVO_4$, are also considered. $BiVO_4$ undergoes a transition from a monoclinic crystal structure to a tetragonal structure at 225 °C and is a "full" ferroelastic material in that there is no coupling to ferroelectric behavior.

Figure 4.24 shows logarithmic plots of the indentation strength behavior of polycrystalline $BaTiO_3$, labeled BT, and $BiVO_4$, labeled BV, from Cook et al. (1983, 1985a) and Baker et al. (1991), using data derived from the published works cited. Symbols with bars represent the means and standard deviation limits of experimental strength measurements at the indentation loads indicated. BT(1) and BT(2) have smaller and larger grain sizes, respectively, relative to BT. Filled symbols represent the results of strength measurements performed at room temperature. The BT and BV materials are in phases of lesser symmetry, tetragonal and monoclinic, respectively, at this temperature. Open symbols represent the results of strength measurements performed at 150 °C, BT, and 300 °C, BV. The BT and BV materials are in phases of greater symmetry, cubic and tetragonal, respectively, at these temperatures. At room temperature, the materials are in ferroelastic phases, and at the elevated temperatures, the materials are in paraelastic phases (i.e., conventional elastic phases). Microstructurally controlled stress distributions arising from ferroelastic effects are present in the materials at room temperature and are relaxed at the elevated temperatures. The fine solid lines represent best fits of Eq. (4.28) to the indentation strength data; the ability of the combined microstructure + lateral crack indentation strength formulation of Section 4.3 to describe the experimental observations of Figure 4.24 is clear, as is the resemblance of the raw data best-fit responses of Figure 4.24 to the universal scaled response of Figure 4.9.

The distinctive feature of Figure 4.24 is that for all four materials, the transition from a ferroelastic state to a paraelastic state was accompanied by a complete or near-complete removal of microstructural effects in the indentation strength response. In more direct terms, the concave strength response at small flaw sizes was eliminated or reduced. In some cases, BT and BT(1), the phase transition lead to a change from a sigmoidal indentation strength response to an ideal response. In some cases, BV and BT(2), the transition lead to change from a microstructure dominated concave response to a lateral crack dominated convex response. In some cases, BT and BV, the transition lead to a uniform decrease in strength, indicated by the arrows in Figure 4.24. Overall, the data provide clear confirmation of the expectation that microstructurally derived internal stress fields in polycrystalline materials contribute significantly to toughening effects in fracture. Close inspection of the relative responses of the $BaTiO_3$ materials in Figure 4.24 shows that the plateau strengths at small flaw sizes decrease in order of increasing grain size, BT(1), BT, BT(2), and that the strengths at large flaw sizes increase in order of increasing grain size, BT(1), BT, BT(2). Both of these trends are consistent with the effects of grain size observed for polycrystalline Al_2O_3 and Si_3N_4 in Sections 4.4.1 and 4.4.2, respectively.

Figure 4.25 plots the results of indentation strength tests of the ferroelastic materials following the format of Figure 4.14. Figure 4.25a shows a logarithmic plot of the scaled experimental indentation strength behaviors of the materials. Symbols represent the means of scaled strengths at the scaled loads indicated; uncertainties are omitted for clarity. Filled and open symbols follow the format of Figure 4.24 for materials in the ferroelastic and paraelastic states. Lines represent the scaled universal strength response and asymptotes. The data all lie on the universal response. The strengths of the ferroelastic group (filled symbols) exhibit clear microstructural effects and the paraelastic group (open symbols) exhibit a clear absence of such effects. The strengths of materials from the ferroelastic group exhibit flaw tolerance at small flaw sizes extending over approximately a factor of 10^2 in indentation load, comparable to the commercial alumina materials of Figure 4.18a. Figure 4.25b shows a logarithmic plot of the scaled effective toughness behaviors of the materials. Symbols are as in Figure 4.25a. Lines represent the scaled universal toughness response and asymptotes. The data all lie on the universal response. The explanation for the flaw tolerance exhibited by the ferroelastic group of materials is apparent in the increase in toughness with flaw size (filled symbols). The effective toughness increases by approximately a factor of 3 as the flaw size increases by a factor of approximately 10^3. In contrast, the toughness of the paraelastic group of materials remains invariant (open symbols). A possible explanation in this case is that the grain boundaries of the materials in the paraelastic state exert very weak perturbation on the crack path and the invariant toughness values reflects predominantly transgranular fracture. Images of BT and BV fracture surfaces from materials in the paraelastic states do indeed exhibit markedly greater proportions of transgranular failure than from those in the ferroelastic states (Pohanka et al. 1978; Baker et al. 1991), consistent with this inference.

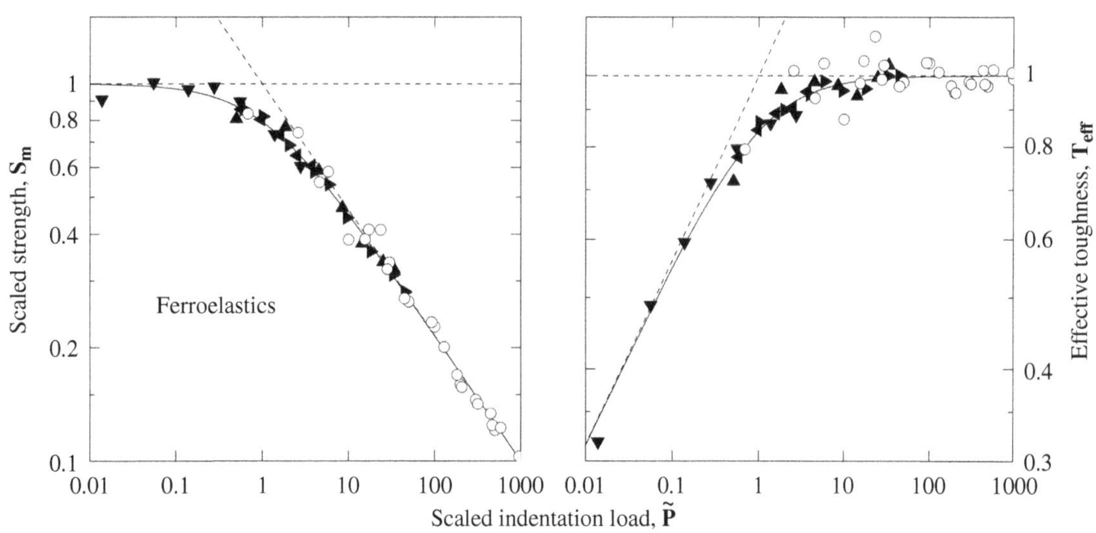

Figure 4.25 Logarithmic plots of (a) scaled Vickers indentation strengths and (b) scaled effective toughness values of ferroelastic materials. Solid lines represent universal responses based on indentation fracture mechanics analysis including microstructural and lateral crack effects. Dashed lines represent small and large indentation load asymptotes. Source: Includes data adapted from Cook et al. (1983, 1985a) and Baker et al. (1991).

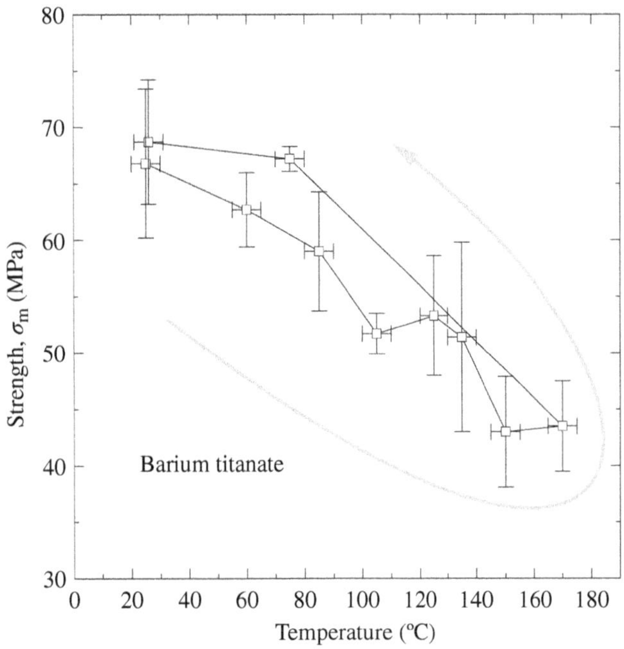

Figure 4.26 Plot of strength variation of polycrystalline barium titanate with temperature. Specimens contained $P = 30$ N Vickers indentations. Temperature was cycled from 25 to 170 °C and reverse, as indicated by arrow.

The reversibility of the toughening processes associated with the temperature-controlled microstructural stress state is illustrated in Figure 4.26. Here, the indentation strengths of the BT material are plotted as a function of temperature for a fixed $P = 30$ N indentation load. Symbols and bars represent the means and standard deviation limits of experimental strength measurements at the temperatures indicated (±5 °C temperature uncertainty). The grayed line is a guide to the eye indicating the direction of the temperature cycle used in the sequential strength measurement experiments. All specimens were equilibrated in silicone oil at the temperatures indicated prior to strength testing. That is, the specimens tested at ≈ 70 °C and ≈ 25 °C on the cooling cycle had been equilibrated at ≈ 170 °C prior to cooling. Although experimental uncertainty precludes definitive assessment of the exact dependence of strength on temperature, it is clear that the strength decreases monotonically with temperature increase and then increases reversibly on temperature decrease. As the changes in homologous temperature are small here relative to the BT processing temperature, it is unlikely that changes in the grain and pore structure could give rise to the strength changes observed. Similarly, stress relaxation effects are also unlikely, and not reversible. Hence, it appears that the cyclic temperature strength measurements reinforce the idea that stress inhomogeneity in polycrystalline structures contributes significantly to microstructural toughening effects. The idea is supported by the observations of Jones and Hoffman (2006) in "resetting" of *R* curve effects in ferroelastic materials by an intermediate annealing step similar to that observed by Knehans and Steinbrech (1982) on renotching of alumina. In the case of ferroelastic materials, it is also likely that annealing removed some "frontal" zone domain transformations that contributed to toughening (see Section 4.4.6), as well as relaxing frictional stresses in the crack wake.

4.4.5 Alumina Composites

It is a clear logical step from the measurements of indentation strength and toughness variation of the polycrystalline alumina and ferroelastic materials to infer that the *deliberate incorporation* of microstructural elements giving rise to internal stress distributions is a potential method for fabricating brittle materials with enhanced flaw tolerance. Thus, this was the step taken in a series of works emanating from Lehigh University in the early 1990s that focused on incorporating aluminum titanate (Al_2TiO_5) and mullite ($3Al_2O_3{\cdot}2SiO_2$) into polycrystalline alumina (Al_2O_3) composite microstructures. Stress distributions arise in polycrystalline Al_2O_3 microstructures from anisotropic thermal expansion effects. Single-crystal Al_2O_3 has hexagonal symmetry, and the basal and axial CTE values differ. On cooling from the processing temperature, the misoriented grains in polycrystalline Al_2O_3 do not strain uniformly and reaction stresses develop. These CTE-based stresses are responsible for the development of microcracks on cooling (see Chapter 1) and the formation and effectiveness of crack

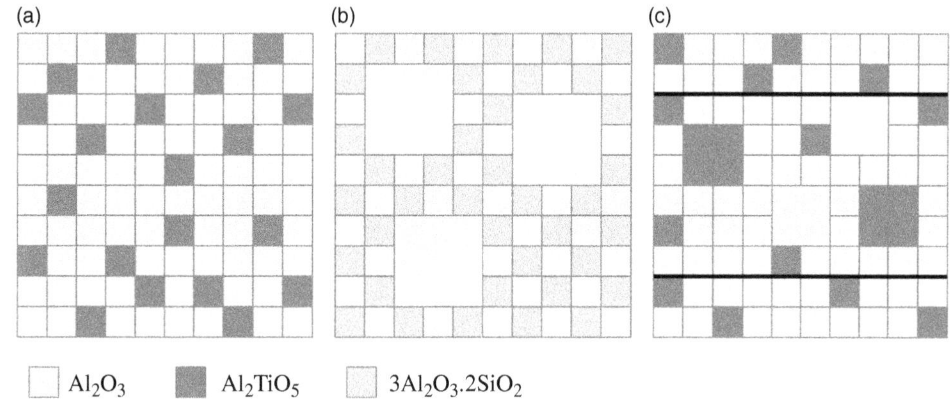

Figure 4.27 Schematic diagrams of alumina (Al_2O_3)-based composite microstructures. (a) Al_2TiO_5 grains dispersed in a polycrystalline Al_2O_3 matrix. (b) Large Al_2O_3 grains distributed in an Al_2O_3–$3Al_2O_3{\cdot}2SiO_2$ (alumina–mullite) matrix. (c) Large Al_2O_3 and Al_2TiO_5 grains distributed in an Al_2O_3 matrix laminated by small-grained Al_2O_3–Al_2TiO_5 layers.

wake bridging tractions on applied loading (see Section 4.2). In the Al_2O_3-based composites, stress distributions arise in the microstructure from inhomogeneous thermal expansion effects. In particular, Al_2TiO_5, although quite anisotropic, has an average CTE much greater than that of Al_2O_3. In contrast, $3Al_2O_3{\cdot}2SiO_2$ has an average CTE less than that of Al_2O_3.

Schematic diagrams of the developed Al_2O_3 composite microstructures are shown in Figure 4.27. The Al_2O_3–Al_2TiO_5 composites (Runyan and Bennison 1991; Bennison et al. 1991; Padture et al. 1993a,b; Lawn et al. 1993) typically contained Al_2TiO_5 volume fractions of 20%, as illustrated in the schematic diagram of Figure 4.27a. Note that on cooling from the stress- and strain-free processing temperature (typically about 1500 °C), the room temperature stress state of the composite places an isolated Al_2TiO_5 grain in hydrostatic tension and the surrounding Al_2O_3 matrix in radial tension and circumferential compression. For many Al_2TiO_5 grains in proximity, these stress states superpose, generating a complicated stress distribution within the microstructure that can greatly affect crack propagation. If the grain sizes become too large, these stresses also cause the material to microcrack (Padture et al. 1993b; Lawn 1993). The Al_2O_3–$3Al_2O_3{\cdot}2SiO_2$ composites (Khan et al. 1998, 2000a,b) typically contained large agglomerates of Al_2O_3 in volume fractions of 30% within a fine-grained matrix of Al_2O_3 + $3Al_2O_3{\cdot}2SiO_2$ in equal proportions, as illustrated in the schematic diagram of Figure 4.27b. Note that the Al_2O_3 + $3Al_2O_3{\cdot}2SiO_2$ matrix has an average CTE less than that of Al_2O_3. As a consequence, on cooling from the processing temperature, the room temperature stress state of the composite places an isolated Al_2O_3 agglomerate in hydrostatic tension and the surrounding mixed-phase matrix in radial tension and circumferential compression. This is identical to that for the Al_2O_3–Al_2TiO_5 composite and leads to the same crack propagation and microcracking effects. A structural variation on the Al_2O_3–Al_2TiO_5 composite microstructure is shown in the schematic diagram of Figure 4.27c (Russo et al. 1992). In this case, a symmetric trilayer laminate structure was formed through tape casting methods. The outer layers were formed of a very fine-grained Al_2O_3–20% Al_2TiO_5 material as described in Figure 4.27a. The inner layer contained larger-grained agglomerates of Al_2O_3 and Al_2TiO_5 within the fine-grained matrix. The idea was to sandwich a flaw tolerant inner layer between two large strength outer layers, thereby maximizing performance of the structure for both large and small flaws.

Figure 4.28 shows logarithmic plots of the indentation strength behavior for the Al_2O_3 composite materials, using data derived from the published works cited: (a) Al_2O_3–Al_2TiO_5, (b) Al_2O_3–$3Al_2O_3{\cdot}2SiO_2$, and (c) and (d) Al_2O_3–Al_2TiO_5 laminar structure. Symbols with bars represent the means and standard deviation limits of experimental strength measurements at the indentation loads indicated. The shaded bands represent the means and standard deviation limits of the intrinsic strengths, σ_i, for each material. Single shaded lines represent individual intrinsic strength measurements. The fine solid lines represent best fits of Eq. (4.28) to the indentation strength data; the ability of the combined microstructure + lateral crack indentation strength formulation of Section 4.3 to describe the experimental observations of Figure 4.28 is clear, as is the resemblance of the raw data best-fit responses of Figure 4.28 to the universal scaled response of Figure 4.9. The grayed lines are guides to the eye and represent the responses of the two smallest grain size Al_2O_3 materials (2.5 and 9 μm) from Figure 4.15, as these are comparable to the matrix grain sizes (about 1–6 μm) of the composites. Even cursory inspection of Figure 4.28 shows that the composite microstructures developed—through significant ceramic processing expertise—exhibited flaw tolerance well in excess of the monolithic analogs. Figure 4.28a shows the indentation strength

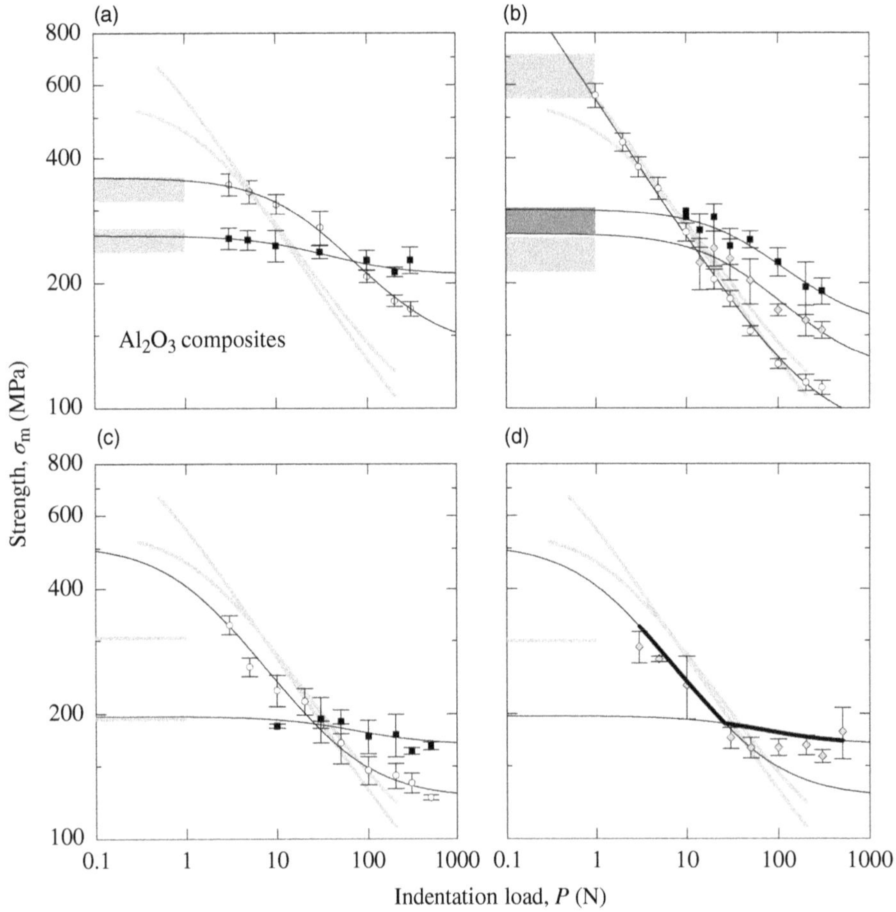

Figure 4.28 Logarithmic plots of Vickers indentation strengths of alumina-based composites. (a) Al_2O_3–Al_2TiO_5 (see Figure 4.27a). (b) Al_2O_3–$3Al_2O_3 \cdot 2SiO_2$ (see Figure 4.27b). (c) and (d) Al_2O_3–Al_2TiO_5 laminar structure (see Figure 4.27c). The grayed lines are repeated in each plot and represent the two smallest grain size materials from Figure 4.15. Source: Data adapted from Bennison et al. (1991), Padture et al. (1993a), Khan et al. (1998, 2000a,b), and Russo et al. (1992).

results for the Al_2O_3–Al_2TiO_5 system from Bennison et al. (1991) (filled symbols) and Padture et al. (1993a) (open symbols). The strength degradation for these systems was reduced by a factor of approximately 0.2 or less in range over the experimental indentation load domain. Figure 4.28b shows the indentation strength results for the Al_2O_3–$3Al_2O_3 \cdot 2SiO_2$ system from Khan et al. (1998, 2000a,b). The open symbols represent the behavior of the Al_2O_3–$3Al_2O_3 \cdot 2SiO_2$ matrix and follow the response of the fine-grained monolithic Al_2O_3 material. The filled and grayed symbols represent the behavior of the Al_2O_3–$3Al_2O_3 \cdot 2SiO_2$ system containing large and small Al_2O_3 agglomerates, respectively. Once again, the strength degradation for these systems was reduced by a factor of approximately 0.2 in range over the experimental indentation load domain. Figure 4.28c shows the indentation strength results for the Al_2O_3–Al_2TiO_5 component materials of the laminar system from Russo et al. (1992). The open symbols represent the behavior of the fine-grained Al_2O_3–Al_2TiO_5 outer layer material and exhibit a slightly more flaw tolerant response than fine-grained monolithic Al_2O_3 material. The filled symbols represent the behavior of the Al_2O_3–Al_2TiO_5 inner layer material containing Al_2O_3 and Al_2TiO_5 agglomerates. The strength degradation for this material was reduced by a factor of approximately 0.25 in range relative to the outer layer material over the experimental indentation load domain. Figure 4.28d shows the indentation strength results for the Al_2O_3–Al_2TiO_5 laminate structure. The best-fit fine and grayed lines are reproduced from Figure 4.28c. The bold line represents a simple maximum strength prediction based on the component strength responses. The observed indentation strength of the laminar structure closely follow the prediction and as intended indeed exhibits large strengths at small flaw sizes and significant flaw tolerance at large flaw sizes.

Figure 4.29 completes assessment of the designed composite materials and plots the results of indentation strength tests following the format of Figure 4.14. Figure 4.29a shows a logarithmic plot of the scaled experimental indentation strength

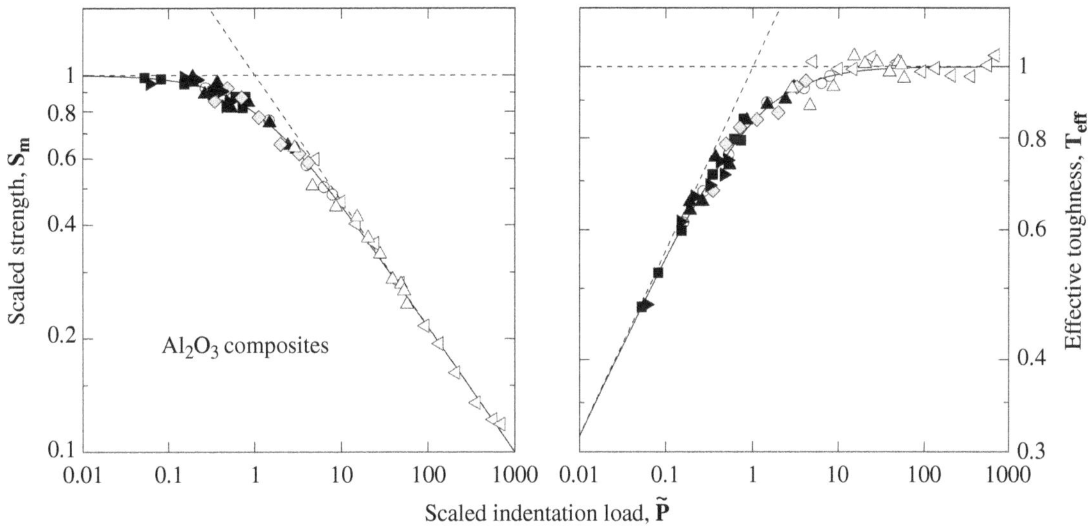

Figure 4.29 Logarithmic plots of (a) scaled Vickers indentation strengths and (b) scaled effective toughness values of alumina-based composites. Solid lines represent universal responses based on indentation fracture mechanics analysis including microstructural and lateral crack effects. Dashed lines represent small and large indentation load asymptotes. Source: Data adapted from Bennison et al. (1991), Padture et al. (1993a), Khan et al. (1998, 2000a,b), and Russo et al. (1992).

behaviors of the materials. Symbols represent the means of scaled strengths at the scaled loads indicated; uncertainties are omitted for clarity. Filled and open symbols follow the format of Figure 4.28a–c for the base and composite materials. Lines represent the scaled universal strength response and asymptotes. The data all lie on the universal response. The strengths of the composites (filled symbols) exhibit clear microstructural effects and the base materials (open symbols) exhibit a clear absence of such effects. The strengths of the composite materials exhibit flaw tolerance at small flaw sizes extending over approximately a factor of 10^2 in indentation load, comparable to the commercial alumina materials of Figure 4.18a. Figure 4.29b shows a logarithmic plot of the scaled effective toughness behaviors of the materials. Symbols are as in Figure 4.29a. Lines represent the scaled universal toughness response and asymptotes. The data all lie on the universal response. The explanation for the flaw tolerance exhibited by the composite materials is apparent in the increase in toughness with flaw size (filled symbols). The effective toughness increases by approximately a factor of 2.5 as the flaw size increases by a factor of approximately 10^3. In contrast, the toughness of the base materials remains invariant (open symbols).

It is instructive to compare the significant flaw tolerance and toughening results presented here for Al_2O_3 composites with the complete lack of flaw tolerance and toughening observed for another set of Al_2O_3 composites with reversed CTE effects. Dispersion of fine (0.1 μm) SiC particles into polycrystalline Al_2O_3 (and MgO) materials to form composite structures was much touted as a strengthening and toughening mechanism (see Zhao et al. 1993 for a brief review). Routine indentation crack length and strength measurements showed that the claims were based on an artifact associated with the machining of test specimens (Zhao et al. 1993) but did reveal some interesting points. First, the addition of the SiC particles altered the Al_2O_3 fracture mode from intergranular to transgranular such that the SiC–Al_2O_3 composite exhibited near-ideal indentation strength behavior as shown in Figure 3.2f. Second, the addition of the SiC particles greatly increased the susceptibility of the material to the incorporation of compressive surface stress on machining as shown in Figure 3.45c. In SiC–Al_2O_3 composites, stress distributions arise in the microstructure from inhomogeneous thermal expansion effects. However, in contrast to the above particle in matrix systems, SiC particles have a CTE much less than the Al_2O_3 matrix. On cooling from the stress- and strain-free processing temperature (1775 °C), the room temperature stress state of the composite placed the nearly completely isolated SiC particles in hydrostatic compression and the surrounding Al_2O_3 matrix in radial compression and circumferential tension—the opposite of the above composites. The particles were too small to initiate matrix microcracking but the stress fields generated in the surrounding matrix certainly aided transgranular fracture and thereby suppressed any crack wake toughening effects.

It is also instructive to compare the flaw tolerance and toughening presented here for Al_2O_3 composites with the much reduced flaw tolerance and toughening observed for polycrystalline oxides with absent CTE effects. Crystals with cubic symmetry may exhibit elastic and fracture anisotropy but cannot exhibit thermal expansion anisotropy.

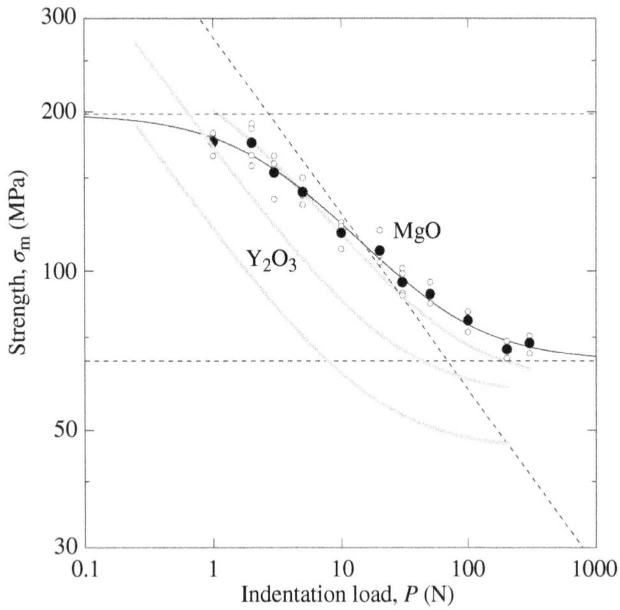

Figure 4.30 Logarithmic plot of Vickers indentation strengths of large grain polycrystalline MgO (symbols). Solid line is fit using a combined microstructural and lateral crack indentation fracture analysis. Dashed horizontal lines indicate strength bounds. Dashed diagonal line indicates ideal indentation response. Upper gray line represents small-grained MgO. Lower gray lines represent polycrystalline Y_2O_3 (see Figure 3.37). Source: Data adapted from Cook et al. (1994b).

As a consequence, polycrystalline cubic materials can display no inherent stress distributions based on anisotropic CTE effects. Microstructural effects on fracture and toughening can only arise from anisotropic deformation and cleavage within grains and weak grain boundary fracture resistance. A clear example of these effects is observed in the indentation strength behavior of polycrystalline MgO. MgO crystals display cubic symmetry. Figure 4.30 shows a logarithmic plot of the indentation strength behavior of polycrystalline MgO from Cook et al. (1994b) using data derived from the published work. Open symbols with bars represent individual strength measurements at the indentation loads indicated for grain size $\lambda = 26\,\mu$m material; filled symbols represent the mean strengths at each load. The fine solid line represents a best fit of Eq. (4.28) to the indentation strength data. The upper and lower horizontal dashed lines represent the asymptotic microstructure controlled and lateral crack controlled strength bounds of σ_m^* and σ_L, respectively. The diagonal dashed line represents the intermediate asymptotic contact controlled strength behavior—note that the complete best-fit response passes clearly through the ideal contact strength asymptote. The ability of the combined microstructure + lateral crack indentation strength formulation of Section 4.3 to describe the experimental observations of Figure 4.30 is clear, as is the resemblance of the raw data best-fit response of Figure 4.30 to the universal scaled response of Figure 4.9. The upper grayed line represents a best fit of Eq. (4.28) to the indentation strength data of the $\lambda = 9\,\mu$m MgO material, restricted to the experimental domain. The absence of significant flaw tolerance and grain size effect in the polycrystalline MgO indentation strength responses in Figure 4.30 is clear. The absence is emphasized by comparison with Figure 4.15 in which significant flaw tolerance and grain size effects in polycrystalline Al_2O_3 are obvious in the same grain size range. The lower gray lines represent best fits of Eq. (4.28) to the indentation strength data of the cubic and cubic + hexagonal polycrystalline Y_2O_3 materials from Figure 3.37e, restricted to the experimental domains. The absence of significant flaw tolerance, except for lateral crack effects at large indentation loads, is clear.

4.4.6 Zirconia

An extremely important class of toughening processes act by dissipating energy during fracture *ahead* of a propagating crack tip. Such processes lead to *frontal zone toughening* and are distinct from the wake zone toughening processes considered in Sections 4.4.1–4.4.5 that dissipate energy behind a crack tip. The most important of the frontal zone toughening processes in brittle materials is the dissipation of energy by irreversible phase transformations. The increase in fracture resistance on crack propagation in this case is known as "transformation toughening." The material predominantly associated with transformation toughening is polycrystalline zirconia, ZrO_2. Details of frontal zone toughening analyses and phase transformations in zirconia are given elsewhere (Heuer 1987; Green et al. 1989; Lawn 1993, Hannink et al. 2000), but a brief description is as follows: The stable equilibrium high-temperature polymorphs of ZrO_2 have cubic and tetragonal crystalline symmetry. The stable equilibrium low-temperature polymorph has monoclinic symmetry. On cooling during processing, from the single-phase tetragonal state or the mixed-phase cubic + tetragonal state, the tetragonal phase can be metastably retained at room temperature by alloying the material with various "stabilizing" metal oxides. For the solely tetragonal state resulting in a tetragonal zirconia polycrystal (TZP), Y_2O_3 and CeO_2 are preferred. For the mixed cubic + tetragonal state resulting in partially-stabilized zirconia (PSZ), CaO and MgO are preferred. The metal oxides stabilize the tetragonal phase against transformation to the monoclinic phase and the associated 5% dilatational strain and 8% shear strain. Under the influence of an applied stress, the retained metastable tetragonal phase can undergo an irreversible tetragonal → monoclinic (t → m) phase transformation. In particular, the t → m transformation can be initiated in the enhanced stress field of a crack tip frontal zone. Energy is thus dissipated on crack propagation in ZrO_2 materials by irreversible deformation adjacent to a crack surface. The dissipated energy is not available for the creation of new fracture surfaces and the creation of the associated surface energy. In the frame of reference of the applied loading, the material thus appears toughened.

A schematic diagram of the transformation toughening process in TZP is shown in Figure 4.31. A polycrystalline specimen consisting of a field of tetragonal ZrO_2 grains is shown in Figure 4.31a. The specimen contains a short crack, indicated in black, that propagates horizontally from the left and terminates in the field. The specimen is loaded by a vertically oriented applied tensile stress indicated by the dark arrows. Under the action of the applied stress, the t → m transformation is initiated in the enhanced stress field of the crack tip at a characteristic transformation stress of σ_T. In TZP, the t → m transformation takes the form of lath-like monoclinic twins that initiate at grain boundaries and extend across grains. Close to the crack tip, the characteristic transformation strain ε_T is reached such that the work/volume performed by the applied stress in fully transforming a grain is $\sim \sigma_T \varepsilon_T$. All the grains in Figure 4.31a have not traversed such a full stress strain cycle of transformation and the zone of transformation has not extended to the full extent allowed by the material. Both of these conditions are attained for long cracks propagating in steady state as shown for the broken specimen in Figure 4.31b. The specimen contains a long crack that has propagated to the right and out of the field of grains and is unloaded. The t → m transformation cycle is completed for the twinned, relaxed grains adjacent to the crack surface, and now in crack wake. The zone of transformation extends to a uniform, maximum width adjacent to each side of the crack surface in the crack wake, w_T. The work/area performed by the applied stress in forming a steady-state transformation zone is thus $\sim 2w_T\sigma_T\varepsilon_T$. Many geometrical factors have been omitted from the above considerations and such factors play a significant role (Green et al. 1989; Lawn 1993) if the transformation process is viewed in a stress framework rather than an energy framework as

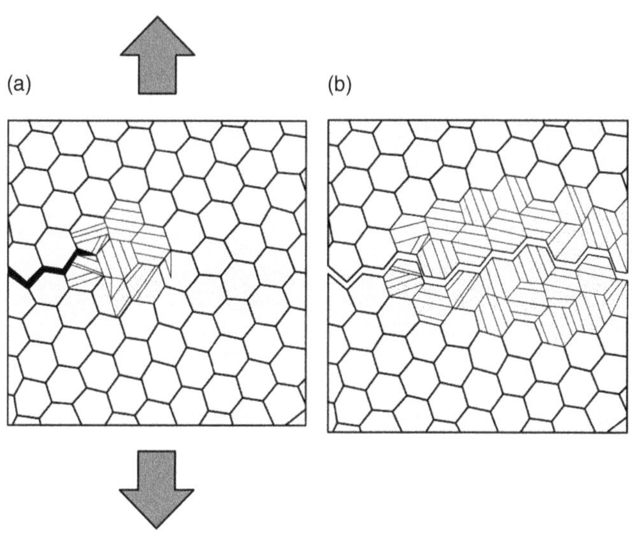

Figure 4.31 Schematic diagram of crack propagation in a phase-transforming tetragonal zirconia polycrystal microstructure. Grain boundaries indicated by fine lines. (a) Initial state with component loaded vertically and initial intergranular crack outlined in bold (left). Stress concentration at crack generates a localized crack front zone of transformation twins. (b) Final state with component separated along the fracture surface. A steady-state zone of transformation twins extends with fixed width along both sides of the crack.

here. The fracture of PSZ is similar to that described with the exception that the t → m transformation and twinning are restricted to tetragonal precipitates within cubic grains. However, in all cases, the underlying physical principles remain the same and in the applied loading reference frame the fracture resistance increases by $\Delta R \sim w_T \sigma_T \varepsilon_T$.

The expression for the energy scaling of ΔR can be used in several ways to make correspondence with the stress framework of fracture. In a "weak" toughening limit, $\varepsilon_T \approx 0$, the SIF formalism and the Irwin relation are assumed to apply (see Chapter 1), such that the steady-state zone size and transformation stress are related by

$$\sigma_T \sim \frac{(R_0 E)^{1/2}}{w_T^{1/2}}, \tag{4.37}$$

and thus

$$T_\infty \sim T_0 \left[1 + \beta E \varepsilon_T / \sigma_T\right]^{1/2}, \tag{4.38}$$

as noted in Section 4.2, where β is a dimensionless geometrical factor. In a "strong" toughening limit, $\varepsilon_T \gg 0$, the conformation of stress and strain at the crack tip is not assumed, although a secant relationship $\sigma_T \sim \varepsilon_T$ between the transformation stress and strain is taken. The Irwin relation of Eq. (4.7) is taken to apply in the limit $\Delta T / T_0 \gg 1$ and thus

$$T_\infty \sim \Delta T \sim \beta E w_T^{1/2} \varepsilon_T, \tag{4.39}$$

where β is again a dimensionless geometrical factor. Note the prominent role of the $E\varepsilon_T$ product in Eqs. (4.38) and (4.39). Steady-state toughening is proportional to the imposed transformation strain and the elastic reaction of the surrounding matrix. Similar scaling ideas apply to toughening by plastic deformation in WC–Co composites (Marshall et al. 1990).

Figure 4.32 shows an example of the steady-state fracture transformation configuration at Vickers indentation cracks in Y-TZP. The square contact impression is visible in the center of the indentation and surrounded by a zone of uplifted transformed material with a somewhat ragged circular perimeter. The surface traces of the cracks emanating from the impression corners are also accompanied by zones of uplifted transformed material. The fracture-related zones are of constant width along the crack wakes. Crack traces with constant width zones of uplifted material are shown in Marshall (1986), Marshall and Swain (1988), and Green et al. (1989). (Short cracks with initial variable width zones as in Figure 4.31a are shown in Yu and Shetty 1989, 1990 and Kaur et al. 2007.) Indentations with surrounding circular outline transformation zones are shown in Marshall (1986), Marshall and Swain (1988), Lai et al. (1989), Green et al. (1989), and Ruiz and Readey (1996). The cracks in these earlier images are either absent (subthreshold indentations) or very short and contained within the transformation zone (trapped, see Cook et al. 1994a and Cook and Braun 1994).

A generalization of Eq. (4.39) is

$$T(c) \sim E w_T^{1/2} \varepsilon_T \beta(c^{1/2} / w_T^{1/2}), \tag{4.40}$$

where $\beta = \beta(c)$ is now a dimensionless function of crack length that is applicable for $c > w_T$ and approaches the limiting value $\beta \to 1$ for $c \gg w_T$. Strictly, β is a function of the ratio $c^{1/2}/w_T^{1/2}$. Many experimental toughness measurements of cracks generated by indentations (Marshall 1986; Marshall and Swain 1988; Anderson and Braun 1990), in edge notched beams (Rose and Swain 1986; Marshall and Swain 1988; Yu and Shetty 1989, 1990), in double cantilever beams (Heuer 1987; Marshall and Swain 1988; Lai et al. 1989), and at pores (Kaur et al. 2007) suggest that β is very close to a linear function of

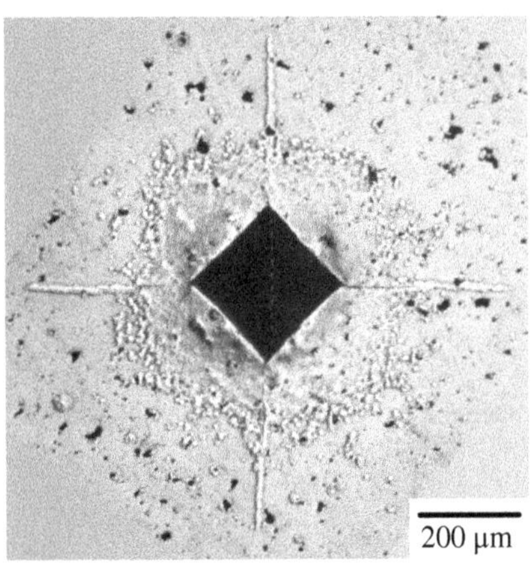

Figure 4.32 Optical image of a Vickers indentation flaw in a tetragonal zirconia polycrystal. The uplifted zones of phase-transformed material surrounding the contact impression and the surface traces of radial cracks are clear.

$c^{1/2}/w_{\mathrm{T}}^{1/2}$. Hence, the T curve used here to consider indentation strength behavior of zirconia is given by the unbounded function

$$T(c) = T_\tau \left(\frac{c}{\delta}\right)^{\tau/2}, \tag{4.41}$$

where the exponent $\tau \leq 1$ and T_τ and δ are empirical parameters describing the toughness behavior over the domain of crack lengths sampled by an indentation strength test. Other bounded functions (arctan, Heuer 1987; exponential, Kaur et al. 2007) are also possible. As will be shown, the advantages of Eq. (4.41) are that it provides analytical tractability and allows broader insight for indentation strength analyses. A form of Eq. (4.41) was used to describe indentation strength observations of alumina in a completely empirically manner (Krause 1988).

Indentation strength analysis for such toughened materials is simply achieved by replacing $T(c)$ from Eq. (4.41) in Eq. (4.20) to gain

$$\psi\sigma_{\mathrm{a}}c^{1/2} + \chi P/c^{3/2} = T_\tau \left(\frac{c}{\delta}\right)^{\tau/2}. \tag{4.42}$$

The equilibrium and stability conditions are then simply written in matrix form similar to Eq. (3.6) as

$$\begin{bmatrix} 1 & 1 \\ 1 & -3 \end{bmatrix} \begin{bmatrix} \mathbf{K_a} \\ \mathbf{K_r} \end{bmatrix} = \begin{bmatrix} 1 \\ \tau \end{bmatrix}, \tag{4.43}$$

where $\mathbf{K_a} = K_{\mathrm{a}}/T(c)$ and $\mathbf{K_r} = K_{\mathrm{r}}/T(c)$ are scaled, generalized, SIF variables with additional crack length dependence. Solution of the simultaneous equations represented by Eq. (4.43) yields the indentation strength σ_{m}:

$$\sigma_{\mathrm{m}} = \frac{(3+\tau)}{4\psi}\left(\frac{T_\tau}{\delta^{\tau/2}}\right)^{4/(3+\tau)}\left(\frac{4\chi P}{(1-\tau)}\right)^{-(1-\tau)/(3+\tau)}. \tag{4.44}$$

The expression for σ_{m} can be written simply as

$$\sigma_{\mathrm{m}} = SP^{(-1/3)(3-3\tau)/(3+\tau)}, \tag{4.45}$$

which reverts to the ideal solution, Eq. (3.9), for $\tau = 0$ characterizing an invariant toughness. Here, the effects of an increasing toughness were approximated by a power law with $\tau > 0$ intended. As a consequence, the inert strength σ_{m} does not decrease with indentation load P as rapidly as the ideal response. The reciprocal nature of the T curve and indentation strength variations are noted that in the general case for any form of strong T curve, the derivatives of the logarithmic strength and toughness variations are related in a negative manner by

$$\frac{\mathrm{d}\ln(\sigma_{\mathrm{m}})}{\mathrm{d}\ln(P)} \approx -\frac{1}{3}\left(1 - 2\frac{\mathrm{d}\ln(T)}{\mathrm{d}\ln(c)}\right). \tag{4.46}$$

Weak variation of strength with indentation load results from strong variation of toughness with crack length. It is noted from Eq. (4.45) that in this case an upper limit is $\tau = 1$, corresponding to no variation of indentation strength with indentation load, i.e., the left side of Eq. (4.46) vanishes. For materials with very steep T curves described by $\tau > 1$, the indentation strength technique is inapplicable and other fracture geometries with greater stabilization, e.g., displacement-controlled double cantilever beams, are required to measure the toughness variation.

Figure 4.33 shows logarithmic plots of the indentation strength behavior of transformation toughened ZrO_2 from Readey et al. (1993), Readey and McCallen (1995), and Marshall (1986), using data derived from the published works cited. Symbols represent the results of individual strength measurements at the indentation loads indicated. Filled and open symbols in the left plot indicate the behavior of fine- and coarse-grained Ce-TZP polycrystals, respectively, from Readey and McCallen. Filled and open symbols in the right plot indicate the behavior of peak-aged and overaged Mg-PSZ polycrystals, respectively, from Marshall. Shaded bands indicate the means and standard deviation limits of intrinsic strengths σ_{i} for each material. The diagonal dashed lines are a guide to the eye repeated in both plots of ideal indentation strength behavior $\sigma_{\mathrm{m}} \sim P^{-1/3}$ drawn to approximate the large load behavior of the overaged Mg-PSZ material. The clear feature of Figure 4.33 is that both TZP materials and the peak-aged PSZ material exhibit significant flaw tolerance: in each of these cases, there is little to no variation of indentation strength out of the intrinsic band for indentation loads varying over a factor of 10^2. The implication is that the t → m transformation was triggered by crack propagation in these materials leading to significant T curves and flaw tolerance. The implication for the overaged PSZ material is that the majority of the t → m transformation was triggered spontaneously by thermal effects *prior* to crack propagation, leading to a weak T curve, weak flaw tolerance, and considerable strength degradation.

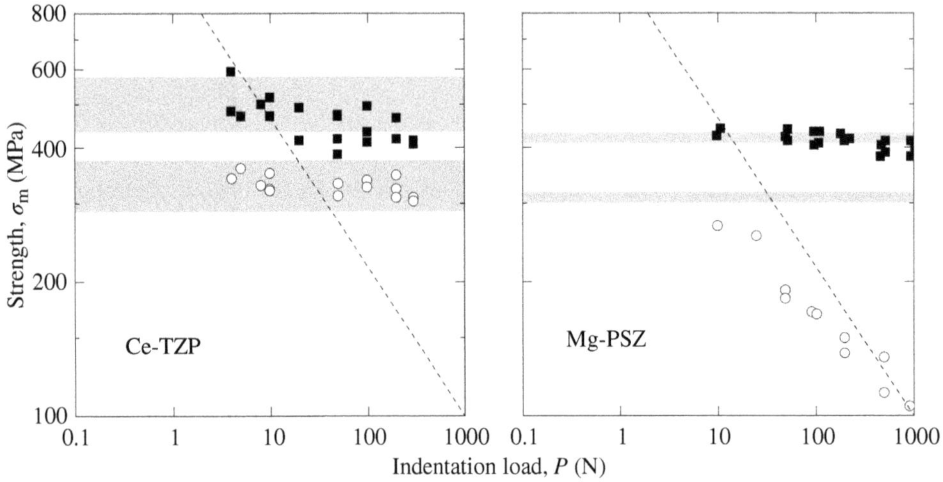

Figure 4.33 Logarithmic plot of Vickers indentation strengths of polycrystalline zirconia. Dashed diagonal lines indicate ideal indentation responses. Filled symbols represent materials exhibiting significant tendency to transformation toughening, leading to almost complete flaw tolerance. Shaded bands represent intrinsic strengths. Source: Data adapted from Readey et al. (1993), Readey and McCallen 1995, and Marshall (1986).

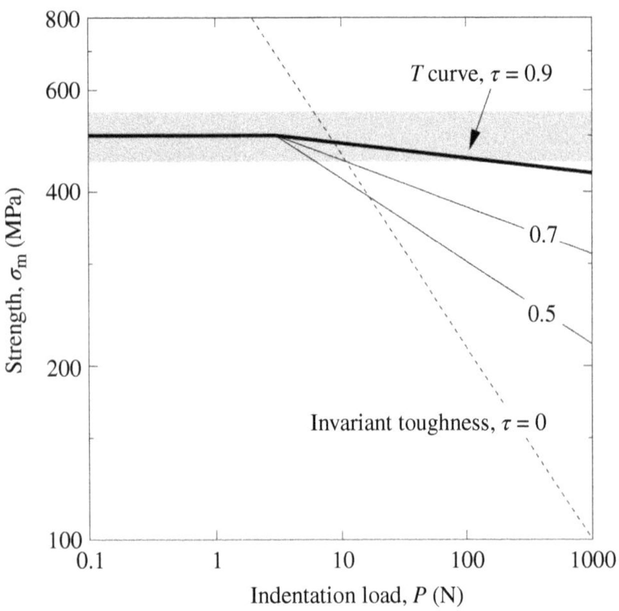

Figure 4.34 Schematic logarithmic indentation strength plot illustrating increased flaw tolerance with increasing slope of T curve. Plot format matches Figure 4.33. Dashed diagonal line indicates ideal indentation response. Shaded band represents intrinsic strength. Bold line approximates transformation toughened zirconia behavior.

The connection between T curve slope and flaw tolerance is shown in the logarithmic indentation strength plot of Figure 4.34, illustrating Eq. (4.45) and replicating aspects of Figure 4.33. The shaded band represents an intrinsic strength of (500 ± 50) MPa, and the dashed line repeats the ideal indentation strength response of Figure 4.33. The solid lines represent the indentation strength response of Eq. (4.45) for the values of τ indicated; the dashed line is given by $\tau = 0$. The important point to note is that as τ is increased the indentation strength response becomes more flaw tolerant. Comparison of Figures 4.34 and 4.33 suggest that the toughened TZP and PSZ materials are well described by T curves with effective slopes given by $\tau \approx 0.9$ over the indentation load and crack length domains measured. The overaged PSZ is described by a T curve with effective slope $\tau < 0.5$ (and the postthreshold strengths of TZP in Figure 3.56 by invariant toughness, a T curve with effective slope $\tau \approx 0$). Indentation strength measurements on an in situ toughened custom SiC material and a zirconia toughened $MoSi_2$ composite exhibited behavior consistent with $\tau \approx 0.7$ (Gilbert et al. 1997; Bhattacharya and Petrovic 1992). The power-law T curve, while empirical, provides insight into toughening effects on strength degradation.

4.4.7 Alumina Impact

A final topic in this survey of microstructural influence on brittle fracture is strength degradation of polycrystalline materials by impact. In particular, a pit-based model of flaw generation, originally developed by Ritter to describe impact strengths (Ritter et al. 1984; Ritter 1985), is developed and applied to polycrystalline alumina data given in Chapter 3. The model incorporates the pervasive idea of weak grain boundaries in the effects of microstructure on polycrystalline impact strength, but does so in a very different way from the crack wake traction formulations developed previously.

The physical principles of the pit model are outlined in the schematic diagrams of Figure 4.35. Figure 4.35a shows a particle of kinetic energy U_K approaching a polycrystalline surface. The flaw created by particle contact with the surface is shown in Figure 4.35b. The flaw consists of "pit" of removed material with characteristic hemispherical radius R. On the periphery of the pit are cracks of size c, a few grain facets in length. Pit formation in a component surface by controlled energy impact may be regarded as Step 1 in the indentation strength sequence illustrated in Figure 3.1. Step 2 of such a sequence is the application of stress to the component and measurement of component strength, indicated by arrows in Figure 4.35b. A relationship between U_K and R may be established as follows. The grain boundary area/volume in the polycrystalline material is $1/\lambda$. λ is the mean intercept length between grain boundaries determined from plane section images. For common equiaxial grain shapes, the "grain size" is approximately 1.25λ. The grain boundary fracture energy/area is 2γ, such that the grain boundary fracture energy/volume is $2\gamma/\lambda$. The volume of the pit created by particle impact is $2\pi R^3/3$. If the First Law of Thermodynamics is invoked such that the kinetic energy of the incoming particle is completely converted into to grain boundary fracture energy within the pit then

$$U_K \sim \frac{\gamma}{\lambda} R^3. \tag{4.47}$$

The cracks at the periphery of the pit are taken to scale with the polycrystalline grain size, $c \sim \lambda$, independent of R. A relative pit size or flaw size ρ may thus be defined by $\rho = R/c$. Combining this term into Eq. (4.47) gives

$$\rho \sim \left(\frac{U_K}{\gamma \lambda^2}\right)^{1/3}, \tag{4.48}$$

similar to the scaling given earlier (Ritter et al. 1984; Ritter 1985). Note that Eqs. (4.47) and (4.48) imply no mechanism for pit formation other than the process is adiabatic and reversible but simply state that energy is conserved.

Equation (4.48) provides a description of Step 1 in the indentation strength sequence. Step 2 requires a simple fracture mechanics analysis of the effects of a stress σ_a applied to a component containing the pit + crack combination constituting a strength-controlling flaw. It is recognized from Figure 4.35b that such a crack propagates in a stress gradient associated with the concentrating effects of the pit: the stress decreases from a significantly enhanced value at the pit periphery $\sigma > \sigma_a$

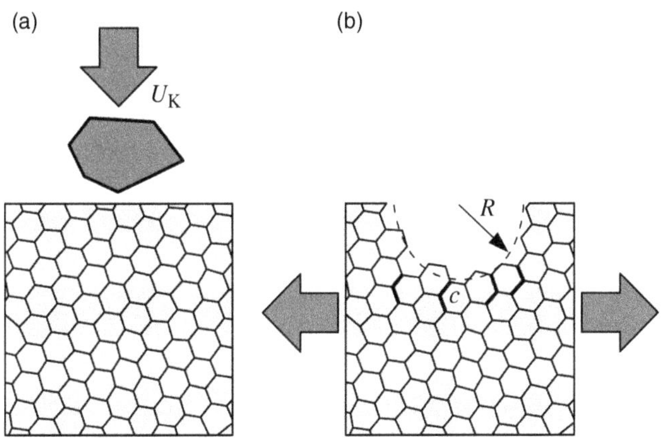

Figure 4.35 Schematic diagram of the pit based impact strength model. (a) A moving particle impacts a polycrystalline surface. (b) The kinetic energy of the particle is transformed by intergranular fracture into grain boundary surface energy, forming a pit with adjacent cracks (shown bold). The pit + cracks is a strength-controlling flaw on component loading (shown as arrows).

to the value in the far field $\sigma = \sigma_a$ associated with the loading of the entire component. The SIF K_ρ for a crack emanating from the pit is

$$K_\rho = \psi(\rho)\sigma_a c^{1/2}, \tag{4.49}$$

where the amplitude term $\psi = \psi(\rho)$ is a function of the relative pit size. In this case, distinct from indentation fracture in a stress gradient considered in Section 3.3.5, K_ρ is here the sole SIF loading the crack. As a consequence, expression of $\psi(\rho)$ as a polynomial does not present great analytical difficulties. The form used is due to Green (1980) and was developed using the integration technique of Eq. (1.82) (see Cook 2023 for other applications). $\psi(\rho)$ scales as

$$\psi(\rho) \sim \left(\frac{1+2\rho}{1+\rho}\right)^{1/2}\left[1+\left(\frac{\rho}{1+\rho}\right)^2+\frac{1}{2}\left(\frac{\rho}{1+\rho}\right)^4\right]. \tag{4.50}$$

The limits are $\psi(0) = 1$ and $\psi(\rho) \to 5/2^{1/2} \approx 3.54$ for $\rho \gg 1$, with intermediate value $\psi(1) = 1.57$. The pit model assumes that cracks generated at the pit periphery by impact are grain boundary intergranular cracks and a fixed number of grain facets in size. Hence, the crack length scales as $c \sim \lambda$ and the toughness value T appropriate to fracture equilibrium scales as $T \sim \gamma^{1/2}$, where γ is the grain boundary fracture energy used above. Physically, $R \geq \lambda$, and thus the relevant domain of Eq. (4.50) is $\rho \geq 1$. Green (1980) and Chao and Shetty (1992) focused on failure from pores and thus on fixed R and variable c, the opposite of the pit model here, and Green thus expanded ψ in terms of $1/\rho$ with no physical domain limitations. Ritter used a somewhat unwieldy arctan + exponential mixed form for ψ. Although $\partial\psi/\partial\rho > 0$ and $\partial\rho/\partial c < 0$, the combination is not enough to disturb the stability result $dK_\rho/dc > 0$, and $K_\rho(c)$ is destabilizing at all crack lengths. The strength of components containing pits is thus obtained by setting equilibrium and simply inverting Eq. (4.49) and using the above scaling relations to obtain

$$\sigma_m \sim \frac{1}{\psi(\rho)}\left(\frac{\gamma}{\lambda}\right)^{1/2}. \tag{4.51}$$

Equations (4.48) and (4.51) form a parametric set for expression of component strength as a function of impact energy, $\sigma_m = \sigma_m(U_K)$, with γ and λ as control variables.

Figure 4.36 is a logarithmic plot of strength variations thus derived, expressed in terms of $\mathbf{S_m}(\mathbf{U})$ scaled to the mean values of σ_m and U_K with base values $\gamma = \lambda = 1$. The mean values are indicated by vertical and horizontal solid lines. The central bold line shows the full $\mathbf{S_m}(\mathbf{U})$ strength response using the base values. The strength varies by a factor of 3.54 between two asymptotic limits shown as dashed horizontal lines. The upper and lower asymptotes correspond to small and large pits, respectively. The diagonal dashed line is a guide to the eye that indicates the ideal contact response $\mathbf{S_m} \sim \mathbf{U_K}^{-2/9}$ (see Chapter 3). The pit model predicts a far weaker dependence of strength on impact energy than does ideal contact. The full strength response passes through an inflection point at $\rho = 1$, which corresponds to the mean strength and impact energy values for the choice of $\gamma = \lambda = 1$ base parameters. Hence, the physically meaningful domain of the central response is $\mathbf{U} \geq 1$. In addition, the experimental domain of impact energies in impact strength experiments is typically a factor of 10^5 (see Chapter 3). Thus, the expected domain of applicability of the pit model (or any other strength model) is reduced further, as indicated by the shaded band in Figure 4.36. The predictive power of the pit model in terms of microstructural variables is indicated by the fine curved lines. The upper line corresponds to parameters that might be used to characterize a glass-bonded, fine-grained alumina, $\gamma = 2$ and $\lambda = 0.1$. The lower line corresponds to parameters that might be used to characterize a glass-free, large-grained alumina, $\gamma = 0.5$ and $\lambda = 10$. The changes in parameters lead to vertical and horizontal shifts of the base response and alterations in strength at a given impact energy within the experimental domain. As expected, strengths of large-grained materials with weak grain boundaries are more susceptible to impact.

Figure 4.37 is a logarithmic plot of impact strength of alumina from Figure 3.41. Symbols with bars represent means and standard deviation limits of measured strengths at the impact energies indicated. Shaded bands represent means and standard deviation limits of intrinsic strengths. The dashed line is a guide to the eye indicating the ideal contact response $\sigma_m \sim U_K^{-2/9}$. The bold solid line is a fit to the data of the strength response from the pit model and highlights both the success and limitations of the pit model. The success is that the weak variation of strength with impact energy is well described. The variation is far weaker than that predicted by ideal contacts and was the main motivations, along with the experimental observations of pits, for the development of the pit model by Ritter. The model also describes the gradually increasing dependence with decreasing impact energy. However, the limitation of the pit model is that the increased dependence is not (and can never be) strong enough, significantly underestimating strengths at small impact energies (whereas the ideal contact model well describes such strengths). The lateral crack based strength model implemented in Figure 3.41 does

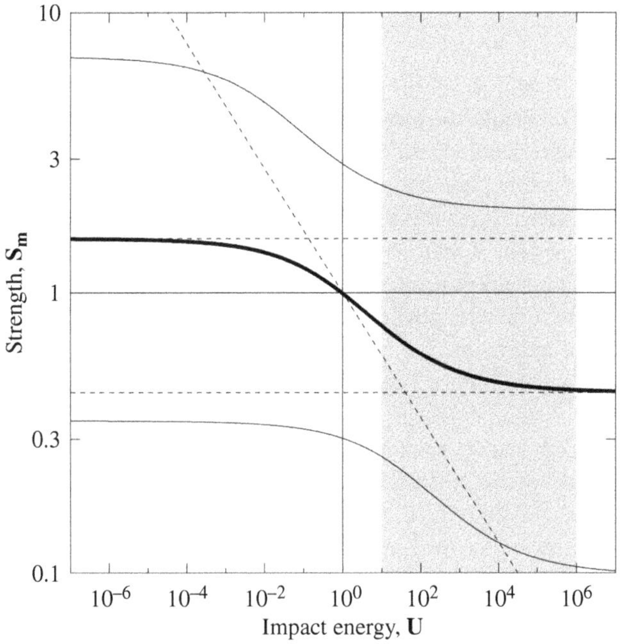

Figure 4.36 Schematic pit-based impact strength plot in scaled logarithmic coordinates. Dashed diagonal line indicates ideal contact response. Shaded band represents typical experimental domain width of impact energies. Bold line represents characteristic pit-based impact strength behavior. Strength bounds indicated by dashed horizontal lines. Fine lines represent the likely range of effects by microstructural variation.

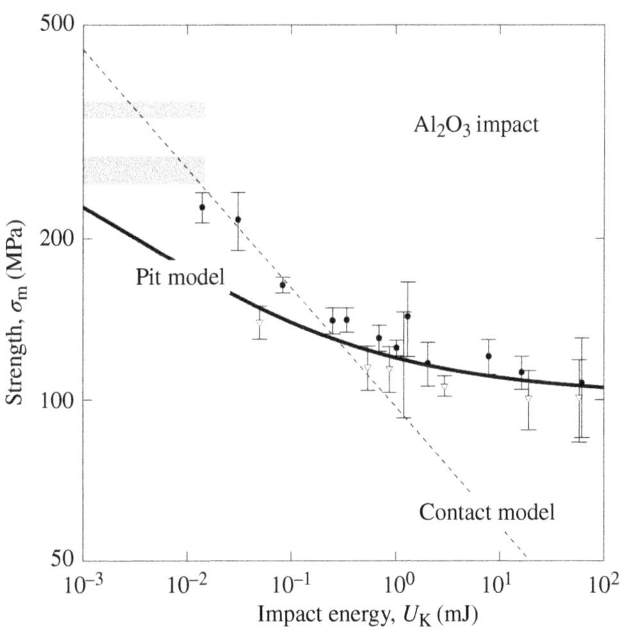

Figure 4.37 Logarithmic plot of impact strength for polycrystalline alumina. Data from Figure 3.41. Dashed diagonal line indicates ideal contact response. Bold line represents a fit to the data of the pit-based impact strength model. The model describes the tendency to strength invariance at large impact energies but underestimates strength increases at small energies.

describe strengths over the entire impact energy domain, probably due to the fact that large, extremely chipped impacts sites look and act much like impact pits. Note that in the lateral crack model the relative residual field is diminished with increasing impact energy and in the pit model the relative crack size is diminished with increasing impact energy. The pit model is explicitly microstructurally based as the pit and crack sizes scale with microstructure. The lateral crack model has no explicit microstructural dependence.

4.5 Summary and Discussion

This chapter has provided an extensive survey and assessment of the indentation strength behavior of brittle materials that exhibit microstructurally-induced toughness variations—*R* curves or *T* curves—in which toughness increases with crack length. Such materials are polycrystalline and include many of the materials encountered in daily life: translucent glass-free alumina, used as the bulb material in the ubiquitous sodium vapor lamp; glass-bonded alumina, used as microelectronic substrates, especially in automobiles; barium titanate, used as the dielectric in the myriad capacitors imbedded in microelectronic devices such as telephones and tablets; glass-ceramics, used in all manner of domestic cookware; and zirconia, used extensively in dental prostheses and in other wear-resistant applications, such as thread guides for fishing rods and sewing machines. The chapter builds on the fundamentals of indentation strength behavior established in Chapter 3. The indentation strength test consists of a two-step sequence: (i) flaw formation on a component surface by a controlled indentation contact followed by (ii) component strength testing. The focus of Chapter 3 was the influence of contact variations (e.g., indentation load and indenter shape) on the resulting strength. Here, in Chapter 4, the focus was on the influence of material microstructure and toughness variations during the strength test. In particular, the effects of fracture resistance and toughness variations arising from crack wake traction zones, pervasive in the fracture of brittle polycrystalline materials, was characterized and analyzed. The analysis made clear connection to the thermodynamics of fracture established in Chapter 1 and showed that toughening processes derive from irreversible dissipation of energy on crack propagation. An expression for toughness variation with crack length was developed for crack wake frictional traction processes and incorporated into an extended indentation strength analysis. A feature of the analysis was superposition onto ideal indentation strength behavior of microstructural effects, dominant at small crack lengths, and lateral crack effects, dominant at large indentations (see Chapter 3). The result was a sigmoidal response that in some form is nearly always observed in indentation strength measurements. The sigmoidal response was fit to the wide variety of polycrystalline materials noted above. A feature of microstructural toughening effects is enhanced material flaw tolerance—the resistance of materials to strength degradation by flaws of increasing size. Some of the materials examined here were deliberately engineered with microstructures to exhibit flaw tolerance, and some impressive results were achieved. As in Chapter 3, the analysis has been directed toward description of indentation strength phenomena in experimentally measurable terms with a minimum of notation and a focus on the *forms* of indentation strength behavior described by scaled variables.

The focus on observed indentation strength behavior has omitted detailed discussion of two aspects of microstructural interactions that pertain at very small, often unobserved, indentation loads. The first aspect concerns the lower-bound base toughness T_0 and the second aspect concerns intrinsic strengths σ_i. These aspects are now addressed. First, terms characterizing the lower bound to material fracture resistance R_0 and the analogous lower bound to toughness T_0 were introduced in comparison to microstructure-based resistance and toughening terms in Eqs. (4.3) and (4.6), respectively. T_0 was recognized as characterizing fracture of the weakest path in a material along which cracks propagated and onto which toughening effects were superposed. Such weak paths were taken as grain boundaries in polycrystals and a mix of interfaces and matrices in composites and glass-ceramics. As indentation strengths depend on both toughness and the crack length derivative of toughness (see Chapter 3), indentation strength formulations $\sigma_m(P)$ including microstructural effects only incorporate information regarding T_0 implicitly, as in Eq. (4.22). Most attention is focused on the indentation load P^* characterizing the microstructural toughening increment, Eq. (4.19), rather than the indentation load P_δ characterizing the base toughness, Eq. (4.18), although it is recognized that (4.22) is only applicable in the domain $P \geq P_\delta$. In the indentation load domain $P < P_\delta$, toughness is given by the invariant value $T = T_0$ and strength is thus given by the applicable ideal indentation response $\sigma = \sigma_0$, where

$$\sigma_0 = 3T_0^{4/3}/4^{4/3}\psi\chi^{1/3}P^{1/3}, \quad P < P_\delta. \tag{4.52}$$

Figure 4.38 illustrates how T_0 and σ_0 appear in plots of indentation toughness and strength using the formats of the experimental results (Section 4.4). (Note that toughness and strength plots are reversed here.) Figure 4.38a shows a logarithmic plot of scaled effective toughness as a function of scaled indentation load for materials exhibiting microstructural toughening effects. The fine curved line represents the universal toughness response, and the dashed lines represent the asymptotes of this response. Effective toughness increases with indentation load until at large loads a maximum steady-state toughness is attained. The experimental data in Section 4.4 all lay on the universal response. The increase in toughness at small indentation loads underlies the flaw tolerance exhibited by many materials. A minimum base toughness appears on this plot as an invariant lower-bound T_0 applicable at very small indentation loads. As an example, fracture of the weakest path in the absence of microstructural effects was characterized here by a lower-bound effective toughness $\mathbf{T_0} = 0.5$, shown

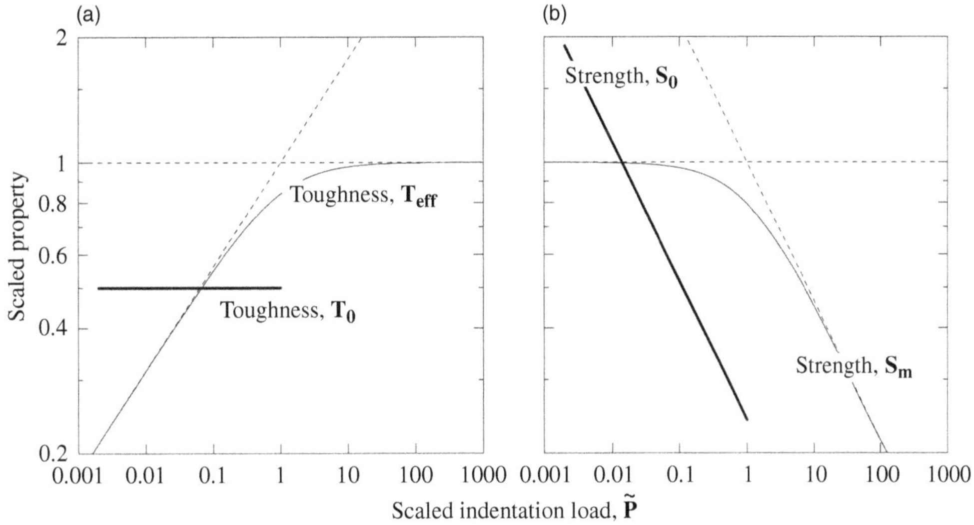

Figure 4.38 Logarithmic plots of (a) scaled effective toughness values and (b) scaled indentation strengths. Fine lines represent unbounded universal responses based on indentation fracture mechanics analysis including microstructural and lateral crack effects. Dashed lines represent small and large indentation load asymptotes. Bold lines indicate the lower-bound toughness value and consequent strengths. Intrinsic strengths intervene to preclude observation of these toughness and strength variations.

by the bold horizontal line. The scaled indentation load for the onset of these effects was thus $\tilde{\mathbf{P}}_\delta \approx 0.1$; for indentation loads less than this, with short attendant cracks, fracture is controlled by $\mathbf{T}_0$. The T curve connects the lower and upper bounds of toughness as crack length increases (in this case, characterized by increasing indentation load).

The resultant indentation strength is shown in Figure 4.38b as a logarithmic plot of scaled strength as a function of scaled indentation load. The fine curved line represents the universal strength response and the dashed lines represent the asymptotes of this response. Strength decreases with indentation load from a small load maximum value until a steady-state strength degradation rate is attained at large loads. The experimental data in Section 4.4 all lay on the universal response. The flaw tolerance at small loads was exhibited by many materials. The effects of a minimum base toughness appear on this plot as a strength degradation response $\mathbf{S}_0$ applicable at very small indentation loads. For the example lower-bound toughness response given in Figure 4.38a, the resulting strength response is shown by the bold diagonal line plotted over the same load domain. Note that the short crack strength response controlled by $\mathbf{T} = \mathbf{T}_0$ is $\mathbf{S}_0 \sim \mathbf{P}^{-1/3}$ and thus parallels the long crack response controlled by $\mathbf{T} = 1$. In this case, the proportionality term connecting strength and toughness was set such that the onset of microstructural effects on strength was $\mathbf{P} \approx 0.01$. For indentation loads less than this value, the strength should increase as indicated by bold line for $\mathbf{S} \geq 1$. There are no reports of experimental observations of such strengths at these small indentation loads; intrinsic strengths always intervene.

The second aspect of strengths at small indentation loads not considered in detail here was the occurrence of intrinsic strengths σ_i. Such strengths are regarded as imbedded in the test specimens, and maybe in the material, and are encountered when intrinsic flaws dominate indentation flaws, usually at small indentation loads, and the intrinsic flaws become strength controlling. Two features of intrinsic strengths are important from a materials engineering perspective: the relationship between intrinsic strength and microstructure and the nature of the transition from strengths controlled by extrinsic flaws (here indentations) to intrinsic strengths (here at small indentation loads). Both of these features are addressed in detail in Chapter 13. However, it is noted that many of the intrinsic strengths measured here (indicated here by shaded bands) in the strength degradation plots for alumina, silicon nitride, glass-ceramics, and alumina composites were all very close to the fitted indentation strength responses extrapolated to small indentation loads. The implication is that in many cases microstructure determines intrinsic strengths and thus intrinsic strengths can be engineered, particularly by grain size modification. A follow on implication is that microstructural effects will thus determine the nature of the transition (e.g., abrupt, gradual) from extrinsic flaws to intrinsic flaws. Another follow on implication is that microstructural effects not included here in the toughening analysis are responsible for truncating strength increases at small flaw sizes such that the $\mathbf{S}_0$ indentation strength branch is not observed. In this regard, it is important to note that subthreshold strengths controlled by ultrasmall crack-free flaws were only observed in Figures 3.51–3.56 for materials with no microstructural effects on fracture: glasses, single crystals, and a fine-grained polycrystal.

Another aspect of "toughening" not addressed in detail in this chapter, although several examples were given in Chapter 3, is toughening associated with reversible processes. As toughening is defined in a selected frame of reference that is not the crack tip (usually that of the applied loading), *any* process, reversible or irreversible, that diverts energy away from the creation of fracture surface can be regarded as toughening. The prime example of such toughening by a reversible process is the effect of a compressive stress field imbedded in a component. The essential physical principle in such toughening is that a SIF K_S associated with a compressive stress field and evaluated in the crack tip frame of reference is negative, $K_S < 0$. When viewed in the applied loading frame of reference, the field appears as a positive contribution to toughness, $T_S = -K_S > 0$, as considered in Eq. (3.91) and following and Eq. (4.10) and following. This idea was considered in detail and implemented by many, including Tandon and Green (1991) and Green et al. (1999) in connection with increasing crack stability and component flaw tolerance in ion-exchanged glasses, by Lange and colleagues in connection with laminated alumina-based ceramics (see Pontin and Lange 2005 for a list of relevant works), and Cook (2005) in connection with laminated cordierite glass-ceramics. Note that in all these cases the materials were not really toughened, simply that components of the materials could be described as tougher.

Finally, the data presented in this chapter (and perhaps those presented in Chapter 3) show that there is a fundamental difference between indentation crack length measurements and indentation strength measurements. Indentation crack length measurements, when correctly and completely performed, almost always conform to the Roesler equation, $c = BP^{2/3}$, where B is the cracking susceptibility, even in systems in which there are clear microstructural, lateral crack, or nonequilibrium effects (Cook 2020). Hence, there is benefit to compiling such measurements and B values to enable materials comparisons (Chapter 2; Cook 2024). Indentation strength measurements, however, very rarely conform completely to the ideal indentation response, $\sigma = SP^{-1/3}$, where S is the contact strength, although may do so if the materials or the indentation load domains are restricted (Chantikul et al. 1981). Hence, a compilation of indentation strength measurements and S values is not so useful for application in a straightforward manner as a compilation of indentation crack length data. Any meaningful comparison of indentation strengths or S values requires additional information regarding the applicable indentation load domain. For example, S only applies at small indentation loads for the majority of untoughened materials; at large loads, lateral cracking shifts indentation strengths away from the ideal response. Chapter 13 considers various methods of comparing indentation strength behavior of toughened and untoughened materials.

The following chapters build on the indentation strength data and analyses here in Chapters 3 and 4 to demonstrate indentation-based strength measurement techniques—to evaluate material toughness, in Chapter 5, and to evaluate nonequilibrium fracture parameters, in Chapter 6.

References

Anderson, R.M. and Braun, L.M. (1990). Technique for the R-curve determination of Y-TZP using indentation-produced flaws. *Journal of the American Ceramic Society* 73: 3059–3062.

Austin, C.R., Schofield, H.Z., and Haldy, N.L. (1946). Alumina in whiteware. *Journal of the American Ceramic Society* 29: 341–354.

Baker, T.L., Faber, K.T., and Readey, D.W. (1991). Ferroelastic toughening in bismuth vanadate. *Journal of the American Ceramic Society* 74: 1619–1623.

Bhattacharya, A.K. and Petrovic, J.J. (1992). R-curve behavior in zirconia-reinforced molybdenum disilicide composites. *Journal of the American Ceramic Society* 75: 23–27.

Bennison, S.J., Chan, H.M., and Lawn, B.R. (1989). Effect of heat treatment on crack-resistance curves in a liquid-phase-sintered alumina. *Journal of the American Ceramic Society* 72: 677–679.

Bennison, S.J., Padture, N.P., Runyan, J.L., and Lawn, B.R. (1991). Flaw-insensitive ceramics. *Philosophical Magazine Letters* 64: 191–195.

Carty, W.M. and Senapati, U. (1998). Porcelain–Raw materials, processing, phase evolution, and mechanical behavior. *Journal of the American Ceramic Society* 81: 3–20.

Cesar, P.F., Rosa, V., Pinto, M.M., Yoshimura, H.N., and Xu, L.R. (2011). Effect of ion exchange on R-curve behavior of a dental porcelain. *Journal of Materials Science* 46: 117–122.

Chantikul, P., Anstis, G.R., Lawn, B.R., and Marshall, D.B. (1981). A critical evaluation of indentation techniques for measuring fracture toughness: II, Strength method. *Journal of the American Ceramic Society* 64: 539–543.

Chantikul, P., Bennison, S.J., and Lawn, B.R. (1990). Role of grain size in the strength and R-curve properties of alumina. *Journal of the American Ceramic Society* 73: 2419–2427.

Chao, L.Y. and Shetty, D.K. (1992). Extreme-value statistics analysis of fracture strengths of a sintered silicon nitride failing from pores. *Journal of the American Ceramic Society* 75: 2116–2124.

Choi, H.J., Cho, K.S., Lee, J.G., and Kim, Y.W. (1997). R-curve behavior of silicon nitride-titanium nitride composites. *Journal of the American Ceramic Society* 80: 2681–2684.

Choi, S.R. and Salem, J.A. (1992). Indentation flaw formation and strength response of silicon nitride ceramics at low indentation loads. *Journal of Materials Science Letters* 11: 1398–1400.

Cook, R.F. (1986). Effect of changes in grain boundary toughness on the strength of alumina. *MRS Online Proceedings Library (OPL)* 78: 199–206.

Cook, R.F. (1990). Segregation effects in the fracture of brittle materials: Ca-Al_2O_3. *Acta Metallurgica et Materialia* 38: 1083–1100.

Cook, R.F. (2005). Toughening of a cordierite glass-ceramic by compressive surface layers. *Journal of the American Ceramic Society* 88: 2798–2808.

Cook, R.F. (2015). Multi-scale effects in the strength of ceramics. *Journal of the American Ceramic Society* 98: 2933–2947.

Cook, R.F. (2017). Fracture mechanics of sharp scratch strength of polycrystalline alumina. *Journal of the American Ceramic Society* 100: 1146–1160.

Cook, R.F. (2020). A critical evaluation of indentation cracks in air. *Journal of the American Ceramic Society* 103: 2278–2295.

Cook, R.F. (2023). *Particle Strengths*. Wiley.

Cook, R.F. (2024). Indentation fracture and deformation v.2. *Zenodo* DOI: 10.5281/zenodo.1149515.

Cook, R.F. and Braun, L.M. (1994). Trapped cracks at indentations: Part II Fracture mechanics model. *Journal of Materials Science* 29: 2192–2204.

Cook, R.F., Braun, L.M., and Cannon, W.R. (1994a). Trapped cracks at indentations: Part I Experiments on yttria-tetragonal zirconia polycrystals. *Journal of Materials Science* 29: 2133–2142.

Cook, R.F., Liniger, E.G., and Pascucci, M.R. (1994b). Indentation fracture of polycrystalline cubic ceramics. *Journal of Hard Materials* 5: 191–212.

Cook, R.F., Liniger, E.G., Steinbrech, R.W., and Deuerler, F. (1994c). Sigmoidal indentation–strength characteristics of polycrystalline alumina. *Journal of the American Ceramic Society* 77: 303–314.

Cook, R.F., Fairbanks, C.J., Lawn, B.R., and Mai, Y.W. (1987). Crack resistance by interfacial bridging: its role in determining strength characteristics. *Journal of Materials Research* 2: 345–356.

Cook, R.F., Freiman, S.W., and Baker, T.L. (1986). Effect of microstructure on reliability predictions for glass ceramics. *Materials Science and Engineering* 77: 199–212.

Cook, R.F., Freiman, S.W., Lawn, B.R., and Pohanka, R.C. (1983). Fracture of ferroelectric ceramics. *Ferroelectrics* 50: 267–272.

Cook, R.F., Lawn, B.R., and Fairbanks, C.J. (1985a). Microstructure-strength properties in ceramics: I, effect of crack size on toughness. *Journal of the American Ceramic Society* 68: 604–615.

Cook, R.F., Lawn, B.R., and Fairbanks, C.J. (1985b). Microstructure-strength properties in ceramics: II, fatigue relations. *Journal of the American Ceramic Society* 68: 616–623.

Cook, R.F. and Michaels, C.A. (2017). Review: coefficients for stress, temperature, and composition effects in fluorescence measurements of alumina. *Journal of Research of the National Institute of Standards and Technology* 122: 1.

Cook, R.F. and Michaels, C.A. (2019). Stress measurements in alumina by optical fluorescence: revisited. *Journal of Research of the National Institute of Standards and Technology* 124: 124020.

Cook, R.F. and Schrott, A.G. (1988). Calcium segregation to grain boundaries in alumina. *Journal of the American Ceramic Society* 71: 50–58.

Dasgupta, R., Hay, J.C., Ortiz-Longo, C.R., White, K.W., and Vipulanandan, C. (1998). Experimental study of the microstructural influence of the strain-softening behavior of cement mortar. *Cement and Concrete Research* 28: 1429–1444.

Dauskardt, R.H., Marshall, D.B., and Ritchie, R.O. (1990). Cyclic fatigue-crack propagation in magnesia-partially-stabilized zirconia ceramics. *Journal of the American Ceramic Society* 73: 893–903.

Dauskardt, R.H., Yu, W., and Ritchie, R.O. (1987). Fatigue crack propagation in transformation-toughened zirconia ceramic. *Journal of the American Ceramic Society* 70: C-248–C-252.

Deuerler, F., Knehans, R., and Steinbrech, R. (1986). Testing-methods and crack resistance behaviour of Al_2O_3. *Journal de Physique Colloques* 47: C1-617–C1-621.

Emery, V.L. (2009). Mud-brick. In *UCLA Encyclopedia of Egyptology* (Ed. Wendrich, W.). UCLA. http://digital2.library.ucla.edu/viewItem.do?ark=21198/zz0026vj53 (accessed 2/22/2024).

Freiman, S.W. and Pohanka, R.C. (1989). Review of mechanically related failures of ceramic capacitors and capacitor materials. *Journal of the American Ceramic Society* 72: 2258–2263.

Gilbert, C.J., Cao, J.J., De Jonghe, L.C., and Ritchie, R.O. (1997). Crack-growth resistance-curve behavior in silicon carbide: small versus long cracks. *Journal of the American Ceramic Society* 80: 2253–2261.

Gonzalez, A.C., Multhopp, H., Cook, R.F., Lawn, B.R., and Freiman, S.W. (1984). Fatigue properties of ceramics with natural and controlled flaws: a study on alumina. In *Methods for Assessing the Structural Reliability of Brittle Materials*, pp. 43–56. ASTM.

Green, D.J. (1980). Stress intensity factor estimates for annular cracks at spherical voids. *Journal of the American Ceramic Society* 63: 342–344.

Green, D.J., Hannink, R.H.J., and Swain, M.V. (1989). *Transformation Toughening of Ceramics*. CRC.

Green, D.J., Tandon, R., and Sglavo, V.M. (1999). Crack arrest and multiple cracking in glass through the use of designed residual stress profiles. *Science* 283: 1295–1297.

Grimes, R.E., Kelkar, G.P., Guazzone, L., and White, K.W. (1990). Elevated-temperature R-curve behavior of a polycrystalline alumina. *Journal of the American Ceramic Society* 73: 1399–1404.

Hannink, R.H., Kelly, P.M., and Muddle, B.C. (2000). Transformation toughening in zirconia-containing ceramics. *Journal of the American Ceramic Society* 83: 461–487.

Hay, J.C. and White, K.W. (1993). Grain-bridging mechanisms in monolithic alumina and spinel. *Journal of the American Ceramic Society* 76: 1849–1854.

Hay, J.C. and White, K.W. (1995). Grain boundary phases and wake zone characterization in monolithic alumina. *Journal of the American Ceramic Society* 78: 1025–1032.

Heuer, A.H. (1987). Transformation toughening in ZrO_2-containing ceramics. *Journal of the American Ceramic Society* 70: 689–698.

Hohlfelder, R.L. and Oleson, J.P. (2014). Roman maritime concrete technology in its Mediterranean context. In *Building for Eternity: The History and Technology of Roman Concrete Engineering in the Sea* (Eds. Oleson, J.P., Jackson, M.D., Hohlfelder, R.L., and Brandon, C.J.) pp. 223–235. Oxbow.

Howell, J.A., Vaudin, M.D., Friedman, L.H., and Cook, R.F., 2020. Microscale mapping of structure and stress in barium titanate. *Journal of Research of the National Institute of Standards and Technology* 125: 125013.

Hu, X. and Wittmann, F.H., 1990. Experimental method to determine extension of fracture-process zone. *Journal of Materials in Civil Engineering* 2: 15–23.

Hübner, H. and Jillek, W. (1977). Sub-critical crack extension and crack resistance in polycrystalline alumina. *Journal of Materials Science* 12: 117–125.

Jona, F. and Shirane, G. (1962). *Ferroelectric Crystals*. Dover.

Jones, J.L. and Hoffman, M. (2006). R-Curve and stress–strain behavior of ferroelastic ceramics. *Journal of the American Ceramic Society* 89: 3721–3727.

Kaur, S., Shetty, D.K., and Cutler, R.A. (2007). R curves and crack-stability map: application to Ce-TZP/Al_2O_3. *Journal of the American Ceramic Society* 90: 3554–3558.

Khan, A., Chan, H.M., Harmer, M.P., and Cook, R.F. (1998). Toughness-curve behavior of an alumina-mullite composite. *Journal of the American Ceramic Society* 81: 2613–2623.

Khan, A., Chan, H.M., Harmer, M.P., and Cook, R.F. (2000a). Toughening of an alumina—mullite composite by unbroken bridging elements. *Journal of the American Ceramic Society* 83: 833–840.

Khan, A., Chan, H.M., Harmer, M.P., and Cook, R.F. (2000b). Alumina agglomerate effects on toughness-curve behavior of alumina–mullite composites. *Journal of the American Ceramic Society* 83: 3089–3094.

Khonsari, M.M. and Amiri, M. (2013). *Introduction to Thermodynamics of Mechanical Fatigue*. CRC.

Kingery, W.D., Bowen, H.K., and Uhlmann, D.R. (1976). *Introduction to Ceramics*. Wiley.

Kovar, D., Bennison, S.J., and Readey, M.J. (2000). Crack stability and strength variability in alumina ceramics with rising toughness-curve behavior. *Acta Materialia* 48: 565–578.

Kovar, D. and Readey, M.J. (1994). Role of grain size in strength variability of alumina. *Journal of the American Ceramic Society* 77: 1928–1938.

Kovar, D. and Readey, M.J. (1996). Grain size distributions and strength variability of high-purity alumina. *Journal of the American Ceramic Society* 79: 305–312.

Knehans, R. and Steinbrech, R. (1982). Memory effect of crack resistance during slow crack growth in notched Al_2O_3 bend specimens. *Journal of Materials Science Letters* 1: 327–329.

Krause, R.F. Jr. (1988). Rising fracture toughness from the bending strength of indented alumina beams. *Journal of the American Ceramic Society* 71: 338–343.

Lai, T.R., Hogg, C.L., and Swain, M.V. (1989). Evaluation of fracture toughness and R-curve behaviour of Y-TZP ceramics. *ISIJ International* 29: 240–245.

Lathabai, S., Rödel, J., and Lawn, B.R. (1991). Cyclic fatigue from frictional degradation at bridging grains in alumina. *Journal of the American Ceramic Society* 74: 1340–1348.

Lathabai, S., Yiu-Wing, M., and Lawn, B.R. (1989). Cyclic fatigue behavior of an alumina ceramic with crack-resistance characteristics. *Journal of the American Ceramic Society* 72: 1760–1763.

Lawn, B.R. (1993). *Fracture of Brittle Solids, 2nd ed.* Cambridge.

Lawn, B.R., Freiman, S.W., Baker, T.L., Cobb, D.D., and Gonzalez, A.C. (1984). Study of microstructural effects in the strength of alumina using controlled flaws. *Journal of the American Ceramic Society* 67: C67–C69.

Lawn, B.R., Marshall, D.B., Chantikul, P., and Anstis, G.R. (1980). Indentation fracture: applications in the assessment of the strength of ceramics. *Journal of the Australian Ceramic Society* 16: 4–9.

Lawn, B.R., Padture, N.P., Braun, L.M., and Bennison, S.J. (1993). Model for toughness curves in two-phase ceramics: I, basic fracture mechanics. *Journal of the American Ceramic Society* 76: 2235–2240.

Lee, S.K., Wuttiphan, S., and Lawn, B.R. (1997). Role of microstructure in Hertzian contact damage in silicon nitride: I, mechanical characterization. *Journal of the American Ceramic Society* 80: 2367–2381.

Li, C.W., Lee, D.J., and Lui, S.C. (1992). R-curve behavior and strength for in-situ reinforced silicon nitrides with different microstructures. *Journal of the American Ceramic Society* 75: 1777–1785.

Lines, M.E. and Glass, A.M. (1977). *Principles and Applications of Ferroelectrics and Related Materials.* Bell Telephone Laboratories.

Littman, R.R., Lorenzon, M., and Silverstein, J. (2014). With & without straw. How Israelite slaves made bricks. *Biblical Archaeology Review* 40: 60–71.

Mai, Y.W. and Lawn, B.R. (1987). Crack-interface grain bridging as a fracture resistance mechanism in ceramics: II, theoretical fracture mechanics model. *Journal of the American Ceramic Society* 70: 289–294.

Marshall, D.B. (1986). Strength characteristics of transformation-toughened zirconia. *Journal of the American Ceramic Society* 69: 173–180.

Marshall, D.B., Morris, W.L., Cox, B.N., and Dadkhah, M.S. (1990). Toughening mechanisms in cemented carbides. *Journal of the American Ceramic Society* 73: 2938–2943.

Marshall, D.B. and Swain, M.V. (1988). Crack resistance curves in magnesia-partially-stabilized zirconia. *Journal of the American Ceramic Society* 71: 399–407.

Mase, G.E. (1970). *Continuum Mechanics.* McGraw-Hill.

Mehta, P.K. and Monteira, P.J.M. (2014). *Concrete: Microstructure, Properties, and Materials, 4th ed.* McGraw-Hill Education.

Michaels, C.A. and Cook, R.F. (2016). Determination of residual stress distributions in polycrystalline alumina using fluorescence microscopy. *Materials & Design* 107: 478–490.

Michaels, C.A. and Cook, R.F. (2018). Residual stress in polycrystalline alumina: comparison of two-dimensional maps and integrated scans in fluorescence-based measurements. *Acta Materialia* 159: 309–319.

Morena, R., Niihara, K., and Hasselman, D.P.H. (1983). Effect of crystallites on surface damage and fracture behavior of a glass-ceramic. *Journal of the American Ceramic Society* 66: 673–682.

Myers, G.A., Michaels, C.A., and Cook, R.F. (2016). Quantitative mapping of stress heterogeneity in polycrystalline alumina using hyperspectral fluorescence microscopy. *Acta Materialia* 106: 272–282.

O'Donnell, H.L., Readey, M.J., and Kovar, D. (1995). Effect of glass additions on the indentation-strength behavior of alumina. *Journal of the American Ceramic Society* 78: 849–856.

Padture, N.P., Bennison, S.J., and Chan, H.M. (1993a). Flaw-tolerance and crack-resistance properties of alumina-aluminum titanate composites with tailored microstructures. *Journal of the American Ceramic Society* 76: 2312–2320.

Padture, N.P., Runyan, J.L., Bennison, S.J., Braun, L.M., and Lawn, B.R. (1993b). Model for toughness curves in two-phase ceramics: II, microstructural variables. *Journal of the American Ceramic Society* 76: 2241–2247.

Pohanka, R.C., Freiman, S.W., and Bender, B.A. (1978). Effect of the phase transformation on the fracture behavior of $BaTiO_3$. *Journal of the American Ceramic Society* 61: 72–75.

Pontin, M.G. and Lange, F.F. (2005). Crack bifurcation at the surface of laminar ceramics that exhibit a threshold strength. *Journal of the American Ceramic Society* 88: 1315–1317.

Rado, P. (1969). *An Introduction to the Technology of Pottery.* Pergamon.

Readey, M.J. and McCallen, C.L. (1995). Microstructure, flaw tolerance, and reliability of Ce-TZP and Y-TZP ceramics. *Journal of the American Ceramic Society* 78: 2769–2776.

Readey, M.J., McCallen, C.L., McNamara, P.D., and Lawn, B.R. (1993). Correlations between flaw tolerance and reliability in zirconia. *Journal of Materials Science* 28: 6748–6752.

Ringwood, T. (1982). Immobilization of radioactive wastes in SYNROC. *American Scientist* 70: 201–207.

Ringwood, A.E., Oversby, V.M., Kesson, S.E., Sinclair, W., Ware, N., Hibberson, W., and Major, A. (1981). Immobilization of high-level nuclear reactor wastes in SYNROC: a current appraisal. *Nuclear and Chemical Waste Management* 2: 287–305.

Ritchie, R.O. and Liu, D. (2021). *Introduction to Fracture Mechanics*. Elsevier.

Ritter, J.E. (1985). Erosion damage in structural ceramics. *Materials Science and Engineering* 71: 195–201.

Ritter, J.E., Strzepa, P., Jakus, K., Rosenfeld, L., and Buckman, K.J. (1984). Erosion damage in glass and alumina. *Journal of the American Ceramic Society* 67: 769–774.

Rosa, V., Yoshimura, H.N., Pinto, M.M., Fredericci, C., and Cesar, P.F. (2009). Effect of ion exchange on strength and slow crack growth of a dental porcelain. *Dental Materials* 25: 736–743.

Rose, L. and Swain, M. (1986). Two R curves for partially stabilized zirconia. *Journal of the American Ceramic Society* 69: 203–207.

Ruiz, L. and Readey, M.J. (1996). Effect of heat treatment on grain size, phase assemblage, and mechanical properties of 3 mol% Y-TZP. *Journal of the American Ceramic Society* 79: 2331–2340.

Runyan, J.L. and Bennison, S.J. (1991). Fabrication of flaw-tolerant aluminum-titanate-reinforced alumina. *Journal of the European Ceramic Society* 7: 93–99.

Russo, C.J., Harmer, M.P., Chan, H.M., and Miller, G.A. (1992). Design of a laminated ceramic composite for improved strength and toughness. *Journal of the American Ceramic Society* 75: 3396–3400.

Sakai, M., Yoshimura, J.-I., Goto, Y., and Inagaki, M. (1988). R-curve behavior of a polycrystalline graphite: microcracking and grain bridging in the wake region. *Journal of the American Ceramic Society* 71: 609–616.

Seymour, L.M., Maragh, J., Sabatini, P., Di Tommaso, M., Weaver, J.C., and Masic, A. (2023). Hot mixing: mechanistic insights into the durability of ancient Roman concrete. *Science Advances* 9: eadd1602.

Shetty, D.K., Rosenfield, A.R., and Duckworth, W.H.F. (1985). Indenter flaw geometry and fracture toughness estimates for a glass-ceramic. *Journal of the American Ceramic Society* 68: C-282–C-284.

Steinbrech, R., Khehans, R., and Schaarwächter, W. (1983). Increase of crack resistance during slow crack growth in Al_2O_3 bend specimens. *Journal of Materials Science* 18: 265–270.

Swain, M.V. (1986). R-curve behaviour in a polycrystalline alumina material. *Journal of Materials Science Letters* 5: 1313–1315.

Swanson, P.L., Fairbanks, C.J., Lawn, B.R., Mai, Y.W., and Hockey, B.J. (1987). Crack-interface grain bridging as a fracture resistance mechanism in ceramics: I, experimental study on alumina. *Journal of the American Ceramic Society* 70: 279–289.

Tandon, R. and Green, D.J. (1991). Crack stability and T-curves due to macroscopic residual compressive stress profiles. *Journal of the American Ceramic Society* 74: 1981–1986.

White, K.W. and Kelkar, G.P. (1991). Evaluation of the crack face bridging mechanism in a $MgAl_2O_4$ spinel. *Journal of the American Ceramic Society* 74: 1732–1734.

Xu, H.H., Wei, L., Padture, N.P., Lawn, B.R., and Yeckley, R.L. (1995). Effect of microstructural coarsening on Hertzian contact damage in silicon nitride. *Journal of Materials Science* 30: 869–878.

Yu, C.S. and Shetty, D.K. (1989). Transformation zone shape, size, and crack-growth-resistance [*R*-Curve] behavior of ceria-partially-stabilized zirconia polycrystals. *Journal of the American Ceramic Society* 72: 921–928.

Yu, C.S. and Shetty, D.K. (1990). Transformation yielding, plasticity and crack-growth-resistance (R-curve) behaviour of CeO_2-TZP. *Journal of Materials Science* 25: 2025–2035.

Zhao, J., Stearns, L.C., Harmer, M.P., Chan, H.M., Miller, G.A., and Cook, R.F. (1993). Mechanical behavior of alumina-silicon carbide "nanocomposites". *Journal of the American Ceramic Society* 76: 503–510.

5

Indentation Strengths: Toughness Estimation

In this Chapter, a modified indentation toughness technique, in which indentation strength and crack length at component failure are combined, is demonstrated. A materials ranking scheme is introduced and compared with "conventional" toughness measurements. The modified indentation technique is used in determination of the variation in toughness with crack length for materials exhibiting significant microstructural effects, including alumina, silicon nitride, silicon carbide, and zirconia. In all, more than twenty materials are assessed to refine the precision and accuracy of the technique and enable toughness estimation for brittle materials at strength relevant length scales. Artifacts associated with conventional indentation toughness methods are largely removed by the modified technique.

5.1 Introduction

The load-bearing performance and reliability of structural ceramic components often depend on the fracture behavior of small flaws. Such flaws are frequently comparable in scale to the ceramic microstructure and consist of cracks associated with pores, inclusions, large grains in polycrystals, or sharp contacts on surfaces. Fracture behavior is determined qualitatively by three factors: (i) the nature of the global loading on the component, e.g., tensile stress generated by thermal shock or by mechanical bending; (ii) the nature of the flaw, including crack geometry and any stress fields locally surrounding the flaw; and (iii) the fracture resistance or toughness characteristics of the ceramic, particularly any variations in toughness at the scale of the flaw. In quantitative fracture mechanics terms, the crack driving force arising from the global loading and local stress field can be expressed functionally as a crack length dependent stress-intensity factor (SIF), $K(c)$, where c is a defined crack length. The crack resistance force, originating in the ceramic bonding and microstructure, is quantified by the toughness, $T(c)$. Fracture equilibrium is given by equating the driving force and the resisting force, $K(c) = T(c)$, making clear that fracture behavior is determined by influences associated with applied loading, $K(c)$, separate from those associated with the material, $T(c)$. Hence, the fracture properties of ceramic materials are most easily made by comparing toughness values. In particular, measurements of toughness at the small length scales associated with strength-controlling cracks and flaws are required to optimize and predict ceramic component load-bearing behavior. Such behavior dictates the performance of ceramic components in diverse applications such as displays in hand-held devices, substrates in automotive microelectronic applications, catalytic particles in chemical reactors, and armor plates to protect people, vehicles, and buildings.

The most common method of assessing toughness of ceramics at small length scales is by indentation crack length measurement (Anstis et al. 1981). In this method, a Vickers diamond pyramid is first brought into contact with a ceramic surface in a controlled load–unload cycle. The surface traces of the resulting indentation cracks are then measured and a simple calibrated power-law formulation involving the crack lengths and the peak indentation load can be used to estimate the toughness. The major issues with this technique are that the calibration assumes that the test environment is inert and that the residual stress field surrounding the contact is well characterized by the ceramic indentation hardness and elastic modulus. Both assumptions lead to inaccuracy, and to a lesser extent, imprecision, in toughness measurement. However, a recent extensive overview of indentation crack length measurements in air—overwhelmingly the most common ceramic fracture test—shows that such tests, although not calibrated toughness measurements, can be used to quantitatively rank ceramic cracking susceptibility (Cook 2020), partially side-stepping environment and residual stress field concerns. A step closer to ceramic toughness measurement at small scales is the two-step indentation strength method (Chantikul et al. 1981). In this method, a controlled indentation flaw is introduced into a ceramic test component as

Indentation Fracture: Strength and Toughness for Brittle Materials Design, First Edition. Robert F. Cook.

above. The strength of the component is then measured under inert conditions and another calibrated power-law formulation involving the strength and the indentation load used to estimate the toughness. Although there are sometimes test environment concerns, there is a much weaker dependence on the nature of the residual field, and toughness measurement by this technique can be quite accurate and precise. Extensions to the technique, quantifying the effects on strength and toughness of multiscale environmental, microstructural, and geometrical influences, have been demonstrated (Cook 2015).

The above points have been considered in detail in previous chapters: The fundamentals of fracture and the importance of toughness were considered in Chapter 1. Indentation crack length measurements were considered in Chapter 2. Indentation strength measurements and the effects of microstructure were considered in Chapters 3 and 4. Here, in Chapter 5, the attributes of both indentation measurement techniques are combined into a modified indentation toughness technique that almost completely removes residual stress field effects from toughness assessment. Indentation crack lengths are still measured, but after inert strength testing of components. The post test crack lengths are combined with the strengths to generate a parameter that is directly proportional to the toughness through the crack geometry alone. The convenience of indentation testing is retained in that the scale and location of the strength-controlling flaw are determined by the experimenter. The modified technique was demonstrated in an earlier work (Cook and Lawn 1983) but involved limited materials and measurements. Here, the technique is revisited with a wider domain of materials and ensuing range of data, including materials in which microstructural effects on toughness are evident and including data from other investigators. In addition, the results of the modified technique are placed in the context of the more recent crack length and strength measurement works (Cook 2015, 2020) and pursue the philosophy of the recent considerations regarding indentation crack lengths in air (Cook 2020) to generate a quantitative ranking scheme for ceramic fracture susceptibility. The work here begins with an overview of the required analysis and experimental method. The results section then follows in three parts: a comparison with conventional test methods; a refinement of the modified technique; and, application of the modified technique to ceramics exhibiting microstructural effects on toughness. A discussion compares the refined modified technique to short and long crack toughness measurement methods.

5.2 Methods

5.2.1 Analysis

This section begins with a brief review of the indentation fracture mechanics developed in Chapters 2 and 3. The review provides a foundation for subsequent development of analysis for toughness estimation by indentation fracture. Analytical development begins with consideration of equilibrium indentation crack lengths in materials with single valued toughness. In inert conditions, the right side of the equilibrium equation above is then $T(c) = T_0$, where T_0 is the material toughness in inert conditions and the "0" subscript indicates no chemical reactions between the material and the environment. For a crack under the sole influence of the indentation residual field, the left side of the equilibrium equation is the SIF $K(c) = \chi_0 P/c^{3/2}$, where the crack length c is measured from the center of the indentation, P is the peak indentation load, and χ_0 is a dimensionless geometry factor appropriate to indentation in an inert environment. Combining these two expressions gives the inert indentation crack length,

$$c_0 = (\chi_0 P/T_0)^{2/3}. \tag{5.1}$$

Under the more common conditions of indentation in air, the right side of the equilibrium equation is $T(c) = T_{air}$, where T_{air} is the toughness in air. Most ceramic materials, especially those containing oxygen, react with atmospheric moisture, such that $T_{air} < T_0$ for an equilibrium crack in the reactive air. The left side of the equilibrium equation is now $K(c) = \chi_{air} P/c^{3/2}$, where χ_{air} is the dimensionless geometry factor appropriate to indentations in air. For most indentations, especially those at large loads, secondary lateral cracks and chips form, relaxing the residual field such that $\chi_{air} < \chi_0$. Combining these latter two expressions gives the equilibrium indentation crack length in air,

$$c_{air} = (\chi_{air} P/T_{air})^{2/3}. \tag{5.2}$$

For most ceramics, $T_{air}/T_0 < \chi_{air}/\chi_0$, such that indentation cracks in air are longer than those in inert conditions, $c_{air} > c_0$. Equation (5.1) was the basis for the original indentation crack length toughness estimation method, calibrated by measurements on materials of known hardness, modulus, and toughness (Anstis et al. 1981). In the recent overview of indentation crack lengths in air (Cook 2020), Eq. (5.2) was the basis for materials fracture susceptibility comparisons using the load-adjusted crack length $c_{air}P^{-2/3}$, noting that this quantity is the invariant cracking susceptibility B for ideal indentations (Chapter 2).

A uniform applied stress σ_a superposed on the indentation crack adds an additional SIF term $K = \psi\sigma_a c^{1/2}$, where ψ is another dimensionless geometry factor, characterizing the effect of the applied stress on the crack. The equilibrium equation for an indentation crack formed in air and subsequently stressed in inert conditions is then

$$K(c) = \chi_{air}P/c^{3/2} + \psi\sigma_a c^{1/2} = T_0. \tag{5.3}$$

The maximum stress σ_m sustainable by the indentation flaw is given by simultaneous solution of Eq. (5.3) and the instability condition, $dK(c)/dc = 0$, such that

$$\sigma_m = 3T_0/4\psi c_m^{1/2}, \tag{5.4}$$

where

$$c_m = (\chi_{air}P/T_0)^{2/3}. \tag{5.5}$$

The stress σ_m is the inert strength of the indented component. The conjugate crack length c_m characterizes the instability configuration and reflects near-equilibrium extension of the indentation crack from c_{air} to c_m under the action of the applied stress. Combining Eqs. (5.4) and (5.5) to eliminate c_m gives $\sigma_m = 3T_0^{4/3}/4^{4/3}\psi\chi_{air}^{1/3}P^{1/3}$, and this is the basis for indentation strength toughness estimation methods (Chantikul et al. 1981; Cook 2015), calibrating the product $\psi\chi_{air}^{1/3}$ by measurements on materials with known properties. It is noted that $\sigma_m P^{1/3}$ and $c_m P^{-2/3}$ are load-adjusted inert instability parameters that are invariant for ideal indentations.

The dependence of inferred toughness on the indentation residual field characterized by χ is reduced in strength measurements relative to crack length measurements by the introduction of the SIF associated with the applied stress: from $T_0 \sim \chi$ (for crack length) to $T_0 \sim \psi^{3/4}\chi^{1/4}$ (for strength) (Anstis et al. 1981; Chantikul et al. 1981). The modified indentation toughness technique eliminates this dependence completely by rearranging Eq. (5.4) to give

$$T_0 = (4\psi/3)\sigma_m c_m^{1/2}. \tag{5.6}$$

Toughness inferred from $\sigma_m c_m^{1/2}$, combined strength–crack length measurements, thus depends entirely on the applied stress related crack geometry factor: $T_0 \sim \psi$ (modified method) (Cook and Lawn 1983).

It is useful to compare the similarities and differences in the application of Eq. (5.6) with the equations and procedures used in toughness determinations using long crack "standard fracture mechanics specimens" (e.g., the double cantilever beam) (Murakami 1987). The major similarity, and most important point, is that in both cases toughness is determined by a product of the peak load or stress supported by the specimen and the crack length at fracture instability, mediated by a crack geometry term. There are, however, three major differences between indentation and long crack toughness assessments. The first major difference is that fracture systems of long crack specimens are usually mechanically unstable, such that fracture instability occurs at peak load with no extension of the crack from an initial length. In contrast, fracture systems for indentation specimens pass from an initial stable configuration, Eq. (5.2), to an unstable configuration at peak load, Eq. (5.5), with attendant crack extension. The second major difference is that the cracks in long crack specimens are usually considered large relative to any microstructural length scale associated with material fracture resistance or toughness variations, such that the toughness measured is regarded as a "steady-state" value (in the context of Chapter 4, $T \to T_\infty$). Again, in contrast, cracks in indentation systems usually span the domain of microstructural length scales associated with toughness variations, such that toughness measured often exhibits a range (in the context of Chapter 4, $T_0 \leq T \leq T_\infty$). The third major difference is that in well-conducted indentation toughness assessments nonreactive environments (e.g. dry N_2, silicone oil, rapid loading) are imposed such that the conditions are inert. In long crack assessments, the test environment is usually air and the specimens are usually considered large enough such that reactive species cannot diffuse to the crack tip during testing. A minor difference is that the geometry term for long crack specimens is usually a function of the initial crack length relative to the specimen dimensions, whereas this is not so in indentation strength tests. Hence, the modified indentation toughness test fits within a recognized toughness measurement framework and combines the ease of indentation crack formation at strength-relevant length scales while bypassing the difficulties associated with significant residual stress field dependence. [Elastic–plastic deformation effects associated with the residual stress field exert geometrical influences on crack formation during the indentation cycle (Cook 2019), but these cause secondary postindentation effects.]

Figure 5.1 Schematic diagram and image of dummy indentation method used for determination of instability crack lengths: central span of bend specimen containing three identical indentations (bending induced tensile stress is horizontal). (a) Failure location, with failure plane extending across specimen; (b) optical micrograph of GC1 ceramic with 20 N indentation and cracks extended perpendicular to stress; (c) intact indentation with extended crack length indicated.

5.2.2 Experiment

The experimental technique for simultaneous determination of inert strength and instability crack length, and thus $\sigma_m c_m^{1/2}$, was primarily the dummy indentation method used earlier (Dabbs et al. 1982; Cook and Lawn 1983). Briefly, three colinear indentations, formed at the same peak load and visually adjudged alike, were generated in air within the central spans of four-point bend specimens of several ceramics and glasses, as shown in Figure 5.1. The specimens were subsequently loaded to failure under inert conditions of flowing dry N_2 gas and rapid loading rate—the gas environment and the resulting short failure times, about 30 ms, minimized diffusion of moisture to the cracks during failure. Failure loads were recorded using a piezoelectric load cell and converted to failure stress using specimen and bend rig dimensions. The failure stresses were consistent with specimens containing single indentations and were thus regarded as σ_m measurements. The indentation loads were selected such that the specimens failed from one of the indentations on a plane perpendicular to the bending direction, Figure 5.1a. The cracks on the remaining, "dummy," indentations were measured after failure using a variety of optical microscopy techniques (Cook 2020) and compared with measurements of indentation cracks on unstressed specimens. The cracks parallel to the failure plane exhibited significant extension and diminished opening while the cracks perpendicular to the failure plane exhibited no significant alteration, see Figure 5.1b. (For crack length observations *during* failure, see figures and measurements in Chapters 3 and 12.) As all indentations experienced the same stress history up to the inert strength, the lengths of the extended cracks were regarded as c_m estimates, see Figure 5.1c.

Using the above technique, the $\sigma_m c_m^{1/2}$ data set used in the earlier work (Cook and Lawn 1983) has been extended significantly here by further measurements on additional specimens and additional materials, and, where possible, by analyzing all data using individual specimen pairs (rather than grouped data as earlier). The data were further extended by including comparable inert or near-inert strength and crack length data obtained by other workers, in particular those of Shetty and colleagues and Bennison and colleagues. Dummy technique measurements on silicon nitride, whisker-reinforced alumina, zirconia, and silicon carbide (Ramachandran and Shetty 1991; Li et al. 1997; Lin et al. 2007; Kaur et al. 2009) using rapid strength measurements in air and dry oil are included. Direct in situ observation measurements on silicon nitride, alumina, and silicon (Marshall 1983; Braun et al. 1992; Li et al. 1992; Kovar et al. 2000; Cook 2006) are also included. In the first, fourth, and fifth of these direct observation studies, the materials do not react with atmospheric moisture, and in the second and third studies, inert conditions were obtained by indentation immersion in dry N_2 and silicone oil, respectively. Direct ex situ measurements on polymer cement composite specimens (Chou et al. 1990) minimized reaction with atmospheric moisture during testing by aging of specimens. The data from other workers were obtained by digitizing and reanalyzing published results.

Overall, data are included from over 25 materials involving over 400 measurements of indentation instability configurations as well as related measurements of initial configurations. The materials included glasses, glass-ceramics,

Table 5.1 Ceramic materials included in modified toughness investigation.

ID	Material	Data	Technique
SLG	Soda-lime silicate glass (Schott-Ruhrglas, Germany)	Dabbs et al. (1982)	Dummy, N_2
BSG	Borosilicate glass (Schott-Ruhrglas, Germany)	Dabbs et al. (1982)	Dummy, N_2
FS	Fused silica (Schott-Ruhrglas, Germany)	Dabbs (1984)	Dummy, N_2
PC	Polymer cement (Custom)	Chou et al. (1990)	Direct, air
GC1	Cordierite glass-ceramic (Pyroceram C9606, Corning, Corning, NY)	Cook (1982)	Dummy, N_2
GC2	Li_2SiO_3 glass-ceramic (Custom, 1.19 μm, Cook et al. 1986)	This work	Dummy, N_2
GC3	Li_2SiO_3 glass-ceramic (Custom, 1.90 μm, Cook et al. 1986)	This work	Dummy, N_2
GC4	Li_2SiO_3-glass ceramic (Custom, 2.30 μm, Cook et al. 1986)	This work	Dummy, N_2
PZT	(Pb, Zr)TiO_3 polycrystal (Plessey, Australia)	This work	Dummy, N_2
BT	$BaTiO_3$ polycrystal (Channel Industries, Santa Barbara, CA)	This work	Dummy, N_2
SYN	Synthetic rock (Synroc B, Australian Atomic Energy Establishment)	This work	Dummy, N_2
A1	Al_2O_3 polycrystal (AD96, Coors, Golden, CO)	This work	Dummy, N_2
A2	Al_2O_3 polycrystal (F99, Friedrichsfeld, Mannheim, Germany)	This work	Dummy, N_2
A3	Al_2O_3 polycrystal (Custom, 2.5 μm, Chantikul et al. 1990)	Braun et al. (1992)	Direct, oil
A4	Al_2O_3 polycrystal (Custom, 4–13 μm)	Kovar et al. (2000)	Direct, N_2
AS	Al_2O_3–SiC (whisker) composite (Ceramatec Inc., Salt Lake City, UT)	Ramachandran and Shetty (1991)	Dummy, air
AT	Al_2O_3–Al_2TiO_5 composite (Custom, Bennison et al. 1991)	Braun et al. (1992)	Direct, oil
TZP	3 mol Y_2O_3–ZrO_2 polycrystal (Tosoh Corporation, Tokyo, Japan)	Li et al. (1997)	Dummy, air
SC1	SiC polycrystal (NC203, Norton Co., Worcester, MA)	This work	Dummy, N_2
SC2	SiC polycrystal (PS-5000, Morgan AM&T, St. Mary's, PA)	Kaur et al. (2009)	Dummy, oil
SC3	SiC polycrystal (Custom, Ceramatec Inc., Salt Lake City, UT)	Kaur et al. (2009)	Dummy, oil
SC4	SiC polycrystal (SiC-N, Cercom, Inc., Vista, CA)	Kaur et al. (2009)	Dummy, oil
SC5	SiC polycrystal (Custom, Ceramatec Inc., Salt Lake City, UT)	Kaur et al. (2009)	Dummy, oil
SN1	Si_3N_4 polycrystal (NC132, Norton Co., Worcester, MA)	Marshall (1983)	Direct, air
SN2	Si_3N_4 polycrystal (SN252, Kyocera, Des Plaines, IL)	Ramachandran and Shetty (1991)	Dummy, air
SN3	Si_3N_4 polycrystal (Custom)	Li et al. (1992)	Direct, air
Si	(001) single-crystal Si	Cook (2006)	Direct, air

polycrystalline ceramics, composites, and a single crystal and are listed by materials class in Table 5.1, along with ID codes for ease of notation, the sources of the materials, the sources of the data (identified by first author and year, often also the material source), and the techniques used to determine the modified toughness parameters (the ID codes are different from those used previously, Cook 2020). The data are reported (Section 5.3) as the mean values of $c_{air}P^{-2/3}$, $c_m P^{-2/3}$, $\sigma_m P^{-1/3}$, and $\sigma_m c_m^{1/2}$ with uncertainties specified as experimental standard deviation limits, using propagation of variance as required. In addition, combining Eqs. (5.2) and (5.5) provides the relative crack extension

$$c_m/c_{air} = (4T_{air}/T_0)^{2/3}. \tag{5.7}$$

The implications of this equation are many-fold: First, the experimentally determined ratio c_m/c_{air} must be greater than 1 to verify stable crack extension during applied stressing to failure. Second, a lower-bound toughness ratio required to generate such a precursor extension is $T_{air}/T_0 \geq 0.25$; this criterion is usually easily met for ceramics reacting with moist air. Third, an upper bound to the relative extension is $4^{2/3} \sim 2.52$ for no reaction, $T_{air}/T_0 = 1$. As the above bounds are fixed, the median values and ranges of c_m/c_{air} are reported here.

5.3 Results

5.3.1 Relation to Conventional Method

Figure 5.2 is a plot of the initial and final indentation crack lengths for the complete materials set examined here, plotted using load-adjusted coordinates $c_{air}P^{-2/3}$ and $c_mP^{-2/3}$, respectively. The plotting scheme is the same as that used in Figure 2.22 and in the recent review of indentation crack lengths in air (Cook 2020), ranking the materials left to right by increasing cracking susceptibility $B = c_{air}P^{-2/3}$ value. The gray band in Figure 5.2 represents the range of $c_{air}P^{-2/3}$ responses from the review for a similarly broad set of ceramic materials (some representative materials are in common, e.g., SC1, A1, SLG). The lower group of small open symbols in Figure 5.2 represents the responses for the set of materials examined here. There is close agreement with the range of the previous observations and with the ranking of materials by susceptibility to indentation cracking. Ceramics regarded as "tough," e.g., SC series and SN series, appear on the left of the plot, ceramics regarded as "not very tough," e.g., SLG and Si, appear on the right side of the plot, and families of materials, e.g., A series and GC series are clustered together. The major ranking irregularity in this scheme is that the "anomalous" glasses, FS and BSG, are not clustered with SLG, and in fact appear in the center of the ranking, implying a much greater toughness. The origin of this observation is that χ is depressed for these materials, leading to shorter than expected crack lengths, Eq. (5.2). Overcoming this effect was a major factor in the initial development of the modified toughness technique (Cook and Lawn 1983).

The upper group of large, solid symbols in Figure 5.2 represents $c_mP^{-2/3}$ responses for the majority of materials examined here. These materials were judged to exhibit load-invariant indentation behavior and are considered here first. The indentation loads used, the number of cracks measured, and the resulting $c_mP^{-2/3}$ mean values and uncertainties are given in Table 5.2. In all cases, $c_mP^{-2/3} > c_{air}P^{-2/3}$, indicating that precursor crack extension occurred during loading and that interpretation of the instability data in ideal indentation terms, Eqs. (5.1)–(5.6), was applicable. In a few cases, notably the SN1 and A3 materials, the relative precursor extension approached the upper bound, ≈ 2.5. In most materials, the precursor extension was much less than this, in fact the lower bound for c_m/c_{air} was 1.2 and the median extension was 1.34, implying $T_{air}/T_0 \approx 0.39$ for these materials, Eq. (5.7) (this small value is discussed in Section 5.4). The trend in increasing $c_mP^{-2/3}$ is similar to that for $c_{air}P^{-2/3}$, although it is clear that the detailed ranking would change if cracking susceptibility $c_mP^{-2/3}$ were used rather than $c_{air}P^{-2/3}$.

The large open symbols in Figure 5.2 represent $c_mP^{-2/3}$ responses of materials that exhibited load-dependent indentation behavior: the responses shown were measured at the largest load applied. These loads are given in Table 5.2, along with the measured $c_mP^{-2/3}$ values and uncertainties where applicable; the loads and measurements are indicated by an asterisk (*).

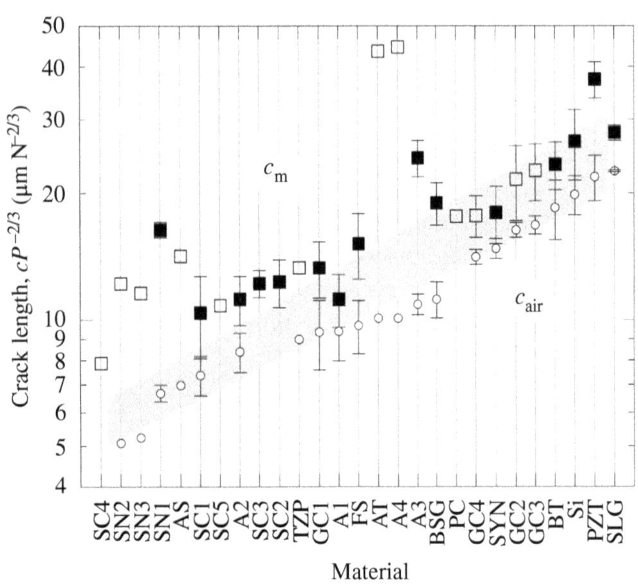

Figure 5.2 Plot of load-adjusted crack length for ceramic materials containing indentations in air, c_{air}, and at instability under applied stress in inert conditions, c_m. Shaded band indicates range of responses from review of conventional, air-based, measurements.

Table 5.2 Ceramic toughness parameters.

Material, *P* (N)	$\sigma_m P^{1/3}$ (MPa N$^{1/3}$)	$c_m P^{-2/3}$ (μm N$^{-2/3}$)	$\sigma_m c_m^{1/2}$ (MPa m$^{1/2}$)
Si, 2–50	133 ± 6 (75)	26.6 ± 5.1 (20)	0.66 ± 0.10
SLG, 0.1–10	137 ± 8 (87)	27.9 ± 1.2 (40)	0.72 ± 0.04
FS, 5–50	190 ± 12 (40)	15.2 ± 2.7 (21)	0.74 ± 0.06
BSG, 5–100	173 ± 19 (129)	19.0 ± 2.2 (35)	0.75 ± 0.09
PZT, 10	126 ± 4	37.4 ± 3.7 (24)	0.77 ± 0.03
PC, 10*	190*	17.6*	0.80*
BT, 10	183 ± 16	23.4 ± 3.0 (24)	0.89 ± 0.08
SYN, 100	281 ± 34	18.0 ± 2.8 (16)	1.2 ± 0.2
GC1, 20	396 ± 16	13.3 ± 2.0 (22)	1.4 ± 0.1
GC2, 100*	361 ± 10* (12)	21.6 ± 4.3* (8)	1.7 ± 0.2*
SC2, 1–30	441 ± 33	12.3 ± 1.6 (30)	1.7 ± 0.1
A1, 5	608 ± 54	11.2 ± 1.6 (21)	2.0 ± 0.2
GC3, 100*	440 ± 25* (11)	22.7 ± 3.5* (8)	2.1 ± 0.2*
GC4, 100*	512 ± 18*	17.7 ± 2.0* (8)	2.2 ± 0.1*
SC3, 1–300	595 ± 74	12.2 ± 0.9 (54)	2.3 ± 0.2
A2, 20	711 ± 36	11.2 ± 1.5 (36)	2.4 ± 0.2
SC1, 10	747 ± 57	10.4 ± 2.3(22)	2.4 ± 0.3
A4, 100*	390*	44.7*	2.6*
A3, 10–200	561 ± 40	24.3 ± 2.4 (8)	2.8 ± 0.2
AS, 322*	993*	14.2*	3.7*
SC4, 300*	1398*	7.9 (6)*	4.0 ± 0.2*
SN1, 20, 50	1045 ± 55	16.4 ± 0.7 (6)	4.2 ± 0.3
TZP, 300*	1666*	13.3*	4.5*
SN3, 294*	1344*	11.6*	4.6*
SC5, 300*	1452*	10.9 (6)*	5.1 ± 0.4*
AT, 300*	876*	43.6*	5.8*
SN2, 374*	2009*	12.2*	7.0*

In many cases, load-dependent responses at indentation instability are an indication of significant microstructural effects giving rise to an increasing toughness with crack length—a *T* curve, *T*(*c*) (Cook 2015). Such toughness increases lead to enhanced stability of a fracture system and elongated stable extension of cracks prior to instability and failure under applied stress (see Chapter 4). An example of elongated extension is noted for the AT material that was deliberately engineered for microstructural influence leading to a *T* curve (Bennison et al. 1991). Other materials that exhibited load-dependent behavior and that are regarded as toughened include the whisker-reinforced AS material and the phase-transforming TZP material. The *T*(*c*) variations determined from application of the modified indentation toughness technique to the complete data sets for the open symbol materials are considered in Section 5.3.3.

5.3.2 Refinement of Modified Technique

Figure 5.3 is a plot of the indentation instability toughness parameter $\sigma_m c_m^{1/2}$ for the complete materials set examined here. The figure uses a plotting scheme similar to that in Figure 5.2, ranking materials from left to right in order of increasing $\sigma_m c_m^{1/2}$ value. The filled symbols represent the mean values and standard deviation limits for $\sigma_m c_m^{1/2}$, determined in most cases by analyzing individual σ_m–c_m pairs. In a few cases, e.g., SLG, the means and standard deviations of grouped $\sigma_m P^{1/3}$ and $c_m P^{-2/3}$ values for a material were used to determine the mean $\sigma_m c_m^{1/2}$ values and standard deviation limits

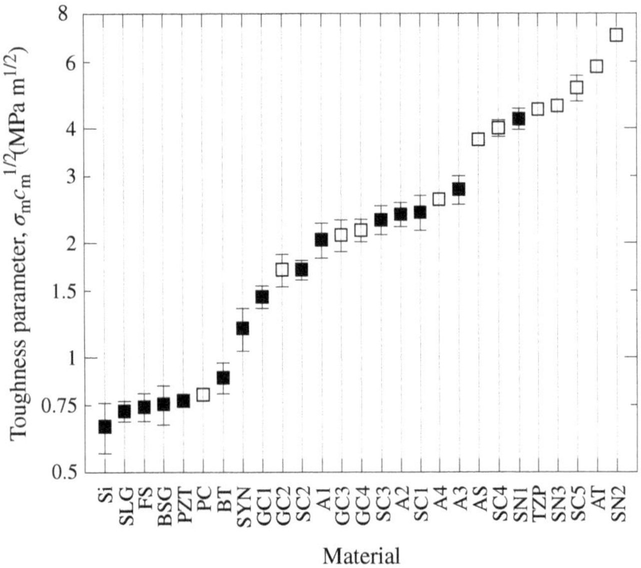

Figure 5.3 Plot of ranked toughness parameter $\sigma_m c_m^{1/2}$ for ceramic materials. The solid symbols represent materials with single valued toughness. The open symbols represent the upper-bound toughness values for materials exhibiting T curves. Tougher materials on right.

were determined by analysis of variance. For all materials, the number of strengths measured, if different from the number of cracks, and the resulting $\sigma_m P^{1/3}$ mean values and uncertainties are given in Table 5.2. As in Figure 5.2, the open symbols in Figure 5.3 represent the $\sigma_m c_m^{1/2}$ large load responses of materials that exhibited load-dependent indentation behavior. The strength values used are given in Table 5.2 with uncertainties where applicable; the strengths are indicated by an asterisk (*). For all materials, the values, and uncertainties where applicable, for $\sigma_m c_m^{1/2}$ are given in Table 5.2, ranked as in Figure 5.3.

There are many points to note in Figure 5.3. First, the toughness parameter $\sigma_m c_m^{1/2}$ values are comparable to accepted long crack toughness values: The toughness of Si and glasses is approximately 0.7 MPa m$^{1/2}$, the toughness of polycrystalline alumina is approximately 2 MPa m$^{1/2}$, and the toughness of microstructurally engineered materials that are toughened by crack-front transformation processes (TZP) or crack wake traction zones (AS, AT) often exceeds 5 MPa m$^{1/2}$. The implication is that the geometrical factor for indentation flaws under applied stress $(4\psi/3) \approx 1$, Eq. (5.6). Second, the $\sigma_m c_m^{1/2}$ values for the glasses, SLG, BSG, and FS, are identical within experimental uncertainty. The artifact of separation arising from crack length or strength measurements alone (Figure 5.2, Table 5.2) is removed. Third, the relative measurement uncertainty for all materials is small: the standard deviation/mean ratio is approximately ±8%, independent of the value of $\sigma_m c_m^{1/2}$ and the material. The first and second points provide support for the accuracy of the modified indentation toughness technique, and the third point provides support for the precision of the technique.

The $\sigma_m c_m^{1/2}$ ranking scheme of Figure 5.3, while extremely useful for product designers selecting brittle materials, is a semiquantitative application of the modified indentation toughness technique. The technique can be rendered more quantitative through a numerical estimate of the crack geometry parameter $(4\psi/3)$ and application of Eq. (5.6). The steady-state toughness values of many of the materials in Figure 5.3 and Table 5.2 have been measured by long crack fracture techniques, e.g. double cantilever beam, single edge-cracked beam, or chevron notched beam and enable such an estimate. For materials with little or no crack–microstructure interactions, e.g., single crystals, glasses, and most glass-ceramics and fine-grained polycrystals, such measurements are crack length invariant and coincide with short crack measurements. Such materials also exhibit indentation load invariant $\sigma_m c_m^{1/2}$ responses and appear predominantly on the left of Figure 5.3. For materials with significant crack–microstructure interactions, e.g., large-grained polycrystals, phase-transforming materials, such measurements provide an estimate of the toughness values applicable for long cracks generated by large indentation loads or by the crack extensions generated on superposed applied loading. Such materials exhibit varying $\sigma_m c_m^{1/2}$ responses and the upper bounds appear predominantly on the right of Figure 5.3. Comparison of $\sigma_m c_m^{1/2}$ values with steady-state toughness measurements T on the same material thus provides a means of estimating ψ, recognizing that invariant and upper-bound $\sigma_m c_m^{1/2}$ values correlate with $T \approx T_0$ and $T \approx T_\infty$, respectively.

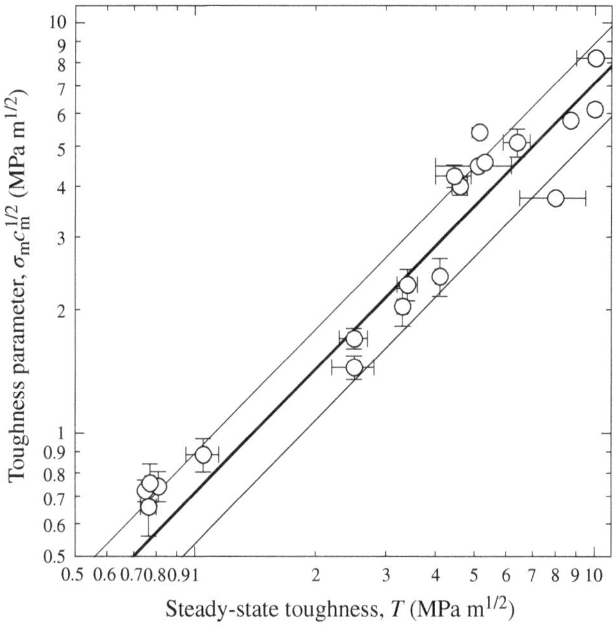

Figure 5.4 Logarithmic plot of indentation toughness parameter $\sigma_m c_m^{1/2}$ as a function of steady-state toughness T. Symbols represent individual materials. Lines indicate linear best fit and bounds, consistent with Eqs. (5.6) and (5.8).

Figure 5.4 is a logarithmic plot of the indentation toughness parameter $\sigma_m c_m^{1/2}$ as a function of steady-state toughness T for materials systems in which both quantities were measured. The plot includes many of the materials from Figure 5.3 and matching toughness values from published works, where available (Wiederhorn 1969; Barker 1973; McHenry et al. 1976; Pohanka et al. 1978; Trantina 1979; Anstis et al. 1981; Marshall 1983; Li et al. 1992; Sherman 1997; Kaur et al. 2009; Lin et al. 2007; DelRio et al. 2015; Xue et al. 2020). The symbols in Figure 5.4 represent the mean responses of individual materials; the bars in $\sigma_m c_m^{1/2}$ represent the experimental standard deviation uncertainties from Table 5.2; the bars in T represent the greater of the experimental uncertainty from a single data source and the range of values encompassing multiple data sources; in some cases, uncertainty could not be estimated. The bold solid line in Figure 5.4 is a visual fit to the data, consistent with the linear relation of Eq. (5.6): the best fit value of the geometry factor used was $\psi = \overline{\psi} = 1.05$. The fine lines in Figure 5.4 indicate bounds of $0.75\overline{\psi}$ and $1.25\overline{\psi}$ and correspond to the oft-cited $\pm 25\%$ relative uncertainty in the analogous ξ_0 factor of Eq. (2.16) used in toughness estimation by indentation crack length measurements (Anstis et al. 1981). The linear fit in Figure 5.4, based on a material-invariant value of $\overline{\psi}$, clearly describes the observations and the bounds include the range of values, 1.00–1.24, assumed in the prior works cited.

The value of $\overline{\psi} = 1.05$ established in Figure 5.4 is used in Section 5.3.3 to estimate toughness variations for a range of materials. However, before doing so, it is important to consider two significant differences between Figure 5.4 and prior, similar appearing, indentation toughness "calibration" plots (Anstis et al. 1981; Chantikul et al. 1981). The first difference is one of censoring. In the prior cited works, data from materials that did not exhibit "ideal" indentation behavior were excluded or censored from analysis. The materials censored were: the "anomalous" glasses that exhibit indentation densification leading to reduced residual stress field amplitudes, cone cracks, and reduced half-penny crack traces; polycrystalline ceramics that exhibit considerable intergranular indentation disruption, lateral cracking, and difficult to observe half-penny crack traces; composites; and phase-transforming materials. No materials or data were excluded from Figure 5.4, which includes representatives of all the prior censored materials. The second difference is one of uncertainty interpretation. In the prior cited works, uncertainties were provided as standard deviations of fitting parameters. Thus, the specified mean $\pm$ uncertainty of the correlation constants, ξ_0 in particular, included about 67% of the (censored) data sets, and mean $\pm 2\times$ uncertainty included most materials, about 90% of the data. More than 90% of the (uncensored) data are included within the range bounds of Figure 5.4, which exhibits a data dispersion about half that of the prior works. Thus, ξ_0 (Anstis et al. 1981), and the related η (Chantikul et al. 1981), should be interpreted as descriptions of designated observations rather than "calibration" constants. $\overline{\psi}$ is much more of a calibration constant as evaluation includes all available material data and is specified as a range. $\overline{\psi}$ can be used with confidence in predicting the toughness of an unknown material from results of a modified indentation technique.

5.3.3 Application of Modified Technique

For materials that exhibit load-dependent $\sigma_m P^{1/3}$ and $c_m P^{-2/3}$ values, it is a simple extension of the above analysis to show that using P as a parameter enables the T curve to be determined via generalization of Eq. (5.6) (Braun et al. 1992; Cook 2015):

$$T(c_m) = (4\overline{\psi}/3)\sigma_m(P)c_m^{1/2}(P). \tag{5.8}$$

Three applications of the modified indentation toughness technique using Eq. (5.8) are demonstrated here. The first application considers the T curves of six engineered polycrystalline Si-based ceramics, Si_3N_4 (SN2 and SN3) and SiC (SC2, SC3, SC4, SC5). These are regarded as very tough ceramics with long T curves. The second application considers the T curves of six engineered polycrystalline oxide ceramics, Al_2O_3 (AS, AT, A4) and ZrO_2 (TZP). These are also regarded as very tough ceramics with long T curves. The third application considers microstructural alteration of a family of Li_2SiO_3 glass-ceramics (GC2, GC3, GC4) and development of the resulting T curves. These are moderately tough ceramics with short T curves. For ease of comparison, the data for all three applications are presented as $T(c_m)$ variations in logarithmic coordinates at the same scale, 8×400. Based on Figure 5.4, the proportionality parameter is taken here as $(4\overline{\psi}/3) = 1.40$.

Figure 5.5 shows the T curves for polycrystalline Si_3N_4 (filled symbols) (Ramachandran and Shetty 1991; Li et al. 1992) and polycrystalline SiC (open symbols) (Kaur et al. 2009), using data derived from the published works cited. The symbols represent toughness values calculated from application of Eq. (5.8) to individual $\sigma_m(P)$–$c_m(P)$ pairs measured at each load P. As used previously, the symbols at the longest crack lengths reflect the data in Table 5.2. The uncertainties in crack length represent those published and the uncertainties in toughness represent those from analysis of variance of strengths and crack lengths. The shaded band represents the range of expected lower-bound toughness values for fine-grained or single crystal Si_3N_4 and SiC as noted by Li et al. (1992). The lines are guides to the eye (recognizing that straight lines in logarithmic coordinates represent power-law behavior, $T \sim c^{\tau/2}$, see Chapter 4). It is clear that the materials exhibited a range of T curve behavior over the crack length domain 20–500 μm, from no apparent increase in toughness for the least tough SiC material to an almost doubling of toughness for the toughest SiC and Si_3N_4 materials. Many of the materials in Figure 5.5 were engineered to exhibit crack wake cohesive traction zones with the necessary crack deflections and formation of frictional interlocks and ligamentary bridges. As detailed in Chapter 4, on crack extension such materials are toughened from base intergranular or transgranular toughness values. The observations here suggest that such toughening varied considerably from material to material in Figure 5.5. However, as the empirical straight lines in Figure 5.5 converge at small crack lengths to the shaded band estimate of untoughened behavior, the observations also suggest that toughening in these materials started from similar base values.

Figure 5.6 shows the T curves for polycrystalline oxides of Al_2O_3 (open symbols) (Ramachandran and Shetty 1991; Braun et al. 1992; Kovar et al. 2000) and ZrO_2 (solid symbols) (Ramachandran and Shetty 1991), using data derived from

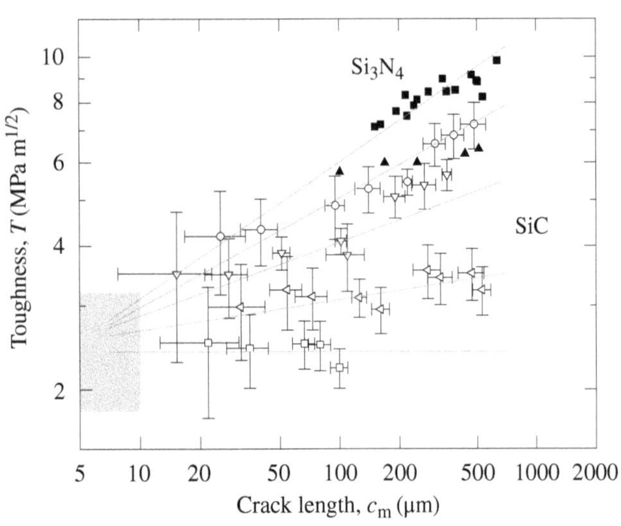

Figure 5.5 Logarithmic plot of toughness T as function of crack length c_m, T curves, for SiC and Si_3N_4 materials. Source: Data adapted from Ramachandran and Shetty (1991), Li et al. (1992), and Kaur et al. (2009).

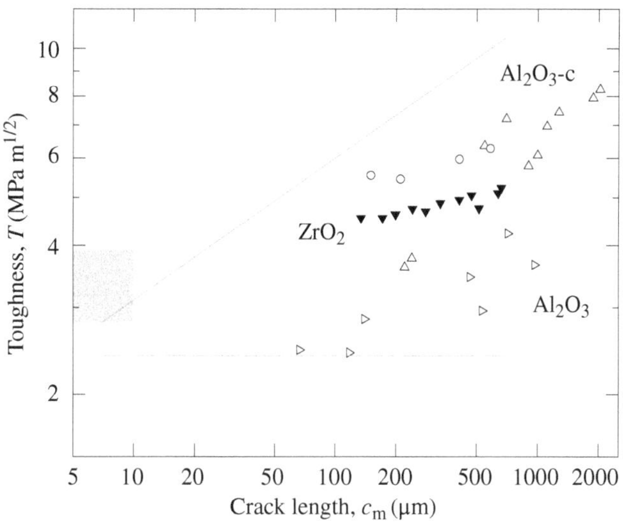

Figure 5.6 Logarithmic plot of toughness T as function of crack length c_m, T curves, for Al_2O_3 and ZrO_2 materials. Source: Data adapted from Ramachandran and Shetty (1991), Braun et al. (1992), and Kovar et al. (2000).

the published works cited. The symbols represent toughness values determined by the same procedures used for the data of Figure 5.5. The shaded band represents the range of expected lower-bound toughness values for fine-grained Al_2O_3 determined from measurements on the A1, A2, and A3 materials. The lines are guides to the eye and repeat the upper and lower extremes from Figure 5.5. The materials exhibited a range of T curve behavior over the crack length domain 50–2000 μm. All the oxides in Figure 5.6 were engineered to exhibit toughening—the Al_2O_3 by crack wake cohesive traction zone formation and the ZrO_2 by crack-front energy consuming phase transformation. The lower open Al_2O_3 symbols in Figure 5.6 represent materials fabricated with grain size distributions that included a significant number of large grains that contributed to cohesive zone formation. The upper open Al_2O_3-c symbols in Figure 5.6 represent alumina composite materials: Al_2O_3–SiC(w) was a composite of a fined-grained alumina matrix containing tough SiC whiskers and Al_2O_3–Al_2TiO_5 was a composite of an alumina matrix containing aluminum titanate particles; the particles were thermally mismatched with the matrix leading to an inhomogeneous internal stress field. The whiskers and stressed particles also contributed to cohesive zone formation. The zirconia material in Figure 5.6 was yttria-stabilized tetragonal zirconia polycrystal and was engineered to exhibit phase-transformation triggered by crack-front stress fields (the monoclinic ZrO_2 phase is stable at ambient conditions and the tetragonal phase dilates and shears on transformation).

Figure 5.7 shows the T curves for oxides of polymer cement (lower filled symbols) (Chou et al. 1990) and lithia silicate glass-ceramics (upper symbols), using data derived from the published work. The symbols and bars represent toughness values determined by the same procedures used for the data of Figure 5.5. The lines are guides to the eye. The materials exhibited a range of T curve behavior over the crack length domain 30–500 μm, but over smaller toughness ranges than the materials in Figures 5.5 and 5.6. The data for the polymer cement exhibit behavior similar to that of the materials in Figures 5.5 and 5.6, albeit at much reduced toughness values and over a much smaller domain of crack lengths. Nevertheless, the straight line slope is similar indicating that toughening processes are not restricted to very tough materials.

Figure 5.8 is a set of optical micrographs of the GC2, GC3, and GC4 materials. The light contrast major matrix phase is the SiO_2 polymorph cristobalite and the dark contrast minor platelet phase is Li_2SiO_3. Both the platelet size and the mean free path in the major phase increase in the order GC2, GC3, GC4. Detailed microstructural morphology and compositional information are given elsewhere (Cook et al. 1986), but it is clear that the materials represent a set of microstructures of increasing scale and from Figure 5.7 that these materials exhibit T curves. For the set of materials, the toughness increases within a crack length domain of 30–500 μm over a range of approximately 1.2–1.6 MPa $m^{1/2}$ to approximately 1.7–2.3 MPa $m^{1/2}$. Larger microstructural scales exhibited greater toughness values at all crack lengths in the order GC2, GC3, GC4. Distinct from the toughness behaviors shown in Figures 5.5 and 5.6, there was a slight tendency to toughness plateaus, at crack lengths longer than approximately 200 μm. Observations of crack extension in these materials showed that microstructurally-induced frictional interlocks and ligamentary bridges formed in the crack wakes, leading to cohesive traction zones (Braun et al. 1992; Cook 2015). Specifically, the platelets appeared more fracture resistant than the matrix, leading to the necessary crack deflection and discontinuous crack propagation required for interlock and bridge formation.

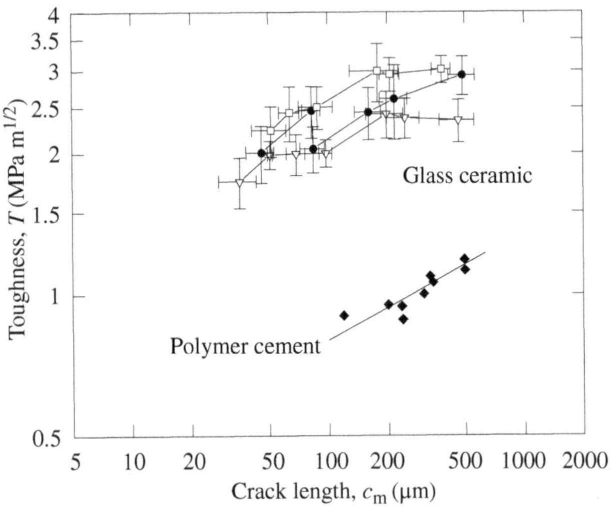

Figure 5.7 Logarithmic plot of toughness T as function of crack length c_m, T curves, for polymer cement and Li_2SiO_3 glass-ceramic materials. Source: Includes data adapted from Chou et al. (1990).

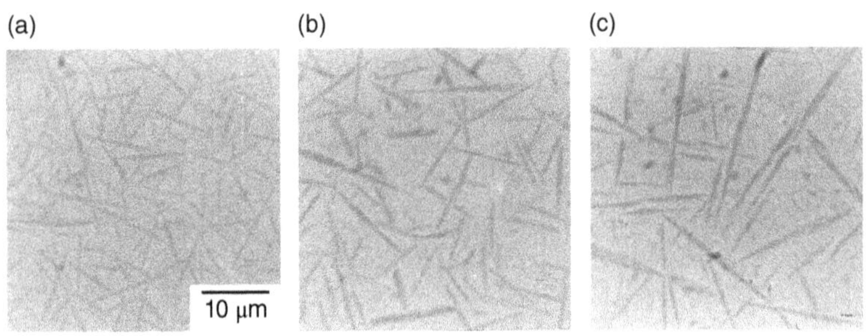

Figure 5.8 Optical micrographs of the (a) GC2, (b) GC3, and (c) GC4 glass-ceramic materials. (Note that material IDs are different from those used earlier, Cook et al. 1986; Cook 2020.)

The implication of Figure 5.7 is that the effects of the cohesive zones, but not necessarily the length scales, increased with the scale of the microstructure. Although detailed discussions of toughening mechanisms are not the focus of this book, it is clear that a multitude of platelet-related elements, of order 10 μm in scale, as shown in Figure 5.8, must be responsible for the observed toughness variations over approximately 100 μm in scale, as shown in Figure 5.7. Such toughness variations could not be discerned by large fracture specimens that typically involve cracks tens of millimeters in length and only sense plateau toughness responses.

There are several quantitative and qualitative differences between the T curves of Figures 5.5 and 5.6 and those of Figure 5.7. First, the toughness and crack length scales describing the responses of the materials in Figures 5.5 and 5.6 are approximately a factor of 2 greater, highlighting the greater load-bearing capacity of these more fracture-resistant materials. Second, the range of toughness values, approximately 2–7 MPa $m^{1/2}$, is much greater for the group of materials in Figures 5.5 and 5.6, indicative of the different materials families represented. Third, with one exception, the responses of individual materials in Figures 5.5 and 5.6 exhibit somewhat weaker relative *increases* of toughness with crack length, although, fourth, the materials in Figures 5.5 and 5.6 do not exhibit tendencies to plateau toughness values at long crack lengths. Again, such toughness variations could not be discerned by large, long crack fracture specimens.

5.4 Discussion

Significant extensions to a modified indentation technique for toughness determination of ceramics and other brittle materials have been provided here. The extensions refine the earlier work (Cook and Lawn 1983) by examining more materials, acquiring new data, including data from other investigators and other techniques (see Tables 5.1 and 5.2), and reanalyzing raw data to better assess uncertainty. As a consequence, the technique can be used with greater confidence in both materials engineering applications and materials science investigations. In particular, for materials engineering, a major new result is the ranking of materials by a direct output of the technique, the toughness parameter $\sigma_m c_m^{1/2}$. The ranking generated here, Figure 5.3, provides a clear foundation and design tool for assessing the relative merits of materials in development and for materials selection. In addition, the level of uncertainty exhibited by the current materials within the ranking provides a quantitative basis for assessing the accuracy and precision of measurements on new materials. Importantly, the ranking is based on the toughness characteristics of materials at length scales relevant to strength-controlling flaws.

As a design tool, implementation of the modified indentation toughness technique requires a little more effort than conventional indentation toughness assessment methods (Anstis et al. 1981; Chantikul et al. 1981; Cook 2020), but exactly the same effort as common long crack toughness measurements using standard fracture mechanics specimens (Murakami 1987). Both modified indentation and standard specimen measurements are based on sound fracture mechanics, both require the creation and measurement of carefully controlled cracks in specimens of known geometry, and both require measurement of the peak sustainable load or stress of the specimen. The long cracks of standard specimen geometries practically eliminate dependence of toughness assessment on the mechanism by which the cracks were initiated and formed. The indentation strength method (Chantikul et al. 1981) significantly reduces dependence on the crack initiation and formation mechanism relative to the indentation crack length method (Anstis et al. 1981), but a weak dependence remains. The modified technique removes this dependence completely, reducing the uncertainty in toughness assessment to that associated with crack geometry, similar to that for long crack measurements. As a consequence of reduced influence of the indentation residual field in application of the modified technique, the materials rankings are considerably different from those arising from conventional crack length measurements, $c_{air}P^{-2/3}$, alone (Figure 5.2), and slightly different from strength measurements, $\sigma_m P^{1/3}$, alone (Table 5.2).

From a materials science perspective, a major new result is the application of the modified indentation toughness technique to determine T curves, $T(c)$, Figures 5.5–5.7. Such determination enables the establishment of structure–properties relations for materials that exhibit microstructural influences on crack propagation. For example, the larger microstructural scale of the GC4 material is correlated with a greater long crack, steady-state toughness, Figure 5.7, whereas the short crack toughness values of GC2, GC3, and GC4 appear to be similar. The implication is that fracture in the matrix of all three glass-ceramics is similar but that larger platelets cause greater energy dissipation during crack extension. Similarly, the enhanced toughness and toughness increase with crack length for Si_3N_4 relative to Al_2O_3–SiC(w) in Figures 5.5 and 5.6 suggests that elongated Si_3N_4 grains are more potent cumulative toughening agents at longer crack length scales than SiC(w) in Al_2O_3. Such observations enable materials science based models of toughening at strength-relevant length scales to be developed. It is noted that some of the T curves reported here differ somewhat from those published earlier (Ramachandran and Shetty, 1991; Braun et al. 1992; Li et al. 1997; Kovar et al. 2000). This is simply a consequence of slightly different numerical factors used in analysis. In addition, T curves determined by direct observation of indentation crack extension from c_{air} to c_m rely on specification of χ, Eq. (5.3) (an example is shown in Chapter 11), that is obviated here in the modified technique by sole reliance on measurements at the c_m instability configuration, Eq. (5.8).

Finally, another materials science insight comes from interpretation of the apparently small value of $T_{air}/T_0 \approx 0.4$. This small value was implied by crack lengths observed in air after indentation relative to those at crack instability in inert conditions under applied stress (Section 5.3.1). For simplicity, the same residual field geometry factor, χ_{air}, was implemented in the strength analysis, Eqs. (5.3) and (5.5), leading to the elimination of this factor on comparison of the initial and final crack lengths, Eq. (5.7). As a consequence, interpretation of the relative crack lengths was made solely in terms of relative toughness values in air and inert conditions, with implications for the chemical reduction of ceramic surface energies on exposure to moisture. The small value of T_{air}/T_0 suggests much greater reduction than more detailed experiments in water imply (Cook 2015); $T_{air}/T_0 \approx 0.7$, placing a lower bound on T_{air}. The resolution to this discrepancy lies in recognizing that the residual field geometry term is changed under the action of applied stress such that $\chi = \chi_{stress}$ in Eqs. (5.3) and (5.5), and Eq. (5.7) becomes

$$c_m/c_{air} = (4\chi_{stress}T_{air}/\chi_{air}T_0)^{2/3}. \tag{5.9}$$

The implication from the combined observations and Eq. (5.9) is that the effect of the residual field is reduced significantly on application of stress to an indentation flaw, such that $\chi_{stress}/\chi_{air} \approx 0.5$. This reduction has wide materials science implications for the behavior of contact flaws under stress.

5.5 Conclusions

Additional measurements and comparisons affirm that the modified indentation toughness technique provides an accurate and precise method for qualitative ranking and quantitative determination of ceramic toughness. The method is applicable to materials that exhibit variations in toughness with crack length, T curves. The modified technique has much in common with standard long crack measurement methods in that a load or stress and crack length at specimen instability are combined to assess material toughness. However, the modified technique has the advantages of ease of indentation crack formation and placement and measurement of material toughness at length scales relevant to strengths of structural ceramics. The artifacts associated with sole (and often limited) use of indentation crack length or indentation strength measurements, particularly for glasses, are largely removed by the modified technique with little additional experimental effort.

References

Anstis, G.R., Chantikul, P., Lawn, B.R., and Marshall, D.B. (1981). A critical evaluation of indentation techniques for measuring fracture toughness: I, direct crack measurements. *Journal of the American Ceramic Society* 64: 533–538.

Barker, L.M. (1973) Short rod K_{Ic} measurements of Al_2O_3. In: *Flaws and Testing. Fracture Mechanics of Ceramics*, vol 3 (Eds. Bradt, R.C., Hasselman, D.P.H., and Lange, F.F.), 483–493. Boston, MA: Springer.

Bennison, S.J., Padture, N.P., Runyan, J.L., and Lawn, B.R. (1991). Flaw-insensitive ceramics. *Philosophical Magazine Letters* 64: 191–195.

Braun, L.M., Bennison, S.J., and Lawn, B.R. (1992). Objective evaluation of short-crack toughness curves using indentation flaws: case study on alumina-based ceramics. *Journal of the American Ceramic Society* 75: 3049–3057.

Chantikul, P., Anstis, G.R., Lawn, B.R., and Marshall, D.B. (1981). A critical evaluation of indentation techniques for measuring fracture toughness: II, strength method. *Journal of the American Ceramic Society* 64: 539–543.

Chantikul, P., Bennison, S.J., and Lawn, B.R. (1990). Role of grain size in the strength and R-curve properties of alumina. *Journal of the American Ceramic Society* 73: 2419–2427.

Chou, Y.S., Mecholsky, J.J., and Silsbee, M. (1990). Fracture toughness of macro-defect-free cement using small crack techniques. *Journal of Materials Research* 5: 1774–1780.

Cook, R.F. (2006). Strength and sharp contact fracture of silicon. *Journal of Materials Science* 41: 841–872.

Cook, R.F. (2015). Multi-scale effects in the strength of ceramics. *Journal of the American Ceramic Society* 98: 2933–2947.

Cook, R.F. (2019). Fracture sequences during elastic-plastic indentation of brittle materials. *Journal of Materials Research* 34: 1633–1644.

Cook, R.F. (2020). A critical evaluation of indentation crack lengths in air. *Journal of the American Ceramic Society* 103: 2278–2295.

Cook, R.F., Freiman, S.W., and Baker, T.L. (1986). Effect of microstructure on reliability predictions for glass-ceramics. *Materials Science and Engineering A* 77: 199–212.

Cook, R.F., Lawn, B.R., and Anstis, G.R. (1982). Fatigue analysis of brittle materials using indentation flaws: part 2 Case study on a glass-ceramic. *Journal of Materials Science* 17: 1108–1116.

Cook, R.F. and Lawn, B.R. (1983). A modified indentation toughness technique. *Journal of the American Ceramic Society* 66: C200–C201.

Dabbs, T.P. (1984) *Fatigue Properties of High Strength Glass with Particular Reference to Optical Fibers.* Ph.D. Thesis. University of New South Wales, Kensington, Australia.

Dabbs, T.P., Lawn B.R., and Kelly P.L. (1982). A dynamic fatigue study of soda-lime silicate and borosilicate glasses using small scale indentation flaws. *Physics and Chemistry of Glasses* 23: 58–66.

DelRio, F.W., Cook, R.F., and Boyce, B.L. (2015). Fracture strength of micro-and nano-scale silicon components. *Applied Physics Reviews* 2: 021303.

Kaur, S., Cutler, R.A., and Shetty, D.K. (2009). Short-crack fracture toughness of silicon carbide. *Journal of the American Ceramic Society* 92: 179–185.

Kovar, D., Bennison, S.J., and Readey, M.J. (2000). Crack stability and strength variability in alumina ceramics with rising toughness-curve behavior. *Acta Materialia* 48: 565–578.

Li, J.-F., Kawasaki, A., and Watanabe, R. (1997). R-curve determination of 3Y-PSZ by the indentation strength-in-bending method. *Journal of the Ceramic Society of Japan* 105: 88–90.

Li, C.W., Lee, D.J.. and Lui, S.C. (1992). R-curve behavior and strength for in-situ reinforced silicon nitrides with different microstructures. *Journal of the American Ceramic Society* 75: 1777–1785.

Lin, G.Y., Lakshminarayanan, R., and Shetty, D.K. (2007). Contact damage initiation in silicon nitride in Hertzian indentation: role of microstructure. *Journal of Materials Science* 42: 3508–3519.

Marshall, D.B. (1983). Controlled flaws in ceramics: a comparison of Knoop and Vickers indentation. *Journal of the American Ceramic Society* 66: 127–131.

McHenry, K.D., Yonushonis, T., and Tressler, R.E. (1976). Low-temperature subcritical crack growth in SiC and Si_3N_4. *Journal of the American Ceramic Society* 59: 262–263.

Murakami, Y. (Ed.) (1987) *Stress Intensity Factors Handbook*. Oxford, UK: Pergamon.

Pohanka, R.C., Freiman, S.W., and Bender, B.A. (1978). Effect of the phase transformation on the fracture behavior of $BaTiO_3$. *Journal of the American Ceramic Society* 61: 72–75.

Ramachandran, N. and Shetty, D.K. (1991). Rising crack-growth-resistance (R-curve) behavior of toughened alumina and silicon nitride. *Journal of the American Ceramic Society* 74: 2634–2641.

Sherman, D. (1997). Fracture toughness evaluation of small thin ceramic specimens. *Journal of the American Ceramic Society* 80: 1904–1906.

Trantina, G.G. (1979). Strength and life prediction for hot-pressed silicon nitride. *Journal of the American Ceramic Society* 62: 377–380.

Wiederhorn, S.M. (1969). Fracture surface energy of glass. *Journal of the American Ceramic Society* 52: 99–105.

Xue, M., Liu, S., Wang, X., and Jiang, K. (2020). High fracture toughness of 3Y-TZP ceramic over a wide sintering range. *Materials Chemistry and Physics* 244: 122693.

6

Indentation Strengths: Reactive Environments

In this chapter the effects of reactive environments on indentation strength are considered. In particular, the influences of exposure to water and slow applied stressing rates during loading of indented components is surveyed and placed in the context of the strengths surveyed in Chapters 3 and 4. The fundamental phenomenon of thermally activated nonequilibrium crack propagation is considered and related to the static and dynamic fatigue of brittle materials. Analyses of strengths in reactive environments are developed, beginning with simple empirical relationships and then including indentation and microstructural effects and the required thermodynamic crack velocity threshold. An extensive survey of dynamic fatigue of brittle materials includes the effects of indentation load and various materials, including glasses, alumina, glass ceramics, and titanates. Static fatigue predictions are demonstrated and verified using dynamic fatigue measurements. The effects of compressive and tensile surface stresses are demonstrated.

6.1 Introduction

This chapter considers indentation strengths of brittle components exposed to reactive environments—reactive strengths. Analyses of such strengths rest on the dual foundations of the thermodynamics of surface energy and isothermal fracture established in Chapter 1 and the mechanics of crack driving forces established in Chapters 3 and 4. The most important difference between the indentation strengths considered here and those in Chapters 3 and 4 is that here surface energy is significantly reduced by the effects of reactive environments, while crack driving forces remain unchanged. This combination of change and no change leads to fracture conditions far from equilibrium and, consequently, time-dependent nonequilibrium crack propagation; conditions and behavior that were negligible in Chapters 3 and 4. The most prevalent group of materials and environments that exhibit such behavior are oxide materials in the presence of water, in either liquid or gaseous form, including atmospheric moisture. The most obvious examples of nonequilibrium brittle fracture behavior are time- or rate-dependent strengths, and these phenomena are the focus of this chapter. Many of the materials considered in Chapters 3 and 4 in terms of inert strengths are also considered here in terms of reactive strengths, including silicate glasses and single-crystal and polycrystalline alumina.

6.1.1 Background Concepts

The starting point for the analysis of brittle fracture strengths in reactive environments is a description of nonequilibrium, irreversible, isothermal fracture, considered in detail in Section 1.6.2. A key element in the description is the variation of equilibrium surface energy γ (expressed as an areal density) in the presence of reactive species. This variation is given by the linked combination of Eqs. (1.71) and (1.72). These two equations form a parametric set for the variation of γ in a gaseous environment containing a partial pressure p of reactive species. Chemical potential μ is used as the linking parameter. Eliminating explicit use of μ gives an expression for the surface energy in experimentally measurable terms as

$$\gamma = \gamma_0 - k_B\theta \int_{-\infty}^{\ln p} \Gamma(p')\mathrm{d}\ln p', \tag{6.1}$$

where p' is a dummy pressure variable, k_B is Boltzmann's constant, and θ is absolute temperature. γ_0 is the upper-bound surface energy in an inert environment, and it is recognized that γ is an equilibrium quantity. Adsorption of reactive species onto a surface is quantified by the coverage Γ, the number of reactive molecules/surface area. Chapters 3 and 4

Indentation Fracture: Strength and Toughness for Brittle Materials Design, First Edition. Robert F. Cook.

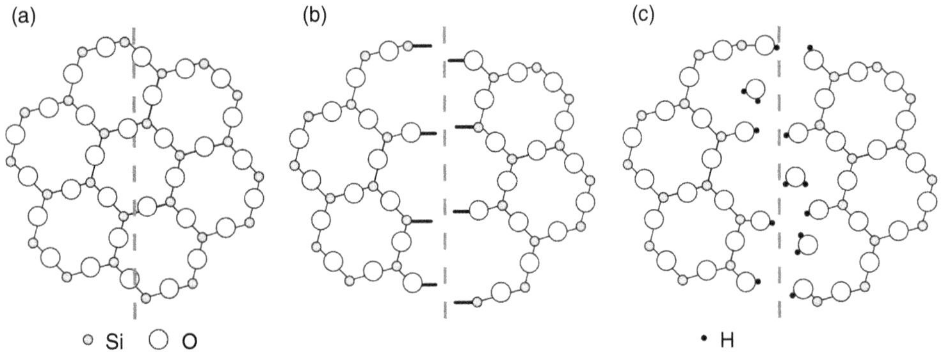

Figure 6.1 Schematic diagrams of fracture of a silicate (Si–O) structure and adsorption of water (H_2O) molecules. Fracture plane is shown as dashed. (a) Initial structure. (b) Structure fractured in inert environment. Broken bonds are shown as bold. (c) Structure fractured in reactive water environment.

considered fracture in inert environments in which reactive species were very sparse (e.g., dry N_2 gas, silicone oil) such that $p \to 0$ leading to $\Gamma \to \Gamma(0) = 0$ and $\ln p \to -\infty$. In this case, the integral of Eq. (6.1) vanishes and the surface energy approaches the inert value, $\gamma \to \gamma_0$. This chapter considers fracture in reactive environments (e.g., moist air) such that $p > 0$ and $\Gamma > 0$ (and $\ln p > -\infty$) such that the integral of Eq. (6.1) is nonzero and the surface energy is reduced from the inert value, $\gamma \leq \gamma_0$.

The macroscopic surface energy reductions associated with the presence of reactive species are significant: Common domestic and laboratory air conditions of 40% relative humidity (RH) lead to $p \approx 0.01$ for water. Experimental values of surface energy of silicate glasses in such conditions are $\gamma/\gamma_0 \approx 0.75$. Other common environments consisting almost entirely of reactive water species (e.g., *very* humid air, liquid water) effectively increase $p \to 1$. Γ thence takes a maximum value, and surface energy takes a minimum value, which for glasses is $\gamma/\gamma_0 \approx 0.5$. This chapter provides experimental assessments of such ratios. Schematic diagrams of the underlying fracture and adsorption processes are shown in the atomic scale sequence of Figure 6.1. Figure 6.1a shows a planar segment of a silicate structure; the Si and O atoms are labeled, and interatomic Si–O bonds are indicated by fine lines. A potential fracture plane perpendicular to the diagram is indicated by a dashed line and intersects several bonds. Figure 6.1b shows the segment parted or fractured in an inert environment. Si–O bonds are ruptured along the dashed line leaving unterminated bonding electrons indicated by the bold lines on both Si and O on the exposed surfaces. Figure 6.1c shows the fractured surfaces on exposure to H_2O molecules. The molecules separate into OH and H and react with the unterminated Si and O surface atoms, respectively, to form SiOH surface terminations. The a→b→c sequence may be written as

$$\text{Si–O–Si} \overset{\text{fracture}}{\rightarrow} \text{Si–} + \text{–O–Si} \overset{\text{H}_2\text{O}}{\rightarrow} \text{Si–OH} + \text{HO–Si}. \tag{6.2}$$

The central and right terms in this sequence are represented by Eq. (6.1). The left and central terms are represented by the thermodynamic considerations of Chapter 1. From a chemical reaction perspective, the left and right terms represent the initial and final states of the reaction and the central term represents the energy barrier separating these states. Under isothermal conditions, the reaction is thermally activated over the barrier leading to a reaction rate. From a mechanical perspective, however, the left and right terms are related to the central term by work and internal elastic energy changes. As a result, the barrier height is determined and biased by the mechanical energy deposited into the reaction. A reaction rate model based on mechanically biased energy barriers has been developed to describe fracture kinetics incorporating Eq. (6.2) (Lawn 1975; Krausz and Krausz 1988) and applied in descriptions of reactive fracture (Cook and Liniger 1993, 1999; Chevalier and Fantozzi 2005; Cook 2015, 2018a, 2019). The model will be applied extensively in this chapter.

The nonnegative entropy constraint of Eq. (1.70) combines mechanical, reaction, and fracture kinetics effects into a condition for isothermal crack propagation. The condition may be written as

$$(\mathscr{G} - 2\gamma)v \geq 0, \tag{6.3}$$

where $\mathscr{G}$ is the mechanical energy release rate acting on the crack and v is the crack velocity. v is measured along a single dimension of the crack, the crack length c, such that $v = dc/dt$ where t is time. Mechanical energy release rates in the domain $\mathscr{G} \geq 2\gamma$ result in isothermal crack velocities in the range $v \geq 0$. Distinct from fracture in inert environments time is a key factor for fracture in reactive environments. Cracks extend with time under sustained applied loading if the condition

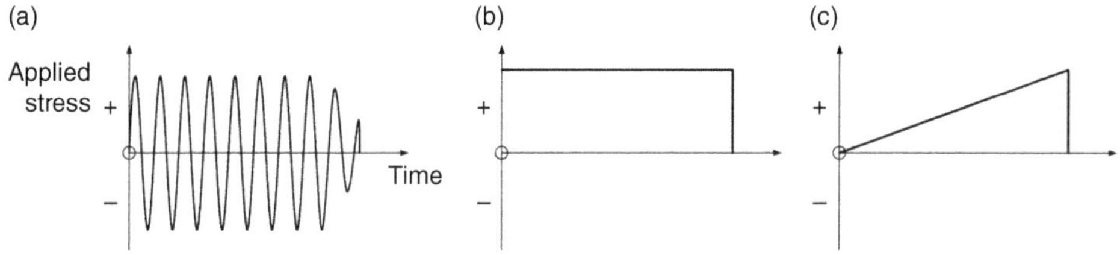

Figure 6.2 Schematic diagrams of applied stress–time behavior leading to component failure (zero support of applied stress). (a) Cyclic fatigue loading commonly associated with ductile material failure. (b) Static fatigue loading associated with brittle material failure. (c) Dynamic fatigue loading associated with brittle material failure and commonly implemented in reactive fracture testing.

of Eq. (6.3) is maintained. However, a common outcome of such extension is that eventually the condition $\mathscr{G} = 2\gamma_0$ is met, leading to the onset of rapid, adiabatic fracture (see Section 1.6.1) and component failure. As a consequence, through comparison with the rapid adiabatic crack extension associated with terminal fracture, the phenomenon of isothermal crack velocity behavior in brittle materials is often termed "slow" or "subcritical" crack growth. More accurately, isothermal brittle fracture is part of the broader phenomenon of *stress corrosion cracking* ($\mathscr{G}$, 2γ, and v represent each term). Such cracking leads to strength degradation of components under sustained loading and is thus part of the even broader phenomenon of mechanical *fatigue*.

When applied in the context of materials engineering, the term "fatigue" usually implies strength degradation and eventual failure of metallic components under sustained, cyclic, loading (Dieter 1986; Hertzberg 1989; Suresh 1998). The loading is often stress controlled and the failure often occurs with little or no warning, as shown in Figure 6.2a. The strength degradation is associated with slow, irreversible crack propagation (Khonsari and Amiri 2013), until a critical condition is reached and rapid failure ensues. All the characteristics associated with the term fatigue applied to metals also apply to strength degradation and eventual failure of *brittle* components under sustained loading: often stress controlled, failure with little warning, strength degradation by slow, irreversible crack propagation, attainment of a critical condition, and rapid failure. The only difference is that the sustained loading need not be cyclic. Steady loading can lead to *static fatigue* failure in brittle materials, Figure 6.2b, and increasing loading can lead to *dynamic fatigue* failure in brittle materials, Figure 6.2c. Note that for most brittle materials, only tensile fatigue loading is of relevance. Early works quantifying static and dynamic fatigue effects in brittle materials (mostly static fatigue in glass) were published in the 1920s and 1930s. In particular, Preston (1935) describes the results of static fatigue experiments and an overview of relevant work at the time. These works were about a century after the quantification of cyclic fatigue effects in metals (Suresh 1998) (although the 1899 work of Grenet noted both static and dynamic fatigue in glass—see partial translation by Preston 1935 and brief historical reviews by Wiederhorn and Bolz 1970; Freiman et al. 2009). The next few decades, 1940s to early 1970s, established most of the phenomena exhibited in fatigue of brittle materials (still with an emphasis on glass). Just as important in this period was that the measured phenomena, mostly quantified by strength–time relationships, were related to the underlying flaw size and crack velocity behavior. Section 6.1.2 details some of the key works from this period in the context of the above reactive fracture concepts. The section also sets the foundation for assessment of indentation strengths in reactive environments, presented later in this chapter.

6.1.2 Background Observations

Figure 6.3 shows semilogarithmic plots of the delayed failure of glasses and porcelains from the work of Baker and Preston (1946a), using data derived from the published work. This 1946 work appears to be the first to introduce the term "static" fatigue to describe observations such as Figure 6.3. Similar observations had been reported in earlier works by Preston (1935, 1942) and were referred to as the effect of the "time element on the apparent strength" (1935) and as "fatigue" and the "decline in strength with the passage of time" (1942). The latter comment was made in reference to unpublished 1939 observations of Baker. Preston (1935) noted that in 1899 Grenet had referred to the "delay in fracturing" regarding similar observations. Soon after, in referring to Preston (1942), Orowan (1944) refers to glass "fatigue" as the observation that "it can be broken by stresses far below its ordinary breaking stress … provided that the load is applied for a sufficiently long time." Also with reference to Preston, Gurney and Pearson (1948) state that "in experiments with glass, 'fatigue' usually denotes deterioration under static loading" and that "static fatigue of glass is due to the spread of cracks." By the late 1940s, fatigue of glass under static load was an accepted experimental fact.

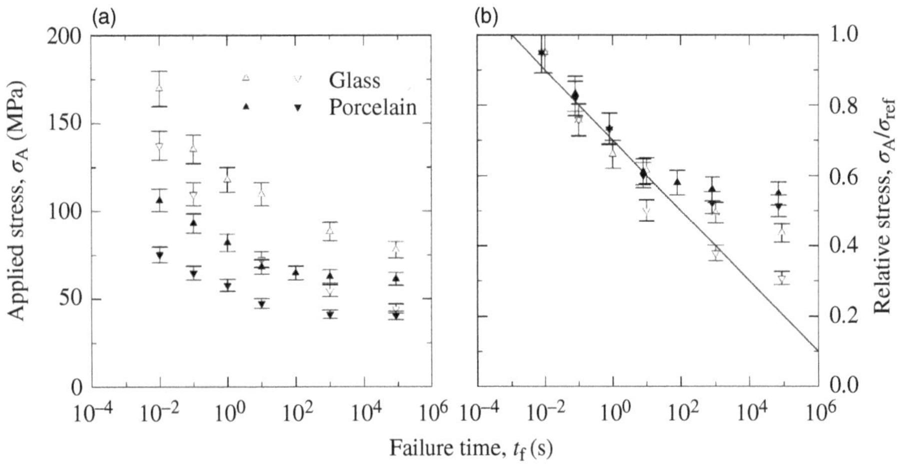

Figure 6.3 Plots of static fatigue behavior, applied stress versus failure time, of glass and porcelain rods in water. (a) Raw data for materials of different compositions (stress converted to MPa). Source: Data adapted from Baker and Preston (1946a). (b) Strength data scaled to relative values by shortest time values in (a).

The static fatigue results of Figure 6.3 illustrate the above statements. The results were obtained using a custom time-controlled electromagnetic apparatus to load glass and porcelain rods in air and water. Refinement of applied load intervals enabled estimates of the applied failure stress σ_A for a prescribed failure time t_f (similar probability based experiments are considered in more detail in Section 6.3.7). Several observations characteristic of static fatigue tests are exhibited by the experimental data of Figure 6.3a. First, the variation in failure time is very large relative to the variation of applied stress. t_f varies, increases, by a factor of 10^8 for variations, decreases, in σ_A of a factor of 3. Second, although there is a material dependence on the value of the applied stress for a specified failure time, the form of the behavior is similar and nearly material independent. Third, the behavior appears to consist of an initial decrease in applied stress as failure time is increased followed by a tendency to an applied stress plateau for very long failure times. The last two characteristics are emphasized in Figure 6.3b, in which the applied stress is shown as a relative value, σ_A/σ_{ref}. Here, σ_{ref} is the stress at the shortest failure time, $t_f = 10^{-2}$ s, adjusted for reported moisture effects (Baker and Preston 1946a). The symbols in Figure 6.3b are as in Figure 6.3a. The diagonal line is a guide to the eye of form $\sigma_A \sim -\ln t_f$. Data for all four systems now fall on a common response. At short times, the response is well described by a logarithmic dependence of the relative applied stress necessary to cause failure at the time imposed. Or, as will be shown, and more physically, the failure time exhibits a negative exponential dependence on the applied stress. However, at long times, the response tends to an invariant relative stress plateau that is about one-third of the short time stress, as also noted by Orowan (1944).

In a series of works in the late 1950s, Charles (1958a,b,c) developed many of the concepts now known to apply to fatigue of brittle materials. In the first two works of the series, Charles showed that the corrosion of glass surfaces *by* water and static fatigue of glass *in* water were thermally activated processes with similar inferred activation energies. The variation of static fatigue failure time with applied stress as a function of temperature for soda-lime glass (SLG) is shown in Figure 6.4a. Symbols represent mean failure conditions in saturated water vapor at the temperatures indicated. Uncertainties (approximately ± 35% relative to mean failure times) are omitted for clarity. Lines are guides to the eye. The shaded band indicates the failure stress at −170 °C and may be regarded as the inert strength. Room temperature failures are indicated in bold. Charles speculated that the corrosion and failure processes were linked by the preferential corrosive dissolution of glass at crack tips, effectively leading to kinetically limited crack propagation perpendicular to the applied stress and thus to delayed failure. Cracks were modeled as propagating elliptical notches using a heuristic power-law dependence of notch propagation velocity on notch length. Analysis based on these assumptions showed agreement with observed static fatigue behavior, especially the sensitivity of lifetime on applied stress. Charles reversed some of the earlier ideas regarding cause and effect in fatigue and regarded static applied stress and lifetime as the independent and dependent variables, respectively.

In the third work of the series, Charles introduced the term "dynamic" fatigue to describe loading in which an applied stress is increased linearly with time as in Figure 6.2c and reported some early experimental results. The variation of dynamic fatigue failure stress, the *dynamic fatigue strength*, with imposed applied displacement rate for SLG is shown in Figure 6.4b in the same semilogarithmic scale as Figure 6.4a. Symbols with bars represent means and standard deviation

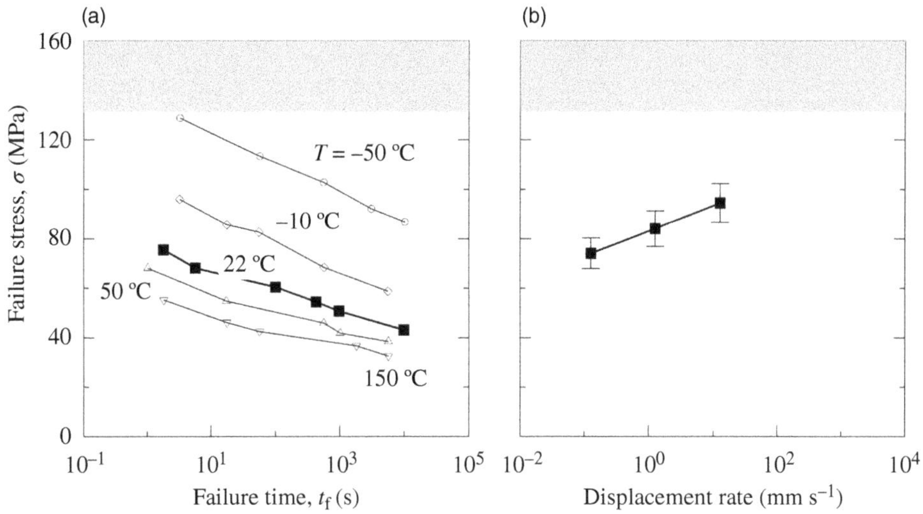

Figure 6.4 Plots of fatigue behavior of abraded glass rods in water. (a) Static fatigue as a function of temperature. (b) Dynamic fatigue at 22 °C. Note the similarity of the magnitudes of the slopes in (a) and (b). (b) Source: Charles (1958a,b,c).

limits of strengths measured in saturated water vapor at room temperature. The bold line is a guide to the eye and the shaded band indicates the inert strength. Analysis of dynamic fatigue tests using the power-law assumption indicated a significant sensitivity of fatigue strength on applied stressing rate, which was both observed in experiments and shown to be in agreement with static fatigue observations: the magnitudes of the slopes of the bold lines are similar. Although dissolution at a blunt notch tip has been shown subsequently to not control fatigue in glass, many of the insights of Charles are the basis for interpretations of fatigue in brittle materials: fatigue is controlled by crack propagation; the crack propagation rate is controlled by a thermally activated chemical reaction—in many brittle materials between an oxide and water; far from the quiescent point of the reaction, the reaction rate is a very sensitive function of stress and crack length; and, dynamic and static fatigue are very closely related. From an experimental perspective, Charles greatly advanced testing efficacy by using controlled flaws in the form of uniform abrasion of the glass test rods such that "the distribution of breaking stresses … [was] narrower" and by introducing the convention of including inert strength measurements in data plots.

It is important to place the work of Charles in context. Many of the ideas developed by Charles had already been established empirically in earlier work at the Preston Laboratories. In particular, that fatigue in glass was thermally activated (Vonnegut and Glathart 1946) and that atmospheric moisture, as well as liquid water, was a key factor in glass fatigue (Baker and Preston 1946b). Both of these ideas were developed analytically and experimentally soon after, not quite independently, by Gurney (1947), Gurney and Pearson (1949), and Stuart and Anderson (1953). Similarly, experiments on fatigue of borosilicate glass (BSG) (Shand 1954) and alumina (Pearson 1956) had extended the earlier detailed fatigue observations on SLG and porcelain (Glathart and Preston 1946) and had affirmed that the key physical process was one of cracking and suggested that the key chemical reaction was one between oxide and water.

In a series of works soon after Charles', Mould and Southwick (1959a,b) and Mould (1960, 1961) of the Preston Laboratories added to the concepts describing fatigue of brittle materials. Mould and Southwick performed extensive experiments regarding lifetimes in static fatigue tests and the effects of flaw type and test environment. In the first work of the series, Mould and Southwick described custom apparatus for (i) generating controlled flaws on glass surfaces by grit blasting and (ii) static fatigue testing of specimens by instrumented electromagnetic and dead weight loading. In the second work of the series, the apparatus were employed in static fatigue testing of glass specimens containing different grit blasted and abrasion flaws. The static fatigue results for tests in water are shown in Figure 6.5a, obtained using a refined load interval approach similar to that of Baker and Preston (1946a,b). Symbols represent mean failure stresses at the failure times indicated; relative uncertainty is approximately $\pm$ 5% of the stress. Severity of surface flaws from abrasion and blasting increases from top to bottom; different symbols indicate different surface treatments. Inert strengths σ_m for each treatment were measured by tests in liquid N_2 and decreased in order of surface flaw severity. The effect of flaw severity on fatigue response was removed by Mould and Southwick through the use of relative stress and time variables, σ_A/σ_{ref} and t_f/t_{ref}. The reference values used were $\sigma_{ref} = \sigma_m$ and $t_{ref} = t_{50}$, the time required for the strength to decrease to $0.5\sigma_m$. The scaled fatigue response is shown in Figure 6.5b. The symbols in Figure 6.5b are as in Figure 6.5a. The diagonal line is a guide to the eye of form $\sigma_A \sim -\ln t_f$.

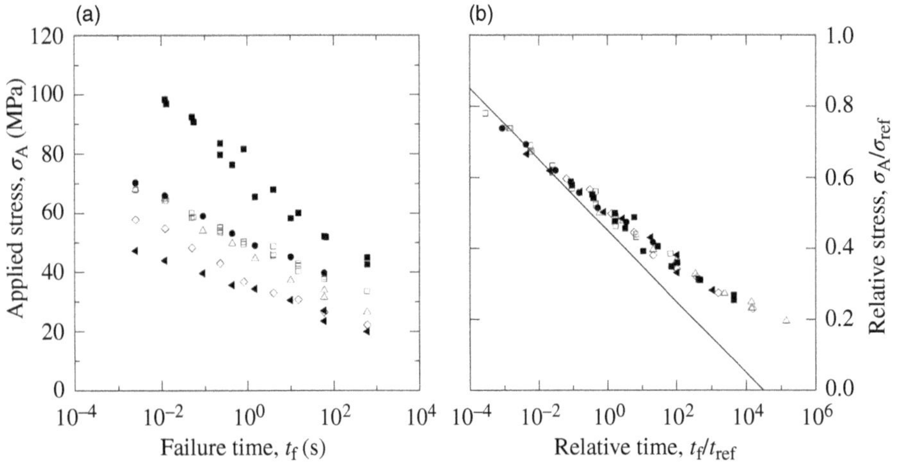

Figure 6.5 Plots of static fatigue behavior, applied stress versus failure time, of glass in water. (a) Raw data for components with different surface treatments (stress converted to MPa). (b) Strength and time data scaled to relative values by reference values of inert strength (not shown) and midrange failure times from (a). (b) Source: Data adapted from Mould and Southwick (1959b).

The data for all flaw systems now fall on a common response. At short times, the response is well described by a logarithmic dependence and at long times the response tends to an invariant relative stress plateau that is about 0.2 of the short time stress. The finding that flaw size simply translated observed stress and lifetime along a universal fatigue response, and that such a response existed, was a major contribution to understanding fatigue behavior. The finding figures prominently in analyses of indentation strengths measured in reactive environments.

In the third and fourth in the series of works, Mould identified the conflicting effects on contact flaw strengths of exposure to reactive environments. Figure 6.6a shows the effects of zero stress aging of abraded components. The filled symbols indicate inert strengths measured after aging in water for the times indicated; the components had identically abraded surfaces. The strengths increase with aging period, consistent with postcontact lateral crack-like growth and residual field relaxation at abrasion flaws, similar to that in Figure 3.19. The open symbols indicate inert strengths measured after aging in various environments for 24 h. The strengths increase with increasing presence of water in the aging environment, from vacuum and 1% RH through 50% RH to 100% RH and liquid water, consistent with water enabling crack extension in the residual field of the abrasion flaw. Figure 6.6b shows the static fatigue responses in water following the 24 h aging steps

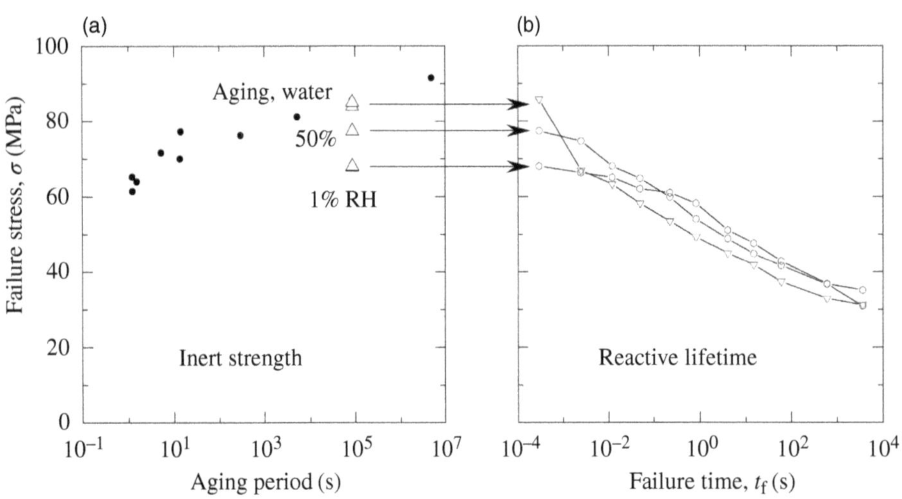

Figure 6.6 Plots demonstrating the effects of aging on strength and lifetime of abraded glass components. (a) Inert strength as a function of aging period. Strengths increase and diverge on aging. (b) Post-aging static fatigue: Failure stresses decrease and converge with failure time. Source: Mould (1960, 1961).

(sequences indicated by arrows). Although there are initial differences in behavior, the static fatigue responses converge on decreasing behavior at extended failure times, suggesting that differences in stress relaxation effects from zero stress aging are largely removed by static fatigue in reactive environments. Note that the plots in Figure 6.6 have identical relative time domains, suggesting that the kinetics of stress field relaxation and strength degradation are similar, consistent with the idea that both phenomena are controlled by crack propagation. The clear quantification of competing effects determining strengths in reactive environments was also a major contribution to the understanding of fatigue behavior. [Similar strengthening followed by weakening behavior for scribed and scratched glass was reported, although less comprehensively, by Shand (1961b) at about the same time, and was explained in much the same way (citing the earlier works of Holland and Turner and Polyakov in support).] Competing effects on strength are examined further in Cook (2018c), in consideration of the influences of moderate corrosion on indentation strengths.

In the late 1960s and early 1970s, Ritter placed dynamic fatigue on a firm basis for evaluating the failure properties of materials in reactive environments (and popularized the term "dynamic fatigue"). Figure 6.7 shows semilogarithmic plots of dynamic fatigue strengths σ_f as a function of applied stressing rate $\dot{\sigma}_a$ for SLG specimens tested in water. The filled symbols in Figure 6.7a show results from Ritter (1969) and Ritter and LaPorte (1975) for abraded rods, using data derived from the published works cited. The open symbols show results from Chandan et al. (1978) for abraded bars, using data derived from the published work. The bar data were measured using a combination of a universal testing machine and an impact machine capable of fast rates, and the shaded band shows the inert strength for the bar specimens. The strength variations with stressing rate in both cases are comparable to those of Figure 6.4. The symbols in Figure 6.7b show results from Ritter (1969) for etched rods, using data derived from the published work. The etched strengths are approximately a factor of 10 greater than the abraded strengths, reflecting the reduction in flaw size on etching, but the dynamic fatigue phenomenon remains the same. The shaded horizontal line indicates the inert strength. In all cases, the diagonal lines are guides to the eye of the form $\sigma_f \sim \ln \dot{\sigma}_a$. Ritter and colleagues demonstrated similar dynamic fatigue behavior for other glasses, ceramics, and test environments (Ritter 1969; Ritter and Sherburne 1971; Ritter and LaPorte 1975; Ritter and Meisel 1976; Ritter and Cavanagh 1976), as did other researchers (Tummala and Foster 1975). In most cases, the data were analyzed using the empirical power-law formulation developed by Charles (1958b,c), sometimes in the absence of supporting observations (as the form of the data was taken as assumed). Dynamic fatigue tests have many experimental advantages over static fatigue tests: Testing machines are usually easier to program and control in application of constant stressing rates rather than in application of constant stresses; the time of specimen failure is better known; and "run outs" are effectively eliminated. Hence, by the mid-1970s dynamic fatigue was the predominant test method for assessing time and rate dependence of strengths of brittle materials. Dynamic fatigue will be the method used in the remainder of this chapter to characterize indentation strengths in reactive environments.

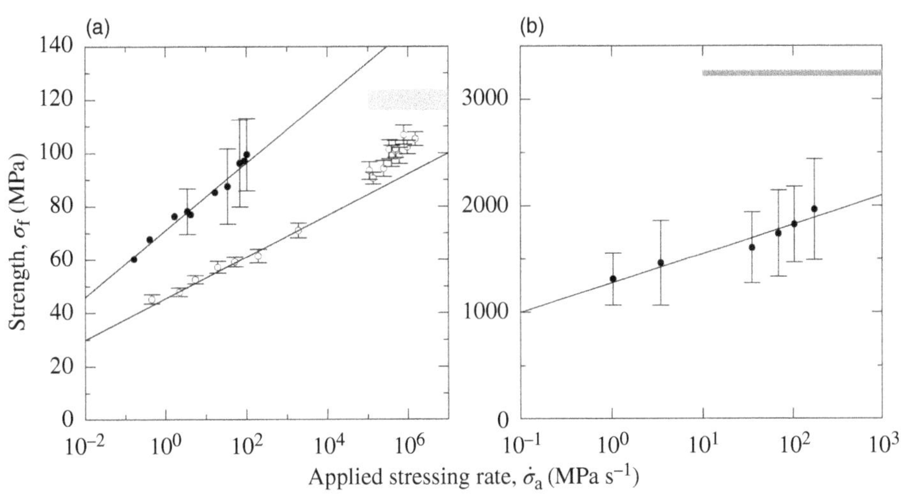

Figure 6.7 Plots of dynamic fatigue behavior, failure stress versus applied stressing rate, for soda-lime silicate glass in water. (a) Abraded rods and bars. (b) Etched rods. Source: Data adapted from Ritter (1969), Ritter and LaPorte (1975), and Chandan et al. (1978).

6.2 Reactive Strength and Lifetime Analyses

This section builds on the concepts and observations of Section 6.1 and develops analyses to describe indentation strengths under dynamic fatigue conditions. In particular, descriptions are sought for the dependence on applied stressing rate of indented component strengths in reactive environments. Complete descriptions include: applied, residual, and microstructural stress fields (Chapter 4); reactive and inert equilibrium surface states, Eq. (6.1); and thermally activated fracture kinetics consistent with the thermodynamic entropy constraint, Eq. (6.3). The section begins with a limited description restricted to a subset of these conditions: applied field only, explicit inclusion of inert surface state only, and fracture kinetics partially consistent with the entropy constraint. Experimentally, the restricted analysis can describe dynamic fatigue data over prescribed domains of applied stress and stressing rate. Perhaps more important, the restricted analysis enables simple interpretation and visualization of fatigue behavior in terms of clearly identifiable fracture parameters and straightforward extension to include indentation and microstructural effects. The section concludes with an outline of a more complete analysis that includes reactive surface energy and physically based fracture kinetics. The latter analysis describes dynamic fatigue data over complete domains of applied stressing rate.

6.2.1 Empirical Crack Velocity: Griffith Flaws

The starting point for analysis is consideration of a simple "Griffith" flaw, a crack with length dimension c in a large component. Prior to loading of the component at time $t = 0$ the crack has initial length c_0. If the component and crack are loaded solely by a uniform applied stress σ_a, the stress-intensity factor (SIF) K acting on the crack is the applied SIF K_a,

$$K = K_a = \psi\sigma_a c^{1/2}, \tag{6.4}$$

where ψ is a dimensionless geometrical term (see Chapter 1). If the component consists of material with Young's modulus E and surface energy γ_0 in an inert environment, the toughness T of the material is the inert toughness T_0,

$$T = T_0 = (2\gamma_0 E)^{1/2}. \tag{6.5}$$

Fracture equilibrium is given by $K = T$ and thus for this system under inert conditions $K_a = T_0$. Hence, combining Eqs. (6.4) and (6.5) at c_0 gives the inert strength of the loaded component σ_0,

$$\sigma_0 = T_0/\psi c_0^{1/2}. \tag{6.6}$$

If the component is loaded under reactive conditions in a nonequilibrium manner, $K_a < T_0$, the crack can extend with time. The fracture kinetics and crack velocity v are determined by the material, environment, and applied loading. Here, the crack velocity is approximated by an empirical power law, similar to that introduced by Charles (1958b,c) and applied subsequently by Ritter and others (Ritter 1969; Ritter and Sherburne 1971; Ritter and LaPorte 1975; Ritter and Meisel 1976; Ritter and Cavanagh 1976; Tummala and Foster 1975; Chandan et al. 1978). The crack velocity is given by

$$v = v_0\left(\frac{K}{T_0}\right)^n, \tag{6.7}$$

where v_0 and n are empirical parameters and Eq. (6.7) bears some resemblance to the Paris "law" often used to describe crack propagation during cyclic fatigue of metals (Dieter 1986; Hertzberg 1989; Suresh 1998). Combining Eq. (6.4) into Eq. (6.7) for a static applied stress $\sigma_a = \sigma_A$ gives a differential equation for crack length

$$\frac{dc}{dt} = v_0\left(\frac{\psi\sigma_A c^{1/2}}{T_0}\right)^n, \tag{6.8}$$

which is simply integrated to give

$$\int_0^{t_f} dt = t_f = \frac{1}{v_0}\left(\frac{\psi\sigma_A}{T_0}\right)^{-n}\int_{c_0}^{c_f} c^{-n/2}dc, \tag{6.9}$$

where c_f is the crack length at time t_f. The critical step at this point in all such analyses is to recognize that the lifetime of a component can be determined by imposing the inert equilibrium condition at the crack length c_f for the current value of the applied stress, in this case σ_A. Hence, extending Eq. (6.6) in this way gives

$$T_0 = \psi\sigma_0 c_0^{1/2} = \psi\sigma_A c_f^{1/2}. \tag{6.10}$$

Equation (6.10) enables T_0 and c_f to be eliminated from Eq. (6.9) in favor of σ_0 and c_0 and the integral to be completed as

$$t_f = \frac{c_0}{v_0}\left(\frac{2}{n-2}\right)\left(\frac{\sigma_A}{\sigma_0}\right)^{-n}\left[1-\left(\frac{\sigma_A}{\sigma_0}\right)^{n-2}\right]. \tag{6.11}$$

Equation (6.11) has two asymptotes: (1) in the reactive kinetic limit, $\sigma_A \ll \sigma_0$, $t_f \sim \sigma_A^{-n}$, and (2) in the inert equilibrium limit, $\sigma_A \to \sigma_0$, $t_f \to 0$, noting that usually $n \gg 2$. The kinetic limit of Eq. (6.11) implies the significant sensitivity of failure time t_f on applied stress $\sigma_A < \sigma_0$ observed in Figures 6.3–6.6.

Combining Eq. (6.4) into Eq. (6.7) for a *dynamic* applied stress $\sigma_a = \dot{\sigma}_a t$, where $\dot{\sigma}_a$ is the applied stressing rate, gives a slightly altered differential equation for crack length

$$\frac{dc}{dt} = v_0\left(\frac{\psi\dot{\sigma}_a t c^{1/2}}{T_0}\right)^n, \tag{6.12}$$

which is integrated in this case to give

$$\int_0^{t_D} t^n dt = \frac{t_D^{n+1}}{n+1} = \frac{1}{v_0}\left(\frac{\psi\dot{\sigma}_a}{T_0}\right)^{-n}\int_{c_0}^{c_f} c^{-n/2}dc, \tag{6.13}$$

where c_f is the crack length at time t_D. The lifetime of a component in this case can be determined by extending Eq. (6.6) as

$$T_0 = \psi\sigma_0 c_0^{1/2} = \psi\sigma_f c_f^{1/2} = \psi\dot{\sigma}_a t_D c_f^{1/2}, \tag{6.14}$$

where $\sigma_f = \dot{\sigma}_a t_D$ is the dynamic fatigue strength of the component at the given applied stressing rate. Equation (6.14) enables T_0 and c_f to be eliminated from Eq. (6.13) in favor of σ_0 and c_0 as before and the integral to be completed as

$$t_D = \frac{c_0}{v_0}\left(\frac{2(n+1)}{n-2}\right)\left(\frac{\sigma_f}{\sigma_0}\right)^{-n}\left[1-\left(\frac{\sigma_f}{\sigma_0}\right)^{n-2}\right], \tag{6.15}$$

or

$$\dot{\sigma}_a^{-1} = \left(\frac{\sigma_0 v_0}{c_0}\right)^{-1}\left(\frac{2(n+1)}{n-2}\right)\left(\frac{\sigma_f}{\sigma_0}\right)^{-(n+1)}\left[1-\left(\frac{\sigma_f}{\sigma_0}\right)^{n-2}\right]. \tag{6.16}$$

Equation (6.16) has two asymptotes: (1) in the reactive kinetic limit, $\dot{\sigma}_a \ll \sigma_0 v_0/c_0$, $\sigma_f \sim \dot{\sigma}_a^{1/(n+1)}$, and (2) in the inert equilibrium limit, $\dot{\sigma}_a \gg \sigma_0 v_0/c_0$, $\sigma_f \to \sigma_0$, noting again that usually $n \gg 2$. The kinetic limit of Eq. (6.16) implies the weak sensitivity of dynamic fatigue strength σ_f on applied stressing rate $\dot{\sigma}_a$ observed in Figures 6.4 and 6.7.

There are several points to note from the above analysis. First, the form of the static fatigue and dynamic fatigue failure time equations, Eqs. (6.11) and (6.15), is identical. Hence, within a given system for identical failure stresses $\sigma_f = \sigma_A$, the failure times are related by a multiplicative constant $t_D = (n+1)t_f$. Second, the failure times are proportional to a characteristic time $t_0 = c_0/v_0$ associated with fatigue of a component in the reactive environment. When combined with the inert strength, a characteristic stress rate σ_0/t_0 is evident. Third, the characteristic parameters enable scaled variables for strength, time, and stress rate to be defined as

$$\begin{aligned}
\mathbf{S} &= \sigma/\sigma_0 \\
\mathbf{t} &= t/t_0 \\
\dot{\mathbf{S}} &= \dot{\sigma}_a t_0/\sigma_0 \\
\Lambda_s &= 2/(n-2);\ \Lambda_d = (n+1)\Lambda_s,
\end{aligned} \tag{6.17}$$

such that Eqs. (6.11) and (6.15) can be written compactly as

$$\mathbf{t}_f = \Lambda_s \mathbf{S}_A^{-n}[1-\mathbf{S}_A^{n-2}], \tag{6.18}$$

and

$$\mathbf{t}_D = \Lambda_d \mathbf{S}_f^{-n}[1-\mathbf{S}_f^{n-2}], \tag{6.19}$$

noting that

$$\mathbf{t}_D = \mathbf{S}_f/\dot{\mathbf{S}}, \tag{6.20}$$

such that Eqs. (6.19) and (6.20) form a parametric set for the function $\mathbf{S}_f(\dot{\mathbf{S}})$. Equations. (6.18)–(6.20) are extremely useful in assessing the effects of changes in n. The kinetic limits of these equations are extremely useful for fitting experimental data as they are simple power laws,

$$\mathbf{t}_f = \Lambda_s \mathbf{S}_A^{-n}, \quad \mathbf{S}_A \ll 1$$
$$\mathbf{S}_f = (\Lambda_d \dot{\mathbf{S}})^{1/(n+1)}, \quad \mathbf{S}_f \ll 1. \tag{6.21}$$

and generate straight lines on logarithmic plots (Charles 1958b,c; Ritter 1969; Ritter and Sherburne 1971; Ritter and LaPorte 1975; Tummala and Foster 1975; Chandan et al. 1978). The physical interpretation of these equations is also straightforward. Failure time decreases significantly with increases in applied stress in static fatigue. Failure stress increases weakly with increases in applied stressing rate in dynamic fatigue. The absolute values of the failure variables are set by the initial flaw size (e.g., Figure 6.7).

6.2.2 Empirical Crack Velocity: Indentation Flaws

The advantages of controlled flaws in fatigue testing were demonstrated by Mould and Southwick (1959a,b) and Mould (1960, 1961) using grit blasting and abrasion. The advantages of controlled indentation flaws in inert strength testing were demonstrated here in Chapters 3 and 4. Analyses for the obvious application of indentation flaws in fatigue testing are enabled by simple modification of Eqs. (6.18)–(6.21). Modification begins with replacement of $K = K_a$ with $K = K_a + K_r$ (see Chapter 3) in Eq. (6.7). Solutions of the resulting differential equation are somewhat lengthy (Fuller et al. 1983) but can be expressed in forms similar to Eqs. (6.18)–(6.21) with changes to accommodate the differences between Griffith flaws and indentation flaws. The characteristic strength and crack length of indentation flaws are the inert strength σ_m attained at the instability crack length c_m. Hence, the characteristic time scale for fatigue of indentation flaws in reactive environments is $t_m = c_m/v_0$. The scaled variables for strength, time, and stress rate are thus

$$\mathbf{S} = \sigma/\sigma_m$$
$$\mathbf{t} = t/t_m$$
$$\dot{\mathbf{S}} = \dot{\sigma}_a t_m/\sigma_m$$
$$\Lambda'_s = (2\pi/n')^{1/2}; \quad \Lambda'_d = (n'+1)\Lambda_s \approx (2\pi n')^{1/2}$$
$$n' = (3n+2)/4, \tag{6.22}$$

where the expressions for Λ' and n' pertain to point flaws (expressions for line flaws are similar, Fuller et al. 1983). The compact forms of the solutions are

$$\mathbf{t}_f = \Lambda'_s \mathbf{S}_A^{-n'}[1 - \mathbf{S}_A^{n'-2}], \tag{6.23}$$

and

$$\mathbf{t}_D = \Lambda'_d \mathbf{S}_f^{-n'}[1 - \mathbf{S}_f^{n'-2}], \tag{6.24}$$

noting that the relation $\mathbf{t}_D = \mathbf{S}_f/\dot{\mathbf{S}}$ remains unaltered. Thus, Eqs. (6.24) and (6.20) form a parametric set for the indentation dynamic fatigue function $\mathbf{S}_f(\dot{\mathbf{S}})$. The kinetic limits of the above equations are also extremely useful for fitting experimental data as they are simple power laws,

$$\mathbf{t}_f = \Lambda'_s \mathbf{S}_A^{-n'}, \quad \mathbf{S}_A \ll 1$$
$$\mathbf{S}_f = (\Lambda'_d \dot{\mathbf{S}})^{1/(n'+1)}, \quad \mathbf{S}_f \ll 1. \tag{6.25}$$

and generate straight lines on logarithmic plots.

It is important to recognize that n' is a perceived crack velocity exponent that describes fatigue in the external frame of reference of the applied stress. The residual stress field does not appear in the applied stress frame of reference, in which fatigue strengths and lifetimes appear to be controlled by Griffith flaws propagating in a material with crack velocity characteristics $v \sim K_a^{n'}$. In the crack tip frame of reference, the crack velocity characteristic are unaltered from $v \sim K^n \sim (K_a + K_r)^n$. Comparison of Eqs. (6.18)–(6.21) and Eqs. (6.23)–(6.25) points to a major advantage and a major difficulty in interpretation of fatigue data in the external component-level frame of reference: The form of the equations is identical. The identical nature of the equations is an advantage as it enables understanding, fitting, and comparison of fatigue responses of indented

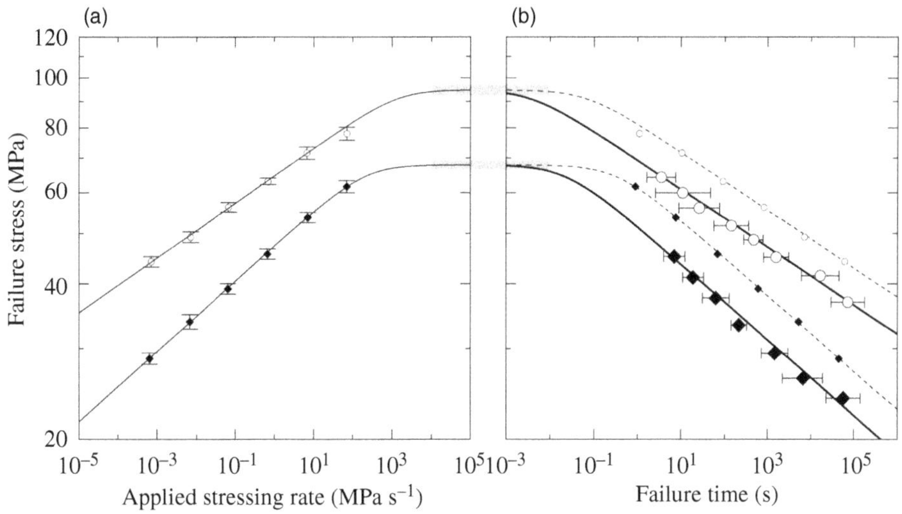

Figure 6.8 Logarithmic plots of fatigue behavior of indented (filled symbols) and indented and annealed (open symbols) disks of soda-lime silicate glass in water. (a) Dynamic fatigue. (b) Static fatigue. Source: Data adapted from Marshall and Lawn (1980) and Chantikul et al. (1981).

components to be readily transferred to and from fatigue responses of components containing Griffith flaws. However, a disadvantage is that without additional information, it can be extremely difficult to assess the residual stress state of fatigue limiting flaws from mean lifetime or strength data alone. The change in logarithmic slope from n to $n' \approx 0.75n$ is often very subtle. However, dispersion and scatter in experimental lifetimes and strengths are often very much reduced if the fatigue limiting flaws are controlled indentations and distinctions may be made.

An example application of controlled indentation flaws in fatigue testing, illustrating the above points and analyses is shown in Figure 6.8 from the work of Marshall and Lawn (1980) and Chantikul et al. (1981), using data derived from the published works cited. Figure 6.8a shows a logarithmic plot of the measured dynamic fatigue responses of SLG disks tested in water and containing $P = 5$ N Vickers indentation flaws. The symbols and bars represent the means and standard deviation limits of dynamic fatigue strengths measured at the applied stressing rates indicated. The shaded bands represent the means and standard deviation limits of inert strengths. The filled symbols represent the strengths of as-indented disks. The open symbols represent the strengths of disks annealed following indentation such that the residual indentation stress field was removed. The lines represent fits of Eqs. (6.18)–(6.21) (open symbols) and Eqs. (6.23)–(6.25) (filled symbols) to the data. The fits clearly describe the data. The advantages of controlled indentation flaw testing are clear in the obvious difference in the dynamic fatigue slopes, characterized by $n = 17.9$ and $n' = 13.9$ and the similar, small, dispersion of strengths for both indented and annealed specimens and reactive and inert tests. Figure 6.8b shows a logarithmic plot of the measured static fatigue responses of the same indented SLG disks. The symbols with bars represent the means and standard deviation limits of static fatigue failure times measured at the applied stresses indicated (dependent and independent variables are reversed relative to Figure 6.8a). Symbols without bars and dashed lines represent dynamic fatigue data repeated from Figure 6.8a in the coordinates of Figure 6.8b using Eq. (6.20). The differences between static fatigue and dynamic fatigue failure times at the same failure stress are clear. The differences are emphasized by the solid lines in Figure 6.8b that represent *predictions* of the static fatigue behavior from the dynamic fatigue behavior using Eqs. (6.18)–(6.21) and Eqs. (6.23)–(6.25). The agreement of the predictions with the observations is clear, as are the effects of residual stress fields in fatigue responses (fits and predictions were also confirmed by numerical solutions of the fatigue equations, Marshall and Lawn 1980; Chantikul et al. 1981; Lawn et al. 1981).

Building on the successful application of indented specimens in dynamic and static fatigue testing, the obvious next step was to incorporate variations in indentation load P. Reversing the scaling of the dynamic fatigue expression in Eq. (6.25) gives

$$\frac{\sigma_f}{\sigma_m} = \left(\Lambda'_d \frac{\dot{\sigma}_a c_m}{\sigma_m v_0}\right)^{1/(n'+1)}. \qquad (6.26)$$

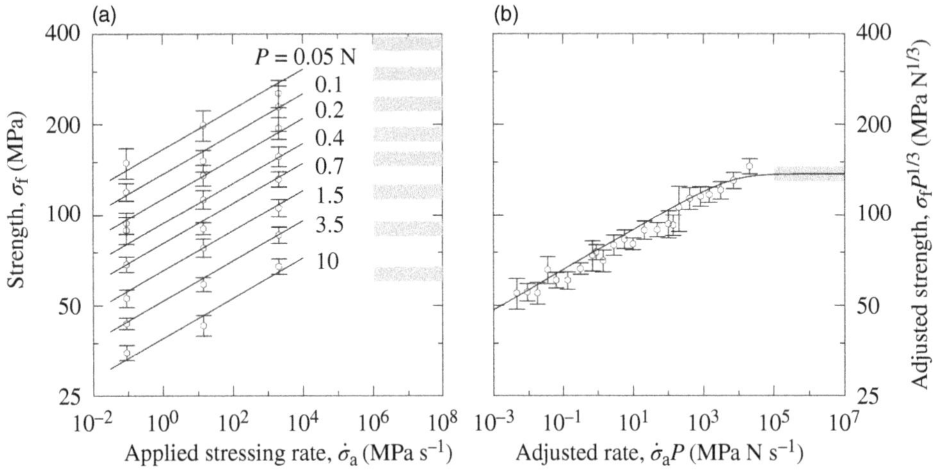

Figure 6.9 Logarithmic plots of dynamic fatigue behavior of soda-lime silicate glass rods in water. (a) Raw data for rods containing indentation flaws formed with different contact loads. (b) Strength and stressing rate data from (a) in load-adjusted coordinates illustrating a universal response. Source: Data adapted from Dabbs et al. (1982).

Multiplying the top and bottom of the left side of Eq. (6.26) by $P^{1/3}$ and top and bottom of the term in parentheses on the right side by P gives

$$\frac{\sigma_f P^{1/3}}{\sigma_m P^{1/3}} = \left(\Lambda'_d \frac{\dot{\sigma}_a P c_m P^{-2/3}}{\sigma_m P^{1/3} v_0}\right)^{1/(n'+1)}. \tag{6.27}$$

$\sigma_m P^{1/3} = S$ and $c_m P^{-2/3} = 4^{2/3} B$ are recognized as indentation load invariant quantities from Chapters 2 and 3. Hence, after gathering terms, Eq. (6.27) becomes

$$\sigma_f P^{1/3} = \left(\frac{4^{2/3}\Lambda'_d S^{n'} B}{v_0}\dot{\sigma}_a P\right)^{1/(n'+1)} = \left(\lambda'_P \dot{\sigma}_a P\right)^{1/(n'+1)}, \tag{6.28}$$

where it is recognized that the first term in parentheses of the central expression consists only of numerical, material, and environmental factors and is therefore indentation load independent. These factors are collapsed into the single load invariant term λ'_P in the expression on the right. The strength and stressing rate parameters $\sigma_f P^{1/3}$ and $\dot{\sigma}_a P$ appearing on the left and right are therefore also load independent, and a plot using these parameters should collapse dynamic fatigue data collected at different indentation loads onto a common response. A similar method can be implemented to arrive at indentation load invariant parameters $t_f P^{-2/3}$ and $\sigma_A P^{1/3}$ for static fatigue, and this is simply a formalized method of arriving at the empirical scaling of Mould and Southwick (1959b) shown in Figure 6.5b.

An example of adjustment of indentation dynamic fatigue parameters so as to remove indentation load effects is shown in Figure 6.9 from the work of Dabbs et al. (1982), using data derived from the published work (see also Cook and Lawn 1984). Figure 6.9a shows a logarithmic plot of the measured dynamic fatigue responses of SLG rods tested in water and containing indentation loads in the domain $P = 0.05$–10 N. The symbols represent the means and standard deviation limits of fatigue strengths measured at the applied stressing rates indicated. The shaded bands represent the means and standard deviation limits of inert strengths. The diagonal lines are best fits to the fatigue data using Eq. (6.25) and a fixed value of $n' = 14.0$. Figure 6.9b shows a logarithmic plot of the fatigue data in the indentation load adjusted coordinates of Eq. (6.28). In these coordinates, the inert strengths collapse onto the load invariant value of S shown as the shaded band. The fatigue strengths collapse onto a common response, indicated by the best fit line obtained using λ'_P as a load invariant fitting parameter and n' as given. The single best fit clearly describes the material response in adjusted coordinates. Dabbs applied the same methodology to demonstrate indentation load invariant dynamic fatigue responses for BSG and fused silica glass (FSG) (Dabbs et al. 1982; Dabbs and Lawn 1985).

6.2.3 Effect of Lateral Cracks and Microstructure

The indentation load adjustment strategy of Eq. (6.28) demonstrated in Section 6.2.2 will be used throughout this chapter in assessment of indentation dynamic fatigue strengths. In particular, the strategy will be extended to account for the effects of lateral cracking and microstructure on crack propagation and fatigue in indented polycrystalline ceramics.

At large indentations lateral cracking and chipping relaxes the indentation residual stress field. The relaxation reduces the driving force on strength controlling indentation half-penny cracks and thus increases the applied stress required to cause failure of an indented component. Increases in the stresses, or component strengths, at large indentation loads to values greater than those extrapolated from small indentation loads were demonstrated extensively in Chapters 3 and 4. Such strength increases were described analytically by the introduction of an effective indentation load $\tilde{P}$,

$$\tilde{P} = \frac{PP_L}{P + P_L}, \tag{6.29}$$

where P_L is a material- and indenter-dependent indentation load at which lateral crack effects become significant in indentation strength testing, see Eqs. (3.62) and (4.26). In dynamic fatigue testing of indented components, the indentation load adjustment strategy of Eq. (6.25) is simply implemented in the presence of lateral cracking by replacing P with $\tilde{P}$.

For all indentation cracks in polycrystalline materials, microstructural effects increase the apparent toughness of the material. Grain scale frictional interlocks and ligamentary bridges in crack wakes act as restraining forces on crack propagation. The restraining forces are not fully developed at small cracks lengths and thus decrease the applied stress required to cause failure of an indented component. Decreases in component strengths at small indentation loads to values less than those extrapolated from large indentation loads were demonstrated extensively in Chapter 4. Such strength increases were described analytically by two changes to the fracture mechanics for indentation strengths of polycrystalline materials. First, the material toughness value was altered to $T_\infty > T_0$, reflecting the enhanced toughness at long crack lengths. Second, a crack length dependent term $\chi P^*/c^{3/2}$ was added to the SIF, reflecting the crack length dependence of the restraining forces, see Chapter 4. In dynamic fatigue testing of indented components, the indentation load adjustment strategy of Eq. (6.25) is simply implemented in the presence of microstructural effects by replacing T_0 with T_∞ and P with $(P + P^*)$. Note that the apparent crack velocity response in an external frame of reference that includes no knowledge of the $(P + P^*)$ term is given by

$$v = v_0\left(\frac{K}{T_\infty}\right)^{n'}. \tag{6.30}$$

In the crack tip frame of reference, the crack velocity characteristics remain unaltered from Eq. (6.7). In the presence of both lateral crack and microstructural effects, indentation load adjustment is accomplished by replacing P with P_{eff},

$$P_{\text{eff}} = \tilde{P} + P^* = \frac{PP_L}{P + P_L} + P^*. \tag{6.31}$$

In fatigue testing Eq. (6.28) becomes

$$\sigma_f P_{\text{eff}}^{1/3} = \left(\lambda'_{\text{P}}\dot{\sigma}_a P_{\text{eff}}\right)^{1/(n'+1)}. \tag{6.32}$$

P_{eff} may use Eq. (6.31) and P_L and P^* calibrated from inert strength tests or be treated simply as an empirical parameter to generate a universal fatigue response for a material. Note that as demonstrated in Chapter 4 $S = \sigma_m P_{\text{eff}}^{1/3}$ and $c_m P_{\text{eff}}^{-2/3}$ are indentation load invariant quantities leading to Eq. (6.32).

An example of indentation load adjusted dynamic fatigue parameters for a polycrystalline material is shown in Figure 6.10 from Cook (2015), using data derived from the published work. The material is a commercial alumina and the inert strength behavior (Figure 4.17) exhibited both lateral crack and microstructural effects (although moderately, shown as a central light gray line) characterized by $P_L = 283$ N and $P^* = 4.24$ N, respectively. Figure 6.10a shows a logarithmic plot of the measured dynamic responses of disks tested in water and containing indentation loads $P = 3$ N (filled symbols) and $P = 100$ N (open symbols). The symbols represent the means and standard deviation limits of fatigue strengths measured at the applied stressing rates indicated. The shaded bands represent the means and standard deviation limits of inert strengths. The diagonal lines are best fits to the fatigue data using Eq. (6.25) and a fixed value of $n' = 24.0$. It is noted that the indentation loads used in the fatigue testing are comparable to the loads characterizing lateral crack and microstructural effects. Hence, it is perhaps not surprising that the effective indentation loads P_{eff}, Eq. (6.31), in fatigue, 7.2 and 78 N, are different from the test loads P, 3 and 100 N. Figure 6.10b shows a logarithmic plot of the fatigue data in the indentation load adjusted coordinates of Eq. (6.32) using these P_{eff} values. In these coordinates, the inert strengths collapse onto the load invariant value of S shown as the shaded band. The fatigue strengths collapse onto a common response, indicated by the best-fit line obtained using λ'_{P} as a load invariant fitting parameter and n' as given. The single best fit clearly describes the material response in lateral and microstructural adjusted indentation coordinates. (Unadjusted coordinates fail to collapse the data and generate responses separated by about three times the uncertainties shown.) The same methodology was applied to demonstrate indentation load invariant dynamic fatigue responses in other polycrystalline ceramics (Cook et al. 1985a,b).

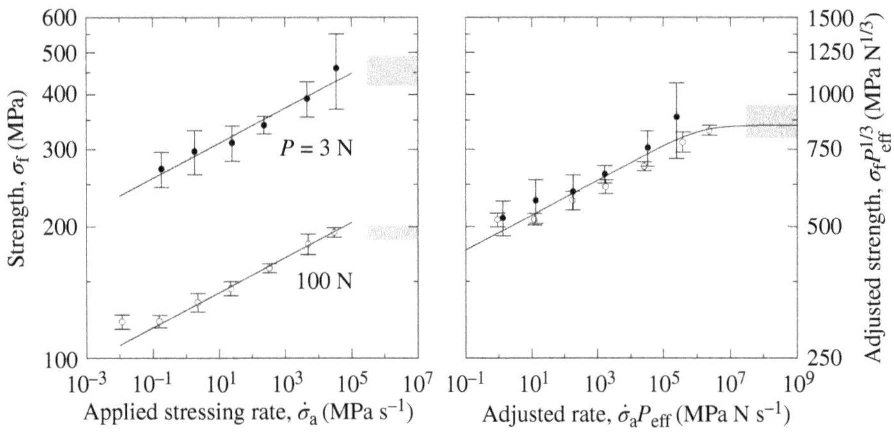

Figure 6.10 Logarithmic plots of dynamic fatigue behavior of polycrystalline alumina disks in water. (a) Raw data for disks containing indentation flaws formed with different contact loads. (b) Strength and stressing rate data from (a) in microstructural and lateral crack load-adjusted coordinates illustrating a universal response. Source: Data adapted from Cook (2015).

6.2.4 Thermally Activated Crack Velocity

The power-law crack velocity expressions Eqs. (6.7) and (6.30) have the advantages of simplicity and ease of manipulation in the derivation of fatigue equations that are then also simple and easily linearized. Although these equations describe major aspects of fatigue behavior, there are serious deficiencies: First, the initial power laws are completely empirical with no physical basis and no reference to the crack propagation process described—fracture in a reactive environment; second, the power laws omit the threshold value of crack driving force at which crack velocity is zero—the threshold is required thermodynamically, see Chapter 1; third, in many cases, the power laws do not well describe experimental observations—at large applied stresses many systems exhibit exponential lifetime dependence and at small nonzero applied stresses many systems exhibit lifetime divergence, e.g., Figures 6.3–6.5, and 6.10.

A crack velocity function that overcomes some of these issues and provides an asymptotic upper bound to a complete description of fatigue behavior is

$$v = v_A \exp\left(\frac{\mathcal{G}}{\eta}\right). \tag{6.33}$$

$\mathcal{G}$ is the mechanical energy release rate and for a static uniform applied stress given by

$$\mathcal{G} = \psi^2 \sigma_A^2 c/E, \tag{6.34}$$

where E is Young's modulus. v_A and η are a characteristic crack velocity and a characteristic thermal energy, $(k_B\theta)$/area, for the fracture system in the reactive environment. (The physical basis of these parameters is shown subsequently.) Fatigue analysis now proceeds as in Section 6.2.1. Combining Eq. (6.34) into Eq. (6.33) gives a differential equation for crack length

$$\frac{dc}{dt} = v_A \exp\left(\frac{\psi^2 \sigma_a^2 c}{E\eta}\right), \tag{6.35}$$

which is simply integrated to give

$$\int_0^{t_f} dt = t_f = \frac{1}{v_A}\int_{c_0}^{c_f} \exp\left(\frac{-\psi^2 \sigma_a^2 c}{E\eta}\right) dc, \tag{6.36}$$

where c_0 is the initial crack length and c_f is the crack length at time t_f. The lifetime of a component can be determined by imposing the inert equilibrium condition at the crack length c_f for the current value of the applied stress σ_A. Hence, extending Eq. (6.6) gives

$$2\gamma_0 = \psi^2 \sigma_0^2 c_0/E = \psi^2 \sigma_A^2 c_f/E. \tag{6.37}$$

Equation (6.37) enables c_f and E to be eliminated from Eq. (6.36) in favor of σ_0 and c_0 and the integral to be completed as

$$t_f = \frac{c_0}{v_A}\left(\frac{\sigma_A}{\sigma_0}\right)^2 \left(\exp\left[-\left(\frac{2\gamma_0}{\eta}\right)\left(\frac{\sigma_A}{\sigma_0}\right)^2\right] - \exp\left[-\left(\frac{2\gamma_0}{\eta}\right)\right]\right). \tag{6.38}$$

The similarity in arithmetic behavior of Eqs. (6.38) and (6.11) is made clear by setting a characteristic time scale for the system $t_A = c_0/v_A$ and characteristic energy and stress ratios $W_0 = 2\gamma_0/\eta$ and $\mathbf{S_A} = \sigma_A/\sigma_0$ to give

$$\frac{t_f}{t_A} = \mathbf{t_f} = \left[\exp(-W_0\mathbf{S_A^2}) - \exp(-W_0)\right]/W_0\mathbf{S_A^2}. \quad (6.39)$$

The two asymptotes of Eq. (6.39) are $\mathbf{S_A} \to 1$, $t_f \to 0$ and $\mathbf{S_A} \ll 1$, $t_f \sim \exp(-W_0\mathbf{S_A^2})$, noting that $W_0 \gg 1$. The first limit is consistent with an upper bound to fatigue of the inert strength. The second, smaller stress, limit is marked by an exponential dependence of failure time on stress and the appearance of a thermal parameter W_0, both consistent with experimental observation. However, a nonzero lower bound to the applied stress at which failure time diverges is still absent.

A crack velocity expression that is physically based and includes the crack propagation limits of a thermodynamic lower-bound threshold and a thermally activated kinetic domain is given by

$$\dot{A} = \dot{A}_0 \sinh\left(\frac{\mathcal{G} - 2\gamma}{\eta}\right), \quad (6.40)$$

where $\dot{A} = dA/dt$ is the rate of change of the crack area, noting that for straight cracks $A \sim c$ and for circular cracks $A \sim c^2$. Consideration shows that $v = dc/dt$, is zero at a *crack velocity threshold* of $\mathcal{G} = 2\gamma$. For $\mathcal{G} \gg 2\gamma$, $v \sim \exp(\mathcal{G}/\eta)$, consistent with Eq. (6.33). The upper bound for applicability of Eq. (6.40) is the inert equilibrium condition $\mathcal{G} = 2\gamma_0$. The macroscopic fracture parameters are related to atomic scale bond rupture parameters by

$$2\gamma = \frac{u_{env}}{a^2}$$
$$\eta = \frac{2k_B\theta}{a^2}$$
$$\dot{A}_0 = 2f_0a^2 \exp\left(\frac{-u_1}{k_B\theta}\right). \quad (6.41)$$

The bonds on the fracture surface occupy area a^2, are separated by energy u_{env} between the bound and fully reacted states in the fracture environment, steps Figure 6.1a → 6.1c, with an intervening energy barrier u_1, step Figure 6.1b. $f_0 \sim k_B\theta/h$ is an intrinsic bond rupture attempt frequency, k_B is Boltzmann's constant, h is Planck's constant, and θ is absolute temperature. The succinct forms of Eqs. (6.40) and (6.41) (and the related inverse relations) reflect a recent, application based, approach to reactive fracture: at indentations (Cook and Liniger 1993), in thin films (Cook and Liniger 1999), in dynamic fatigue (Cook 2015), at interfaces (Hughey et al. 2004), and for reliability predictions (Cook 2018a, 2019). [Planar approximations of Eq. (6.40) for indentation cracks were used by Lathabai et al. (1989a) in dynamic fatigue and Wan et al. (1990) in crack propagation.]

Crack propagation equations similar in form to Eq. (6.40) have a long analytical history, extending back to the rate theory of Glasstone et al. (1941). The early works of Gibbs and Cutler (1951) and Stuart and Anderson (1953) established the essential physical principle that macroscopic crack velocity reflected the biasing of atomic scale energy barrier u_1 by applied loading. An important implication of these works was that a nonzero force or stress was required to maintain fracture in quiescent equilibrium states. The forces or stresses maintained stationary cracks and the nonzero loading thus reflected the crack velocity threshold. Gibbs and Cutler considered a stabilizing field and Stuart and Anderson considered a destabilizing field. Stuart and Anderson integrated an equation similar to Eq. (6.40) assuming planar symmetry and a series method to fit the static fatigue failure time data of Baker and Preston (1946a), Figure 6.3, including both the exponential and divergent limits. Shand (1961a) presented a diagram of nonequilibrium crack velocity variation that showed the predominant features: an exponential kinetic region separating divergent and vanishing crack velocities at inert and reactive equilibria, respectively. Shand also showed the consequent static fatigue diagram for glass components containing flaws of different sizes. The predicted behavior was entirely consistent with the observations of Baker and Preston, Charles, and Mould and Southwick. (Shand provided few supporting references for these diagrams however.) Gibbs and Cutler, Stuart and Anderson, and Shand all considered the physical mechanism of fatigue to be activated bond rupture. In contrast, Charles and Hillig (1962) developed a model for fatigue that was based on thermally activated preferential dissolution of material at assumed blunt crack tips. The predictions of the model had many qualitative features in common with those of the earlier models, including that of the static fatigue behavior predicted by Shand. Although blunt crack tips were never observed, the corrosion dissolution views of Charles and Hillig were prevalent in the interpretations and reviews of crack velocity behavior by Wiederhorn in the early 1970s (Wiederhorn and Bolz 1970; Wiederhorn 1972a,b, 1975). (In a somewhat contradictory manner, as Wiederhorn's experiments used a sharp crack based SIF approach.) Lawn (1975) established nonequilibrium crack velocity equations in forms similar to Eqs. (6.40) and (6.41) by biasing the energy barriers to bond rupture by deviations of a fracture system from equilibrium in the reactive environment, that is by $(\mathscr{G} - 2\gamma)$. (The Lawn analysis was apparently independent of the almost identical earlier approaches of Gibbs and Cutler, Stuart and Anderson, and Shand.) Similar crack velocity functions were also developed by Brown (1979) and Krausz and Krausz (1988) and repeated by Lawn (1993). The analyses of Gibbs and Cutler, Stuart and Anderson, Charles and Hillig, Brown, Krausz and Krausz, and, to a lesser extent, that of Lawn are complicated, with many variables that are not simple to measure or interpret. The clear forms of Eqs. (6.40) and (6.41) reflect the analytical approach of Lawn but are directed toward application. Detailed considerations of the physical principles underlying Eq. (6.41) and comparisons with power law and dissolution based velocity formulations are given elsewhere (Cook and Liniger 1993; Cook 1986, 2018a—a close repetition of these analyses is given in Vlassak et al. 2005.)

For experimental application, Eq. (6.40) is expressed in a form that separates material-dependent crack velocity parameters from component geometry. This expression is

$$dc/dt = \upsilon = \upsilon_0 \sinh\left[(\mathscr{G} - 2\gamma)/\eta\right]/b, \tag{6.42}$$

where υ_0 is a material-dependent characteristic crack velocity with Arrhenius form,

$$\upsilon_0 = 2f_0 a \exp\left(-u_1/k_B\theta\right), \tag{6.43}$$

and $b = dA/adc$ is a dimensionless factor that depends on component and crack geometry (the ratio of the crack front length to the bond length). For planar systems b is invariant. An example of a planar $\upsilon(\mathscr{G})$ curve is shown in Figure 6.11a using typical brittle material parameters of $2\gamma = 1\,\mathrm{J\,m^{-2}}$, $\upsilon_0 = 10^{-3}\,\mathrm{m\,s^{-1}}$, $\eta = 0.4\,\mathrm{J\,m^{-2}}$, $2\gamma_0 = 7\,\mathrm{J\,m^{-2}}$, and $b = 10^6$ (with associated typical atomic scale variables for fracture at 25 °C, Cook 2018a). For all component geometries, $\upsilon = 0$ at the crack velocity threshold $\mathscr{G} = 2\gamma$. Or, expressed in SIF terms, $\upsilon = 0$ at $K = T$. For planar crack geometries far from the threshold, $\upsilon \sim \exp(\mathscr{G})$, consistent with Eq. (6.33), and such a trend is evident in Figure 6.11a. In the seminal double cantilever beam experiments of Wiederhorn and Bolz, expressed in terms of K, $\upsilon \sim \exp(K^2)$ is expected and is consistent with their

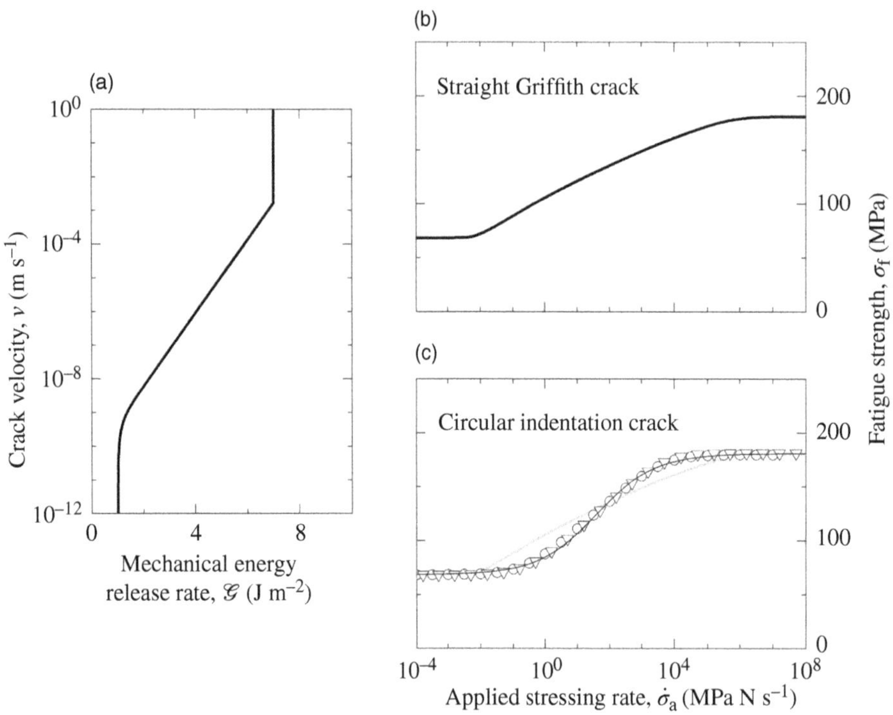

Figure 6.11 Thermally activated crack velocity function and consequent dynamic fatigue behavior. Semilogarithmic coordinates. (a) Crack velocity $\upsilon(\mathscr{G})$ function for typical material fracture parameters at room temperature. Note bounded exponential variation. (b) Dynamic fatigue behavior approximated for a planar Griffith flaw using the integrated $\upsilon(\mathscr{G})$ function of (a). Note bounded near-logarithmic variation. (c) Dynamic fatigue behavior for a circular indentation flaw obtained by numerical integration of the $\upsilon(\mathscr{G})$ function of (a) (symbols and bold line). Note bounded logarithmic variation.

observations (Wiederhorn and Bolz 1970; Cook and Liniger 1993). The crack velocity curve in Figure 6.11a is identical in form to that presented by Shand (1961a) and later by Cook (1986) (the latter including advocacy for a reactive bond rupture view of the threshold in fatigue).

For all component geometries consisting of crack length c_0 in a static uniform stress field σ_A, the reactive equilibrium or threshold condition is given by

$$\mathscr{G} = \psi^2\sigma_{\min}^2 c_0/E = 2\gamma. \tag{6.44}$$

$\sigma_{\min}$ is the minimum applied stress required to cause component failure in the reactive environment. $\sigma_{\min}$ is referred to as the "Orowan" strength in reference to the work of Orowan (1944), in which the concept of a reduced strength in reactive environments arising from a decrease in surface energy was suggested. The maximum applied stress required to cause component failure in an inert environment is the Griffith strength $\sigma_{\max}$,

$$\mathscr{G} = \psi^2\sigma_{\max}^2 c_0/E = 2\gamma_0. \tag{6.45}$$

Static and dynamic fatigue occur between these two strength bounds. Estimation of the ratio $2\gamma/2\gamma_0$ is thus extremely important in brittle materials design and a major numerical output of fatigue tests (also in cyclic fatigue of metals, Dieter 1986; Hertzberg 1989; Suresh 1998). The characteristic time, stress, and energy parameters for fatigue of components between the strength bounds in the reactive environment are referenced to those in Eqs. (6.42) and (6.44),

$$\begin{aligned}
t_0 &= c_0/v_0\\
\mathbf{t} &= t/t_0\\
\mathbf{S} &= \sigma/\sigma_{\min}\\
\dot{\mathbf{S}} &= \dot{\sigma}_a t_0/\sigma_{\min}\\
W &= 2\gamma/\eta,
\end{aligned} \tag{6.46}$$

noting that in these scaled coordinates the bounds are $1 \leq \mathbf{S} \leq \mathbf{S}_{\max}$. Static fatigue behavior is described by integration of Eq. (6.42). Assuming planar geometry, the failure time in the scaled coordinates is given by (Cook 2018a)

$$\mathbf{t}_f = \frac{b}{W\mathbf{S}_A^2}\ln\left(\frac{e^{W(\mathbf{S}_{\max}^2-1)}-1}{e^{W(\mathbf{S}_{\max}^2-1)}+1}\cdot\frac{e^{W(\mathbf{S}_A^2-1)}+1}{e^{W(\mathbf{S}_A^2-1)}-1}\right). \tag{6.47}$$

For $\mathbf{S}_A \to \mathbf{S}_{\max}$, the failure time vanishes, $\mathbf{t}_f \to 0$, and for $\mathbf{S}_A \to 1$, the failure time diverges, $\mathbf{t}_f \gg 1$, both effects consistent with many experimental observations. The kinetically limited central asymptote, $1 \ll \mathbf{S}_A \ll \mathbf{S}_{\max}$, is

$$\mathbf{t}_f \to \frac{2b}{W\mathbf{S}_A^2}e^{-W(\mathbf{S}_A^2-1)}, \tag{6.48}$$

consistent with Eq. (6.39): an exponential increase in failure time with decreasing applied stress, also consistent with many experimental observations. Hence, in all respects, the physically based crack velocity expression describes the fundamental characteristics of static fatigue. Details are discussed elsewhere (Cook 2018a).

As noted in Section 6.1, dynamic fatigue tests are usually performed more easily and thus more often than static fatigue tests. However, direct extension of the integration methodology to generate a dynamic fatigue description analogous to Eq. (6.47) is not possible. Even for the kinetic limit, Eq. (6.33), of Eq. (6.42), a closed form solution to the implied dynamic fatigue integral is not possible. However, a guide to the form of dynamic fatigue behavior arising from Eq. (6.42) can be gained by following the simple multiplicative relationship between failure times in static and dynamic fatigue observed in the empirical power-law analyses, Eqs. (6.18), (6.19), (6.23), and (6.24). Hence, substituting $\mathbf{t}_f \to \mathbf{t}_D$ and $\mathbf{S}_f \to \mathbf{S}_A$ in Eq. (6.47) and using Eq. (6.20), a dynamic fatigue response for planar Griffith flaws, $\mathbf{S}(\dot{\mathbf{S}}_a)$, may be generated. Such a relationship is shown in the semilogarithmic plot of Figure 6.11b using the $v(\mathscr{G})$ behavior of Figure 6.11a and an initial crack length of $c_0 = 10$ μm. This approximated response exhibits many of the experimental characteristics of dynamic fatigue behavior: the applied stressing rate domain and fatigue strength range are typical for brittle materials; the strength approaches the upper-bound inert Griffith strength at fast stressing rates and the lower-bound Orowan strength at slow rates; and the strengths follow a near-logarithmic trend at intermediate rates. Two points preclude the direct application of responses similar to that shown in Figure 6.11b to dynamic fatigue of components containing indentation flaws in polycrystalline materials: (i) the mechanical energy release rate is perturbed from that for a Griffith flaw, Eq. (6.34), by the addition of residual stress and microstructural terms $\sim c^{-3/2}$ and (ii) the cracks are circular, such that the geometry term is not invariant but increases with crack size, $b \sim c$.

All the complicating analytical factors are handled in a straightforward manner by numerical integration techniques. Several approaches are possible. The most direct is to estimate all fracture parameters for a material, e.g., (ψ, χ, P^*) and $(\upsilon_0, 2\gamma, \eta, 2\gamma_0)$, specify the test parameters P and $\dot{\sigma}_a$, and integrate to determine σ_f as a function of $\dot{\sigma}_a$ on a case-by-case basis. This was the approach in Cook (2015). A more flexible, but somewhat less exact, approach is to determine solutions for a broad domain of material and test parameters and develop a parameterized function to describe the solutions. The function can then be fit to experimental data through variation of the function parameters. This was the approach in earlier work (Marshall and Lawn 1980; Chantikul et al. 1981; Lawn et al. 1981) and in Figures 6.8–6.10. In these cases, the functions were framed in scaled coordinates as in Sections 6.2.1–6.2.3. A similar parameterized function approach will be used here. Figure 6.11c shows a semilogarithmic plot of dynamic fatigue responses in the same coordinates as Figure 6.11b; the gray line reproduces the planar Griffith flaw response. The symbols represent full numerical solutions of Eq. (6.42) for indentation flaws, matching the bounding Orowan and Griffith strengths, Eqs. (6.44) and (6.45), to the bounding indentation counterparts,

$$\sigma_{min} = 3(2\gamma E)^{2/3}/4^{4/3}\psi(\chi P_{eff})^{1/3}, \tag{6.49}$$

and

$$\sigma_{max} = 3(2\gamma_0 E)^{2/3}/4^{4/3}\psi(\chi P_{eff})^{1/3}, \tag{6.50}$$

noting that these are simply variants of Eq. (3.9), and matching the applied stressing rate at the median strength. It is clear that the approximated response does not quite match the numerical response. However, a functional form that does match the numerical response and that is amenable to fitting dynamic fatigue data is shown by the solid line. This function is obtained by noting that for the power-law crack velocity approximation the relationship between static and dynamic responses is one of inverse functions, $\mathbf{t_f} \sim \mathbf{S_A}^{-n}$ and $\mathbf{S_f} \sim \dot{\mathbf{S}}_\mathbf{a}^{1/n}$. For the physically based crack velocity expression, the static fatigue response varies as $\mathbf{t_f} \sim \ln[\tanh(\mathbf{S_A})]$ (Cook 2018a), which suggests a dynamic fatigue inverse of $\mathbf{S_f} \sim \tanh[\ln(\dot{\mathbf{S}}_\mathbf{a})]$. The fitting function used in Figure 6.11c follows this form and is given by

$$\sigma_f = \sigma_{min} + [\sigma_{max} - \sigma_{min}]\left[1 + \tanh[m_{med}\ln(\dot{\sigma}_a/\dot{\sigma}_{med})]\right]/2, \tag{6.51}$$

where σ_{min} and σ_{max} are the strength bounds as given above and m_{med} and $\dot{\sigma}_{med}$ characterize the slope and stressing rate of the median strength response. In fitting Eq. (6.51) to experimental data, σ_{max} is fixed and σ_{min}, m_{med}, and $\dot{\sigma}_{med}$ are fitting parameters. Consideration of Eqs. (6.51) and (6.42) indicates that m_{med} is related to $1/\eta$ and $\dot{\sigma}_{med}$ is related to $1/\upsilon_0$. These relationships are not pursued here. Attention is focused on the parameter σ_{min}, which through σ_{max} provides an estimate of the surface energy ratio $2\gamma/2\gamma_0$. Combining Eqs. (6.49) and (6.50) gives $2\gamma/2\gamma_0 = (\sigma_{min}/\sigma_{max})^{3/2}$. The surface energy ratio is critical in designing with brittle materials as it allows a means to overcome fatigue effects. Equation (6.51) provides a clear method for estimating this ratio by enabling an estimate of $\sigma_{min}/\sigma_{max}$.

6.3 Reactive Strengths Observations

This section presents experimental measurements of indentation strengths of glasses and ceramics under reactive dynamic fatigue loading conditions. The reactive conditions are exposure to water and the dynamic applied stressing rates vary from 10^{-3} to 10^3 MPa s^{-1}, leading to fatigue failure times from 24 h to 10 ms. Inert strengths are included with all fatigue measurements. Indentation loads vary from $P = 3$ N to $P = 100$ N and in some cases intrinsic flaws, $P = 0$ N, are included. Although the measurements draw heavily from previously published works, many have been supplemented with unpublished data and all have been reanalyzed for this book using Eq. (6.51). The section begins with consideration of dynamic fatigue behavior of groups of specimens containing a single indentation load and is followed by consideration of multiple groups tested using multiple indentation loads. Example applications of the measurements are given as dynamic and static fatigue predictions and comparisons with independent measurements. Surface stress effects in reactive indentation fracture, for both compressive and tensile stresses, are then examined. Finally, the section considers probabilistic methods in fatigue and illustrates how such methods can be used in reliability predictions.

6.3.1 Single Contact Loads

Figure 6.12 shows a semilogarithmic plot of the dynamic fatigue behavior of a glass-bonded polycrystalline alumina (AD90) containing $P = 10$ N Vickers indentations in water at room temperature. The symbols represent the means and standard

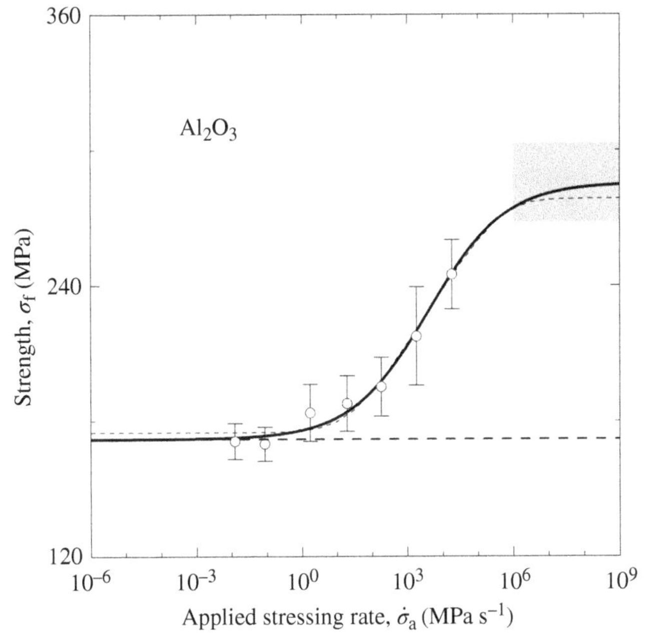

Figure 6.12 Semilogarithmic plot of dynamic fatigue behavior of polycrystalline alumina disks in water. Note lower-bound threshold fatigue strength. Lines are fits based on physically based crack velocity behavior. Source: Includes data adapted from Cook et al. (1985b) and Cook (2015).

deviation limits of fatigue strengths measured at the applied stressing rates indicated. The shaded band represents the mean and standard deviation limits of the inert strength. Data are included from prior works (Cook et al. 1985b; Cook 2015). The solid line represents a visual best fit to the fatigue data of Eq. (6.51) constrained by the inert strength. The fine dashed line indicates a prior regressed best fit to the data obtained by integration (Cook 2015); there is no significant difference between the fits. The bold dashed line indicates the best-fit reactive strength. It is clear that the fits, both of which include inert and reactive upper and lower bounds separated by a kinetically limited region, describe the data. In particular, the lower bound is marked by a tendency at slow applied stressing rates to a strength plateau that is indicative of crack velocity threshold effects. Estimation from the plot shows that the upper and lower bounds are approximately $\sigma_{max} \approx 300$ MPa and $\sigma_{min} \approx 180$ MPa, such that the surface energy ratio for this alumina in water is $2\gamma/2\gamma_0 \approx (180/300)^{3/2} \approx 0.46$. This value will be shown to be typical for oxides in water; details are given in the discussion. The plotting scheme used here, including the factor of 3 in strength and the factor of 10^{15} in applied stressing rate, will be used throughout this section.

Figures 6.13 and 6.14 show semilogarithmic plots of the dynamic fatigue behavior of a range of indented materials in water at room temperature, in approximate order of increasing strength. Polycrystals, glass-ceramics, and a single crystal are included. The symbols, lines, and bands are as described for Figure 6.12. Different indentation loads were used for each material and data are included from prior works. In Figure 6.13: (a) $Pb(Zr)TiO_3$ ($P = 10$ N, Cook and Lawn 1984); (b) "Zerodur" glass-ceramic ($P = 10$ N, tested in air, Viens 1990); (c) Bi–$BaTiO_3$ ($P = 5$ N, Cook 2015); and (d) Cordierite glass-ceramic ($P = 20$ N, Cook 2015). In Figure 6.14: (a) "Pyroceram" glass-ceramic ($P = 20$ N, Cook et al. 1981; Multhopp et al. 1983; Cook and Lawn 1984); (b) polycrystalline SiC ($P = 10$ N, Multhopp et al. 1983; Cook and Lawn 1984); (c) polycrystalline Al_2O_3 ($P = 20$ N, Cook and Lawn 1984); and (d) single-crystal Al_2O_3, Sapphire ($P = 3$ N, Cook 1986, 2015). It is clear that the fits (both integrated and parameterized) describe the data and that plateau "threshold" strengths at slow stressing rates are prominent in many cases—and also discernible in Figures 6.8–6.10. Some materials (e.g., sapphire) exhibit substantial decreases in fatigue strength and some (e.g., SiC) exhibit very small decreases. The use of controlled indentation flaws enables such trends and differences between materials to be discerned clearly.

6.3.2 Multiple Contact Loads

The effects of variations in indentation load on fatigue responses were discussed in the context of the power-law approximation to crack velocity in Eqs. (6.26)–(6.28) and demonstrated for SLG in Figure 6.9. Considerations of the effects were

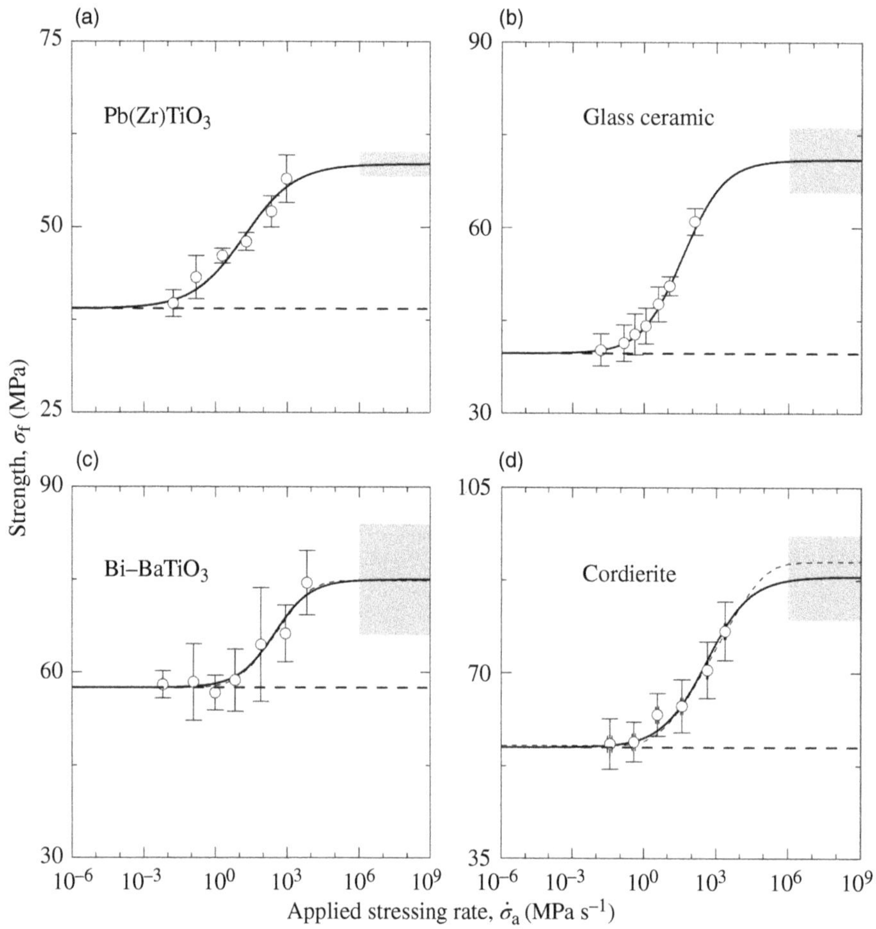

Figure 6.13 Semilogarithmic plots of dynamic fatigue behavior of polycrystalline ceramics in water and (b) air. Single indentation loads used. Note lower-bound threshold fatigue strengths. Lines are fits based on physically based crack velocity behavior. Source: Includes data adapted from Cook and Lawn (1984), Viens (1990), and Cook (2015).

extended to include lateral crack and microstructural influences in Eqs. (6.31) and (6.32) and demonstrated for polycrystalline alumina in Figure 6.10. Consideration of these equations and Eqs. (6.47) and (6.51) imply that such scaling of fatigue variables to incorporate indentation load effects is applicable generally. Hence, in this section, the experimental dynamic fatigue data for materials containing a range of test indentation loads P have been adjusted with the effective indentation load P_{eff} as in Figure 6.10, $\sigma_f \rightarrow \sigma_f P_{eff}^{1/3}$ and $\dot{\sigma}_a \rightarrow \dot{\sigma}_a P_{eff}$. Similarly, the fitting parameters in Eq. (6.51) have been modified to their load invariant adjusted counterparts as $\sigma_{min} \rightarrow S_{min}$ and $\dot{\sigma}_{med} \rightarrow \dot{S}_{med}$, noting that $S = \sigma_{max} P_{eff}^{1/3}$ is an invariant inert strength constraint parameter for a material (see Section 4.3). The dynamic fatigue behavior in load-adjusted coordinates is thus given by extending Eq. (6.51) as

$$\sigma_f P_{eff}^{1/3} = S_{min} + [S - S_{min}]\left[1 + \tanh[m_{med} \ln(\dot{\sigma}_a P_{eff}/\dot{S}_{med})]\right]/2. \tag{6.52}$$

Figures 6.15 and 6.16 show semilogarithmic plots of the dynamic fatigue behavior of a range of indented materials in water at room temperature, in approximate order of increasing adjusted strength. Glasses, glass-ceramics, and polycrystals are included. The symbols, lines, and bands are as described for Figure 6.12 with the addition that the parameters are all indentation load adjusted. Different indentation load domains were used for each material and data are included from prior works. In Figure 6.15: (a) SLG (P = 0.05 N–P = 10 N, Dabbs et al. 1982); (b) BSG (P = 5 N–P = 100 N, Dabbs et al. 1982); (c) FSG, amorphous SiO_2 (P = 5 N–P = 50 N, Dabbs and Lawn 1985); and (d) Synroc (P = 10 N–P = 100 N, Cook et al. 1982b; Cook and Lawn 1984). These materials exhibited either ideal indentation strength behavior or behavior influenced by lateral cracking. (For Figure 6.15a–c, see also Cook and Lawn 1984.) In Figure 6.16: (a) polycrystalline $BaTiO_3$ (P = 3 N, 30 N, Cook et al. 1985b); (b) "Macor" glass-ceramic (P = 5 N, 20 N, Cook 2015); (c) polycrystalline glass-bonded Al_2O_3 (AD96) (P = 0–20 N, Gonzalez et al. 1984); and (d) polycrystalline Al_2O_3 (Vistal 1) (P = 0–100 N, Cook et al. 1985b; Cook

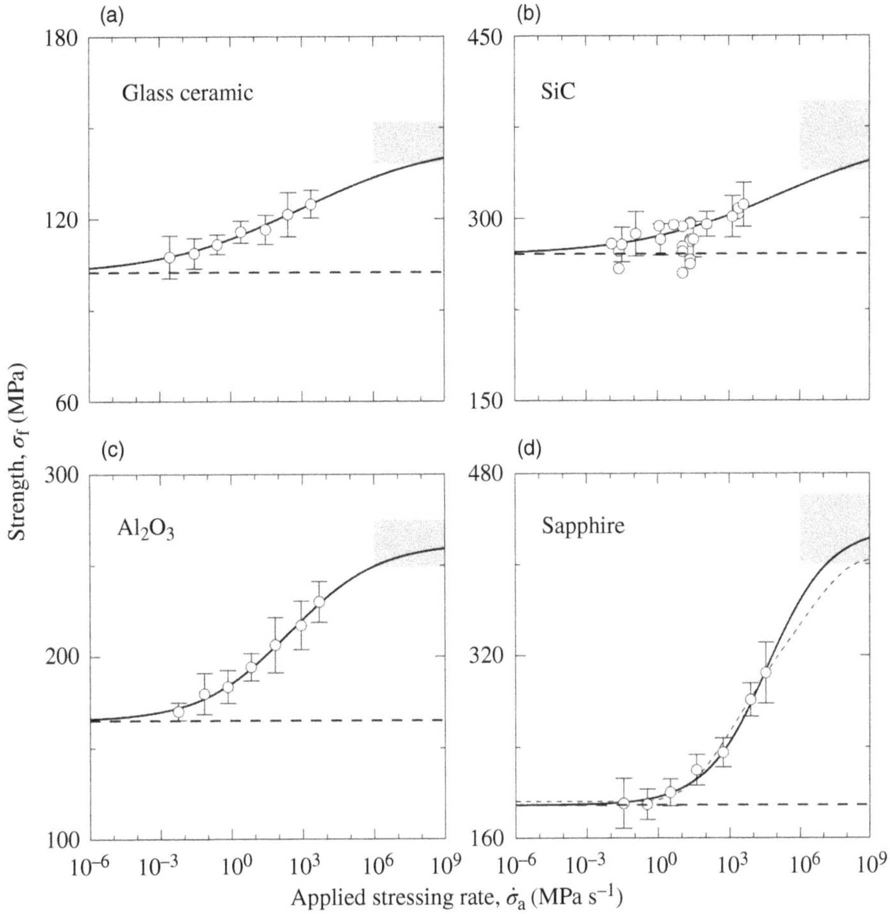

Figure 6.14 Semilogarithmic plots of dynamic fatigue behavior of polycrystalline and (d) single-crystal ceramics in water and air. Single indentation loads used. Note lower-bound threshold fatigue strengths. Lines are fits based on physically based crack velocity behavior. Source: Includes data adapted from Cook et al. (1981), Cook and Lawn (1984), Multhopp et al. (1983), Viens (1990), and Cook (1986, 2015).

2015). These materials exhibited indentation strength behavior influenced by both lateral cracking and microstructural effects. In all cases, $P \neq P_{eff}$. The use of the indentation load adjusted fatigue parameters converged the data to common load invariant material responses well described by the load-adjusted fitting function given by Eq. (6.52). The fine dashed line in Figure 6.15a represents the power-law fit of Figure 6.9b recast into semilogarithmic coordinates. Over the domain of the data, the power-law approximation describes the fatigue response but diverges from the thermally activated response at slow stressing rates and approaches an asymptote of zero strength. In load-adjusted coordinates, all materials exhibited fatigue responses that included a tendency to a plateau threshold strength S_{min} at slow stressing rates. It is notable that the responses for the alumina materials included the fatigue responses of specimens with strengths controlled by intrinsic flaws (i.e., no indentation, $P = 0$ N). These responses fall onto the common responses for the materials. The convergence provides further support for the implication from inert strength tests (see Chapter 4) that such flaws are determined by microstructural influences and amenable to strength prediction from controlled flaw tests—in both inert and reactive environments.

6.3.3 Predictions

One of the major attributes of Eq. (6.52) is that it enables fatigue predictions. It is noted that the strength parameters S and P_{eff} in Eq. (6.52) are determined from inert strength tests in which indentation load P is varied, as described in Chapter 4. The fatigue parameters S_{min}, m_{med}, and $\dot{S}_{med}$ in Eq. (6.52) are determined from fatigue tests in which the test indentation load P is fixed. Hence, the fatigue response $\sigma_f(\dot{\sigma}_a)$ at any other designated test load can be predicted simply by determining

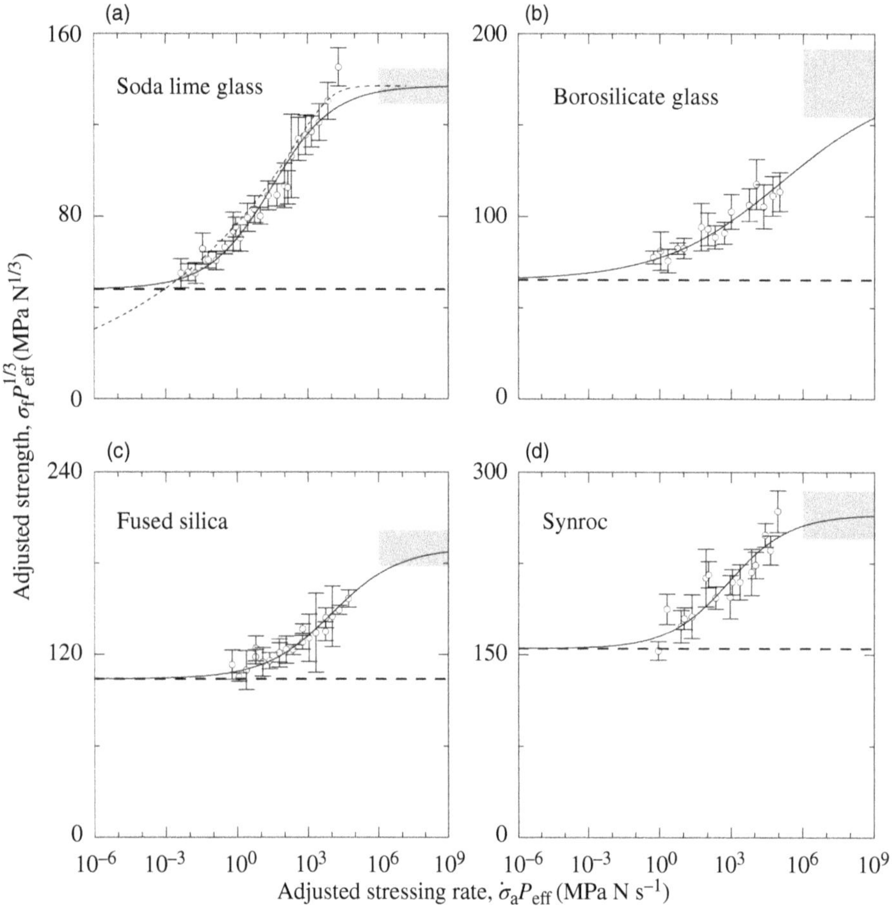

Figure 6.15 Semilogarithmic plots of dynamic fatigue behavior of glasses and (d) a polycrystalline ceramic in water. Multiple indentation loads used. Note lower-bound threshold fatigue strengths. Lines are fits based on physically based crack velocity behavior. Source: Includes data adapted from Dabbs et al. (1982), Dabbs and Lawn (1985), Cook et al. (1982b), and Cook and Lawn (1984).

a new value of P_{eff} and applying Eq. (6.52). This was the approach in Cook (2015), in which dynamic fatigue measurements of components containing $P = 100$ N indentations were used to predict the dynamic fatigue responses of components containing $P = 3$ N indentations. The predictions were applied successfully to two alumina materials, one exhibiting near-ideal indentation strength behavior and the other exhibiting significant perturbations from ideal. As noted, the latter predictions required extrapolation of fatigue responses from large flaws affected by lateral cracking and microstructural toughening to small flaws weakly affected by such influences, and thus required correct evaluation of P_{eff}. Figures 6.10, 6.16c and d demonstrate this predictive ability in a different way by illustrating universal responses in load-adjusted coordinates of fatigue responses determined at different values of P. Two further examples of the predictive ability of indentation dynamic fatigue measurements are illustrated here. The first example is the prediction of dynamic fatigue responses of materials that are closely related to test materials—reflecting a common step in microstructural development. The second example is the use of dynamic fatigue data from components containing indentation flaws to predict the static fatigue response of components containing intrinsic flaws—this example reflects a conventional reliability procedure.

The first example of fatigue prediction concerns changes in microstructural scale. Figure 6.17 shows dynamic fatigue plots in load-adjusted coordinates for (a) the Al_2O_3 material of Figure 6.16d, heat treated so as to increase the grain size from approximately 20 to 25 μm (Cook et al. 1985a,b) and (b) SiO_2-based glass-ceramic materials containing precipitated lithium silicate platelets, with different mean free paths between the platelets, 2.3 μm (open symbols) and 1.2 μm (filled symbols) (Cook et al. 1986). Fracture in the Al_2O_3 materials is predominantly intergranular, and fracture in the glass-ceramic is predominantly between platelets. In both cases, no significant change in fracture surface chemistry, and thus no change in fracture kinetics, is expected. In such cases, the fatigue strength simply scales with the inert strength, or, in load-adjusted coordinates with S. The bold solid lines in Figure 6.17 represent such fatigue predictions for the two materials systems. In

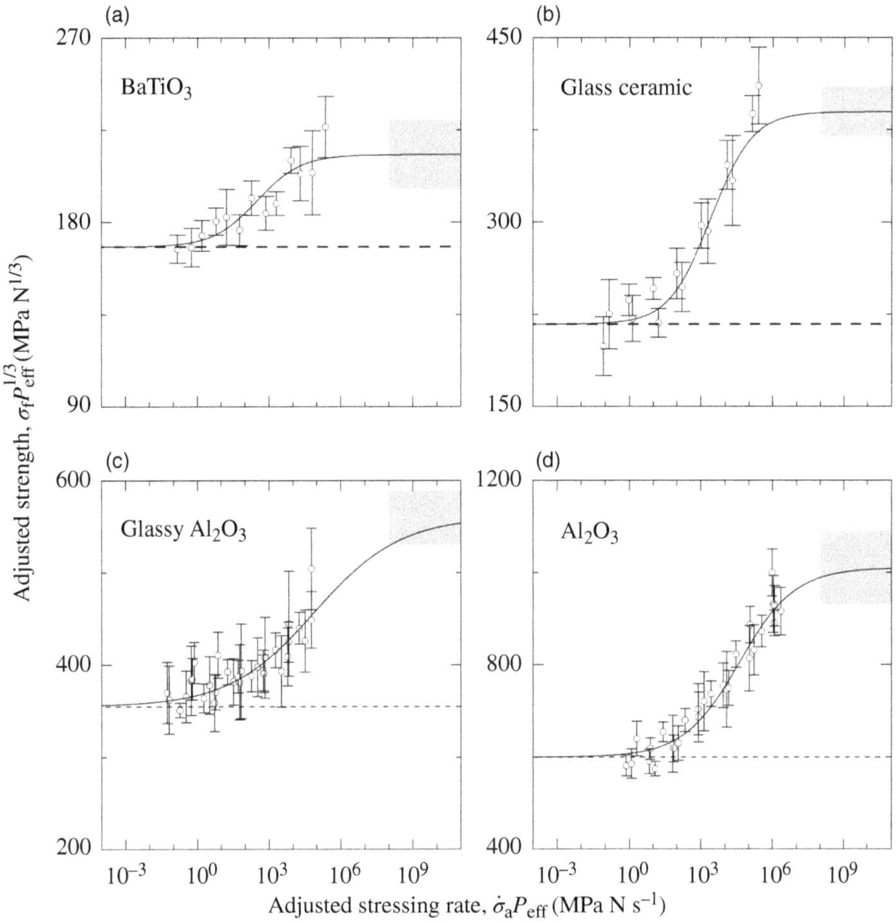

Figure 6.16 Semilogarithmic plots of dynamic fatigue behavior of polycrystalline ceramics in water. Multiple indentation loads used. Note lower-bound threshold fatigue strengths. Lines are fits based on physically based crack velocity behavior. Source: Includes data adapted from Gonzalez et al. (1984), Cook et al. (1985b), and Cook (2015).

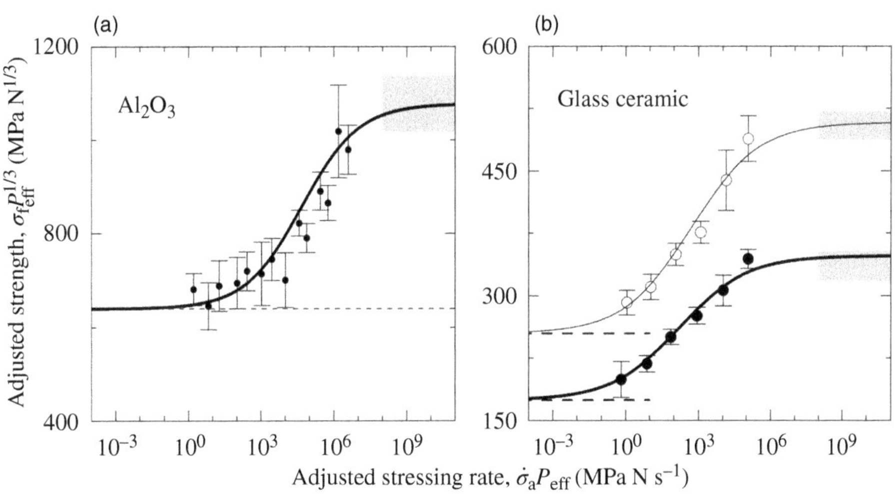

Figure 6.17 Semilogarithmic plots of dynamic fatigue behavior of polycrystalline ceramics in water. Multiple indentation loads used. Bold lines are predictions based on inert strength calibrations and physically based crack velocity behavior. Source: Includes data adapted from Cook et al. (1985a,b) and Cook (1986).

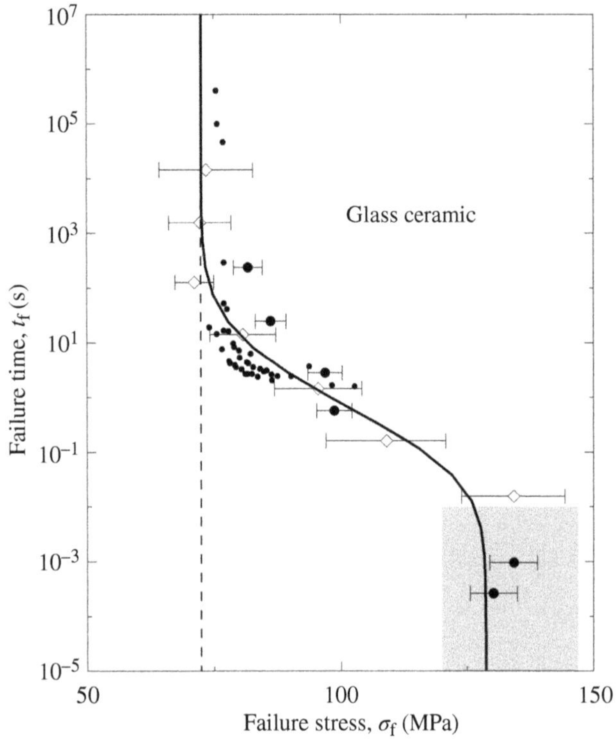

Figure 6.18 Semilogarithmic plot of static fatigue behavior of glass-ceramic in water. Indentation and intrinsic flaws used. Open symbols and bold line represent dynamic fatigue indentation calibration measurements and fit based on physically based crack velocity behavior. Filled symbols are intrinsic flaw dynamic and static fatigue intrinsic flaw measurements. Source: Includes data adapted from Smyth and Magida (1983), Heslin et al. (1986), and Cook (2015).

each case, fatigue behavior and S_{base} were both determined for the base microstructures and S_{var} alone determined for the variants. Predictions of fatigue behavior for the microstructural variants were then determined by scaling the base fatigue response by S_{var}/S_{base}. For the Al_2O_3 material, the base response is shown in Figure 6.16d. For the glass-ceramic material, the base response is shown by the open symbols. The filled symbols in each plot represent the measured fatigue response of the microstructural variant. The agreement with the predictions is clear.

The second example of fatigue prediction concerns changes in both loading type, dynamic and static fatigue, and flaw type, indentation and intrinsic. For both changes, materials evaluation is more easily performed using dynamic fatigue and indentation flaws. However, assessments of reliability inevitably require predictions for static fatigue of components containing intrinsic flaws. Two cases are considered. Figure 6.18 shows a static fatigue plot for the machinable glass-ceramic "Macor." The load-adjusted dynamic fatigue response is shown in Figure 6.16b. The open symbols in Figure 6.18 represent the means and standard deviation limits of dynamic fatigue failure times t_D at the failure stresses σ_f indicated for specimens containing 20 N indentation flaws. The shaded band represents the mean and standard deviation limits of the 20 N inert strength. The solid line represents the best fit to the data using Eq. (6.52) and setting $P = 20$ N. The lower-bound threshold stress is indicated by the dashed line. The filled symbols with bars represent the means and standard deviation limits of dynamic fatigue and inert strength data for specimens containing intrinsic flaws (Smyth and Magida 1983). The agreement between the data sets suggests that the intrinsic flaws in the components are well approximated by 20 N indentations. The comparison of static and dynamic fatigue in Sections 6.2.1 and 6.2.2 implies that the solid line should be interpreted as an upper bound to static fatigue failure times. The consideration of thermally activated crack velocity in Section 6.2.4 implies that the dashed line should be interpreted as the lower-bound stress at which fatigue failure times diverge. The individual filled symbols in Figure 6.18 represent static fatigue failure times measured at the applied stresses indicated for components containing intrinsic flaws (Heslin et al. 1986). The agreement with the predictions is clear.

Figure 6.19 shows a static fatigue plot for a polycrystalline alumina (AD96). The load-adjusted dynamic fatigue response is shown in Figure 6.16c. The shaded band represents the mean and standard deviation limits of the inert strength

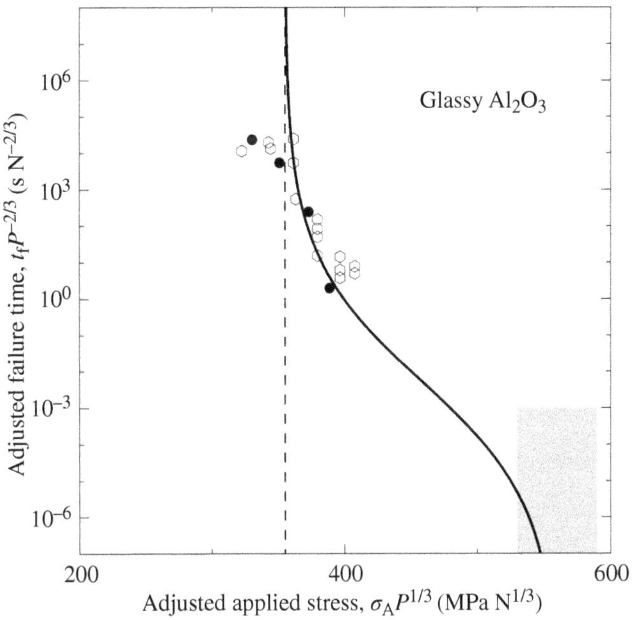

Figure 6.19 Semilogarithmic plot of static fatigue behavior of polycrystalline in water. Indentation and intrinsic flaws used. Bold line represents prediction from dynamic fatigue (Figure 6.16c) using physically based crack velocity behavior. Filled and open symbols are intrinsic and indentation flaw, respectively, static measurements. Source: Includes data adapted from Gonzalez et al. (1984).

parameter S. The solid line represents the best fit to the dynamic fatigue data using Eq. (6.52) and transformed into static fatigue load-adjusted coordinates of $t_f P^{-2/3}$ and $\sigma_A P^{1/3}$. The lower-bound threshold stress is indicated by the dashed line. As previously, the solid line should be regarded as an upper bound to failure time and the dashed line as a lower-bound stress. The open symbols indicate static fatigue measurements of specimens containing 5 N indentations. The filled symbols indicate static fatigue measurements of specimens containing intrinsic flaws ($\approx$ 2 N equivalent) (Gonzalez et al. 1984). The agreement between the static fatigue behavior of the intrinsic and indentation flaws is clear. However, in this case, the overall agreement with the predicted behavior is weaker, although well within experimental uncertainty if variation in the inert strength is considered.

6.3.4 Surface Stress Effects

The effects of surface stresses on the indentation strengths of components in inert environments were considered in detail in Chapter 3. Attention was focused on the strength-enhancing effects of surface compression and the potentially deleterious effects of compensating bulk tension. Compressive effects induced by chemical means (ion infiltration or exchange) and physical means (lamination, machining, or tempering) were examined. Here, the effects of surface stresses on the fatigue behavior of components in reactive environments are considered. Three systems are examined. Two systems are ceramics previously studied in Chapter 3 that included compressive surface stresses. Indentation fatigue strengths are examined for these systems. A new thin film on substrate system is introduced that includes a tensile surface stress. Indentation cracks are used as part of direct crack velocity measurements for this tensile film system.

Figure 6.20 shows semi-logarithmic plots of dynamic fatigue behavior of components of two glass-ceramic materials. Both sets of components contained strength controlling $P = 20\,N$ indentation flaws. Figure 6.20a shows the behavior of the laminated glass-ceramic considered in Figure 3.46. Figure 6.20b shows the behavior of the machined glass-ceramic considered in Figure 3.45a. In all cases, symbols with bars represent the means and standard deviation limits of experimental measurements of strengths at the applied stressing rates indicated. Shaded bands represent the means and standard deviation limits of inert strengths. Open symbols and gray lines represent the base component measurements and best fits from Figure 6.13d and Figure 6.14a. The filled symbols in Figure 6.20a represent the fatigue strengths of components containing indentation cracks small with respect to the thickness of the uniformly compressed laminated surface layer (Cook 2005). As a consequence, during fatigue loading the cracks grew but remained contained within the layer, including at the instability failure state. The fatigue strengths are thus predicted from the base response by simple translation through

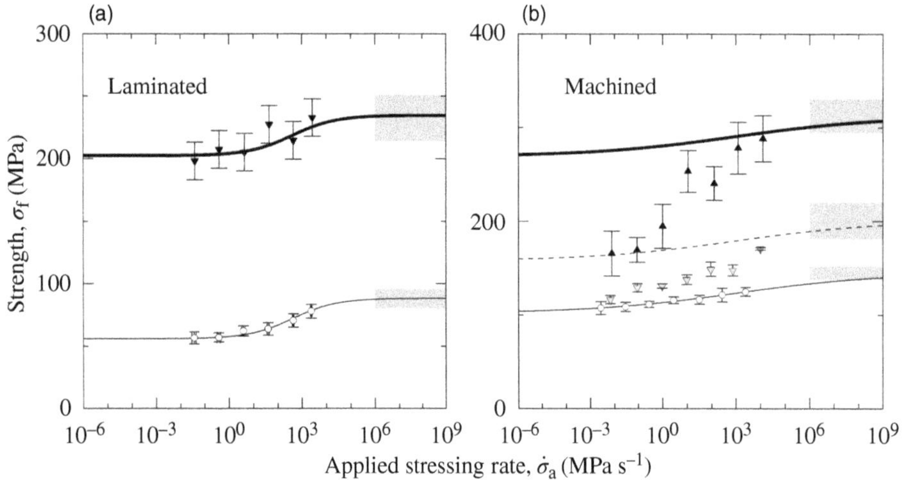

Figure 6.20 Semilogarithmic plots of dynamic fatigue behavior of compressive surface stressed glass-ceramics in water. Single indentation loads used. Fine lines represent fits to base unstressed components. Bold and dashed lines are predictions based on inert strength calibrations of stressed components. (a) Laminated composite. (b) Machined components. Source: Includes data adapted from Cook et al. (1982a) and Cook (2005).

the magnitude of the compressive stress represented by the difference in inert strengths, approximately 140 MPa. Such a prediction is indicated by the bold line. The agreement with the experimental observation is clear. The filled symbols in Figure 6.20b represent the fatigue strengths of components containing indentation cracks comparable in scale to the thickness of the inhomogeneously compressed machined surface layer (Cook et al. 1981). The magnitude of surface compression induced by machining is a maximum at the surface and on proceeding into the bulk decreases to near zero, typically on a length scale of ≈ 100 μm. As a consequence, during fatigue loading in this case, the cracks grew and extended through the layer of decreasing compression. As such, only an upper bound to the fatigue strength can be predicted by simple translation through the difference in inert strengths, here about 180 MPa. This expectation is borne out by the experimental data, which decrease markedly with decreasing stressing rate, and thus with increasing crack extension. The base response was obtained by removal of the machined surface by careful fine polishing. The results of an intermediate polishing stage are indicated by the grayed symbols and dashed line. Here too a prediction based on the inert strength provides an upper bound to fatigue behavior. Note that in both machined cases the data converge on the base response at slow applied stressing rates, implying that bulk tension is negligible in the machined components.

A surface stressed system that can in many respects be regarded as the complete opposite of the ceramic systems just considered is shown in the schematic diagrams of Figure 6.21. The diagrams show cross sections of indentation contacts of increasing load on a film on substrate system. The substrate is much tougher and harder than the film such that all fracture and irreversible deformation generated on indentation is restricted to the film. The progression of indentations in Figure 6.21 is (a) a small crack-free subthreshold indentation with irreversible deformation contained in the film, (b) a midsized postthreshold indentation with a half-penny like crack and deformation and fracture contained in the film,

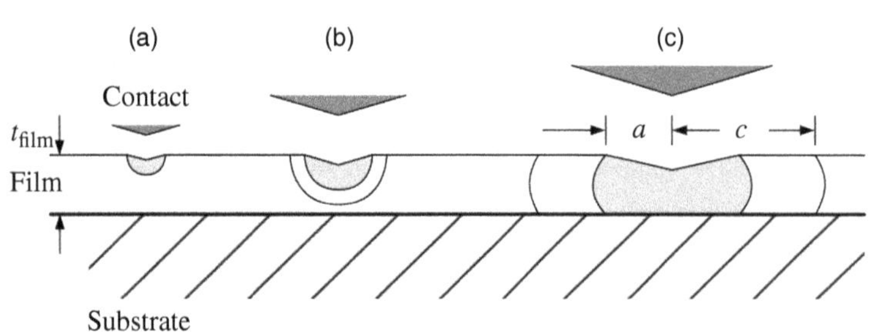

Figure 6.21 Schematic cross-section diagrams of indentation flaws in a film on substrate system. (a) A small, subthreshold flaw consisting of a plastic deformation zone (pdz). (b) A midsized postthreshold flaw consisting of a pdz + half-penny crack contained within the film. (c) A large flaw consisting of a pdz + channel cracks extending to the substrate.

and (c) a large indentation in which the deformation and fracture extend to the film–substrate interface and cracks take on a "channel" like morphology. The progression is similar to that of impact flaws of increasing velocity shown in Figure 3.10, here omitting lateral cracks at the large indentation. The progression is different from that of indentation flaws of increasing size shown in Figure 3.24 as here the cracks at the large indentation do not extend into the bulk. Another, greater, difference is that in Chapter 3 attention was focused on compressive surface layers or films and the shortening of indentation cracks, as shown in Figure 3.25. Here, attention is focused on tensile films and the failure mechanism of interest is unchecked film splitting or channel cracking leading to $c \gg t_{\text{film}}$. Such channel cracking can be triggered by surface contacts, characterized by contact dimension a.

The fracture mechanics of postthreshold, channel cracked, and lateral cracked and delaminated indentations for film on substrate systems, with an emphasis on compressive films, is considered in detail elsewhere (Thurn and Cook 2004a,b). For tensile films here, the contact conditions for film failure by subthreshold and postthreshold flaws are simply obtained by reversing the emphasis of the fracture mechanics analyses of Chapter 3. Replacing the failure stress with the film stress σ_{film} and the contact load with the contact dimension a^2 in Eqs. (3.122) and (3.8) gives the scaling of the critical contact sizes for film failure, $a_{\text{sub}} \sim \sigma_{\text{film}}^{-2}$ and $a_{\text{post}} \sim \sigma_{\text{film}}^{-3/2}$. The SIF for a crack in the channel geometry is given by (Thurn and Cook 2004b)

$$K = \lambda(E_f H_f^2)^{1/3} a^2 / t_{\text{film}} c^{1/2} + \psi \sigma_{\text{film}} t_{\text{film}}^{1/2}, \tag{6.53}$$

where the first term on the right side is an indentation residual stress term and the second term is a film stress term. E_f and H_f are film elastic modulus and hardness, respectively. λ and ψ are dimensionless crack geometry terms. For indentations in air, the equilibrium condition is $K = T_{\text{air}}$, where T_{air} is the film toughness in air. Setting equilibrium and inverting Eq. (6.53) gives an expression for the stable equilibrium channel crack length in air

$$c_{\text{air}} = \left[\frac{\lambda(E_f H_f^2)^{1/3}}{t_{\text{film}}(T_{\text{air}} - \psi \sigma_{\text{film}} t_{\text{film}}^{1/2})} \right]^2 a^4. \tag{6.54}$$

This equation is applicable for sufficiently large indentation contacts, $a \geq a_{\text{sub}}, a_{\text{post}}$. There is then a much stronger dependence of equilibrium crack length on contact size, a^4, than conventional indentation cracks, $a^{4/3}$. The tendency to long stable cracks at contacts in tensile oxide films is increased if the humidity of the air is increased such that T_{air} is decreased. In this case, as $(T_{\text{air}} - \psi \sigma_{\text{film}} t_{\text{film}}^{1/2}) \to 0$, the crack length diverges. The limiting case is for immersion of the thin film system in the fully reactive environment of water such that $T = T_{H_2O}$, $(T_{H_2O} - \psi \sigma_{\text{film}} t_{\text{film}}^{1/2}) < 0$, and unchecked nonequilibrium channel cracking occurs.

An example of nonequilibrium channel cracking in a tensile oxide film containing an indentation and exposed to water is shown in Figure 6.22. The film is a 900 nm thick organosilicate film containing a 5 N Vickers indentation (Cook and Liniger 1999). The film stress is $\sigma_{\text{film}} \approx 70$ MPa (tensile). Figure 6.22a shows the as-indented film in air. Figure 6.22b and c shows the indentation exposed to water for 3 and 10 s, respectively. The nonequilibrium crack extension is apparent (in this case, the *initial* indentation and cracks did penetrate the Si substrate—the effects on *final* crack propagation in the film are negligible). The crack velocity behavior of the channel cracks can be measured in a straightforward manner using this

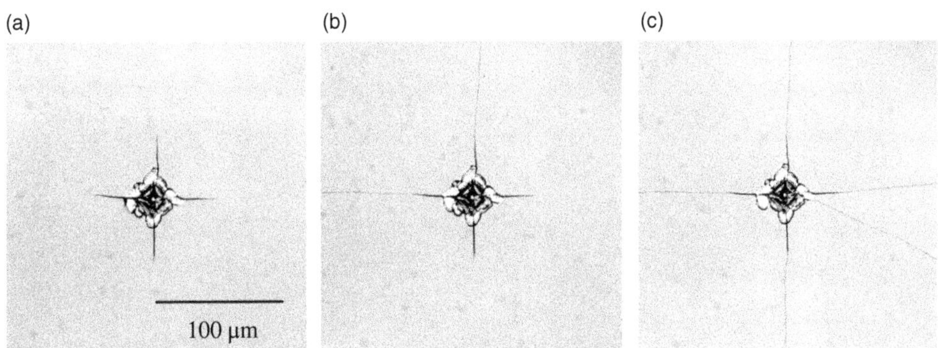

Figure 6.22 Optical images of nonequilibrium channel cracking in a tensile organosilicate thin film on silicon substrate. Cracks originate from an indentation flaw. (a) Initial indentation in air. (b) After 3 s exposure to water. (c) After 10 s exposure to water. Source: Robert F. Cook.

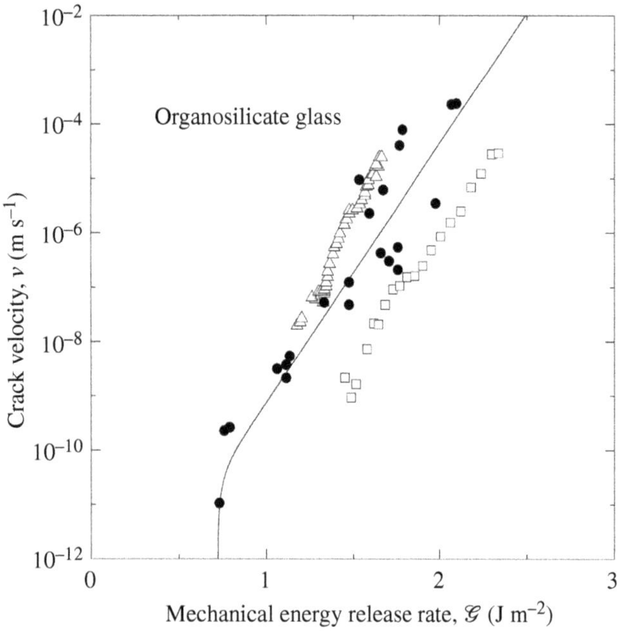

Figure 6.23 Semilogarithmic plot of crack velocity $v(\mathscr{G})$ for organosilicate films in water using data from observations such as Figure 6.22 (filled symbols) and interfacial split beam measurements (open symbols). Line represents fit using physically based crack velocity behavior. Source: Data adapted from Cook and Liniger (1999), Guyer and Dauskardt (2004), and Vlassak et al. (2005).

geometry as control over indentation location enables reproducible crack length measurements and the mechanical energy release rate $\mathscr{G} = \psi^2\sigma^2_{\text{film}}t_{\text{film}}$ is crack length independent (although σ_{film} and t_{film} must be measured). The $v(\mathscr{G})$ behavior of the organosilicate film is shown in the semi-logarithmic plot of Figure 6.23. The filled symbols indicate the results of experiments similar to those shown in Figure 6.22, using film thickness to vary $\mathscr{G}$. The line represents a visual fit to these data of Eq. (6.42). Note that the crack velocity threshold for FSG in water is $2\gamma \approx 1.5\ \text{J m}^{-2}$ (Cook 2019); organosilicate films are not very fracture resistant. The open symbols in Figure 6.23 represent subsequent crack velocity measurements on similar films obtained using a macroscopic split beam apparatus (Hughey et al. 2004; Guyer and Dauskardt 2004; Vlassak et al. 2005). The agreement with the exponential crack velocity trend is clear. It is noted that the crack velocity approaches 1 mm s^{-1} and thus cracks in a film covering a 300 mm diameter Si wafer in water could traverse the full wafer diameter in approximately 5 min if the film was too thick or the wafer was flexed. Multiple crack initiation points, including preexisting channel cracks, could lead to fragmentation of the entire wafer under such circumstances. An example is shown in Figure 6.24 of fragmentation of a 1 μm thick organosilicate film containing a 5 N indentation. Figure 6.24a shows the as-indented film in air. Figure 6.24b shows the indentation site after 16 h exposure to water. In the terms of this chapter, reactive fracture lead to static fatigue failure of the film from the indentation flaw.

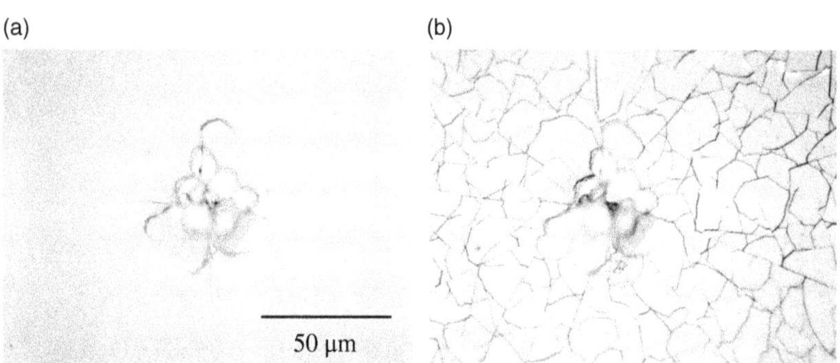

Figure 6.24 Optical images of nonequilibrium fragmentation in a tensile organosilicate 1 μm thin film on silicon substrate. Fragmentation originates from an indentation flaw. (a) Initial indentation in air. (b) After 16 h exposure to water. Source: Robert F. Cook.

6.3.5 Probabilistic Methods

Probabilistic methods have been used to characterize the fatigue characteristics of brittle components since the earliest studies. In these methods, the probability of failure is measured as a function of time for components exposed to similar loading conditions. For brittle components, the probabilistic element is assumed to arise from variability in flaw sizes in components on initial loading or from variability in flaw size development during loading. In either case, there is variability in failure time. Sections 6.3.1–6.3.4 in this chapter have considered deterministic methods, in which the median or mean failure time is estimated as a function of variations in applied loading, assuming that flaw sizes and development are invariant. Use of controlled indentation flaws reduces this assumption to near certainty. Some early works reversed the dependent and independent variables such that failure load was estimated as a function of variations in time of loading (Baker and Preston 1946a; Mould and Southwick 1959a,b), but the deterministic principle remained the same. Later in this section, a third option is considered in which loading time is fixed and failure load is estimated as a function of variations in initial flaw size.

One of the earliest works that used a probabilistic method to assess fatigue of brittle components did not measure failure time explicitly (Glathart and Preston 1946). Instead, the number of loading cycles applied prior to component failure was measured (somewhat akin to cyclic loading of metals, Figure 6.2a, without the stress reversals). The components were glass drinking tumblers and the cycles included filling, handling, and passing the tumblers through a commercial dishwasher. Two experiments were performed. One in which the tumblers were used in a restaurant and the number of cycles tracked and one in which restaurant use was simulated in a laboratory and the number of cycles tracked. Many hundreds of tumblers were used. Figure 6.25a shows a semilogarithmic plot of failure probability as a function of the number of cycles prior to failure in each test environment. The median number of cycles to failure (0.5 failure probability) is identified easily in the plot, about 10 cycles for each test, although it is clear that a restaurant cycle was approximately twice as damaging as a laboratory cycle, a trend that continued until all tumblers had failed (1.0 failure probability), about 30 cycles. The comparison of median and complete failure parameters values points to a near ubiquitous phenomenon in fatigue: failure times are heavily skewed. There are typically many early fails followed by a broad distribution of later fails, including components that do not fail in the duration of a test ("run outs"). Hence, the use of semi-logarithmic coordinates to display fatigue data.

A later work used a conventional static loading approach to assess the probabilistic fatigue properties of annealed glass rods (Ritter and Sherburne 1971). A custom dead weight loading apparatus was used and the failure probability measured for groups of rods exposed to different applied stress levels for specified periods of time in a controlled humidity chamber. Figure 6.25b shows a semilogarithmic plot of failure probability as a function of loading time for the groups of rods. The groups are indicated as symbols connected by lines and the applied stress level increases from bottom to top in the plot. The behaviors of the rods in the groups exposed to the greatest and least applied stresses are indicated by filled symbols. For all groups, the failure probability increases with increasing failure time. Rods in the group exposed to the least applied stress exhibited no fails at the shortest exposure time and rods in the group exposed to the greatest applied stress had all failed by the longest exposure time. Note that for most groups not all specimens failed in the specified exposure time; there were significant numbers of run outs. The failure times at the median failure probability 0.5, along a horizontal line in Figure 6.25b, were estimated for each applied stress level and a conventional static fatigue plot created. A scaled universal static fatigue response was thence developed for the glass rods and compared, favorably, with the earlier response (Mould and Southwick 1959b) and with dynamic fatigue behavior.

About the same time, a static loading approach was used to assess the probabilistic fatigue properties of disks of dental porcelain (Southan and Jørgensen 1973, 1974). A custom dead weight loading apparatus was used and the failure probability measured for groups of disks exposed to different applied stress levels for specified periods of time in water. Figure 6.25c shows a semilogarithmic plot of failure probability as a function of loading time for the groups of disks. The symbols and lines are as in Figure 6.25b. Although the measurements are not as comprehensive as those on glass, the results are clearly similar. Subsequent rapid fracture testing of surviving specimens showed no effects of the level of prior sustained loading, suggesting that crack initiation as well as crack propagation was an important part of two-step brittle static fatigue behavior, a conclusion also reached by Glathart and Preston (1946).

As part of a comparative study of crack propagation and brittle component lifetime an extensive series of conventional static fatigue tests were performed on rods of electrical porcelain (Soma et al. 1980). The rods were loaded in four-point bending to different applied stress levels and the failure time measured for exposure on loading to water, air, and kerosene. Figure 6.25d shows a semilogarithmic plot of failure probability as a function of failure time for the groups of rods. Different symbols indicate different groups; for ease of comparison, run outs in the kerosene group have been omitted. The applied

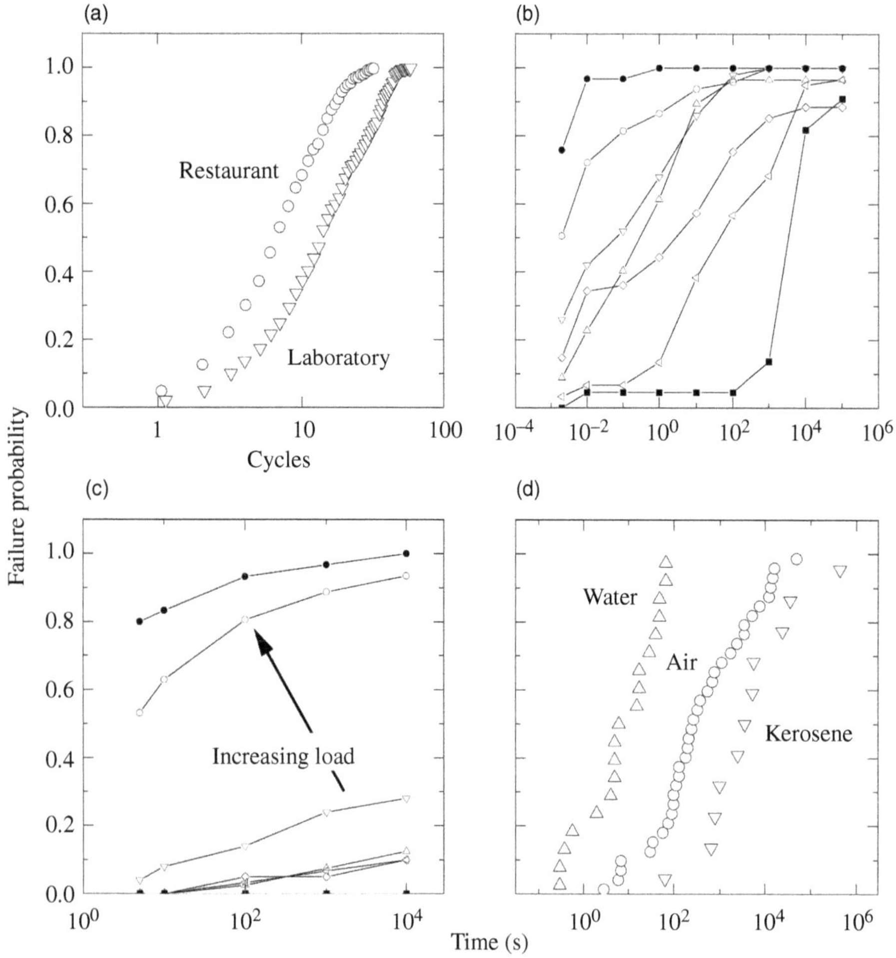

Figure 6.25 Semilogarithmic plots of failure probability as a function of time for loaded brittle components. (a) Glass drinking tumblers exposed to use and dishwashing cycles. (b) Glass rods exposed to static loads in controlled humidity. (c) Dental porcelain disks exposed to static loads in water. (d) Electrical porcelain rods exposed to static loads in the environments indicated. For plots bcd, loading increases top to bottom as indicated.

stress levels were kerosene 85 MPa, air 90 MPa, and water 95 MPa. Hence, the fracture enhancing effects of the fatigue environments (stress + water activity) increase from right to left. The failure times at fixed probability level reflect these effects and decrease in the same order. Predictions of failure probability as a function of static fatigue failure time were generated using an empirical power-law crack velocity function and shown to be in agreement with experimental observations, at least for short failure times. At long failure times, the crack velocity based predictions underestimated failure times (and the significant number of run outs). The underestimates were probably as a consequence of the threshold omitted from the assumed power-law crack velocity function.

The experimental variables in a conventional static fatigue test method are usually assigned as constrained (flaw size), independent (applied stress), and dependent (failure time). The probabilistic studies outlined here point to the possibility that other test methods and assignments might exchange the nature of the variables in an experiment and generate the same information. A few examples of limited changes in assignment have been noted in this chapter. In early works, failure time was treated as the independent variable and applied stress treated as the dependent variable and evaluated by experimental load refinement (Baker and Preston 1946a; Mould and Southwick 1959b). The later probabilistic works also treated failure time as the independent variable and applied stress as the dependent variable that was evaluated by analytical interpolation (Ritter and Sherburne 1971; Southan and Jørgensen 1974). More significant changes of variable assignments in static fatigue experiments have been demonstrated (Liniger and Cook 1993; Cook 2018b). The changes take advantage of the controlled nature of indentation flaws and assign failure time as the constrained variable, flaw size as the independent variable, and applied stress as the dependent variable. The estimation of applied stress is by a probabilistic

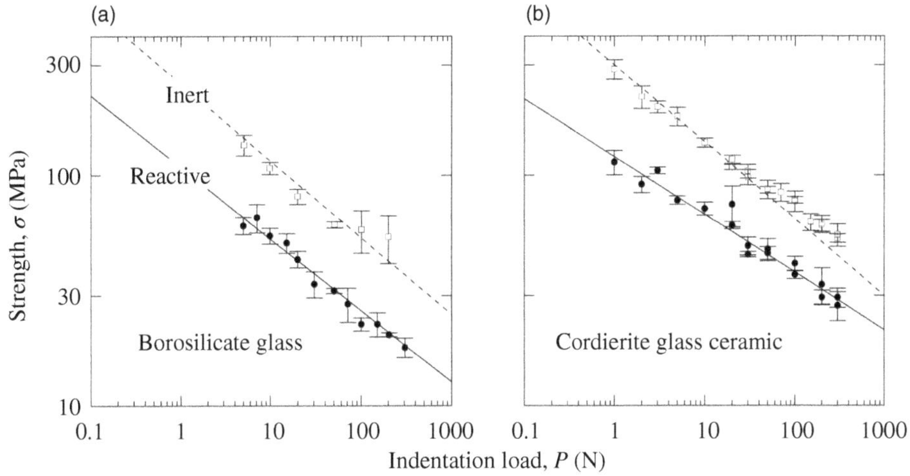

Figure 6.26 Logarithmic plots of inert (open symbols) and reactive (filled symbols) indentation strengths, illustrating the strength degrading effects of a reactive environment. (a) Borosilicate glass. Loaded in water for 10 min. (b) Cordierite glass-ceramic. Loaded in water for 1 h.

interpolation method. The output is the variation of a "reactive strength" as a function of indentation load, similar to the inert indentation strength degradation behavior studied extensively in Chapters 3 and 4. The experimental procedure is described briefly as follows: Controlled flaws of single indentations are introduced into test specimens over a domain of indentation loads, similar to inert strength measurements. A group of identically indented specimens is then loaded in a controlled environment for a specified length of time; each specimen at a selected separate applied stress. At the end of the specified time, the pass (run out) and fail state for each specimen is noted. The process is repeated with new groups of specimens using refined applied stress domains until the probability of failure as a function of applied stress is established. The median stress is an estimate of the reactive strength at the chosen indentation load.

Figure 6.26 shows logarithmic plots of the inert and reactive strengths of (a) BSG and (b) cordierite glass-ceramic, using data derived from the published works cited. The open symbols and bars represent the means and standard deviation limits of inert strengths measured at the indentation loads P indicated. The filled symbols represent the medians and 0.17 and 0.83 quantile bounds for reactive strengths at the indentation loads indicated for loading times in water of (a) 10 min and (b) 1 h. The dashed lines are guides to the eye of form $\sigma \sim P^{-1/3}$ consistent with ideal inert indentation strength. The solid lines are guides to the eye of form $\sigma \sim P^{-x}$, where $x \leq 1/3$. The diminished reactive strengths and diminished sensitivity of reactive strength to indentation load are consistent with nonequilibrium crack velocity effects during static loading in the reactive environment (Liniger and Cook 1993; Cook 2018b). The experimental advantages of such a testing procedure are that the duration of loading segments are known in advance and that every test counts as run outs are incorporated into the probabilistic estimate of reactive strength. The analytical advantages of the procedure are that the susceptibility of a material to strength degradation by fatigue is readily apparent by comparison with inert strengths and that the probability of failure can be specified in the plotting scheme. Note that the plotting scheme of Figure 6.26 also emphasizes the fact that small flaws are proportionately more susceptible to fatigue than large flaws,

6.4 Summary and Discussion

This chapter has provided an extensive survey and assessment of the indentation strength behavior of brittle materials in reactive environments. The brittle materials studied have been predominantly oxides, in the form of glasses, polycrystalline ceramics, and glass-ceramics. The reactive environments have been either gaseous in the form of moist air or liquid; in both cases, the reactant was H_2O. The indentation strength methodology was identical to that used in Chapters 3 and 4: emplacement of a controlled indentation flaw in the surface of a test component followed by strength testing of the component. The important difference here was that the rate or time of loading in the strength test was a key variable in the procedure. The majority of reactive indentation strength data presented here were the result of dynamic fatigue tests, in which strength is measured as a function of applied stressing rate. At fast applied stressing rates, the strengths approach an upper bound of the inert or Griffith strength that was the focus of Chapters 3 and 4. At slow applied stressing rates, the strengths approach a lower bound of the reactive or Orowan strength. The two bounds are separated at intermediate rates by fatigue strengths that are limited by the kinetics of crack growth and that increase with increasing applied stressing rate. Analyses were developed from fundamental physically based crack velocity behavior to describe the entire dynamic fatigue response. The analyses were extended to include the effects of indentation load and applied in description of a multitude of experimental measurements from a wide range of materials. Simple fatigue predictions were demonstrated for changes in microstructural scale, application of static loading, and the effects of surface and thin film stresses.

The major focus of the historically early experiments on fatigue of brittle materials was quantifying the kinetics of strength degradation—the decreases in failure stress or strength with increases in failure time in both static and dynamic fatigue. As the presence of water was demonstrated as a necessary factor in such fatigue, a localized chemical dissolution rate oriented view of the phenomenon developed. The subsequent development of fracture mechanics showed that chemical reaction with water was indeed necessary for fatigue but that the phenomenon was driven by a global mechanical energy gradient. It is now known that if this gradient is greater than the surface energy of a material in the reactive environment of interest, nonequilibrium fracture and fatigue result from thermally activated bond rupture. The materials science focus of fatigue is then on the details of nonequilibrium fracture kinetics. The materials engineering focus is to design and manufacture components that exhibit adequate reliability in service while incorporating materials susceptible to fatigue. There are several approaches to this problem. One approach is to assume that fatigue will occur and design the component such that the failure time is greater than the service life. A more conservative approach is to design the component such that fatigue does not occur. In practice, this latter approach requires the surface energy of the material in the service environment to always exceed the applied mechanical energy gradient. In fracture mechanics terms, $2\gamma > \mathscr{G}$. Reactive indentation strength tests enable this design approach through determination of the ratio $2\gamma/2\gamma_0$, where $2\gamma_0$ is the surface energy in an inert environment (and usually known).

Figure 6.27 applies the results of this chapter and a few other studies (Scholz 1972; Cook 2006) in a plot of the ranked ratio $2\gamma/2\gamma_0$ for a range of materials exposed to water-containing environments. The ratio was evaluated from dynamic fatigue and inert strength results using the fitted parameters σ_{min} and σ_{max} or S_{min} and S in Eqs. (6.49) and (6.50) to give $2\gamma/2\gamma_0 = (\sigma_{min}/\sigma_{max})^{3/2} = (S/S_{min})^{3/2}$. The acronyms used are: SLG, soda-lime glass; BSG, borosilicate glass; GC, glass-ceramic; PZT, lead zirconate titanate; and BT, barium titanate. Filled symbols represent surface energy ratios determined from the indentation dynamic fatigue results of Sections 6.3.1 and 6.3.2. Open symbols represent ratios determined from other dynamic and static fatigue tests, including from Section 6.1.2. Figure 6.27 has several applications. A first point to note is that for many brittle materials the ratio $2\gamma/2\gamma_0$ is considerably less than 1. Designers using brittle materials must therefore

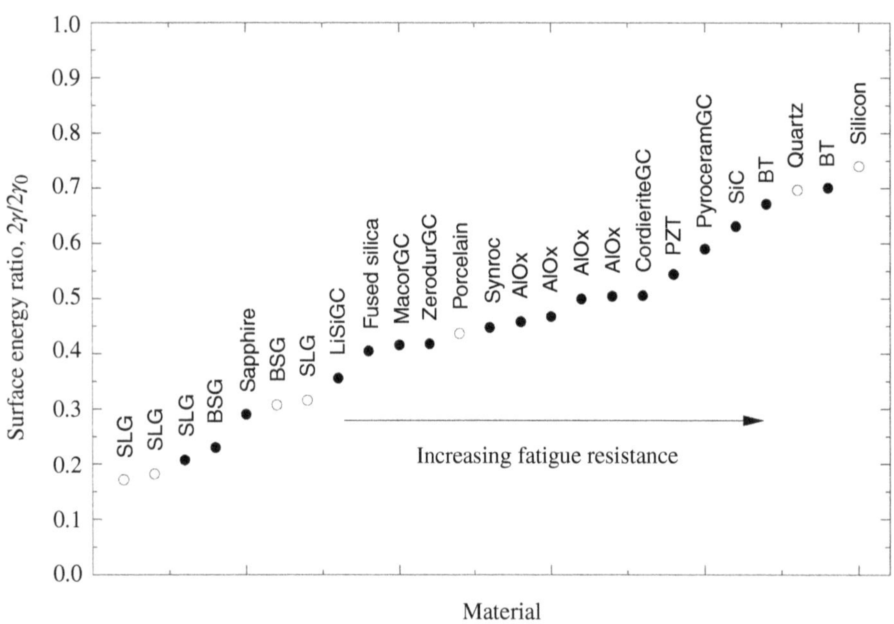

Figure 6.27 Plot of ranked surface energy ratio (water/inert) determined from fatigue tests on brittle materials. Materials with large ratios, e.g., Si, SiC, exhibit greater fatigue resistance than materials with small ratios, e.g., soda-lime silicate glass, sapphire.

be mindful that many of these materials will exhibit significant fatigue effects and adjust component dimensions (larger) and allowed stresses (smaller) accordingly. A second point is that the plot provides a benchmark for developers and manufacturers of brittle materials. The relative fatigue behavior of a new material or of batches of materials in manufacturing can be compared against the behavior of preexisting materials.

Further points regarding Figure 6.27 bear on the materials science of fatigue. The first of these is that it is apparent in the plot that there are groups of materials with similar surface energy ratios. In order of increasing resistance to fatigue, the groups are: the modified glasses SLG and BSG; the amorphous SiO_2-based materials, FSG, some glass-ceramics, and porcelain; the polycrystalline Al_2O_3 materials; the cordierite- and titanate-based materials; and the Si-based materials. There are some materials apparently "out of order" (sapphire, Synroc, and quartz) but the overall trend is clear. Materials with weak bond polarity, e.g., Si and SiC, exhibit greater fatigue resistance than materials with polar bonds, e.g., SLG and BSG. A second materials science point is that Figure 6.27 describes what is thermodynamically possible in fatigue—no information regarding the kinetics of fatigue is included. For example, Pyroceram and SiC are regarded as displaying reduced susceptibility to fatigue in the sense that the variation of dynamic fatigue strength with applied stressing rate is reduced relative to SLG. However, the perceived reduced susceptibility could arise simply from the decreased surface energy ratio, with no change in the kinetics of bond rupture (see Section 6.2.4 and Cook 2018a). A third materials science point is that Figure 6.27 was determined from strength ratios based on derivatives of strengths with respect to applied stressing rates. The same materials science points could be inferred from direct $v(\mathscr{G})$ crack velocity measurements, providing the derivative $\mathrm{d}\ln(v)/\mathrm{d}\mathscr{G}$ is taken as the indicator of the fatigue threshold (not the absolute value of v). For example, FSG can exhibit some very slow crack velocities (e.g., Muraoka and Abé 1996; Wiederhorn private communication 2006) but that does not mean it does not exhibit a crack velocity threshold (Cook 2019).

Finally, it is important to place the emphasis of this chapter, indentation strength in reactive environments, in the context of related work. If the environment is extremely reactive, corrosion and dissolution of a material can occur in the long periods of exposure associated with fatigue loading. Such corrosion and dissolution occurs preferentially at surface flaws, such that, from an indentation fracture mechanics perspective, the effect of a corrosive environment is to cause a time-dependent decrease in ψ and χ. These decreases lead to reduced SIF and can lead to strength *increases* in fatigue tests. The extension of indentation fracture from reactive to corrosive environments is considered elsewhere (Cook 2018c). Corrosive effects were noted in the earliest fatigue works (Mould 1960, 1961). If fracture specimens are extremely large, transport of reactive species *to* rupturing crack tip bonds can limit crack velocity rather than the reaction rate of reactive species *with* crack tip bonds. Such effects are observed in planar specimens (e.g., double cantilever beam) at fast crack velocities if the transverse (not crack length) dimension of the specimen is large and unreactive species are present in the test environment (Soma et al. 1980; Chantikul et al. 1983). The length of the transport path, the crack dimension, is usually short enough in indentation fracture tests in reactive environments that transport limitations are not noticeable (Chantikul et al. 1983). In indentation strength tests in which an enclosure of dry N_2 is formed about the specimen or drop of dry silicone oil is used, transport of moisture is impeded on length scales of the specimen or oil drop and inert conditions are obtained. The fatigue performance of components in service depends on the behavior of intrinsic flaws. Hence, valid indentation-based reliability predictions for such components depend on empirical verification of the behavior of unindented components (e.g., Figures 6.18 and 6.19). However, materials comparisons using unindented components are limited: usually the experimental flaw size scatter is too great and the nature of the flaws is unknown. This was the case in the early static fatigue experiments on porcelain (Baker and Preston 1946a;) and is the case for the more recent dynamic fatigue studies of dental ceramics (e.g., Morena et al. 1986; Fairhurst et al. 1993). Hence, although there are a large number of such fatigue studies, they are not included here as most use uncontrolled intrinsic flaws. Finally, the behavior of brittle materials under cyclic loading, Figure 6.2a, has been studied since the early works on fatigue (Gurney and Pearson 1948), although the studies are not numerous. The reason for this lies, at least partially, in the observation that for many brittle materials cyclic fatigue behavior was not "special." That is the effects of cyclic fatigue could be reduced to a static fatigue response by integrating the cyclic loading of Figure 6.2a to obtain the equivalent static loading, Figure 6.2c (Gurney and Pearson 1948; Evans and Fuller 1974). However, more recent studies have shown that this is not true for materials that exhibit R curves or T curves on monotonic loading. The cyclic fatigue behavior of such materials is significantly affected by the degradation of toughening mechanisms on cyclic loading: reverse plasticity impedes phase transformations (Dauskardt et al. 1987, 1990) and reverse displacement destroys frictional interlocks (Lathabai et al. 1989b, 1991). Hence, such behavior is not considered here as it is not strictly a phenomenon that depends on the reactive environment. Similarly, it is not apparent that the small cyclic fatigue effect observed in crack velocity measurements of BSG (Dill et al. 1997) is strictly a reactive influence.

References

Baker, T.C. and Preston, F.W. (1946a). Fatigue of glass under static loads. *Journal of Applied Physics* 17: 170–178.

Baker, T.C. and Preston, F.W. (1946b). The effect of water on the strength of glass. *Journal of Applied Physics* 17: 179–188.

Brown, S.D. (1979). Multibarrier kinetics of brittle fracture: I, stress dependence of the subcritical crack velocity. *Journal of the American Ceramic Society* 62: 515–524.

Chandan, H.C., Bradt, R.C., and Rindone, G.E. (1978). Dynamic fatigue of float glass. *Journal of the American Ceramic Society* 61: 207–210.

Chantikul, P., Lawn, B.R., and Marshall, D.B. (1981). Micromechanics of flaw growth in static fatigue: influence of residual contact stresses. *Journal of the American Ceramic Society* 64: 322–325.

Chantikul, P., Lawn, B.R., Richter, H., and Freiman, S.W. (1983). Relation between multiregion crack growth and dynamic fatigue of glass using indentation flaws. *Journal of the American Ceramic Society* 66: 515–518.

Charles, R.J. (1958a). Static fatigue of glass. I. *Journal of Applied Physics* 29: 1549–1553.

Charles, R.J. (1958b). Static fatigue of glass. II. *Journal of Applied Physics*, 29: 1554–1560.

Charles, R.J. (1958c). Dynamic fatigue of glass. *Journal of Applied Physics* 29: 1657–1662.

Charles, R.J. and Hillig, W.B. (1962). The kinetics of glass failure by stress corrosion. In *Proceedings of the Symposium sur la Resistance Mecanique du Verre et les Moyens de L'Ameliorer*. pp 511–527. Union Scientifique Continentale du Verre.

Chevalier, J. and Fantozzi, G. (2005). Slow crack propagation in ceramics at the nano- and micro-scale: effect of the microstructure. In *Fracture Mechanics of Ceramics* (Eds. Bradt, R.C., Munz, D., Sakai, M., White, K.W.) pp. 173–190. Springer.

Cook, R.F. (1986). Crack propagation thresholds: a measure of surface energy. *Journal of Materials Research* 1: 852–860.

Cook, R.F. (2005). Toughening of a cordierite glass-ceramic by compressive surface layers. *Journal of the American Ceramic Society* 88: 2798–2808.

Cook, R.F. (2006). Strength and sharp contact fracture of silicon. *Journal of Materials Science* 41: 841–872.

Cook, R.F. (2015). Multi-scale effects in the strength of ceramics. *Journal of the American Ceramic Society* 98: 2933–2947.

Cook, R.F. (2018a). Long-term ceramic reliability analysis including the crack-velocity threshold and the "bathtub" curve. *Journal of the American Ceramic Society* 101: 5732–5744.

Cook, R.F. (2018b). A simple method of short-term mechanical reliability prediction for ceramics in reactive environments. *Journal of the American Ceramic Society* 101: 2727–2731.

Cook, R.F. (2018c). Strength of brittle materials in moderately corrosive environments. *Journal of the American Ceramic Society* 101: 1684–1695.

Cook, R.F. (2019). Thermal activation effects in crack propagation and reliability of fused silica. *Journal of the American Ceramic Society* 102: 7575–7583.

Cook, R.F., Freiman, S.W., and Baker, T.L. (1986). Effect of microstructure on reliability predictions for glass ceramics. *Materials Science and Engineering* 77: 199–212.

Cook, R.F. and Lawn, B.R. (1984). Controlled indentation flaws for the construction of toughness and fatigue master maps. *Journal of Research of the National Bureau of Standards* 89: 453–465.

Cook, R.F., Lawn, B.R., and Anstis, G.R. (1982a). Fatigue analysis of brittle materials using indentation flaws: Part 2 Case study on a glass-ceramic. *Journal of Materials Science* 17: 1108–1116.

Cook, R.F., Lawn, B.R., Dabbs, T.P., Reeve, K.D., Ramm, E.J., and Woolfrey, J.L. (1982b). A comparative study of the fatigue properties of Synroc and borosilicate glass. *Journal of the American Ceramic Society* 65: C-172–C-173.

Cook, R.F., Lawn, B.R., Dabbs, T.P., and Chantikul, P. (1981). Effect of machining damage on the strength of a glass-ceramic. *Journal of the American Ceramic Society* 64: C-121–C-122.

Cook, R.F., Lawn, B.R., and Fairbanks, C.J. (1985a). Microstructure-strength properties in ceramics: I, effect of crack size on toughness. *Journal of the American Ceramic Society* 68: 604–615.

Cook, R.F., Lawn, B.R., and Fairbanks, C.J. (1985b). Microstructure-strength properties in ceramics: II, fatigue relations. *Journal of the American Ceramic Society* 68: 616–623.

Cook, R.F. and Liniger, E.G. (1993). Kinetics of indentation cracking in glass. *Journal of the American Ceramic Society* 76: 1096–1105.

Cook, R.F. and Liniger, E.G. (1999). Stress-corrosion cracking of low-dielectric-constant spin-on-glass thin films. *Journal of the Electrochemical Society* 146: 4439–4448.

Dabbs, T.P. and Lawn, B.R. (1985). Strength and fatigue properties of optical glass fibers containing microindentation flaws. *Journal of the American Ceramic Society* 68: 563–569.

Dabbs, T.P., Lawn, B.R., and Kelly, P.L. (1982). Dynamic fatigue study of soda-lime silicate and borosilicate glasses using small scale indentation flaws. *Physics and Chemistry of Glasses* 23: 58–66.

Dauskardt, R.H., Marshall, D.B., and Ritchie, R.O. (1990). Cyclic fatigue-crack propagation in magnesia-partially-stabilized zirconia ceramics. *Journal of the American Ceramic Society* 73: 893–903.

Dauskardt, R.H., Yu, W., and Ritchie, R.O., 1987. Fatigue crack propagation in transformation-toughened zirconia ceramic. *Journal of the American Ceramic Society*, 70 (10): C-248–C-252.

Dieter, G.E. (1986). *Mechanical Metallurgy, 3rd ed.* McGraw-Hill.

Dill, S.J., Bennison, S.J., and Dauskardt, R.H. (1997). Subcritical crack-growth behavior of borosilicate glass under cyclic loads: evidence of a mechanical fatigue effect. *Journal of the American Ceramic Society* 80: 773–776.

Evans, A.G. and Fuller, E.R. (1974). Crack propagation in ceramic materials under cyclic loading conditions. *Metallurgical and Materials Transactions B* 5: 27–33.

Fairhurst, C.W., Lockwood, P.E., Ringle, R.D., and Twiggs, S.W. (1993). Dynamic fatigue of feldspathic porcelain. *Dental Materials* 9: 269–273.

Freiman, S.W., Wiederhorn, S.M., and Mecholsky, J.J. Jr. (2009). Environmentally enhanced fracture of glass: a historical perspective. *Journal of the American Ceramic Society* 92: 1371–1382.

Fuller, E.R. Jr., Lawn, B.R., and Cook, R.F. (1983). Theory of fatigue for brittle flaws originating from residual stress concentrations. *Journal of the American Ceramic Society* 66: 314–321.

Gibbs, P. and Cutler, I.B. (1951). On the fracture of glass which is subjected to slowly increasing stress. *Journal of the American Ceramic Society* 34: 200–206.

Glathart, J.L. and Preston, F.W. (1946). The fatigue modulus of glass. *Journal of Applied Physics* 17: 189–195.

Glasstone, G., Laidler, K.J., and Eyring, H. (1941). *The Theory of Rate Processes*. McGraw-Hill.

Gonzalez, A.C., Multhopp, H., Cook, R.F., Lawn, B.R., and Freiman, S.W. (1984). Fatigue properties of ceramics with natural and controlled flaws: a study on alumina. In *Methods for Assessing the Structural Reliability of Brittle Materials*, ASTM STP 844 pp. 43–56. ASTM.

Gurney, C. (1947). Delayed fracture in glass. *Proceedings of the Physical Society* 59: 169–185.

Gurney, C. and Pearson, S. (1948). Fatigue of mineral glass under static and cyclic loading. *Proceedings of the Royal Society of London, Series A: Mathematical and Physical Sciences* 192: 537–544.

Gurney, C. and Pearson, S. (1949). The effect of the surrounding atmosphere on the delayed fracture of glass. *Proceedings of the Physical Society, Section B* 62: 469–476.

Guyer, E.P. and Dauskardt, R.H. (2004). Fracture of nanoporous thin-film glasses. *Nature Materials* 3: 53–57.

Hertzberg, R.W. (1989). *Deformation and Fracture Mechanics of Engineering Materials, 3rd ed.* Wiley.

Heslin, T.M., Magida, M.B., and Forrest, K.A. (1986). A method for developing design diagrams for ceramic and glass materials using fatigue data. NASA report No. NASA-RP-1174. NASA.

Hughey, M.P., Morris, D.J., Cook, R.F., Bozeman, S.P., Kelly, B.L., Chakravarty, S.L., Harkens, D.P., and Stearns, L.C. (2004). Four-point bend adhesion measurements of copper and permalloy systems. *Engineering Fracture Mechanics* 71: 245–261.

Khonsari, M.M. and Amiri, A. (2013). *Introduction to Thermodynamics of Mechanical Fatigue*. CRC.

Krausz, A.S. and Krausz, K. (1988). *Fracture Kinetics of Crack Growth*. Kluwer.

Lathabai, S. and Lawn, B.R. (1989a). Fatigue limits in noncyclic loading of ceramics with crack-resistance curves. *Journal of Materials Science* 24: 4298–4306.

Lathabai, S., Mai, Y.-W., and Lawn, B.R. (1989b). Cyclic fatigue behavior of an alumina ceramic with crack-resistance characteristics. *Journal of the American Ceramic Society* 72: 1760–1763.

Lathabai, S., R odel, J., and Lawn, B.R. (1991). Cyclic fatigue from frictional degradation at bridging grains in alumina. *Journal of the American Ceramic Society* 74: 1340–1348.

Lawn, B.R. (1975). An atomistic model of kinetic crack growth in brittle solids. *Journal of Materials Science* 10: 469–480.

Lawn, B.R. (1993). *Fracture of Brittle Solids, 2nd ed.* Cambridge.

Lawn, B.R., Marshall, D.B., Anstis, G.R., and Dabbs, T.P. (1981). Fatigue analysis of brittle materials using indentation flaws: Part 1 General theory. *Journal of Materials Science* 16: 2846–2854.

Liniger, E.G. and Cook, R.F. (1993). A controlled flaw technique for lifetime characterization. *Journal of the American Ceramic Society* 76: 2123–2126.

Marshall, D.B. and Lawn, B.R. (1980). Flaw characteristics in dynamic fatigue: the influence of residual contact stresses. *Journal of the American Ceramic Society* 63: 532–536.

Morena, R., Beaudreau, G.M., Lockwood, P.E., Evans, A.L., and Fairhurst, C.W. (1986). Fatigue of dental ceramics in a simulated oral environment. *Journal of Dental Research* 65: 993–997.

Mould, R.E. (1960). Strength and static fatigue of abraded glass under controlled ambient conditions: III, aging of fresh abrasions. *Journal of the American Ceramic Society* 43: 160–167.

Mould, R.E. (1961). Strength and static fatigue of abraded glass under controlled ambient conditions: IV, effect of surrounding medium. *Journal of the American Ceramic Society* 44: 481–491.

Mould, R.E. and Southwick, R.D. (1959a). Strength and static fatigue of abraded glass under controlled ambient conditions: I, general concepts and apparatus. *Journal of the American Ceramic Society* 42: 542–547.

Mould, R.E. and Southwick, R.D. (1959b). Strength and static fatigue of abraded glass under controlled ambient conditions: II, effect of various abrasions and the universal fatigue curve. *Journal of the American Ceramic Society* 42: 582–592.

Multhopp, H., Cook, R.F., and Lawn, B.R. (1983). Universal fatigue curves for ceramics using indentation flaws. *Journal of Materials Science Letters* 2: 683–684.

Muraoka, M. and Abé, H. (1996). Subcritical crack growth in silica optical fibers in wide range of crack velocities. *Journal of the American Ceramic Society* 79: 51–57.

Orowan, E. (1944). The fatigue of glass under stress. *Nature* 154: 341–343.

Pearson, S. (1956). Delayed fracture of sintered alumina. *Proceedings of the Physical Society. Section B* 69: 1293–1296.

Preston, F.W. (1935). The time factor in the testing of glassware. *Journal of the American Ceramic Society* 18: 220–224.

Preston, F.W. (1942). The mechanical properties of glass. *Journal of Applied Physics* 13:623–634.

Ritter, J.E. Jr. (1969). Dynamic fatigue of soda-lime-silica glass. *Journal of Applied Physics* 40: 340–344.

Ritter, J.E. Jr. and Cavanagh, M.S. (1976). Fatigue resistance of a lithium aluminosilicate glass-ceramic. *Journal of the American Ceramic Society* 59: 57–59.

Ritter, J.E. Jr. and LaPorte, R.P. (1975). Effect of test environment on stress-corrosion susceptibility of glass. *Journal of the American Ceramic Society* 58: 265–267.

Ritter, J.E. Jr. and Meisel, J.A. (1976). Strength and failure predictions for glass and ceramics. *Journal of the American Ceramic Society* 59: 478–481.

Ritter, J.E. Jr. and Sherburne, C.L. (1971). Dynamic and static fatigue of silicate glasses. *Journal of the American Ceramic Society* 54: 601–605.

Scholz, C.H. (1972). Static fatigue of quartz. *Journal of Geophysical Research* 77: 2104–2114.

Shand, E.B. (1954). Experimental study of fracture of glass: II, experimental data. *Journal of the American Ceramic Society* 37: 559–572.

Shand, E.B. (1961a). Fracture velocity and fracture energy of glass in the fatigue range. *Journal of the American Ceramic Society* 44: 21–26.

Shand, E.B. (1961b). Correlation of strength of glass with fracture flaws of measured size. *Journal of the American Ceramic Society* 44: 451–455.

Smyth, K.K. and Magida, M.B. (1983). Dynamic fatigue of a machinable glass-ceramic. *Journal of the American Ceramic Society* 66: 500–505.

Soma, T., Matsui, M., Oda, I., and Yamamoto, N. (1980). Applicability of crack propagation data to failure prediction in porcelain. *Journal of the American Ceramic Society* 63: 166–169.

Southan, D.E. and Jørgensen, K.D. (1973). An appropriate mechanical test for dental porcelain. *Australian Dental Journal* 18: 246–250.

Southan, D.E. and Jørgensen, K.D. (1974). The endurance limit of dental porcelain. *Australian Dental Journal* 19: 7–11.

Stuart, D.A. and Anderson, O.L. (1953). Dependence of ultimate strength of glass under constant load on temperature, ambient atmosphere, and time. *Journal of the American Ceramic Society* 36: 416–424.

Suresh, S. (1998). *Fatigue of Materials, 2nd ed.* Cambridge.

Thurn, J. and Cook, R.F. (2004a). Mechanical and thermal properties of physical vapour deposited alumina films: Part I. Thermal stability. *Journal of Materials Science* 39: 4799–4807.

Thurn, J. and Cook, R.F. (2004b). Mechanical and thermal properties of physical vapour deposited alumina films: Part II, elastic, plastic, fracture, and adhesive behaviour. *Journal of Materials Science* 39: 4809–4819.

Tummala, R.R. and Foster, B.J. (1975). Strength and dynamic fatigue of float glass surfaces. *Journal of the American Ceramic Society* 58: 156–157.

Viens, M.J. (1990). *Fracture Toughness and Crack Growth of Zerodur*, Vol. 4185. pp. 1–30. National Aeronautics and Space Administration, Office of Management, Scientific and Technical Information Division.

Vlassak, J.J., Lin, Y., and Tsui, T.Y. (2005). Fracture of organosilicate glass thin films: environmental effects. *Materials Science and Engineering A* 391: 159–174.

Vonnegut, B. and Glathart, J.L. (1946). The effect of temperature on the strength and fatigue of glass rods. *Journal of Applied Physics* 17: 1082–1085.

Wan, K.T., Lathabai, S., and Lawn, B.R. (1990). Crack velocity functions and thresholds in brittle solids. *Journal of the European Ceramic Society* 6: 259–268.

Wiederhorn, S.M. (1972a). A chemical interpretation of static fatigue. *Journal of the American Ceramic Society* 55: 81–85.

Wiederhorn, S.M. (1972b). Environmental stress corrosion cracking of glass. In *Corrosion Fatigue*. pp. 731–742. NACE.

Wiederhorn, S.M. (1975). Crack growth as an interpretation of static fatigue. *Journal of Non-Crystalline Solids* 19: 169–181.

Wiederhorn, S. and Bolz, L.H. (1970). Stress corrosion and static fatigue of glass. *Journal of the American Ceramic Society* 53: 543–548.

7

Dynamic Fragmentation Patterns in Flexed Ceramic Disks

In this chapter fragmentation arising from dynamic fracture of flexed ceramic disks is considered at three different length scales—the number of fragments generated, the mass distribution of the fragments, and the branching distance of propagating cracks. Indentation-strength techniques are used to control the strain energy at failure in polycrystalline ceramic specimens, thereby enabling clear measurements of fragmentation phenomena at all three scales. Data from these tests are combined with data determined from an extensive survey of published works, encompassing over forty glasses and ceramics, leading to quantitative comparisons of the parameters used to characterize fragmentation patterns. Assessments include suppression of the number of fragments generated on failure of ceramics relative to glasses, the substantial decrease in width of fragment mass distributions generated by disk flexure relative to explosively generated fragments, and the correlation of crack branching resistance with material fracture resistance. Innovative data presentation schemes are introduced to aid in these assessments. Examples of practical application of the data are included and the fragmentation patterns are compared with those generated by desiccation of saturated powders ("mud flat" cracking) and thermal expansion mismatch.

7.1 Introduction

Fragmentation patterns often develop during brittle failure when crack velocities are extremely large and fracture is rapid. An example of a fragmentation pattern that developed in a glass-ceramic disk fractured in biaxial flexure is shown in Figure 7.1a. The pattern consists of two cracks that propagated in opposite directions from a strength-controlling flaw at the disk center before a series of sequential crack branching events formed a radiating fragmentation pattern composed of fragments with angular points. The entire pattern developed in less than 100 μs with crack velocities greater than 1 km s^{-1}. Such crack velocities are comparable to the speed of bulk acoustic waves, leading to dynamic fracture conditions in which the kinetic energy of wave motion in the material is comparable to the mechanical and surface energies that control quasi-static fracture (Chapter 1; see also Lawn 1993). The nature of fragmentation patterns formed by biaxial flexure of ceramic disks such as in Figure 7.1a is the focus of this chapter.

Fragmentation of biaxially loaded disks into predominantly radial patterns was noted in the earliest brittle fracture studies. In considerations of "the propagation of fissures in glass ...," Preston hypothesized that cracks propagate in brittle materials so as to maintain directions at right angles to the principal tension (Preston 1931) and that the direction of this tension may change as the crack advances. In particular, if an instability splits the "wave-front" (forming what is now termed "hackle") then the crack will fork or form "radiants" as the tension experienced by each crack front segment will be in a different direction. Preston noted that the tendency to forking increased with increasing glass strength (Preston 1935). Schematic diagrams of fragmentation patterns formed in glass under uniaxial or equibiaxial tension, identical to that shown in Figure 7.1a, were presented (Preston 1931, 1935). Preston also noted that the crack branching phenomenon produced "dagger-shaped fragments ... that makes glass dangerous" (Preston 1935). Clearly, the number, size, shape, and kinetic energy of fragments formed when a brittle material fractures determine the capacity for injury, and hence the original focus of the manufacturers of glass products for domestic use, and more recently for architectural and automotive applications, in minimizing the capacity for fragmentation or to contain the fragments once formed (Bennison et al. 1999).

Images of fragmentation patterns similar to Figure 7.1a and as described by Preston have been reported for biaxially stressed disks of many materials, but principally glass (Gorham and Rickerby 1975; Matthewson and Field 1980; Shetty et al. 1983a; Nakasa and Nakatsuka 1990; Quinn 1999; Quinn et al. 2005; Salem 2013; DeLellis et al. 2020). In addition, Terao (1978) describes fragmentation patterns in glass similar to Figure 7.1a, generated by localized heating and consequent

Indentation Fracture: Strength and Toughness for Brittle Materials Design, First Edition. Robert F. Cook.

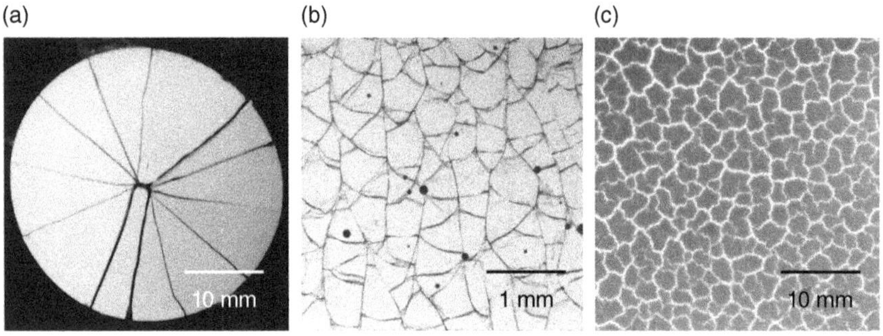

Figure 7.1 Optical images of three fragmentation patterns. (a) Glass-ceramic disk fractured in biaxial flexure—the topic of this work; (b) region of polymeric film fractured by thermal expansion mismatch with substrate; (c) region of wet starch film fractured by desiccation.

biaxial stress fields, from the extensive and very early, 1930s, work of Hirata. Images of fragmentation patterns similar to Figure 7.1a formed in biaxially stressed disks of other materials include: ion-exchanged glass (Donald and Hill 1988; Hill and Donald 1989; Donald et al. 1993; Kooi et al. 2008; Tandon and Glass 2015); polycrystalline alumina (Al_2O_3, Shetty et al. 1983b; Nakasa and Nakatsuka 1990; Borger et al. 2002); porous alumina (Sapozhnikov et al. 2015); polycrystalline MgF_2 (Mecholsky et al. 1998, 2020); polycrystalline Si_3N_4 (Quinn and Wirth 1989, 1990); polycrystalline ZrO_2 (Nakasa and Nakatsuka 1990, 1991); acryl (Nakasa and Nakatsuka 1991); single-crystal and polycrystalline Si (Cook 2006); and polycrystalline spinel ($MgAl_2O_4$) (Tokariev et al. 2012; Salem 2013). Images of fracture in biaxially loaded square plates of glass-ceramic and alumina (Shetty et al. 1981; Choi et al. 2000) are similar to Figure 7.1a. Laser-induced cracks in single-crystal Si disks (Murr and Szilva 1975) are similar to those formed by localized heating studied by Hirata.

Fragmentation patterns generated by uniaxial loading, also similar to those described by Preston, have been reported many times for glass rods and bars (Terao 1953; Bowden et al. 1967; Congleton and Petch 1967; Blauel and Richter 1969; Field 1971; Sundaram and Tippur 2018; Nakamura et al. 2020) and appear as one half of the pattern shown in Figure 7.1a. Mixed-mode loading, uniaxial + shear, such as occurs in pressurized glass bottles or tubes (Aoki and Sakata 1980; Bradt 2011; Quinn 2020), or shear loading of glass lathes (Quinn 2020) have also been observed to give rise to patterns similar to those of Figure 7.1a. Figure 7.2 shows an enlarged view of the central region of Figure 7.1a at the length scale of primary crack branching. The strength-controlling flaw is located between the two initial cracks and subsequent crack bifurcation and trifurcation events leading to specimen-level fragmentation. Localized images of biaxially-loaded specimens, similar to Figure 7.2, showing cracking at the length scale of initial branching have also been reported for Si_3N_4 and SiC (Choi and Gyekenyesi 1998) and glass (Nakasa and Nakatsuka 1991; Tang et al. 2014; Sundaram and Tippur 2018). Pervasive crack branching, Figure 7.2, and consequent fragmentation pattern formation, Figure 7.1a, in brittle material fracture has become so well accepted that the phenomena are sometimes studied quantitatively with no accompanying image (Kirchner 1978; Deobald and Kobayashi 1992; Oakley 1996).

In examining images such as Figures 7.1a or 7.2, care must be taken to recognize that fragmentation of brittle materials also occurs under the localized loading of sharp contacts or indentations or impacts. If the locally contacted or impacted material is in the form of a disk or plate, the resulting fragmentation patterns can often appear superficially similar to those generated by global biaxial loading. In fragmentation on both local and global loading, there are predominant radiating "primary" cracks, as observed, for example, in impact experiments on Al_2O_3 (Jones et al. 2017) and porcelain (Fragassa et al. 2018). However, contact or impact generated fragmentation patterns differ from biaxially generated patterns in that there are

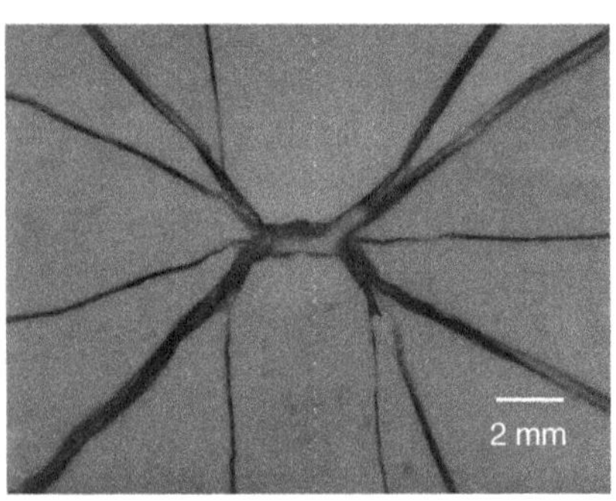

Figure 7.2 Enlarged view of failure location of glass-ceramic disk from Figure 7.1a, showing initial flaw location and crack branching events.

many closely separated primary cracks that rarely branch and many secondary circumferential "secondary" cracks that form between the primary cracks. These observations are particularly common in tempered or exchanged glasses that contain central zones of tension (Barstow and Edgerton 1939; Hillig and Charles 1961; Zijlstra and Burggraaf 1969; Sakai et al. 1991; Glass et al. 2007; Kooi et al. 2008; Lee et al. 2012; Davydova and Uvarov 2013; Tandon and Glass 2015; Karlsson 2017), but are also observed in SiC (Zinszner et al. 2012, 2015; Forquin et al. 2018) and other materials, as noted in a review of impact by Subhash et al. (2008).

Fracture systems in other materials also form fragmentation patterns, as shown in Figure 7.1b and c. Thin films with elastic, plastic, or thermal expansion properties different from a substrate often exhibit fragmentation patterns if the substrate is mechanically loaded or the temperature of the system is altered. The patterns that develop in these cases frequently consist of arrays of long, parallel, primary cracks with shorter, secondary cracks spanning the primary cracks, so as to form "ladders." In many cases, the patterns resemble cracks generated in ion-exchanged glasses far from an impact or contact site. The pattern in Figure 7.1b exhibits these features and was formed by cooling a polymeric film between two quartz plates. The pattern in Figure 6.24b is similar and was formed in a condensed organosilicate film on a silicon substrate. Other systems exhibiting similar patterns include metallic or oxide films on stretched polymeric substrates (Frank et al. 2009; Wu et al. 2015; Ruoho et al. 2020). Films formed of solid particles held together by a liquid often exhibit fragmentation patterns when the liquid evaporates. Such patterns are formed by desiccation cracks that often generate planar arrays of cells with convoluted edges, forming "mud-flat" patterns (Goehring et al. 2015). The pattern in Figure 7.1c exhibits these features and was formed by drying a water saturated starch film on a glass substrate. Often such patterns form within larger fragments with straight-sides in both drying and stretching geometries (Gao et al. 1998; Qin et al. 2014; Seghir and Arscott 2015). It is clear that the desiccation cracks of Figure 7.1c do not resemble those of Figure 7.1a, as noted earlier in experiments on alumina and agar powders (Nakasa and Nakatsuka 1994).

This chapter is divided into a further three major sections, each focusing on an experimental aspect of fragmentation of brittle materials, largely ceramics, in biaxial flexure. The sections proceed in order of decreasing length scale: First, the scale of the specimen (number of fragments), second, the scale of the fragment (mass distributions), and third, the scale of the crack (branching distances). Although each of these subjects has been considered before in various contexts, they have not been linked or simultaneously considered in a single study. Here, original experimental results are presented in each of the three sections and combined with previously published results from other works where appropriate. Fragmentation patterns generated by impacts or contacts, thin film constraints, or desiccation are not considered.

7.2 Number of Fragments

Results from a variety of polycrystalline ceramic materials were generated for this investigation and combined with results on amorphous (glassy), single-crystalline, and polycrystalline materials published by other researchers and the current author. The polycrystalline materials examined here were AlN, Ca-doped Al_2O_3, Cr-doped Al_2O_3, cordierite ($2MgO–2Al_2O_3–5SiO_2$) glass-ceramic (GC), MgO, and Y_2O_3. All samples except AlN were used in previously published ceramic strength studies (Cook 1990, 2005, 2015; Cook et al. 1994a,b) and considered in Chapters 3–6. A range of grain-sizes was included for the Ca–Al_2O_3 and MgO materials (Cook 1990; Cook et al. 1994b), cubic phase and larger grain-sized mixed cubic + hexagonal phase materials were included for the Y_2O_3 materials (Cook et al. 1994b), and base and compressive surface stressed materials were included for the glass-ceramic (Cook 2005). The grain-size d, Young's modulus E, and fracture resistance R of the materials are given in Table 7.1 along with the dimensions, diameter × thickness, of the disk specimens used for strength testing. The AlN and Al_2O_3 materials were tested in their as-sintered form, the MgO was polished to the 6 μm level and then annealed, and the glass-ceramic and Y_2O_3 materials were polished to the 1 μm level. Full specimen details are given in the earlier works (Cook 1990, 2005, 2015; Cook et al. 1994a,b).

The testing consisted of a two-step procedure, here briefly outlined. First, the disk specimens were indented with a Vickers diamond pyramid in the center of the prospective tensile test face, using controlled peak contact loads, P, in the domain 0.2–300 N. The indentation contact generated a square plastic deformation impression and stable cracks emanating from the impression corners. The contact load determined the indentation crack size and thus controlled the dominant flaw size in a disk. (Details of indentation cracking are discussed in Chapter 2.) Figure 7.3 shows indentation flaws formed at two different indentation loads in the glass-ceramic material and demonstrates how the flaw size was controlled. As the flaw size determines the maximum sustainable stress or strength of the specimen, the strength and hence stored elastic strain energy in the specimen at failure, U_E, are easily controlled through the indentation load. Second, the specimens were

Table 7.1 Ceramic properties, test specimen, and loading rig dimensions.

Material	Grain Size *d* (μm)	Young's Modulus *E* (GPa)	Fracture Resistance *R* (J m^{-2})	Specimen Dimensions (mm × mm)	Rig Dimensions (mm × mm)
Cordierite GC	3	128	18	32 × 3.3	28 × 4
Y_2O_3	48	180	8	20 × 2	18 × 1.5
MgO	10–26	305	8	32 × 3.5	28 × 4
AlN	10	340	12	25 × 2	20 × 3.5
Ca–Al_2O_3	15–60	400	20–100	25 × 1.1	20 × 2
Cr–Al_2O_3	15–25	400	20–100	27 × 2	20 × 2

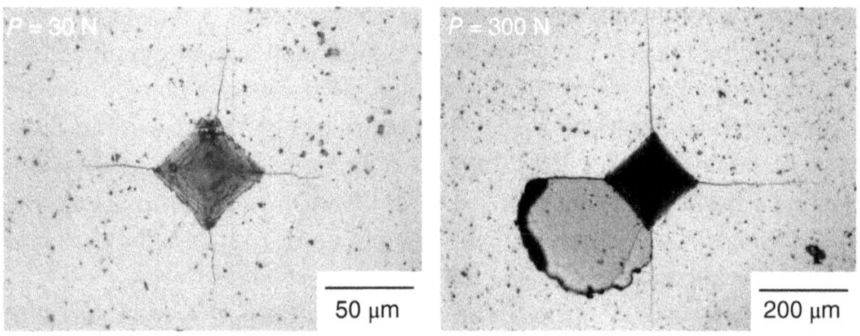

Figure 7.3 Optical micrographs of Vickers indentation flaws in cordierite glass-ceramic at the indentation loads *P* indicated. The residual contact impression and crack lengths increase with load, enabling control of specimen strengths. A chip of material separated from the material is visible at the larger indentation load.

broken in biaxial flexure using a flat-on-three-ball test rig (Marshall 1980). Most failure times were less than 50 ms, using a piezoelectric load cell to monitor the failure loads. Such times ensure "inert" conditions by not allowing diffusion of reactive species such as water to the crack tip during failure. Failure stresses were calculated from plate theory (Roark and Young 1983) and the testing rig dimensions, diameters of three-ball circular support × circular loading flat, given in Table 7.1. Stressing rates for these tests were approximately 10^4 MPa s^{-1}. To examine the effect of "reactive" conditions on the strength and subsequent fragmentation behavior, some tests on the glass-ceramic were conducted in water at slower stressing rates extending to 10^{-2} MPa s^{-1}.

Figure 7.4 shows examples of strength behavior for the glass-ceramic. Base and surface stressed specimen strengths (Cook 2005) are indicated by the open and closed symbols, respectively. Solid lines are indentation fracture mechanics-based best fits to the data. Decreasing the indentation load increased the size of the failure-causing flaw and hence the inert strength decreased with increasing load, Figure 7.4a. At very small loads, the specimens failed from flaws intrinsic to the specimen, and "intrinsic strength" is indicated by the left bands. At large loads, chipping led to strength increase above the ideal response. Similar indentation and intrinsic strength behavior were observed for the other materials (Chapter 3; see also Cook 2015). Glass-ceramic specimens containing a compressive surface stress exhibited greater strengths at a given indentation load than the base material. Decreasing the applied stressing rate in water provides more time for activated crack growth and hence the reactive strength decreased with decreasing stressing rate, Figure 7.4b. At very fast stressing rates, the reactive strengths approached the "inert" level, indicated by the right bands. Similar inert and reactive strength behaviors were observed for the other materials (Chapter 6; see also Cook 2015). Specimens with compressive surface stress exhibited greater strengths at a given stressing rate. Extensive details of compressive stress effects on strength are given elsewhere (Cook 2005).

Figure 7.4 highlights the advantages of the controlled flaw indentation-strength technique. By controlling the size of the dominant flaw in a specimen, the stress at failure can be varied over a wide range, leading to a wide variation in the consequent fragmentation pattern. Without this control, the strengths and resulting patterns would only vary within the limits set by the intrinsic flaw distribution. (Details of variations in indentation strength measurements are discussed in Chapters 3–6.) Figure 7.5 shows a sequence of fragmentation patterns that developed in the glass-ceramic disks on failure at the range of strengths shown in Figure 7.4. The strengths are indicated above the images. At small strengths, the specimens broke into

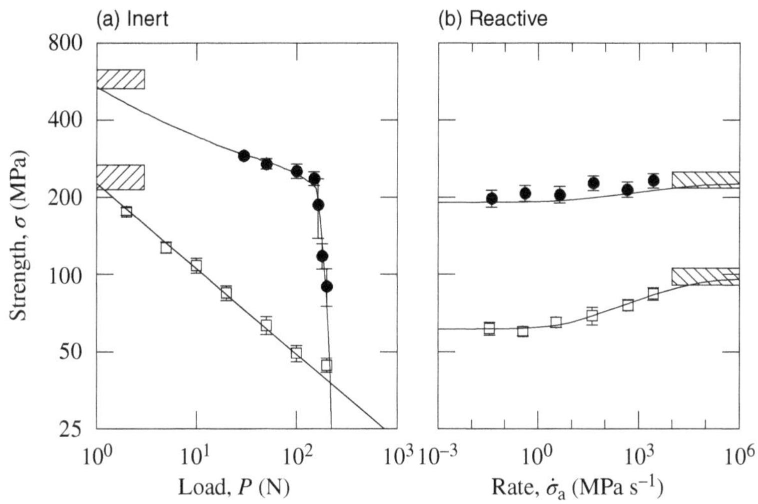

Figure 7.4 Plots of indentation strength for cordierite glass-ceramic disks. (a) Inert strength, in which indentation load, and thus flaw size, is varied and stressing rate is fixed at a rapid value that excludes chemical reaction; (b) reactive strength, in which indentation load is fixed and stressing rate, and thus extent of chemical reaction with the environment, is varied.

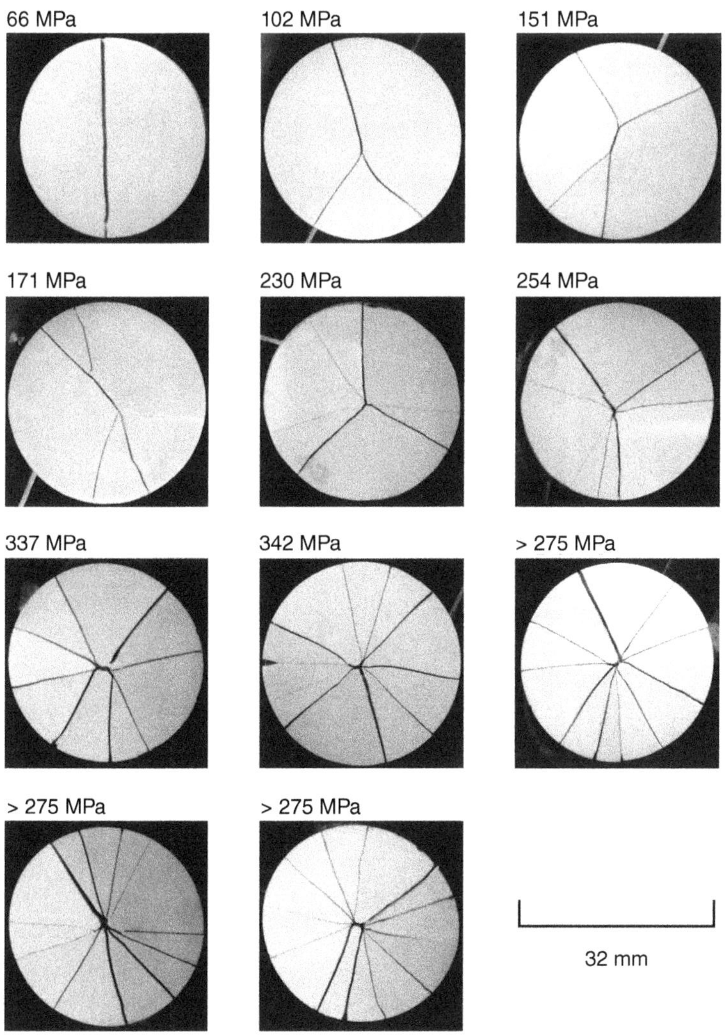

Figure 7.5 Sequence of images of glass-ceramic disks fragmented in biaxial flexure at the strengths indicated. The number of fragments increases with strength and with surface compression (final three disks with strengths > 275 MPa contained surface compression and central tension).

two or three fragments, and as the strength increased, the number of fragments increased. The fragments are wedge-shaped, formed by cracks radiating from the central flaw. Figure 7.1a appears as the end member of the sequence. Fragmentation patterns and sequences were similar for the other polycrystalline materials examined here, and similar sequences also appear in many of the works regarding fragmentation patterns cited above, most notably the work considering polycrystalline Si (Cook 2006). No correlation was observed here between the form of the fragmentation pattern and the method of strength control (indentation flaw size in inert tests or applied stressing rate in reactive tests). Strength appeared to be the sole determiner of the number and shape of fragments, consistent with the works cited above, most notably those in which indentation load and environment (Shetty et al. 1983a) and stressing rate (Salem 2013) were varied in a similar manner to here. At fixed strength, surface stress increased the number of fragments relative to the base material, also consistent with earlier observations (Shetty et al. 1983a).

Observations such as Figure 7.5 are placed on a quantitative basis in Figure 7.6, which plots the number of fragments observed in strength tests of the six polycrystalline materials examined here. All disks were analyzed by optical microscopy to determine the number of fragments and only those fragments that had become physically separated from their neighbors were counted. In some cases, it was clear that cracks had initiated within fragments but had failed to propagate to the edges of the disk. These cracks were not included in computing the number of fragments.

There are a number of clear features in Figure 7.6: The first feature is a definite trend of more fragments generated by specimens with greater strengths. The second feature is that such a trend is extremely material dependent and individual material variations exhibit significant dispersion (e.g., large strength range with small change in fragment number for the

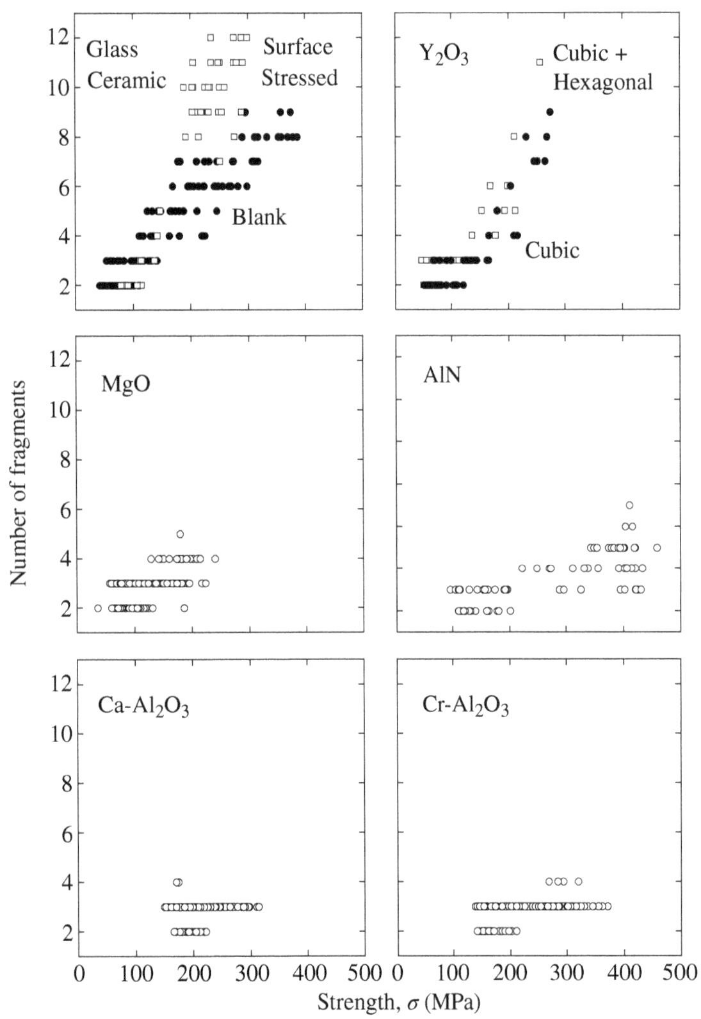

Figure 7.6 Plots of the number of fragments versus strength for disks of six polycrystalline materials fractured in biaxial flexure. Symbols represent between 100 and 200 observations of single disks for each material.

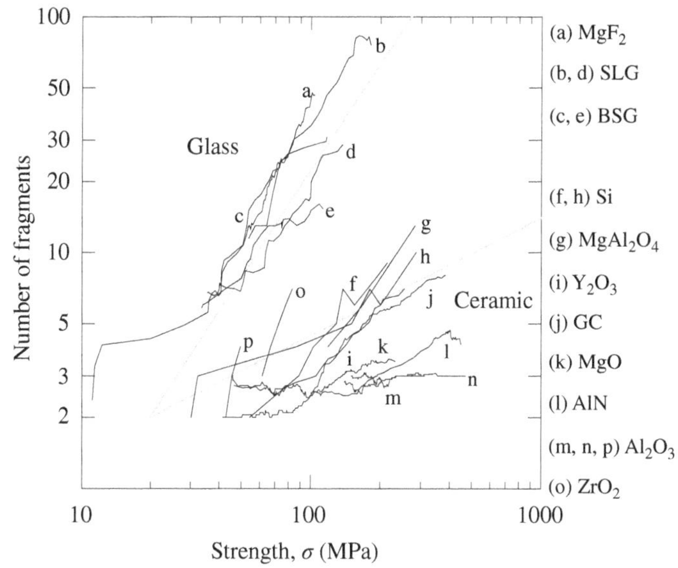

Figure 7.7 Composite plot of the number of fragments formed versus strength for fracture of biaxially flexed disks of glasses, ceramics, and semiconductor materials. SLG: soda-lime silicate glass. BSG: borosilicate glass. GC: cordierite glass-ceramic. All materials polycrystalline except (b, c, d, e) glasses and (f) single crystal.

Al_2O_3 materials and the opposite for the glass-ceramic). Although the indentation load domain was invariant, the range of strengths generated differed considerably from material to material, indicating a third feature of significant microstructural and hence fracture resistance effects on strength. Microstructural effects include: (i) the strength range covered by the AlN and Cr–Al_2O_3 materials is similar yet more fragments were generated in AlN; (ii) the Y_2O_3 material containing hexagonal phase generated more fragments at the same strength than that consisting of cubic phase alone; and (iii) the glass-ceramic material containing a compressive surface stress generated more fragments at the same strength than the base material.

Figure 7.7 shows a composite logarithmic plot of the responses of the polycrystalline materials in Figure 7.6 along with responses for similarly-sized specimens of other materials, including polycrystalline MgF_2 Figure 7.7a (Mecholsky et al. 1998), soda-lime silicate glasses (SLG) Figure 7.7b (Quinn et al. 2005) and Figure 7.7d (Nakasa and Nakatsuka 1990), borosilicate glasses (BSG) Figure 7.7c (DeLellis et al. 2020) and Figure 7.7e (Nakasa and Nakatsuka 1990), single-crystal Figure 7.7f and polycrystalline Figure 7.7h Si (Cook 2006), polycrystalline spinel Figure 7.7g (Salem 2013), polycrystalline partially stabilized ZrO_2 Figure 7.7o (Nakasa and Nakatsuka 1990), and polycrystalline Al_2O_3 Figure 7.7p (Nakasa and Nakatsuka 1990). The responses were determined from published numerical, graphical, or pictorial data from the cited sources. In most cases, the raw data in Figure 7.7 were smoothed using a multipoint moving average to remove the dispersion obvious in Figure 7.6. The bold solid lines in Figure 7.7 are guides to the eye illustrating empirical power-law responses of $\sigma^{1.5}$ and $\sigma^{0.5}$.

It is clear that the responses in Figure 7.7 fall into two groups dominated by amorphous glasses or crystalline ceramics and semiconductors. The solid line guides to the eye are approximate descriptions of the two groups. The diverse responses are in accord with common experience: a dropped drinking glass usually shatters into many pieces and is largely irrecoverable. A dropped ceramic coffee cup frequently breaks cleanly into a few pieces and is often repairable ("glue the handle back on") or even transformed into the art of Kintsugi (Richman-Abdou 2019). On closer examination and comparison with Table 7.1, it is clear that there is a tendency to generate more fragments as the Young's modulus E decreases, consistent with the idea that the fragmentation process is the conversion of strain energy into surface energy: greater strain energy generates more fragments. Of course, this is only the driving force element of the thermodynamic balance and does not consider how that conversion is to take place. That is, whether the fragmentation or branching criterion is met, or if the conversion is possible: greater fracture resistance R requires greater strain energy to balance the new surfaces created during fragmentation of tougher, more fracture resistant materials. For instance, the alumina materials may exhibit few fragments because the strain energy at failure is small, the criterion for branching is not met, or the fracture resistance is great. The exceptions to the groups in Figure 7.7 highlight the sensitivity of fragmentation to materials details: polycrystalline MgF_2 is grouped with glasses although exhibiting a greater modulus and Si has similar toughness to glass and is grouped with ceramics.

7.3 Distribution of Fragment Sizes

The images of Figure 7.5 show that fragments sizes are not identical within a fractured specimen and that there is a distribution of fragment masses. Such size or mass distributions have been considered since early fragmentation studies concerned with exploding armaments (Mott 1947) and are well reviewed by Grady and Kipp (1985) and Grady (2010), including descriptions of more recent fragmentation distributions generated by controlled blasts and impacts. The original interest in distributions of fragment masses was directed toward estimating the probability of generating large fragments able to cause maximal damage in an armament detonation. This application of fragmentation remains of current interest (Gold and Baker 2008; Pike and Zuo 2014). However, there are many additional areas of application, including rock fragmentation in mining (Hogan et al. 2012, 2014), designing safe structures in architectural and transportation applications (Bennison et al. 1999), ceramic armor (to protect against the large fragments or penetrators) (O'Donnell 1991; Woodward et al. 1994), forensic evidence investigations (Locke and Unikowski 1991, 1992; Hicks et al. 1996), and environmental and geological assessments (Matthews 1983; Novák-Szabó et al. 2018; Pioli et al. 2019). Fragment mass distributions also provide fundamental information regarding different mechanisms and material dependencies of the fragmentation process (Åström 2006). Here, the distributions of fragment masses resulting from the above glass-ceramic strength tests were determined and compared with distributions resulting from the more frequently reported, predominantly impact, tests.

Glass-ceramic specimens with identical numbers of fragments between 2 and 12 were grouped together, independent of strength, and an electric balance used to determine the individual fragment masses, m, constituting a group. Each group consisted of $N = 40$–100 fragments. As the mass of an intact specimen was approximately 6 g, the number average mass within a group was (6 g/number of fragments). An empirical distribution function (edf), $\mathrm{Pr}(m)$, was formed from the fragment masses within each group. The masses were ranked from smallest to largest as $m(i)$ with rank index i varying from 1 to N. The quantity $(i - 0.5)/N$ was generated for each value of the index and the discrete function pairing $\mathrm{Pr}(m) = [m(i), (i - 0.5)/N]$ formed the edf for each group. Again, only those fragments that were physically distinct were counted in determining the fragment mass distributions. The function $\mathrm{Pr}(m)$ gives the probability that a fragment selected at random from a group has mass $\leq m$. Figure 7.8 shows plots of the edf mass variations for each group of glass-ceramic fragments for (a) blank components, 2–8 fragments, and (b) surface stressed components, 9–12 fragments. For ease of visualization, individual fragment masses are shown as symbols for the two-fragment group and as solid lines for the other groups. The distributions are weakly sigmoidal and move to smaller masses as the number of fragments increases. There was a tendency in some groups of the blank material toward a bimodal distribution of large and small fragments. This was particularly evident in the edf of the six-fragment group but also in the distributions of the 4-, 5-, and 7-fragment groups. Visual support for the observation is clear in Figure 7.5 and no doubt reflects the inhomogeneity of the flat on three-balls test geometry (Börger et al. 2002, 2004). For historical reasons associated with estimating the probability of forming large, destructive, exploding shell fragments, fragment mass distribution data are usually presented as the complementary edf, $\overline{Pr}(m) = 1 - \mathrm{Pr}(m)$ (Grady and Kipp 1985; Grady 2010). The data in Figure 7.8 would then decrease

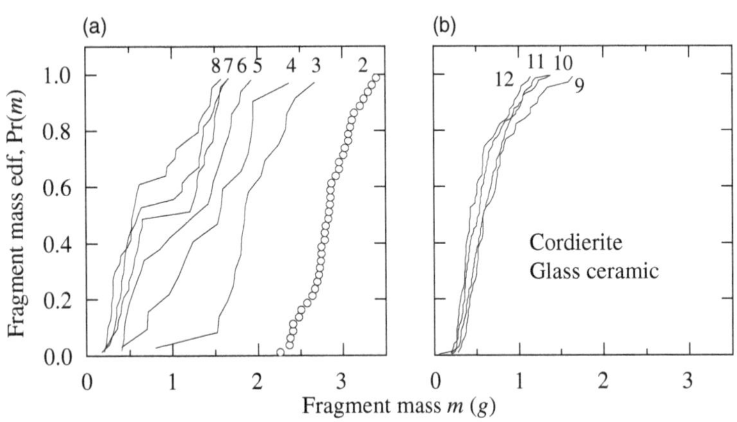

Figure 7.8 Empirical distribution function (edf) plots of fragment mass for glass-ceramic disk specimens fractured in biaxial flexure and exhibiting the different numbers of fragments indicated. (a) Blank specimens; (b) stressed specimens containing surface compression and central tension. Stressed specimens exhibit greater numbers of fragments and different shaped edf responses.

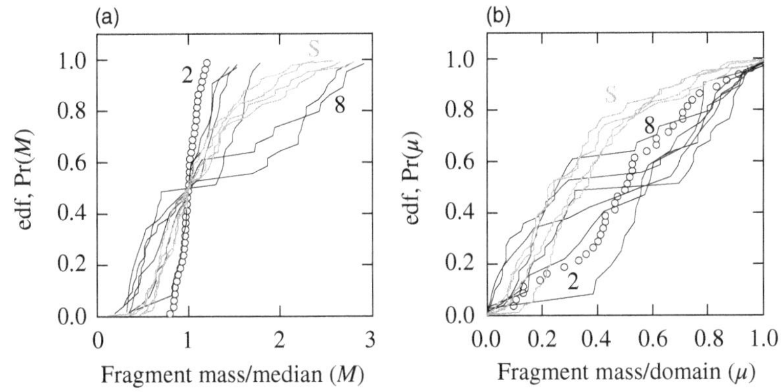

Figure 7.9 Plots of fragment mass distributions from Figure 7.8 using scaled mass axes. (a) Distributions scaled by median mass illustrating increased relative mass dispersion with greater number of fragments; (b) distributions scaled by mass domain width illustrating near-linear distributions.

with increasing mass as $\overline{Pr}(m)$ gives the probability that a fragment selected at random from a group has mass $> m$. Greater insight into the relative forms of the distributions in Figure 7.8 is gained however by scaling the mass axes to common values.

Figure 7.9a is a composite distribution plot with the fragment mass domain in each distribution scaled by the specific distribution median mass, $M = m/m_{\text{median}}$. Identical notation to Figure 7.8, is used and all distributions in the scaled coordinates pass through (1, 0.5). It is clear that the relative width of the distributions, dispersion in M, increases as the number of fragments increases (from 2 to 8), i.e., the relative separation between the lightest and heaviest fragments increases as the number of fragments and, see Figure 7.7, the strength increases. An alternative interpretation is thus that the fragmentation process becomes more disordered as the strength increases. Figure 7.9a also shows that this disorder is suppressed by a compressive surface stress. Although the strengths and numbers of fragments are large for the surface-stressed (S) materials, the variability in M is less than the trend in strength would suggest.

Figure 7.9b is a composite distribution plot with fragment masses scaled by relative position in a distribution mass domain, $\mu = (m - m_{\text{min}})/(m_{\text{max}} - m_{\text{min}})$. In this case, all distributions in the scaled coordinates pass through (0, 0) and (1, 1). The similarities in edf shapes within the blank and surface stressed material groups are clear—the blank materials exhibit weakly perturbed linear variations and the surface stressed materials exhibit concave variations. The nature of the blank material edf perturbation, from linear to bimodal, noted above, is observed in the many horizontal segments. The small-mass branches of the bimodal responses appear to be in common and overlie the surface stressed responses, suggesting that fragment mass disorder is suppressed at large strengths, generating a common edf.

The behavior of the cordierite glass-ceramic is most easily compared with that of other fragmentation systems using a plotting scheme similar to Figure 7.9a. Such a comparison is shown in Figure 7.10: Fragment mass edf plots for several "typical" fragmentation systems were determined by digitizing published works and scaling to a common format. The mass axes were scaled to distribution median values as M. The median thus appears as $M = 1$, and median values are shown as vertical lines. All plots in Figure 7.10 extend over 10^5 in M. The edf axes were scaled to distribution maxima as 1.0. The median is thus $\Pr(M) = 0.5$, and median values are shown as horizontal lines. The resulting symbols in Figure 7.10 indicate digitized values for: (a) an impacted thin SiO_2 plate sandwiched between polymer layers (Grady and Kipp 1985); (b) an impacted granite block (Hogan et al. 2012); (c) a dropped clay tablet (Meibom and Balslev 1996); (d) an impacted ZrO_2 cylinder (Davydova et al. 2018); and (e) an impacted egg shell (predominantly $CaCO_3$) (Wittel et al. 2004). The cordierite glass-ceramic flexed disks appear as (f). The great extent of the relative fragment mass domains in systems (a)–(e) necessitated the use of a logarithmic scale, which also enables direct comparison of edf shapes independent of fragment mass magnitude.

The most obvious feature of Figure 7.10 is the commonality of the relative fragment mass domains observed in the typical fragmentation experiments (a)–(e), about 10^4. These domains are quite different from the extent of the disk fragmentation system (f), about 10^1. Flexure generated fragments clearly exhibit distribution widths different from typical impact generated fragments. A second clear feature of Figure 7.10 is that there is no common distribution shape. Although the shapes differ considerably, most typical fragmentation systems exhibit a convex tendency, which is distinct from the near-linear tendency of the disk fragmentation system, Figure 7.9b.

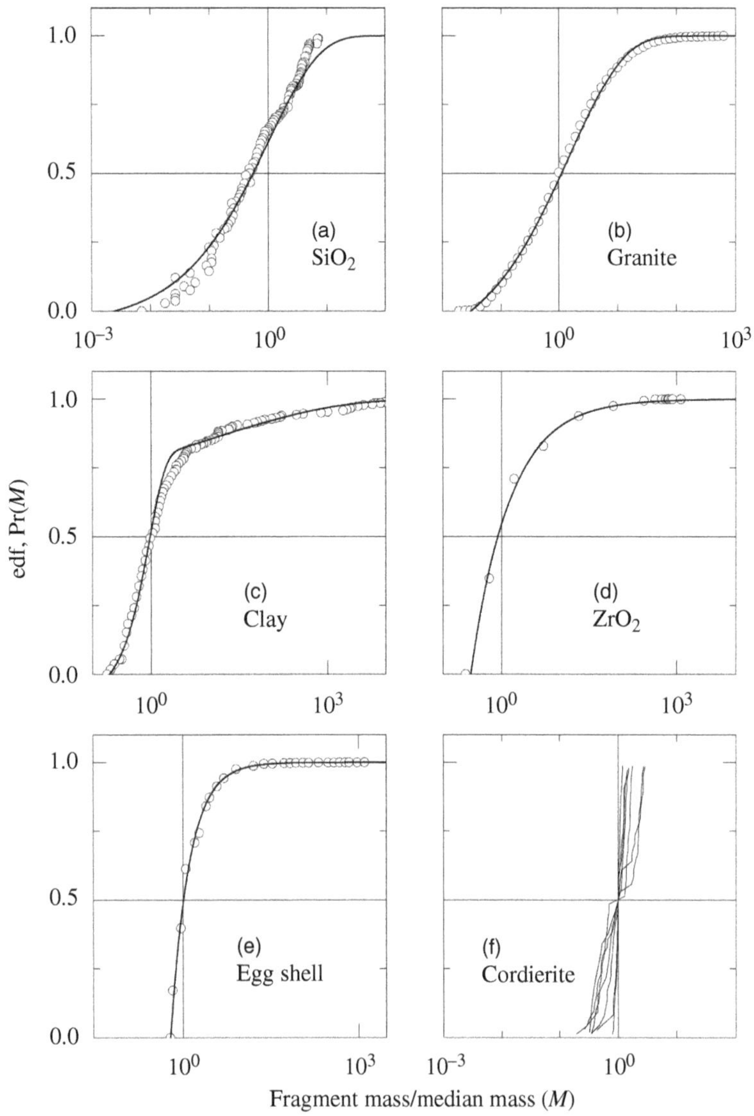

Figure 7.10 Plots of experimental edf behavior for five typical fragmentation systems and the glass-ceramic disk data. (a)–(e) Symbols represent experimental data, lines represent empirical best fits; (f) lines represent experimental data from Figure 7.8 and Figure 7.9. Source: Includes data adapted from Grady and Kipp (1985), Hogan et al. (2012), Meibom and Balslev (1996), Davydova et al. (2018), and Wittel et al. (2004).

In more detail, the solid lines in Figure 7.10a–e represent best fits to the data using empirical equations frequently implemented to describe fragmentation distributions. In accord with the usual plotting scheme, the equations are usually given as complementary cumulative distribution functions, $\overline{H}(m)$, fit to $\overline{Pr}(m)$ data. Here, cumulative distribution functions $H(M) = 1 - \overline{H}(M)$, in scaled coordinate M, are fit to Pr(M) data. In Figure 7.10a, $\overline{H} = A\exp(-M^{1/2})$, now known as the Mott distribution (Grady and Kipp 1985), is used, where A is fitting parameter. This behavior is exhibited by cylindrical steel shells (Weimer and Rogers 1979). A more flexible equation, $\overline{H} = A\exp(-M^{\alpha})$, the powered exponential, is illustrated in Figure 7.10b, where A and α are fitting parameters. Such an equation describes most sigmoidal appearing data; $\alpha = 1$, the exponential distribution, describes fragmented expanded rings (Grady and Kipp 1985). Of even greater flexibility is the bilinear exponential equation, $\overline{H} = A\exp(-M^{\alpha}) + B\exp(-M^{\beta})$, illustrated in Figure 7.10c, where A, α, B, and β are fitting parameters. Such an equation describes many explosion generated fragment distributions (Grady and Kipp 1985). A final equation is the simple power-law $\overline{H} = AM^{-\alpha}$, illustrated in Figure 7.10d and e, where A and α are fitting parameters. This equation has been used extensively, and examples include impact fragments of glass and ferroelectric spheres (Gilvarry and Bergstrom 1961; Grady 2008); gypsum disks and spheres (Oddershede et al. 1993); Si plates (Huang et al. 2016); a large variety of glass and ceramic plates and rods, including Al_2O_3, SiC, ZrO_2, and granite (O'Donnell 1991; Woodward et al. 1994; Davydova and Uvarov 2013; Davydova et al. 2015, 2017, 2018; Zinszner et al. 2015), and meteoritic fragments (Halliday et al. 1989). In contrast, the disk data are described by a simple linear relationship $H = 0.5 + A(1 - M)$, where A is a fitting parameter, Figure 7.9. It is reemphasized that these equations are all empirical descriptions. An extensive survey of over 1000 fragmentation data sets and 17 distributions (Sanchidrián et al. 2014) concluded that "No general quantitative indications can be given on what characteristics a data set must have to be best fitted by one or another distribution."

7.4 Crack Branching Distances

The number and size distribution of the fragments generated on specimen fracture depends on the distance between crack branching events and the angles between crack branches. The major factor is the crack branching distance, and branching angles will not be considered here. However, it is noted that the angles observed in the ceramics studied in this work and elsewhere were typically about 45°, although variation was considerable; both factors recognized in early work (Preston 1935). The value is less than that determined for distortion of the crack tip stress field during dynamic fracture. As first calculated by Yoffe (1951), crack velocities are large during dynamic fracture and the maximum crack tip tensile stress does not occur directly ahead of the tip but in two lobes at approximately ± 60° to the crack plane. A recent review of crack branching angle observations in specimens of various geometries suggests that the angle and its variation are sensitive functions of the shear/tension mode of loading (Quinn 2020). Here, the loading will be primarily restricted to equibiaxial tension in loaded disks.

Figure 7.11 is a schematic diagram of the branching phenomenon observed in failure of a disk containing a strength-controlling indention flaw, similar to Figure 7.2. Branching occurs at radii r_i measured from the flaw center, and the "branching distance" is taken as the average of the two initial radii, $\lambda_b = (r_1 + r_2)/2$. These radii are typically the shortest in a branching pattern and frequently indistinguishable. Figure 7.12 shows a logarithmic plot of the variation in λ_b with σ_f for the blank glass-ceramic disk specimens shown in Figures 7.4, 7.6, and 7.8. Individual symbols represent measurements from single specimens. The branching distances, of order millimeters for strengths of hundreds of

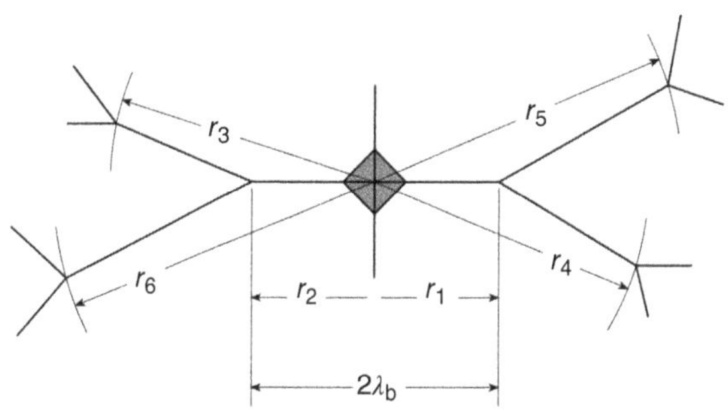

Figure 7.11 Schematic diagram of crack branching at fracture from an indentation flaw. The central impression of the indentation is visible along with two cracks propagating left and right leading to branching and fragmentation. The branching radii r_i and principle branching distance λ_b are illustrated.

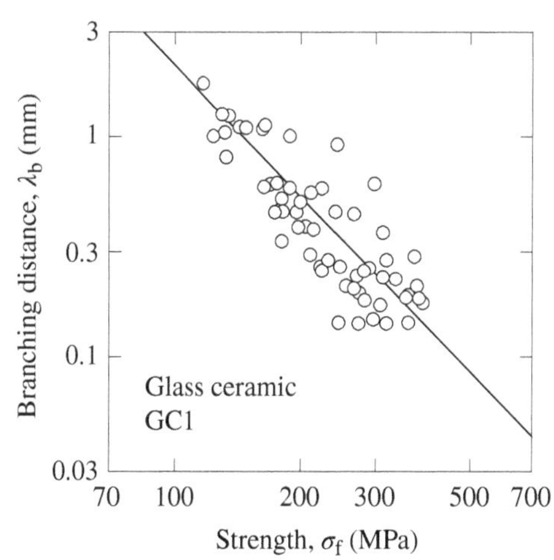

Figure 7.12 Plot of branching distance versus strength for cordierite glass-ceramic disks. The symbols represent individual specimen measurements, and the solid line represents an empirical fit to the data of the form $\lambda_b \sim \sigma_f^{-2}$.

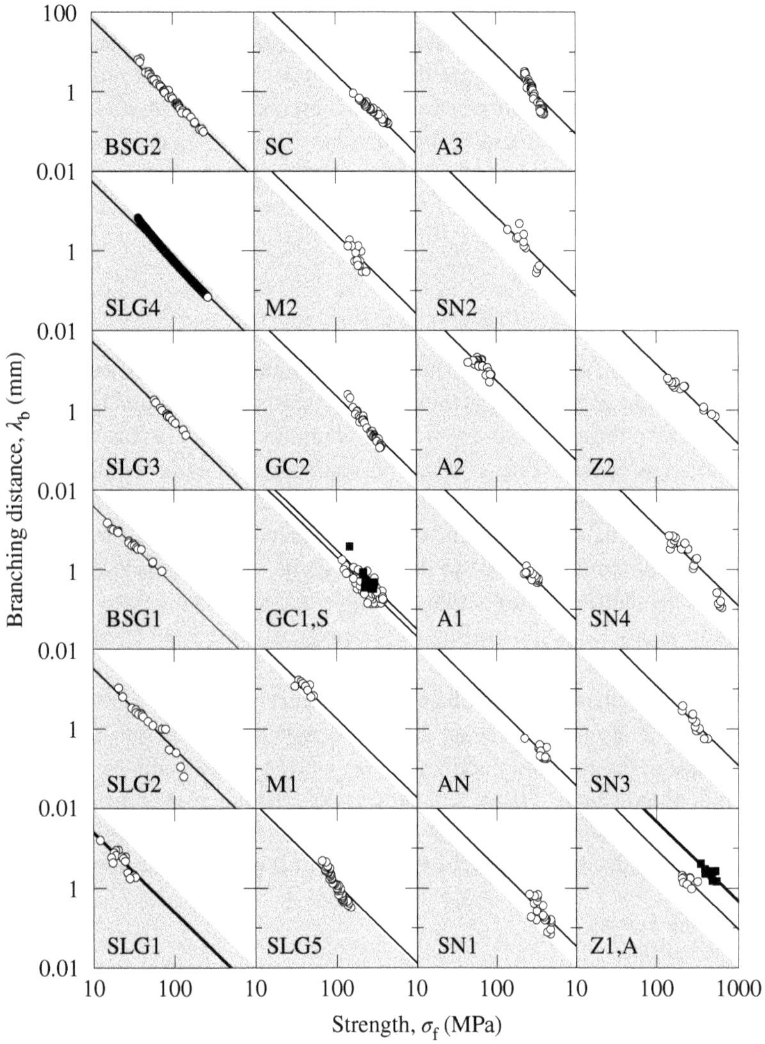

Figure 7.13 Plot of branching distance versus strength for a variety of glass and ceramics. The symbols represent individual specimen measurements, and the solid lines represent empirical fits to the data of the form. The shaded regions are a guide to the eye and are invariant.

megapascals, are typical for ceramics and glasses. Also typical is that the branching distance decreased significantly with increased strength. The solid line in Figure 7.12 is a best fit to the data of a commonly observed and reported description (see below) for branching of $\lambda_b = S_B\sigma_f^{-2}$, where S_B here is an empirical "branching constant." The data in Figure 7.12 are well described by this equation over the domain of strengths measured.

Figure 7.13 shows the results of an extensive survey of branching distances, including the polycrystalline AlN, MgO, and glass-ceramic materials examined here along with glasses and other polycrystalline ceramics. The families of materials are labeled for ease of reference and include soda-lime silicate glass (SLG), borosilicate glass (BSG), MgO (M), glass-ceramic (GC), SiC (SC), Si_3N_4 (SN), AlN (AN), Al_2O_3 (A), and ZrO_2 (Z). For example, the cordierite glass-ceramic material shown in Figure 7.12 is labeled GC1. The results from other researchers include those from Congleton and Petch (1967) (SLG1, M1, A2), Aoki and Sakata (1980) (SLG2, BSG1), Shetty et al. (1983a) (SLG3, GC2, A3), Oakley (1996) (SLG4), Quinn (1999) (BSG2), Choi and Gyekenyesi (1998) (SLG5, SC, SN1, SN2, SN3, SN4, Z2), Choi et al. (2000) (A1), and Kirchner et al. (1981) (Z1). (Note that the labels differ from those in Chapter 2.)

The results from other researchers in Figure 7.13 were obtained by digitizing published data and scaling to a single format. Some commonly cited results, in which branching distances were inferred from fracture surface features, e.g., Kirchner (Kirchner 1978) and Kirchner and Kirchner (Kirchner and Kirchner 1979), are not included. The format of Figures 7.12 and 7.13 has the independent failure variable, the strength σ_f, as the abscissa and the dependent variable,

Table 7.2 Crack branching parameters of glasses and ceramics.

Data Source	Material	S_B (MPa^2 m)	N (Observations)	Identification
Congleton and Petch (1967)	Soda-lime glass	2.48 ± 0.76	17	SLG1
Aoki and Sakata (1980)	Soda-lime glass	3.35 ± 1.16	17	SLG2
Aoki and Sakata (1980)	Borosilicate glass	3.98 ± 0.54	17	BSG1
Shetty et al. (1983a)	Soda-lime glass	5.34 ± 0.30	14	SLG3
Oakley (1996)	Soda-lime glass	5.66 ± 0.85	540	SLG4
Quinn (1999)	Borosilicate glass	6.75 ± 1.14	46	BSG2
Choi and Gyekenyesi (1998)	Soda-lime glass	13.0 ± 4.8	43	SLG5
Congleton and Petch (1967)	MgO	18.8 ± 4.2	10	M1
This work	Cordierite	21.2 ± 9.2	59	GC1
Shetty et al. (1981)	Cordierite	23.4 ± 7.5	37	GC2
Mecholsky et al. (1998)	MgF_2	24.5 ± 3	2	
This work	MgO	27.4 ± 11.2	15	M2
Choi and Gyekenyesi (1998)	SiC	29.4 ± 3.6	34	SC
This work	Cordierite	31.7 ± 12.7	21	GC1S
Choi and Gyekenyesi (1998)	Si_3N_4	34.3 ± 16.9	20	SN1
This work	AlN	36.7± 10.8	14	AN
Choi et al. (2000)	Al_2O_3	54.6 ± 8.7	25	A1
Congleton and Petch (1967)	Al_2O_3	56.4 ± 13.0	22	A2
Choi and Gyekenyesi (1998)	Si_3N_4	74.1 ± 43.4	12	SN2
Shetty et al. (1983b)	Al_2O_3	90.1 ± 38.8	41	A3
Kirchner et al. (1981)	ZrO_2	91.1 ± 28.5	9	Z1
Choi and Gyekenyesi (1998)	Si_3N_4	122.2 ± 34.3	12	SN3
Choi and Gyekenyesi (1998)	Si_3N_4	124.1 ±59.0	20	SN4
Choi and Gyekenyesi (1998)	ZrO_2	143.9 ±30.8	14	Z2
Kirchner et al. (1981)	ZrO_2	465 ± 116	10	Z1A

the branching distance λ_b, as the ordinate, and uses the same logarithmic scale (1:2) in all cases. The format enables an unbiassed assessment of the σ_f–λ_b relationship for a given material and easy between-material comparisons. (A common format reverses the independent and dependent variables and linearizes the data assuming a particular relationship; see many of the works cited above.) Between-material comparisons are further enhanced by the inclusion of an invariant shaded triangular contour along the value $S_B = 10\,\text{MPa}^2$ m in each subplot. Contours of S_B in the shaded region are < 10 and those in the unshaded region > 10. The solid lines in Figure 7.13 are best fits to the data of the form $\lambda_b = S_B\sigma_f^{-2}$. The materials, identification labels, number of observations, and the best fit values of S_B are given in Table 7.2, along with the data source first author and year. The values given in Table 7.2 are means and standard deviations of the experimental quantity $\sigma_f^2\lambda_b$ evaluated for each material. The lines and values should be interpreted as representative contours of branching behavior assuming the quadratic form given, rather than unbiassed descriptions of the observed behavior. The subplots in Figure 7.13 are arranged so that materials with smaller values of S are represented in the lower left of the figure and materials with larger values are represented in the upper right (the extremes of behavior are indicated by bold lines).

There are many clear trends apparent in Figure 7.13 and Table 7.2. First, for all materials examined, crack branching distance decreased with increasing strength, independent of the nature of the strength-controlling flaw (indentation, machining abrasion, incidental contact), the strength measurement environment (inert, reactive), or the measurement geometry (ball on ring, ring on ring, flat on balls). In some cases, the strength domain was limited by a limited domain of uncontrolled flaws (e.g., A1, M1, Z1) such that a clear trend is difficult to detect, although large numbers of specimens containing

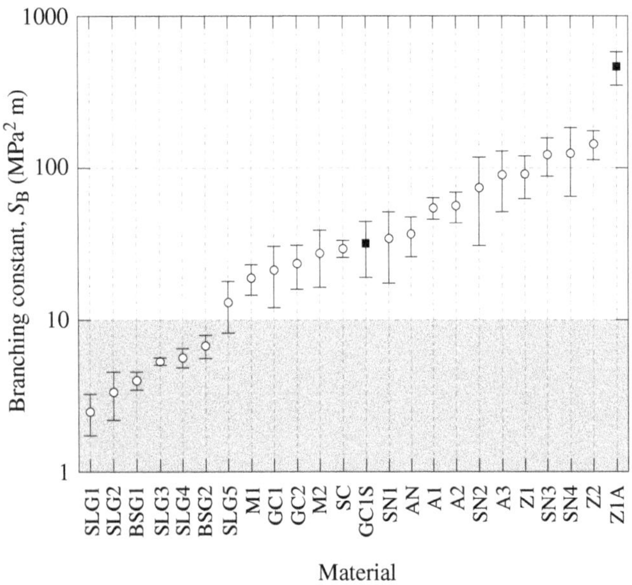

Figure 7.14 Plot of material branching constant S_B ranked by mean value. Glasses have S_B values < 10 MPa^2 m (shaded region), and ceramics have S_B values > 10 MPa^2 m. More fragmentation resistant materials on right of plot. (Note that 10 MPa^2 m = 10^4 MPa^2 mm.)

uncontrolled flaws can give rise to significant trend (e.g., SLG4). In other cases, controlled indentation flaws were used to extend the flaw size domains with consequent extended strength domains (e.g., SLG3, BSG2, GC1) and clear and significant branching trends were observed. Second, glasses tend to exhibit behavior on contours of S_B <10 MPa^2 m and ceramics on contours of S_B > 10 MPa^2 m. That is, glass observations (all SLG, BSG) appear within the shaded region and ceramic observations (all others) appear within the unshaded region. Third, in more detail regarding the last point, there is a clear gradation of behavior from optical glasses (e.g., SLG1), through microelectronic glass-ceramics and ceramics (e.g., GC1), to structural ceramics (e.g., Z2) in terms of increasing fragmentation resistance as quantified by movement from lower left to upper right in Figure 7.13 or top to bottom in Table 7.2 (MgF_2 now groups with ceramics). Intuitively, greater fragmentation resistance is associated with greater fracture resistance (and, coincidentally, usually greater elastic modulus and contact hardness). This association extends to within material comparisons: the solid symbols in the Z1A subplot indicate peak aged material with greater transformation toughening and clearly greater fragmentation resistance; the solid symbols in the GC1S subplot indicate surface compressed material with somewhat greater fragmentation resistance. An opposite case is the SN1 material tested at 1100 °C with less fragmentation resistance than the other SN materials. Figure 7.14 is a plot of all the S_B values determined here, using the same notation as Figure 7.13 and ranked in increasing order making the above material dependencies clear.

A final observation from Figure 7.13 is that the branching behavior of some materials is not completely described by $\lambda_b = S_B\sigma_f^{-2}$. The most common deviation is that the λ_b response has a greater than quadratic dependence on strength. In cases in which the deviation is small, the empirical relation can be modified to describe the observed behavior as $\lambda_b = S_B(\sigma_f - \sigma_0)^{-2}$, where σ_0 is an empirical shift parameter, e.g., SLG4, BSG2. In cases in which the deviation is large an empirical power-law relation $\lambda_b = S_B\sigma_f^{-x}$ is required, where $x > 2$ is an empirical exponent, e.g., SLG5, GC2, A3. These factors should be taken into account when interpreting Figure 7.14.

7.5 Application Examples

In this section, three sets of results are presented, without analysis, as examples of potential applications of the work above. Two of the examples concern the number of fragments generated during component failure—Si die fragmentation and ceramic microstructure effects on fragmentation—and one example concerns crack branching—the variation of branching radii in a fragmentation sequence. The first two examples are applicable in component and materials engineering and the third example is applicable in developing a fundamental basis for fragmentation.

7.5.1 Si Die Failure

Figure 7.15 shows an array of rectangular single-crystal Si dies used in microelectronic devices. The dies are approximately 15 × 10 mm and bonded with a polymeric adhesive to a composite polymer–glass fiber substrate such that the active face is down and the inactive (001) die face is visible. The die edges are aligned along [110] directions. The die active faces, substrates, and adhesive layers are not visible. The dies and substrates are approximately 0.7 and 1 mm thick, respectively. The entire die–substrate assembly was cooled to approximately −70°C and then warmed to room temperature. The greater coefficient of thermal expansion of the substrates relative to the Si dies caused bending of the assemblies on cooling, placing the back faces of the dies in biaxial tension. The images of Figure 7.15 were recorded after the cooling cycle and warming the assemblies to room temperature.

The dies have fragmented along [110] directions as a consequence of the cooling cycle, similar to [100] branching observed in MgO (Bowden et al. 1967). The number of fragments for each die in Figure 7.15 is indicated; the number of fragments increases from top to bottom. The bottom right image in Figure 7.15 is a schematic diagram generated by close examination of the most fragmented die, illustrating the 27 fragments. The practical application of Figure 7.15 is in establishing the strength distribution of dies and propensity to failure in an assembly. The stress developed in a die is determinable from substrate, adhesive, and die thermomechanical properties, assembly geometry, and cooling temperature. The strength of a die is determinable from fragmentation data such as Figures 7.6 and 7.7. Comparison of the stress and strength enables model verification and engineering of die edge finishing and strength-controlling flaws, adhesive

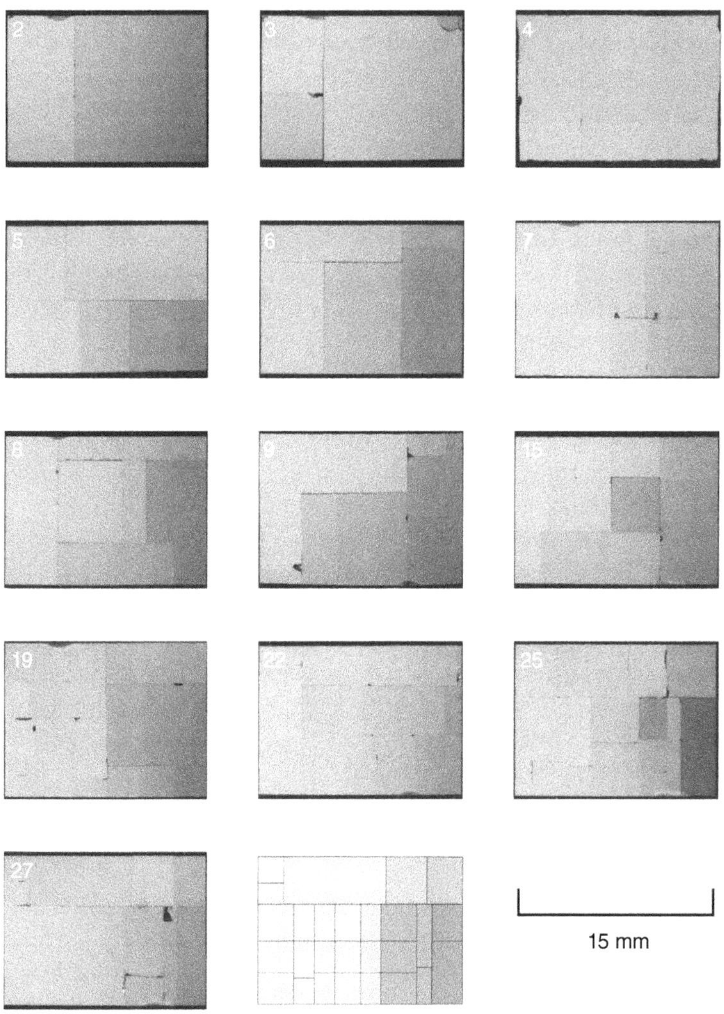

Figure 7.15 Sequence of optical images of Si dies fragmented in biaxial flexure generated by thermal expansion mismatch with an underling polymeric substrate (not visible). The number of fragments is indicated, along with a schematic diagram of the most fragmented die.

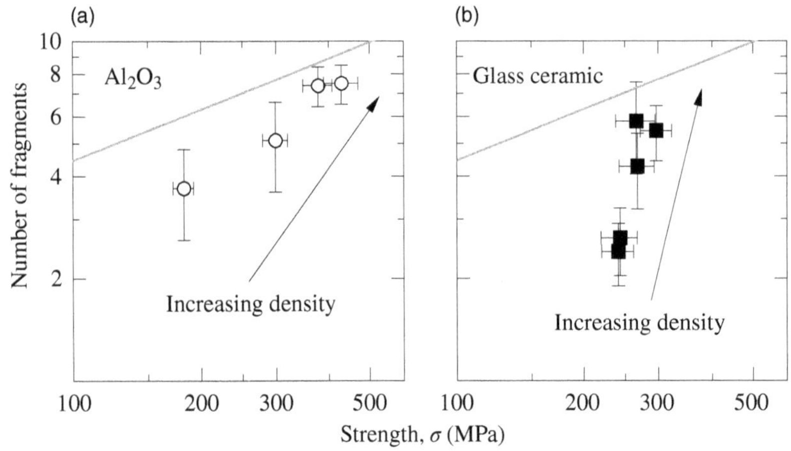

Figure 7.16 Plots of the number of fragments as a function of strength for two ceramics of varying relative density, (a) Al_2O_3 and (b) cordierite glass-ceramic. As density increases, strength and the number of fragments generated on failure increase.

fixing temperature, and substrate materials and geometry. Even in the absence of calculation, empirical comparison of Figures 7.7 and 7.15 enables the peak stress reached by the dies during the cooling cycle to be estimated as 500 MPa. The advantage of fragmentation data in engineering trouble shooting is clear as they address the question of "over stressed or under strength?"

7.5.2 Ceramic Microstructure Development

Figure 7.16 examines the tendencies to fragmentation as ceramic microstructures develop and, in particular, how the tendency to fragmentation changes during ceramic processing and is controlled by relative density or porosity. The subplot of Figure 7.16a shows the number of fragments formed as a function of strength for Al_2O_3 from the work of Sapozhnikov et al. (2015). The relative density of fired Al_2O_3 disks was varied by controlling the pressure used to form the disks in the unfired ("green") state, thereby altering the fired relative density from 0.77 to 0.96 with consequent increase in strength. The number of fragments increased with increasing strength (and density) toward the representative response of dense ceramics reproduced from Figure 7.7 as the solid line.

The subplot in Figure 7.16b shows the number of fragments formed as a function of strength for cordierite glass-ceramic. The relative density of fired glass-ceramic disks was varied by controlling the volume fraction of Si_3N_4 whiskers added to the disks in the unfired state from 0.15 to 0, thereby altering the fired relative density from 0.85 to 0.99 with consequent increase in strength. As above, the number of fragments increased with increasing strength (and density) toward the representative response of dense ceramics. The sensitivity of the number of fragments to strength is significant for both materials in Figure 7.16 and reinforces the observations of Figure 7.7. However, the interpretation is somewhat different. In Figure 7.7, variations in strength and the number of fragments were caused by variations in the size of strength-controlling flaws for a given material. In Figure 7.16, variations in strength are caused by variations in flaw size (probably pore size) *and* variations in material (different density and microstructure). The practical application of Figure 7.16 is in designing and engineering ceramic fabrication methods and microstructures with optimized fragmentation behavior and in reducing diagnosis of fabrication state to a semiquantitative measurement (number of fragments, not strength).

7.5.3 Branching Radius Variation

The number and mass distribution of fragments generated on failure of a brittle component depends on the detailed variations of the branching geometry (angle, radii) with branching node. In particular, a major factor is the variation of branching radius, r_i, with node index i, Figure 7.11. The early schematic diagrams of Preston (1935) make clear that a nonlinear increase in r_i with i was observed in uniaxial and biaxial loading of glass plates and pressure loading of bottles. Somewhat later images of uniaxial failure of glass plates by Clark and Irwin (1966), Bowden et al. (1967), and Field (1971) indicate similar nonlinear behavior (and also, in agreement with Preston, very small, $\approx 20°$, uniaxial branching angles). Figure 7.17 places the branching radius behavior on a quantitative basis in plots of r_i versus i for glasses and ceramics. The plot of

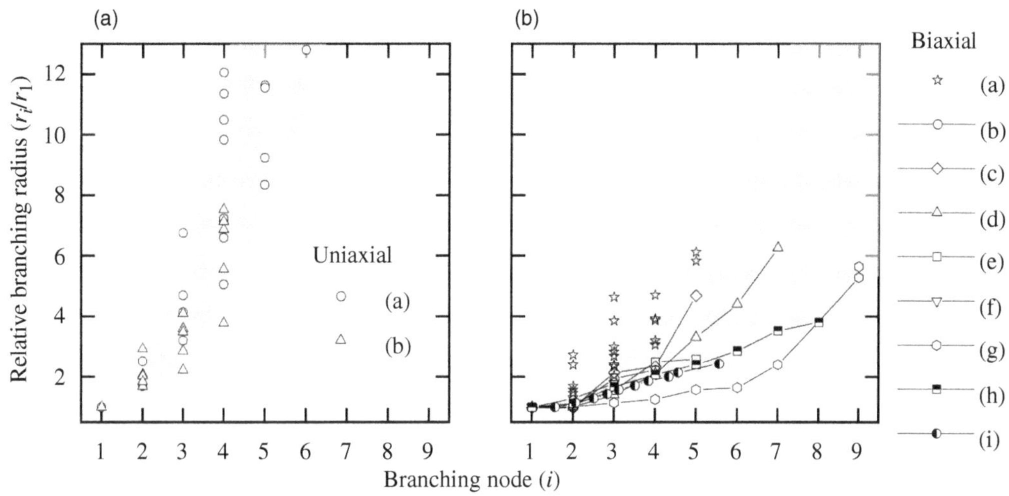

Figure 7.17 Plots of branching node radii for uniaxial and biaxial loading and fragmentation of glasses and ceramics. Symbols represent measurements of (normalized) radii at the crack branching nodes indicated. The shaded bands are guides to the eye of quadratic variation.

Figure 7.17a shows data, digitized from published images of glass uniaxial fracture, from (a) Clark and Irwin (1966) and (b) Bowden et al. (1967) and Field (1971). Individual symbols represent measurements at individual branching nodes. Comparison of different systems is enabled by normalizing the branching radii as r_i/r_1 noting that for these unidirectional fragmentation patterns $r_1 = \lambda_b$ and r_2 is distinct from r_1. A clear feature is that the branching radii at a given node exhibit considerable dispersion such that there is considerable overlap between radii at adjacent nodes. (An alternative interpretation is that the radii are quantized and that the nodes exhibit dispersion.) The shaded band highlights the nonlinear variation of radius with node and is of the form $(r_i/r_1 - 1) \sim (i - 1)^2$, clearly describing the median behavior and ranges of radii.

The plot of Figure 7.17b shows data from glass and ceramic biaxial fracture, digitized from published sources where necessary, for (a) soda-lime silicate glass (Terao 1978), (b) ZrO_2 (Nakasa and Nakatsuka 1990), (c) AlN (AN, this work), (d) (350 ± 35) MPa strength cordierite glass-ceramic (GC1, this work), (e) (245 ± 25) MPa GC1, (g) (140 ± 25) MPa GC1, (f) BSG (Nakasa and Nakatsuka 1990), (h) (245 ± 25) MPa surface stressed GC1S, and (i) laminated and thermally tempered glass (Sakai et al. 1991). As above, individual symbols represent measurements at individual branching nodes. Symbols with lines represent measurements at branching nodes averaged over many specimens; the lines are guides to the eye. The same normalization and scale is used as in the previous plot, noting that for these multidirectional fragmentation patterns $r_1 + r_2 = 2\lambda_b$ and $r_2 \approx r_1$. Once again there is considerable overlap between radii at adjacent nodes and the shaded band highlights the nonlinear variation of radius with node of the form$(r_i/r_1 - 1) \sim (i - 1)^2$, although with smaller amplitude than the previous plot. Systems (h) and (i) are regarded as toughened and yet are not clearly distinguished from the other biaxial systems, and no significant distinction between the strength levels of the GC1 system, (d)–(f), is observed.

Overall, Figure 7.17 demonstrates the quadratic "stretching out" during fragmentation pattern development, the dispersion in node radii (or vice verse), the difference in uniaxial and biaxial patterns, and the weak material dependence of normalized pattern shape. These features form the fundamental basis for fragmentation model development as demonstrated, for example, by probabilistic sequential algorithms (Quinn et al. 2005) or described by fractal-like dimensions (Nakasa and Nakatsuka 1994; Mecholsky et al. 1998).

7.6 Conclusions

An extensive examination of experimental data, both new and previously published, describing fracture of biaxially stressed brittle disks has placed many fragmentation phenomena on a firm quantitative basis. Data, many requiring new digitization from published sources, from over 40 materials fragmentation systems and hundreds of specimens, were examined. The phenomena have been characterized at three length scales—(i) that of the disks, the number of fragments, (ii) that of the fragments, the distribution of fragment masses, and (iii) that of the propagating fracture, the variation of crack branching distances. For the first phenomenon, a distinction between ceramics and glasses is observed. Ceramic failures generate fewer fragments at the same failure strength than glasses and the number of fragments generated have a weaker dependence on strength. No dependence on the number of fragments with flaw type, specimen environment, or testing protocol is observed. Failure of ceramic and glass disks containing small specimen-level stresses generates far fewer fragments, tens at most, than chemically stressed or physically tempered glass specimens, which typically number hundreds, suggesting a different fracture mechanics basis. For the second phenomenon, a clear distinction exists between disk fragment mass distributions and impact or explosion fragment mass distributions. Although, the dispersion of disk fragment mass distributions increases with strength, the relative domain width, about a factor of 10, is very small compared with impact or explosion generated fragment mass distributions, about a factor of 10^4, suggesting a different probabilistic basis. For the third phenomenon, no dependence of crack branching distance on flaw type, specimen environment, or testing protocol is observed. To first order, an inverse quadratic dependence of crack branching distance on strength is observed, and the resulting branching coefficient can be used to rank the fragmentation resistance of materials. For all three phenomena, the indentation–strength method provides a great advantage in expanding and determining the domains of flaws and thus strengths responsible for the strain energies in specimens at failure and thus range of fragmentation behavior. The data here provide a quantitative basis for simulation of fundamental fragmentation phenomena and qualitative examples of applications in materials and component development.

References

Åström, J.A. (2006). Statistical models of brittle fragmentation. *Advances in Physics* 55: 247–278.

Aoki, S. and Sakata, M. (1980). Crack bifurcation under hydrostatic pressure. *Engineering Fracture Mechanics* 13: 491–499.

Barstow, F.E. and Edgerton, H.E. (1939). Glass-fracture velocity. *Journal of the American Ceramic Society* 22: 302–307.

Bennison, S.J., Jagota, A., and Smith, C.A (1999). Fracture of glass/poly(vinyl butyral) (Butacite®) laminates in biaxial flexure. *Journal of the American Ceramic Society* 70: 1760–1770.

Blauel, J.G. and Richter, H. (1969). Thermisch induzierte Innenbrüche in Glasstäben (in German). *Glastechnishe Berichte* 42: 136–139.

Börger, A., Supancic, P., and Danzer, R. (2002). The ball on three balls test for strength testing of brittle discs: stress distribution in the disc. *Journal of the European Ceramic Society* 22: 1425–1436.

Börger, A., Supancic, P., and Danzer, R. (2004). The ball on three balls test for strength testing of brittle discs-Part II: analysis of possible errors in the strength determination. *Journal of the European Ceramic Society* 24: 2917–2928

Bowden, F.P., Brunton, J.H., and Field, J.E., Heyes, AD (1967). Controlled fracture of brittle solids and interruption of electrical current. *Nature* 216: 38–42.

Bradt, R.C. (2011). The fractography and crack patterns of broken glass. *Journal of Failure Analysis and Prevention* 11: 79–96.

Choi, S.R. and Gyekenyesi, J.P. (1998). Crack branching and fracture mirror data of glasses and advanced ceramics. Nasa Report. NASA/TM-1998-206536.

Choi, S.R., Powers, L.M., and Nemeth, N.N. (2000). Slow crack growth behavior and life/reliability analysis of 96 wt % alumina at ambient temperature with various specimen/loading configurations. Nasa Report. NASA/TM-2000-210206.

Clark, A.B.J. and Irwin, G.R. (1966). Crack-propagation behaviors. *Experimental Mechanics* 6: 321–330.

Congleton, J. and Petch, N.J. (1967). Crack-branching. *Philosophical Magazine* 16: 749–760.

Cook, R.F. (1990). Segregation effects in the fracture of brittle materials: Ca-Al_2O_3. *Acta Metallurgica* 38: 1083–1100.

Cook, R.F. (2005). Toughening of a cordierite glass-ceramic by compressive surface layers. *Journal of the American Ceramic Society* 88: 2798–2808.

Cook, R.F. (2006). Strength and sharp contact fracture of silicon. *Journal of Materials Science* 41: 841–872.

Cook, R.F. (2015). Multi-scale effects in the strength of ceramics. *Journal of the American Ceramic Society* 98: 2933–2947.

Cook, R.F., Liniger, E.G., Steinbrech, R.W., and Deuerler, F. (1994a). Sigmoidal indentation-strength characteristics of polycrystalline alumina. *Journal of the American Ceramic Society* 77: 303–314.

Cook, R.F., Liniger, E.G., and Pascucci, M.R. (1994b). Indentation fracture of polycrystalline cubic materials. *Journal of Hard Materials* 5: 190–212.

Davydova, M., Pantelev, I., and Naimark, O. (2017). Fragmentation of Mansurov granite under quasi-static compression. *Frattura ed Integrità Strutturale* 42: 170–180.

Davydova, M. and Uvarov, S. (2013). Fractal statistics of brittle fragmentation. *Frattura ed Integrità Strutturale* 24: 60–68.

Davydova, M., Uvarov, S., and Chudinov, V. (2015). Statistical laws of dynamic fragmentation of ZrO_2 ceramics. *Applied Mechanics and Materials* 784: 468–475.

Davydova, M., Uvarov, S., and Naimark, O. (2018). The effect of porosity on fragmentation statistics of dynamically loaded ZrO_2 ceramics. *Frattura ed Integrità Strutturale* 43: 106–112.

DeLellis, D.P., Mecholsky, N.A., Mechosky, J.J. Jr., and Quinn, G.D. (2020). A fractal analysis of crack branching in borosilicate glass. *Journal of the American Ceramic Society* 103: 5283–5290.

Deobald, L.R. and Kobayashi, A.S. (1992). Dynamic fracture characterization of Al_2O_3 and SiC_w/Al_2O_3. *Journal of the American Ceramic Society* 75: 2867–2870.

Donald, I.W. and Hill, M.J.C. (1988). Preparation and mechanical behaviour of some chemically strengthened lithium magnesium alumino-silicate glasses. *Journal of Materials Science* 23: 2797–2809.

Donald, I.W., Hill, M.J.C., Metcalfe, B.L., Bradley, D.J., and Bye, A.D. (1993). The mechanical properties and fracture behavior of some chemically strengthened silicate and borate glasses. *Glass Technology* 34: 114–119.

Field, J.E. (1971). Brittle fracture: its study and application. *Contemporary Physics* 12: 1–31.

Forquin, P., Rossiquet, G., Zinszner, J.-L., and Erzar, B. (2018). Microstructure influence on the fragmentation properties of dense silicon carbides under impact. *Mechanics of Materials* 123: 59–76.

Fragassa, C., Camargo, F.V., Pavlovic, A., de Figueiredo Silveira, A.C., Minak, G., and Bergmann, C.P. (2018). Mechanical characterization of gres porcelain and low-velocity impact numerical modeling. *Materials* 11: 1082.

Frank, S., Handge, U.A., Olliges, S., and Spolenak, R. (2009). The relationship between thin film fragmentation and buckle formation: synchrotron-based in situ studies and two-dimensional stress analysis. *Acta Materialia* 57: 1442–1453.

Gao, S., Nakasa, K., and Kato, M. (1998). Analysis of repeating-division and delamination processes of sprayed WC-Co coating under biaxial tension (in Japanese). *Journal of the Society of Materials Science, Japan* 47: 1151–1157.

Gilvarry, J.J. and Bergstrom, B.H. (1961). Fracture of brittle solids. II. Distribution function for fragment size in single fracture (Experimental). *Journal of Applied Physics* 32:400–410.

Glass, S.J., Tandon, R., and Heller, D. (2007). Characterization of crack branching and fragmentation patterns for ion-exchanged glass. In *Fractography of Glasses and Ceramics V: Ceramic Transactions*, Vol. 199 (Eds. Varner, J.R., Quinn, G.D., and Wightman, M.) pp. 407–420. Wiley.

Goehring, L., Nakahara, A., Dutta, T., Kitsunezaki, S., and Tarafdar, S. (2015). *Desiccation Cracks and their Patterns: Formation and Modelling in Science and Nature.* Weinheim, Germany: Wiley.

Gold, V.M. and Baker, E.L. (2008). A model for fracture of explosively driven metal shells. *Engineering Fracture Mechanics* 75: 275–289.

Gorham, D.A. and Rickerby, D.G. (1975). A hydraulic strength test for brittle samples. *Journal of Physics E: Scientific Instruments* 8: 794–796.

Grady, D.E. (2008). Fragment size distributions from the dynamic fragmentation of brittle solids. *International Journal of Impact Engineering* 35: 1557–1562.

Grady, D.E. (2010). Length scales and size distributions in dynamic fragmentation. *International Journal of Fracture* 163: 85–99.

Grady, D.E. and Kipp, M.E. (1985). Geometric statistics and dynamic fragmentation. *Journal of Applied Physics* 58: 1210–1222.

Halliday, I., Blackwell, A.T., and Griffin, A.A. (1989). The flux of meteorites on the Earth's surface. *Meteoritics* 24: 173–178.

Hicks, T., Vanina, R., and Margot, P. (1996). Transfer and persistence of glass fragments on garments. *Science & Justice* 36: 101–107.

Hill, M.J.C. and Donald, I.W. (1989). Stress profile characteristics and mechanical behaviour of chemically strengthened lithium magnesium aluminosilicate glasses. *Glass Technology* 30:123–127.

Hillig, W.B. and Charles, R.J. (1961). Concerning dislocations in glass. *Journal of Applied Physics* 32: 123–124.

Hogan, J.D., Rogers, R.J., Spray, J.G., and Boonsue, S. (2012). Dynamic fragmentation of granite for impact energies of 6–28 J. *Engineering Fracture Mechanics* 79: 103–125.

Hogan, J.D., Vincet, G., Rogers, R.J., Spray, J.G., and Schneider, M. (2014). Debris field kinetics during the dynamic fragmentation of polyphase natural ceramic blocks-Ejecta measurements and damage modes. *Experimental Mechanics* 54: 211–228.

Huang, J.Y., Huang, J.W., Sun, T., Fezzaa, K., Xu, S.L., and Luo, S.N. (2016). Dynamic deformation and fracture of single crystal silicon: fracture modes, damage laws, and anisotropy. *Acta Materialia* 114: 136–145.

Jones, T.L., Meredith, C.S., and Becker, B. (2017). First static and dynamic analysis of 3-D-printed sintered ceramics for body armor applications. *American Ceramic Society Bulletin* 96: 36–41.

Karlsson, S. (2017). Spontaneous fracture in thermally strengthened glass – a review and outlook. *Ceramics-Silikàty* 61: 188–201.

Kirchner, H.P. (1978). The strain intensity criterion for crack branching in ceramics. *Engineering Fracture Mechanics* 10: 283–288.

Kirchner, H.P., Gruver, R.M., Swain, M.V., and Garvie, R.C. (1981). Crack branching in transformation-toughened zirconia. *Journal of the American Ceramic Society* 64: 529–533.

Kirchner, H.P. and Kirchner, J.W. (1979). Fracture mechanics of fracture mirrors. *Journal of the American Ceramic Society* 62: 198–202.

Kooi, J.E., Tandon, R., Glass, S.J., and Mecholsky, J.J. Jr. (2008). Analysis of macroscopic crack branching patterns in chemically strengthened glass. *Journal of Materials Research* 23: 214–225.

Lawn, B.R. (1993). *Fracture of Brittle Solids, 2nd ed.* Cambridge, UK: Cambridge.

Lee, H., Cho, S., Yoon, K., and Lee, J. (2012). Glass thickness and fragmentation behavior in stressed glasses. *New Journal of Glass and Ceramics* 2: 138–143.

Locke, J. and Unikowski, J.A. (1991). Breaking of flat glass-part 1: size and distribution of particles from plain glass windows. *Forensic Science International* 51: 251–262.

Locke, J. and Unikowski, J.A. (1992). Breaking of flat glass-part 2: effect of pane parameters on particle distribution. *Forensic Science International* 56: 95–106.

Marshall, D.B. (1980). An improved biaxial flexure test for ceramics. *American Ceramic Society Bulletin* 59: 551–553.

Matthews, E.R. (1983). Measurements of beach pebble attrition in Palliser Bay, southern North Island, New Zealand. *Sedimentology* 30: 787–799.

Matthewson, M.J. and Field, J.E. (1980). An improved strength-measurement technique for brittle materials. *Journal of Physics E: Scientific Instruments* 13: 355–359.

Mecholsky, J.J. Jr., DeLellis, D.P., and Mecholsky, N.A. (2020). Relationship between fractography, fractal analysis and crack branching. *Journal of the European Ceramic Society* 40: 4722–4726.

Mecholsky, J.J. Jr., Linhart, R., Kwitkin, B.D., and Rice, R.W. (1998). On the fractal *Nature* of crack branching in MgF_2. *Journal of Materials Research* 13: 3153–3159.

Meibom, A. and Balslev, I. (1996). Composite power laws in shock fragmentation. *Physical Review Letters* 76: 2492–2494.

Mott, N.F. (1947). Fragmentation of shell cases. *Proceedings of the Royal Society London A: Mathematical and Physical Science* 189: 300–308.

Murr, L.E. and Szilva, W.A. (1975). Laser-induced fracture in silicon. *Journal of Materials Science* 10: 1536–1548.

Nakamura, T., Kawabata, T., Takashima, Y., and Yanagimoto, F. (2020). Effect of the stress field on crack branching in brittle material. *Theoretical and Applied Fracture Mechanics* 108: 102583.

Nakasa, K. and Nakatsuka, J.-I. (1990). Crack propagation pattern in a disk of brittle material under axisymmetrical tension (in Japanese). *Journal of the Society of Materials Science, Japan* 39: 687–693.

Nakasa, K. and Nakatsuka, J.-I. (1991). Crack initiation, propagation and branching in a disk of brittle material under axisymmetrical tension. *Engineering Fracture Mechanics* 39: 661–670.

Nakasa, K. and Nakatsuka, J.-I. (1994). Analysis of crack branching morphology in a disk of brittle material under axisymmetric tension by using branching dimension. *Engineering Fracture Mechanics* 47: 403–415.

Novák-Szabó, T., Sipos, A.Á., Shaw, S., Bertoni, D., Pozzebon, A., Grottoli, E., Sarti, G., Ciavola, P., Domokos, G., and Jerolmack, D.J. (2018). Universal characteristics of particle shape evolution by bed-load chipping. *Science Advances* 4: eaao4946.

Oakley, D.R. (1996). Crack branching in float glass subjected to biaxial loading. *Journal of Non-Crystalline Solids* 196: 139–143.

Oddershede, L., Dimon, P., and Bohr, J. (1993). Self-organized criticality in fragmenting. *Physical Review Letters* 71: 3107–3110.

O'Donnell, R.G. (1991). An investigation of the fragmentation behaviour of impacted ceramics. *Journal of Materials Science Letters* 10: 685–688.

Pike, A.W. and Zuo, Q.H. (2014). Geometric design consideration for controlled fragmentation of metallic shells. *Finite Elements in Analysis and Design* 91: 59–67.

Pioli, L., Bonadonna, C., and Pistolesi, M. (2019). Reliability of total grain-size distribution of tephra deposits. *Scientific Reports* 9: 10006.

Preston, F.W. (1931). The propagation of fissures in glass and other bodies with special reference to the split wave front. *Journal of the American Ceramic Society* 14: 419–427.

Preston, F.W. (1935). The angle of forking of glass cracks as an indicator of the stress system. *Journal of the American Ceramic Society* 18: 175–176.

Qin, Z., Pugno, N.M., and Buehler, M.J. (2014). Mechanics of fragmentation of crocodile skin and other thin films. *Scientific Reports* 4: 4966.

Quinn, J.B. (1999). Extrapolation of fracture mirror and crack-branch sizes to large dimensions in biaxial strength tests of glass. *Journal of the American Ceramic Society* 82: 2126–2132.

Quinn, G.D. (2020). On crack branching angles in glasses and ceramics. *Journal of the European Ceramic Society* 40: 4711–4721.

Quinn, G.W., Quinn, J.B., Mecholsky, J.J. Jr., and Quinn, G.D. (2005). Computer modeling of crack propagation using fractal geometry, in *Ceramic Engineering and Science Proceedings* 26: 77–84.

Quinn, G.D. and Wirth, G. (1989). Biaxial stress rupture of silicon nitride. *Materials Science and Engineering* A 109: 147–152.

Quinn, G.D. and Wirth, G. (1990). Multiaxial strength and stress rupture of hot pressed silicon nitride. *Journal of the European Ceramic Society* 6: 169–177.

Richman-Abdou, K. (2019). Kintsugi: the centuries-old art of repairing broken pottery with gold. https://mymodernmet.com/kintsugi-kintsukuroi/ (accessed 1/12/21).

Roark, R.J. and Young, W.C. (1983). *Formulas for Stress and Strain, 5th ed.* Tokyo, Japan: McGraw-Hill.

Ruoho, M., Niemelä, J.P., Guerra-Nunez, C., Tarasiuk, N., Robertson, G., Taylor, A.A., Maeder, X., Kapusta, C., Michler, J., and Utke, I. (2020). Thin-film engineering of mechanical fragmentation properties of atomic-layer-deposited metal oxides. *Nanomaterials* 10: 558.

Sakai, T., Ramulu, M., Ghosh, A., and Bradt, R.C. (1991). Cascadating fracture in a laminated tempered safety glass panel. *International Journal of Fracture* 48: 49–69.

Salem, J.A. (2013). Transparent armor ceramics as spacecraft windows. *Journal of the American Ceramic Society* 96: 281–289.

Sanchidrián, J.A., Ouchterlony, F., Segarra, P. and Moser, P. (2014). Size distribution functions for rock fragments. *International Journal of Rock Mechanics and Mining Sciences* 71: 381–394.

Sapozhnikov, S.B., Kudryavtsev, O.A., and Dolganina, N.Yu. (2015). Experimental and numerical estimation of strength and fragmentation of different porosity alumina ceramics. *Materials and Design* 88: 1042–1048.

Seghir, R. and Arscott, S.,(2015). Controlled mud-crack patterning and self-organized cracking of polydimethylsiloxane elastomer surfaces. *Scientific Reports* 5: 14787.

Shetty, D.K., Rosenfield, A.R., Bansal, G.K., and Duckworth, W.H (1981). Biaxial flexure studies of a glass-ceramic. *Journal of the American Ceramic Society* 64: 1–4.

Shetty, D.K., Rosenfield, A.R., and Duckworth, W.H. (1983a). Crack branching in ceramic disks subjected to biaxial flexure. *Journal of the American Ceramic Society* 66: C10–C12.

Shetty, D.K., Rosenfield, A.R., Duckworth, W.H., and Held, P.R. (1983b). A biaxial-flexure test for evaluating ceramic strengths. *Journal of the American Ceramic Society* 66: 36–42.

Subhash, G., Maiti, S., Geubelle, P.H., and Ghosh, D. (2008). Recent advances in dynamic indentation fracture, impact damage and fragmentation of ceramics. *Journal of the American Ceramic Society* 91: 2777–2791.

Sundaram, B.M. and Tippur, H.V. (2018). Dynamic fracture of soda-lime glass: a full-field optical investigation of crack initiation, propagation and branching. *Journal of the Mechanics and Physics of Solids* 120: 132–153.

Tandon, R. and Glass, S.J. (2015). Fracture initiation and fragmentation in chemically tempered glass. *Journal of the European Ceramic Society* 35: 285–295.

Tang, Z., Abrams, M.B., Mauro, J.C., Venkataraman, N., Meyer, T.E., Jacobs, J.M., Wu, X., and Ellison, A.J. (2014). Automated apparatus for measuring the frangibility and fragmentation of strengthened glass. *Experimental Mechanics* 54: 903–912.

Terao, N. (1953). Sur une relation entre la résistance à la rupture et le foyer d'eclatement du verre (in French). *Journal of the Physical Society of Japan* 8: 545–549.

Terao, N. (1978). Experimental study of crack propagation in brittle amorphous solids-bifurcation of a crack in glass. *Physica Status Solidi A* 47: 565–575.

Tokariev, O., Steinbrech, R.W., Schnetter, L., and Malzbender, J. (2012). Micro- and macro-mechanical testing of transparent $MgAl_2O_4$ spinel. *Journal of Materials Science* 47: 4821–4826.

Weimer, R.J. and Rogers, H.C. (1979). Dynamic fracture phenomena in high–strength steels. *Journal of Applied Physics* 50: 8025–8030.

Wittel, F., Kun, F., Herrmann, H.J., and Kröplin, B.H. (2004). Fragmentation of shells. *Physical Review Letters* 93: 035504.

Woodward, R.L., Gooch, W.A. Jr., O'Donnell, R.G., Perciballi, W.J., Baxter, B.J., and Pattie, S.D. (1994). A study of fragmentation in the ballistic impact of ceramics. *International Journal of Impact Engineering* 15: 605–618.

Wu, K., Zhang, J.Y., Li, J., Wang, Y.Q., Liu, G., and Sun, J. (2015). Length-scale-dependent cracking and buckling behaviors of nanostructured Cu/Cr multilayer films on compliant substrates. *Acta Materialia* 100: 344–358.

Yoffe, E.H. (1951). The moving Griffith crack. *The London, Edinburgh, and Dublin Philosophical Magazine and Journal of Science* 42: 739–750.

Zijlstra, A.L. and Burggraaf, A.J. (1969). Fracture phenomena and strength properties of chemically and physically strengthened glass. II. Strength and fracture behavior of chemically strengthened glass in connection with the stress profile. *Journal of Non–Crystalline Solids* 1: 163–185.

Zinszner, J.-L., Forquin, P., and Rossiquet, G. (2012). Design of an experimental configuration for studying the dynamic fragmentation of ceramics under impact. *The European Physical Journal Special Topics* 206: 107–115.

Zinszner, J.L., Forquin, P., and Rossiquet, G. (2015). Experimental and numerical analysis of the dynamic fragmentation in a SiC ceramic under impact. *International Journal of Impact Engineering* 76: 9–19.

8

Edge Chipping at Small Scales and Strengths of Diced Components

In this chapter the morphology of cracks, chips, and scallops formed by sharp particle contacts adjacent to edges of brittle materials is investigated, along with the resulting component strengths. Indentations are used as model sharp contacts and the edge-chipping behavior of indentation flaws is studied using variations in contact load and edge proximity. The load and length scales studied are newtons and micrometers, respectively. The materials focus is on the behavior of single crystal Si and composite Al_2O_3–TiC. The variation of indentation crack lengths parallel to edges is studied along with the critical condition for chip initiation from such cracks. The ensuing variation of chip size as a function of contact load and edge proximity is examined. The small-scale chip initiation and size responses are in agreement with large-scale chip measurements. The small-scale chip initiation and size measurements are combined with indentation-strength measurements to form upper and lower bound strength predictions as a function of edge-chip scallop size. Comparison with commercially-diced and inspected Si components validates the approach.

8.1 Introduction

A recent overview has highlighted the pervasive nature of *edge chipping* by sharp particle contacts (Lawn 2021). The overview included qualitative examples describing the detriments and applications of chipping in anthropological, artistic, biomedical, geological, and industrial contexts. Quantitative attention was focused on the conditions for chipping in terms of proximity of a contacting particle to a component edge and load on the particle. The measurements considered were primarily macroscopic, with length scales of order millimeters (mm) and load scales of order kilonewtons (kN). This chapter will extend the overview and present experimental measurements, also in an industrial context, of edge chipping related to sawing of components to size. Specifically, such sawing is known as "dicing" in the manufacture of semiconductor and magnetic storage devices, and edge damage during or subsequent to dicing is the focus here. Large wafers are used as substrates during the fabrication of such devices, and the wafers are sawn or diced into many rectangular dies at fabrication completion, each die containing a single semiconducting or magnetic device. Avoidance of defects associated with edge chipping during wafer dicing is thus a critical aspect of maximizing device manufacturing yield (Lei et al. 2012). As a consequence of the industrial importance of wafer dicing, the chapter here will focus on the chipping behavior of the two most important wafer materials in the semiconductor and magnetic storage industries: single-crystal silicon (Si) and the polycrystalline ceramic composite Al_2O_3–TiC. These materials were not considered earlier. In addition, the chapter will extend the earlier work in two other ways: (i) the length and load scales for edge chipping will be smaller, micrometers (μm) and newtons (N), respectively; and (ii) the experiments and analyses will include the resulting fracture strengths of chipped components. Finally, the chapter will demonstrate another of the many applications of indentation fracture techniques in the science of, and engineering with, brittle materials.

The major characteristics of an edge chip are shown in Figure 8.1. A sharp particle impinges on a material surface near a component edge and undergoes a load-unload contact cycle with peak load P, indicated by dark shading. The distance of the contact point from the edge is h. If the peak load P is large enough or the contact distance h is small enough, an edge chip, indicated by light shading, is formed by fracture on a curved surface adjacent to the edge. The chip fracture surface extends from the contact, predominantly parallel to the edge, before curving and exiting the material on the contact surface, the edge, and the free surface, predominantly perpendicular to all three. The distance from the fracture exit on the free surface to the component edge is s (cf Chai and Lawn 2007a), taken as the characteristic size of the chip and most often $s > h$. The separation of the fracture exits on the component edge is D (Chai and Lawn 2007a). Often, but not always,

Indentation Fracture: Strength and Toughness for Brittle Materials Design, First Edition. Robert F. Cook.

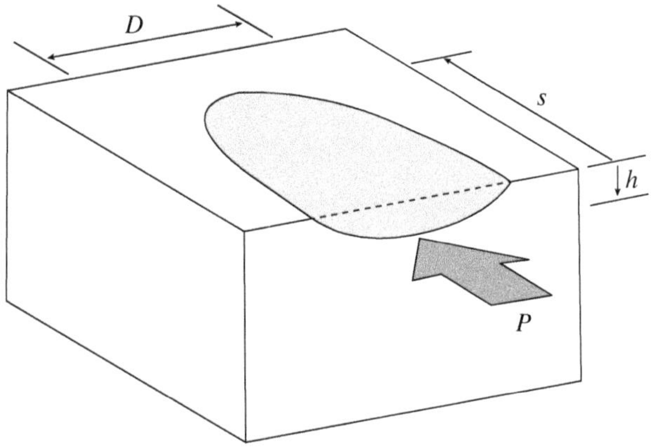

Figure 8.1 Schematic diagram of an edge chip (shaded) formed by a sharp contact, load *P*, distance *h* from the edge (dashed). The chip dimensions perpendicular and parallel to the edge are *s* and *D*, respectively.

the chip outline on the free surface is approximately semicircular such that $D \approx 2s$. If P is too small or h too large, fracture may still occur along a curved surface at the contact but if the cracks formed do not exit the material no chip is generated. The development of edge chips from contact cracks as P and h are varied is a focus of this work, along with the subsequent variations in s.

Three aspects of edge chipping are major concerns in dicing and are illustrated in Figure 8.2. The first concern is the removal of chipped material from the die to form very small loose particles of wafer material that lead to device defects in subsequent manufacturing processes. An example of an edge chip leading to a free particle is shown in Figure 8.2a, which shows a micrometer-scale chip formed during dicing of Al_2O_3–TiC. The striations of the diced surface are apparent in the upper section of the image, the polished surface of the front face of the wafer is apparent in the lower section of the image, and the grains of the polycrystalline ceramic are apparent on the curved chip surface. The contact that caused the chip was on the diced surface leading to a nearly semi-circular chip outline on the polished surface, similar to that shown in Figure 8.1. A second example of an edge chip leading to a free particle is shown in Figure 8.2b, which shows a hundred

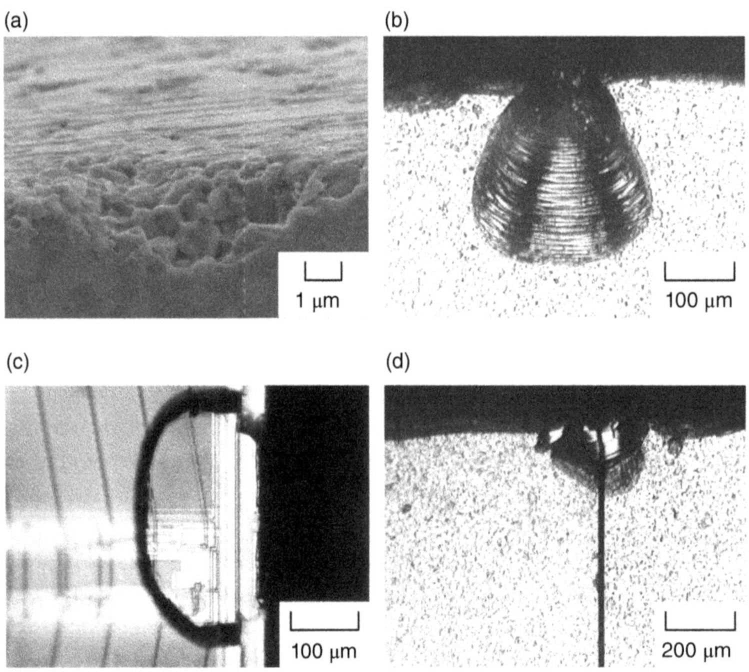

Figure 8.2 Images of edge chips illustrating deleterious effects of dicing. (a) Chip in Al_2O_3–TiC polished polycrystal. (b) Chip in single-crystal Si on die back face. (c) Chip in single-crystal Si on die front face with material attached. (d) Chip in single-crystal Si on die back face and initiated failure crack.

micrometer scale chip formed during dicing of Si. In this case, the diced edge is apparent at the top of the image, remnant film deposition features on the back face of the wafer are apparent in the lower section of the image, and ripple markings associated with fracture in single-crystal Si are apparent on the curved chip surface. The contact that caused the chip was on the diced surface, not shown, leading to a typical *scallop-shaped chip* outline on the wafer back face and $D < s$. Edge-chip "scallops" and "scalloping" during dicing are extremely common.

The second concern arising from edge chipping during dicing is the direct generation of defects in devices fabricated on the wafer front face. An example of an edge chip leading to a device defect is shown in Figure 8.2c, which shows a 100-μm scale chip formed by die handling subsequent to dicing of Si. Here, the diced edge is apparent on the right of the image, active device features on the wafer front face are apparent in the left of the image, and the displaced chip associated with subsurface fracture in single-crystal Si is apparent (the chip is held in place by electrostatic effects). The contact that caused the chip was on the diced surface, leading to a scallop-shaped chip outline on the wafer front face. The third concern arising from edge chipping during dicing is the generation of strength-controlling flaws leading to complete die failure in subsequent die handling and, particularly, during semiconductor device packaging. The common packaging geometry of active front face attachment of a die to a polymer substrate places a tensile stress in the die back face. The stress is a driving force for brittle fracture of Si from back face cracks associated with edge chips. An example of complete die failure caused by crack propagation from an edge chip is shown in Figure 8.2d, which shows a Si die back face similar to that in Figure 8.2b. The crack extending across the die from top to bottom is apparent. The variation of die strength σ with the edge-chip parameters P, h, and s is another focus of this chapter.

This chapter is divided into a further four sections. In Section 8.2, the overall development of an edge chip from a sharp contact flaw is described as P is increased or h is decreased. The experimental variations of crack lengths at contacts influenced by an edge but in the pre-chip state are presented. This section also presents experimental measurements of the threshold conditions for chipping in terms of contact load and edge proximity relations and compares the small-scale results here with the earlier larger-scale observations (Chai and Lawn 2007a). Section 8.3 presents and compares experimental measurements of the most obvious manifestation of chipping, the variation of characteristic chip size s in the post-chip state in terms of contact conditions. In Sections 8.2 and 8.3, Vickers indentations (Chapter 2) are used as model sharp contacts to generate controlled edge chips. Section 8.4 presents experimental measurements of die strengths controlled by edge chips, including those generated by dicing flaws and model indentation flaws. Section 8.5 summarizes Chapter 8 in context and suggests some directions for future work.

8.2 Sharp Contact Cracking at Edges

8.2.1 Contact Morphology

Model sharp contact flaws are formed by Vickers diamond pyramid indentations. Such flaws are often used in contact and fracture testing (Chapters 2–7), were used in earlier edge-chipping studies (Chai and Lawn 2007a; Lawn 2021), and are also used here. An advantage of indentation flaws in chip testing is that the contact load and location can be controlled, thereby controlling chip formation. Indentation contacts also have the advantage of mimicking the essential aspects of "natural" sharp contact flaws, including the formation of a plastic deformation zone (pdz) and cracks localized to the contact (Lawn and Cook 2012; Marshall et al. 2015). The following qualitative description of edge-chip development is based on extensive observations of controlled indentation deformation and fracture adjacent to component edges in many materials. The greatest experimental concern is variation in the indentation distance to an edge, h, although behavior is also greatly influenced by variations in the indentation load, P. The description is summarized in Figure 8.3. The sequence Figure 8.3a–d illustrates indentation behavior at fixed P as h is decreased from very large values for indentations remote from an edge to very small values for indentations close to an edge.

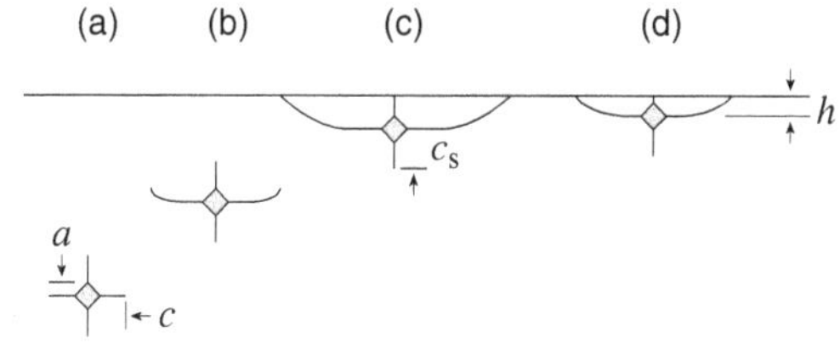

Figure 8.3 Schematic plan-view diagram of edge-chip development at Vickers indentation contacts. Contact impression is shown as shaded. (a) Remote from the edge: standard indentation. (b) Near the edge: cracks elongated parallel to edge. (c) Critical distance to edge: cracks elongated and curved to reach edge and form chip. (d) Close to edge: cracks curved and truncated by edge to form small chip.

Figure 8.3a shows a schematic plan view diagram of a "standard" Vickers indentation formed far from a component edge. The indentation contact generates a residual square surface impression, shown shaded, of semidiagonal a. The diagonal dimension is fixed at peak load, such that the mean contact pressure or hardness is $H = P/2a^2$. Inverting this equation gives the variation of a with P for materials of invariant hardness:

$$a = (1/2H)^{1/2}P^{1/2}. \tag{8.1}$$

The contact also generates an approximately hemispherical subsurface pdz (not shown) of characteristic dimension $b > a$ (Chapter 2), and the ratio b/a is a load-invariant material parameter. During the contact cycle, cracks are initiated in the pdz, propagate into the surrounding elastic matrix, and are stabilized at lengths in equilibrium with the indentation residual stress field set by b/a. The most important of these cracks for chip formation and strength control are half-penny cracks with semicircular outlines centered on the pdz (Cook and Pharr 1990; Cook 2019). At a standard indentation, two identical, perpendicular, half-penny cracks are formed, aligned parallel to the indenter diagonals and extending into the contacted surface. The surface traces of the half-penny cracks are of length c. The variation of c with P is given by the Roesler equation

$$c = BP^{2/3}, \tag{8.2}$$

where B is the material- and indenter-dependent cracking susceptibility (Chapter 2). B is predominantly dependent inversely on the material toughness T in the indentation environment and weakly dependent on the b/a ratio set by the indenter and material (b/a can be expressed in terms of E/H, where E is the material Young's modulus). Hence, for materials with invariant toughness contacted by a single indenter, B is invariant. Combining Eqs. (8.1) and (8.2) to eliminate P leads to

$$c/a^{4/3} = (2H)^{2/3}B, \tag{8.3}$$

which implies that the adjusted crack length ratio $c/a^{4/3}$ should be load invariant for standard indentations far from an edge. Extensive measurements over extended load domains on many materials are well described by Eqs. (8.1)–(8.3) (Chapter 2; see also Cook 2020). Deviations in the ratio $c/a^{4/3}$ should thus be a sensitive measure of edge effects.

Figure 8.3b shows a schematic view of a Vickers indentation close to an edge. The indentation impression and thus perceived hardness are unaffected by the edge, as are the surface traces of the half-penny cracks perpendicular to the edge. However, the surface traces of the half-penny cracks parallel to the edge are elongated and curved toward the edge. In the presence of an edge, the compliance introduced by the free surface beyond the edge significantly perturbs the indentation stress field adjacent to the contact. As a consequence, the driving force for the half-penny crack parallel to the free surface is considerably enhanced relative to the perpendicular crack, which experiences little to no compliance change and is thus relatively unaffected. Figure 8.4 shows examples of Vickers indentations on various materials with enhanced cracking adjacent to edges. An indentation on soda-lime silicate glass is shown in Figure 8.4a, and the distance to the edge (top of image) is indicated. The significantly elongated and curved cracks, initially parallel to the edge are apparent, as is the slightly elongated crack perpendicular to and truncated by the edge. The bright areas are enhanced reflections from lateral cracks parallel to the indented surface (Cook 2019; Lawn 2021; Chapter 2). The shaded areas are diminished reflections from the half-penny cracks initially perpendicular to the indented surface and curving toward the free surface (Chai and Lawn 2007a) (note the lack of shading and curvature on the perpendicular half-penny cracks). This overall contact morphology is extremely common. At greater a/h ratios, the crack curvature is greater, as shown in Figure 8.4b by an indentation on a diced Si surface (the striations were formed in dicing). The crack perpendicular to the edge is slightly longer and truncated by the edge as in Figure 8.4a. At smaller a/h ratios, the crack curvature is less, as shown in Figure 8.4c by an indentation on a polished Si surface. The cracks perpendicular to the edge are both shorter in this case. At very small a/h ratios, no cracks perpendicular to the edge may form, as shown in Figure 8.4d by an indentation in an Al_2O_3 coating on an Al_2O_3–TiC substrate. In this case, the indentation is too small (subthreshold) to initiate cracks when placed remote from an edge. However, near an edge, the enhanced driving force arising from the adjacent free surface initiates cracks parallel to the edge.

8.2.2 Crack Lengths

Figure 8.4 makes clear that edge proximity primarily affects contact-induced cracks parallel to an edge. This effect is demonstrated quantitatively in Figure 8.5, which shows the variation of crack length c as a function of distance from the edge h

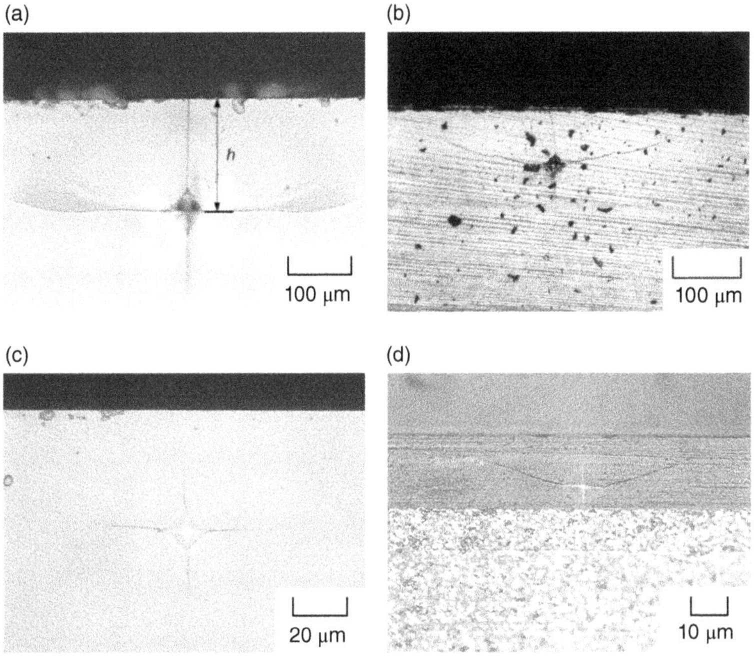

Figure 8.4 Images of indentation flaws adjacent to component edges illustrating enhanced cracking parallel to the edge (top in each image): (a) soda-lime silicate glass; (b) diced Si surface; (c) polished Si surface; and (d) diced cross-section of Al_2O_3 film on Al_2O_3–TiC ceramic composite substrate.

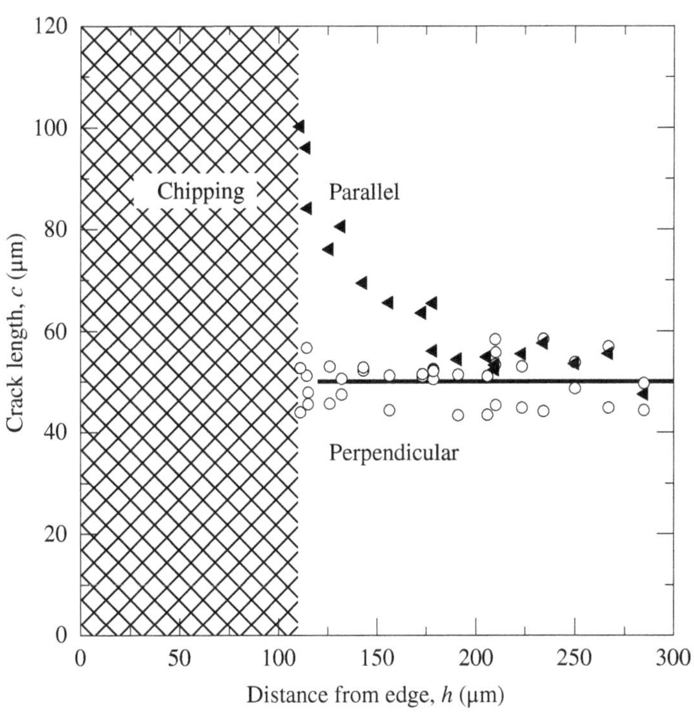

Figure 8.5 Crack length as a function of distance to an edge for indentation cracks in Al_2O_3–TiC. The cracks parallel to an edge increase in length significantly as the edge is approached before chipping occurs. Cracks perpendicular to the edge are unaffected.

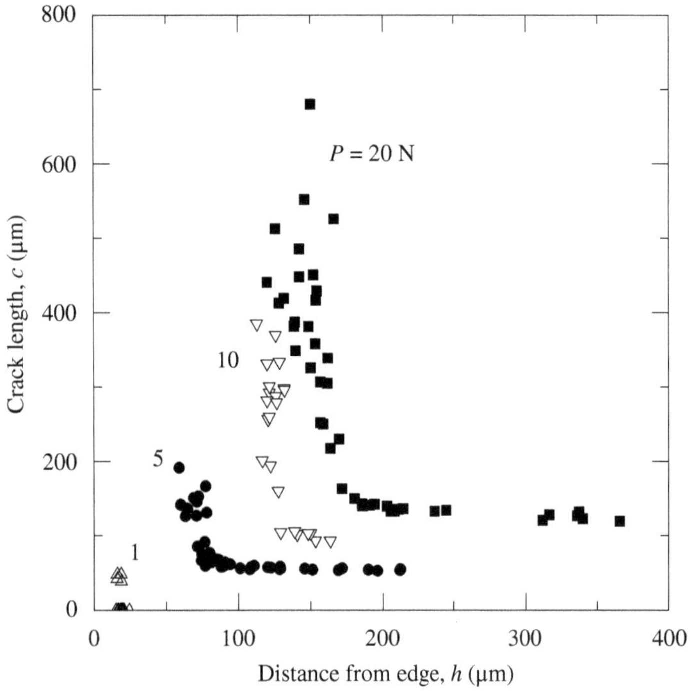

Figure 8.6 Lengths of cracks parallel to an edge for indentations in soda-lime silicate glass at the indentation loads indicated. Crack lengths increase with increasing load and edge proximity.

for indentation cracks in Al_2O_3–TiC. The indentation load was $P = 25$ N, resulting in a contact dimension of $a \approx 30$ μm. The open and closed symbols represent measurements of individual crack lengths, perpendicular and parallel to the edge, respectively. Remote from the edge, $h \approx 10a$, 300 μm, both sets of crack are similar, $c \approx 50$ μm, indicated by the solid horizontal line. As h was decreased, the cracks perpendicular to the edge did not vary appreciably in length. However, the cracks parallel to the edge increased significantly in length as h was decreased until the critical distance for chipping was reached, $h \approx 3a$, and the cracks exited the material in the configuration shown in the schematic diagram of Figure 8.3c. At indentation locations closer to the edge than this, chips formed, indicated by the cross-hatched region. Figure 8.6 shows measurements as a function of indentation load of crack lengths parallel to an edge in soda-lime silicate glass. At each load indicated, symbols represent individual crack lengths and chipping occurred at all distances less than the smallest indicated. The behavior is similar to that of Figure 8.5. Remote from the edge, cracks lengths are small and tend to an invariant length. Near the edge, the crack lengths increase significantly. At a critical distance, decreasing with decreasing indentation load, chipping and material removal occurred.

Materials comparisons of edge effects on indentation crack lengths, e.g., between Figures 8.5 and 8.6, and assessment of indentation load effects, e.g., of the various indentation load responses in Figure 8.6, is accomplished using the adjusted crack length $c/a^{4/3}$. Figure 8.7a replots the crack length $c(h)$ data of Figure 8.6 as an adjusted crack length ordinate $c/a^{4/3}$, following Eq. (8.3) and using the measured a values for each load. For simplicity, the abscissa in Figure 8.7a also follows Eq. (8.3) to generate an adjusted edge distance $h/a^{4/3}$. As expected, the adjusted crack length data lie on a load-independent response, indicated as a guide to the eye by the solid line. At large adjusted distances to the edge, the adjusted crack lengths tend to an invariant value. At small adjusted distances to the edge, $h/a^{4/3} \approx 1\ \mu\ m^{-1/3}$, the adjusted crack lengths increase significantly, before chipping occurs at very small adjusted distances (some small, subthreshold, indentations have no cracks). Figure 8.7b and c shows similar plots of adjusted edge cracking parameters for Al_2O_3–TiC and Si, respectively. The behavior is qualitatively similar to that of Figure 8.7a: invariant adjusted crack lengths far from an edge and significant increases in crack length near an edge prior to the onset of chipping. Material dependencies are revealed in quantitative comparisons of invariant adjusted crack lengths remote from an edge and in the adjusted distances to an edge at which crack lengths increase.

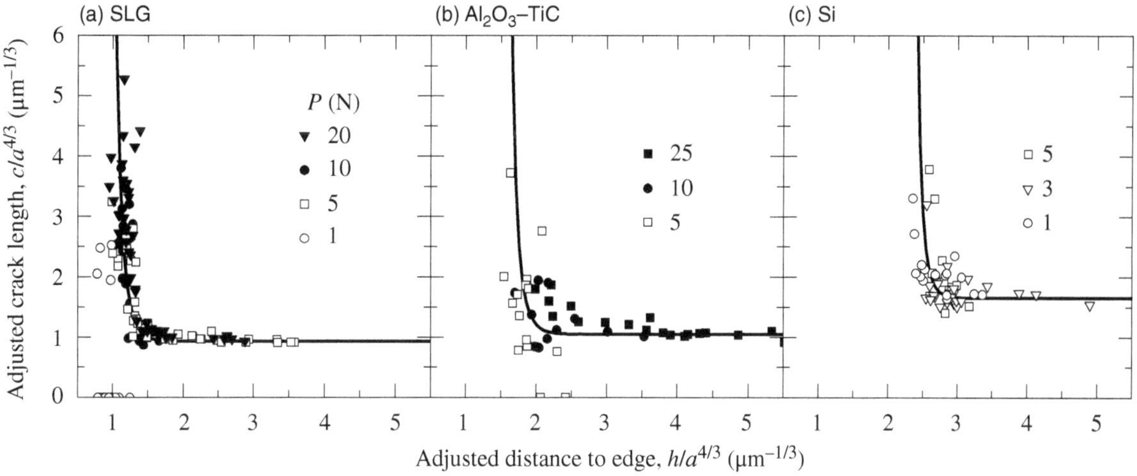

Figure 8.7 Plots of adjusted indentation crack length as a function of adjusted distance to an edge for three materials. (a) Soda-lime silicate glass. (b) Al_2O_3–TiC. (c) Si. In all three materials, use of adjusted parameters removes indentation load dependence and illustrates the increase in crack length adjacent to edges.

8.2.3 Chipping Threshold

In Figure 8.7, some dispersion is exhibited in the edge distances at which crack lengths increase significantly, about a factor of 0.3 in relative adjusted edge distance coordinates $h/a^{4/3}$. The dispersion implies sensitivity in the crack driving force to the details of contact and edge geometry. Similar dispersion is also observed in the critical condition for chipping at small edge distances, implying a similar sensitivity in the chip initiation conditions to contact and edge geometry. An example of dispersion in the critical condition for chipping is shown in Figure 8.8, which is a semilogarithmic plot of the probability of chip formation in Si as a function of adjusted edge distance for four indentation loads. As in Figure 8.7, the use of the adjusted edge distance removes the load dependence. Consistent with Figure 8.7c, chipping was observed at adjusted distances of approximately 2 (<2.5, see Figure 8.7c) with dispersion about a relative factor of 0.5. The implication is that chip initiation is a probabilistic phenomenon if the macroscopic parameter of $h/a^{4/3}$ is the sole controlled variable.

The data of Figure 8.8 were obtained by the technique of performing a series of indentations in the configurations of Figure 8.3b–d, and determining probabilities within the tested (P, h) domain. A second technique is to perform a series of static indentations in configuration Figure 8.3b and halt investigation when configuration Figure 8.3c is encountered. This approach provides an upper bound to the (P, h) conditions defining the domain of configuration Figure 8.3c; uncertainty in h is generated as P is controlled. Yet a third technique is to perform a single dynamic indentation and observe the indentation in situ as P is increased and halt the experiment when configuration Figure 8.3c is encountered. This approach provides a lower bound to the (P, h) conditions defining the domain of configuration Figure 8.3c; uncertainty in P is generated as h is controlled. This latter technique was that implemented in a previous macroscale study (Chai and Lawn 2007a), although uncertainties were not given.

Figure 8.9 shows estimations of the initiation conditions for indentation edge-chip formation in (P, h) coordinates for several materials. Chip initiation was taken as the configuration of Figure 8.3c and recorded using the three techniques above. The open symbols represent observations from the present microscale study, with bars representing the estimated uncertainties. The solid symbols represent observations from a macroscale study (Chai and Lawn 2007a) digitized from published results, recognizing these as lower bounds. The solid line is the published $P \sim h^{3/2}$ empirical fit to the macro scale glass data. The data from tests at the microscale (about 100 μm edge distance) and macroscale (about 1000 μm edge distance) are in agreement both qualitatively and quantitatively. Tougher materials (alumina, zirconia) exhibit greater chip initiation loads (relative to glass) at the macro-scale and the micro-scale (Al_2O_3–TiC exhibits greater initiation loads than glass). Over a factor of 10^2 in edge distance, and therefore 10^3 in load, the $P \sim h^{3/2}$ trend is observed. Extrapolation of the glass behavior from the macroscale to the microscale underestimates the critical edge distance by a relative factor of about 0.3. As the extrapolation does not take into account the lower-bound nature of the macroscale data, the underestimation is consistent with the probabilistic data of Figure 8.8.

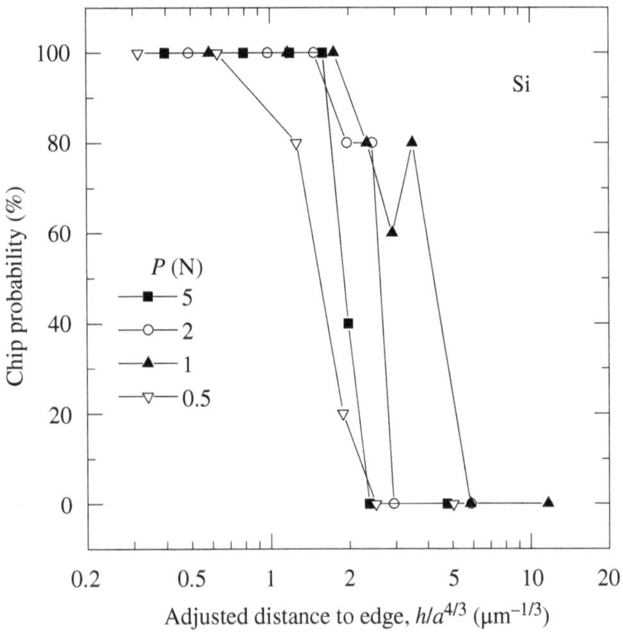

Figure 8.8 Plot of chip probability as a function of adjusted edge distance for several indentation loads in Si. Remote from the edge, the probability of indentation-induced chipping is zero. Close to the edge, the probability increases to certainty over a range of adjusted edge distances. Approximately 10 indentations/point.

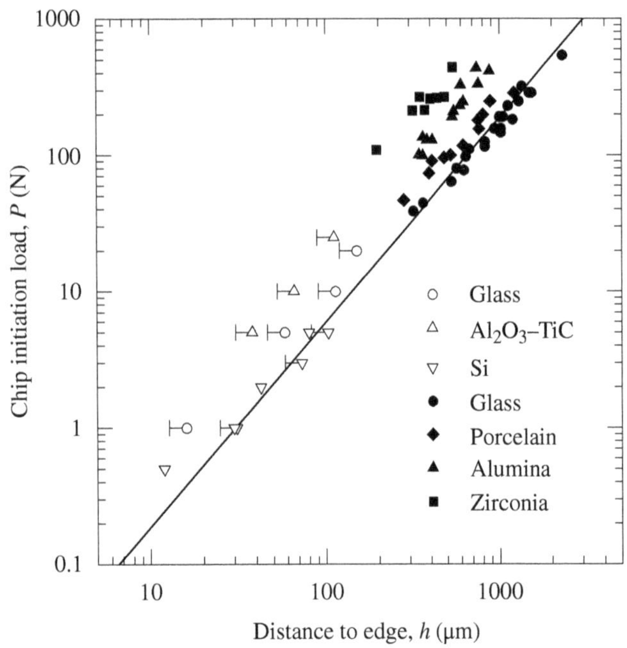

Figure 8.9 Plot of indentation edge-chip initiation conditions expressed in terms of contact load and edge distance. Macroscale and microscale observations are in agreement. Source: Includes macro-scale data adapted from Chai and Lawn (2007a).

8.3 Edge Chip Size Variation

Figures 8.5–8.7 considered variations in the chip-free configuration, Figure 8.3b. Figures 8.8 and 8.9 considered variations in the chip initiation configuration, Figure 8.3c. In this section, variations in the well-developed chip configuration, Figure 8.3d, are considered. An example of chip size variation for indentation flaws in Al_2O_3–TiC is shown in Figure 8.10.

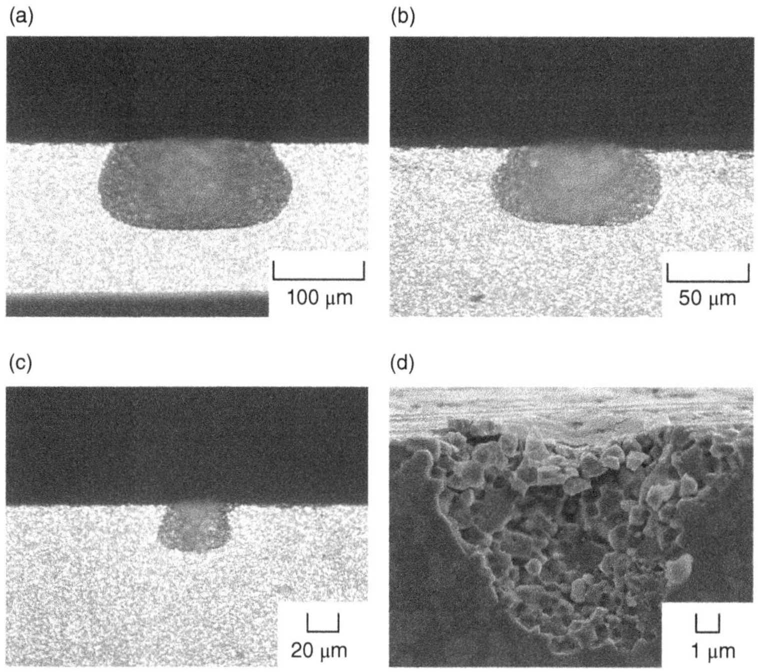

Figure 8.10 Images of edge chips in Al_2O_3–TiC generated by indentations on the contact face perpendicular to the chip. Edge distances of 20–3 μm were used to form the chips from largest to smallest, (a)–(d).

A particular focus of edge-chip characterization is the variation in chip size on the noncontacted, free surface, Figure 8.1, as a function of contact load P and edge proximity h. In contrast to Sections 8.1 and 8.2, in which indentation crack lengths c were observed to became longer and chipping prevalence was observed to increase as edge proximity h decreased, chip size s will be seen to decrease as h decreases. An example of this behavior is shown in Figure 8.10, in which the flaws were created from largest to smallest, Figure 8.10a–d, using indentation loads from $P = 3$ to 0.25 N with associated impression dimensions of approximately 4–2 μm. Edge distances were $h = 20$–3 μm. The scallop size s decreased from approximately 100–10 μm—the potency of the free edge to generate chips at sharp contacts is clear as $s > h$ and $s > a$. The commonly observed variation in chip morphology with size is also noticeable in Figure 8.10: larger chips, Figure 8.10a and b, are oblate ($D > s$, see Figure 8.1 for notation), moderate chips, Figure 8.10c, are near semicircular ($D \approx s$), and small chips, Figure 8.10d, are slightly prolate ($D < s$). The tilt of the specimen showing the smallest chip in Figure 8.10d enables the intact half of the contact impression to be observed on the indented face. In this case, the impression size and chip size are comparable to the scale of the microstructure. The similarity between the small-scale artificial chip of Figure 8.10d and the dicing related natural chip of Figure 8.2a is clear.

Figure 8.11 shows logarithmic plots of the variations in chip size s with edge distance h for small-scale chipping: in Figure 8.11a of two Al_2O_3–TiC systems and in Figure 8.11b of Si. Also shown in Figure 8.11 (small open symbols) are the responses observed for large-scale chipping of zirconia, alumina, porcelain, and glass digitized from the published results of an earlier study (Chai and Lawn 2007a), along with a characteristic linear response for large chips, $s = 5h$ (solid line). For the Al_2O_3–TiC and Si measurements presented here, a range of indention loads P was used, leading to a range of contact dimensions a. In the case of the Al_2O_3–TiC systems, all indentation loads used were subthreshold. That is, remote from an edge no indentation generated an indentation half-penny crack. In the case of Si, all indentation loads used were postthreshold. That is, remote from an edge all indentations generated indentation half-penny cracks. The impression dimensions a for the Al_2O_3–TiC systems varied from 2 to 14 μm, those for Si varied from 4 to 16 μm. The edge distances for the Al_2O_3–TiC systems varied from 3 to 50 μm, those for Si varied from 2 to 80 μm. For each value of h, five to eight indentations were made and the symbols with bars in Figure 8.11 represent the means and standard deviations of the experimentally observed dimensions. For both Al_2O_3–TiC and Si, only a few indentations at the larger h values exhibited chipping and measurements in Si for these cases are represented as isolated symbols. Different symbols in Figure 8.11 represent different P values. Optical microscopy was used for all measurements.

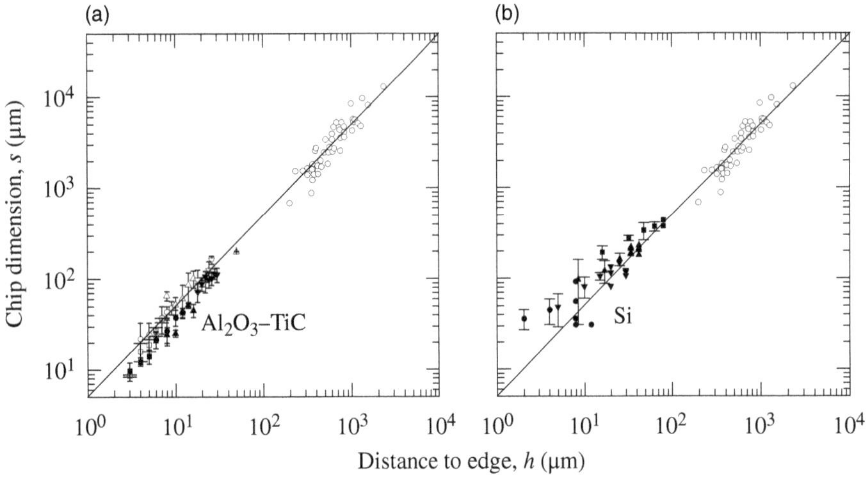

Figure 8.11 Plots of chip dimensions as a function of edge distances for (a) Al_2O_3–TiC and (b) Si. Different symbols represent different indentation loads. Source: Includes macro-scale data adapted from Chai and Lawn (2007a).

The solid symbols in Figure 8.11a represent the Al_2O_3–TiC system shown in the images of Figures 8.2a and 8.10. The chipping data for this material are slightly overestimated by a linear response extrapolated from larger-scale data associated with earlier work (Chai and Lawn 2007a). The small *s* open symbols in Figure 8.11a represent the Al_2O_3 film–Al_2O_3–TiC substrate system shown in the image of Figure 8.4d. The data for this material exhibit slightly larger chip dimensions for a given edge distance than the bare Al_2O_3–TiC system. Observations of chipping in the film on substrate system suggest a weak tendency to interfacial fracture that may explain the slightly greater chip dimensions. The data for the Al_2O_3 film on Al_2O_3–TiC system are well described by the extrapolated linear response. The solid symbols in Figure 8.11b represent the Si system shown in the image of Figure 8.2b. In this case, the chipping data at the largest *h* values for the material are well described by the extrapolated linear response, but the smaller *h* value data are clearly underestimated. In more detail, the Si data deviate systematically from the extrapolated linear response for edge distances less than the largest distance to give rise to anomalous chipping at a given indentation load. Specifically, the Si responses at each load exhibit greater chip sizes and a weaker dependence of chip size on edge distance, approximately $s \sim h^{1/2}$.

Two major points may be drawn from Figure 8.11. The first point, supporting the conclusion reached in the earlier work (Chai and Lawn 2007a), is that chip size *s* as a function of edge distance *h* is predominantly material invariant. The results here add three more material responses in support of this conclusion and extend the domain of applicability by a factor of 100 to smaller scales. The second point builds on the discussion above regarding the determination of different bounds on chipping parameters by different test methods. The macroscale tests provided lower bounds on the chip sizes *s* at each edge distance *h*, as the reported *s*(*h*) responses (small, open symbols) reflected the minimum contact load chip initiation condition, Figure 8.3c. The microscale tests provide the upper bounds and greater values of *s* at each *h* value, as loading continues beyond the minimum initiation condition to greater loads if the minimum condition does not generate a chip. As a consequence of the overload, longer transverse cracks, Figures 8.3b and 8.7, are generated prior to chipping and thus larger chips are eventually generated from these larger precursors. As a consequence, materials that generate many chip initiation features during indentation contact will tend to the solid line *s*(*h*) lower bound, such as Al_2O_3–TiC in Figure 8.11a. Materials that generate chip initiation features in a gradual, probabilistic manner during contact will exhibit a range of responses greater than the lower bound, such as Si in Figure 8.11b. Hence, the first point above should probably be modified to state that chipping behavior is material invariant for the lower-bound contact conditions pertaining to chip initiation.

Another way of looking at this second point regarding bounds on chipping is to replot the Si data of Figure 8.11b as chip aspect ratio, s/h, as a function of the adjusted impression size, a/h, as shown in Figure 8.12. At a given edge distance, as the impression size, and thus peak contact load, increases, the chip aspect ratio increases from the minimum value of 5 to almost 20. In physical terms, small contacts far from the edge generate "stubby" chips and large contacts close to the edge generate "slender," shard-like chips. The minimum aspect ratio stubby chips reflect a material-invariant lower bound on chip initiation. That is, the probability of chip initiation is increased if the load and thus impression dimension are increased beyond the threshold at a given edge distance, or if the edge distance is decreased at the threshold load. In either case, the chip aspect ratio becomes more slender.

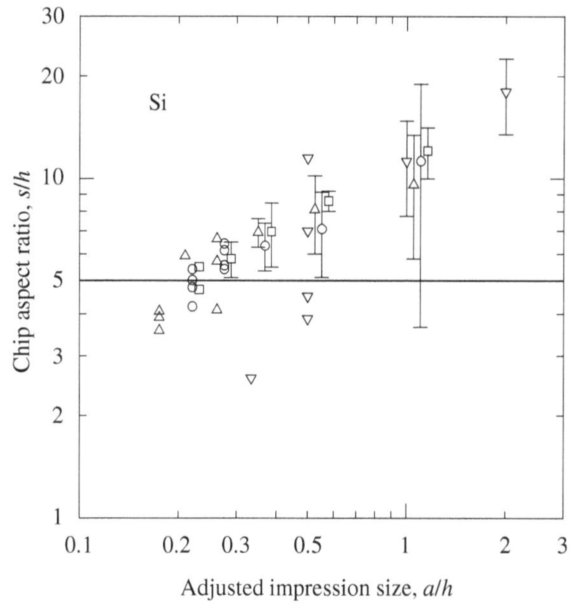

Figure 8.12 Plot of chip aspect ratio as a function of adjusted impression size for Si. Different symbols represent different indentation loads and are displaced for clarity. An aspect ratio of 5 observed for chips generated by relatively small impressions reflects a material-invariant lower bound on edge chipping.

8.4 Edge Chip Strength Variation

In addition to damaging active devices and creating loose particles of material, Figure 8.2a–c, edge chips formed during or subsequent to dicing decrease the failure load of a component by generating strength-controlling flaws, Figure 8.2d. The most visible feature of such flaws is the scallop shaped-chip of material removed from the component edge, and the connection between a visible scallop and a crack propagating across a component is often clear. Hence, assessments of "edge quality" usually quantify the number and size of scallops visible on diced edges, with the expectation that such parameters will correlate not just with device damage and particle creation, but with component strength.

There are two major reasons why such a scallop size–component strength correlation might not be straightforward. First, the predominant plane of the scallop crack is parallel to the free surface. The cracks that control component strength are perpendicular to the free surface, however, as is the tensile stress applied to the component subsequent to scallop formation (often by bending). Although the size of a scallop is related to the size of a strength-controlling crack by the contact load, edge proximity has a much greater effect on scallop size than on the lengths of cracks perpendicular to the edge, c_s, Figure 8.3c. Plan view inspection of scallop dimensions does not provide information regarding the aspect ratio of the scallop or the likely contact load or dimensions. Estimating the perpendicular crack length is thus difficult. The second reason for uncertainty in strength estimation from scallop size measurements is that an edge contact removes a substantial, but unknown fraction, of the pdz beneath a contact, along with the scallop-shaped particle of material. pdz formations at contacts generate significant residual stress fields and thus driving forces for crack propagation, including cracks perpendicular to edges. The driving force for crack propagation is thus reduced in a variable manner by scallop formation.

Hence, visual inspections can certainly enable assessments of edge "roughness" and provide clear predictions for the direct effects of chipping, such as generation of device defects and free particles. However, as edge chipping is an indirect measure of strength-limiting flaws, edge inspections can only provide bounds on the strengths of components limited by near-edge contacts. The rest of Section 8.4 builds on Sections 8.2 and 8.3 to provide such bounds and interpret measurements of edge chipping *and* strength on diced Si components.

8.4.1 Strength Measurements

Figure 8.13 shows logarithmic plots of the strength σ of Si dies containing controlled indentation flaws. The dies were approximately 15 mm × 8 mm × 0.7 mm and were strength tested in four-point bending perpendicular to the long axis of a die. Flaws were introduced on the 15 mm × 0.7 mm die face at specified distances h from the prospective tensile die edge. The same set of indentation loads P used in Figures 8.11 and 8.12 were used to introduce the flaws. The strengths of dies

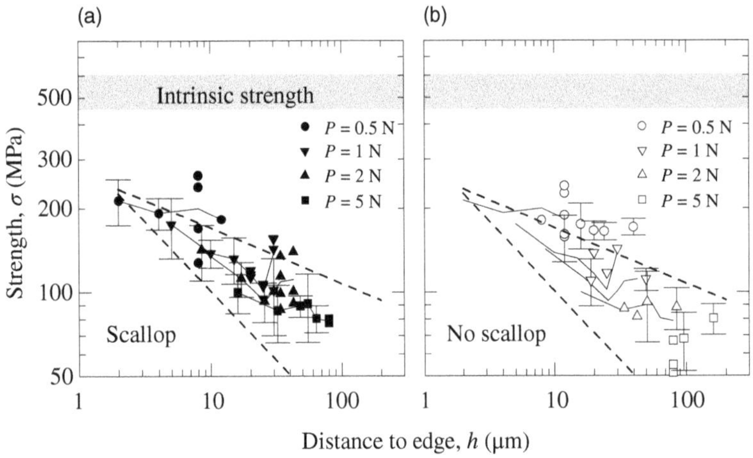

Figure 8.13 Logarithmic plots of strength for Si dies containing indentation flaws located distances from die edges. Indentation loads are indicated. Gray bars indicate indention-free intrinsic strengths. (a) Closed symbol data are for indentation flaws with edge scallops. (b) Open symbol data are for indentation flaws with no scallop.

containing no such flaws and considered by optical inspection to exhibit no significant edge scallop (active 15 mm × 8 mm die faces, defects <1 μm) are shown as the gray bands, (529 ± 77) MPa, and this strength is referred to as the intrinsic strength of the dies, σ_I. The strengths of dies containing introduced edge flaws were all less than this value. At least eight dies for each (P, h) combination were indented, examined for the formation of a scallop, and strength tested.

The filled symbols in Figure 8.13a represent the strengths of dies containing indentation flaws that exhibited an edge chip in the form of a removed scallop of material, Figures 8.2b, 8.3c and d, and 8.10. The symbols with bars represent the means and standard deviations of at least five strength measurements at the (P, h) combinations indicated. Other symbols represent individual strength measurements. As in Figures 8.11 and 8.12, smaller indentation loads P, and thus smaller contact impressions a, enabled smaller h values to be examined. Also as in Figures 8.11 and 8.12, at larger edge distances the probability of scallop formation decreased, indicated by the increasing proportion of individual symbols at large h values for a given P value. The solid lines are guides to the eye and connect the median strengths at each h value for a specified value of P. There is a clear decrease in strength with increasing edge distance, both within a fixed P set and between P values. The dashed lines are upper and lower bounds to the data. The strength lower bound is of the form $\sigma \sim h^{-0.5}$ and corresponds to the material-invariant lower bound for scallop initiation indicated in Figure 8.11. The strength upper bound is of the form $\sigma \sim h^{-0.20}$ and corresponds to the deviations to larger scallop sizes exhibited by Si from the material-invariant bound of Figure 8.11. The bounds are considered in greater detail in Section 8.4.2.

The open symbols in Figure 8.13b represent the strengths of dies containing indentation flaws that exhibited no scallop, Figures 8.3b and 8.4. The symbols with bars represent the means and standard deviations of at least five strength measurements at the (P, h) combinations indicated. Other symbols represent individual strength measurements. As in Figure 8.6, smaller indentation loads P, and thus smaller contact impressions a, enabled smaller h values to be examined. As in Figure 8.8, at smaller edge distances, the probability of scallop formation increased, indicated by the increasing proportion of individual symbols at small h values for a given P value. The solid and dashed lines are reproduced from Figure 8.13a and are the guides to the eye and bounds for the scallop strength data, respectively. As in Figure 8.13a, the scallop-free strength data exhibit a clear decrease in strength with increasing edge distance, both within a fixed P set and between P values. The scallop-free strength data are only marginally greater than the scallop median strength trends for selected indentation loads and for all loads are well within the upper and lower bounds determined for the scallop strength data. The similarity in strengths supports the expectation above that information regarding scallop size alone, or, as here, even scallop presence or absence, may be an insufficient indicator of strength. Section 8.4.2 illustrates how the data of Figure 8.13, when combined with those from Figures 8.9 and 8.11, can provide practical strength prediction bounds for direct comparison with industrially-based measurements.

8.4.2 Strength Analysis and Prediction

In industrial microelectronics production, Si devices are always inspected visually after dicing to assess edge quality, usually to estimate electrical yield loss due to edge chip or scallop formation and disruption of the active device, Figure 8.2c.

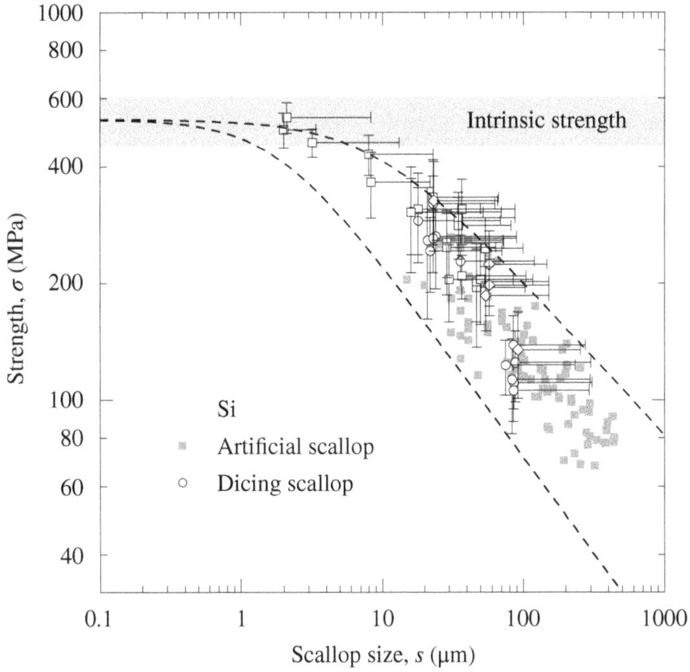

Figure 8.14 Plot of Si die strength as a function of edge-chip scallop size. Solid symbols: artificial indentation-induced scallops. Open symbols: dicing-induced scallops. Gray bar: maximum, intrinsic strength. Dashed lines: strength bounds predicted from separate tests.

The parameters most often reported are the active face scallop density expressed as number of scallops/edge length, mean scallop size, and maximum observed scallop size. The devices are sometimes inspected visually on the (diced) side face, and rarely, if ever, inspected mechanically by strength testing. Hence, yield loss and reliability truncation due to handling and packaging stresses that are large relative to die strength must be estimated from visual inspection data. Such estimates require dedicated experiments in which the same sets of dies undergo both the customary visual inspection and strength testing.

Here, three different form factors of fully fabricated commercial dies were sawn to size, all approximately 15 mm × 8 mm, in 38 different dicing runs. After the completed production sequences, the dies were inspected visually using standard protocols and the scallop values reported for each dicing run as the average scallop size and the maximum observed scallop size. Approximately 10 dies were sampled randomly from each run after visual inspection and subsequently strength tested in bending. In a separate control experiment, approximately 80 dies were sampled randomly from the dicing runs and indented so as to generate an edge scallop and the scallop size measured. The strengths of these dies were then also tested in bending.

Figure 8.14 is a logarithmic plot of the strength σ of the Si dies containing both artificial and dicing scallops as a function of scallop size s. Note that h, P, and a are not known or not specified for these dies. The shaded bar represents the intrinsic strength σ_I of pristine dies, repeated from Figure 8.13. The open symbols with bars represent measurements of the dies containing dicing scallops. Different open symbols represent different die form factors; there was no significant distinction. An open symbol position indicates the mean die strength and the mean scallop size for a dicing run. The vertical bars indicate the standard deviations of the strength measurements. The horizontal bars indicate the maximum scallop dimension observed in a dicing run. The solid symbols represent measurements of individual dies containing the artificial indentation scallops. The strength data for the dicing and artificial scallops overlap considerably and exhibit a clear decrease in strength with increasing scallop size. The dashed lines in Figure 8.14 are lower- and upper-bound predictions of the strength, based on the separate strength results of Figure 8.13 and determined as follows.

The lower-bound strength prediction is based on the material-invariant scallop (chip) initiation load response of Figure 8.9, which shows that the contact load for scallop initiation varies with edge distance as

$$P \sim h^{3/2}. \tag{8.4}$$

Comparison with Eq. (8.2) implies that the (unobserved) crack perpendicular to the edge at scallop initiation, Figure 8.3c, notated here as c_s, should thus vary linearly with edge distance,

$$c_s \sim h. \tag{8.5}$$

This crack controls component strength, but is not involved in scallop formation and only observed in section view, Figures 8.3 and 8.4, but not plan view, Figure 8.2. Approximating the strength variation with crack length to be described by the Griffith equation (Chapter 2; see also Lawn 1993), gives

$$\sigma \sim c_s^{-1/2}, \tag{8.6}$$

and thus for strength-controlling cracks formed at scallop initiation, using Eq. (8.5),

$$\sigma \sim h^{-1/2}. \tag{8.7}$$

The final step is to replace the unknown edge distance with the measured scallop size using the material-invariant lower-bound response of Figure 8.11,

$$s \sim h. \tag{8.8}$$

Hence, from Eqs. (8.6) and (8.8), the variation of strength with scallop size is

$$\sigma \sim s^{-1/2}. \tag{8.9}$$

Equation (8.9) is lower bound to the strength as Figure 8.11b shows that similar-sized scallops may be generated over a range of contact loads, if the smaller loads are further from the edge. A smaller load gives rise to a smaller crack and therefore a greater strength. As edge distance, contact load, and strength-controlling crack size are all unknown, Eq. (8.9) is thus a lower-bound variation.

In practical terms, the lower-bound calibration strength σ_L measured for known values of edge distance, Eq. (8.7) and the lower dashed lines in Figure 8.13, was given by $\sigma_L = 320h^{-1/2}$, where $[\sigma_L]$ = MPa and $[h]$ = μm. The lower bound for scallop size as a function of edge distance, Eq. (8.8) and the solid lines in Figure 8.11, was given by $s = 5h$. The full lower-bound prediction was obtained by forming the smooth p-norm of the intrinsic strength, σ_I, and the scallop-controlled strength, $\sigma_L(s)$:

$$\sigma^q = \sigma_I^q + \sigma_L^q, \tag{8.10}$$

where q is an empirical smoothing exponent and here $q = -4$ was used. The lower dashed line in Figure 8.14 is the lower-bound strength prediction from Eq. (8.10).

The upper-bound strength prediction is based on empirical Si-specific strength and scallop size responses. The upper bound of the Si strength variation with edge distance for the indentation flaws, Figure 8.13, is given by

$$\sigma \sim h^{-x}, \tag{8.11}$$

where x is an empirical exponent and here $x = 0.20$ was used. The variation of the Si small scallop dimensions with edge distance for indentation flaws, Figure 8.11b, is given by

$$s \sim h^{0.5}, \tag{8.12}$$

Combining Eqs. (8.11) and (8.12) gives

$$\sigma \sim s^{-2x}. \tag{8.13}$$

In more detail, the proportionality terms in Eqs. (8.11) and (8.12) are decreasing and increasing functions, respectively, of indentation load. As a consequence, the proportionality term in Eq. (8.13) is nearly load invariant and thus explicit significant indentation load effects are not expected if strength is expressed as a function of scallop size.

In practical terms here, the upper-bound calibration strength σ_U measured for known values of edge distance, Eq. (8.11) and the upper dashed lines in Figure 8.13, was given by $\sigma_L = 270h^{-0.20}$, where $[\sigma_U]$ = MPa and $[h]$ = μm. The upper bound for scallop size as a function of edge distance, Eq. (8.12) and the solid symbols in Figure 8.11b, was given by $s = 50h^{0.5}$, where $[s] = [h]$ = μm. The full upper-bound prediction was obtained by forming a p-norm as above:

$$\sigma^q = \sigma_I^q + \sigma_U^q, \tag{8.14}$$

where again $q = -4$. The upper dashed line in Figure 8.14 is the upper-bound strength prediction from Eq. (8.14).

The upper- and lower-bound strength predictions of Figure 8.14 approach the appropriate asymptotic limits, $\sigma \sim s^{-0.4}$ and $\sigma \sim s^{-0.5}$, respectively, at large scallop dimensions. Both predictions approach the intrinsic strength limit, $\sigma_I = 529$ MPa at small scallop dimensions. The bounds are derived from controlled flaw indentation scallop and strength measurements. Hence, it is expected that the self-consistent analysis above should describe independently measured strengths as a function of scallop size. This expectation is realized by the fact that the strength prediction bounds enclose the control, solid symbol, indentation data of Figure 8.14. The similarity in appearance of dicing-induced die fabrication scallops and indentation-induced artificial scallops, Figures 8.2 and 8.10, leads to the expectation that the scallop-controlled strengths of both groups would be similar. This expectation is realized by the overlap of the open and closed symbol experimental data of Figure 8.14 and the fact that the strength prediction bounds enclose all the strength data and match the upper-bound dicing-derived data.

8.5 Summary

Small-scale (micrometer) chips generated by light sharp contacts (newtons) adjacent to the edges of brittle material components behave similarly to large-scale chips (millimeters) formed by heavy contacts (kilonewtons). A critical contact combination of peak load P and distance from the edge h generates a chip of characteristic size s on the free, noncontacted surface. The work here has extended to smaller scales by more than a factor of 100 the quantitative relations between P, h, and s established in a previous large-scale chip study (Chai and Lawn 2007a) and extended the materials set to include the chipping characteristics of two industrially-relevant materials, Si and Al_2O_3–TiC. The work here has also identified and quantified the behavior of nonchip cracks prior to the onset of chipping and quantified chip size variations for conditions beyond the critical (P, h) chipping threshold. In particular, the material-invariant chipping conditions identified earlier (Chai and Lawn 2007a) are shown to be lower-bound chipping probability conditions.

In distinction to the focus on chip morphology in previous works (Chiu et al. 1998; Chai 2011, 2017; Chai and Lawn, 2007a,b) and the recent overview (Lawn 2021), a major goal here has been the prediction of the strengths of chipped components. In particular, the strengths of diced Si components generated in semiconductor microelectronics fabrication have been considered. A feature of Si dicing is the formation of chips on sawn edges in a characteristic scallop morphology. Estimation of device yield and reliability using predictions of die strength from the customary postprocessing scallop quantification was thus a key motivation. Experiments and analyses here demonstrated that characterization of indentation- and dicing-induced scallop morphology, combined with targeted indentation–strength tests, can be used to predict and verify dicing-controlled die strengths. In this regard, the work here should be viewed as complementary to experiments and analyses regarding strengths of dies controlled by back grinding (Chapter 10). In the work here, the dies were full thickness and strengths were controlled by edge flaws arising from dicing. In the complementary work, back face ground dies were thinned prior to dicing and strengths were controlled by surface flaws arising from grinding.

Just as the work on grinding flaws builds on the fundamental materials science of sharp linear contacts, the scallop work here builds on the fundamental materials science of strength of sharp "point" contacts. Specifically, as noted in the recent overview (Lawn 2021), the chip crack generated by a sharp contact adjacent to an edge is directly analogous to the lateral crack generated by a sharp contact on a material surface far from edges. (In fact, lateral cracking is often referred to simply as "chipping," Cook 2015.) In the case of contacts far from edges, lateral cracks remove material and weaken indentation residual stress fields (Swain and Lawn 1976; Marshall et al. 1982; Chen et al. 2005), thereby increasing the strengths associated with the two sets of half-penny cracks (Chapters 3 and 4, see also Cook and Roach 1986; Cook 2015). In the case of contacts adjacent to edges, one set of half-penny cracks removes material (the "scallop") and weakens the residual field, thereby increasing the strength associated with the remaining truncated half-penny crack. The work here is a first step in quantifying this phenomenon.

From a materials engineering perspective, the dicing flaws studied here lie in the domain of one-dimensional arrays of flaws, in this case arrayed along edges, and contrast with two- and three-dimensional arrays of flaws in areas (e.g. grinding striations) and volumes (e.g. pores), respectively. The framework to analyze the size distributions of such flaws and the resulting strength distributions is well established and has been applied to sawn polycrystalline Al_2O_3 components among others (Cook and DelRio 2019). The framework has also been applied in analysis of the flaw distributions and resulting strength distributions of microscale microelectromechanical systems (MEMS) components. MEMS components are typically submillimeter in scale and formed from layers of polycrystalline Si. The strength-limiting flaws in MEMS are usually nanometer-scale grain boundary grooves on component side-walls (Cook et al. 2019). Such flaws thus form one-dimensional arrays with a characteristic linear flaw density, just as dicing-induced flaws. As the grooves are formed chemically, the flaws are smaller than the physically formed dicing edge chips. Hence, MEMS strengths are much greater, a few GPa, compared with die strengths, a few hundreds of MPa. The strength framework is of course still applicable, and a potential next step in edge-chip analysis is to apply the strength framework to an array of edge chips.

References

Chai, H. (2011). On the mechanics of edge chipping from spherical indentation. *International Journal of Fracture* 169: 85–95.

Chai, H. (2017). Modelling edge chipping in flint knapping, cutting tools and sharp teeth using a trapezoidal prism structure. *International Journal of Solids and Structures* 104: 1–7.

Chai, H. and Lawn, B.R. (2007a). A universal relation for edge chipping from sharp contacts in brittle materials: a simple means of toughness evaluation. *Acta Materialia* 55: 2555–2561.

Chai, H. and Lawn, B.R. (2007b). Edge chipping of brittle materials: effect of side-wall inclination and loading angle. *International Journal of Fracture* 145: 159–165.

Chen, X., Hutchinson, J.W., and Evans, A.G. (2005). The mechanics of indentation induced lateral cracking. *Journal of the American Ceramic Society* 88: 1233–1238.

Chiu, W.C., Thouless, M.D., and Endres, W.J. (1998). An analysis of chipping in brittle materials. *International Journal of Fracture* 90: 287–298.

Cook, R.F. (2015). Multi-scale effects in the strength of ceramics. *Journal of the American Ceramic Society* 98: 2933–2947.

Cook, R.F. (2019). Fracture sequences during elastic-plastic indentation of brittle materials. *Journal of Materials Research* 34: 1633–1644.

Cook, R.F. (2020). A critical evaluation of indentation crack lengths in air. *Journal of the American Ceramic Society* 103: 2278–2295.

Cook, R.F. and DelRio, F.W. (2019). Determination of ceramic flaw populations from component strengths. *Journal of the American Ceramic Society* 102: 4794–4808.

Cook, R.F. and Pharr, G.M. (1990). Direct observation and analysis of indentation cracking in glasses and ceramics. *Journal of the American Ceramic Society* 73: 787–817.

Cook, R.F. and Roach, D.H. (1986). The effect of lateral crack growth on the strength of contact flaws in brittle materials. *Journal of Materials Research* 1: 589–600.

Cook, R.F., DelRio, F.W., and Boyce, B.L. (2019). Predicting strength distributions of MEMS structures using flaw size and spatial density. *Microsystems & Nanoengineering* 5: 49.

Lawn, B.R. (1993). *Fracture of Brittle Solids, 2nd ed.* Cambridge.

Lawn, B.R. (2021). Chipping: a pervasive presence in nature, science and technology. *Journal of Materials Science* 56: 8396–8405.

Lawn, B.R. and Cook, R.F. (2012). Probing material properties with sharp indenters: a retrospective. *Journal of Materials Science* 47: 1–22.

Lei, W.S., Kumar, A., and Yalamanchili, R. (2012). Die singulation technologies for advanced packaging: a critical review. *Journal of Vacuum Science & Technology B* 30: 040801.

Marshall, D.B., Cook, R.F., Padture, N.P., Oyen, M.L., Pajares, A., Bradby, J.E., Reimanis, I.E., Tandon, R., Page, T.F., Pharr, G.M., and Lawn, B.R. (2015). The compelling case for indentation as a functional exploratory and characterization tool. *Journal of the American Ceramic Society* 98: 2671–2680.

Marshall, D.B., Lawn, B.R., and Evans, A.G. (1982). Elastic/plastic indentation damage in ceramics: the lateral crack system. *Journal of the American Ceramic Society* 65: 561–566.

Swain, M.V. and Lawn, B.R. (1976). Indentation fracture in brittle rocks and glasses. *International Journal of Rock Mechanics and Mining Sciences & Geomechanics Abstracts* 13: 311–319.

9

Scratches and Lateral Cracking at Linear Sharp Contacts

In this chapter a fracture mechanics analysis is developed for lateral cracking at linear sharp contacts. Such contacts are commonly encountered as surface scratches formed by loaded wedges, rolled particles, or dragged asperities. The most visible features of a scratch are the lateral cracks extending parallel to the surface from the linear contact. The analysis here takes an energy-based approach to predict lateral crack length as functions of contact load and geometry and material modulus, hardness, and toughness, independent of the scratch origin. The basis of the analysis is the relaxation of residual elastic energy associated with contact deformation and conversion into surface energy associated with lateral cracking. The analysis is supported by experimental measurements of cracks from dragged and rolled contacts. The analysis is applied in derivation of the mineralogical Mohs scale, providing a fundamental basis for the scale in terms of material properties and a clear definition of scratch resistance supported by experiment. The need for more measurements in this field is highlighted.

9.1 Introduction

Sharp contacts on the surfaces of brittle materials generate two major fracture systems. As noted in Chapters 2 and 3, these are the half-penny crack system, often termed "median-radial," in which cracks extend perpendicular to the surface from a contact (Lawn et al. 1980); and the lateral crack system, in which cracks extend parallel to the surface from a contact (Marshall et al. 1982). As the half-penny crack system is the predominant flaw in determining the failure behavior of brittle components, half-penny cracks have received the most attention, as noted in Chapters 3–6. Sharp contact flaws in brittle materials such as ceramics, glasses, semiconductors, rocks, and minerals are characterized by localized plastic deformation zones (pdz) and extended elastic deformation fields remaining after sharp contact events. These residual fields both initiate half-penny cracks and significantly enhance crack propagation under the action of superposed applied stress. As a consequence, if a contact on a component is large enough, the ensuing flaw, consisting of half-penny cracks + residual field, is strength-controlling. Deliberately introduced sharp indentation flaws, such as those formed by Vickers diamond pyramid probes, thus act as model flaws for studying the fracture behavior of structural ceramics and other brittle materials. Such studies include estimation of the cracking susceptibility through the measurement of surface traces of half-penny cracks (Chapter 2; see also Anstis et al. 1981; Cook, 2020) and determination of the effects of microstructure and environment on toughness through indentation strength measurements (Chapters 3–6; see also Chantikul et al. 1981; Cook 2015).

Observations of sharp contact flaws, however, particularly indentation flaws generated in the conventional load ranges of most indenters, 10–100 N, quickly reveal that it is the second crack system above, consisting of *lateral cracks* extending almost parallel to the contacted surface, that is the most visible. The air gap and concomitant surface uplift associated with lateral cracks generate significant contrast on optical observation of sharp contacts, particularly in bright- and dark-field optical microscopy, rendering lateral cracks the largest and most obvious feature (Chapter 2; see also Marshall et al. 1982). This is particularly so at large contact loads when lateral cracks curve and extend to the material surface, removing conchoidal chips of material from the surface adjacent to the contact (Chapter 3; see also Glaesemann et al. 1987; Cook et al. 1990). Such lateral cracking or chipping degrades the optical properties of surfaces, is responsible for material removal under erosive conditions, and significantly affects half-penny crack-based strength and toughness analyses through modification of the residual stress field (Chapters 3–6; see also Cook and Roach 1986; Cook 2015).

The sharp contacts considered in Chapters 1–8, largely Vickers indentations, have been restricted to a single location and are considered *point flaws*. Here, lateral cracks at sharp *linear flaws*, in which a contact location is extended in a line,

Indentation Fracture: Strength and Toughness for Brittle Materials Design, First Edition. Robert F. Cook.

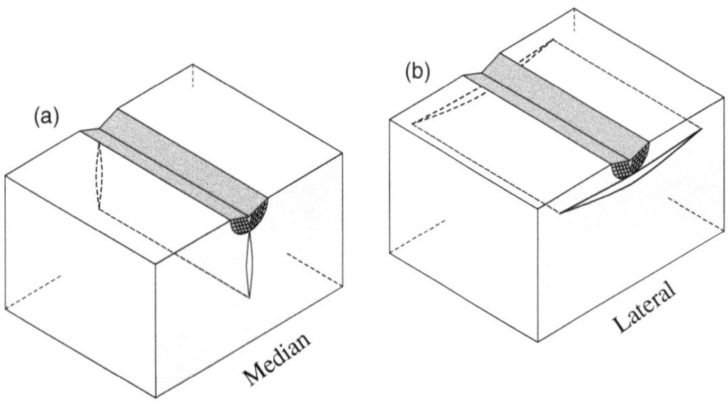

Figure 9.1 Schematic cross-section and surface diagrams of crack morphologies generated at linear sharp contact flaws. (a) Median; (b) lateral.

are considered. Such cracks are often the most visible evidence of *accidental* contacts leading to "scratches" by static, dragged, or rolled sharp objects or particles. The analogous *deliberate* contacts leading to linear flaws are those generated by (i) static sharp edges or corners on load-bearing surfaces, (ii) dragged sharp scribes used to scratch markings or score lines into surfaces, or (iii) rolling beveled disk cutters used to score glass or tile for breaking. Schematic diagrams of cracking at linear sharp contact flaws are shown in Figure 9.1, in direct analogy to Figure 2.5c and d. Figure 9.1a shows a median crack, or "median vent," formed beneath a linear pdz generated on a brittle material surface (pdz indicated by dark shading). The median crack is perpendicular to the surface and straight-fronted beneath the pdz. The residual stress fields arising in the elastic matrices of material surrounding the pdz of linear flaws are similar to those arising at point flaws. As a consequence, the fracture mechanics of linear median cracks is nearly identical to that of point half-penny cracks and is discussed in detail elsewhere (Fuller et al. 1983; Symonds et al. 1983; Cook 1985, 1994, 2017, 2018). Figure 9.1b shows lateral cracks formed beneath the linear pdz. The lateral cracks are also straight-fronted but extend to both sides of the flaw beneath the pdz and parallel to the surface. The focus of this chapter is the fracture mechanics of linear lateral cracks. The mechanics is somewhat similar to that developed earlier for point lateral cracks (Marshall et al. 1982). As in the consideration of edge chipping (Chapter 8), the fracture analysis of lateral cracking at sharp contacts must take into account the lack of constraint and thus increased compliance of the free surface parallel to the crack plane. Such lack of constraint is absent for half-penny cracks, which are well approximated by propagation in unbounded spaces (Chapters 1 and 2).

A shared assumption of the linear lateral crack analysis developed here and the point lateral crack analysis developed earlier is that lateral cracking is determined solely by the postindentation form of the pdz. This assumption is consistent with extensive observations and analysis (Cook and Pharr 1990; Cook 2019; references therein) showing that lateral cracking is maximized at, or occurs after, peak contact load. As such, the detailed mechanism of linear pdz generation need not be specified and lateral cracking at many types of linear contact events can be described uniformly by a single analysis. Schematic diagrams showing examples of linear pdz generation at sharp linear contact flaws are given in Figure 9.2. Figure 9.2a shows a long straight wedge loaded perpendicular to a material surface, forming a linear pdz along the length of

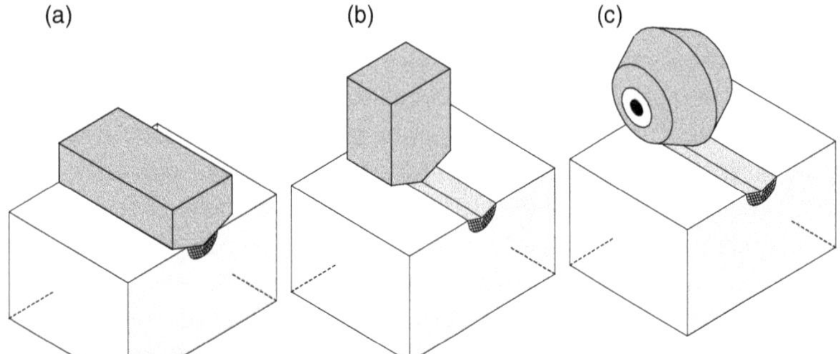

Figure 9.2 Schematic diagrams of plastic deformation zone (pdz) formation at straight linear flaw contacts. (a) Elongated static wedge loading; (b) short dragged wedge loading; (c) circular biconical rolling loading. The resulting pdz is the same in all cases.

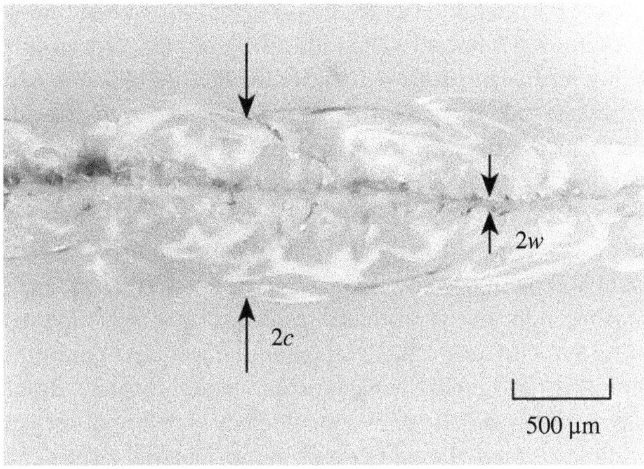

Figure 9.3 Reflected light plan view image of a surface flaw in glass generated by linear sharp contact with a rolled disk cutter. The central contact track and extensive lateral cracks are visible.

the wedge. Figure 9.2b shows a short wedge loaded normal to a material surface and translated or dragged in a straight line parallel to and along the surface, forming a linear pdz along the length of the translated path. Figure 9.2c shows a biconical roller loaded normal to a material surface and rolled in a straight line parallel to and along the surface, forming a linear pdz along the length of the rolled path. In all cases, the cross section of pdz formed is shown shaded. These three geometries are the most prevalent in forming linear contact flaws and are all considered here.

An example of a sharp linear contact flaw on a brittle surface is given in Figure 9.3, which shows an optical bright-field plan view of a flaw generated by a disk cutter rolled manually on soda-lime silicate glass (SLG). The residual surface *track*, indicating plastic deformation of the glass at the cutter contact, is visible as the straight line at the center of the flaw, width $2w$; track contrast is associated with residual surface deformation. The subsurface lateral cracks, indicating fracture of the glass, are visible as bright undulating areas on either side of the track, width $2c$; contrast is associated with the air gap and reflection from the lower cracked surface. In this chapter, crack length c refers to *lateral* crack length. The lateral cracks are clearly much larger than the track. The subsurface median crack is not visible. Similar images have been reported in applications by "fresh" SiC grit paper (Lawn and Wilshaw 1975), a commercial roller (Pan et al. 2008), and in recent works using commercial or custom scratch testers to generate surface damage in glass by dragged sharp probes (Subhash et al. 2002; Gu and Yao 2011; Gu et al. 2011; Bensaid et al. 2018; Suratwala et al. 2019; Wang et al. 2020; Zakiev et al. 2020).

Greater detail of a sharp linear contact flaw generated by a rolling disk cutter is shown in Figure 9.4. Figure 9.4a shows a flaw in SLG similar to Figure 9.3 but in a reflected light bright-field cross section. The location on the glass surface of the contact track and associated pdz is visible at the top of the image. A median crack extends perpendicular to the surface from the base of zone. Lateral cracks extend approximately parallel to the surface from near the base of the zone. Figure 9.4b is a transmitted cross-polarized light image of the same flaw. The large areas of reversed contrast are indicative of birefringence induced by the extended residual stress and strain fields about the flaw. Cross-section images very similar to Figure 9.4a,

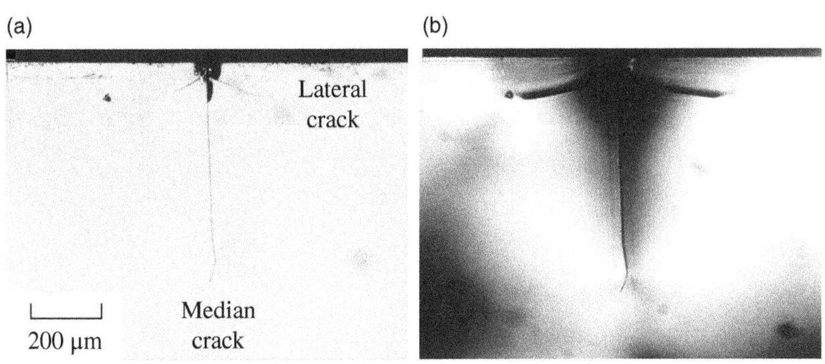

Figure 9.4 Cross-section images of a linear sharp contact flaw in glass formed by a disk roller. (a) Reflected light image; (b) transmitted cross-polarized light image.

showing a contact track, pdz, median crack, and lateral cracks have been reported since the early studies of linear flaws, including: scratches formed by dragged Vickers probes on glass (Peter 1964; Evers 1967; Swain 1979); scratches formed by dragged conical probes on zirconia (Lee et al. 2000, 2001); scratches formed by single-point grinding of Al_2O_3 (Gruver and Kirchner 1973), glass (Swain 1979; Kirchner 1984), and Si_3N_4 (Kirchner and Isaacson 1982); scratches formed by multipoint grinding of glass (Kirchner 1984) and Si (Pei et al. 1999); line flaws formed by static sharp wedge loading of glass and rock (Swain and Lawn 1976); and line flaws formed by rolled disk cutters on glass (Howes and Szameitat 1984; Tomei et al. 2018). Cross sections of line flaws formed by sharp wedge microindentation of film-on-substrate systems are also similar to Figure 9.4a, showing contact tracks and pdz in the film, median cracks in the substrate, and lateral delamination cracks at the film–substrate interface (Yeap et al. 2007) along with associated film uplift (De Boer and Gerberich 1996). Polarized light cross sections similar to Figure 9.4b, showing significant birefringence indicative of residual stress fields, have been reported for linear flaws in glass formed by rolled disk cutters (Swain 1981a; Symonds et al. 1983). Longitudinal sections of multiple indentation arrays and grinding tracks on Si_3N_4 show great similarity and exhibit behavior consistent with a significant residual field (Marshall et al. 1983). The implication from the above works is that the phenomenology of deformation and cracking at linear sharp contacts is well established, typified by Figures 9.3 and 9.4, and extends across material classes and contact types, thereby providing a firm foundation for analysis.

Early fracture mechanics analyses of linear flaws generated by sharp contacts were performed in the 1980s, primarily by Swain in the context of rolled contacts during glass cutting and Kirchner in the context of dragged contacts during ceramic grinding and machining. The focus of these researchers was the formation of median cracks during contact and subsequent crack propagation to failure under the action of applied stress. Prediction of component strength was a predominant goal. The contributions of Swain and Kirchner are summarized and reviewed by the current author in two subsequent works considering deformation and fracture at linear flaws generated by rolled (Cook 1994) and dragged (Cook 2017) contacts on glass and Al_2O_3, respectively. Although these latter works advanced the fracture mechanics significantly, the emphasis was still on median cracks and strength. However, a broadly applicable central feature of these works was the separation of the factors leading to track and pdz formation—the load and geometry of rolled, dragged, or wedge contacts—from the influence of the track and zone in fracture mechanics analyses. As a consequence, the fracture mechanics framework developed in these later works self-consistently describes and predicts track dimensions, median crack lengths, and component strengths, including measurements from the earlier and other (Cook 2018) works, along with the effects of material microstructure and lateral crack related stress relaxation at large contact loads. Ono and Tanaka (2001) performed a roller-based study of median crack lengths and strengths of scribed flat panel display glass and showed the clear influence of lateral cracks on scratch strengths. (Curiously, Ono and Tanaka do not compare their results with either the earlier experiments of Swain or the earlier analyses of Cook.) The above works, including those of Swain, Kirchner, and Ono and Tanaka, are extended in Chapter 10 in the context of scratching controlled strengths, through a brief review and consideration of median crack lengths at linear flaws.

The matrix of phenomena: sharp point contact, sharp linear contact, median crack length, and lateral crack length is then complete except for the last element—the lengths of lateral cracks at sharp linear contacts. This chapter develops the fracture mechanics to address this omission. As in the earlier works (Cook 1994, 2017), a key element of the analysis will be separation of contact details from fracture details. A primary focus is to enable estimates of surface and edge quality and material removal by linear contacts and scratches. A secondary focus is more philosophical and is concerned with understanding scratch tests and the related Mohs scale. The chapter begins with an analysis section that first considers linear contacts and then considers linear lateral cracks. The results section then follows and compares behavior observed for dragged and rolled contacts with predictions from the analysis. Further predictions from the analysis are then used in interpretation of the Mohs scale, cited extensively—but not without debate—in mineralogy and other materials fields. The work here narrows that debate by application of the developed scratch analysis to experimental observations. A discussion places the work here in the context of previous lateral crack observations and analyses and considers some engineering applications.

9.2 Analysis

Lateral cracks at linear sharp contacts are described here by fracture mechanics analysis based on energy changes associated with cracking after contact. The analysis leads to results analogous to those obtained earlier based on force balances at point contacts on monolithic materials (Marshall et al. 1982) and at film on substrate systems (Marshall and Evans 1984;

Rossington et al. 1984; Ritter et al. 1989; Rosenfeld et al. 1990; Thurn and Cook 2004). The energy approach provides insight into the origin of the force balances by recognizing three interrelated internal energy changes on cracking after contact; two associated with elastic deformation and one associated with surface formation. The analysis begins with consideration of the contact deformation process and is followed by consideration of the energy changes associated with subsequent lateral cracking.

9.2.1 Contact Deformation

Figure 9.5 shows a cross section of a deformation and fracture sequence at a linear sharp contact. In Figure 9.5a, the section is at peak load, characterized by a contact or track width of $2w$ and a sharp contact- or probe-included angle of $2\phi_p$. The contact extends perpendicular to the image, consisting of many sections along the contact length. For static contact by a long sharp wedge as in Figure 9.2a, all cross sections along the wedge and thus the contact are identical. The wedge length is taken as L. For a peak load P imposed on such a wedge, the peak *line load*, load/length, imposed on the material is $P_L = P/L$ and is identical at all sections, as are w and ϕ_p. (Throughout, all line quantities are indicated by subscript L, e.g., "X_L.") Contact between the probe and the surface is characterized by the mean supported contact stress, the hardness H, given by

$$H = P_L/2w. \tag{9.1}$$

An example is a $2w \times L = 0.1$ mm × 10 mm contact on a $H = 10$ GPa material, for which $P = 10$ kN; wedge loads are large (Swain and Lawn 1976). For a rigid probe, the contact depth d_p at peak load is given by

$$d_p = w \cot \phi_p. \tag{9.2}$$

Deformation of the material by the probe is both elastic and plastic. For brittle materials, plastic deformation is localized to a small zone, the pdz, bounded at the surface by the contact (in this case the track width) and in the subsurface by a

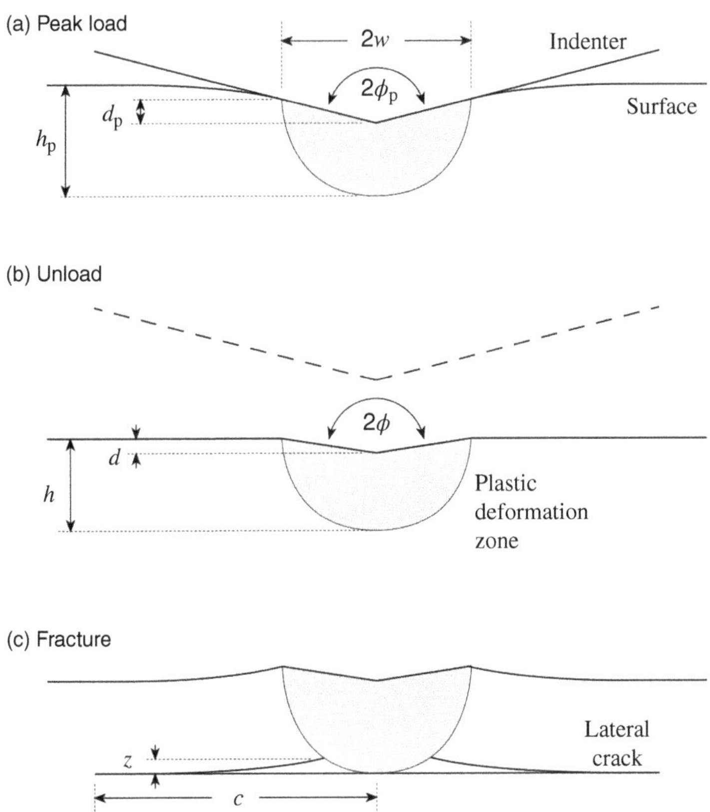

Figure 9.5 Schematic cross sections of a sharp contact deformation and fracture sequence in the formation of a linear flaw. (a) Peak load, consisting of elastic–plastic deformation; (b) unload, consisting of elastic recovery; (c) fracture, consisting of stress relaxation by lateral crack formation.

characteristic depth, h_p. Elastic deformation is extended, including surface deformation exterior to the contact and bulk deformation interior and exterior to the pdz boundary. At peak load, the transverse stress on the median plane bisecting the contact, is tensile, leading to the formation of a median vent or median crack on this plane directly beneath the contact. These cracks can control the strength of components containing linear contact flaws and are used by glaziers, tilers, and thin film display manufacturers to break glass and ceramics from scored surfaces. Median cracks are considered in Chapter 10 and elsewhere (Cook 1994, 2017, 2018).

For dragged or rolled contacts, Figure 9.2b and c, respectively, all sections along the contact are not identical. In these more common cases, the contact width, included angle, and line load vary along the length of the contact according to the longitudinal contact profile (e.g., triangular for dragged indenters, circular for a glass-cutting disk rollers). For such contacts, the cross section of Figure 9.5a, and w, ϕ_p, and P_L, pertain to the location of maximum contact depth associated with the moving profile. Variations in these quantities along the length of the contact lead to nonlinear relationships between the value of P_L effective during fracture at a linear flaw and the physical load P applied in flaw formation. The simplest of these relationships is that for a dragged fixed profile probe such as an indenter or the short sharp wedge of Figure 9.2b. In these cases, the contact length and width are comparable such that the contact area $\sim w^2$. For materials of invariant hardness the relationship $P_L \sim P^{1/2}$ results (Symonds et al. 1983; Cook 2017). For simplicity, the wedge picture is borne in mind throughout the following, but the lateral crack analysis developed is general and describes linear flaws formed by static, dragged, or rolled contacts. The focus of analysis is on the residual geometry after contact and P_L, rather than the loads active during contact. Connection with observation or experiment and with the physical load P is made subsequent to the fracture analysis and is considered here in Section 9.3.

On unloading, Figure 9.5b, elastic contact deformation is recovered, leaving a residual contact impression, the track, associated with plastic deformation. Here, obtuse probes ($2\phi_p > 120°$) are considered, for which there is very little elastic deformation transverse to the contact. The recovered surface is characterized by a track of residual depth $d < d_p$ given by

$$d = w \cot \phi, \tag{9.3}$$

where $2\phi > 2\phi_p$ is the residual included angle and the track width w is taken as invariant from peak load to the unloaded state. An example is $\phi \approx 76°$ such that $\cot \phi \approx 1/4$, typical for a scratch generated by a dragged Vickers indenter. The scale of the residual pdz can be estimated by modeling the contact as a pressurized cylindrical cavity in an elastic–perfectly plastic medium (Hill 1950), assuming that the scaling extends to the half-space of the contacted material. At peak load, the (half-space) cross-sectional area of the cavity is taken to be the contact cross section, wd_p, and the pressure exerted on the surrounding material is H. H is also taken as the measure of material resistance to plastic yielding (consistent with the lack of transverse recovery of the track width) such that an (approximately hemicylindrical) pdz forms about the contact. Recovery on unloading decreases the cross-sectional areas of both the track, to its residual value of wd, and the pdz, to a characteristic value $h^2 < h_p^2$, such that mechanical equilibrium is established over the cross section between the hydrostatically compressed pdz and the surrounding elastically restraining matrix. The pressurized cylindrical cavity analysis implies variation of h with material properties and track geometry as (Cook 1985),

$$h \sim w(E'/H)^{2/3}(\cot \phi)^{1/2}, \tag{9.4}$$

where E' is the material plane strain modulus (Chapter 1). Equation (9.4) is similar in form to that developed earlier for an analogous spherical cavity model applied to point contacts (Lawn et al. 1980), and, in comparison to Eq. (9.3), introduces materials dependence and diminished contact geometry dependence, consistent with the elastic–plastic deformation configuration. (See Chapter 2; the dimensions a and b characterizing point flaws are analogous to the dimensions w and h used here to characterize line flaws.)

Independent of the formation mechanism and any material influences, for a flaw of length L, the dilatational strain, Δ_r, developed in the residual pdz is given by

$$\Delta_r \sim -wdL/h^2L \sim -w^2 \cot \phi/h^2. \tag{9.5}$$

The (volumetric) elastic energy density, $\mathcal{U}_E$, in a cross section of the flaw varies as

$$\mathcal{U}_E \sim \Delta_r^2 \tag{9.6}$$

such that the total residual elastic energy in the system, U_r, is given by

$$U_r = (M/2)(w^4 \cot^2 \phi/h^4)h^2L$$

and the elastic line energy, $U_{\rm rL}$, of the flaw is thus

$$U_{\rm rL} = (M/2)(w^4 \cot^2\phi/h^4)h^2. \tag{9.7}$$

The parameter M is an elastic modulus that takes into account the exact geometries of the pdz and the stress and strain fields in a cross section of the flaw (pdz + matrix). Equation (9.7) represents elastic potential energy stored in the material as a result of work performed by the contact during flaw formation. The energy is analogous to that stored in a one-dimensional deformed linear spring. Rewriting Eq. (9.7) as

$$U_{\rm rL} = (M/E^2)(Ew^2 \cot\phi/h)^2/2$$

suggests the definition of two spring-like line quantities associated with the residual pdz,

$$\lambda_{\rm rL} = M/E^2 \tag{9.8}$$

and

$$F_{\rm rL} = Ew^2 \cot\phi/h \tag{9.9}$$

as a line compliance and line force, respectively. M absorbs any ν dependence on modulus, see Chapter 1, and introduction of the plane stress Young's modulus E is recognition of the one-dimensional nature of the spring. The elastic line energy of Eq. (9.7) is then expressed as

$$U_{\rm rL} = \lambda_{\rm rL} F_{\rm rL}^2/2 \tag{9.10}$$

or, alternatively, defining a displacement $u_{\rm rL}$ conjugate to the force by

$$u_{\rm r} = -\lambda_{\rm rL} F_{\rm rL}, \tag{9.11}$$

the elastic line energy is expressed as

$$U_{\rm rL} = u_{\rm r}^2/2\lambda_{\rm rL}. \tag{9.12}$$

It is noted that

$$u_{\rm r} = -Mw^2 \cot\phi/Eh < 0$$

and thus it is convenient to define the displacement magnitude $\delta = |u_{\rm r}|$. In arithmetic terms, Eqs. (9.10)–(9.12) have familiar forms describing a spring, in which $u_{\rm r}$ is the displacement of the spring from its natural length, and $F_{\rm rL}$ is the force exerted by the displaced spring on its surroundings. Here, $u_{\rm r}$ is negative—the spring is deformed to a shorter length—and $F_{\rm rL}$ is positive—relative to the spring, the force is outward. In physical terms, the spring is the pdz exerting a force to push the zone out of the constraining matrix. Relaxing the constraint reduces the elastic line energy deriving from the contact imposed dilatational strain, Eq. (9.5). Lateral crack formation associated with such relaxation is considered in Section 9.2.2.

9.2.2 Lateral Cracking

Figure 9.5c shows the configuration of a linear sharp contact with lateral cracks. An implicit coordinate axis is fixed in the material and positive along the material outward normal (vertical in the image). Three changes are apparent relative to the uncracked configuration of Figure 9.5b. First, crack opening displacement at the crack mouths extends the pdz, leading to surface uplift relative to the coordinate fixed point at the intersection of the zone and the matrix. Second, the displacement and surface uplift cause bending deformation of material in the matrix adjacent to the pdz. Third, crack extension leads to the creation of new surfaces in the matrix. The effects of these changes on system energy are considered in turn.

Crack opening displacement leads to expansion of the pdz and thus relaxation of the compressed spring and reduction in the residual elastic line energy. Rewriting Eq. (9.9) to include a small extension of the zone $h \to h(1 + z/\delta)$ by crack opening displacement $z \geq 0$, Figure 9.5, leads to modification of Eq. (9.10), $U_{\rm rL} \to U_{\rm rL}/(1 + z/\delta)^2 \approx U_{\rm rL}(1 - z/\delta)^2$. Equation (9.12) is thus generalized to

$$U_{\rm EL} = (\delta - z)^2/2\lambda_{\rm rL}. \tag{9.13}$$

Crack opening displacement also leads to deformation in the matrix that is well approximated by the bending of two cantilevers. The cantilevers are formed by the material and lateral crack surfaces with fixed ends built into the matrix and free ends displaced by the extending pdz. The line energy of the deformed cantilevers is given by

$$U_{\rm CL} = z^2/2\lambda_{\rm cL}, \tag{9.14}$$

where λ_{cL} is the line compliance of the two cantilever system,

$$\lambda_{cL} = 2c^3/Eh^3 \tag{9.15}$$

and c is the lateral crack length that defines the length of an individual cantilever.

The formation of the lateral cracks generates new fracture surfaces in the material with surface energy areal density 2γ. As there are two cracks, the crack area/length is $2c$, such that the surface energy/length of the lateral cracks is

$$U_{SL} = 4c\gamma. \tag{9.16}$$

The total line energy of the lateral crack fracture system is the sum of Eqs. (9.13), (9.14), and (9.16),

$$\begin{aligned} U_L &= U_{EL} + U_{CL} + U_{SL} \\ &= (\delta - z)^2/2\lambda_{rL} + z^2/2\lambda_{cL} + 4c\gamma. \end{aligned} \tag{9.17}$$

It is noted that the energy U_L of Eq. (9.17) depends on two extensive system variables, z and c, and hence there are two forces associated with two different derivatives of the energy. The first of these is

$$\begin{aligned} F_L(z) = -\partial U_L/\partial z &= F_{EL} + F_{CL} \\ &= (\delta - z)/\lambda_{rL} - z/\lambda_{cL}, \end{aligned} \tag{9.18}$$

which expresses the net crack opening force F_L as the competing sum of the spring force associated with the residual pdz attempting to open the crack and the cantilevers attempting to close the crack, noting that the third, surface energy, term plays no role. Mechanical equilibrium is determined by setting $F_L = 0$ in Eq. (9.18) to give

$$z = \delta/(1 + Mh^3/2Ec^3) \tag{9.19}$$

using Eqs. (9.8) and (9.15) and noting that as a function of c Eq. (9.19) describes a range of equilibrium configurations. Consideration of Eq. (9.18) shows that $\partial F_L/\partial z < 0$ such that the equilibria are stable.

The second force associated with a derivative of the total energy is, using Eq. (9.15),

$$\begin{aligned} F_L(c) = -\partial U_L/\partial(2c) &= \mathcal{G} - R \\ &= 3Eh^3z^2/2c^4 - 2\gamma, \end{aligned} \tag{9.20}$$

which expresses the net crack extension force as the difference between the mechanical energy release rate $\mathcal{G}$ associated with the cantilevers attempting to extend the crack and the fracture resistance R associated with the surface energy attempting to retract the crack. $\mathcal{G}$ and R are configurational forces associated with crack area changes in the system and are the conventional terms used in fracture mechanics (Chapter 1; see also Lawn 1993). Note the absence of the minus sign in the definition of R. Fracture equilibrium is determined by setting $\mathcal{G} = R$ in Eq. (9.20), noting that such a condition here describes a range of equilibrium configurations as it is a function of z.

Specification of an equilibrium crack configuration begins by combining z from Eq. (9.19) into the definition of $\mathcal{G}$ from Eq. (9.20) to give, after some manipulation,

$$\mathcal{G} = 6\xi(U_{rL}/h)\left[\frac{(c/h)}{\xi + (c/h)^3}\right]^2, \tag{9.21}$$

where $\xi = (M/2E)$ is a dimensionless geometry factor of order unity. Equation (9.21) shows that $\mathcal{G}$ is the product of an energy/area, (U_{rL}/h), and a dimensionless function of relative crack length, (c/h). Building on the fracture mechanics concepts established in Chapter 1, Eq. (9.21) has several interpretations and applications. The first interpretation is that Eq. (9.21) embodies the physics of the First Law of Thermodynamics by describing energy associated with surface creation in terms of energy associated with elastic deformation (which was in turn generated by work performed by the contact). The second interpretation is to recognize that $\mathcal{G}(c)$ in Eq. (9.21) passes through a maximum at $(c/h) = (2/\xi)^{1/3}$ and thus Eq. (9.21) represents a destabilizing to stabilizing transition and embodies the fracture mechanics of crack initiation (Sections 1.7.2 and 2.4). Such fracture mechanics has also been well studied for half-penny cracks at indentation point contacts in ceramics (Lawn and Evans 1977; Lawn and Marshall 1979), cracks at phase-transforming expanding inclusions in glass (Swain 1981b), and cracks at constrained conducting lines in microelectronic dielectrics (Thurn and Cook 2002).

The first application of Eq. (9.21) is then to assess lateral crack initiation at linear flaws, building on the crack initiation and threshold analyses developed in Chapters 1, 2, and 8 and extending the range of crack initiation geometries considered in the earlier works by Lawn, Swain, and Thurn. Threshold effects at linear flaws are commonly observed in domestic and

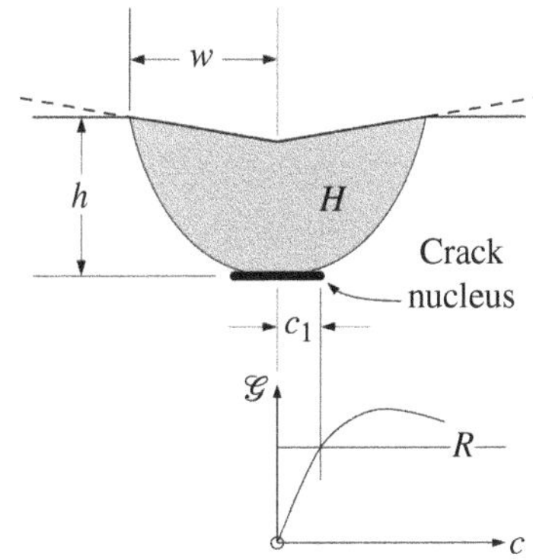

Figure 9.6 Schematic cross-section diagram of a linear flaw contact impression at lateral crack initiation. The plastic deformation zone (pdz) is indicated by shaded region. A lateral crack nucleus on the zone periphery is indicated by bold line. The inset graph shows the variation of the mechanical energy release rate at initiation.

industrial contexts and are of great engineering importance in designing scratch-resistant and mar-free surfaces. Following the discussion in Section 2.4, it is recognized that imposition of the equilibrium condition $\mathcal{G} = 2\gamma$ at the maximum of $\mathcal{G}(c)$ leads to a minimum value of U_{rL} and hence a lower-bound track width w^* required to generate lateral cracks at linear sharp contacts. This lower bound is very rarely encountered in practice (e.g., Figure 2.24). Hence, also following Section 2.4, attention here will focus on the estimation of the upper-bound track width w_Ω associated with the imposition of unstable fracture equilibrium on small lateral crack nuclei at the pdz boundary.

In developing a description of lateral crack initiation, attention shifts to the behavior of small unstable crack nuclei, $c < w, h$. Figure 9.6 shows a schematic cross-section diagram of a post-contact linear flaw and lateral crack initiation. The diagram is similar to a rotated version of Figure 2.22. The peak load sharp probe location is indicated by dashed lines, and the track semiwidth w is indicated. The pdz is shown shaded, and the pdz depth h is indicated. The characteristic pressure within the pdz is H. Attention here is focused on the small lateral crack nucleus of dimension c_1 exterior to the pdz. In this case, the nucleus is formed by the intersection of shear faults within the pdz adjacent to the pdz base. The nucleus is in a state of radial tension (relative to the contact). In the limit of very small cracks, $(c/h) \ll 1$, the behavior of the mechanical energy release rate $\mathcal{G}(c)$ in Eq. (9.21) is destabilizing:

$$\mathcal{G} = (6/\xi)(U_{rL}/h)(c/h)^2. \tag{9.22}$$

If deformation within the pdz associated with the linear flaw maintains geometrical similarity on loading, the scale c_1 of a lateral crack nucleus formed at the base of the pdz is

$$c_1 = \Omega_\phi w \tag{9.23}$$

following Eq. (2.22). $\Omega_\phi = \Omega(\phi) < 1$ is a proportionality factor that depends on the included contact angle. Imposing the equilibrium condition, as in the inset of Figure 9.6, $\mathcal{G} = R = 2\gamma = T^2/E$ where T is the material toughness in the linear flaw initiation environment and combining Eqs. (9.4), (9.7), (9.22), and (9.23) gives, after collecting terms and rearranging, the scaling of the track width w_Ω at lateral crack initiation:

$$w_\Omega \sim \frac{T^2}{H^2}\left(\frac{E}{H}\right)^{4/3}\left(\frac{\cot\phi}{\Omega_\phi^4}\right)^{1/2}. \tag{9.24}$$

For linear flaws formed by dragged sharp contacts, the normal contact load at lateral crack initiation is given by $P_\Omega \sim Hw_\Omega^2$ and thus

$$P_\Omega \sim \frac{T^4}{H^3}\left(\frac{E}{H}\right)^{8/3}\left(\frac{\cot\phi}{\Omega_\phi^4}\right). \tag{9.25}$$

The primary T and H scaling behaviors of Eqs. (9.24) and (9.25) are in agreement with those for initiation of half-penny cracks at point flaws, Eqs. (2.24) and (2.25), and with the lower-bound contact load for lateral crack initiation given by Marshall et al. (1982). For linear flaws formed by other mechanisms (wedging, rolling), the $P(w)$ behavior is different and hence so is the scaling of the initiation load [Eq. (9.25)]. In all cases, the secondary scaling behaviors of the E, H, $\cot\phi$, and Ω parameters depend on the loading geometry and differ slightly from those given previously. An important implication from Eqs. (9.24) and (9.25) is that Ω_ϕ must be a decreasing function of ϕ such that overall $dP_\Omega/d\phi > 0$ and smaller loads initiate lateral cracks at more acute (smaller ϕ) contacts.

The second, and perhaps more important, application of Eq. (9.21) is to consider the behavior of $\mathcal{G}(c)$ on the stabilizing branch well away from the maximum, at $(c/h) \gg 1$, such that

$$\mathcal{G} = 6\xi(U_{\mathrm{rL}}/h)/(c/h)^4. \tag{9.26}$$

For fixed profile contacts (ϕ is invariant with contact depth), such as by static wedges, dragged indenters, or conical rollers, deformation on a track cross section maintains geometrical similarity, such that $h \sim w$ and thus $U_{\mathrm{rL}} \sim Mw^2$, Eq. (9.7), and Eq. (9.26) is simply expressed as

$$\mathcal{G} \sim w^5/c^4, \tag{9.27}$$

which is indeed stabilizing as $\partial\mathcal{G}/\partial c < 0$. The importance of Eq. (9.27) is that imposing the equilibrium condition gives the variation of lateral crack length in terms of track width in experimentally testable terms, $c \sim w^{5/4}$. Further, the mechanism by which the track was generated was not specified, such that Eq. (9.27) can be expressed in terms of the most commonly controlled contact parameter, the normal load P. As discussed above, with reference to Figure 9.2, for static wedges $w \sim P$, Eq. (9.1), such that $c \sim P^{5/4}$. For dragged indenters, $w \sim P^{1/2}$ (Symonds et al. 1983) such that $c \sim P^{5/8}$. For conical rollers, $w \sim P^{2/3}$ (Cook 1994) such that $c \sim P^{5/6}$. These latter two relations, pertaining to the most common experimental conditions, will be examined in Section 9.3.

9.3 Results

Observations of lateral crack lengths are considered here, beginning with those formed by sharp point contacts, conical and Vickers indenters, followed by those formed by sharp linear contacts, a dragged Vickers indenter and rolled biconical wheels. The point contact flaw observations provide background and context for the linear flaw observations and loosely follow the data presentation scheme of Chapter 2. The two sets of linear sharp contact observations make use of the analyses of Section 9.2 regarding lateral crack lengths. The first set of linear flaws considers a Vickers pyramidal indenter dragged in two orientations on two glasses to generate lateral cracks at different contact loads. The second set considers multiple biconical glass cutters rolled on a single glass to generate lateral cracks with different contact geometries at different contact loads.

9.3.1 Point Contacts

The line flaw lateral crack analysis developed in Section 9.2 is similar to that developed by Marshall et al. (1982) for lateral cracks at indentation point flaws. The point flaw analysis includes expressions for the threshold indentation load associated with initiation of lateral cracks (see Section 9.2) and for the postthreshold stable equilibrium lateral crack lengths. The expression for the equilibrium crack lengths, applicable at large indentation loads, can be summarized as

$$c \sim (\cot\phi)^{5/6}(E^{3/8}H^{-1/2}T^{-1/2})P^{5/8}. \tag{9.28}$$

This expression includes dependencies on an indenter characteristic ($\cot\phi$), material characteristics (E, H, and T), and the applied contact load P. The load dependence for the point contacts considered is identical to that derived in Section 9.2 for dragged contacts, $P^{5/8}$. The angular and material dependencies are similar, but not identical, to those developed in Section 9.2. The different dependencies are discussed in Sections 9.3.3, 9.4, and 9.5.

Tests of Eq. (9.28) are made most simply by measuring the load dependence of lateral crack length, $c(P)$, for variations in either indenter or material. Swain and Lawn (1976) performed the first test by measuring the size of lateral crack surface chips as a function of indentation load in SLG, using various tungsten carbide (WC) conical indenters. Figure 9.7 encapsulates the results in a logarithmic plot of lateral crack length as a function of indentation load, using data derived from the published work. The series of WC cones had included angles in the domain $\phi = 30°$ to $\phi = 70°$. (Some of the related half-penny crack data were presented in Figure 2.14.) Symbols represent individual crack length measurements. Results from the extremes of cone-included angle, 30° and 70°, are shown as filled symbols; intermediate angles are shown as open symbols. The solid line represents the $P^{5/8}$ dependence of Eq. (9.28) fit to the upper bounds of the $\phi = 30°$ data (straight line of slope 5/8 in logarithmic coordinates). The dashed line represents the same $P^{5/8}$ dependence for the $\phi = 70°$ data, predicted from the solid line fit using the angular scaling of Eq. (9.28), $c_{70}/c_{30} = (\cot 70°/\cot 30°)^{5/6} \approx 0.27$.

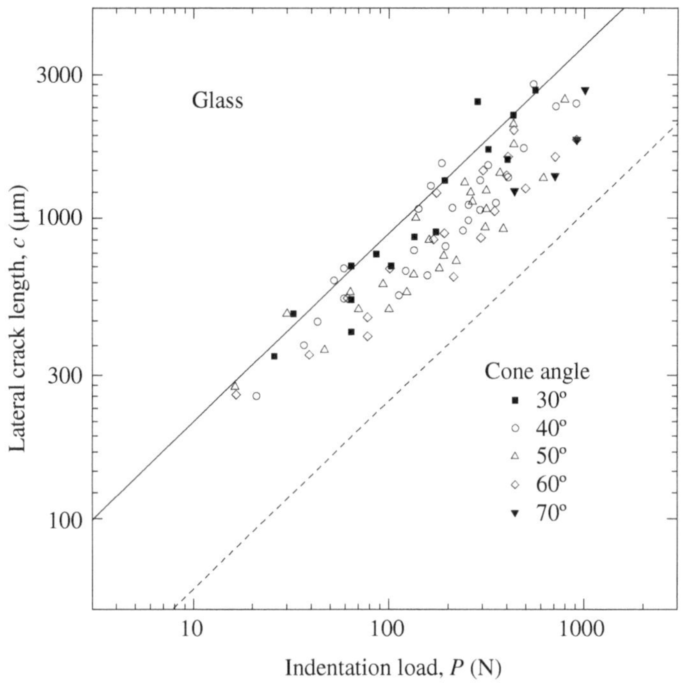

Figure 9.7 Logarithmic plot of lateral crack length as a function of indentation load for conical indentations of soda-lime glass. Semi-included cone angles are indicated. Lines represent predicted data bounds. Source: Adapted from Swain and Lawn (1976).

There are two major points to note in Figure 9.7. First, although lateral crack size does increase with indentation load as expected, a dependence on indenter-included angle is not very clear and the data appear quite scattered. The scatter is in spite of the fact that the angular dependence for indentation lateral cracks, $(\cot \phi)^{5/6}$, Eq. (9.28), is stronger than that for indentation half-penny cracks, $(\cot \phi)^{4/9}$, Eq. (2.20). This point is emphasized by the dashed line in Figure 9.7 that, notwithstanding the scatter in the experimental data, clearly overestimates the reduction in lateral crack length with increasing indenter-included angle. In contrast, the related plot of variation in indentation half-penny cracking susceptibility, Figure 2.18, although exhibiting some scatter, shows clear agreement with the predicted, but reduced, angular dependence. The second point to note is that there does not appear to be a significant trend in threshold indentation load with indenter-included angle. Equation (9.25) suggests that this variation is $P_\Omega \sim \cot \phi$ such that a factor of ≈ 4.8 should be observed for the indenters used. Such a factor is observed for the extremes of cone-included angle, 30° and 70°, but there is no trend for the intermediate angles. Overall, observations of included angle effects on lateral cracking at indentation point flaws are only in weak agreement with pdz-based residual field models: Cracks are longer and thresholds are smaller as indenters become more acute but the effects are not as pronounced as predicted.

The second test of Eq. (9.28) is made by measuring the load dependence of lateral crack length, $c(P)$, using a single indenter and variations in material. Such measurements, including measurements derived from the published work of Marshall et al. (1982), are shown in Figure 9.8 as a logarithmic plot of lateral crack length as a function of indentation load. The data represent Vickers indentations in soda-lime and As_2S_3 glasses, Y_2O_3, ZnS, and MgF_2 polycrystals, and cordierite glass-ceramic. Symbols with bars represent the means and standard deviations of at least four measurements at each indentation load indicated. The solid line represents the $P^{5/8}$ dependence of Eq. (9.28) fit to the mid-points of the As_2S_3 glass data. The dashed line represents the same $P^{5/8}$ dependence for the cordierite glass-ceramic data, predicted from the solid line fit using the elastic modulus, hardness, and toughness scaling of Eq. (9.28) and materials properties values taken from Marshall et al. (1982) and Chapters 3 and 11. The As_2S_3 and cordierite materials exhibit the extremes of mechanical properties, and the predicted lateral crack responses differ by ≈ 0.37. The agreement between observation and prediction is clear, as is the gradation in lateral cracking behavior for materials with mechanical properties intermediate between the extremes. It is noted that the countervailing effects of decreased toughness and decreased stiffness lead soda-lime glass to exhibit a similar lateral cracking response to cordierite glass-ceramic. Similar agreement and countervailing effects were demonstrated earlier by Marshall et al. (1982). An overall conclusion is that observations of material effects on lateral cracking at indentation point flaws are in good agreement with pdz-based residual field models.

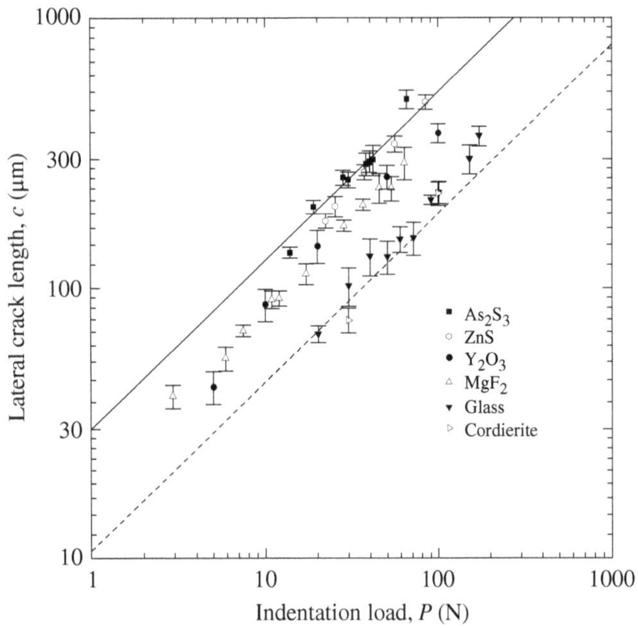

Figure 9.8 Logarithmic plot of lateral crack length as a function of Vickers indentation load for the materials indicated. Lines represent predicted data bounds. Source: Includes data adapted from Marshall et al. (1982).

9.3.2 Dragged and Rolled Contacts

For linear sharp contacts formed by dragging, the variation of lateral crack length with normal contact load is similar to that for point contacts, Eq. (9.28). The linear contact variation is obtained by expanding Eq. (9.27) to include contact angle and material properties and replacing track width with contact load. Hence, combining Eqs. (9.4), (9.7), and (9.26), imposing the equilibrium fracture condition $\mathcal{G} = T^2/E$, and using the dragged contact deformation relation $w \sim (P/H)^{1/2}$ gives, after collecting terms and rearranging, the scaling of lateral crack size with normal contact load:

$$c \sim (\cot \phi)^{5/8}(E^{2/3}H^{-5/6}T^{-1/2})P^{5/8}, \tag{9.29}$$

where T is the material toughness in the lateral cracking environment. As in Eq. (9.28), this expression includes dependencies on an indenter characteristic ($\cot \phi$), material characteristics (E, H, and T), and the applied contact load P. The angular and material dependencies are similar, but not identical, to those presented in Eq. (9.28).

Tests of Eq. (9.29) are made most simply by measuring the load dependence of lateral crack length, $c(P)$, for variations in indenter and material, as illustrated previously for point contacts. Here, both variations are considered, although the distinctions are not as well defined as those considered for point contacts. Ahn et al. (2003) and Gu and Yao (2011) performed scratch tests using dragged contacts formed by Vickers diamond pyramids on soda-lime glass with indenter diagonal forward and borosilicate glass with indenter face forward, respectively. Figure 9.9 illustrates the key phenomena in a logarithmic plot of track width w and lateral crack length c measured as a function of normal contact load P, using data derived from the published works cited. The filled symbols represent the results of Ahn et al., and the open symbols represent the results of Gu and Yao. In each case, the lower symbols indicate the track widths, and the upper symbols indicate the lateral crack lengths. The lines are fits to the data of Gu and Yao. The lower diagonal line represents a $P^{1/2}$ fit (straight line, slope 1/2) to the track width data using invariant hardness H_{glass}. The upper diagonal line represents a $P^{5/8}$ fit (slope 5/8) to the crack length data using Eq. (9.29) and invariant elastic modulus E_{glass} and toughness T_{glass}. Despite some scatter, the data are consistent with the expected trends (1/2 and 5/8), and the data of Ahn et al. are described similarly.

Not all normal loads used in the scratch tests of Ahn et al. and Gu and Yao gave rise to lateral cracks. Definite lateral crack threshold loads and related threshold track widths were observed, consistent with Eqs. (9.24) and (9.25), indicated by the vertical lines in Figure 9.9. The observed lateral cracks were formed at postthreshold loads, and the track widths formed at loads less than indicated by the vertical lines were subthreshold. Further examples of subthreshold scratches include those formed by a dragged Berkovich indenter in polycrystalline AlN using a 15 mN load (Gao et al. 2021) and single-crystal $LiNbO_3$ using a 10 mN load (Gruber et al. 2022), both comparable to the subthreshold observations of Gu and

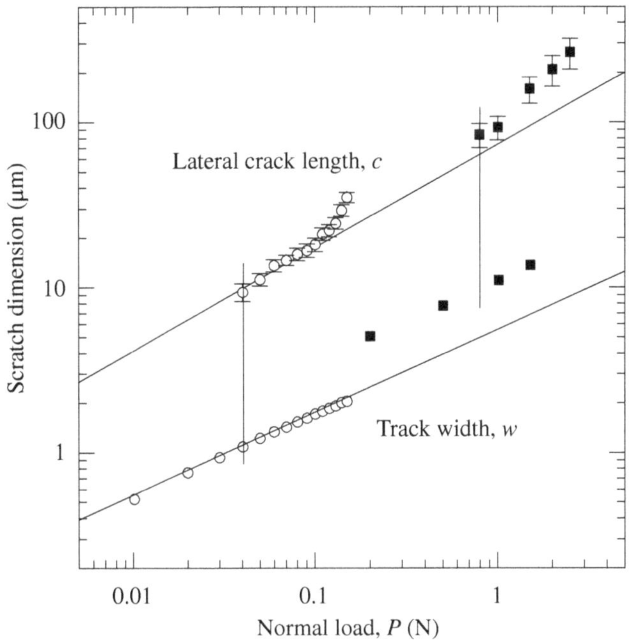

Figure 9.9 Logarithmic plot of scratch dimensions as a function of normal contact load for linear flaws formed by dragged Vickers indenters. Lines indicate behavior consistent with constant hardness and toughness. Source: Adapted from Ahn et al. (2003) and Gu et al. (2011).

Yao (open symbols), and those formed by a dragged Vickers indenter in polycrystalline Al_2O_3 (Cook 2017), using loads up to 3 N comparable to the observations of Ahn et al. (filled symbols).

There is a further qualitative point to be gained from Figure 9.9: Dragged scratch loads are small, approximately 0.1–1 N, far less than the loads typically required to generate point contact flaws, e.g., as in Figure 9.7, which are approximately 100 N. This difference in scale, pointing out the potency of small contact loads at linear contacts to cause surface damage, was noted previously in work considering sharp and blunt scratches on polycrystalline Al_2O_3 (Cook 2017, 2018) and to a lesser extent in work considering rolled tracks on glass (Symonds et al. 1983; Cook 1994). The damaging ability of "light" contacts to cause scratch damage is well known in domestic and industrial contexts, although sometimes this is an advantage: a human typically generates 10–100 N by static arm force, which is more than enough to score glass for breaking and shaping (see also Ono and Tanaka 2001 for examples of similar machine-applied forces).

Another test of Eq. (9.29) is made by measuring the load dependence of lateral crack length, $c(P)$, for a single material and variations in contact geometry, in this case for various biconical rollers. Lateral crack lengths at roller tracks are compared here for rollers of different conical geometry, radius R and included angle ϕ, contacting a single soda-lime glass material. Hence, Eq. (9.27) is reexpressed as

$$\mathcal{G} \sim w^5 \cot\phi / c^4 \qquad (9.30)$$

by including a geometrical rather than a material dependence from Eq. (9.7). Track width for roller flaws including a geometrical dependence is (Cook 1994)

$$w \sim P^{2/3}(R\cot\phi)^{-1/3}. \qquad (9.31)$$

Imposing equilibrium, combining Eqs. (9.30) and (9.31), and inverting leads to

$$c \sim P^{5/6}[R/(\cot\phi)^{1/2}]^{-5/12}, \qquad (9.32)$$

which shows the predominant $P^{5/6}$ dependence for lateral cracks generated at roller tracks, with lesser influences of roller radius and included angle.

Following Eq. (9.32), Figure 9.10 shows a logarithmic plot of lateral crack length c as a function of the adjusted normal contact load parameter $P(R/\cot^{1/2}\phi)^{-1/2}$ for conical roller flaws formed on soda-lime glass. The flaws were generated by nine different hard metal rollers with radii between 2.75 and 1.85 mm and included semi-angles between 75° and 42.5° rolled at approximately 0.5 mm s^{-1} on the glass surface. The raw data are from the unpublished results of Swain (M. Swain, personal communication 1985). The line in Figure 9.10 is a guide to the eye of slope 5/6, consistent with the response from Eq. (9.32).

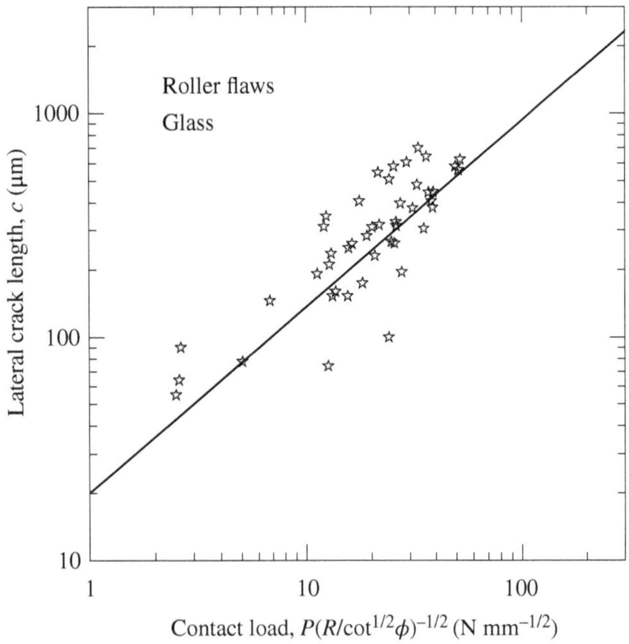

Figure 9.10 Logarithmic plot of lateral crack lengths at linear flaws generated by disk roller contacts using ranges of normal contact loads, roller radii, and included roller angles. Abscissa coordinate adjusts the contact load to compare different roller shapes. Line indicates predicted trend. Source: Adapted from Swain (private communication 1985).

The data are consistent with this expected trend. However, as in Figure 9.7, although lateral crack size does increase with indentation load as expected, a convergence of data to remove dependencies on roller radius and included angle is not very clear and the data appear quite scattered. However, the dependence of lateral crack length on normal contact load at roller tracks is much stronger than for point contacts (logarithmic slope 5/6 vs 5/8), and the dependence of lateral crack length on contact geometry is much weaker than for point or dragged contacts [e.g., $(\cot \phi)^{1/4}$ vs $(\cot \phi)^{5/8}$]. Hence, the scatter in Figure 9.10 likely reflects inherent scatter in lateral cracks at roller tracks. Such a conclusion is supported by the observation that *median* cracks formed at roller tracks in the same contact load and roller geometry domains are very well described, with minimal scatter, by pdz-based residual field models (Cook 1994). A pictorial example of lateral crack variability leading to crack length data scatter is shown in Figure 9.3.

Comparison of the dragged and rolled lateral crack data of Figures 9.9 and 9.10, and other scratch systems can be made by direct application of Eq. (9.27) ($c \sim w^{5/4}$). Figure 9.11 shows a logarithmic plot of lateral crack length c as a function of track width w for scratches formed on a range of materials by a range of dragged and rolled contacts. The plot includes the Vickers scratch data of Ahn et al. (2003) and Gu and Yao (2011) on glass, and Swain (1979) on glasses and sapphire, the Knoop scratch data of Matsuoka et al. (2017) on glass, the conical scratch data of Subhash and Bandyo (2005) on a series of polycrystalline Al_2O_3 materials, and the roller scratch data of Swain (1985) [modified as $w(\cot \phi)^{1/4}$, Eq. (9.30)], using data derived from the published works cited. The data symbols are identified by first author. The shaded region indicates the subthreshold lateral crack domain, $w \leq c$. The line is a guide to the eye of form $c \sim w^{5/4}$. The diversity of lateral crack responses reflects the diversity of scratch systems, although it does appear that the ideal slope 5/4 response is an upper bound for lateral cracks long relative to the track width, and linear slope 1 response describes short, barely postthreshold, lateral cracks. Subthreshold dragged tracks (within shaded region) were observed in polycrystalline Al_2O_3 (Cook 2017) at approximately the same track widths as postthreshold tracks were observed in single-crystal Al_2O_3 (sapphire, Swain and Lawn 1976). The explanation probably lies primarily with an increased value of T for the polycrystalline alumina, such that P_Ω was increased and a similar value of H led to unchanged w.

Overall, it is clear from Figures 9.7–9.11 that lateral cracking at sharp contacts—point, dragged, or rolled—is well described by pdz-based residual field models. Functionally and empirically for a given contact load, which can be quite light, "scratching" of surfaces is far more influenced by material variations than by contact geometry variations. These observations are reflected in a major application of scratch testing, the distinction of rocks and minerals via the Mohs scale in geological and mineralogical field testing. Section 9.4 considers sharp contacts dragged so as to form scratches on the 10

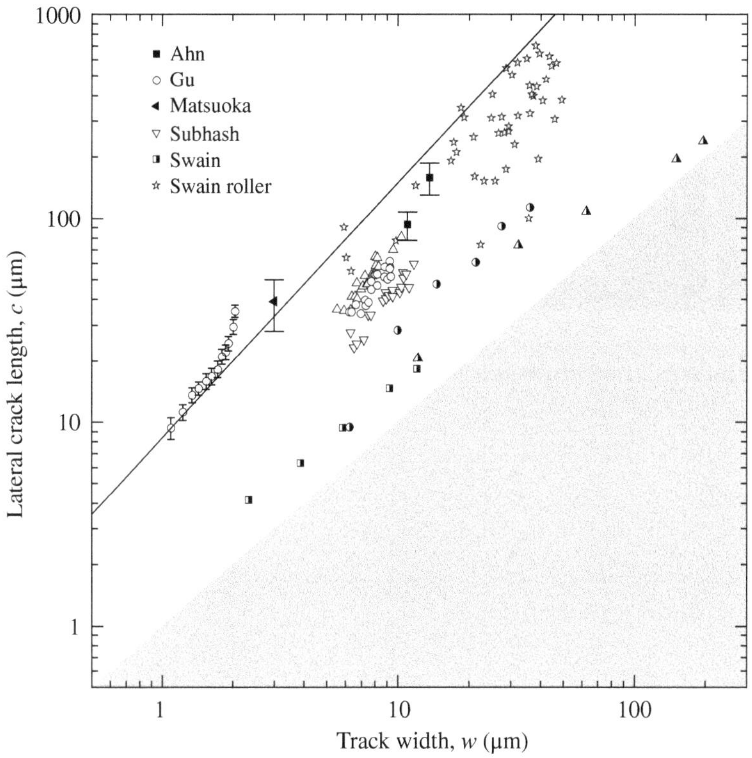

Figure 9.11 Logarithmic plot of lateral crack length c as a function of track width w for scratches formed on a range of materials by a range of dragged and rolled contacts. Symbols identify the first author. Shaded region indicates subthreshold conditions. Line is a guide to the eye of ideal asymptotic form $c \sim w^{5/4}$. Source: Adapted from Ahn et al. (2003), Gu and Yao (2021), Swain (1979, 1985), Matsuoka et al. (2017), and Subhash and Bandyo (2005).

different minerals that constitute the archetypes of the Mohs scale. Lateral crack analysis is applied to evaluate resistance to scratching in terms of plastic, fracture, and wear properties of the Mohs minerals.

9.4 Mohs Scale Minerals

The *Mohs scale* was originally developed as a method to guide mineral identification. The method semiquantitatively ranks 10 standard minerals through increasing resistance to scratching by a metal file or a standard mineral. Scratch resistance is assessed by visual inspection of the severity of an empirical residual scratch, leading to a ranking by a number, the *Mohs number*, on a scale from 1, least scratch resistant, archetype talc, to 10, most scratch resistant, archetype diamond. The intermediate archetype minerals include calcite ($CaCO_3$), 3, and quartz (SiO_2), 7.

An application of the Mohs scale is illustrated in Figure 9.12, which shows specimens of (a) a calcite single crystal, (b) a polycrystalline steel file, and (c) a quartz single crystal. All surfaces of the specimens were flat and terminated in points shown on the left of the images. The single crystals were terminated by reflective crystallographic planes. The points of all specimens were dragged by hand for approximately 5 mm on the surfaces of all other specimens with contact loads of approximately 10 N. The resulting scratches are indicated by arrows in the images. Visible residual scratches were formed on the calcite surface by both the quartz crystal (upper arrow and scratch in Figure 9.12a) and the steel file (lower arrow and scratch in Figure 9.12a). A visible residual scratch was formed on the steel surface by the quartz crystal (arrow and scratch in Figure 9.12b), but not by the calcite crystal. No scratches could be formed on the quartz crystal by either the steel file or the calcite crystal. Figure 9.12 provides an example of field-scale use of scratching in developing the Mohs scale. Laboratory-scale assessment of the scratch damage generated is shown in Figure 9.13, which shows an enlarged view of the quartz-generated scratch in calcite of Figure 9.12a (upper). This scratch has much in common with the conical roller scratch in glass of Figure 9.3: There is a central linear plastic deformation track of characteristic width $2w$. There are subsurface lateral cracks that follow the track and turn to exit the surface in irregular patterns and which exhibit characteristic lateral

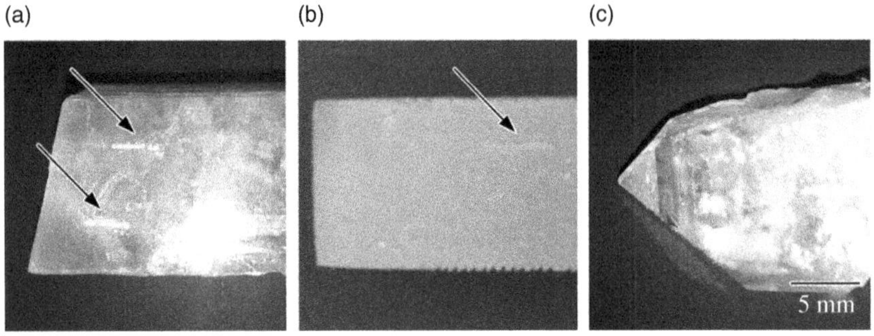

Figure 9.12 Images of material surfaces used in the assessment of the Mohs scale. (a) Calcite single crystal; (b) steel file; (c) quartz single crystal. Points used to form scratches by dragged contacts shown to left. Resulting scratches indicated by arrows.

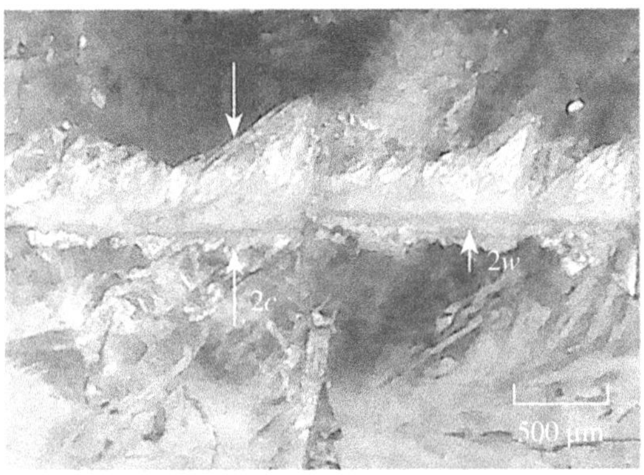

Figure 9.13 Enlarged image of scratch in calcite formed by dragged quartz point (from Figure 9.12a) (cf. Figure 9.3).

crack length $2c$. Unlike the undulating lobe-like lateral crack outlines in glass shown in Figure 9.3, the lateral cracks in calcite are angular and clearly affected by crystallographic cleavage effects. The extent of the visible scratch in Figure 9.12a is clearly formed by the surface terminations of the lateral cracks shown in Figure 9.13.

The Mohs scale is often referred to as a "hardness" scale, although it has long been known that the assessments of Mohs number for many materials do not vary linearly or even monotonically with conventional hardness measurements. The lack of clear understanding of the Mohs scale stems in large part from the lack of clear definition of what constitutes "scratch resistance," what the severity (width? depth? optical contrast?) of a residual scratch represents—plasticity, fracture, or wear—or how this combination of phenomena varies from material to material. This issue was addressed in an earlier comprehensive work that used indentation techniques to measure modulus, hardness, and toughness of the Mohs scale minerals at scratch-relevant length scales (Broz et al. 2006). This work also included a brief history of the Mohs scale and its historical relationship to hardness. Although recognizing the importance of material hardness in the development of the Mohs scale, the work concluded that "other properties such as toughness and modulus are involved in the phenomenon of scratch resistance" and "Until the physical processes of cracking associated with scratching versus indentation are studied in more detail, the best explanation is that scratch resistance is fundamentally different from hardness, and that fracture toughness may dominate in some minerals. There is no simple way to predict the relationship of material properties and Mohs number." Section 9.4 uses the lateral crack analysis developed in Section 9.2 to address the problems raised in these conclusions.

The approach here is to examine three candidate phenomena that give rise to a visually apparent scratch when a sharp contact is dragged on a surface. These phenomena are plastic deformation, quantified by the track width w, fracture, quantified by the lateral crack length c, and the line wear volume, V_L, a new parameter quantifying the volume of material removed per length of scratch. The variations of the spatial parameters, w, c, and V_L are predicted from combinations of the material properties of modulus, hardness, and toughness, E, H, and T, using the values for Mohs minerals developed by Broz (Broz et al. 2006). Scratch resistance is predicted by the inverse comparison of parameters, for example, shorter lateral crack lengths imply a greater relative scratch resistance. Predicted scratch resistance is then compared with the Mohs number.

The simplest spatial parameter to quantify is the track width w, which for dragged contacts of fixed included angle varies with contact load and material hardness as

$$w \sim (P/H)^{1/2}. \tag{9.33}$$

Contacts at which scratch damage is determined by plastic deformation alone, i.e., no cracks are formed, are termed "subthreshold" and occur at small loads. Under these conditions, the relative track width and thus scratch resistance of a material relative to a reference is given by

$$\text{Scratch resistance (plasticity)} = (w_{\text{ref}}/w) \sim (H_{\text{ref}}/H)^{-1/2}. \tag{9.34}$$

The assumption in this prediction is that the contact details of load and shape are invariant such that the test material and the reference material are scratched in the same way. The hardness of the reference material is H_{ref} such that scratching produces a plastic deformation track of width w_{ref}. If the hardness of the test material is greater than that of the reference material, the track width is smaller and the scratch resistance of the test material by this measure, Eq. (9.34), is greater.

At greater "post-threshold" contact loads, lateral cracks are formed and the characteristic spatial parameter of the scratch is the lateral crack length c. Combining Eqs. (9.27) and (9.33) and imposing the equilibrium condition expressed in terms of toughness and modulus, replacing 2γ with the more conventionally used toughness T using $T^2 = 2\gamma E$ (Chapter 1; see also Lawn 1993), give

$$\mathcal{G} \sim E(P/H)^{5/2}/c^4 \sim T^2/E, \tag{9.35}$$

where information regarding contact geometry has again been suppressed. Inverting this expression gives the contact load and material dependence of lateral crack lengths at dragged contacts,

$$c \sim (P/H)^{5/8}E^{1/2}T^{-1/2} \tag{9.36}$$

noting that Eq. (9.36) is somewhat similar in form to Eq. (9.33) but is extended by the inclusion of elastic and fracture property dependencies. The form of Eq. (9.36) is typical for expressions developed in indentation fracture mechanics, in that material influences are represented by sequences of material properties raised to fractional powers as shown in Sections 9.2 and 9.3 (see also Lawn et al. 1980; Marshall et al. 1982). The relative crack length and thus scratch resistance of a material relative to a reference is given by

$$\text{Scratch resistance (fracture)} = (c_{\text{ref}}/c) \sim (H_{\text{ref}}/H)^{-5/8}(E_{\text{ref}}/E)^{1/2}(T_{\text{ref}}/T)^{-1/2}. \tag{9.37}$$

Once again, the assumption is that the contact load and shape are invariant. Scratching of the reference material produces a lateral crack of length c_{ref}. If the combination of properties given by Eq. (9.36) for the test material is greater than that for the reference material, the lateral crack is smaller and the scratch resistance of the test material by this measure, Eq. (9.37), is greater.

At even greater contact loads, the lateral cracks curve toward and intersect the material surface, removing chips of material such that the spatial parameter characterizing the scratch is the volume of material removed/length of scratch V_{L}. This parameter has the dimensions of area and varies as the cross section of the pdz-lateral crack complex, Figure 9.5b and c, such that

$$V_{\text{L}} \sim hc \sim w(h/w)c. \tag{9.38}$$

The last term in Eq. (9.38) has been written to reflect the fact that the parameter V_{L}, the line wear volume, is the product of three factors that exhibit different material dependencies of increasing complexity: plastic, elastic–plastic, elastic–plastic–fracture. Combining the requisite Eqs. (9.4), (9.33), and (9.36) into Eq. (9.38), suppressing geometrical and load dependencies and retaining only material dependencies, gives

$$V_{\text{L}} \sim (1/H)^{1/2}(E/H)^{2/3}(1/H)^{5/8}E^{1/2}T^{-1/2}. \tag{9.39}$$

The relative line wear volume and thus scratch resistance of a material relative to a reference is thus given by

$$\text{Scratch resistance (wear)} = (V_{\text{Lref}}/V_{\text{L}}) \sim (H_{\text{ref}}/H)^{-1.8}(E_{\text{ref}}/E)^{1.2}(T_{\text{ref}}/T)^{-0.5} \tag{9.40}$$

using decimal notation for simplicity. Once again, the assumption is that the contact load and shape are invariant. Scratching of the reference material produces a line wear volume of V_{Lref}. If the combination of properties given by Eq. (9.39) for the test material is greater than that for the reference material, the wear volume is smaller and the scratch resistance of the test material by this measure, Eq. (9.40), is greater. [Note that if instead of Eq. (9.38), $V_{\text{L}} \sim w^2$ is assumed and Eq. (9.33) is used, Archard's law for *plasticity*-controlled wear results.]

The materials properties scaling of Eqs. (9.39) and (9.40) can be assessed against the extensive single-pass, single-point linear contact experiments of Subhash and colleagues (Subhash et al. 2002, 2006; Subhash and Bandyo 2005; Subhash and Klecka 2007) and the multipoint, multipass grinding experiments of Marshall et al. (1987). The Subhash group used

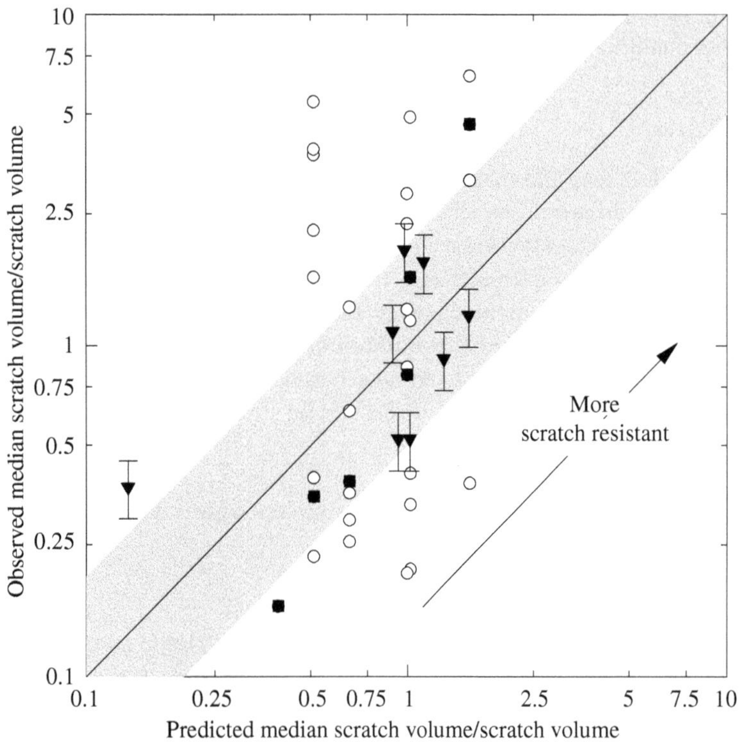

Figure 9.14 Logarithmic plot of observed scratch volume for various materials versus scratch volume predicted using linear flaw lateral crack analysis. Line indicates ideal agreement, shaded band indicates a factor of 2 variation. Source: Adapted from Subhash et al. (2002) and Subhash and Bandyo (2005).

a pendulum-based custom-instrumented scratching apparatus and subsequent profilometer measurements to determine the volume displaced as a function of normal force by dragged 90° conical contacts. The materials examined were SLG and the polycrystalline ceramics Al_2O_3, ZrO_2, Si–Al–O–N, Si_3N_4, and SiC. The experiments by Marshall et al. used an instrumented grinding machine to determine the normal force required to remove a set point of cut by very small dragged particle contacts. The materials were single-crystal and polycrystalline Al_2O_3 and some glass-ceramics. Figure 9.14 shows a logarithmic plot of scratch volumes observed for the materials examined as a function of volumes predicted by the wear analysis above, using data derived from the published works cited. Open symbols indicate individual measurements by Subhash et al. at different contact forces, and filled symbols indicate the median values for each material; filled symbols with bars indicate the median values from Marshall et al. and approximate ranges for the medians. All values are relative to the overall median quantities. The solid line indicates ideal agreement, and the shaded band indicates a factor of 2 variation in predicted volume about this line.

It is clear that the data in Figure 9.14 are restricted to a small domain of predicted relative scratch volumes, about a factor of 4 (the exception is a "machinable" glass-ceramic). The restriction is a consequence of the focus in the original works on highly engineered wear-resistant ceramics with, thus, similar E, H, and T values. The same restriction also applies to many, more conventional, multipass, multipoint, "wear" studies (Hsu and Shen 1996, 2004; Tewari et al. 2009). However, it is also clear that the relative ranges of observed scratch volumes for the materials in Figure 9.14 are considerable. Although peak contact force for each material varied by a factor of approximately 4, scratch volumes varied about the median values by a factor of approximately 40. Restricted domains of materials properties and extended ranges of observed removed volumes are common features of most wear studies. This experimental difficulty has significantly impeded the definitive assessment of materials scaling of wear volumes, as in Eq. (9.39) for example (Hsu and Shen 1996, 2004; Tewari et al. 2009). Notwithstanding this caveat, the median experimental data are consistent with the predicted materials trend. The force sensitivity observed in these well-controlled single-pass, single-point tests illustrates the difficulties in predicting the material dependencies of wear volumes in the less controlled, traditional, multipass, multipoint wear tests (Hsu and Shen 1996, 2004; Tewari et al. 2009). The force sensitivity in Figure 9.14 also illustrates the potential difficulties in predicting the Mohs number from material properties.

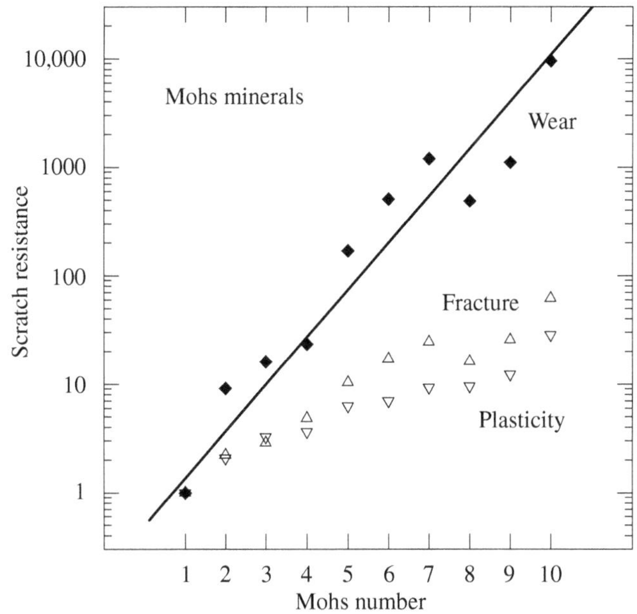

Figure 9.15 Variation of predicted scratch resistance for the Mohs minerals assuming dominance by various phenomena and the lateral crack analysis developed here. The predicted volume of the wear track shows the greatest variation with Mohs number and is the most likely candidate for the basis of the Mohs scale.

Equations (9.34), (9.37), and (9.40) provide a means of assessing the relationship between material properties and Mohs number. Figure 9.15 shows a semilogarithmic plot of predicted scratch resistance as a function of Mohs number, assuming dominance by plastic, fracture, or wear effects using the above relationships and the material properties of the Mohs minerals. The ratios of the left sides of Eqs. (9.34), (9.37), and (9.40) were determined using apatite (Mohs number 5) as the reference and then normalizing the resulting values by those of talc (Mohs number 1) to generate a relative resistance for each parameter. The lower open symbols show the relative scratch resistance based on residual track width as determined by plasticity. The central open symbols show the relative scratch resistance based on lateral crack length as determined by fracture. The upper filled symbols show the relative scratch resistance based on volume of material removed as determined by wear.

It is instructive to consider Figure 9.15 from the perspective of Mohs scale development. Relative scratch resistance determined by plastic effects (lower open symbols) varies by about a factor of 30 over the Mohs scale, and just over 10 if diamond is excluded. This in itself makes plastic track width a weak indicator of scratch resistance. In addition, to obtain experimentally viable track width measurements, it is better to use small, subthreshold contact loads such that the tracks are not deformed by cracking effects, e.g., as in Figures 9.3 and 9.13. Such tracks are small and difficult to observe and measure, even with an optical microscope in the laboratory. The data of Figure 9.15 thus suggest that field identification and distinction of minerals by plastic track width comparison would be extremely difficult and thus it is unlikely that the Mohs scale was based on this phenomenon. Relative scratch resistance determined by fracture effects (central open symbols) varies by about a factor of 60 over the Mohs scale, and just over 20 if diamond is excluded. This makes crack length a better indicator of scratch resistance than track width, but not much. However, crack length measurements are also difficult, especially if the cracks are subsurface in opaque or translucent materials. Figures 9.3 and 9.13 are exceptions. The data of Figure 9.15 thus also suggest that field identification and distinction of minerals by comparison of lateral crack lengths at scratches would be difficult and thus it is unlikely that the Mohs scale was based on this phenomenon either.

In contrast to the above considerations, relative scratch resistance determined by wear effects (upper filled symbols) varies by about a factor of 10,000 over the Mohs scale, and just over 1000 if diamond is excluded, making line wear volume a robust indicator of scratch resistance. Further, wear tracks, in which surface material is removed to form a linear flaw, are easy to observe and assess, e.g., Figure 9.12. That is, the lateral cracks exit the surface as in Figures 9.3 and 9.13. The created subsurface texture and orientation at the flaw are usually very different from that of the original surface allowing easy visual assessment of the existence and scale of scratches in the field, even in opaque or inhomogeneous materials. Hence, identification and distinction of minerals by line wear volume at scratches is relatively easy and thus it is likely that the

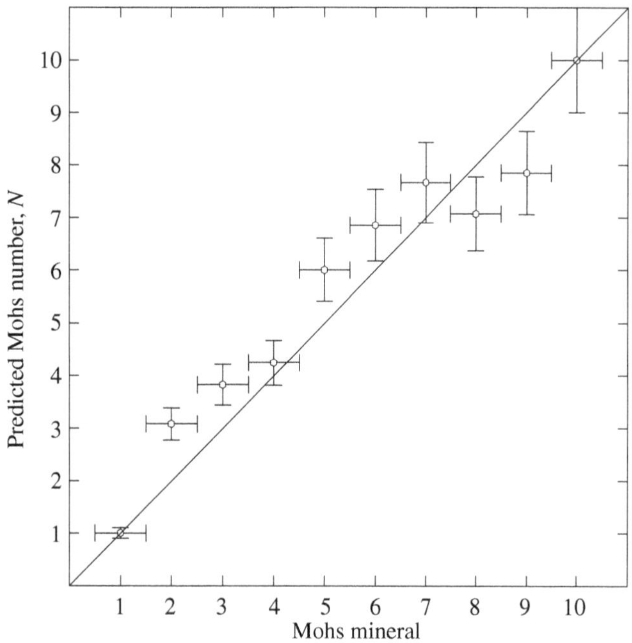

Figure 9.16 Predicted Mohs number N based on independently measured hardness, toughness, and elastic modulus of the Mohs minerals and a physically based linear contact flaw lateral crack analysis. Variation normalized to talc (1) and diamond (10).

Mohs scale was based on this phenomenon. The solid line in Figure 9.15 is an empirical fit to the predicted scratch resistance of the form $\exp(N-1)$, where N is the Mohs number and the use of $(N-1)$ in the exponential emphasizes that the Mohs scale is *relative*, in this case relative to mineral 1, talc. The exponential description also emphasizes that the material properties responses of different minerals are exaggerated by the scale and that scratch volume might vary considerably as in Figure 9.14.

If the material properties dependence of scratch resistance by wear, Eqs. (9.39) and (9.40), is simplified and approximated as $H^2T^{1/2}/E$, the connection between Mohs number and conventional hardness, toughness, and modulus is

$$N \sim 2\ln H + (1/2)\ln T - \ln E - 1. \tag{9.41}$$

Equation (9.41) provides a physical basis for the empirical description of the Mohs scale detailed over 60 years ago by Tabor (Tabor 1954, 1956) and often repeated graphically in textbooks (Callister 1997), and emphasizes the dominance, but not exclusivity (Broz et al. 2006), of hardness in determining Mohs number. Figure 9.16 shows a prediction of Mohs number N for the Mohs minerals using Eq. (9.41). The uncertainties represent $\pm$ 0.5 for the integer mineral order and $\pm 0.1N$ for the numerical prediction, assuming typical E, H, and T uncertainties (Broz et al. 2006). The line indicates ideal agreement. As might be anticipated, agreement in the predicted trend is greater for materials in the center of the scale and less so for the extreme materials in which scratches are less well defined (talc, 1, simply deforms without fracture on sharp contact and sapphire, 9, is barely damaged by most contacts). The uncertainties also suggest that specification of a "Mohs number" for a mineral with a value any more precise than ± 1 is unlikely to lead to accurate comparison of the mechanical properties that determine scratch resistance or wear, or that permit mineral identification.

9.5 Summary and Discussion

The analyses developed here describing lateral cracks at linear sharp contacts considerably extend indentation fracture mechanics. Such mechanics has been well established for the half-penny crack system at point contacts (Chapter 2; see also Lawn et al. 1980; Anstis et al. 1981; Cook 2020), the lateral crack system at point contacts, Eq. (9.28) (Marshall et al. 1982), and its interaction with half-penny cracks (Cook et al. 1990), the strengths of components controlled by half-penny cracks (Chantikul et al. 1981), and the effects of lateral cracks at point contacts on such strengths (Chapter 3; see also Cook and Roach 1986; Cook 2015). In addition, the fracture mechanics of median cracks formed at linear wedge, roller, and dragged contacts, and the resulting strengths is are also well established (Swain and Lawn 1976; Cook 1994, 2017, 2018). Here, analysis is developed to describe the lengths of lateral cracks at linear flaws. The flaws are considered to be formed by static, rolled, or dragged sharp contacts, and the primary focus is on cracks that are "well developed," i.e., much larger than the contact, Eqs. (9.29), (9.32), and (9.36). These descriptions are important for predicting engineering applications and control of linear flaws, including the optical quality of scratched surfaces, strengths resulting from scored lines, and edge quality of broken components. In addition, analyses of lateral cracks that are small with respect to contacts and in "initiation" states have also been considered here. Such analyses are important in the engineering contexts of defining maximum allowed energies, loads, or contact geometries required to avoid scratches. The initiation analyses follow the fracture mechanics developed in Chapter 1 for half-penny cracks at indentation contacts (see also Lawn and Evans 1977; Lawn and Marshall 1979).

It is important to distinguish the emphasis on lateral crack lengths in this chapter from prior, complementary, emphases on lateral crack appearance and lateral crack effects. In this chapter, attention was focused on specifying the stable equilibrium lengths of lateral cracks as functions $c(P)$ of the normal contact loads applied in different contact geometries to different material surfaces (c is the crack length, P is the load). In considering indentation strengths in Chapter 3, two other parameters specifying lateral crack behavior were introduced, both separate from, but related to, the function $c(P)$. The first parameter is the lateral crack development, L_D, that quantifies (visually) the formation of lateral cracking patterns at Vickers indentations. L_D is determined as a function of indentation load, $L_D = L_D(P)$, and varies from $L_D = 0$ for no visible lateral cracking, to $L_D = 0.67$ for subsurface lateral cracks visible in all four indentation quadrants, to $L_D = 1$ for surface chips visible in all four indentation quadrants. The importance of L_D is that increasing values of L_D, particularly for $L_D \rightarrow 0.67$, are correlated with relaxation of the indentation residual stress field that controls component strengths. The measurements of lateral crack lengths by Marshall et al. (1982) were performed in this region, $L_D \leq 0.67$. The measurements of lateral crack or chip lengths by Swain and Lawn (1976) were performed in the saturated region $L_D = 1$. Extending the L_D concept to linear flaws suggests $L_D \approx 0.67$ for both the roller scratch of Figure 9.3 and the dragged scratch of Figure 9.13. Clearly, $L_D(P)$ can be correlated with the amplitude of $c(P)$, noting that $L_D = 0$ at the lateral crack threshold load P_Ω (see Section 9.2.2) at which the $c(P)$ response vanishes. The advantage of L_D is that it provides a measure of what a sharp contact flaw "looks like" and thus provides complementary information to $c(P)$ that provides a measure of sharp contact flaw size. The second parameter is the lateral crack relaxation load, P_L, that quantifies the effects of lateral cracking on indentation strength responses. P_L is a material parameter (much as the threshold load P_Ω) and is determined from indentation strength degradation measurements, $\sigma(P)$. For small values of indentation load, $P \ll P_L$, indentation strengths approach ideal, decreasing, behavior. For large values of indentation load, $P \gg P_L$, indentation strengths approach an invariant strength plateau. For most materials, $P_L \approx 100$ N, such that nearly all indentation strengths are affected by mesoscopic lateral crack stress relaxation effects (Chapters 3 and 4; see also Cook and Roach 1986; Cook 2015). Clearly, P_L can be correlated with a load on the lateral crack $c(P)$ response (e.g., $c \approx 10a$, say, or $P_L \approx 100P_\Omega$) (and can also be estimated from half-penny crack length measurements, Cook et al. 1990; Cook and Liniger 1992). The advantage of P_L is that it provides a direct measure of how much a sharp contact flaw is relaxed by lateral cracking; a value that is difficult to gauge from $c(P)$ size measurements alone.

A useful extension of the analysis here would be to recast the earlier point contact lateral crack analysis (Marshall et al. 1982) in energy terms, rather than as a force balance based on an equation similar to Eq. (9.4). Such recasting would highlight the similarities and differences of the two geometries and provide insight in cases for which the distinction is unclear, for instance slow dragged or rolled contacts. An example of such convergence is the identical load dependence of well-developed lateral crack lengths predicted for point, Eq. (9.28) (Marshall et al. 1982), and dragged, Eq. (9.29), contacts. Another extension of the analysis would be to consider crack lengths far from initiation adjacent to the deformation zone and at propagation termination by the surface. Geometric similarity of the contact would suggest that such exiting linear lateral cracks should scale as $c \sim w$, an idea forwarded earlier in the context of wedge loading (Swain and Lawn 1976), for which, thus, $c \sim P$. Geometric similarity of point contacts suggests that exiting point lateral cracks should scale as $c \sim a$ and thus, $c \sim P^{1/2}$, independent of indenter-included angle. The data of Figure 9.7 are consistent with this assumption and may explain the limited success of Eq. (9.28) in this context. The geometric similarity idea should be testable by the determination of stress trajectories and bears directly on a major application of the analysis here—the prediction of wear volumes, scratch resistance, and interpretation of the Mohs scale in terms of material properties.

The underlying physical and analytical basis for the Mohs scale was provided by the analysis developed here, along with support for that basis through application of experimentally determined material properties and displayed in Figure 9.15. The analysis would appear to answer many of the questions raised earlier (Broz et al. 2006): the physical process of cracking associated with scratching is well defined, Figure 9.5c, leading to a clear definition of scratch resistance, Eq. (9.40), that is different from hardness and involves other material properties of toughness and modulus; a simple predictive relationship between material properties and Mohs number is provided, Eq. (9.41) and the resulting Figure 9.16 is the first of its type.

The physical basis of experimental estimation of Mohs number—naked eye visual observation—would have been available to Mohs as a macroscopic test, implementable in the field. It is unclear that microscopic tests, based on specialized instrumentation in the laboratory (Gu and Yao 2011; Bensaid et al. 2018; Suratwala et al. 2019; Wang et al. 2020; Zakiev et al. 2020), or nanoscale considerations including atomistic computation (Gerberich et al. 2015) truly bear on the origin of the Mohs number, as Mohs characterizes a macroscopic phenomenon. Both the microscopic tests and the nanoscale considerations bear on the onset of fracture at elastic–plastic scratching (and thus have more to do with ductile–brittle transitions at contacts than visible scratch resistance, see Cook 2023). Such tests and considerations are therefore important in assessing the mechanisms of crack initiation and applicability of the Mohs number to characterize a particular scratch. The fundamental basis of Mohs number appears to be not microscopic but macroscopic: hardness, toughness, and modulus combined to result in macroscopic scratch resistance defined as reciprocal line wear volume. Although there is no "Mohs hardness" or "scratch hardness," hardness does play a primary role in determining Mohs number, as recognized by Tabor (1954, 1956). Strictly, hardness is not a material property, as it depends on the shape of the probe used to assess the mean contact pressure at a contact. Thus, geometry and properties of the wearing asperity must play a secondary role in determining scratch resistance, a conclusion also recognized by Tabor.

A clear conclusion from this work is that much more experimental measurement of lateral cracking at linear sharp contacts is required. The survey work of Swain and the analyses here in the context of indentation fracture provide a robust framework for such measurements. The variation of flaw dimensions, track width and lateral crack length, with load for static contact by sharp wedges (Swain and Lawn 1976), dragged contacts, Figure 9.9 (Swain 1979), and rolled contacts, Figure 9.10 (M. Swain, personal communication 1985) should be extended to a variety of materials and contact geometries. The variation of wear volume and scratch resistance, Figure 9.15, with material and the nature of the contacting asperity, in the contexts of the Mohs scale and engineering application, should be determined.

References

Ahn, Y., Cho, N.G., Lee, S.H., and Lee, D. (2003). Lateral crack in abrasive wear of brittle materials. *JSME International Journal Series A Solid Mechanics and Material Engineering* 46: 140–144.

Anstis, G.R., Chantikul, P., Lawn, B.R., Marshall, and D.B. (1981). A critical evaluation of indentation techniques for measuring fracture toughness: I, direct crack measurements. *Journal of the American Ceramic Society* 64: 533–538.

Bensaid, N., Benbahouche, S., Roumili, F., Sangleboeuf, J.-C., Le Cam, J.-B., and Rouxel, T. (2018). Influence of the normal load of scratching on cracking and mechanical strength of soda-lime-silica glass. *Journal of Non-Crystalline Solids* 483: 65–69.

Broz, M.E., Cook, R.F., and Whitney, D.L. (2006). Microhardness, toughness and modulus of Mohs scale minerals. *American Mineralogist* 91: 135–142.

Chantikul, P., Anstis, G.R., Lawn, B.R., and Marshall, D.B. (1981). A critical evaluation of indentation techniques for measuring fracture toughness: II, strength method. *Journal of the American Ceramic Society* 64: 539–543.

Callister, W.D. Jr. (1997). *Materials Science and Engineering: An Introduction, 4th ed.* New York, NY: Wiley.

Cook, R.F. (1985). *Strength Characterisation of Ceramics using Controlled Indentation Flaws*. Ph.D. thesis. Kensington, Australia: University of New South Wales.

Cook, R.F. (1994). Deformation and fracture by sharp rolling contacts. *Journal of the American Ceramic Society* 77: 1263–1273.

Cook, R.F. (2015). Multi-scale effects in the strength of ceramics. *Journal of the American Ceramic Society* 98: 2933–2947.

Cook, R.F. (2017). Fracture mechanics of the scratch strength of polycrystalline alumina. *Journal of the American Ceramic Society* 100: 1146–1160.

Cook, R.F. (2018). Blunt scratch strength of polycrystalline alumina. *Journal of the American Ceramic Society* 101: 16–19.

Cook, R.F. (2019). Fracture sequences during elastic-plastic indentation of brittle materials. *Journal of Materials Research* 34: 1633–1644.

Cook, R.F. (2020). A critical evaluation of indentation crack lengths in air. *Journal of the American Ceramic Society* 103: 2278–2295.

Cook, R.F. (2023) *Particle Strengths*. New York, NY: Wiley.

Cook, R.F. and Liniger, E.G. (1992). Grain-size effects in the indentation fracture of MgO. *Journal of Materials Science* 27: 4751–4761.

Cook, R.F., Pascucci M.R., and Rhodes W.H. (1990). Lateral cracks and microstructural effects in the indentation of Y_2O_3. *Journal of the American Ceramic Society* 73: 1873–1878.

Cook, R.F. and Pharr, G.M. (1990). Direct observation and analysis of indentation cracking in glasses and ceramics. *Journal of the American Ceramic Society* 73: 787–817.

Cook, R.F. and Roach D.H. (1986). The effect of lateral crack growth on the strength of contact flaws in brittle materials. *Journal of Materials Research* 1: 589–600.

de Boer, M.P. and Gerberich, W.W. (1996). Microwedge indentation of the thin film fine line–II. Experiment. *Acta Materialia* 44: 3177–3187.

Evers, M. (1967). Plastic deformation of glass with diamond indenters (in German). *Glasstechnische Berichte* 40: 41–43.

Fuller, E.R. Jr., Lawn, B.R., and Cook, R.F. (1983). Theory of fatigue for brittle flaws originating from residual stress concentrations. *Journal of the American Ceramic Society* 66: 314–321.

Gao, S., Li, H., Kang, R., Zhang, Y., and Dong, Z. (2021). Effect of strain rate on the deformation characteristic of AlN ceramics under scratching. *Micromachines* 12: 77.

Gerberich, W.W., Ballarini, R., Hintsala, E.D., Mishra, M., Molinari, J.-F., and Szlufarska, I. (2015). Toward demystifying the Mohs hardness scale. *Journal of the American Ceramic Society* 98: 2681–2688.

Glaesemann, G.S., Ritter, J.E. Jr., and Jakus, K. (1987). Mixed-mode fracture in soda-lime glass using indentation flaws. *Journal of the American Ceramic Society* 70: 630–636.

Gruber, M., Leitner, A., Kiener, D., Supancic, P., and Bermejo, R. (2022). Effect of crystal orientation on the hardness and strength of piezoelectric $LiNbO_3$ substrates for microelectronic applications. *Materials & Design* 213: 110306.

Gruver, R.M. and Kirchner, H.P. (1973). Effect of surface damage on the strength of Al_2O_3 ceramics with compressive surface stresses. *Journal of the American Ceramic Society* 56: 21–24.

Gu, W. and Yao, Z. (2011). Evaluation of surface cracking in micron and sub-micron scale scratch tests for optical glass BK7. *Journal of Mechanical Science and Technology* 25: 1167–1174.

Gu, W., Yao, Z., and Li, K. (2011). Evaluation of subsurface crack depth during scratch test for optical glass BK7. *Proceedings of the Institution of Mechanical Engineers, Part C: Journal of Mechanical Engineering Science* 225: 2767–2774.

Howes, V.R. and Szameitat, A. (1984). Morphology of sub-surface cracks below the scratch produced on soda lime glass by a disc cutter. *Journal of Materials Science Letters* 3: 872–874.

Hill, R. (1950). *The Mathematical Theory of Plasticity*. Oxford, UK: Oxford.

Hsu, S.M. and Shen, M.C. (1996). Ceramic wear maps. *Wear* 200: 154–175.

Hsu, S.M. and Shen, M. (2004). Wear prediction of ceramics. *Wear* 256: 867–878.

Kirchner, H.P. (1984). Comparison of single-point and multipoint grinding damage in glass. *Journal of the American Ceramic Society* 67: 347–353.

Kirchner, H.P. and Isaacson, E.D. (1982). Residual stresses in hot-pressed Si_3N_4 grooved by single-point grinding. *Journal of the American Ceramic Society* 65: 55–60.

Lawn, B.R. (1993). *Fracture of Brittle Solids, 2nd ed.* Cambridge, UK: Cambridge.

Lawn, B.R. and Evans AG (1977). A model for crack initiation in elastic/plastic indentation fields. *Journal of Materials Science* 12: 2195–2199.

Lawn, B.R., Evans A.G., and Marshall D.B. (1980). Elastic/plastic indentation damage in ceramics: the median/radial crack system. *Journal of the American Ceramic Society* 63: 574–581.

Lawn, B.R. and Marshall D.B. (1979). Hardness, toughness, and brittleness: an indentation analysis. *Journal of the American Ceramic Society* 62: 347–350.

Lawn, B. and Wilshaw, R. (1975). Indentation fracture: principles and applications. *Journal of Materials Science* 10: 1049–1081.

Lee, S.K., Jensen, R.P., and Readey, M.J. (2001). Effect of grain size on scratch damage in Y-TZP ceramics. *Journal of Materials Science Letters* 20: 1341–1343.

Lee, S.K., Tandon, R., Readey, M.J., and Lawn, B.R. (2000). Scratch damage in zirconia ceramics. *Journal of the American Ceramic Society* 83: 1428–1432.

Marshall, D.B. and Evans, A.G. (1984). Measurement of adherence of residually stressed thin films by indentation. I-Mechanics of interface delamination. *Journal of Applied Physics* 56: 2632–2638.

Marshall, D.B., Evans, A.G., Khuri-Yakub, B.T., Tien, J.W., and King, G.S. (1983). The nature of machining damage in brittle materials. *Proceedings of the Royal Society London A* 385: 461–475.

Marshall, D.B., Lawn, B.R., and Cook, R.F. (1987). Microstructural effects on grinding of alumina and glass-ceramics. *Journal of the American Ceramic Society* 70: C-139–C-140.

Marshall, D.B., Lawn, B.R., and Evans, A.G. (1982). Elastic/plastic indentation damage in ceramics: the lateral crack system. *Journal of the American Ceramic Society* 65: 561–566.

Matsuoka, J., Guo, D., and Yoshida, S. (2017). Cross-section morphology of the scratch-induced cracks in soda-lime-silica glass. *Frontiers in Materials* 4: 8.

Ono, T. and Tanaka, K. (2001). Effect of scribe-wheel dimensions on the cutting of AMLCD glass substrate. *Journal of the Society for Information Display*, 9: 87–94.

Pan, C.T., Hsieh, C.C., Su, C.Y., and Liu, Z.S. (2008). Study of cutting quality for TFT-LCD glass substrate. *The International Journal of Advanced Manufacturing Technology* 39: 1071–1079.

Pei, Z.J., Billingsley, S.R., and Miura, S. (1999). Grinding induced subsurface cracks in silicon wafers. *International Journal of Machine Tools and Manufacture* 39: 1103–1116.

Peter, K. (1964). Brittle fracture and microplasticity of glass in indentation experiments (in German). *Glasstechnische Berichte* 37: 333–345.

Ritter, J.E., Lardner, T.J., Rosenfeld, L., and Lin, M.R. (1989). Measurement of adhesion of thin polymer coatings by indentation. *Journal of Applied Physics* 66: 3626–3634.

Rosenfeld, L.G., Ritter, J.E., Lardner, T.J., and Lin, M.R. (1990). Use of the microindentation technique for determining interfacial fracture energy. *Journal of Applied Physics* 67: 3291–3296.

Rossington, C., Evans, A.G., Marshall, D.B., and Khuri-Yakub, B.T. (1984). Measurements of adherence of residually stressed thin films by indentation. II. Experiments with ZnO/Si. *Journal of Applied Physics* 56: 2639–2644.

Subhash, G. and Bandyo, R. (2005). A new scratch resistance measure for structural ceramics. *Journal of the American Ceramic Society* 88: 918–925.

Subhash, G. and Klecka, M. (2007). Ductile to brittle transition depth during single-grit scratching on alumina ceramics. *Journal of the American Ceramic Society* 90: 3704–3707.

Subhash, G., Loukus, J.E., and Pandit, S.M. (2002). Application of data dependent systems approach for evaluation of fracture modes during a single-grit scratching. *Mechanics of Materials* 34: 25–42.

Subhash, G., Marszalek, M.A., and Maiti, S. (2006). Sensitivity of scratch resistance to grinding-induced damage anisotropy in silicon nitride. *Journal of the American Ceramic Society* 89: 2528–2536.

Suratwala, T., Steele, R., Shen, N., Ray, N., Wong, L., Miller, P.E., and Feit, M. (2019). Lateral cracks during sliding indentation on various optical materials. *Journal of the American Ceramic Society* 103: 1343–1357.

Swain, M.V. (1979). Microfracture about scratches in brittle solids. *Proceedings of the Royal Society London* A 366: 575–597.

Swain, M.V. (1981a). Median crack initiation and propagation beneath a disc glass cutter. *Glass Technology* 22: 222–230.

Swain, M.V. (1981b). Nickel sulphide inclusions in glass: an example of microcracking induced by a volumetric expanding phase change. *Journal of Materials Science* 16: 151–158.

Swain, M.V. and Lawn B.R. (1976). Indentation fracture in brittle rocks and glasses. *International Journal of Rock Mechanics and Mineral Science and Geomechanics Abstracts* 13: 311–319.

Symonds, B.L., Cook R.F., and Lawn B.R. (1983). Dynamic fatigue of brittle materials containing indentation line flaws. *Journal of Materials Science* 18: 1306–1314.

Tabor, D. (1954). Mohs's hardness scale—a physical interpretation. *Proceedings of the Physical Society B* 67: 249–257.

Tabor, D. (1956). The physical meaning of indentation and scratch hardness. *British Journal of Applied Physics* 7: 159–166.

Tewari, A., Basu, B., and Bordia, R.K. (2009). Model for fretting wear of brittle ceramics. *Acta Materialia* 57: 2080–2087.

Thurn, J. and Cook, R.F(2002). Stable dielectric fracture at interconnects from electromigration stresses. *Acta Materialia* 50: 2627–2637.

Thurn, J. and Cook, R.F. (2004). Mechanical and thermal properties of physical vapor deposited alumina films: II, elastic, plastic, fracture and adhesive behavior. *Journal of Materials Science* 39: 4809–4819.

Tomei, N., Murakami, K., Fukunishi, T., Yoshida, S., and Matsuoka, J. (2018). Direct observation of crack propagation in a liquid crystal display glass substrate during wheel scribing. *International Journal of Applied Glass Science* 9: 105–113.

Wang, W., Wan, Z., Yang, S., Feng, J., Dong, L., and Lu, L. (2020). Experimental and numerical investigation on the effect of scratch direction on material removal and friction characteristic in BK7 scratching. *Materials* 13: 1842.

Yeap, B.K., Zeng, K., Jiang, H., Shen L., and Chi D. (2007). Determining interfacial properties of submicron low-k films on Si substrate by using wedge indentation technique. *Journal of Applied Physics* 101: 123531.

Zakiev, I., Gogotsi, G.A., Storchak, M., and Zakiev, V. (2020). Glass fracture during micro-scratching. *Surfaces* 3: 211–224.

10

Strengths of Silicon Devices Determined by Wafer Backgrinding

In this chapter the effects of wafer backgrinding on the strength of silicon devices in the form of dies are investigated experimentally and analytically. In particular, the scale and orientation of striations generated by in-feed grinding on wafer and die back faces are shown to be critical factors in determining the spatial arrangement of die strengths over a wafer and the strength distributions of dies sampled from a wafer. The scale of the striations is considered in the context of controlled indentation and scratch flaw experiments, leading to the assessment that strength-controlling grinding striations are formed by translated sharp contacts with approximately 0.1 N normal load. The effect of striation orientation is described by a simple expression fit to experimental die strength measurements. An analytical formulation of striation orientation based on in-feed grinding geometry parameters implements the expression to generate wafer strength maps and sample strength distributions. The maps and distributions are of direct engineering applicability.

10.1 Introduction

Semiconductor devices are the key components of electronic systems in the 21st century. The devices gather, process, and store information in compact and economical forms and enable systems as pervasive and diverse as telephones and display tablets, washing machines, automobiles, credit cards, satellites, and ballistic missiles—in fact any system or "thing" that requires any sort of computational capability. The overwhelmingly predominant material used in semiconductor devices is silicon (Si) in single-crystal form. Ge, GaAs, and SiC are used far less frequently for specialized electronic applications. Polycrystalline Si is also used, again less frequently, in photovoltaic applications and in microelectromechanical systems (MEMS). As thus might be anticipated, considerable attention has been given to the electrical properties of Si and to the fabrication of Si components so as to obtain the desired electrical performance (e.g., Sze and Lee 2012; Campbell 2012). As also might be anticipated, attention has been given more recently to the worldwide economic and geopolitical effects of the manufacture and application of such pervasive systems (Miller 2022). This chapter considers a critical, but often overlooked, step in Si semiconductor device fabrication—that of grinding the electrically finished devices to size for use in application. The grinding process controls Si device strength through the formation of flaws in the form of striations on the backfaces of devices and thus controls device and system mechanical performance. This chapter is then complementary to Chapter 8, in which the dicing process controlled Si device strength through the formation of chips on the edges of devices. The chapter here is also a follow on to Chapter 9, as grinding striations are a form of scratch damage. This chapter begins with a brief overview of semiconductor device fabrication, with an emphasis on the mechanical aspects of manufacture. This is followed by sections considering the nature of grinding striations and the consequent effects on device strength. A model of grinding controlled strength is then developed, based on these observations of striations and strengths, and model applications are demonstrated.

10.1.1 Semiconductor Device Fabrication

Semiconductor device fabrication may be divided into four sequential stages: two electrical, followed by two mechanical. In the first, electrical, stage, known as front end of line (FEOL), dopants are implanted into or deposited onto the surface of a large circular wafer of semiconducting material—usually single-crystal Si. The dimensions of wafers used currently in fabrication are predominantly 300 mm diameter × 775 μm thick or, less commonly, 200 mm × 725 μm, and all are oriented such that the wafer face is the Si(001) plane (Si exhibits a cubic crystal structure). The wafer is the substrate for the many

Indentation Fracture: Strength and Toughness for Brittle Materials Design, First Edition. Robert F. Cook.

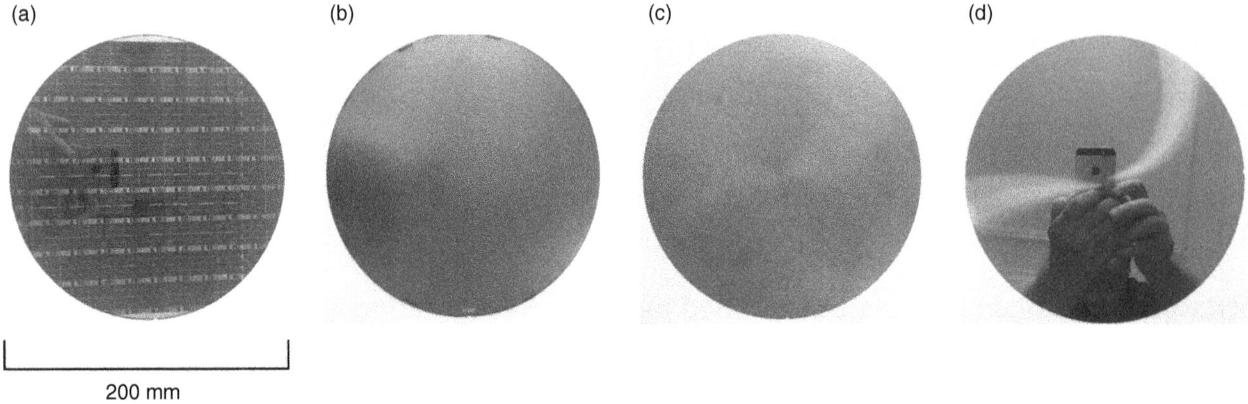

Figure 10.1 Frontface and backface of a 200 mm Si wafer at various stages during semiconductor device fabrication. (a) Frontface of full 725 μm thickness wafer after electrical circuit formation into a rectangular array of VLSI devices; (b) backface of full thickness wafer; interference shadings indicate films residual from VLSI processing; (c) backface of 420 μm thick wafer after first, Z1, coarse stage of backgrinding; wafer surface is matte and grinding striations are evident; (d) backface of 380 μm thick wafer after second, Z2, fine stage of backgrinding; wafer surface generates specular reflection.

billions of discrete electrical circuit elements formed in the Si during FEOL processing. In the second electrical stage, known as back end of line (BEOL), successive layers of conductors and insulators are deposited and patterned over the wafer surface to form a three-dimensional array interconnecting the individual circuit elements. BEOL processing generates very large-scale integrated (VLSI) circuits that form individual devices over the wafer surface. The devices are aligned in a rectangular array, along Si⟨110⟩ directions, as shown in Figure 10.1a. After completion of the FEOL and BEOL stages, the VLSI devices are tested for quality and a "wafer map" of electrical performance of the devices is generated.

The third, mechanical, stage is known as wafer finishing. In this stage, material is removed from the Si wafer substrate in two substages: backgrinding and dicing. In the backgrinding substage, material is removed uniformly from the back (nondevice) face of the wafer, Figure 10.1b, parallel to the wafer surface, to generate a substrate of appropriate thickness and backface surface finish. Usually the majority of material is removed by a coarse rotating grinding wheel operation, known as Z1, leaving residual surface and subsurface damage in the wafer backface as shown in Figure 10.1c. Hence, backface surface treatments are often completed by fine grinding, known as Z2, shown in Figure 10.1d, or chemically assisted etching or polishing. In the next substage, dicing, material is locally removed from the front (device) side of the wafer, perpendicular to the wafer surface, to generate dies of appropriate length and width. Usually material is removed in this substage by a rotating dicing saw blade that penetrates the entire thinned wafer perpendicular to the wafer surface. The blade is translated along Si⟨110⟩ directions, between the devices defined in the FEOL and BEOL stages (see Chapter 8 for a discussion of dicing damage on the resulting device edges). After dicing, the now individual or singulated devices are sorted and picked using the electrical wafer map generated after the BEOL stage. In the fourth stage, known as packaging, the thinned, rectangular, sorted and picked devices in the form of dies—commonly referred to as "chips"—are attached mechanically and electrically to larger metallic, polymeric, or ceramic substrates and encapsulated to form packaged devices suitable for use in electronic applications, for example in hand-held tablets, cell phones, and smart cards.

10.1.2 Wafer Backgrinding and Die Strength

The focus of the work here is mechanical performance of devices determined by the third and fourth fabrication stages above, rather than the more commonly studied electrical performance (Sze and Lee 2012; Campbell 2012) determined by the first two stages. The mechanically oriented focus is a consequence of the significant commercial driving force for fabrication and use of thinned dies. Smaller and thinner dies enable greater density of devices in applications and thus greater application functionality, e.g., through more chips/cell phone. In addition, thinned devices enable reduced power consumption through fewer parasitic electrical and thermal effects and are critical to new high-density packaging schemes, such as stacked dies. Hence, mechanical backgrinding is likely to remain an essential part of the semiconductor fabrication process: the diversity of consequent challenges generated by thinned wafers and dies in semiconductor fabrication and application is reviewed by Marks et al. (2015). The current work will consider a specific challenge in detail—the mechanical strength of dies limited by backgrinding damage, Figure 10.1c and d. Such strengths determine device yield in the fabrication stages of dicing, sorting, picking, and packaging and determine packaged device reliability in operation. During fabrication, die mechanical strengths limit the loads that can be applied to dies (e.g., by robotic handling devices). During operation, die

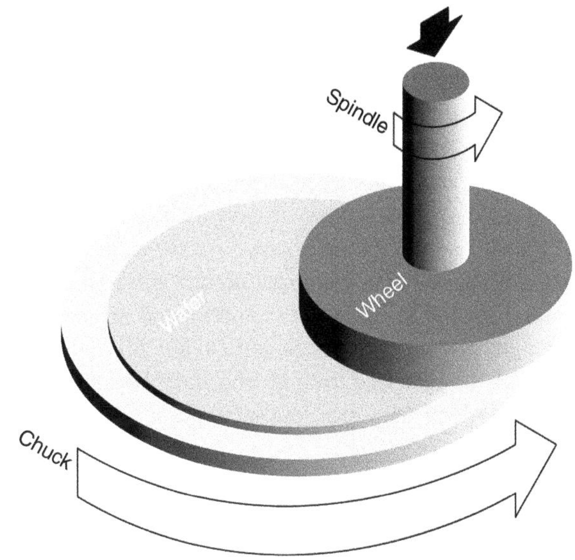

Figure 10.2 Schematic diagram of wafer backface in-feed grinding process. The rotations of the wafer, large disc fixed to underlying chuck, and grinding wheel, small disk fixed to spindle, about their respective axes are indicated by open arrows. The linear "feed" motion of the grinding wheel center "in" toward the wafer is indicated by the solid arrow.

mechanical strengths limit the thermal expansion mismatch stresses that can be applied to dies by packaging materials, (e.g., polymer circuit boards) and limit the stresses that can be applied to dies by users (e.g., by bending of a credit card). Hence, to optimize semiconductor device fabrication a wafer map of die mechanical strength is required, similar to that generated for die electrical performance.

A common grinding process that determines die strength is "in-feed" grinding. In this process, the abrasive face of a rapidly rotating circular wheel is used to remove material from the backface of a semiconductor wafer. Material removal is facilitated by small, sharp, hard particles (diamonds, alumina, silicon carbide) imbedded in the abrasive face of the grinding wheel. The wheel diameter is less than that of the wafer, and the wheel and wafer axes, although parallel, are not colinear. A schematic diagram is shown in Figure 10.2. In operation, the wafer is rotated slowly about the wafer axis while the rapidly rotating grinding wheel is simultaneously translated slowly "into" the wafer parallel to the wheel axis. In this way, the grinding wheel center describes a helical path relative to the wafer center, and the grinding wheel face removes material from the entire wafer backface, Figure 10.1c.

In-feed grinding generates surface damage as an array of striations in the form of circular arcs radiating from the wafer center. As a consequence of the grinding geometry and chuck and wheel motion, the striations exhibit a systematic variation of orientation over the backface of the wafer, Figure 10.3, and thus from die to die. The striations are residual plastic deformation and fracture tracks generated by individual abrasive particles translated in contact with the wafer surface.

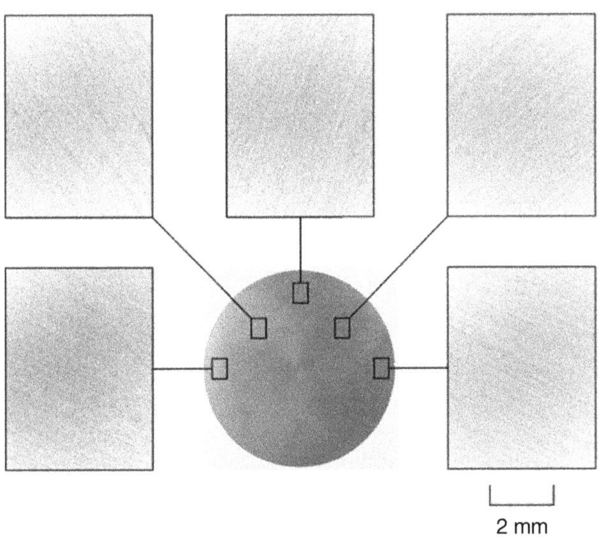

Figure 10.3 Optical micrographs of the backface of a Z1 ground wafer. The overall wafer view from Figure 10.1c is shown along with five images illustrating different orientations of near-parallel grinding striations in different locations on the wafer backface. Z1 + Z2 grinding striations are similarly oriented.

The strength-controlling nature of these striations in terms of orientation relative to a die edge and subsequent package orientation is the focus of this chapter.

Specifically, Sections 10.2–10.4 demonstrate the generation of wafer-scale maps of die mechanical strength, and consequent strength distributions, as controlled by in-feed grinding. First, brief overviews are given of prior observations of linear flaws in Si surfaces, single scratches and multiple striations. Next, experimental measurements of strengths of Si dies are presented, using controlled flaws to characterize strengths. The observations and measurements are placed in the context of linear sharp contact flaw analyses (Chapter 9; see also Cook 2017, 2018) and provide background for the principal section of the current work that follows—the development of a model to predict die strength maps and distributions using in-feed grinding parameters and experimental strength measurements. The basis of the model is outlined in two earlier patents (Mendelson et al. 1999, 2001) and several recent works considering the linkages between flaw population and sampled strength distributions (Cook and DelRio 2019a,b; Cook et al. 2020; Cook 2023; DelRio et al. 2020). The strength distributions of dies resulting from in-feed wafer backgrinding are compared with those resulting from strength measurements of Si MEMS, bulk ceramics, and small particles (Cook and DelRio 2019a,b; Cook et al. 2020; Cook 2023; DelRio et al. 2020).

10.2 Linear Flaws in Si

10.2.1 Single Scratches

The fracture mechanics of strength-controlling linear flaws—scratches—is reviewed and developed in Chapter 9, with a focus on lateral crack formation, and in two recent works (Cook 2017, 2018), with a focus on superposed applied stress effects and the resulting scratch strength. Chapter 9 and the recent works consider large-scale features formed by dragged or rolling contacts on ceramics, glasses, semiconductors, and other brittle materials and make a clear distinction between such linear flaws and the more frequently studied point-like sharp indentation flaws. An important finding to emerge from these considerations is that extended striations or scratches are much more potent as strength-degrading surface flaws than localized contacts. For a typical brittle material, the normal load required to generate a strength-controlling striation or scratch flaw by a translated sharp contact is about 0.1 N or less, whereas the load required to generate a strength-controlling impact or indentation flaw by the same sharp contact at normal incidence is about 1 N or greater. The nature of the damage leading to strength-controlling striations is shown in the schematic diagram of Figure 10.4, which depicts two sharp particles translated over a brittle material surface at small, slightly different, normal loads. In both cases, the translated particle generates a striation in the form of a residually deformed, linear surface impression (shown as light shading). Beneath each striation is

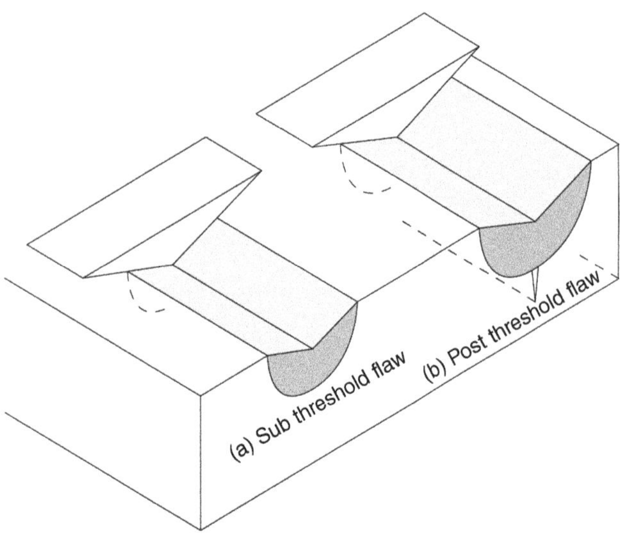

Figure 10.4 Schematic diagrams of linear flaws generated in surfaces by translated sharp contacts. The top view shows the contact and the ensuing plastic deformation track (light shading). The cross-section view shows the subsurface plastic deformation zone (dark shading). (a) Small, subthreshold contact load; (b) large, postthreshold contact load exhibiting larger plastic deformation track and zone and subsurface median crack (dashed line).

a localized, near semicylindrical, zone of plastic deformation (shown in section as dark shading). Compressive strain in the localized plastic deformation zone generates a tensile residual stress field in the surrounding elastically deformed matrix. If the striation is of sufficient size, the residual field can initiate and stabilize cracks in the matrix adjacent to the plastic deformation zone. In particular, a crack is shown beneath the zone for the slightly larger striation, Figure 10.4b. Such a crack, formed centrally beneath the zone, perpendicular to the surface, is known as a *median crack*, and the striation–crack combination is termed a "postthreshold flaw" as the threshold conditions for crack initiation, such as normal load and particle acuity, are exceeded. The slightly smaller striation, Figure 10.4a, is termed a "subthreshold flaw" as the threshold conditions for crack initiation are not exceeded. Under the action of a superposed applied stress, a large striation forming a postthreshold flaw is often strength controlling through the ensuing instability and propagation of the median crack leading to component failure. Small striations can also be strength controlling, as an applied stress can "pop-in" a median crack at a subthreshold flaw, leading to immediate instability, crack propagation, and failure.

As will be shown below, the strength-controlling damage generated by backgrinding Si wafers most probably consists of multiple subthreshold linear contact flaws. That is, multiple, curved versions of Figure 10.4a. Early, 1980s, research on linear damage in Si, however, considered single postthreshold flaws generated by translated contacts with large normal loads, typically greater than 0.25 N. The focus was on damage morphology: Such flaws exhibited significant lateral cracking, chipping, and material removal parallel to the surface, in addition to the surface impression and subsurface median crack of Figure 10.4b, and were more aptly described as scribe lines or scratches than as striations. A cross section illustrating the impression and distinct lateral and median cracks at a scribe line in Si (Misra and Finnie 1979) is very similar to that of a glass-cutting roller track (Howes and Szameitat 1984). Much more surface damage and material removal is visible in images of scratches in Si generated by a sharp probe (Lim and Danyluk 1985, 1988; Danyluk et al. 1988). An early study of damage generated by translated blunt probes on Si (Leu and Scattergood 1988) showed that much greater contact loads were required, $\approx$100 N, and that the damage morphology was significantly different and dominated by near-surface circular chatter cracks (similar to those on glass, Lawn and Wilshaw 1975; Lawn et al. 1984). The observations of flaws formed on Si surfaces by large-load scribing or scratching provide extreme upper bounds to the extent of damage generated by backgrinding.

10.2.2 Multiple Striations

Subsequent research, in the early 1990s–2000s, on linear damage in Si did consider the effects of typical backgrinding flaws on the strength of Si wafers and dies. The twofold motivation for this consideration was (i) the introduction of larger wafers that were more susceptible to both failure during handling and grinding-induced dimensional nonuniformities (Bawa et al. 1995; Mendelson et al. 1996); and (ii) the introduction of polymeric-based packaging schemes that placed the die in tension and thus generated a greater likelihood of package-induced device failure (Michaelides and Sitaraman 1999; Yu et al. 2002). A review of sharp contact damage and fracture strength of Si summarized the research findings regarding the backgrinding-strength linkage (Cook 2006): After frontface processing, the backfaces of Si wafers, Figure 10.1b, are reasonably strong ($\approx$200 MPa), although not as strong as MEMS devices ($\approx$2 GPa, DelRio et al. 2015); coarse backgrinding, Figure 10.1c, reduces this strength for Si dies considerably ($\approx$50 MPa); subsequent fine backgrinding, Figure 10.1d, restores the strength to the wafer level or greater ($\approx$300 MPa); final stage etching can increase the strengths even more, nearly to MEMS levels. The qualitative trends in Si strength during wafer finishing were identified in an early work addressing the backgrinding–strength linkage (Hudson and Perrin 1990).

Practical experience and further research led to the quantitative findings above and to the recognition of the different effects on the Si surface of the various wafer finishing steps. As summarized by Marks et al. (2015): Coarse grinding, designated Z1, performs the greatest reduction in wafer thickness (and thus removes the majority of wafer material), hundreds of micrometers at approximately 5 μm s^{-1}, and generates an optically rough surface, root mean square (rms) roughness of several micrometers, and significant subsurface damage, tens of micrometers in depth. Fine grinding, designated Z2, reduces the surface roughness, removing material at approximately 1 μm s^{-1}, to generate an optically smooth surface, rms roughness of several nanometers, and removes sub surface damage. An optional polishing step, designated Z3, continues these processes at a finer level. Final etching is designed to reduce the surface roughness to subnanometer levels and remove even more subsurface damage. Both polishing and etching involve significant chemical effects.

More detailed research regarding Si die strength, in the late 1990s–late 2000s, focused on Z1 backgrinding parameters. The Z1 step generates the greatest surface roughness and largest amount of subsurface damage and is therefore the most strength controlling. Two extensive studies in 1997 identified most of the important phenomena regarding the geometrical

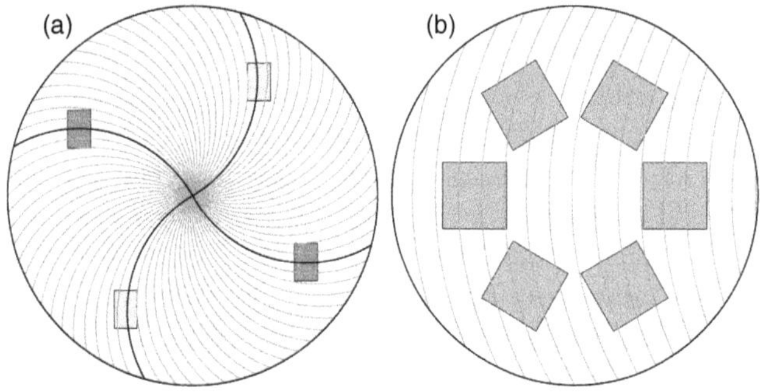

Figure 10.5 Schematic diagrams of wafer backfaces with representative examples of grinding striations and dies. (a) In-feed striations and rectangular dies. Uniaxial strength tests exhibit a significant effect of die location; (b) creep-feed striations and square dies. Biaxial strength tests exhibit no significant effect of die orientation.

linkages between wafer backgrinding and die strength. Lee et al. (1997) found that: (i) the striations generated by in-feed backgrinding formed an array of curved arcs radiating from the center of the wafer; (ii) a die picked from the wafer exhibited nearly parallel striations, oriented relative to the die axes in a direction that depended on die location within the wafer, Figure 10.3; and (iii) the most significant factor in determining die strength was striation orientation relative to the die axes. In particular, for uniaxial flexure tests in which the long axis of a rectangular die was in tension, strength was significantly greater for dies exhibiting striations more closely aligned with the long axis. A schematic diagram of an in-feed ground wafer backface and example grinding striations and die locations are shown in Figure 10.5a.

McGuire et al. (1997) used a Z1 variation known as "creep-feed" grinding, in which the wafer is translated, rather than rotated, beneath the grinding wheel, and found that: (i) the striations generated formed a parallel array of curved arcs across the wafer backface; (ii) a die picked from the wafer exhibited parallel striations oriented relative to the die axes in a direction that depended most strongly on wafer orientation relative to the creep-feed direction; and (iii) in equibiaxial flexure tests, in which the entire face of a square die was placed in isotropic tension, there was no significant effect on strength of striation orientation relative to the die axes, implying that the strength-controlling striation damage was not location dependent. Cross-section measurements showed the grinding damage to extend 15 μm beneath the surface and an inverse correlation was demonstrated between damage depth and strength. A schematic diagram of a creep-feed ground wafer backface and example grinding striations and die locations are shown in Figure 10.5b.

Two patents by Mendelson et al. (1999, 2001) placed the backgrinding–strength linkage on sound quantitative experimental and analytical foundations, and generated clear design bounds on grinding parameters to meet specified strength requirements. Mendelson et al. determined analytical expressions for the shape of striation arrays formed by in-feed, creep-feed, and other backgrinding geometries and provided expressions for experimentally determined variations of strength with striation orientation. Combining the analyses with the measurements enabled quantitative strength bounds to be established along with clear qualitative descriptions of strength behavior as systems deviated from optimum grinding parameters (e.g., grinding wheel diameter relative to die size). A partial intent of this chapter is to extend the patents of Mendelson et al. by providing (i) greater insight into, and better descriptions of, strength behavior with regard to grinding-induced flaws; and (ii) complete descriptions of strength behavior beyond the bounding values, including spatial variations and distributions.

An extensive series of studies by Pei and colleagues (Pei et al. 1999, 2005, 2008; Pei and Strasbaugh, 2001, 2002; Pei, 2002; Chidambaram et al. 2003a,b; Sun et al. 2004, 2005a,b), while not considering strength, explored the technology and damage phenomenology of in-feed grinding in detail. In particular, Pei et al. (1999) showed striation cross sections with median cracks extending to about 20 μm beneath the surface, similar in form to those in Figure 10.4b and in the earlier observations of somewhat greater damage by Misra and Finnie (1979) and Howes and Szameitat (1984); Pei (2002), and Pei and Strasbaugh (2002) used Makyoh ("magic mirror") techniques to show images of backgrinding striation patterns on wafers as an array of radiating circular arcs, similar to those described by Lee et al. (1997), Figure 10.5a; Chidambaram et al. (2003b) developed a mathematical model to describe and predict the shape of the grinding striation arrays as a function of grinding parameters. The model was in agreement with the earlier observations (Lee et al. 1997; Pei 2002; Pei and Strasbaugh 2002) and similar to that developed earlier by Mendelson et al. (1999, 2001); Sun et al. (2005b) developed a model for the depth variation of the striations, resulting in depths between 0.2 and 0.4 μm. This model was in agreement with single-point ductile turning observations that showed that the transition in Si from the subthreshold flaw configuration, Figure 10.4a, to the postthreshold configuration, Figure 10.4b, occurred at an impression or striation depth between 0.2 and 1 μm (Chao et al. 2002). In addition, the values were consistent with Raman measurements of a heavily deformed surface layer developed on Si backgrinding, thickness of approximately 0.5 μm (Chen and De Wolf 2003), etching studies of removal of a heavily stressed layer (relieving Si wafer bow), thickness of approximately 0.6 μm (Haapalinna et al. 2004), and measurements and modeling of grinding striation depths of the related SiC ceramic, between 0.1 and 0.5 μm (Agarwal and Rao 2010).

Many additional explicit strength studies reinforced the idea that uniaxial die strengths depended on backgrinding striation orientation (Wu et al. 2003; Chong et al. 2004; McLellan et al. 2004; Dries et al. 2008; Zhao et al. 2009; Chae et al. 2010). Specifically, striations parallel to the tensile axis applied in strength testing (usually by bending along the long axis of a die) led to the greatest strengths, and striations perpendicular to the tensile axis led to the weakest strengths, as illustrated in Figure 10.5a. As the die location on the wafer determined the striation orientation, there was a clear correlation between strength and location, e.g., as shown in Chae et al. (2010ab). In addition, many studies reinforced the idea that die strengths increased significantly if the striations were diminished or removed by fine grinding or etching of the wafer backface, or if the front face was tested (Liu et al. 2002; Cotterell et al. 2003; Funke et al. 2004; Jeong et al. 2004; Yoshikawa et al. 2012; Watanabe et al. 2014). In these cases, the correlation between die strength and location was diminished or lost, as observed earlier by McGuire et al. (1997). Measurements of the total damage depth created by backgrinding were in the range of 15–30 μm (Jeong et al. 2004; Yoshikawa et al. 2012; Watanabe et al. 2014), and it was recognized that not all damage was removed by etching.

In summary, the above observations provide physical bases and semiquantitative descriptions of backgrinding controlled die strength. However, a fully predictive strength model requires experimental measurements and quantitative descriptions of the variations in die strength with flaw size and striation orientation. Section 10.3 presents such strength measurements and descriptions.

10.3 Strength of Si

This section begins with an overview of fracture mechanics analysis required for the examination of indentation strength and grinding strength measurements of Si. The analysis is based on several prior considerations in this book: (i) crack initiation from Section 1.7; (ii) residual stress field amplitude from Section 2.4; (iii) indentation fracture mechanics from Section 3.2, and (iv) linear flaw geometry from Section 9.2. The initial goal of the analysis here is the specification of the strength of Si components containing linear sharp contact flaws—scratches and consequent median cracks—in terms of experimental scratch parameters. Subthreshold (uncracked) and postthreshold (cracked) flaws are considered. Subsequent to this specific analysis, a general framework for contact flaw controlled strengths, linear and point, subthreshold and postthreshold, is developed and illustrates the commonality of terms and behavior. The analyses and framework enable the interpretation of strength measurements in Sections 10.3.3, 10.3.4, and 10.4.

10.3.1 Indentation Fracture Mechanics–Linear Contact Flaws

The development of strength analyses begins with an extension of the flaw geometry descriptions of Figure 10.4 (cf. Figures 2.22, 9.5, and 9.6). Figure 10.6a shows a cross section of the surface of a brittle component containing a subthreshold flaw formed by a linear sharp contact. The residual width of the surface contact is $2w$ and the residual included angle is 2ϕ. A localized residual plastic deformation zone (pdz) of compacted and sheared material is formed in the subsurface beneath the contact and has a characteristic depth h. A crack nucleus of dimension c_1 is formed by shear fault intersections at the pdz depth periphery. Plastic strain in the pdz generates a residual stress field in the surrounding matrix of elastically deformed material. The combination of pdz, crack nucleus, and residual field is termed a "subthreshold flaw," as there is no well-defined crack. The initiation of a median crack forms a "postthreshold flaw" or, potentially, leads to failure of a component containing the flaw. The subthreshold crack initiation and failure processes are described by a stress-intensity factor (SIF), now developed, characterizing the effects of the residual stress field on the crack nucleus at a scratch.

The characteristic residual pressure p_w exerted by the pdz on the surrounding matrix scales as

$$p_w \sim -\kappa\left(\frac{\Delta V}{V}\right), \tag{10.1}$$

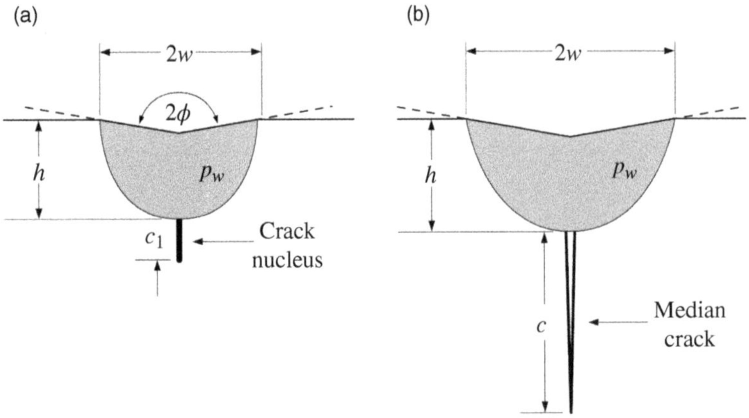

Figure 10.6 Schematic cross sections of linear scratch flaws formed in brittle material surfaces. Contact induced localized plastic deformation zones (pdz) shown shaded. (a) Sub threshold flaw consisting of pdz and small crack nucleus c_1; (b) postthreshold flaw consisting of pdz and well-developed median crack c.

where κ is the material bulk modulus, V is the initial volume of material forming the pdz, and ΔV is the imposed change in volume as a consequence of the scratch track formation. In this case

$$V \sim whL \tag{10.2}$$

and

$$\Delta V \sim -w^2 \cot\phi L, \tag{10.3}$$

where $L \gg w, h$ is the length of the flaw into the plane of the diagram of Figure 10.6a. Combining Eqs. (10.2) and (10.3) into Eq. (10.1) and omitting terms in material Poisson's ratio give

$$p_w \sim (w/h)E\cot\phi, \tag{10.4}$$

where E is the material Young's modulus. Using the pressurized cylindrical cavity analysis of Eq. (9.4) to express the aspect ratio (w/h) in terms of material and indenter parameters modifies both the material and angular dependencies to give

$$p_w \sim H(E/H)^{1/3}(\cot\phi)^{1/2}, \tag{10.5}$$

where H is the material hardness.

For crack nuclei small relative to the pdz ($c_1 < h$), the residual stress acting on a nucleus is approximately uniform, $\approx p_w/2$, and thus the SIF K_n acting on a crack nucleus is given as before (Chapter 2) by

$$K_n = \Omega_1 H c^{1/2}, \tag{10.6}$$

where Ω_1 is a material- and indenter-dependent geometry constant characterizing the coupling of the residual field and the scale of the median crack nucleus. The characteristic contact stress H appears as the amplitude of K_n acting on the crack nucleus. Note that Eq. (10.6) is the asymptotic destabilizing SIF response applicable to small cracks and the onset of initiation in the complete crack domain (Chapter 1).

For geometrically similar scratch contacts (invariant ϕ), deformation within the pdz associated with the linear flaw also maintains geometrical similarity. The scale c_1 of a median crack nucleus formed at the base of the linear pdz is then a direct generalization of that for a half-penny crack nucleus at the base of a point pdz (Chapter 2) and a lateral crack nucleus formed at the base of the linear pdz (Chapter 9), and given by

$$c_1 = \Omega_2 w, \tag{10.7}$$

following Eq. (2.22), where $\Omega_2 < 1$ is a proportionality factor that depends on the included contact angle. Imposing the equilibrium condition $K_n = T$ for a crack nucleus $c = c_1$, where T is the material toughness in the median crack initiation environment, and combining Eqs. (10.6) and (10.7) give the now familiar scaling of the track width w^* at median crack initiation:

$$w^* = \frac{1}{\Omega^2}\frac{T^2}{H^2}, \tag{10.8}$$

where $\Omega = \Omega_1 \Omega_2^{1/2}$. For Si, $T \approx 1$ MPa m $^{1/2}$, $H \approx 10$ GPa, suggesting that $\Omega \approx 0.06$ for $w^* \approx 3\,\mu$m consistent with Figure 10.3. For linear flaws formed by such dragged sharp contacts $P = Hw^2$ and the normal contact load at median crack initiation P^* thus scales as derived previously

$$P^* = \frac{1}{\Omega^2} \frac{T^4}{H^3}. \tag{10.9}$$

Using the values above implies $P^* \approx 0.3$ mN—individual contacts forming the striations are likely very light.

Failure of backface ground dies containing subthreshold striation flaws is determined by extension of Eq. (10.6) to include the effects of a superposed uniform applied tensile stress σ_a. The SIF characterizing the applied stress is $K_a = \psi \sigma_a c^{1/2}$ (Chapters 1 and 3), where ψ is a geometry constant characterizing the coupling of the applied field and the scale of the median crack nucleus. The total SIF K acting on a subthreshold striation flaw is

$$K = K_n + K_a = \Omega_1 H c^{1/2} + \psi \sigma_a c^{1/2}. \tag{10.10}$$

Imposing the equilibrium condition $K = T$ for $c = c_1$ and now combining Eqs. (10.10) and (10.7) gives the failure strength σ_f for components with subthreshold flaws

$$\sigma_f = \frac{\Omega_1 H}{\psi} \left[\left(\frac{P^*}{P} \right)^{1/4} - 1 \right], \quad P < P^*, \tag{10.11}$$

similar to that derived in Chapter 3 and earlier (Cook 2023; Jung 2004a,b). Note that the K_a term in Eq. (10.10) is taken to be sufficiently greater than the K_n term that any deviations in K_n from destabilizing asymptotic behavior are not significant in the unstable dynamic propagation of the crack and eventual component failure (Chapter 1). The coupling of the tensile stress applied to the crack nucleus is anticipated to be greater than the coupling of the resolved stress from within the pdz, such that $\Omega_1 < \psi$. Nevertheless, the characteristic strength of subthreshold flaws is set by H—subthreshold strengths are large. The hardness of Si is $H \approx 10$ GPa and subthreshold indentation strengths of 2–3 GPa have been observed (Jung 2004ab), consistent with values of $\Omega_1/\psi \approx 0.1$ and $P^* \approx 30$ mN. The difference between applied indentation loads and perceived striation loads is addressed in Section 10.3.2.

Figure 10.6b shows a cross section of the surface of a brittle component containing a postthreshold flaw formed by a linear sharp contact scratch. The form and dimensions of the flaw are as in Figure 10.6a with an important exception: A median crack of dimension c is formed beneath the pdz such that the flaw is "postthreshold." The postthreshold crack propagation and failure processes are also described by an SIF, distinct from K_n above. The SIF K_s characterizing the effects of the residual stress field on the median crack at the scratch scales as

$$K_s \sim F_L / c^{1/2}, \tag{10.12}$$

where F_L is the line force acting to open the crack (Murakami 1987). Equation (10.12) is the asymptotic stabilizing SIF response applicable for large cracks and termination of the initiation process in the complete crack domain (Chapter 1). The crack opening line force exerted by the pdz scales as

$$F_L \sim p_w h \sim E \cot \phi w, \tag{10.13}$$

using Eq. (10.4). For linear flaws formed by dragged sharp contacts $P = Hw^2$ and thus

$$\begin{aligned} F_L &\sim E \cot \phi \left(\frac{P}{H} \right)^{1/2} \\ &\sim E^{1/2} \cot \phi \left(\frac{E}{H} \right)^{1/2} P^{1/2} \\ &= \chi_s P^{1/2}, \end{aligned} \tag{10.14}$$

where χ_s is a factor with dimensions of stress$^{1/2}$ that absorbs all material and geometrical effects. Note that the leading material term in χ_s is $E^{1/2}$ reflecting the fact that the crack driving force characterized by K_s derives from elastic energy stored in the residual field. Combining Eq. (10.14) into Eq. (10.12) and setting equilibrium, $K_s = T$, gives the stable equilibrium length c_{air} for median cracks at scratches formed by dragged sharp contacts in air:

$$c_{air} = P \left(\frac{\chi_s}{T} \right)^2. \tag{10.15}$$

The linear response predicted by Eq. (10.15) for median cracks at scratches is distinct from that of the nonlinear $P^{2/3}$ response for half-penny cracks at indentations observed and discussed in great detail (Chapter 2 and references

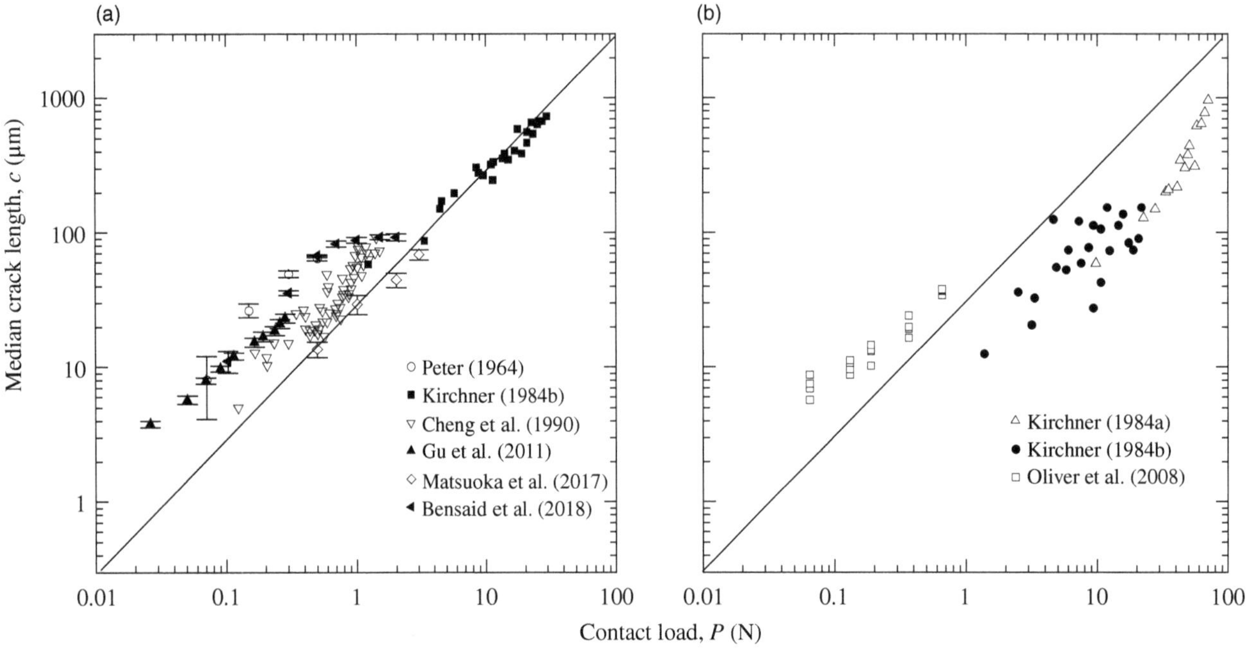

Figure 10.7 Logarithmic plots of median crack lengths as a function of normal contact load at postthreshold linear flaws formed by dragged contacts. (a) Single-point, single-pass flaws formed in glass. Source: Data adapted from Peter (1964), Kirchner (1984a), Cheng et al. (1990), Gu et al. (2011), Matsuoka et al. (2017), and Bensaid et al. (2018). (b) Single-point and multi-point flaws formed in Si, Si_3N_4, and glass. Source: Data adapted from Kirchner (1984a,b) and Oliver et al. (2008).

therein). There are significantly fewer observations of scratch cracks than of indentation cracks, a consequence of the greater difficulty of experimentation and lack of surface trace in the former case. However, cross-sectional scratch crack measurements consistent with Eq. (10.15) have been made since the earliest studies. Figure 10.7 is a logarithmic plot of median crack length versus contact load for several scratch systems. Figure 10.7a shows the response of glasses containing single-point scratches generated by dragged Vickers (Peter 1964; Cheng et al. 1990; Gu et al. 2011; Bensaid et al. 2018), Knoop (Matsuoka et al. 2017), and conical (Kirchner 1984b) indenters. The glasses were all soda-lime silicate except those of Gu, which were borosilicate. The contacts were all single-pass except those of Kirchner which were multipass. Symbols represent individual measurements at the normal contact loads indicated. The solid line is a guide to the eye of slope 1 and well describes the observations. No significant trend with contact geometry, including the effects of face or edge orientation for Vickers indenters, is apparent. Figure 10.7b repeats the straight line guide of Figure 10.7a for comparison with the responses of some other systems: single-point, single-pass scratching of single-crystal Si (Oliver et al. 2008), single-point multipass scratching of polycrystalline Si_3N_4 (Kirchner 1984a), and multipoint, multipass scratching of glass (Kirchner 1984b). Again, the observations are well described by a linear trend. Si exhibited single-point scratch crack lengths comparable to those of glass and Si_3N_4 exhibited single-point scratch crack lengths shorter than those of glass, consistent with similar and greater toughness values, respectively. The shorter multipoint crack lengths exhibited by glass are probably attributable to greater surface wear effects.

The cracking observations of Figure 10.7 provide support for the pdz-based fracture mechanics approach to backface grinding effects. The failure of backface ground dies containing postthreshold striation flaws is determined by extension of Eq. (10.12) to include the effects of superposed stress σ_a characterized by the applied SIF $K_a = \psi\sigma_a c^{1/2}$. The total SIF K acting on a postthreshold striation flaw is then

$$K = K_s + K_a = \chi_s P^{1/2}/c^{1/2} + \psi\sigma_a c^{1/2}, \tag{10.16}$$

which has an identical mathematical structure (the addition of independent stabilizing and destabilizing SIF terms) to that considered in Chapter 3 for the effects of applied stress on indentation point flaws. The method of solution and interpretation are identical. Imposing the equilibrium condition $K = T$ and the instability condition $dK/dc = dT/dc = 0$ and solving the resulting two equations simultaneously gives the crack length c_m and stress σ_m at equilibrium instability:

$$c_m = 4P\left(\frac{\chi_s}{T}\right)^2 \tag{10.17}$$

and

$$\sigma_{\rm m} = \frac{T}{2\psi c_{\rm m}^{1/2}} = \frac{T^2}{4\psi \chi_{\rm s} P^{1/2}} \tag{10.18}$$

Again, the $P^{-1/2}$ trend of Eq. (10.18) for dragged scratch flaws is distinct from the $P^{-1/3}$ trend much studied for indentation flaws (Chapter 3 and references therein). Combining the $\sigma \sim c^{-1/2}$ strength responses of dragged scratches from Figure 1.10c with the linear $c \sim P$ dragged scratch crack length responses of Figure 10.7 indicates that Eq. (10.18) is well obeyed. A strength calculation for roller-induced flaws, beginning with an equation similar to Eq. (10.16), leads to a $P^{-2/3}$ trend for the strength of rolled scratch flaws that is also well demonstrated experimentally (Cook 1994). [Apparent exceptions to these observations (Levengood 1958; Ono and Tanaka 2001) are probably the results of unaccounted-for specimen geometry effects, e.g., those of thin, wide plates.] Section 10.3.2 develops a unified framework for contact flaw controlled strengths that de-emphasizes the load-specific nature of the above analyses.

10.3.2 Indentation Fracture Mechanics—General Contact Flaws

The fracture mechanics analyses of Section 10.3.1 and Chapters 2 and 3 were based on the specification of four SIF acting on contact flaws: $K_{\rm a}$, characterizing the effects of a global applied stress field; $K_{\rm n}$, characterizing the effects of a localized initiating field; and $K_{\rm r}$ and $K_{\rm s}$ characterizing the effects of distributed residual fields associated with point and line flaws, respectively. The similarities of, and analogies between, these fields in generalized contact flaw strength analyses are observed most easily when $K_{\rm n}$, $K_{\rm r}$, and $K_{\rm s}$ are reexpressed in terms of flaw contact impression dimensions, a for point and w for line flaws.

The functional form of the SIF $K_{\rm a}$ describing the effects of a global, uniform, applied stress $\sigma_{\rm a}$ on a crack at a contact flaw in a component has no inherent scales,

$$K_{\rm a} = \psi \sigma_{\rm a} c^{1/2},$$

as c is an *arbitrary* crack length. ψ is a dimensionless geometrical factor that depends on crack shape. The amplitude term of the SIF, $\sigma_{\rm a}$, is also arbitrary and is usually the controlled variable in strength tests.

The SIF $K_{\rm n}$ describing the effects of stress localized to a small crack, a crack nucleus $c \ll a$ or $c \ll w$, at a contact flaw has similar functional form to $K_{\rm a}$:

$$K_{\rm n} = \Omega H c^{1/2}.$$

In this case, however, the amplitude term is not arbitrary and is given by the characteristic contact stress, the hardness H, of the contacted material. Ω is a dimensionless factor that depends on crack shape, contact geometry (through the included angle ϕ of the contact), and material elastic and plastic properties (through the elastic modulus/hardness ratio E/H of the material).

The total SIF K acting on a crack nucleus at a contact flaw under the superposed action of an applied stress is thus

$$K = K_{\rm a} + K_{\rm n} = \psi \sigma_{\rm a} c^{1/2} + \Omega H c^{1/2}.$$

K characterizes a destabilizing SIF field, $dK/dc > 0$, as both terms are destabilizing. The applied failure stress of the small crack, the strength of the component $\sigma_{\rm a} = \sigma_{\rm f}$, is thus given by imposing the equilibrium condition $K = T$ and simply inverting the above equation:

$$\sigma_{\rm f} = \frac{\Omega H}{\psi}\left[\frac{T}{\Omega H c^{1/2}} - 1\right].$$

Crack nuclei generated at contact flaws are not of arbitrary length, however, and for geometrically similar contacts, the crack nucleus length is proportional to the contact impression dimension, $c \sim a$ or $c \sim w$. Hence, defining a characteristic contact flaw strength $\sigma^* = (\Omega H/\psi)$ and characteristic contact flaw dimensions $a^* = w^* = (T/\Omega H)^2$ enables the above equation to be written as

$$\sigma_{\rm f} = \sigma^* \left[\left(\frac{a^*}{a}\right)^{1/2} - 1\right]$$

and

$$\sigma_{\rm f} = \sigma^* \left[\left(\frac{w^*}{w}\right)^{1/2} - 1\right].$$

These equations describe subthreshold strengths for point and line flaws. As expressed here in terms of contact dimensions, the similarity of subthreshold strength variation with contact flaw scale, independent of geometry, is clear. The physical interpretation of these equations is that an applied stress field aids a contact stress field to pop-in a well-developed crack at a contact and subsequently dominates the crack driving force, leading to continued crack propagation and thus component failure. In the limit of spontaneous pop-in, $w \to w^*$ for example, the subthreshold strength vanishes, $\sigma_f \to 0$.

Flaw geometry effects on subthreshold strengths are assessed most easily by the dependence of flaw scale on the normal contact load P. For point flaws and dragged linear flaws $a \sim (P/H)^{1/2}$ and $w \sim (P/H)^{1/2}$. Setting characteristic loads P^* by $a^* \sim (P^*/H)^{1/2}$ or $w^* \sim (P^*/H)^{1/2}$ gives Eq. (10.11):

$$\sigma_f = \sigma^* \left[\left(\frac{P^*}{P} \right)^{1/4} - 1 \right].$$

However, for rolled linear flaws $w \sim P^{2/3}$, such that

$$\sigma_f = \sigma^* \left[\left(\frac{P^*}{P} \right)^{1/3} - 1 \right]$$

and for wedge linear flaws $w \sim P$ and

$$\sigma_f = \sigma^* \left[\left(\frac{P^*}{P} \right)^{1/2} - 1 \right],$$

where P^* is defined appropriately in each case.

The SIF K_r and K_s describing the effects of stress distributed over well-developed cracks at point and line contact flaws have similar functional forms, different from K_a:

$$K_r = \Lambda H a^2 / c^{3/2}.$$

and

$$K_s = \Lambda H w / c^{1/2},$$

respectively. In these cases, similar to K_a, the crack length is arbitrary. The amplitude terms, similar to K_n, include a dependence on the hardness H of the contacted material. In addition, the amplitude terms include dependencies on the scale of the contact, the area of the point flaw and the area/length of the line flaw. Λ are dimensionless factors similar to Ω that depend on crack shape, contact geometry, and material elastic and plastic properties. Distinct from K_a and K_n, the functional forms of K_r and K_s, although slightly different, are both stabilizing as $dK_r/dc < 0$ and $dK_s/dc < 0$.

The total SIF K acting on well developed cracks at point and line contact flaws under the superposed actions of applied stress are thus

$$K = K_a + K_r = \psi \sigma_a c^{1/2} + \Lambda H a^2 / c^{3/2}$$

and

$$K = K_a + K_s = \psi \sigma_a c^{1/2} + \Lambda H w / c^{1/2}.$$

In these cases, K characterizes stabilizing-to-destabilizing SIF fields and the applied failure stress of well-developed cracks, the strengths of the components $\sigma_a = \sigma_{max}$, is given by imposing simultaneously equilibrium, $K = T$, and instability, $dK/dc = dT/dc = 0$. Solving these equations (Chapter 3; see also Cook 2023) gives

$$\sigma_{max} = \frac{3T^{4/3}}{4^{4/3} \psi (\Lambda H a^2)^{2/3}}$$

and

$$\sigma_{max} = \frac{T^2}{4\psi(\Lambda H w)}.$$

Defining a characteristic postthreshold contact flaw strength $\sigma^\dagger = (\Lambda H/\psi)$ and contact flaw dimensions $a^\dagger = w^\dagger = (T/\Lambda H)^2$ as previously enables the above equations to be written as

$$\sigma_{max} = (3/4^{4/3}) \sigma^\dagger \left(\frac{a^\dagger}{a} \right)^{2/3}$$

and

$$\sigma_{\max} = (1/4)\sigma^{\dagger}\left(\frac{w^{\dagger}}{w}\right).$$

These equations describe postthreshold strengths for point and line flaws. As expressed here in terms of contact dimensions, the different dependencies of postthreshold strength variations with contact flaw scale are clear.

As noted previously, flaw geometry effects on postthreshold strengths are also assessed most easily by the dependence of strength on the normal contact load P. For point flaws and dragged linear flaws $a \sim (P/H)^{1/2}$ and $w \sim (P/H)^{1/2}$. Setting postthreshold characteristic loads $P^{\dagger}$ by $a^{\dagger} \sim (P^{\dagger}/H)^{1/2}$ or $w^{\dagger} \sim (P^{\dagger}/H)^{1/2}$ as before gives

$$\sigma_{\max} = \sigma^{\dagger}\left(\frac{P^{\dagger}}{P}\right)^{1/3}$$

and

$$\sigma_{\max} = \sigma^{\dagger}\left(\frac{P^{\dagger}}{P}\right)^{1/2}$$

as shown in Chapter 3 and in Cook (2017). For rolled linear flaws $w \sim P^{2/3}$, such that

$$\sigma_{\max} = \sigma^{\dagger}\left(\frac{P^{\dagger}}{P}\right)^{2/3}$$

as shown in Cook (1994), and for wedge linear flaws $w \sim P$, and

$$\sigma_{\max} = \sigma^{\dagger}\left(\frac{P^{\dagger}}{P}\right),$$

where, as before, $P^{\dagger}$ is defined appropriately in each case.

This section has highlighted the similarities and differences in general strength behavior of point contact and line contact flaws, including both subthreshold and postthreshold contacts. In particular, in SIF-based strength formulations flaw size effects were shown to appear in crack length variables in subthreshold analyses and in amplitude variables in postthreshold analyses. Differences in flaw geometry were reflected in differences in the power laws describing the crack length and amplitude dependencies. The similarities and differences were framed in terms of the flaw contact impression dimensions and then reexpressed in terms of the experimentally more accessible normal contact load. An issue beyond the current scope is the effect of increasing toughness $T(c)$ on the contact strength behavior. This point was addressed in detail in Chapter 4 and showed that such T curve effects lead to much reduced dependence of strength on point indentation loads. Linear scratch flaws are expected to behave in much the same way, and an example is shown in Lee et al. (2000) of near invariant scratch strengths measured as a function of contact load in transformation toughened zirconia.

In practical terms regarding linear striation flaws formed by backgrinding of Si wafers, the relevant contacts are those made by dragged or rolled abrasive particles pressed between the grinding wheel and the wafer. For these cases, the general analyses above show that the power-law exponents controlling the load dependence of subthreshold flaws are 1/4 and 1/3, respectively, and those controlling the load dependence of postthreshold flaws are 1/2 and 2/3. Postthreshold strengths are less than subthreshold strengths and occur at greater loads. Using the above analyses, Figure 10.8 shows a logarithmic plot of strength as a function of contact load for striation flaws. The plot is a generalization of Figure 3.2 in two ways: First, the contact flaws represented are linear rather than point, and second, two different contact geometries, dragged and rolled, are represented. Identical parameters, typical of Si strengths, were used for both geometries: the threshold contact load $P^* = 10$ mN, the threshold strength $\sigma^* = 20$ GPa, the postthreshold contact load $P^{\dagger} = 1$ N, and the postthreshold strength $\sigma^{\dagger} = 50$ MPa. The postthreshold region ($P \geq P^*$) is indicated hatched in Figure 10.8, the solid and dashed lines indicate the dragged and rolled responses, respectively, and the shaded regions indicate the ranges of strength between the two contact geometries. As in Figure 3.2, within the postthreshold contact domain, there is only one strength response. Within the subthreshold domain, two strength responses are possible, as both postthreshold and subthreshold strength responses appear, although the fraction of observed postthreshold responses usually decreases as the contact load is reduced to values much less than the threshold load ($P \leq P^*$). The major points of Figure 10.8 are that striation flaws formed by linear contacts are similar in strength behavior to point flaws and that the exact nature of the contact details only weakly perturbs the strength response. Section 10.3.3 examines experimental strength responses of indented, scratched, and ground Si surfaces and interprets the responses in terms of the analyses here.

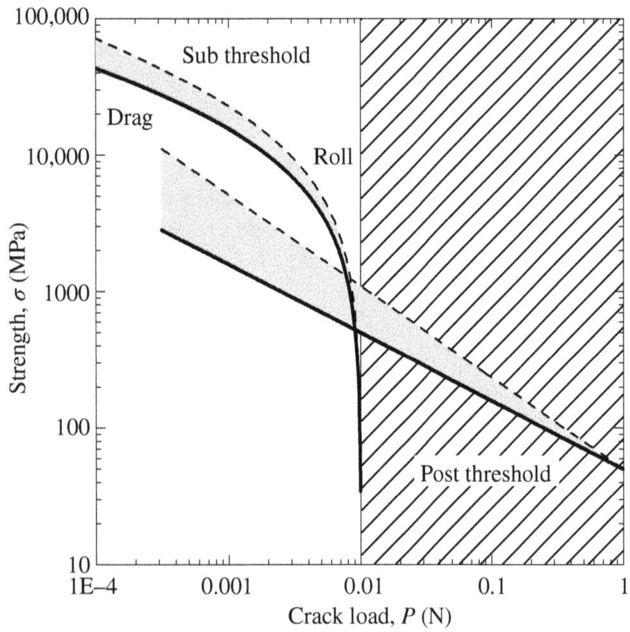

Figure 10.8 Logarithmic plot of strength of linear flaws as a function of contact load. Dragged contacts are indicated by solid lines; rolled contacts are indicated by dashed lines. Postthreshold loads are indicated by the hatched region. Strengths can exhibit bimodal behavior in the subthreshold region. The shaded regions indicate likely grinding flaw responses.

10.3.3 Observations—Controlled Flaws

Figure 10.9 shows optical microscope images of the backface of a Si wafer processed by in-feed Z1 + Z2 grinding and containing subsequently introduced controlled flaws. The flaws were generated by a sharp Vickers indenter, either by conventional indentation contact perpendicular to the wafer surface and are indicated by arrows in Figure 10.9a and b, or by Vickers contact and scratching parallel to the surface adjacent to a broken edge (left) of the wafer, Figure 10.9c and d. Detailed comparisons of the forms and fracture mechanics of indentations and scratches can be found in Chapters 2 and 9 (see also Cook 2017). The backface parallel grinding striations, approximately 2 μm in width, are clear and oriented slightly diagonally in the images.

The indentation in Figure 10.9a was formed with a contact load of $P = 0.1$ N and is barely visible within the striation array at this magnification. The indentation scale is consistent with earlier measurements (Cook 2006) that showed the dimensions of the surface impression at this load are approximately 2–3 μm and the surface traces of any associated cracks are approximately 3–5 μm. As the indentation surface damage in Figure 10.9a is comparable in size to the visible striations, the $P = 0.1$ N indentation load thus provides a likely lower bound to the contacts required to generate strength-controlling flaws on a ground surface. The indentation in Figure 10.9b was formed with an indentation load of $P = 1$ N and the square indentation contact impression and surface traces of cracks radiating from the impression corners are clearly visible; the dimensions are consistent with the earlier measurements of 10–15 μm for the impression dimensions and 20–30 μm for the cracks. The indentation surface damage in Figure 10.9b is larger than the grinding striations and could thus most probably constitute a strength-controlling flaw and reduce the wafer strength from the level set by the grinding striations. In an engineering context, the indentation loads used to generate the flaws in Figure 10.9a and b bracket the gravitational force exerted by the mass of the wafer. The mass of a 200 mm diameter wafer is about 50 g, leading to about 0.5 N weight; about 1.2 N weight for a 300 mm wafer. Hence, simply placing a wafer on a sharp particle under its own weight is enough to generate a flaw greater in size than the grinding damage.

The linear flaw in Figure 10.9c was formed with a normal contact load of 0.1 N, as in Figure 10.9a, but the indenter was translated at approximately 0.5 mm s^{-1} from right to left over the wafer surface and off the wafer edge. The resulting flaw consists of a linear impression with adjacent fine debris and closely resembles the ideal linear contact flaws of Figure 10.4 and the observations of Danyluk et al. (1988), Misra and Finnie (1979), and Chao et al. (2002). No surface cracks are visible and as the subsurface is not visible it is not possible to ascertain if the flaw exhibits a subthreshold or postthreshold configuration. The edge appears to only weakly perturb the flaw as indicated by the appearance of slightly more debris. The width of

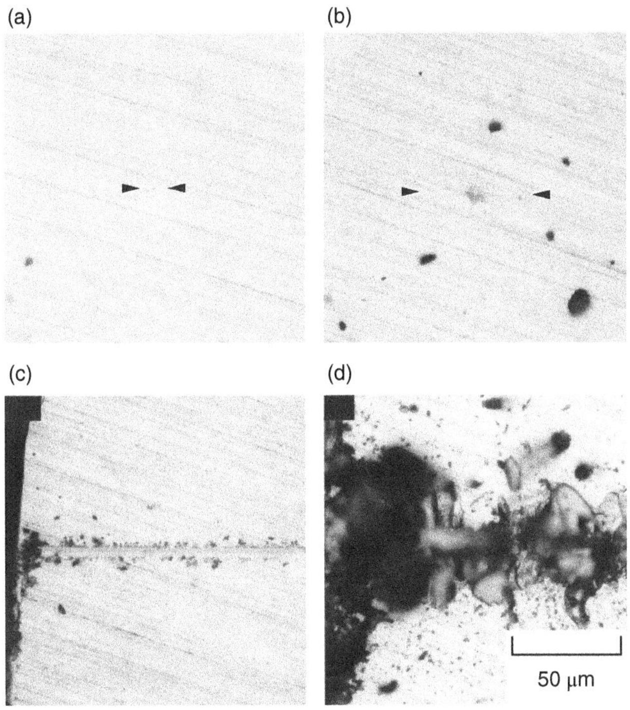

Figure 10.9 Optical micrographs of Z1 + Z2 ground wafer backfaces showing parallel arrays of grinding striations and introduced indentations (indicated by arrows) and scratch flaws. (a) 0.1 N normal Vickers indentation; (b) 1 N normal Vickers indentation; (c) 0.1 N Vickers indentation scratch; (d) 1 N Vickers indentation scratch.

the track in Figure 10.9c is approximately 10 μm and it is clear that the damage generated is far greater than that generated in Figure 10.9a and much larger than the grinding striations and could thus constitute a strength-controlling flaw. In fact, the implication of Figure 10.9c is that linear flaws formed by much smaller normal contact loads, as the grinding striations appear to have been, could be strength controlling. The flaw in Figure 10.9d was formed with a normal contact load of 1 N, as in Figure 10.9c, but the indenter was translated over the wafer surface and off the wafer edge as in Figure 10.9c. The resulting flaw consists of a barely discernible plastic deformation track covered by significant amount of coarse debris formed from fragments and cracked material. Surface lateral cracks adjacent to the track are clear suggesting the existence of a subsurface median crack as in Figure 10.4b. The flaw resembles the scratch observations of Lim and Danyluk (1985, 1988). The edge significantly perturbs the flaw as indicated by the generation of more cracking and debris. The scratch damage in Figure 10.9d is approximately 50 μm in extent, much larger than that generated in Figure 10.9b and the grinding striations and thus easily constitutes a strength-controlling flaw.

Figure 10.10 shows the strength of backface ground Si dies, 15.5 mm length × 6.9 mm width × 370 μm thick, tested in four-point bending along the length. Details of strength testing can be found elsewhere (Cook 2017). The die backfaces were Z1 + Z2 in-feed ground and many contained an introduced controlled indentation or scratch flaw as described above, Figure 10.9. Figure 10.10 is thus a strength degradation plot in logarithmic coordinates (Chapter 3): filled symbols represent indentation strength measurements at the loads P indicated; open symbols represent scratch strength measurements at the normal contact loads indicated; individual symbols represent individual strength measurements; uncertainty bars represent the mean and standard deviation limits for at least five measurements; and the shaded band represents the mean and standard deviation limits of approximately 35 as-ground dies selected randomly from several wafers.

The maximum strength level in Figure 10.10 is set by the as-ground dies, 282 ± 52 MPa. The strength is decreased or degraded by the introduction of controlled flaws, more so by the linear scratches than by the point indentations. The upper line in Figure 10.10 is of slope −1/3, describing a $P^{-1/3}$ dependence of strength on indentation load characteristic of indentation flaws, and quantified by $\sigma P^{1/3} = 130$ MPa $N^{1/3}$ typical for single-crystal Si (Chapter 3, see also Cook 2006). This line intersects the as-ground strength level at an indention load of $P = 0.1$ N, suggesting that indentation-like sharp contact flaws of this magnitude could describe the strength-controlling element of the grinding damage. This assessment is consistent with the observations of Figure 10.9 that showed that 0.1 N indentation flaws were comparable in scale to the grinding

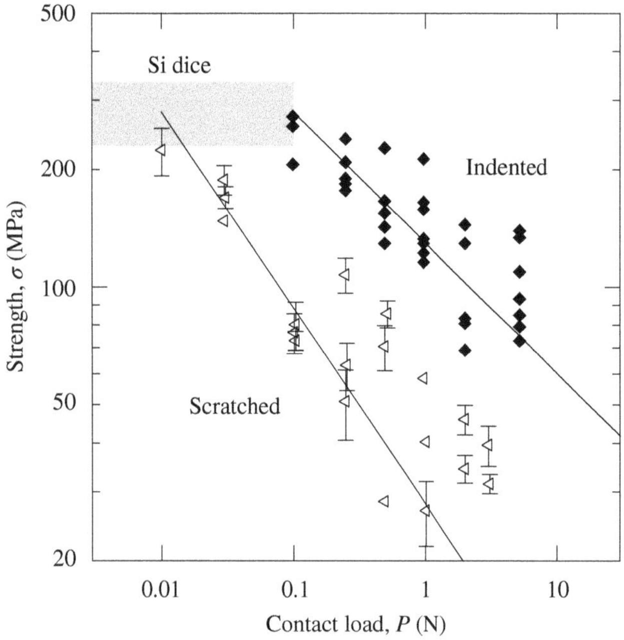

Figure 10.10 Indentation strength degradation plot for Si dies containing introduced indentation or scratch flaws at the normal contact loads indicated. Symbols represent strength measurements, lines represent ideal contact mechanics responses appropriate to Si, and shaded band indicates intrinsic strength range determined by grinding.

striations (a) and that 1 N indentation flaws were larger than the grinding striations (b). The lower line in Figure 10.10 is of slope −1/2, describing a $P^{-1/2}$ dependence of strength on indentation load characteristic of scratch flaws (Cook 2017), and quantified by $\sigma P^{1/2} = 28\ \text{MPa N}^{1/2}$. This line intersects the as-ground strength at an indentation load of $P = 0.01$ N, suggesting that scratch-like sharp contact flaws of this magnitude could describe the strength-controlling element of the grinding damage. This assessment is consistent with the observation of Figure 10.9c that showed that 0.1 N scratch flaws were larger in scale than the grinding striations. The assessment is also consistent with earlier work on alumina that showed that linear flaw strengths were shifted by approximately a factor of 10 to smaller contact loads relative to indentation strengths (Cook 2017).

Figure 10.10 implies that the Z1 + Z2 resultant grinding flaws can be characterized as sharp contacts by loads between 0.01 N, for translated contacts as in Figure 10.4, and 0.1 N, for static contacts as by indentations. These values are in quantitative agreement with earlier work on Si (Cook 2006) and linear flaws (Cook 2017). Another point of quantitative agreement is with independent measurement of the toughness of Si, $T = 0.75\ \text{MPa m}^{1/2}$ (Cook 2006; DelRio et al. 2015). Assuming the flaws are simple Griffith cracks, a characteristic crack length c can be estimated from $(T/\psi\sigma)^2$, where ψ is a geometrical constant, here taken as $\psi = 1.5$ appropriate to linear surface cracks. Using these values gives $c \approx 3\ \mu\text{m}$, consistent with the observations, Figure 10.9, of the grinding striations. Overall, in the context of strength-controlling flaws, Figures 10.9 and 10.10 provide a self-consistent quantitative description of the scale of grinding damage that is also consistent with independent measurements of Si properties and behavior. Section 10.3.4 extends consideration of Si die strength beyond the *scale* of the strength-controlling flaws to flaw *orientation*.

10.3.4 Observations—Grinding Flaws

Figure 10.11 shows the strength of Z1 or Z1 + Z2 backface ground Si dies measured in four-point bending along the length of the die. A range of die dimensions similar to those stated above was measured. The strength test geometry is characterized by the angle θ between the long edge of the die and the direction of striations on the die backface, defined in the schematic diagram in Figure 10.11. During testing, the backface of a die is placed in uniaxial tension along the die length, and therefore θ is also the angle between the striations and the tensile axis. Symbols and bars in Figure 10.11 represent the experimental means and standard deviation limits of at least five strength measurements under the conditions specified at the angle θ indicated. Individual symbols represent individual strength measurements; for clarity, some symbols are displaced.

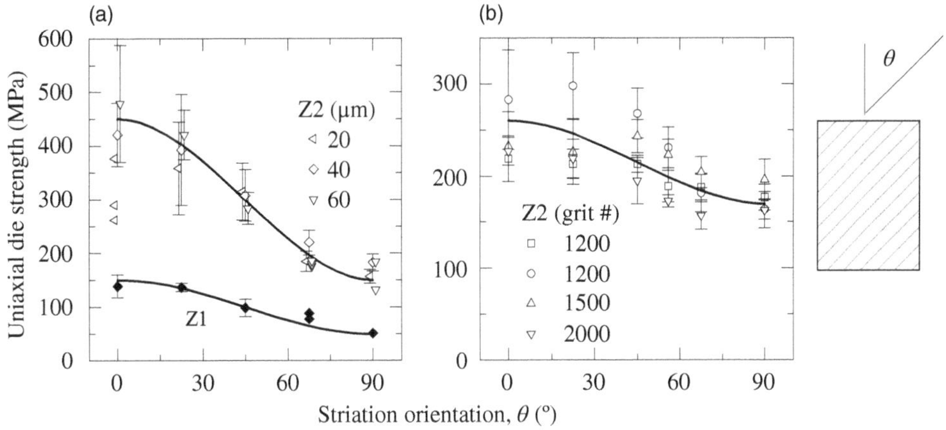

Figure 10.11 Variations of uniaxial strength with striation orientation for Si dies. Symbols represent strength measurements for dies with striations at the orientation angle indicated. Lines represent best fits of empirical polynomial between maximum and minimum strength bounds, Eq. (10.19). (a) Z1 grinding and Z1 + Z2 depth study; (b) Z1 + Z2 grit study. Schematic diagram shows definition of θ.

Figure 10.11a shows the variation of strength with striation orientation for two different backface grinding conditions, Z1 and Z1 + Z2. The strength measurements are consistent with the observations noted in Section 10.2: The strengths of Z1 ground dies are considerably less than those of Z1 + Z2 ground dies, for all values of θ. For both conditions, the strength is a maximum for striations parallel to the long edge of the dies, $\theta = 0°$, and significantly reduced to a minimum for striations perpendicular to the long edge, $\theta = 90°$. The observed variation of strength with orientation can be described by

$$\sigma = \sigma_{max} - (\sigma_{max} - \sigma_{min})[2(\theta/90)^3 - 3(\theta/90)^2], \quad (10.19)$$

where θ is expressed in degrees and σ_{max} and σ_{min} are empirical fitting parameters characterizing upper and lower bounds of strength, respectively. Equation (10.19) provides a slightly better smooth continuous description of the experimental observations and is more easily implemented analytically than the piece-wise description used earlier (Mendelson et al. 1999, 2001; Cook 2017). The solid lines in Figure 10.11 are visual best fits of Eq. (10.19) to the strength data: $\sigma_{max} = 150$ MPa and $\sigma_{min} = 50$ MPa for the Z1 ground dies in Figure 10.11a. Comparison with Figure 10.10 suggests that the Z1 grinding flaws in Figure 10.11a correspond to striations formed with normal contact loads of approximately 0.3 N.

Three sets of Z1 + Z2 ground die strengths are also shown in Figure 10.11a. All sets were identically Z1 ground but were subsequently Z2 ground so as to remove three different thicknesses of material, 20, 40, and 60 μm. There was no significant effect on measured strength of Z2 thickness reduction. The strength invariance has two implications: that the initial 20 μm reduction removed the Z1 surface damage layer, consistent with the earlier estimates cited above, and that the initial reduction introduced a layer of Z2 damage that remained invariant with increasing Z2 grinding. For all thickness reductions, the Z2 strengths exhibited a clear increase in dispersion with decreasing θ, implying that there was greater variability in the strength-controlling nature of the grinding flaws as striations approached orientations more closely aligned with die edges. The best fit of Eq. (10.19) to the Z1 + Z2 strength data in Figure 10.11a provides $\sigma_{max} = 450$ MPa and $\sigma_{min} = 150$ MPa.

Figure 10.11b shows many of the same trends in strength variation with striation orientation for four different Z1 + Z2 backface grinding sets. As above, the sets were identically Z1 ground; in this case, subsequent Z2 grinding used different abrasive grit sizes. Similar to the above, the strengths are considerably greater than those of Z1 ground dies for all values of θ, with maxima for striations parallel to the long edge of the dies and minima for striations perpendicular to the long edge. The strengths are described by Eq. (10.19) using $\sigma_{max} = 260$ MPa and $\sigma_{min} = 90$ MPa as shown by the solid line. There was no significant effect on strength of Z2 grit size implying that all grits used removed the Z1 surface damage approximately equally, although perhaps not as effectively as in Figure 10.11a. In addition, there was an increase in the strength dispersion with decreasing θ for all grits. Comparison with Figure 10.10 suggests that the Z2 grinding flaws in Figure 10.11a and b correspond to striations formed with normal contact loads between 0.03 and 0.1 N.

The uniaxial strength measurements shown in Figure 10.11 are in agreement with previous observations of the effects of backgrinding striation severity and orientation (Section 10.2) and with the current flaw size estimations and strength degradation measurements (Figures 10.9 and 10.10). The quantitative description of striation orientation effects on uniaxial strength, Eq. (10.19) and solid lines in Figure 10.11, is a key element of model development in Section 10.4. Before such

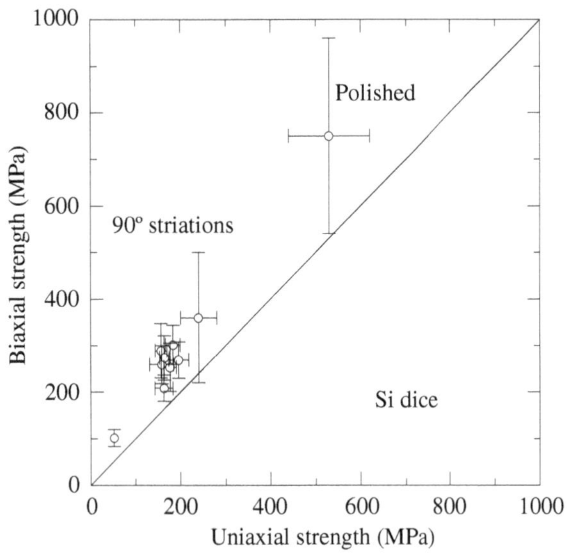

Figure 10.12 Plot of biaxial strength versus uniaxial strength for Si dies measured using square and rectangular specimens with a range of backface surface finishes. Biaxial strengths are systematically greater than uniaxial strengths.

development, additional biaxial strength measurements will be considered, providing both further agreement with previous observations and insight into model interpretation.

Figure 10.12 shows measurements of Si biaxial strength as a function of die uniaxial strength for a variety of backgrinding conditions. Symbols represent the means and standard deviation limits of at least 10 strength measurements for each geometry in each grinding condition. The uniaxial strengths were measured for dies in the minimum strength orientation with striations perpendicular to the long edge, $\theta = 90°$. The biaxial strengths were measured with 32 mm × 32 mm square coupons cut from the same wafers as the uniaxial test dies. The backfaces of the coupons exhibited parallel arrays of grinding striations; the group of test coupons from a selected grinding condition included all array orientations relative to the square edges. The square coupons were tested in flat-on-three-ball flexure, which placed the ground backface center of the coupon in equibiaxial stress, significantly greater at the center of the coupon than the edges and thereby largely preventing edge flaws from controlling coupon strength. As a consequence, biaxial flexure testing enables an assessment of greater strengths associated with smaller surface flaws in the center of the coupon. Details of biaxial strength testing can be found in Chapter 3 (see also Cook 2017).

To minimize interference from edge flaws in the creation of strength degradation plots, early studies of brittle ceramic strengths (Cook et al. 1985) took advantage of the biaxial flexure geometry to extend the domain of introduced strength-controlling indentation flaws to very small contact loads. Similarly motivated application of the biaxial geometry was used in subsequent indentation strength measurements of Si (Cook 2006). The geometry also enabled crystallographic striation controlled Si strength measurements (McGuire et al. 1997; Chae et al. 2010) and strength measurements of entire wafers (Funke et al. 2004; Jeong et al. 2004). Other studies have focused on the ability of *combined* application of uniaxial and biaxial strength testing geometries to separate edge and surface flaw populations (Cotterell et al. 2003; Zhao et al. 2009) and this is the intent here in Figure 10.12. The mean biaxial strengths in Figure 10.12 are consistently about 1.5 times greater than the matching mean uniaxial strengths, with somewhat greater biaxial strength dispersion. Strength in the equibiaxial test coupon geometry is controlled by the largest surface flaw in the stressed central region, independent of flaw or striation orientation relative to the coupon edge. Hence, the implication of Figure 10.12 is that strength-controlling surface flaws generated by a variety of grinding conditions are smaller and slightly more variable in size than $\theta = 90°$ edge flaws. This finding extends to polished dies and coupons, in which visible grinding striations are absent. The small increase in mean strength and increased strength dispersion for biaxial strengths relative to uniaxial strengths imply that clear distinction between surface and edge flaw populations is difficult for $\theta = 90°$ or random θ uniaxial die orientations, consistent with previous observations (Cotterell et al. 2003; Zhao et al. 2009). Comparison of Figures 10.11 and 10.12 suggest that distinction between surface and edge flaw populations for the much greater strength $\theta = 0°$ dies is likely although $\theta = 0°$ dies also exhibit great strength dispersion.

As applied stress conditions in the center of a loaded square coupon are equibiaxial, the strength-controlling nature of surface flaws associated with striations should not depend significantly on striation orientation relative to coupon edges. Hence, biaxial strength should be invariant with test coupon position on a wafer. Figure 10.13 shows a plot of relative biaxial

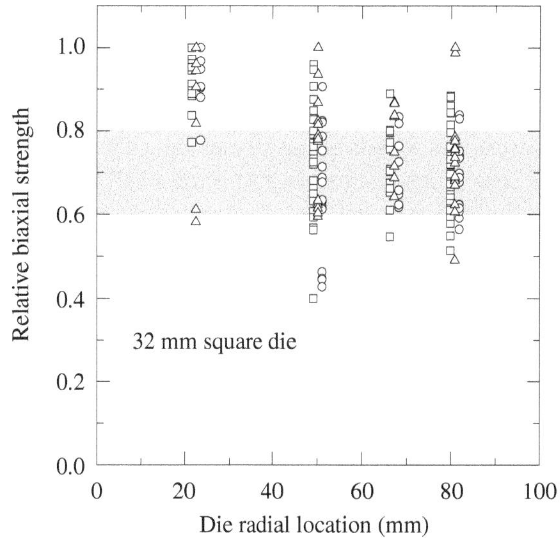

Figure 10.13 Plot of biaxial strengths of backface ground Si dies as a function of die radial location relative to wafer center. Different symbols represent dies sampled from different ground wafers. The strengths are normalized to the greatest strength observed for each wafer. The shaded band is a guide to the eye.

strength as a function of coupon radial location for several Z1 + Z2 ground 200 mm wafers. To compare radial variations between grinding conditions, the strengths are normalized by the maximum observed for each condition. Symbols represent individual strength measurements. Different symbols represent different wafers. Radial location is measured from wafer center. Although there is considerable dispersion, consistent with Figure 10.12, the majority of strengths fall between 0.6 and 0.8 of the maximum observed for each wafer, as indicated by the shaded band and consistent with the earlier observations (McGuire et al. 1997). Slightly greater strengths were observed for coupons from locations closer to the wafer center, suggesting a slight radial dependence to the grinding process, perhaps associated with chuck shape or grinding wheel or spindle tilt (Chidambaram et al. 2003a; Sun et al. 2005a,b).

A major motivation for the work here, considered in a final experimental observation, is that of grinding striation orientation effects on the loads supported by packaged dies. Figure 10.14 shows an empirical distribution function (edf) of the failure loads of ground packaged dies. Packaging consisted of die attachment to a metal lead frame and sealing in a polymer encapsulant. The failure loads were determined by deforming the packaged dies at constant rate in uniaxial three-point bending with the die ground faces in tension and recording the peak supported loads. The edf is bimodal, consisting of a number of small-load failures, separation, and a number of large-load failures. Example experimental load–displacement responses are shown in the inset. The large-load failures (bold line) were characterized by linearly increasing load–displacement responses until peak loads were reached at which the supported loads abruptly decreased to

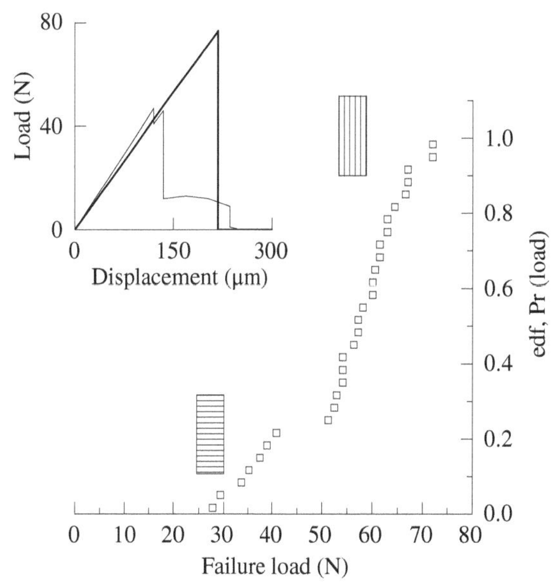

Figure 10.14 Empirical distribution function (edf) of failure loads for packaged devices containing backface ground Si dies. The edf is bimodal, consisting of weak packages containing dies with backface grinding striations perpendicular to the die long axis and strong packages containing dies with backface grinding striations parallel to the die long axis. The inset shows the observed load–displacement behavior.

zero and the packages and included dies were separated into two pieces. The small-load failures (fine line) were characterized by similarly sloped increasing load–displacement responses until smaller peak loads were reached that were followed by small-load drops and re-loads until large load drops to a nonzero loads. Displacement continued at these intermediate loads until eventual slow load drops to zero load at which point the packages were intact but plastically deformed and the internal dies were broken. All dies were depackaged after mechanical testing by dissolving the polymer encapsulant and the backfaces subsequently inspected. In all cases, the small-load failures were associated with striations perpendicular to the die long axes and the large-load failures were associated with striations parallel to the die long axes, indicated in the schematic diagrams.

Figure 10.14 makes clear that the above considerations regarding wafer backgrinding and the consequent effects of striation orientation on die strength are critical in determining finished device performance and reliability. Section 10.4 develops a model to predict die strength as a function of wafer grinding parameters.

10.4 Grinding Controlled Strength Model

10.4.1 Analysis

Figure 10.15 is a plan view diagram of in-feed grinding of a wafer backface, similar to the diagram shown in earlier works (Mendelson et al. 1999, 2001) and slightly different from that used in a later model (Chidambaram et al. 2003b). The wafer is fixed in an XY coordinate system with radius w and frontface edge exclusion e. The rectangular dies are of edge lengths s_x and s_y, taken such that the long edge s_y is parallel to Y. A specific die is characterized by center coordinates (c_x, c_y) in XY such that

$$(|c_x| + s_x/2)^2 + (|c_y| + s_y/2)^2 \leq (w - e)^2. \tag{10.20}$$

An example is indicated by the open dot symbol in Figure 10.15.

A grinding wheel, radius g, is passed over the wafer such that the grinding wheel center, indicated by the solid dot, describes a circular path relative to the wafer center (the grinding wheel is rotating about its own center). The grinding wheel center path is constrained to radius g such that the grinding wheel periphery passes through the wafer center, as shown in Figure 10.15. The constraint leads to an invariant isosceles triangle of points formed by the wafer center, the grinding wheel center, and the intersection of the wafer and grinding wheel peripheries. The triangle thus rotates with the grinding wheel and an example position is shown in Figure 10.15, along with the invariant angles θ_g and θ_w characterizing the triangle, given by

$$\cos\theta_g = w/2g \tag{10.21}$$

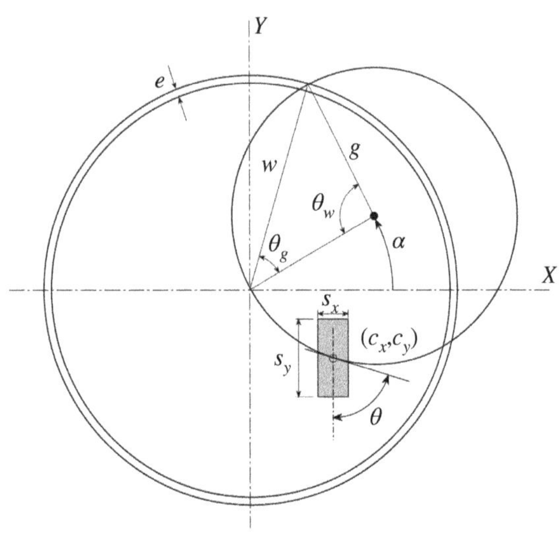

Figure 10.15 Plan view diagram of an in-feed wafer backgrinding process, illustrating the key dimensions and angles used in developing a die strength model. The wafer is the large circle, the grinding wheel is the small circle, and a representative die is shown shaded.

and $\theta_w = \pi - 2\theta_g$. The grinding wheel center path is described by the variable angle α, such that the grinding wheel center coordinates (g_x, g_y) in XY are given by

$$\begin{aligned} g_x &= g\cos\alpha \\ g_y &= g\sin\alpha. \end{aligned} \tag{10.22}$$

The angle α, measured in a conventional anticlockwise sense as shown in Figure 10.15, increases linearly with time as the grinding wheel center moves in the circular path about the wafer center. The grinding wheel trailing periphery thus rotates about the wafer center, generating circular striation arcs radiating from the wafer center as shown in Figure 10.5a. The points (x, y) along such arcs, for general values of α describing the grinding wheel position, are described by

$$(x - g_x)^2 + (y - g_y)^2 = g^2, \tag{10.23}$$

where (g_x, g_y) are given by Eq. (10.22).

The α values of interest here (strictly, $\alpha \bmod 2\pi$) are those for which a striation arc passes through the center of a specific die (c_x, c_y). As the dies are small relative to the wafer and grinding wheel dimensions, the striations on the backface of a die are well approximated by a parallel array of striations tangent to such a center striation. The discrete value of α characterizing the striation passing through a specific die center is obtained by setting $(x, y) = (c_x, c_y)$ and determining α in Eq. (10.22) such that Eq. (10.23) is satisfied. Such a determination is easily made by first defining the angle $\theta_c = \tan^{-1}(c_y/c_x)$ characterizing the die position [note that this definition differs from that used earlier (Mendelson et al. 1999, 2001)] and second, by recognizing from inspection of Figure 10.15 that the upper bound to α is given by $\theta_c + \pi/2 > \theta_g + \alpha$. Simple numerical search can then determine α and thus (g_x, g_y) from Eq. (10.22) and thence the (x, y) arc from Eq. (10.23). The derivative along the arc, dy/dx from Eq. (10.23), is $dy/dx = -(x - g_x)/(y - g_y)$. Hence, the orientation of the strength-controlling striations on the die backface is given by

$$\tan\theta = (c_y - g_y)/(c_x - g_x). \tag{10.24}$$

Once θ is determined for a die from implementation of Eqs. (10.21)–(10.24), the die strength can be determined from Eq. (10.19).

10.4.2 Results

Figure 10.16 is a diagram of an in-feed ground wafer simulated using the above analysis. The system dimensions were: wafer radius $w = 100$ mm, edge exclusion $e = 3$ mm, die size $s_x \times s_y = 10\text{ mm} \times 16\text{ mm}$, and grinding wheel radius $g = 70$ mm. The diagram is to scale with dimensions given in mm. The wafer edge and edge exclusion boundary are shown as the large bold circles. The edges of the 156 dies meeting the constraint of Eq. (10.20) are shown as the solid rectangular outlines; the centers of such dies are shown as solid dotted symbols. The circular arcs that pass through the centers of the dies, determined from Eqs. (10.21)–(10.24), are shown as fine lines.

Figure 10.17 shows the edf $\Pr(\theta)$ for the striation angles at the center of each die resulting from Figure 10.16. The edf forms a straight line across the entire angle domain, indicating that the striation angles result from a uniform probability density. Other simulations (not shown) using smaller die dimensions confirmed this behavior and showed that the deviations from ideal linear behavior in Figure 10.17 resulted from discrete nonzero die size. The key implication from Figure 10.17 is that as the striation orientation distribution density is uniform, any consequent strength distribution is dependent entirely on the strength–orientation relationship and any perturbations from that relationship arising from experimental or sampling effects.

Figure 10.18 shows whole-wafer strength edf predictions using the orientation edf of Figure 10.17, the strength–orientation function of Eq. (10.19), and the strength parameters describing the Z1 and Z1 + Z2 strengths of Figure 10.11. As anticipated above, the strength edf responses are simply rotated, discrete versions of the solid lines shown in Figure 10.11. Each response consists of 156 points, each point representing the uniaxial strength of a single die. The edf responses increase in strength from Z1 (left) to Z1 + Z2, g (center, grit study, Figure 10.11b) to Z1 + Z2, d (right, depth study, Figure 10.11a). In all cases, the edf variations exhibit near-linear responses at the center of the strength domain and near-vertical responses adjacent to the upper and lower strength bounds.

The ground die strength edf behavior is quite distinct from the strength edf behavior observed for most other materials. Bulk ceramics, Si MEMS components, brittle particles, glass fibers, and Si nanowires (Cook and DelRio, 2019a,b) overwhelmingly exhibit near-horizontal responses adjacent to the upper and lower strength bounds. That is, for most other

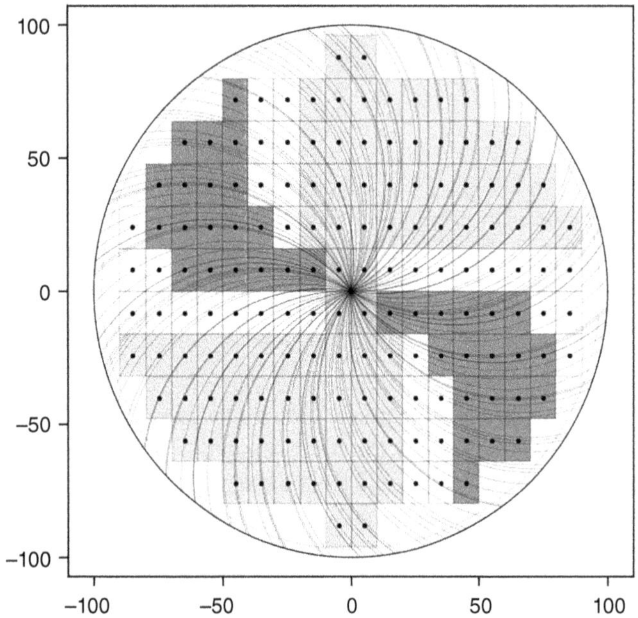

Figure 10.16 Mechanical strength map resulting from simulation of a back ground wafer. The frontface edge exclusion and die outlines are shown as light gray lines. The back face grinding striations that pass through the centers of dies are shown as dark gray lines. Strong dies are shown as light gray shaded rectangles, weak dies are shown as dark gray shaded rectangles. Dimensions in mm.

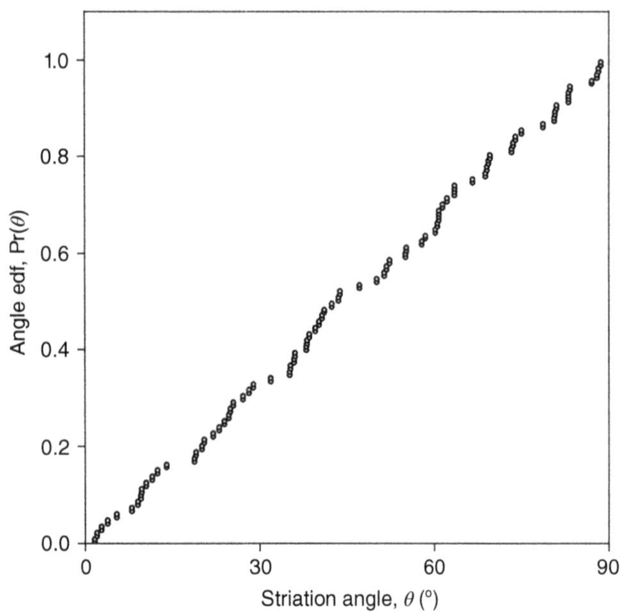

Figure 10.17 Empirical distribution function (edf) of center striation orientation angles for the backface ground wafer and dies in Figure 10.16. The linear behavior indicates a uniform distribution of angles.

materials the strength edf is classically sigmoidal. Particles can often exhibit concave strength edf responses, but the slopes at the bounds are clearly finite (Cook 2023). In addition, the strength edf domains in Figure 10.18 for Si die failure are wider than those of conventional materials, such as ceramics, glasses, and MEMS, which typically exhibit ratios of approximately 2. Particles can exhibit ratios of 4 or greater (Cook 2023). The strength domains in Figure 10.18 (derived from Figure 10.11) exhibit ratios of approximately 3. Although detailed consideration is beyond the scope of this work, it appears that the distributions of die strengths, reflecting a single type of flaw/die, have much in common with particle strength distributions, reflecting a single flaw/particle. Multiple flaws/specimen in conventional materials give rise to extreme value effects that

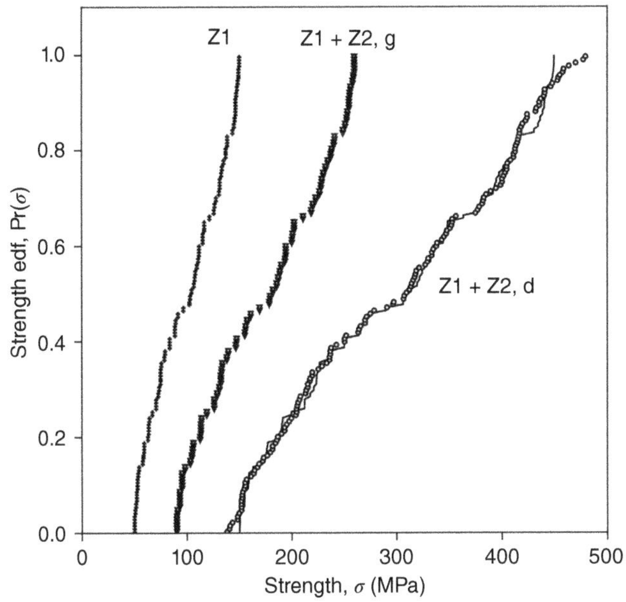

Figure 10.18 Empirical distribution functions (edf) for strength behavior of dies from backface ground wafers. Z1 grinding generates the weakest strength dies. Z1 + Z2 grinding generates dies with greater strengths. The edf shapes are atypical for brittle materials.

contract strength distribution domains and lead to sigmoidal features in strength edfs. Here and in particles, the single flaw/specimen removes extreme value effects giving rise to broad distributions that reflect the entire flaw population (here, the entire wafer). Further inspection of Figure 10.18 shows that the discrete deviations from the overall response are identical for the three grinding conditions (symbols for Z1 and Z1 +Z2, g; solid line for Z1 + Z2, d). As the same wafer-level orientation edf (Figure 10.17) was used in strength edf determinations, the deviations from smooth responses arising from the set of discrete orientations are the same for all grinding conditions.

A final point to note from Figure 10.18 is the effect of systematic uncertainty in strength measurement. Figure 10.11 shows dispersions in strength about the mean behavior given by Eq. (10.19) reflecting such uncertainty. The solid line in Figure 10.18 shows the mean Z1 + Z2, d strength edf determined assuming no dispersion; it is identical in shape to the Z1 and Z1 +Z2, g edf curves indicated by the solid symbols. The open symbols in Figure 10.18 show a Z1 + Z2, d strength edf including dispersion. This edf was calculated by assuming a 10% uniform variation about the mean value and reranking the resultant strengths. The process mimics experimental uncertainty or "scatter." The effect on the edf is minor; the randomized edf is slightly broader than the original and exhibits small deviations, but is otherwise not very noticeably different. Other simulations (not shown) indicate that even greater uncertainty leads to similar small effects. The reranking process smooths out most of the deviations by interchanging opposing strength deviations, leading to little change from the mean response.

The die strengths from the Z1 + Z2, d simulation of Figure 10.18 are replotted in Figure 10.19 as a function of die radial position relative to the wafer center. The full domain of strengths is observed at all radii and there is no variation in strength with radial position. The same die strength information is replotted in Figure 10.16 using radial and angular position—light gray for the strongest 1/3 of dies and dark gray for the weakest 1/3 of dies. It is clear that there is a significant variation in strength with angular position. Quadrants of the wafer alternate between strong dies (quadrants 1 and 3) and weak dies (quadrants 2 and 4). Figure 10.18 provides the mechanical strength wafer "map" advocated in Section 10.1.

The angular effect is considered in greater quantitative detail in Figure 10.20, which shows strength edf behavior for dies selected using the wafer map information of Figure 10.16. The solid line in Figure 10.20 is the strength edf behavior of the entire ensemble of dies from the wafer (reproducing the open symbol response from Figure 10.18). The open and filled symbols in Figure 10.20 form strength edf responses of dies selected from the same ensemble but restricted to quadrants 1 and 3 or quadrants 2 and 4, respectively. There are clear differences arising from angular selection. The dies selected from the odd numbered quadrants are stronger, and the strength distribution is convex. In contrast, dies selected from the even numbered quadrants are weaker, and the strength distribution is concave. Dies selected from quadrants 1 and 3 have a median strength about twice that of dies selected from quadrants 2 and 4.

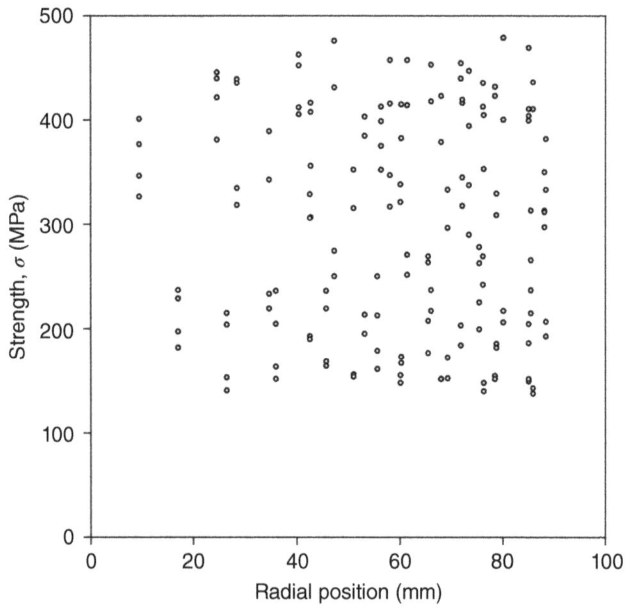

Figure 10.19 Plot of uniaxial strength of backface ground dies versus position on wafer quantified in terms of radius from wafer center. There is no significant effect of radial position.

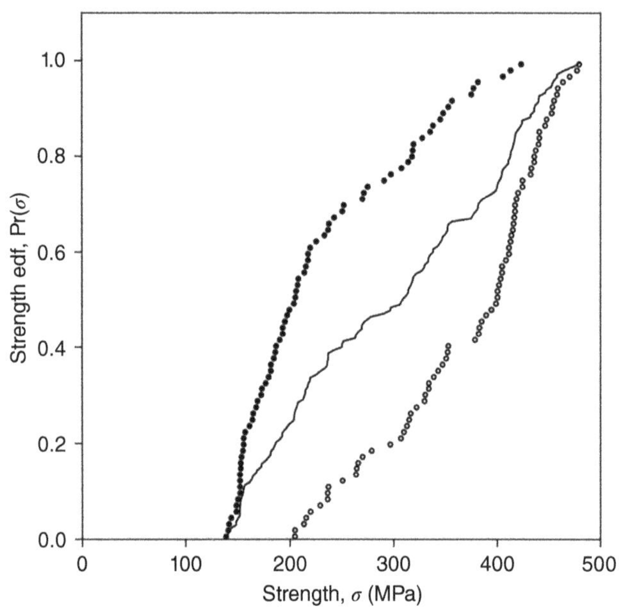

Figure 10.20 Empirical distribution functions (edf) for strength of dies from a Z1 + Z2 backface ground wafer. The solid line is the strength edf for the entire wafer. The solid and open symbols are edf responses for dies sampled from different angular quadrants (same shading scheme as Figure 10.16) illustrating the significant effect on strength.

A clear implication from Figure 10.20, reinforcing the mechanical map concept, is that die selection by angular position from a backface ground wafer provides an opportunity for optimized mechanical performance of packaged devices. Selected high-strength dies could be placed in greater stress packages, e.g., on polymer substrates rather than ceramic substrates, or in applications requiring greater reliability, e.g., in automobiles rather than game consoles. A related implication is that die selection from a mechanical map provides an opportunity for optimized commercial performance, as dies from an entire wafer can be utilized in appropriate packages and applications. Conversely, Figure 10.20 provides a clear warning that angularly selected dies could provide a misleading representation of the mechanical performance of a wafer. The greatest danger is that a few angularly selected (e.g., quadrant 1) dies are selected and tested, exhibiting strengths greater than

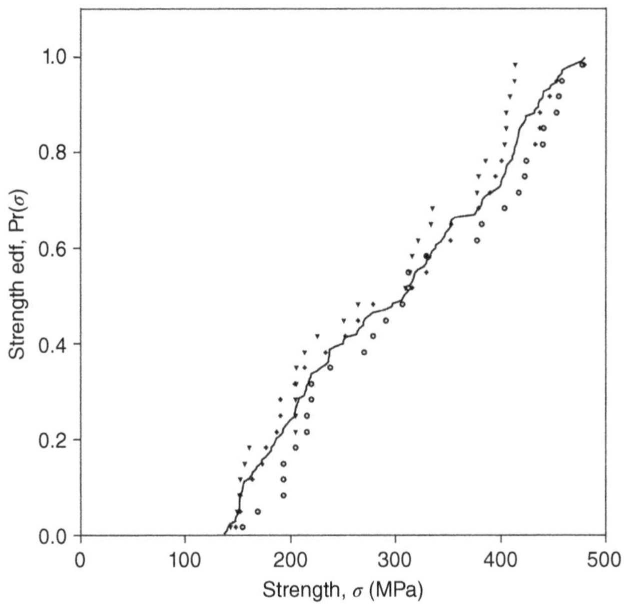

Figure 10.21 Empirical distribution functions (edf) for strength of dies from a Z1 + Z2 backface ground wafer. The solid line is the strength edf for the entire wafer. The symbols are edf responses for 30 dies sampled at random from the wafer; different symbols represent separate samples.

the wafer as a whole. Device processing decisions regarding backgrinding and packaging would then be made based on erroneous, overestimated, strengths. Conversely, random selection from the entire wafer, even in the presence of strength uncertainty, would provide an accurate, if less precise, estimate of die behavior. Figure 10.21 reproduces the solid line from Figure 10.20 as the strength edf behavior of the entire ensemble of 156 dies from the wafer. The symbols in Figure 10.21 form the strength edf responses of 30 dies sampled at random from the ensemble; the three different symbols represent three different samples. Although there are deviations from the ensemble response, the edf responses of the samples exhibit the same domains and medians as the ensemble and would provide accurate guidance for grinding and packaging optimization.

10.5 Summary

A method for generating wafer maps of semiconductor device die strengths was developed. The strength maps are in close analogy with the maps of device electrical performance generated after wafer frontface processing. In contrast, the strength maps are based on wafer backface mechanical processing parameters, specifically those characterizing wafer grinding. The backface parameters and frontface parameters are related through the variables of die size and location. A common method of wafer backface processing—in-feed grinding—was considered in detail, and a brief overview of prior observations and analyses of in-feed grinding of Si was presented. In-feed grinding generates wafer backface striations in the form of radiating circular arcs that lead to parallel arrays of striations on die backfaces. For the common loading configuration of bending a rectangular die, the striation array orientation relative to the die long axis is a major factor in controlling die strength. Considerable physical insight and support for the map generation process was provided by an extensive consideration of controlled indentation flaw and sharp scratch behavior. Observations and strength measurements of Si showed that strength-controlling striations were most likely subthreshold linear contact flaws formed by abrasive particles contacting the wafer backface with normal loads less than 0.1 N, less than the weight of a Si wafer. Strength measurements of ground dies showed a significant dependence on striation scale and orientation and with loading geometry; for the most common rectangular die loading configuration, uniaxial strength decreased by a factor of 3 for grinding striations perpendicular to the die long axis relative to striations parallel to the axis. Coarse grinding decreased strengths at all orientations, and strengths measured in biaxial loading were greater than those measured in uniaxial loading (as in bending). A simple cubic expression was used to describe the smooth variation of die mean strength with striation orientation. This expression and the strength bounds were used in a model of die strength based on grinding geometry parameters. The model was implemented numerically and used to generate wafer mechanical maps and die strength distributions—both the maps and the distributions are practical for direct engineering application. The model provides a rare example of "forward" strength distribution analysis, in which the flaw distribution, here the orientation distribution, and the strength dependence, here the cubic expression, are known. Such knowledge enables sample strength distributions to be predicted from population properties, here wafer properties. Examples of the effects of grinding severity, experimental uncertainty, radial position, and angular selection were demonstrated. ("Backward" strength distribution analyses, in which population characteristic are inferred from sample strength distributions, are far more common.) The model is easily extended to other grinding geometries and provides a foundation for consideration of other strength-controlling flaws, such as those generated by dicing of full thickness wafers.

References

Agarwal, S. and Rao, P.V. (2010). Modeling and prediction of surface roughness in ceramic grinding. *International Journal of Machine Tools and Manufacture* 50: 1065–1076.

Bawa, M.S., Petro, E.F., and Grimes, H.M. (1995). Fracture strength of large diameter silicon wafers. *Semiconductor International* 18: 115–118.

Bensaid, N., Benbahouche, S., Roumili, F., Sangleboeuf, J.-C., Le Cam, J.-B., and Rouxel, T. (2018). Influence of the normal load of scratching on cracking and mechanical strength of soda-lime-silica glass. *Journal of Non-Crystalline Solids* 483: 65–69.

Campbell, S.A. (2012). *Fabrication Engineering at the Micro- and Nanoscale, 4th ed*. Oxford.

Chae, S.-H., Zhao, J.-H., Edwards, D.R., and Ho, P.S. (2010a). Effect of backside scratch direction on the Si die strength. *Inter Society Conference on Thermal and Thermomechanical Phenomena in Electronic Systems*, 1–6.

Chae, S.H., Zhao, J.H., Edwards, D.R., and Ho, P.S. (2010b). Effect of dicing technique on the fracture strength of Si dies with emphasis on multimodal failure distribution. *IEEE Transactions on Device and Materials Reliability* 10: 149–156.

Chao, C.L., Ma, K.J., Liu, D.S., Bai, C.Y., and Shy, T.L. (2002). Ductile behaviour in single-point diamond-turning of single-crystal silicon. *Journal of Materials Processing Technology* 127: 187–190.

Chen, J. and De Wolf, I. (2003). Study of damage and stress induced by backgrinding in Si wafers. *Semiconductor Science and Technology* 18: 261–268.

Cheng, W., Ling, E., and Finnle, I. (1990). Median cracking of brittle solids due to scribing with sharp indenters. *Journal of the American Ceramic Society* 73: 580–586.

Chidambaram, S., Pei, Z.J., and Kassir, S. (2003a). Fine grinding of silicon wafers: a mathematical model for the chuck shape. *International Journal of Machine Tools and Manufacture* 43: 739–746.

Chidambaram, S., Pei, Z.J., and Kassir, S. (2003b). Fine grinding of silicon wafers: a mathematical model for grinding marks. *International Journal of Machine Tools and Manufacture* 43: 1595–1602.

Chong, D.Y.R., Lee, W.E., Lim, B.K., Pang, J.H.L., and Lew, T.H. (2004). Mechanical characterization in failure strength of silicon dice. *Intersociety Conference on Thermal Phenomena and Thermomechanical Phenomena In Electronic Systems*, volume 2, 203–210.

Cook, R.F. (1994). Deformation and fracture by sharp rolling contacts. *Journal of the American Ceramic Society* 77: 1263–1273.

Cook, R.F. (2006). Strength and sharp contact fracture of silicon. *Journal of Materials Science* 41: 841–872.

Cook, R.F. (2017). Fracture mechanics of the scratch strength of polycrystalline alumina. *Journal of the American Ceramic Society* 100: 1146–1160.

Cook, R.F. (2018). Blunt scratch strength of polycrystalline alumina. *Journal of the American Ceramic Society* 101: 16–19.

Cook, R.F. (2023). *Particle Strengths*. New York, NY: Wiley.

Cook, R.F. and DelRio F.W. (2019a). Material flaw populations and component strength distributions in the context of the Weibull function. *Experimental Mechanics* 59: 279–293.

Cook, R.F. and DelRio, F.W. (2019b). Determination of ceramic flaw populations from component strengths. *Journal of the American Ceramic Society* 102: 4794–4808.

Cook, R.F., DelRio, F.W., and Boyce B.L. (2020). Predicting strength distributions of MEMS structures using flaw size and spatial density. *Microsystems & Nanoengineering* 5: 1–12.

Cook, R.F., Lawn, B.R., and Fairbanks, C.J. (1985). Microstructure-strength properties in ceramics: I-effect of crack size on toughness. *Journal of the American Ceramic Society* 68: 604–615.

Cotterell, B., Chen, Z., Han, J.-B., and Tan, N.-X. (2003). The strength of the silicon die in flip-chip assemblies. *Journal of Electronic Packaging* 125: 114–119.

Danyluk, S., Lee, S.-W., Ahn, J.-H., and Kahn, A. (1988). Resistivity of linear scratches in doped (100) n-type single-crystal silicon. *Journal of Applied Physics* 63: 4568–4571.

DelRio, F.W., Boyce, B.L., Benzing, J.T., Friedman, L.H., and Cook, R.F. (2020). Shoulder fillet effects in strength distributions of microelectromechanical system components. *Journal of Micromechanics and Microengineering* 30: 125013.

DelRio, F.W., Cook, R.F., and Boyce, B.L. (2015). Fracture strength of micro-and nano-scale silicon components. *Applied Physics Reviews* 2: 021303.

Dries, B., Kingsbury, J., and Mohamed, E. (2008). A methodology for improving granularity of wafer strength testing to characterize and qualify a 150 micron DFM wafer thinning process. *Advanced Semiconductor Manufacturing Conference*, 94–99.

Funke, C., Kullig, E., Kuna, N., and Möller, H.J. (2004). Biaxial fracture test of silicon wafers. *Advanced Engineering Materials* 6: 594–598.

Gu, W., Yao, Z., and Li, K. (2011). Evaluation of subsurface crack depth during scratch test for optical glass BK7. *Proceedings of the Institution of Mechanical Engineers, Part C: Journal of Mechanical Engineering Science* 225: 2767–2774.

Haapalinna, A., Nevas, S., and Pähler, D. (2004). Rotational grinding of silicon wafers—sub-surface damage inspection. *Materials Science and Engineering B* 107: 321–331.

Howes, V.R. and Szameitat, A. (1984). Morphology of sub-surface cracks below the scratch produced on soda-lime glass by a disc cutter. *Journal of Materials Science Letters* 3: 872–874.

Hudson, B. and Perrin, D. (1990). The effects of wafer processing on silicon die strength. *International Symposium for Testing and Failure Analysis (ASM International*, Pittsburgh, PA, 1990) p. 515

Jeong, S.-M., Park, S.-E., Oh, H.-S., and Lee, H.L. (2004). Evaluation of damage on silicon wafers using the angle lapping method and a biaxial fracture strength test. *Journal of Ceramic Processing Research* 5: 171–174.

Jung, Y.G., Pajares, A., and Lawn, B.R. (2004a). Effect of oxide and nitride films on strength of silicon: A study using controlled small-scale flaws. *Journal of Materials Research* 19: 3569–3575.

Jung, Y.G., Pajares, A., Banerjee, R., and Lawn, B.R. (2004b). Strength of silicon, sapphire and glass in the subthreshold flaw region. *Acta Materialia* 52: 3459–3466.

Kirchner, H.P. (1984a). Damage penetration at elongated machining grooves in hot-pressed $Si_3 N_4$. *Journal of the American Ceramic Society* 67: 127–132.

Kirchner, H.P. (1984b). Comparison of single-point and multipoint grinding damage in glass. *Journal of the American Ceramic Society* 67: 347–353.

Lawn, B. and Wilshaw, R. (1975). Indentation fracture: principles and applications. *Journal of Materials Science* 10: 1049–1081.

Lawn, B.R., Wiederhorn, S.M., and Roberts, D.E. (1984). Effect of sliding friction forces on the strength of brittle materials. *Journal of Materials Science* 19: 2561–2569.

Lee, S.-M., Sim, S.-M., Chung, Y.-W., Jang, Y.-K., and Cho, H.-K. (1997). Fracture strength measurement of silicon chips. *Japanese Journal of Applied Physics* 36: 3374–3380.

Lee, S.K., Tandon, R., Readey, M.J., and Lawn, B.R. (2000). Scratch damage in zirconia ceramics. *Journal of the American Ceramic Society* 83: 1428–1432.

Leu, H.J. and Scattergood, R.O. (1988). Sliding contact fracture on glass and silicon. *Journal of Materials Science* 23: 3006–3014.

Levengood, W.C. (1958). Effect of origin flaw characteristics on glass strength. *Journal of Applied Physics* 29: 820–826.

Lim, D.S. and Danyluk, S. (1985). Surface damage of single-crystal silicon abraded in ethanol and deionized water. *Journal of Materials Science* 20: 4084–4090.

Lim, D.S. and Danyluk, S. (1988). Correlation of dynamic friction and the dislocation etch pit density surrounding annealed scratches in (111) p-type silicon. *Journal of Materials Science* 23: 2607–2612.

Liu, D.S., Shih, Y.S., Ni, C.Y., Chiou, D.Y., Chung, C.L., and Huang, M.L. (2002). Influences of material source, wafer process and location on silicon die strength. *Experimental Techniques* 6: 29–35.

Marks, M.R., Hassan, Z., and Cheong, K.Y. (2015). Ultrathin wafer pre-assembly and assembly process technologies: a review. *Critical Reviews in Solid State and Materials Sciences* 40: 251–290.

Matsuoka, J., Guo, D., and Yoshida, S. (2017). Cross-section morphology of the scratch-induced cracks in soda-lime-silica glass. *Frontiers in Materials* 4: 8.

McGuire, K., Danyluk, S., Baker, T.L., Rupnow, J.W., and McLaughlin, D. (1997). The influence of backgrinding on the fracture strength of 100 mm diameter (111) p-type silicon wafers. *Journal of Materials Science* 32: 1017–1024.

McLellan, N., Fan, N., Liu, S., Lau, K., and Wu, J. (2004). Effects of wafer thinning condition on the roughness, morphology and fracture strength of silicon die. *Journal of Electronic Packaging* 126: 110–114.

Mendelson, R.L., Cook, R.F., Diefenderfer, D.F., and Liniger E.G. (1999). Method and apparatus for preventing chip breakage during semiconductor manufacturing using wafer grinding striation information. United States Patent, 5,888,838.

Mendelson, R.L., Cook, R.F., Diefenderfer, D.F., and Liniger E.G. (2001). Method and apparatus for preventing chip breakage during semiconductor manufacturing using wafer grinding striation information. United States Patent, 6,171,873 B1.

Mendelson, R., Diefenderfer, D.F., Gumbert, C.M., and Singh, B. (1996). Optimized backgrind wafer process minimizes stress effects. *Semiconductor International* 19: 221–225.

Michaelides, S. and Sitaraman, S.K. (1999). Die cracking and reliable die design for flip-chip assemblies. *IEEE Transactions on Advanced Packaging* 22: 612–613.

Miller, C. (2022). *Chip War: The Fight for the World's Most Critical Technology*. Scribner.

Misra, A. and Finnie, I. (1979). On the scribing and subsequent fracturing of silicon semiconductor wafers. *Journal of Materials Science* 14: 2567–2574.

Murakami, Y. (Ed.) (1987) *Stress Intensity Factors Handbook*. Pergamon.

Oliver, A.D., Wallner, T.A., Tandon, R., Nieman, K., and Bergstrom, P.L. (2008). Diamond scribing and breaking of silicon for MEMS die separation. *Journal of Micromechanics and Microengineering* 18: 075026.

Ono, T. and Tanaka, K. (2001). Effect of scribe-wheel dimensions on the cutting of AMLCD glass substrate. *Journal of the Society for Information Display* 9: 87–94.

Pei, Z.J. (2002). Study on surface grinding of 300 mm silicon wafers. *International Journal of Machine Tools and Manufacture* 42: 385–393.

Pei, Z.J., Billingsley, S.R., and Miura, S. (1999). Grinding induced subsurface cracks in silicon wafers. *International Journal of Machine Tools and Manufacture* 39: 1103–1116.

Pei, Z.J., Fisher, G.R., and Liu, J. (2008). Grinding of silicon wafers: a review from historical perspectives. *International Journal of Machine Tools and Manufacture* 48: 1297–1307.

Pei, Z.J., Fisher, G.R., Bhagavat, M., and Kassir, S. (2005). A grinding–based manufacturing method for silicon wafers: an experimental investigation. *International Journal of Machine Tools and Manufacture* 45: 1140–1151.

Pei, Z.J. and Strasbaugh, A. (2001). Fine grinding of silicon wafers. *International Journal of Machine Tools and Manufacture* 41: 659–672.

Pei, Z.J. and Strasbaugh, A. (2002). Fine grinding of silicon wafers: designed experiments. *International Journal of Machine Tools and Manufacture* 42: 395–404.

Peter, K. (1964). Brittle fracture and microplasticity of glass in indentation experiments (in German). *Glasstechnische Berichte* 37: 333–345.

Sun, W., Pei, Z.J., and Fisher, G.R. (2004). Fine grinding of silicon wafers: a mathematical model for the wafer shape. *International Journal of Machine Tools and Manufacture* 44: 707–716.

Sun, W., Pei, Z.J., and Fisher, G.R. (2005a). Fine grinding of silicon wafers: machine configurations for spindle angle adjustments. *International Journal of Machine Tools and Manufacture* 45: 51–61.

Sun, W., Pei, Z.J., and Fisher, G.R. (2005b). Fine grinding of silicon wafers: effect of chuck shape on grinding marks. *International Journal of Machine Tools and Manufacture* 45: 673–686.

Sze, S.M. and Lee, M-K. (2012). *Semiconductor Devices: Physics and Technology, 3rd ed.* Wiley.

Watanabe, N., Miyazaki, T., and Yoshikawa, K. (2014). Damage evaluation of wet-chemical Si-wafer thinning/backside via exposure process. *IEEE Transactions on Components, Packaging Manufacturing Technology* 4: 741–747.

Wu, J.D., Huang, C.Y., and Liao, C.C. (2003). Fracture strength characterization and failure analysis of silicon dies. *Microelectronics Reliability* 43: 269–277.

Yoshikawa, K., Miyazaki, T., Watanabe, N., and Aoyagi, M. (2012). Wet-chemical silicon wafer thinning process for high chip strength. *ECS Transactions*, 45: 141–150.

Yu, F., Sang, W., Pang, E., Liu, D., and Teng, J. (2002). Stress analysis in silicon die under different types of mechanical loading by finite element method (FEM). *IEEE Transactions on Advanced Packaging* 25: 522–527.

Zhao, J.-H., Tellkamp, J., Gupta, V., and Edwards, D.R. (2009). Experimental evaluations of the strength of silicon die by 3-point-bend versus ball-on-ring tests. *IEEE Transactions on Advanced Packaging* 32: 248–255.

11

Strength and Toughness of Cordierite Glass-Ceramic Composites

In this chapter the toughness and strength of a series of glass ceramic-whisker composites are investigated in an extensive series of indentation fracture-based tests. The tests include measurements of hardness from indentation deformation, measurements of relative toughness in air from indentation crack lengths and in situ measurements of crack extension in stressed specimens, and measurements of toughness and assessments of flaw size from inert indentation- and intrinsic-strength tests. The extensive tests illustrate the power of indentation techniques to evaluate the fracture properties of structural ceramics at small length scales. In the case here, whisker addition led to significant increases in composite toughness, from 1.4 MPa $m^{1/2}$ to 2.6 MPa $m^{1/2}$, and increases in intrinsic flaw size, such that composite strength remained invariant at approximately 250 MPa. An Appendix considers the contrasting behavior of a series of glass ceramic-particulate composites.

11.1 Introduction

The development and application of advanced structural ceramics require knowledge of mechanical properties of ceramic materials and of the load-bearing capacity of ceramic components. In particular, ceramics are typically brittle and fail by crack propagation with little accompanying plastic deformation—measurements of toughness and fracture strength are thus required for structural ceramic design and development. As fracture strengths of structural ceramics are typically controlled by cracking from small, microscopic flaws, toughness at short crack lengths, and distributions of flaw sizes are critical mechanical properties for comparing materials and predicting component performance. As demonstrated throughout this book, controlled-flaw indentation techniques are ideally suited for evaluating these properties and many other small-scale fracture and deformation phenomena (see also Lawn and Cook 2012; Marshall et al. 2015). The basis of such techniques is the controlled loading and unloading of a sharp indenter onto a ceramic surface to form a residual contact impression and cracks. Early works in this area (Lawn et al. 1980; Anstis et al. 1981; Chantikul et al. 1981) showed that measurements of indentation crack lengths or of strengths of components containing strength-controlling indentation flaws can be used to estimate toughness and, when the measurements are combined, estimate intrinsic flaw sizes. Subsequent advances in fracture properties evaluation based on these techniques are summarized in overview works considering the now pervasive method of indentation crack length measurements in air (Cook 2020) and sophisticated indentation strength methods that enable quantification of multiscale effects in fracture, including contact damage, material microstructure, and environmental reaction kinetics (Cook 2015). The crack length and strength methods are the focus of Chapters 2–6 in this book.

Further advances in ceramic toughness, strength, and flaw size determination are enabled by recent works of experimental fracture measurement in two areas. The first fracture measurement takes advantage of indentation techniques and enables direct assessment of variations in toughness with crack length (Braun et al. 1992; Kovar et al. 2000). The measurement is based on in situ observations of crack propagation during indentation strength tests. (Ex situ crack propagation measurements have also been implemented but do not always provide a complete assessment of toughness variation. Detailed in Chapter 5, see also Cook and Lawn 1983.) The second fracture measurement implements probabilistic methods and enables direct assessment of flaw size distributions (Cook and DelRio 2019a,b). The measurement makes distinction between a population of flaws and a sampled distribution and has been applied to machining flaws, indentation flaws, and pores in bulk ceramics, to ceramic particles (Cook 2023), and to silicon microelectromechanical systems components (Cook et al. 2019; DelRio et al. 2020).

Indentation Fracture: Strength and Toughness for Brittle Materials Design, First Edition. Robert F. Cook.

This chapter will apply the above two more recent techniques in a comprehensive set of fracture measurements on an advanced structural ceramic, a composite of cordierite glass-ceramic reinforced with silicon nitride whiskers. Specifically, a group of microstructural variants of the ceramic composite with different whisker contents will be examined using both the earlier indentation crack length and strength measurement techniques and the more recent toughness variation and flaw size distribution measurement techniques. The chapter begins with a description of the experimental methods, emphasizing materials characterization and indentation-based testing and provides background observations in these areas. This is followed by a summary of the required analysis methods, including indentation fracture mechanics and the probabilistic failure framework. For both the experimental and analysis methods, extensive details are given in the chapters and works cited above, and the focus in this chapter is to enable measurement interpretation. The results follow, considering toughness and strength measurements. The aim of the chapter is not to find the "best" composite, but to demonstrate how a comprehensive set of measurements can be used to guide advanced structural ceramic development and application. The Appendix provides a brief overview of the indentation fracture properties of a series of cordierite glass-ceramic composites reinforced with chromia particles. The whisker and particle composites exhibit contrasting effects of microstructural manipulation on strength and toughness.

11.2 Experimental Methods

11.2.1 Materials

The glass-ceramic composites were formed from tapecast greensheets (Tummala, 1991; Knickerbocker 1992; Knickerbocker et al. 1993) that contained cordierite glass powder and up to 35% by relative volume (vol%) 1 μm × 10 μm β-phase Si_3N_4 whiskers (Ube Industries Ltd, Ube City, Japan; see Homeny et al. 1990), along with binders and dispersants. The volume fractions of whiskers, V_f, were determined relative to densified 0% "blank" glass-ceramic and analyses here predominantly focused on $V_f = 0\%$, 5%, 10%, and 15% materials. Figure 11.1a shows a group of whiskers and the glass particles of a greensheet. During mixing of the glass, whiskers, binder, and dispersant, the initially longer whiskers were broken down to their 10 μm length. On tape casting, there was some preferential alignment of the whiskers parallel to the greensheet faces, as shown in the cross section of a fractured greensheet in Figure 11.1b. The porosity of the greensheet is visible in both Figures 11.1a and b. The 150 μm thick greensheets were laminated, free-sintered, and then diced to produce 5.5 mm × 7.0 mm × 55 mm bars, or punched before sintering to produce 28.6 mm × 2.3 mm disks. The edges of the bars were chamfered and polished, and the surfaces of the disks were polished. All compositions underwent identical sintering and polishing treatments.

The density of the sintered composites was measured using the Archimedes' displacement technique on several bars of each composition. Comparisons with the theoretical densities revealed that the addition of whiskers significantly reduced the relative densities, ρ_r, of the sintered glass-ceramic composites such that there were significant volume fractions of porosity, $V_p = (1 - \rho_r)$. The variation of ρ_r with V_f is shown in Figure 11.2. The uncertainty in relative density measurement was approximately 0.1% and is smaller than the symbol size. The mean values of ρ_r are given in Table 11.1. The nature of the porosity in the sintered composites was assessed using mercury porosimetry measurements. For materials with $V_f < 10\%$, the open porosity was <0.3% (compared with total porosity $V_p = 6.3\%$ for the $V_f = 10\%$ material), whereas for

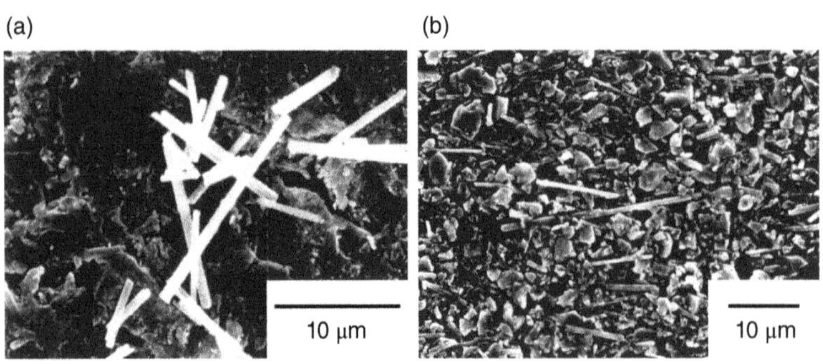

Figure 11.1 Scanning electron microscope (SEM) images of (a) a group of Si_3N_4 whiskers imbedded in the cordierite glass particles of a composite greensheet and (b) a cross section of a fractured greensheet showing the partial alignment of the whiskers.

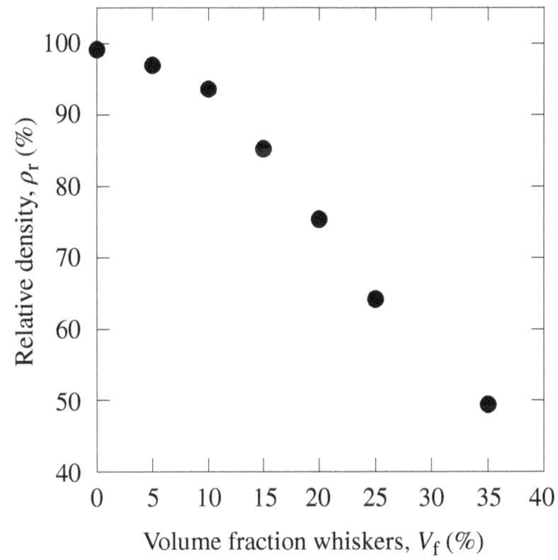

Figure 11.2 Plot of relative density of sintered composites as a function of volume fraction of Si_3N_4 whiskers added to cordierite glass-ceramic. The whiskers clearly retard densification.

Table 11.1 Bulk and contact properties of glass-ceramic whisker composites.

Material and Property							
Volume Fraction of Whiskers, V_f (%)	0	5	10	15	20	25	35
Relative Density, ρ_r (%)	99.2	97.0	93.7	85.3	75.4	64.2	49.4
Young's Modulus, E (GPa)	120 ± 10	136 ± 12	133 ± 11	104 ± 8	—	—	—
Coefficient of Restitution, e	0.82	0.80	0.77	0.70	—	—	—
Hardness, H (GPa)	8.3 ± 0.5	8.2 ± 0.6	6.7 ± 0.8	3.1 ± 0.4	—	—	—
Cracking Susceptibility, B (μm $N^{-2/3}$)	13.4 ± 0.9	11.3 ± 0.7	11.1 ± 0.7	11.3 ± 0.5	—	—	—

$V_f = 15\%$, the open porosity was 7.3% (compared with $V_p = 14.7\%$). These measurements suggest a change in the microstructure from a porous solid to an open foam in the range $10\% < V_f < 15\%$.

Fracture cross sections of sintered specimens with $V_f = 5\%$ and $V_f = 15\%$ are shown in Figure 11.3 (the same magnification is used for both specimens). The image of the $V_f = 5\%$ material, Figure 11.3a, shows porosity at two different scales. At the smaller scale, near spherical (circular in section) isolated pores, approximately 1 μm in diameter, are distributed throughout the material and are referred to here as "micropores." Such micropores are typical in the closed porosity microstructures that develop during final stage sintering of monolithic ceramics and constitute the majority of the porosity. At the larger scale, a single lenticular (elliptical in section) pore, approximately 30 μm in length, is shown. The pore contains a cluster of entangled whiskers and very little sintered glass-ceramic and is referred to here as a "macropore." Such macropores are a consequence of incomplete mixing of constituents in the greensheet and persist throughout sintering of composites. Although large, such macropores constitute a minority of sintered material porosity and, as is typical, were observed here to be distributed sparsely throughout the composites. Macropores were observed more frequently here in the materials with greater whisker volume fractions.

The image of the $V_f = 15\%$ material, Figure 11.3b, differs from Figure 11.3a in several ways. First, the fracture surface is much more rough in the $V_f = 15\%$ material. Second, the distributed porosity (dark regions) in this material is angular and irregular in shape, much more pervasive, and larger, up to 5 μm in size, than the micropores in Figure 11.3a. Such porosity, comparable in scale to the densified glass-ceramic regions, is typical of the open porosity microstructures that develop during early stage or arrested sintering of monolithic ceramics. Close inspection of the densified regions in Figure 11.3b reveals cross sections of very small isolated spherical pores and broken whiskers (small bright disks, also visible in Figure 11.3a). Nevertheless, the implication from Figure 11.3b is that the irregular porosity is open and constitutes the majority, leading to the rough fracture surface. Clearly the whiskers retarded densification of the composite by both inhibiting the viscous sintering process (Knickerbocker et al. 1993) where the whiskers were homogeneously distributed and by forming undensifiable agglomerates where heterogeneously distributed.

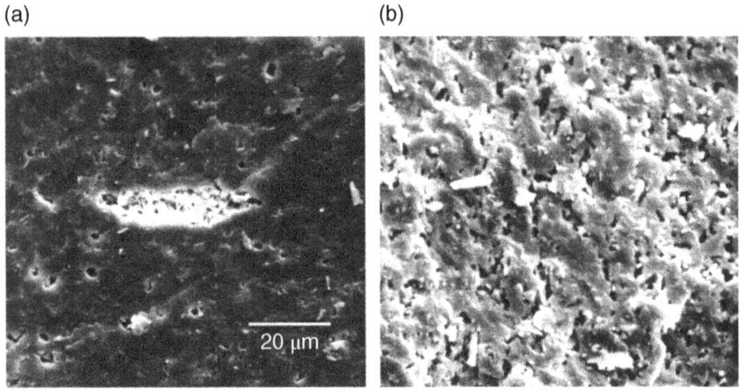

Figure 11.3 SEM images of fracture cross sections of sintered cordierite glass-ceramic silicon nitride whisker composites. (a) $V_f = 5\%$ volume fraction of whiskers, (b) $V_f = 15\%$.

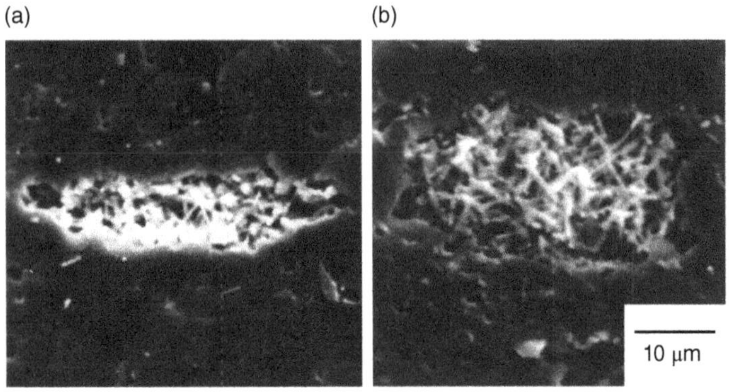

Figure 11.4 SEM images of isolated macropores in cordierite glass-ceramic silicon nitride whisker composites. (a) $V_f = 5\%$ volume fraction of whiskers, (b) $V_f = 15\%$.

Magnified views of macropores in $V_f = 5\%$ and $V_f = 15\%$ composites are shown in Figure 11.4 (the 5% specimen is that from Figure 11.3a; the same magnification is used for both specimens). In both images, the macropores contain almost no glass-ceramic particles and consist of entangled whisker agglomerates with many whiskers perpendicular to the greensheet tapecasting and lamination directions (horizontal in the images). The macropore in the $V_f = 15\%$ material, Figure 11.4b, is larger than that in the $V_f = 5\%$ material, Figure 11.4a. The observations of Figures 11.3 and 11.4 were typical of many images examined for the composites.

Young's modulus E was determined from the load–displacement responses of small bars 1.0 mm × 5.0 mm × 35 mm cut from the larger test bars. A stiff, three-point bend fixture under displacement control, instrumented with a resistance displacement gauge and a resistance load cell were used to measure the responses. The values of E are given in Table 11.1. The uncertainties reflect test and between-specimen variability. E was slightly greater for the 5 and 10% compositions than that of the blank 0% glass-ceramic, but significantly less for the 15% composition. Typical values of the Young's moduli for dense cordierite glass-ceramic and silicon nitride are 121 and 300 GPa, respectively. These values and the observations of Figures 11.3 and 11.4 imply that there were competing influences of stiffer whiskers and compliant porous glass-ceramic regions in determining the moduli of the composites. The initial increase in modulus at small V_f implies good bonding between at least some of the whiskers and the glass-ceramic matrix before a decrease in modulus at large V_f associated with porosity.

11.2.2 Contact Properties

A Vickers diamond pyramid indenter was used for all indentation tests in this study. The indentation load–displacement, P–h, behavior of the composite materials was determined using an instrumented indenter with capacitance displacement gauges and resistance load cell (Cook and Pharr 1990; Thurn et al. 2002). Polished disks of material were indented to a peak depth of 20 μm at a penetration rate of 1 μm s^{-1} and the resulting P–h data recorded. Complete indentation cycles are

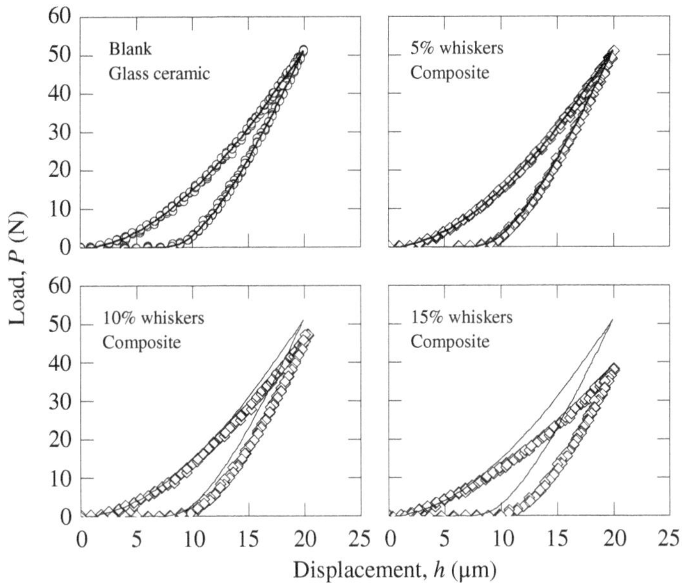

Figure 11.5 Indentation load–displacement plots for cordierite glass-ceramics containing Si_3N_4 whiskers. The symbols represent individual load–displacement measurements, and the solid lines represents the response of the blank material. Data shown are every second point for single replicated cycles.

shown in Figure 11.5. The open symbols are data for the designated material; the response of the blank glass-ceramic is shown as the repeated solid lines and provides a reference.

The load–displacement behavior of the composites was typical of that for ceramics: combined elastic and plastic deformation during loading (leading to a quadratic dependence of P on h); elastic recovery during unloading (leaving a remnant displacement and contact impression). Comparison of the composite indentation responses with that of the blank glass-ceramic shows that addition of the whiskers (with the concomitant increase in porosity) leads to smaller sustained peak loads at peak displacement and larger remnant displacements. That is, the composites were less contact resistant and exhibited greater irreversible deformation relative to reversible deformation during contact. An indication of the relative amounts of irreversible and reversible deformation during contact is the coefficient of restitution e, which gives a measure of the prospective relative rebound velocity of an impacting particle (Tabor 1951). The value of e may be calculated from the load–displacement data by $e = (U_r/U)$, where U is the total amount of energy expended by the indenter during the loading cycle and U_r is the amount of energy recovered during the unloading cycle, represented by the areas under the upper and lower P–h curves, respectively. The values of e are given in Table 11.1 and indicate that the addition of whiskers significantly decreased e, implying that small particle impacts would be more likely to generate deformation damage in a composite surface relative to a blank glass-ceramic surface. The likely mechanism for the decrease in e is the enhanced pore collapse and crushing of whisker entanglements during indentation contact of the composites. The smooth responses in Figure 11.5 imply that the 20 μm indentation depth sampled many pores (Cook 2018a), consistent with Figure 11.3.

Figure 11.6a is a schematic plan diagram of a Vickers indentation flaw. The central square is the residual contact impression indicating remnant plastic deformation, with depth formed as in Figure 11.5. The residual impression semidiagonal is the dimension indicated. Figure 11.6b and c shows optical micrographs of indentations in the blank glass-ceramic and 15 vol% whisker composite, respectively. Both indentations were formed with a peak load of $P = 100$ N. In both micrographs, the square contact impression is evident, and, as expected from the indentation load–displacement tests, the remnant impression in the composite is larger implying a smaller contact resistance. A measure of contact resistance is the mean contact pressure associated with plastic deformation, the hardness H, defined by (Tabor 1951)

$$H = P/2a^2. \tag{11.1}$$

Inverting this equation gives

$$a = (1/2H)^{1/2}P^{1/2}, \tag{11.2}$$

showing that the indentation impression dimension a should increase as $P^{1/2}$ in a material of constant hardness.

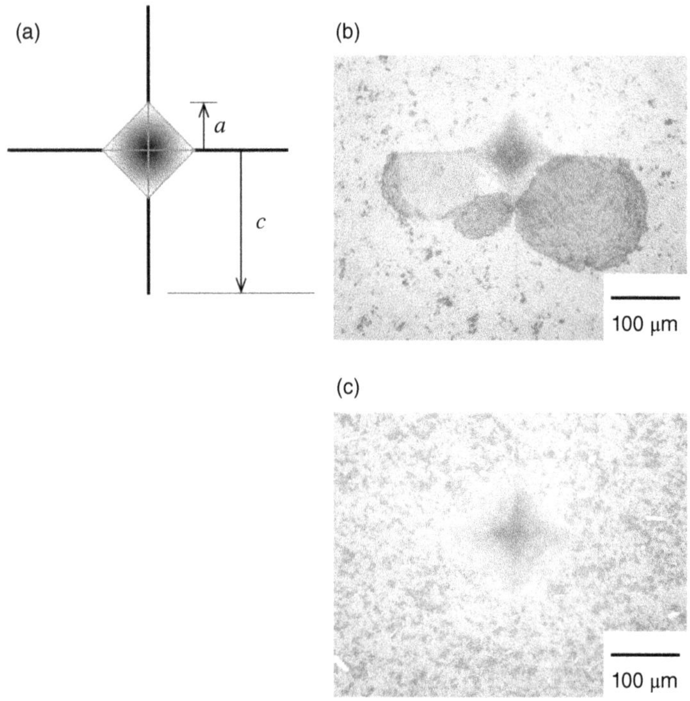

Figure 11.6 Vickers indentation flaws. (a) Schematic plan diagram. Optical micrographs of 100 N indentations in (b) blank cordierite glass-ceramic and (c) glass-ceramic containing 15 vol% Si_3N_4 whiskers. The smaller contact impression but more extensive cracks indicate that the blank glass-ceramic is harder but less tough than the composite.

Indentations were made on the polished disks in the load range P = 2–300 N and the resulting contact impressions measured by optical microscopy. The results are shown in Figure 11.7. Individual symbols represent the means of eight measurements at each load (uncertainty is smaller than the symbol size), and the lower solid lines are slope-1/2 best fits to the impression data consistent with Eq. (11.2). Over the load domain tested, all materials were described by constant hardness and the values are given in Table 11.1. (Here and following, values are given as means and standard deviations of experimental measurements, using propagation of variance as required.) Hardness decreased significantly on whisker addition and the decrease is probably explained by the same pore collapse mechanism suggested for the decreasing coefficient of restitution. The values are consistent with the idea that the variation of hardness with whisker volume fraction is almost completely dominated by material porosity.

The micrograph of the indentation in the blank glass-ceramic, Figure 11.6b, shows the two dominant crack systems generated at sharp contacts (Cook and Pharr 1990). Surface traces of cracks are visible emanating from the contact impression corners. These cracks propagate perpendicular to the material surface and control the maximum stress—the strength—sustainable by the indentation (Chantikul et al. 1981). Opposing surface traces most likely reflect a single crack system that is connected beneath the impression to form a semicircular half-penny crack, although at small loads this morphology is less common and indentations are typified by less connected surface-localized radial cracks. The curved traces and surfaces of lateral cracks are visible emanating from the contact impression edges. These cracks propagate approximately parallel to the material surface and are responsible for material removal—chipping—at the contact and exert a secondary influence on strength by modifying the indentation residual stress field (Cook and Roach 1986; Cook 2015).

The whisker composites exhibited a greater resistance to both half-penny and lateral cracking at contacts, as shown in the micrograph of Figure 11.6b. Both indentation crack initiation (whether a crack forms at a contact or not) and indentation cracking susceptibility (how long cracks are) were impeded in the composites. The increased resistance to indentation cracking was determined by measuring the lengths of surface traces of half-penny cracks c_{air} (from the contact impression center) as a function of indentation load P over the load range of 2–300 N. The results are shown in Figure 11.7, noting that cracks were not observable in the 15% composite at small indentation loads as a consequence of suppressed crack initiation. Individual symbols are the means of eight measurements at each load (uncertainty is smaller than the symbol size). Indentation cracking susceptibility is quantified by recognizing that the stress-intensity factor (SIF) K acing on the

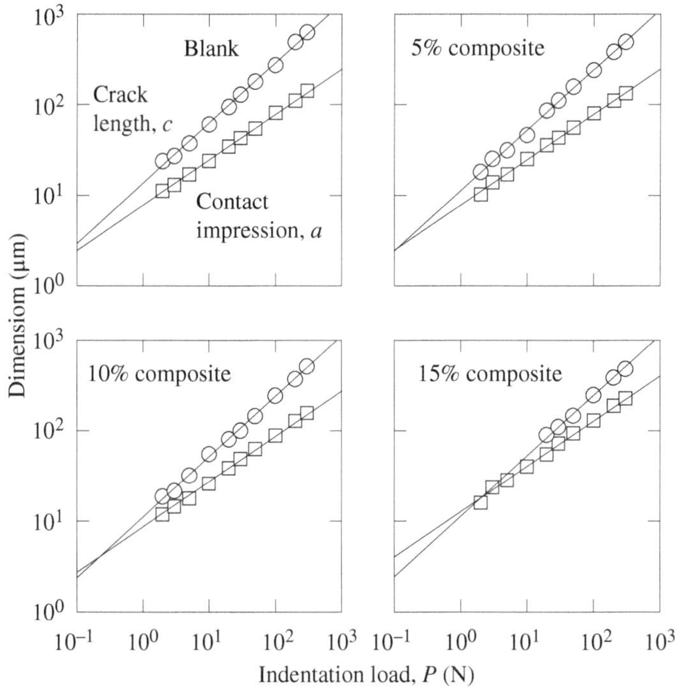

Figure 11.7 Plots of the indentation half-penny crack length and contact impression dimension as a function of indentation load for cordierite glass-ceramics containing silicon nitride whiskers. The solid lines are best fits to the data consistent with constant hardness and toughness, Eqs. (11.2) and (11.4).

indentation half-penny cracks is given by

$$K = \chi P/c^{3/2}, \tag{11.3}$$

where χ is a dimensionless load-invariant factor (see Chapter 2 for details). The equilibrium condition for the indentation cracks is $K = T_{air}$, where T_{air} is the toughness of the material in air. Imposing this equilibrium condition and inverting Eq. (11.3) gives (Cook 2020)

$$c_{air} = BP^{2/3}, \tag{11.4}$$

where B is the cracking susceptibility. Equation (11.4) shows that the equilibrium indentation crack length should increase as $P^{2/3}$ in a material of constant toughness, noting that the experimental measurable $B = (T_{air}/\chi)^{-2/3}$. More details are given below and in the analysis section. The parallel construction of Eqs. (11.1)–(11.2) and Eqs. (11.3)–(11.4) is noted.

The upper solid lines in Figure 11.7 are slope-2/3 best fits to the crack length data consistent with Eq. (11.4). Over the load domain tested, all materials were described by constant cracking susceptibility and the values are given in Table 11.1. The composites exhibited smaller values of B, consistent with greater, microstructure-invariant values of T_{air}/χ. To further investigate fracture behavior, the dependence of χ on the elastic and plastic properties of an indented material must be considered. χ is a dimensionless, material-dependent parameter characterizing the intensity of the residual stress field acting on indentation cracks. The residual field arises from the strain mismatch between the remnant plastic contact impression and the surrounding elastic matrix (Lawn and Cook 2012). The material dependence of χ is given by

$$\chi = \xi(E/H)^{1/2}, \tag{11.5}$$

where ξ is a dimensionless geometry term that is invariant for ideal indentations. Combining Eq. (11.5) into the expression for B gives

$$T_{air} = \xi(E/H)^{1/2}B^{-3/2}, \tag{11.6}$$

which permits an absolute estimate of T_{air} (Anstis et al. 1981; Cook 2020). Of perhaps greater use here is a relative estimate that allows for comparison of microstructural effects and avoids explicit knowledge of ξ:

$$T_{air}/T^0_{air} = [(E/H)^{1/2}B^{-3/2}]/[(E/H)^{1/2}B^{-3/2}]^0, \tag{11.7}$$

where X^0 indicates quantity X evaluated for the blank $V_f = 0\%$ material, noting that E, H, and B are given in Table 11.1.

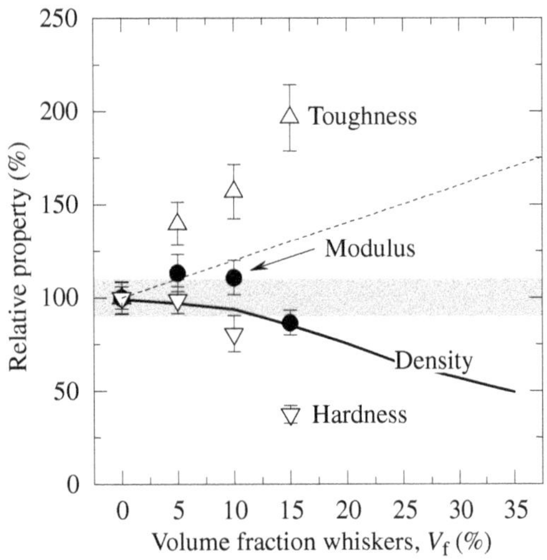

Figure 11.8 Plot of relative density and relative mechanical properties of cordierite glass-ceramic composites as a function of silicon nitride whisker volume fraction. The resistance to localized plastic deformation measured by the relative hardness depends critically on porosity of the composites. Conversely, the resistance to fracture measured by relative toughness depends on whisker addition. Relative modulus is influenced by both and exhibits little variation.

Using Eq. (11.7), Figure 11.8 plots T_{air}/T^0_{air} along with H/H^0, E/E^0, and ρ_r as a function of V_f for the blank and composite materials. As guides to the eye: the shaded band indicates no change in property relative to the base material, consistent with the average uncertainty; the upper dashed line indicates rule of mixtures change in property of a factor of 3 between cordierite and silicon nitride end members. As discussed above, the relative modulus E/E^0 does not vary much as compliant porosity and stiff whiskers offset. Relative hardness H/H^0 decreases with increase in V_f at a greater rate than does relative density. This observation implies that the *mechanism* of plastic deformation at contacts is altered in the composites by the addition of whiskers and the consequent porosity from that in dense cordierite glass-ceramic. As discussed, the tendency toward more foam-like deformation localized to the contact by pore collapse and crushing of whisker entanglements is most likely (Cook 2018a). Relative toughness T_{air}/T^0_{air} increases at a greater rate with increase in V_f than does the rule of mixtures estimate. This observation implies that the *mechanism* of fracture is altered in the composites from that in dense cordierite glass-ceramic. In particular, the addition of the whiskers toughens the composites beyond any deleterious effects of the associated porosity. The toughening probably derives from cohesive bridging zones formed behind the tips of propagating cracks (Cook 2015). The zones are associated with whisker sliding and friction caused by crack-opening displacements. This toughening mechanism is obviously not available to either monolithic cordierite or silicon nitride and hence the relative toughness exceeds a rule of mixtures estimate.

11.2.3 Toughness

Relative toughness and toughness variations can be estimated in greater detail from in situ observations of indentation crack extension under superimposed applied stress (Braun et al. 1992; Kovar et al. 2000). During the application of stress, indentation half-penny cracks extend stably from their initial length to a final length at the onset of instability and failure. Monitoring this stable extension is an effective method of gauging the relative toughness of materials.

Bars of each composition were indented at their centers ($P = 100$ N) and loaded in a stiff four-point bending fixture mounted on an optical microscope (Tandon and Cook 1993; Hughey et al. 2004). Loading was controlled by a stepping motor and monitored by a resistance load cell, and the indentation crack extension measured optically. Applied stress was calculated from the bending load using standard beam theory. Care was taken to increment the load so that crack velocities averaged over a few load increments were <0.01 μm s^{-1} and the system remained as close as possible to equilibrium. The experiments were performed in air so the toughness value was that appropriate to equilibrium in a moist environment, T_{air}. Two bars, four cracks, were measured for each composition except 15%, for which one bar was measured.

11.2.4 Strength

Inert strength tests of specimens containing dominant indentation flaws were performed on the composites. Such tests carry many advantages: the flaw size can be predetermined via the easily controlled indentation load and a wide range of flaw sizes across the length scale of interest can be used; the geometry of the flaws is known from previous studies (Cook and Pharr 1990) and the influence of lateral cracking in altering the flaw geometry detected in the strength response (Cook and Roach 1986); as strengths at a given flaw size can be compared between materials, toughness properties of materials can be compared (Chantikul et al. 1981) and the variation of toughness with flaw size inferred from the strength response (Cook 2015); and finally, comparison with the intrinsic (unindented) strength can be used to infer the intrinsic flaw size (Lawn et al. 1980).

Bar specimens were indented at their centers with the indentation diagonals aligned perpendicular to the bar axes. The specimens were then broken with the indented face in tension in a flexible four-point bend fixture (to prevent torque on the specimens). Rapid loading rates were used such that specimen failure occurred in ≈50 ms and thus reactive species (primarily water vapor) were not able to access the crack tip during propagation to failure. The fixture was instrumented with a piezoelectric load cell and peak, failure, loads were recorded electronically. Failure stress was calculated from peak load using standard beam theory. Approximately five specimens were broken at each indentation load in the range $P = 2\text{–}300$ N.

To gauge the intrinsic strengths, tests were performed on specimens containing no introduced indentation flaws. Polished disk specimens were used so that the failure distributions were representative of flaws intrinsic to the materials (and not an artifact of edge preparation). The disks were tested in biaxial flexure under inert conditions using a flat-on-three-ball test rig (Marshall 1980) instrumented with the piezoelectric load cell. Strength was determined from peak load using standard plate theory. Approximately 50 specimens for each of the 5%, 8%, 10%, and 15% composites were tested and compared with 142 specimens of the blank 0% material (Cook and DelRio 2019b).

11.3 Analysis Methods

The SIF arising from uniform stress σ_a applied to a crack of length c is

$$K = \psi \sigma_a c^{1/2}, \tag{11.8}$$

where ψ is a dimensionless, stress-invariant, geometrical factor. The net SIF acting on a stressed indentation half-penny crack is then the sum of Eqs. (11.3) and (11.8):

$$K = \chi P/c^{3/2} + \psi \sigma_a c^{1/2}. \tag{11.9}$$

Equation (11.9) is the basis for the toughness and strength analyses that follow, in both reactive (moist) air and inert environments.

11.3.1 Toughness Variation

If the equilibrium condition $K = T_{air}$ is imposed for a crack in air loaded as described by Eq. (11.9), measurement of a $\sigma_a(c)$ trajectory for a specified P enables determination of material toughness variation $T_{air}(c)$ (Braun et al. 1992; Kovar et al. 2000). Such a determination depends on knowledge of the geometrical factors χ and ψ. The evaluation of these factors to within a common multiplicative factor may be performed through an experiment on a material with single valued toughness (which does not need to be known numerically). Here, the base glass-ceramic was used for geometrical factor evaluation, assuming invariant toughness in air $T_{air} = T^0_{air}$—an assumption supported here by Figure 11.7. Imposing the equilibrium condition for this material and rearranging Eq. (11.9) gives

$$(\psi/T^0_{air})\sigma_a c^{1/2} = 1 - (\chi/T^0_{air})P/c^{3/2}. \tag{11.10}$$

Simple linear fitting of the experimental quantities $P/c^{3/2}$ and $\sigma_a c^{1/2}$ suggested by Eq. (11.10) thus provides the related calibration factors. Here, $(\psi/T^0_{air}) = (1/1.12)$ and $(\chi/T^0_{air}) = (1/25.67)$ were determined, both with dimensions of $(\text{MPa m}^{1/2})^{-1}$. For materials with crack geometry assumed identical to that of the calibration material, these factors enable determination of relative toughness. Combining Eqs. (11.5) and (11.9) and imposing equilibrium gives

$$T_{air}/T^0_{air} = (\psi/T^0_{air})\sigma_a c^{1/2} + (\chi/\chi^0)(\chi/T^0_{air})P/c^{3/2}, \tag{11.11}$$

where the relative residual stress term (χ/χ^0) is

$$(\chi/\chi^0) = (E/H)^{1/2}/(E^0/H^0)^{1/2}. \tag{11.12}$$

It is noted that Eq. (11.11) is an extension of Eq. (11.7) for nonzero applied stress and is identical to Eq. (11.7) for $\sigma_a = 0$.

11.3.2 Indentation Strength Variation

If the equilibrium condition $K = T_{inert}$ is imposed for a crack described by Eq. (11.9) in an inert environment, the maximum sustainable applied stress is

$$\sigma_m = 3T_{inert}^{4/3}/4^{4/3}\psi(\chi P)^{4/3}. \tag{11.13}$$

The stress σ_m is the strength of a component containing a dominant indentation flaw and thus depends on P. Equation (11.13) pertains for ideal indentations at moderate indentation loads and can be reexpressed similarly to Eq. (11.4) as

$$\sigma_m = SP^{-2/3}, \tag{11.14}$$

where S is an experimentally determined parameter with dimensions of MPa $N^{1/3}$ (introduced as $\hat{S}$ in Chapter 3; for simplicity the caret is omitted here). Equation (11.14) shows that the inert indentation strength σ_m should decrease as $P^{-1/3}$ in a material of constant toughness, noting that the experimental measurable $S = 3(T_{inert}/4)^{4/3}/\psi\chi^{1/3}$. Combining Eq. (11.5) into the expression for S gives

$$T_{inert} = \eta(E/H)^{1/8}S^{3/4}, \tag{11.15}$$

permitting an absolute estimate of T_{inert}, where η is a dimensionless geometry term that is invariant for ideal indentations (Chantikul et al. 1981). Of perhaps greater use here is a relative estimate that allows comparison of microstructural effects and avoids explicit knowledge of η:

$$T_{inert}/T_{inert}^0 = [(E/H)^{1/8}S^{3/4}]/[(E/H)^{1/8}S^{3/4}]^0. \tag{11.16}$$

similar to Eqs. (11.7) and (11.11). At large indentation loads, lateral cracking as in Figure 11.6 may relax χ thereby increasing the strength (Cook and Roach, 1986; Cook 2015)—this effect was small here and is not considered in detail. At small indentation loads, flaws intrinsic to the material become dominant and strength controlling, leading to an upper bound on strength σ_f. The intrinsic flaws can be defined by a characteristic contact load P_f using Eq. (11.14)

$$\sigma_f = SP_f^{-1/3}. \tag{11.17}$$

The inert strength over the complete large-to-small indentation load domain (indentation-to-intrinsic strength range) can be described by a smooth p-norm function as in an earlier analysis (Cook and DelRio 2018b). Using an empirical exponent $\alpha < -1$ to connect the strengths by

$$\sigma^\alpha = \sigma_m^\alpha + \sigma_f^\alpha, \tag{11.18}$$

and combining Eqs. (11.14), (11.17), and (11.18) gives

$$\sigma = \sigma_f[1 + (P/P_f)^{-\alpha/3}]^{1/\alpha}, \tag{11.19}$$

Equation (11.19) describes the full experimental strength range, once the limiting experimental asymptotic values of S and σ_f are determined. (σ_i and P_i were used to characterize intrinsic strengths in Chapters 3 and 4; subscript "f" is used here to avoid confusion with i introduced in Section 11.3.3. Setting $\sigma_f = \sigma^*$, $P_f = P^*$, and $\alpha = -1$ generates the indentation strength function identical to that used extensively in describing microstructural effects in Chapter 4. This completely different interpretation of indentation strength variation is considered in the Appendix.)

11.3.3 Intrinsic Strength Variation

The intrinsic strength data were analyzed as detailed in recent works considering strength variations and distributions of silicon components and nanowires (Cook and DelRio 2019a; Cook et al. 2019), ceramics (Cook and DelRio 2019b), indentation crack lengths (Cook 2020), and particles (Cook 2023). A brief account of application of the analysis is given here.

An empirical distribution function (edf) of intrinsic strengths, $Pr(\sigma)$, was formed for each material. The strengths were ranked from smallest to largest, $\sigma(i)$, where i is an index that spans the domain $i = 1$ to $i = N$, where N was the number of intrinsic strength measurements in the sample group for each material. The quantity $(i - 0.5)/N$ was formed for each index value and the discrete parametric function of the pairing $Pr(\sigma) = [\sigma(i), (i - 0.5)/N]$ formed the edf for each material.

A smoothing function was fit visually to each edf by the continuous function $H(\mu)$. The smoothing function is based on a nonlinear perturbation of an incomplete Beta function (Cook and DelRio 2019b) and is given by

$$H(\mu) = 30(\mu^{3p}/3 - \mu^{4p}/2 + \mu^{5p}/5), \quad (11.20)$$

where μ is a relative strength parameter,

$$\mu = (\sigma - \sigma_{\min})/(\sigma_{\max} - \sigma_{\min}).$$

Equation (11.20) describes a sigmoidal variation from 0 to 1 between empirical strength fitting bounds of $\sigma_{\min}$ and $\sigma_{\max}$; p is an empirical fitting factor of order unity controlling sigmoid symmetry. $H(\mu)$ is an estimator of the cumulative distribution function (cdf) of strengths $H(\sigma)$ set by the population of flaws, the sampling procedure, and the specimen size (for details see works cited above).

The strength-controlling intrinsic flaw distributions in the samples are determined from the strength measurements through specification of $\sigma(c)$. Here, the Griffith relation,

$$\sigma = Dc^{-1/2}, \quad (11.21)$$

was used, where D is a function of material fracture resistance and intrinsic crack geometry (D is similar to T/ψ, Chapter 1). A key feature of Eq. (11.21) is that $\mathrm{d}\sigma/\mathrm{d}c < 0$, i.e., strength decreases as flaw size increases. Other, more complicated, strength-flaw size relations could be selected (e.g., related to cracks at pores, Chao and Shetty 1992), but all would be characterized by this negative derivative. As a consequence of the derivative, the sample crack length cdf, $H(c)$, is related to the sample strength cdf by $H(c) = 1 - H(\sigma)$. The crack length probability density function (pdf), $h(c)$, is related to the cdf by

$$h(c) = \mathrm{d}H(c)/\mathrm{d}c. \quad (11.22)$$

The determination of $h(c)$ for each material enables comparison of the strength-controlling flaws, provided values for D are known. Here, D/D^0 was taken to vary as $T_{\mathrm{inert}}/T^0_{\mathrm{inert}}$.

11.4 Results

11.4.1 Toughness Measurements

Figure 11.9 shows the measured applied stress–crack length responses for indentation cracks in air for each composition. Symbols represent individual stress–crack length observations. The solid line represents the best fit to the blank glass-ceramic data as described by Eq. (11.10) and the stated geometrical factors and is reproduced for reference in each subplot. In all cases, there was stable crack extension before failure. The stress to cause failure increased with whisker volume fraction and the stable extension, although less for the 5 and 10% materials, was considerably greater for the 15% material. The enhanced failure stresses and increased extension are both indicative of toughening. However, direct evidence of the ability of the whiskers to impede crack propagation was observed during stressing of the composites: frequently the crack tip would stop movement and become blocked at a whisker for many load increments before passing through or around the whisker and extending a considerable distance into the ceramic matrix in a single increment. Often these erratic crack tip motions would be associated with discrete changes in the crack-opening displacement near whiskers behind the crack tip, supporting the idea that frictional interlocks behind the tip were influencing overall crack behavior. The "staircase" motion in the stress-extension responses of the composites indicates this erratic crack tip behavior. Observations on other sets of indentation cracks were similar.

Figure 11.10 is a plot of the relative toughness of the composite materials using the data from Figure 11.9 and Eqs. (11.11) and (11.12). Although there is some scatter, it is clear that the toughness of the composites is greater than that of the blank glass-ceramic and increases with increasing volume fraction of whiskers. It is also clear that the toughness does not increase dramatically for cracks longer than approximately 200 μm, consistent with the length scale for the onset of steady-state toughness estimated from the indentation strength data, approximately 20–50 μm, as shown in Figure 11.7. The relative toughness values are also consistent with those inferred from the indentation crack length data. The values determined from the direct long-crack measurements in Figure 11.10 are 1.35 (5% composite), 1.6 (10%), and 1.9 (15%), consistent with those from the static indentation short-crack measurements in Figure 11.8.

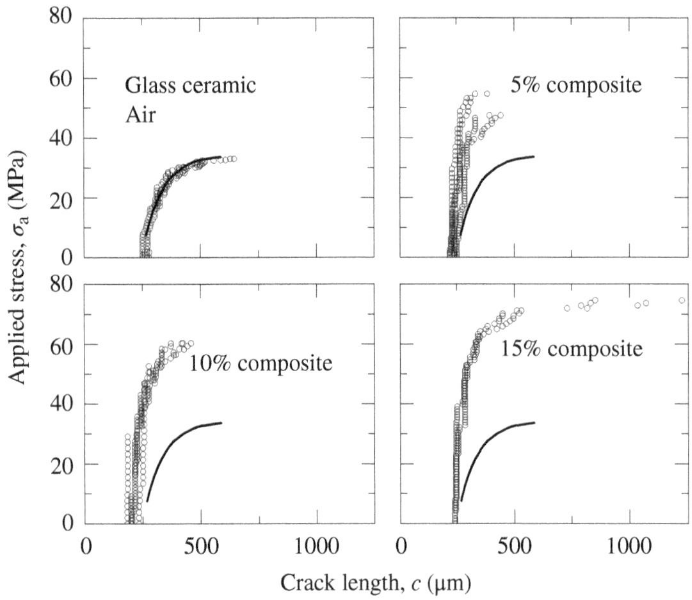

Figure 11.9 Plots of equilibrium indentation crack length under the action of applied stress for glass-ceramic silicon nitride whisker composites in air. All materials exhibit stable crack extension, and the enhanced composite responses indicate whisker toughening.

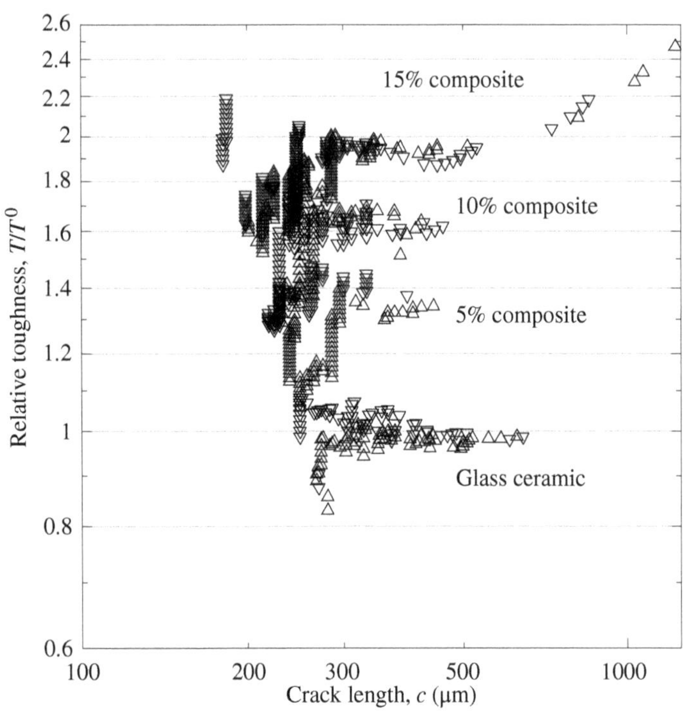

Figure 11.10 Plot of relative toughness variation with crack length for glass-ceramic silicon nitride whisker composites in air. Toughness increases significantly with whisker volume fraction (indicated) at long crack lengths and exhibits considerable scatter at short crack lengths.

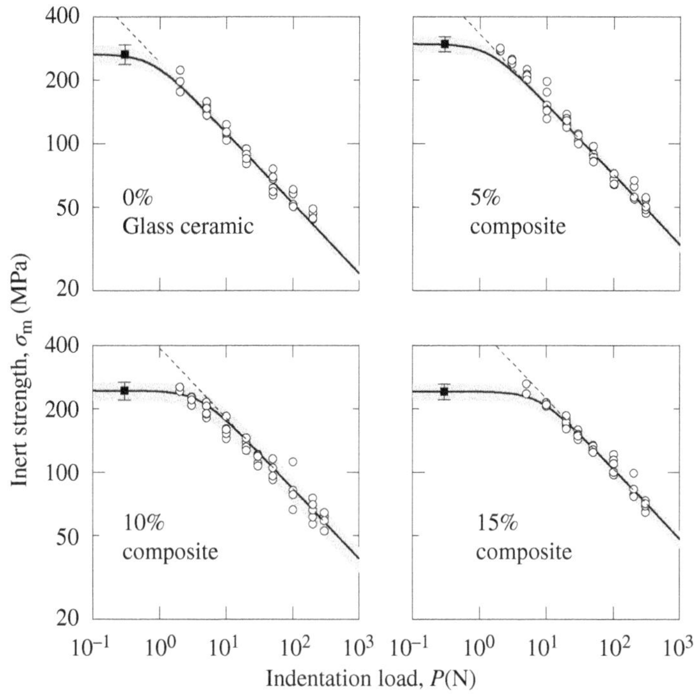

Figure 11.11 Plots of indentation (open symbols) and intrinsic (solid symbols) strengths of glass-ceramic silicon nitride whisker composites as a function of indentation load. Lines and bands represent best fits consistent with indentation flaw dominance at large loads and intrinsic flaw dominance at small loads.

11.4.2 Strength Measurements

Figure 11.11 is a plot of inert indentation and intrinsic strengths of the materials. Open symbols represent individual indentation strength measurements and, as expected, decrease with increasing indentation load and flaw size. The dashed lines represent best fits to the data consistent with Eq. (11.14). For the blank glass-ceramic, the fit was performed for $P < 100$ N (at loads greater than 100 N, the strength data show a slight increase above the $P^{-1/3}$ dependence, indicative of lateral crack influences that reduce the residual stress field, Cook and Roach 1986; Cook 2015). Over this load range, the data for the blank material are well described by Eq. (11.14), consistent with constant toughness (Chantikul et al. 1981). For the composites, the fits were performed for $P > 100$ N (at loads less than 100 N, the strength data show slight decreases below the $P^{-1/3}$ dependence, indicative of small toughness decreases, Cook 2015). Over this load range, the data for the composites are also well described by Eq. (11.14), consistent with constant toughness. Table 11.2 gives the means and standard deviations of the best fit S parameters describing the dashed lines. Solid symbols represent the means and standard deviations of the intrinsic strengths. Table 11.2 gives the means and standard deviations of the σ_f and P_f parameters describing the solid symbols. Using these values, the solid lines in Figure 11.11 represent mean best fits to the entire data ranges of strengths using Eq. (11.19) and the shaded bands represent standard deviations of these means.

Table 11.2 Strength and toughness properties of glass-ceramic whisker composites.

Material and Property					
Volume Fraction of Whiskers, V_f (%)	0	5	8	10	15
Indentation Strength Parameter, S (MPa N$^{1/3}$)	242 ± 19	331 ± 26	—	389 ± 51	482 ± 42
Intrinsic Strength, σ_f (MPa)	266 ± 29	297 ± 25	268 ± 26	243 ± 24	241 ± 21
Intrinsic Load, P_f (N)	0.75 ± 0.30	1.38 ± 0.48	—	4.1 ± 2.0	8.0 ± 2.9
Toughness, T_{inert} (MPa m$^{1/2}$)	1.40 ± 0.08	1.81 ± 0.11	≈1.95	2.09 ± 0.21	2.62 ± 0.18

Two major features appear in Figure 11.11. The first is that the indentation strengths, and therefore associated S parameters, Table 11.2, increase with increasing volume fraction of whiskers, indicating increasing toughening. Using the values in Tables 11.1 and 11.2 gives the relative inert toughness of the composites $T_{\text{inert}}/T^0_{\text{inert}}$ from Eq. (11.16) as 1.3 (5% composite), 1.5 (10%), and 1.9 (15%), consistent with those from the reactive environment short-crack measurements of Figure 11.8 and the long-crack measurements of Figure 11.10. The absolute values of the toughness T_{inert} can be approximated by a characteristic value of $\eta = 0.61$ (Cook et al. 1987) leading to the toughness values shown in Table 11.2. The second major feature in Figure 11.11 is that the intrinsic strengths σ_f do not vary appreciably with the volume fraction of whiskers. The combination of the two features leads to a significant increase in the characteristic indentation load P_f with whisker volume fraction. As P_f marks the transition from indentation flaw control to intrinsic flaw control, the increase suggests that additional whiskers increase both the toughness and the intrinsic flaw size in the composites.

The variation of the intrinsic flaw sizes in the materials is investigated further through analysis of the intrinsic strength distributions. Figure 11.12 shows plots of edf measurements for composite materials. The symbols represent individual strength measurements. The solid lines represent best fits to the measurements using Eq. (11.20). The dashed line represents a best fit to measurements on the blank glass-ceramic (Cook et al. 2019) and is reproduced as a reference in each subplot. The distributions exhibit sigmoidal behavior and, as anticipated from Figure 11.11, vary little with whisker volume fraction. The pdf describing the flaws underlying the strength variations of Figure 11.12 were determined using the above analysis and identifying D with the values given in Table 11.2. Figure 11.13 shows the resulting crack size pdf (interpolating the 8% D value as 1.95 MPa m$^{1/2}$). Consistent with the single-valued characterization, the more detailed assessment of flaw size distribution in Figure 11.13 shows that the flaw sizes controlling intrinsic strengths increase significantly with increasing whisker volume fraction. In all cases, the inferred flaw size distributions extend about a factor of 2 from the smallest to largest flaw and are skewed toward larger flaws. In the case of blank glass-ceramic, the flaws are probably surface machining or polishing flaws. In the case of the composites, the flaws are probably associated with the bulk pores shown in Figures 11.3 and 11.4. The flaw sizes in Figure 11.13 are probably upper bounds for two reasons: The increase in toughness with crack length in the composites was not taken into account and the intensifying influence on applied stress adjacent to pores was not taken into account. Nevertheless, the sizes of the flaws inferred from self-consistent application of indentation and intrinsic strength measurements are comparable to those observed.

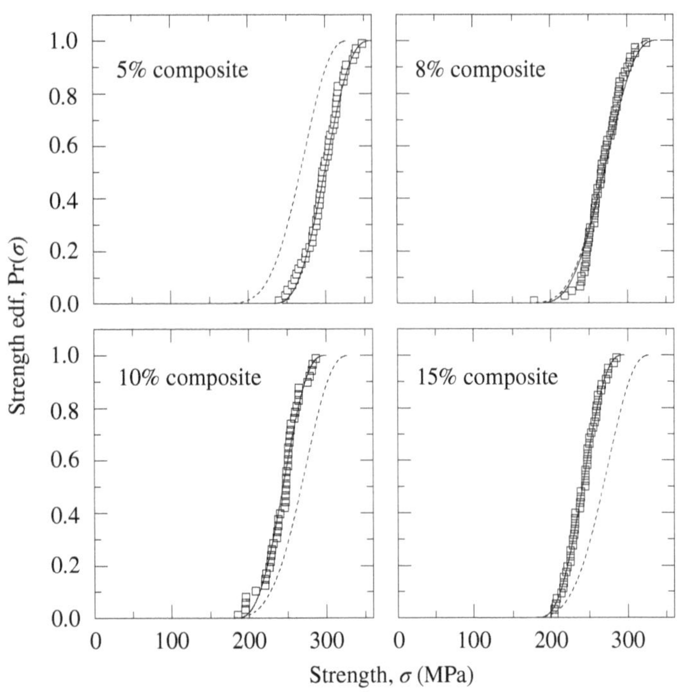

Figure 11.12 Plots of intrinsic strength edf measurements for glass-ceramic silicon nitride whisker composites. Dashed line shows response of blank glass-ceramic. Strength varies little with whisker content.

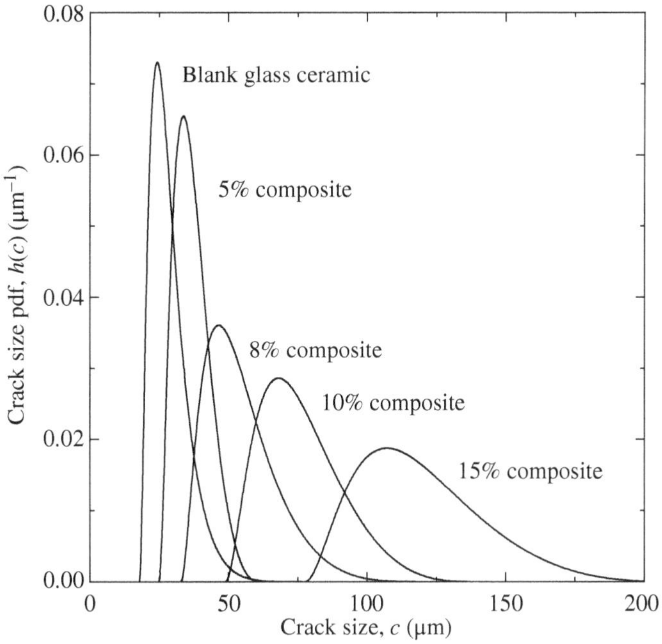

Figure 11.13 Plot of intrinsic flaw pdf responses for glass-ceramic silicon nitride whisker composites. Flaw size increases significantly with whisker content.

11.5 Conclusions

Composites of cordierite glass-ceramics containing silicon nitride whiskers were fabricated by free sintering. The introduction of whiskers retarded the densification of the composites such that for whisker volume fractions greater than 15 vol%, significant open porosity along with entangled whisker agglomerates were generated. Material properties that depended on the composite behavior as a whole were well described by volume-averaged estimates. In particular, Young's modulus was invariant and exhibited a trade-off with whisker addition between the influence of stiff whiskers and the counteracting effects of increased porosity generated by the whiskers on sintering. Material properties that depended on the local response of a constituent element and for which the response mechanism was altered (hardness, toughness, and strength) were not well described by volume-based estimates. Hardness exhibited a significant decrease on whisker addition and was dominated by the enhanced porosity. Toughness exhibited a significant increase on whisker addition, although this increase was not reflected in the strength. In each case, the composites responded differently from the bulk constituents: pore collapse controlled the contact deformation leading to the softening; whisker pullout controlled the crack propagation leading to the toughening; and, porous whisker entanglements increased the dominant flaw size, counteracting the toughening, leading to an invariance in the strength with the volume fraction of whiskers added. Quantitatively, toughness increased monotonically with whisker volume fraction to give a steady-state toughness at 15 vol% of approximately 2.6 MPa $m^{1/2}$, compared with that of the blank glass-ceramic of 1.4 MPa $m^{1/2}$.

References

Anstis, G.R., Chantikul, P., Lawn, B.R., and Marshall, D.B. (1981). A critical evaluation of indentation techniques for measuring fracture toughness: I, direct crack measurements. *Journal of the American Ceramic Society* 64: 533–538.

Braun, L.M., Bennison, S.J., and Lawn, B.R. (1992). Objective evaluation of short-crack toughness curves using indentation flaws: Case study on alumina-based ceramics. *Journal of the American Ceramic Society* 75: 3049–3057.

Chantikul, P., Anstis, G.R., Lawn, B.R., and Marshall, D.B. (1981). A critical evaluation of indentation techniques for measuring fracture toughness: II, strength method. *Journal of the American Ceramic Society* 64: 539–543.

Chao, L.Y. and Shetty, D.K. (1992). Extreme-value statistics analysis of fracture strengths of a sintered silicon nitride failing from pores. *Journal of the American Ceramic Society* 75: 2116–2124.

Cook, R.F. (2015). Multi-scale effects in the strength of ceramics. *Journal of the American Ceramic Society* 98: 2933–2947.

Cook, R.F. (2018a). Model for instrumented indentation of brittle open-cell foam. *MRS Communications* 8: 1267–1273.

Cook, R.F. (2018b). A simple method for short-term mechanical reliability predictions for ceramics in reactive environments. *Journal of the American Ceramic Society* 101: 2727–2731.

Cook, R.F. (2020). A critical evaluation of indentation crack lengths in air. *Journal of the American Ceramic Society* 103: 2278–2295.

Cook, R.F. (2023). *Particle Strengths.* Wiley.

Cook, R.F. and DelRio, F.W. (2019a). Material flaw populations and component strength distributions in the context of the Weibull function. *Experimental Mechanics* 59: 279–293.

Cook, R.F. and DelRio, F.W. (2019b). Determination of ceramic flaw populations from component strengths. *Journal of the American Ceramic Society* 102: 4794–4808.

Cook, R.F., DelRio, F.W., and Boyce, B.L. (2019). Predicting strength distributions of MEMS structures using flaw size and spatial density. *Microsystems & Nanoengineering* 5: 49.

Cook, R.F., Fairbanks, C.J., Lawn, B.R., and Mai Y.-W. (1987). Crack resistance by interfacial bridging: its role in determining strength characteristics. *Journal of Materials Research* 2: 345–356.

Cook, R.F. and Lawn, B.R. (1983). A modified indentation toughness technique. *Journal of the American Ceramic Society* 66: C200–C201.

Cook, R.F. and Pharr, G.M. (1990). Direct observation and analysis of indentation cracking in glasses and ceramics. *Journal of the American Ceramic Society* 73: 787–817.

Cook, R.F. and Roach, D.H. (1986). The effect of lateral crack growth on the strength of contact flaws in brittle materials. *Journal of Materials Research* 1: 589–600.

DelRio, F.W., Boyce, B.L., Benzing, J.T., Friedman, L.H., and Cook, R.F. (2020). Shoulder fillet effects in strength distributions of microelectromechanical system components. *Journal of Micromechanics and Microengineering* 30: 125013.

Homeny, J., Neergaard, L.J., Karasek, K.R., Donner, J.T., and Bradley, S.A. (1990). Characterisation of β–silicon nitride whiskers. *Journal of the American Ceramic Society* 73: 102–105.

Hughey, M.P., Morris, D.J., Cook, R.F., Bozeman, S.P., Kelly, B.L., Chakravarty, S.L.N., Harkens, D.P., and Stearns, L.C. (2004). Four-point bend adhesion measurements of copper and permalloy systems. *Engineering Fracture Mechanics* 71/2: 245–261.

Knickerbocker, J.U. (1992). Overview of the glass ceramic/copper substrate-A high performance multilayer package for the 1990s. *American Ceramic Society Bulletin* 71: 1393–1401.

Knickerbocker, S.H., Kumar, A.H., and Herron, L.W. (1993). Cordierite glass ceramics for multilayer ceramic packaging. *American Ceramic Society Bulletin* 72: 90–95.

Kovar, D., Bennison, S.J., and Readey, M.J. (2000). Crack stability and strength variability in alumina ceramics with rising toughness-curve behavior. *Acta Materialia* 48: 565–578.

Lawn, B.R. and Cook, R.F. (2012). Probing material properties with sharp indenters: A retrospective. *Journal of Materials Science* 47: 1–22.

Lawn, B.R., Marshall, D.B., Chantikul, P., and Anstis, G.R. (1980). Indentation fracture: applications in the assessment of strength of materials. *Journal of the Australian Ceramic Society* 16: 4–9.

Marshall, D.B. (1980). An improved biaxial flexure test for ceramics. *American Ceramic Society Bulletin* 59: 551–553.

Marshall, D.B., Cook, R.F., Padture, N., Oyen, M.L., Pajares, A., Bradby, J., Reimanis, I., Tandon, R., Page, T., Pharr, G.M., and Lawn, B.R. (2015). The case for indentation as an enduring and exploratory characterization tool. *Journal of the American Ceramic Society* 98: 2671–2680.

Tabor, D. (1951) *The Hardness of Metals.* Oxford.

Tandon, R. and Cook, R.F. (1993). Indentation crack initiation and propagation in tempered glass. *Journal of the American Ceramic Society* 76: 85–889.

Thurn, J., Morris, D.J., and Cook, R.F. (2002). Depth-sensing indentation at macroscopic dimensions. *Journal of Materials Research* 17: 2679–2690.

Tummala, R.R. (1991). Ceramic and glass ceramic packaging in the 1990s. *Journal of the American Ceramic Society* 74: 895–908.

Appendix

The indentation fracture properties of a series of glass-ceramic + particle composites are described in this Appendix, providing a contrasting microstructural effect to the glass-ceramic + whisker composites considered in the main chapter. The glass-ceramic matrix was identical to that used for the whisker composites. Volume fractions between 0.05 and 0.13 of Cr_2O_3 particles with median sizes of 5 and 11 μm were mixed into the matrix material prior to sintering. The tape casting, disk specimen formation, and sintering procedures for the particle composites were identical to those used for the whisker composites. Finished densities varied between 93% and 99% of the theoretical value for each composition. Mean free paths between particles in the matrix varied from 5 to 16 μm. An example microstructure and indentation crack for the composite is shown in Figure 11.A1; the composite contains 0.07 by volume of particles and is 98% theoretical density. The matrix is gray, the bright areas are particles, and dark areas are pores. A 50 N indentation crack proceeds right to left. The paths of cracks in the composites were noticeably deflected relative to those in the base glass-ceramic.

Figure 11.A2 shows a logarithmic plot of the Vickers indentation contact dimensions, impression semidiagonal a and crack length c_{air}, for the composites. Symbols represent the mean values for several indentations at each indicated contact

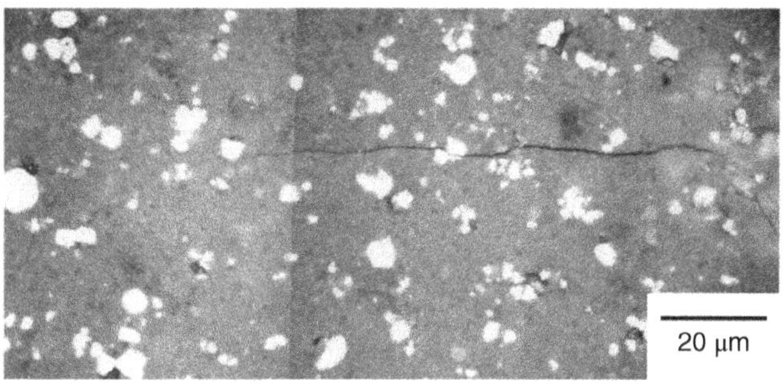

Figure 11.A1 Optical microscope image of a polished surface of ceramic composite consisting of cordierite glass ceramic matrix (gray) containing Cr_2O_3 particles (bright) and pores (dark). An indentation crack proceeds through the microstructure (right to left). The crack path in the composite is deflected relative those in the base glass ceramic (Figures 2.6a and 2.21).

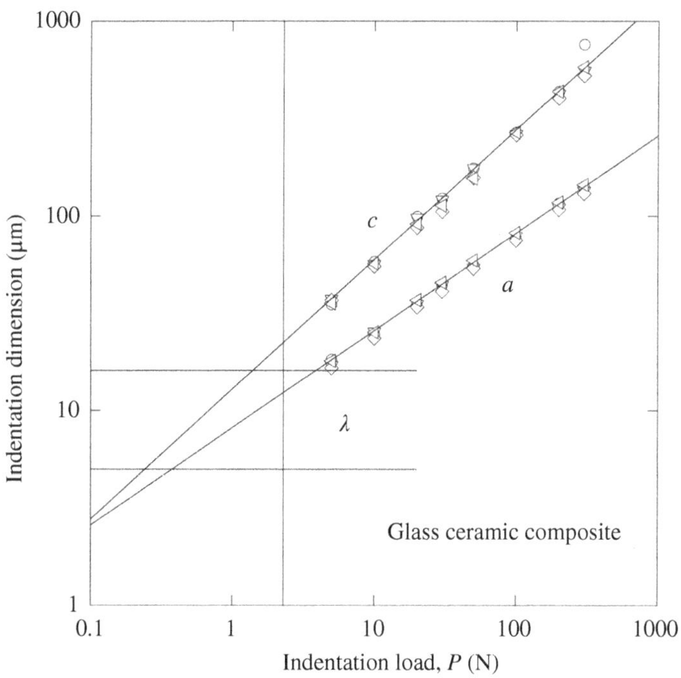

Figure 11.A2 Logarithmic plot of contact dimensions for the cordierite–Cr_2O_3 composites. Horizontal lines indicate the range of microstructural scales λ. Vertical line indicates indentation load P^* characterizing the onset of microstructural effects. Diagonal lines represent the response of the base material.

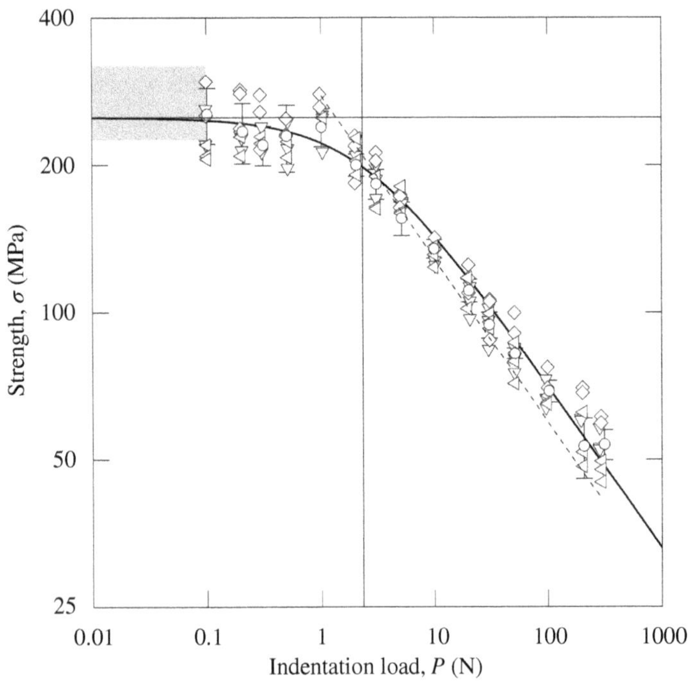

Figure 11.A3 Logarithmic plot of indentation strength behavior for the cordierite–Cr_2O_3 composites. Different symbols represent mean strengths for different materials. Shaded band represents intrinsic strength. Solid line indicates best fit to the data taking microstructural effects into account.

load P. Different symbols represent different composites. The diagonal lines represent the $P^{1/2}$ and $P^{2/3}$ responses of the base glass-ceramic. For the domain of indentation loads used, the indentation impression dimensions and crack lengths were not significantly different from those of the base material and exhibited no effect of changes in microstructure. The horizontal lines represent the bounds of the characteristic scales λ in the microstructure, from particle size to interparticle spacing. It is clear that the indentation cracks generated were in most cases considerably larger than these length scales and thus might be expected to reflect the saturated effects of any microstructural interactions. In any event, indentation crack lengths nearly always exhibit $P^{2/3}$ behavior and weak microstructural sensitivity (Cook 2020).

Figure 11.A3 shows a logarithmic plot of the indentation strength σ of the composites tested in biaxial flexure. Every composite specimen was inspected visually after strength measurement to verify indentation failure. Symbols with bars represent the means and standard deviation limits of at least five strength measurements at the indentation loads P indicated. Individual symbols represent the mean values for three to five strength measurements at each indicated contact load. Different symbols represent different composites. No significant variation of indentation strength with microstructure is apparent. The shaded band represents the mean and standard deviation limits, (272 ± 42) MPa, of the measured intrinsic strengths for the combined measurements for all composites. No significant variation of intrinsic strength with microstructure was observed. The dashed line indicates the ideal $P^{-1/3}$ behavior extended over the domain of observations for the base glass-ceramic material from Figure 3.2c. The bold solid line indicates a visual best fit to the strength data of the microstructurally influenced $\sigma(P)$ strength response of Eq. (4.24),

$$\sigma = \sigma^*(1 + P/P^*)^{-1/3}.$$

The value of $P^* = 2.3$ N is indicated by the vertical line and repeated in Figure 11.A2. The horizontal line in Figure 11.A3 indicates the small P asymptotic strength response of $\sigma^* = 250$ MPa. For small indentation loads $P \leq P^*$, the indentation strength is nearly invariant and approaches this strength level. As noted in Chapter 4, materials that exhibit indentation strengths with clear microstructural influences also exhibit characteristic strength values σ^* that well approximate intrinsic strengths, as observed here.

Note that the strength equation above, Eq. (4.24), is very different in interpretation from the similar function Eq. (11.19) used to describe the strength of the whisker composites. Equation (11.19) is an empirical heuristic expression used to interpolate between indentation strengths and intrinsic strengths. The physical interpretation of Eq. (11.19) is that the ideal indentation strength response is truncated by an independent intrinsic strength level. There are no indentation strength data leading to the truncated strength level. This was the case in Figure 11.11. Equation (4.24) is a physically based expression that describes the variation of indentation strengths over the full indentation domain. The interpretation of Eq. (4.24) is that the ideal indentation strength response is perturbed by microstructural effects. Indentation strength observations extend to the intrinsic strength, as shown in Figure 11.A3. If intrinsic strengths are determined by microstructurally controlled flaws, the low load indentation strengths encompass the intrinsic strengths. This is the case in Figure 11.A3.

Insight into the origin of the microstructural effects in the particle composites is gained by noting that the Cr_2O_3 particles have a greater elastic modulus and greater coefficient of thermal expansion than the cordierite matrix. On cooling from the sintering temperature, the particles thus develop significant, several hundred MPa, hydrostatic tension. The stress field in the matrix adjacent to a particle is one of radial tension and circumferential compression of the same magnitude, although equilibrium requires a mean stress level in the matrix of several tens of MPa of compression. The implication from the strength data is that at long crack lengths the composite is weakly toughened by the near zero global compressive stress on the crack plane. At short crack lengths, the toughness of the composite decreases as significant local tensile stresses act on a crack—recognizing that in biaxial flexure that any of the four indentation half-penny cracks may cause failure.

12

Crack Propagation in Toughened Materials

In this chapter indentation crack propagation under the action of applied stress is studied in three brittle material systems that exhibit toughening: polycrystalline alumina, phase transforming zirconia, and surface stressed ion exchanged glass. Based on the ideal indentation crack extension behavior established in Chapter 3, perturbed crack extension analyses for toughened materials are developed. The variations include changes to residual field and crack geometry, similar and dissimilar crack driving and resistance fields, and piecewise continuous resistance fields. The analyses provide the foundation for interpreting the experimental observations, which include extended and contracted crack advance, erratic crack advance, and trapped cracks.

12.1 Introduction

This chapter considers indentation crack propagation in brittle materials that exhibit crack length-dependent toughening. Such toughening arises from the dissipation or redirection of energy by interactions between a propagating crack and material microstructures or component structures. The effects of toughening on indentation fracture were considered in detail in Chapter 4 with an emphasis on component failure and indentation strengths, $\sigma_m(P)$, where σ_m is the strength of a component containing an indentation flaw formed by contact load P. In this chapter, the emphasis is on indentation crack propagation trajectories prior to failure, $\sigma_a(c)$, where σ_a is the applied stress and c is the indentation crack length. The applied stress is bounded by $0 \leq \sigma_a \leq \sigma_m$, and the crack length is bounded by $c_{air} \leq c \leq c_m$, where c_{air} is the initial crack length formed on indentation and c_m is the final crack length. The ratio c_m/c_{air} is a key parameter in indentation crack propagation measurements. Observations and measurements of $\sigma_a(c)$ were described and analyzed in earlier Chapters 3, 5, and 11 in this book. In Chapter 3, $\sigma_a(c)$ behavior was examined for materials of invariant toughness. Images of crack propagation in cordierite glass-ceramic and silicate glass were shown, and the near-ideal indentation crack propagation behavior of cordierite and silicon nitride was demonstrated. The behavior of cracks at residual stress-free annealed indentations was also examined, along with the behavior of partially stress relaxed indentations in glass and silicon that exhibited the pervasive phenomenon of lateral cracking. In Chapter 5, observations and measurements of the end point of strength tests (σ_m, c_m) were described and analyzed for a range of materials, leading to quantitative estimates of material toughness, including variations of toughness with crack length. An important finding of this chapter was that the average ratio c_m/c_{air} for the range of materials examined was ≈ 1.34, significantly less than the ideal value of 2.52, although there was considerable material-to-material variation. In Chapter 11, measurements of $\sigma_a(c)$ were described and analyzed for whisker-reinforced cordierite glass-ceramic composite materials, demonstrating almost invariant toughness that correlated well with indentation strength measurements.

Here, indentation crack propagation $\sigma_a(c)$ behavior is considered for three toughened materials: polycrystalline alumina, tetragonal zirconia polycrystals (TZP), and ion-exchanged glass. The indentation strength behaviors were considered in Chapters 3 and 4. The toughening mechanisms include crack wake toughening (alumina), crack front toughening (zirconia), and component surface compression (ion-exchanged glass). An example of the crack propagation behavior to be analyzed is shown in Figure 12.1, which shows stable crack extension under applied tensile stress at a $P = 300$ N indentation flaw in zirconia in air (similar to the images in Figures 3.5 and 3.21). The stress direction is horizontal in the images: Figure 12.1a shows the as-indented zero stress state, Figure 12.1c shows the near maximum stress state, just prior to failure, and Figure 12.1b shows an intermediate state. In Figure 12.1a, all surface traces of the indentation cracks are of similar length. On application of the applied stress, as in Figure 12.1b and c, the vertical cracks extend and the horizontal

Indentation Fracture: Strength and Toughness for Brittle Materials Design, First Edition. Robert F. Cook.

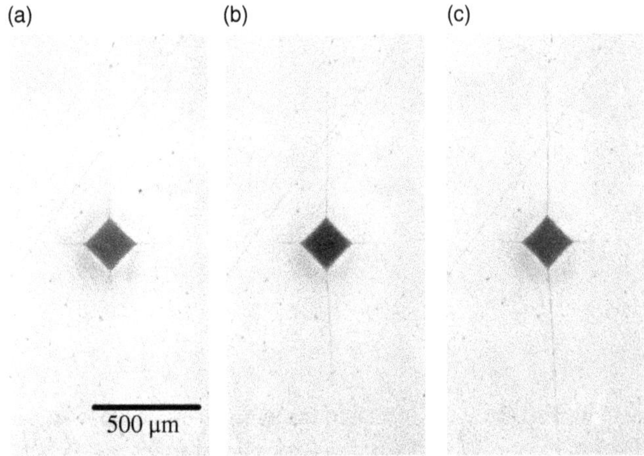

Figure 12.1 In situ indentation crack propagation observations for tetragonal zirconia polycrystal in air. Applied tensile stress direction is horizontal. (a) Initial symmetric indentation flaw. (b) Partially loaded flaw. Vertical cracks are extended stably. (c) Flaw loaded immediately prior to failure. Vertical cracks are extended to the limit of stability.

cracks remain at the as-indented lengths. Figure 12.1b and c depicts stable equilibrium configurations: Unless the applied stress was increased, the vertical cracks remained at stable extended lengths. The stable indentation crack extension in this toughened material was a factor of approximately 3.1. The (σ_a, c) trajectory resulting from such measurements is shown in Figure 12.2. The trajectory proceeds from left to right from $(0, c_{air})$ to (σ_m, c_m). Symbols represent individual crack length measurements at the applied stresses indicated; different symbols represent different indentation tests. The solid line is a fit to the data based on indentation fracture mechanics analyses and well describes the data, including the enhanced stable crack extension.

This chapter surveys experimental and analytical indentation crack propagation phenomena similar to those illustrated in Figures 12.1 and 12.2. The survey of experimental behavior will focus on the three toughened material systems noted above. The survey of analytical behavior will focus on developing physically based explanations for crack propagation observations rather than developing formulations to be fit to experimental data. The reasons for this latter approach are twofold:

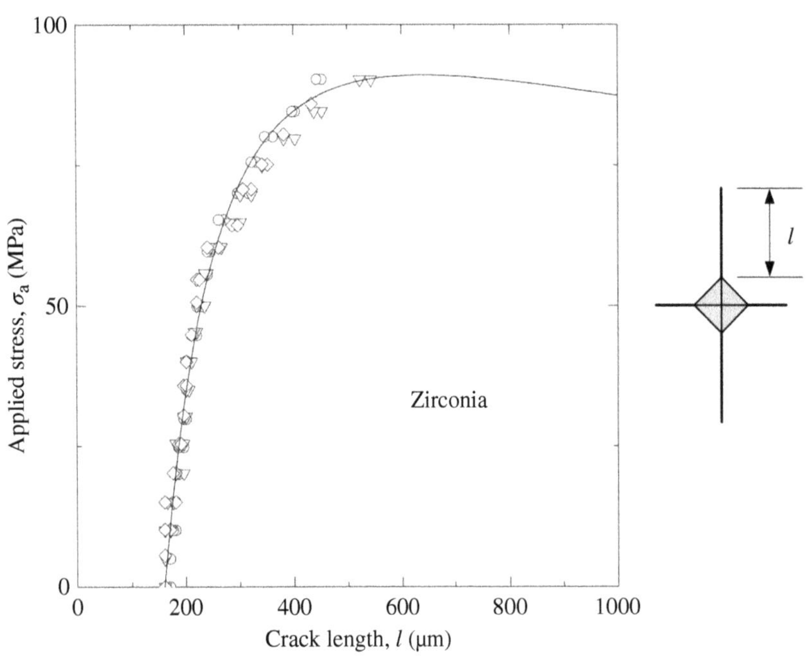

Figure 12.2 Trajectory of applied stress versus crack length for propagation in air of an indentation crack in tetragonal zirconia polycrystal similar to Figure 12.1. Note schematic diagram defining crack length.

microstructural influences introduce parameters into fracture analyses that are usually not well quantified and that are usually stochastic. Hence, although comparison of experimental crack propagation data with limiting analytical forms provides physical insight, fitting of such forms to experimental data is not usually informative. The crack propagation analyses developed here are based on the fracture mechanics of idealized indentations of Chapter 3. Simple perturbations reflecting microstructural influences are superposed on the base analysis. Many of the perturbations were introduced in the context of indentation strength in Chapters 3 and 4.

12.2 Indentation Crack Extension Analyses

12.2.1 Residual Field Variation

The simplest perturbation to the idealized indentation fracture analysis is to vary the power-law dependence of the residual stress term, reflecting geometrical changes in the indentation stress field and indentation cracks. The prevalent indentation crack geometry in tough materials such as the TZP of Figure 12.1 is the radial geometry of Figure 2.5a (Anderson and Braun 1990; Cook et al. 1994). Such cracks are also prevalent in tough WC–Co composites (Shetty et al. 1985) and are usually referred to as exhibiting the "Palmqvist" geometry (see Chapter 2). The surface traces are characterized by the length l from the contact impression corner to the crack tip, as shown in the plan view diagram of Figure 12.2. The residual stress-intensity factor (SIF) K_r acting on the cracks is given by (Cook et al. 1994)

$$K_r = \chi Ea/l^{1/2}, \tag{12.1}$$

where E is the elastic modulus of the material, a is the contact impression semidiagonal, and χ is a dimensionless indentation geometry factor. T curve toughness variation effects in TZP are very short scale (Anderson and Braun 1990; see also Figure 4.32) such that the appropriate toughness for the indentation cracks is the steady-state value T_∞. The stable equilibrium indentation crack length l_0 is then given by setting $K_r = T_\infty$ and inverting to gain

$$l_0 = \left(\chi Ea/T_\infty\right)^2 \sim P, \tag{12.2}$$

recognizing that $P = 2Ha^2$, where H is the hardness. (For simplicity in this chapter, the initial crack length will be specified in the same inert environment as the strength tests, hence the "0" subscript. Cracks at indentations in air are usually longer than in inert environments. See Chapters 2 and 3.) Equation (12.2) is similar to that characterizing median crack lengths at dragged scratch flaws, Eq. (10.15). Under the action of a superposed applied stress σ_a, the applied SIF is $K_a = \psi\sigma_a l^{1/2}$. The total SIF on such a crack is then

$$K = K_a + K_r = \psi\sigma_a l^{1/2} + \chi Ea/l^{1/2}, \tag{12.3}$$

and is an example of the generalized indentation fracture SIF of Eq. (3.27) with $r = 1$. The combined conditions of equilibrium and instability for this system are then given by a matrix expression similar to Eq. (3.116) as

$$\begin{bmatrix} 1 & 1 \\ 1 & -1 \end{bmatrix} \begin{bmatrix} \mathbf{K_a} \\ \mathbf{K_r} \end{bmatrix} = \begin{bmatrix} 1 \\ 0 \end{bmatrix}, \tag{12.4}$$

where $\mathbf{K_a} = K_a/T_\infty$ and $\mathbf{K_r} = K_r/T_\infty$ are scaled SIF variables. Solution of the simultaneous equations represented by Eq. (12.4) yields the indentation strength σ_m,

$$\sigma_m = \frac{T_\infty^2}{4\psi\chi Ea} = SP^{-1/2}, \tag{12.5}$$

and the radial crack length at instability l_m,

$$l_m = \left(2\chi Ea/T_\infty\right)^2 = 4l_0. \tag{12.6}$$

Equation (12.5) is similar to that characterizing strengths of components containing dragged scratch flaws, Eq. (10.18) and such indentation strength, $\sigma_m(P)$, behavior was a focus of Chapters 3, 4, and 11. Here attention is focused on the radial crack propagation trajectory $\sigma_a(l)$ for various TZP materials. It is convenient to introduce scaling relations similar to Eq. (3.24) as

$$\begin{aligned} \mathbf{C} &= l/l_0 \\ \mathbf{S_a} &= \sigma_a/\sigma_m, \end{aligned} \tag{12.7}$$

Setting fracture equilibrium $K = T_\infty$ and inverting Eq. (12.3) using Eqs. (12.5)–(12.7) gives

$$\mathbf{S_a} = 4\mathbf{C}^{-1/2}[1 - \mathbf{C}^{-1/2}]. \tag{12.8}$$

Equation (12.6) shows that ideal radial crack extension at failure is a factor of 4, greater than the 2.52 characterizing ideal half-penny crack extension, and Eq. (12.8) shows that $\mathbf{S_a} = 1$ at $\mathbf{C} = 4$. Equation (12.8) was a fit to the crack extension data of Figure 12.2 and shown as the solid line. The fit accommodates the extended radial crack propagation although, in common with similar plots in Chapter 3 for half-penny crack propagation, the data do not quite extend to the peak of the fit.

The analysis above is simply generalized for variations in impression size and material, both of which can alter the geometry of the indentation residual field. Different materials are characterized by different steady-state toughness values T'_∞, and different residual field geometries are characterized by variations in the exponent r, typically between 1 and 3. The residual SIF is then given by

$$K_r = \chi E a / l^{r/2}. \tag{12.9}$$

Setting the revised equilibrium $K_r = T'_\infty$ and using Eq. (12.9) gives the postindentation crack length as

$$l_0 = \left(\chi E a / T'_\infty\right)^{2/r}. \tag{12.10}$$

Observations on a range of TZP materials with different T'_∞ values showed that l_0 at large indentation loads was approximately invariant with respect to material, suggesting that $\chi E / T'_\infty$ was invariant for the TZP materials considered (Cook et al. 1994; Cook and Braun 1994). Under the influence of applied stress, the combined conditions of equilibrium and instability for the TZP materials are then given by

$$\begin{bmatrix} 1 & 1 \\ 1 & -r \end{bmatrix} \begin{bmatrix} \mathbf{K_a} \\ \mathbf{K_r} \end{bmatrix} = \begin{bmatrix} 1 \\ 0 \end{bmatrix}, \tag{12.11}$$

where $\mathbf{K_a} = K_a / T'_\infty$ and $\mathbf{K_r} = K_r / T'_\infty$. Solution of the simultaneous equations represented by Eq. (12.11) yields

$$\sigma_m = \frac{\beta T'_\infty}{\psi} \left(\frac{T'_\infty}{E\chi}\right)^{1/r} \frac{1}{a^{1/r}}, \tag{12.12}$$

where $\beta = r/(r+1)^{(r+1)/r}$ is a numerical coefficient used earlier ($\beta = 1/4$ for $r = 1$ and $\beta = 3/4^{4/3}$ for $r = 3$). Equation (12.12) is an obvious generalization of Eq. (12.5). Note that now $l_m/l_0 = (1+r)^{2/r}$, reflecting the increase in precursor crack extension with decreasing r ($l_m/l_0 = 4$ for $r = 1$ and $l_m/l_0 = 2.52$ for $r = 3$). Introducing scaling relations of

$$\mathbf{a} = a / a_{\text{ref}}$$
$$\mathbf{T} = T'_\infty / T_{\text{ref}}, \tag{12.13}$$

where a_{ref} and T_{ref} are reference impression sizes and toughness values, and setting fracture equilibrium $K = T'_\infty$ leads to a generalization of Eq. (12.8) as

$$\mathbf{S_a} = \frac{\mathbf{T}\mathbf{C}^{-1/2}}{\beta}\left[1 - \mathbf{a}\mathbf{C}^{-r/2}\right]. \tag{12.14}$$

Equation (12.14) will be used in Section 12.3.2 to describe crack propagation in TZP materials, using the data of Figure 12.2 as the reference configuration.

12.2.2 Similar Driving and Resistance Fields

The variation in the precursor crack extension considered in Section 12.2.1 arose from a change in crack geometry. The geometrical change was characterized by a change in the functional nature of the residual stress field term K_r, as shown in Eq. (12.9). A change in the *amplitude* of the residual stress field term was used in Chapter 4, Eq. (4.21), to characterize the effects of microstructure using $K_r \to K'_r$, where

$$K'_r = \chi (P + P^*) / c^{3/2}. \tag{12.15}$$

The indentation load like term P^* characterizes the effects of microstructural crack wake tractions. The matching toughness is the steady-state value T_∞. Half-penny cracks were considered, with length c measured from the indentation contact impression center, and $r = 3$ in Eq. (12.15). The stable equilibrium indentation crack length c_0 is then given by setting $K_r' = T_\infty$ and inverting to gain

$$c_0 = \left[\chi(P + P^*)/T_\infty\right]^{2/3}. \tag{12.16}$$

Under the action of a superposed applied stress σ_a, the applied SIF is $K_a = \psi\sigma_a c^{1/2}$ as used previously. The combined conditions of equilibrium and instability for this system are then given by the matrix expression

$$\begin{bmatrix} 1 & 1 \\ 1 & -3 \end{bmatrix} \begin{bmatrix} \mathbf{K_a} \\ \mathbf{K_r'} \end{bmatrix} = \begin{bmatrix} 1 \\ 0 \end{bmatrix}, \tag{12.17}$$

where $\mathbf{K_a} = K_a/T_\infty$ and $\mathbf{K_r'} = K_r'/T_\infty$ are scaled SIF variables. The solution of the simultaneous equations represented by Eq. (12.17) yields the indentation strength σ_m,

$$\sigma_m = \frac{\beta T_\infty^{4/3}}{\psi[\chi(P + P^*)]^{1/3}}, \tag{12.18}$$

and the crack length at instability c_m,

$$c_m = \left[4\chi(P + P^*)/T_\infty\right]^{2/3} = 4^{2/3}c_0 \approx 2.52c_0. \tag{12.19}$$

In Chapter 4, Eq. (12.18) was applied extensively and successfully in description of the indentation strengths of polycrystalline ceramics that exhibited microstructural effects on fracture. The important point to note here is that the associated precursor crack extension ratio c_m/c_0 is unaltered by such microstructural effects.

The physical implication of Eq. (12.19) is that the ratio c_m/c_0 is invariant with indentation load P and thus invariant with the relative indentation and microstructural scales. The mathematical reason for the invariance arises from the similar functional forms of the residual SIF and crack length dependent toughness terms: $K_r(c) = \chi P/c^{3/2}$ and $T(c) = T_\infty - \chi P^*/c^{3/2}$. However, as noted in Chapter 4, although Eq. (12.18) is successful in providing a description of the mean indentation strength response for many materials, there is often significant dispersion of strengths about the mean. Such dispersion was particularly prevalent at small indentation loads in large grain size materials—cases in which the crack length and microstructural scales were significantly different (see also Cook et al. 1994). Hence, a refined implication of Eq. (12.19) arising from the consideration of strength data is that the invariant ratio $c_m/c_0 \approx 2.52$ is the expected *mean* response for a set of indentations in a material. For specific indentations, microstructurally induced local variations in P^* along the path of a propagating crack will lead to scatter in the $\sigma_a(c)$ trajectory and dispersion about the mean c_m/c_0 ratio. This interpretation of Eq. (12.19) will be used in Section 12.3.1 in consideration of crack propagation in polycrystalline Al_2O_3.

12.2.3 Dissimilar Driving and Resistance Fields

A clear implication of Sections 12.2.1 and 12.2.2 is that the extent of precursor crack growth on loading an indentation flaw is a function of the relative gradients of the active SIF and toughness. Such gradients are encapsulated in the power laws used to quantify these terms and thus far attention has focused on the values $r = 1$ and $r = 3$ for residual SIF and toughness terms of the form $K, T \sim c^{-r/2}$. Greater insight into precursor crack growth can be gained by examining fracture behavior as a function of the toughness gradient. In particular, a convenient measure of toughness variation is the logarithmic derivative of the toughness T with respect to crack length c, given by

$$\frac{\mathrm{d}\ln(T)}{\mathrm{d}\ln(c)} = \frac{\tau}{2}. \tag{12.20}$$

If τ is invariant over a crack length domain of interest, integration of Eq. (12.20) gives rise to the power law of Eq. (4.41)

$$T(c) = T_\tau\left(\frac{c}{\delta}\right)^{\tau/2}, \tag{12.21}$$

where the lower-bound T_τ and length scale δ are empirical parameters and the exponent is constrained by $0 \le \tau \le 1$ in description of indentation fracture (see Section 4.4). The SIF for an indentation half-penny crack is $K_r(c) = \chi P/c^{3/2}$. Setting the equilibrium condition $K_r(c) = T(c)$ and inverting thus gives the stable equilibrium crack length c_0 as

$$c_0 = \left(\frac{\chi P \delta^{\tau/2}}{T_\tau}\right)^{2/(3+\tau)}, \tag{12.22}$$

which devolves to the expression for untoughened materials at $\tau = 0$. It is important to recognize that although Eq. (12.21) can be used to describe toughness measurements and indentation crack length and strength variations, application of the expression on an unbounded domain is empirical.

The equilibrium and stability conditions for an indentation half-penny crack under applied stress in such a toughness field are written in matrix form as, Eq. (4.43),

$$\begin{bmatrix} 1 & 1 \\ 1 & -3 \end{bmatrix} \begin{bmatrix} \mathbf{K_a} \\ \mathbf{K_r} \end{bmatrix} = \begin{bmatrix} 1 \\ \tau \end{bmatrix}, \tag{12.23}$$

where $\mathbf{K_a} = K_a/T(c)$ and $\mathbf{K_r} = K_r/T(c)$ are scaled SIF variables with crack length dependence that depends on both SIF and toughness variation. In Chapter 4, attention was focused on the strength determined by Eq. (12.23). Here, attention is focused on the crack length c_m at instability,

$$c_m = \left(\frac{4\chi P \delta^{\tau/2}}{(1-\tau)T_\tau}\right)^{2/(3+\tau)}, \tag{12.24}$$

such that

$$c_m/c_0 = \left(\frac{4}{(1-\tau)}\right)^{2/(3+\tau)}. \tag{12.25}$$

It is clear from Eq. (12.25) that precursor crack extension increases significantly as $\tau \to 1$ or as $T \to c^{1/2}$. A moderate T curve, $\tau = 0.5$, within a crack length domain of interest leads to $T \sim c^{0.25}$ and $c_m/c_0 \approx 3.3$. This value of τ describes the toughness responses of the toughened Si_3N_4 and Al_2O_3 composite of Figures 5.5 and 5.6. A weak T curve, $\tau = 0.2$, leads to $T \sim c^{0.1}$ and $c_m/c_0 \approx 2.7$ and this value of τ describes the less tough Si_3N_4, SiC, Al_2O_3, and glass-ceramics of Figures 5.5–5.7 (see also Lin et al. 2007). An absent T curve, invariant toughness, $\tau = 0$, leads to $c_m/c_0 = 2.52$. Care needs to be exercised however in inferring T curve characteristics solely from c_m/c_0 ratios. If the T curve and residual indentation field vary over the same crack length domain, the equilibrium and instability matrices of Eqs. (12.11) and (12.23) are combined as

$$\begin{bmatrix} 1 & 1 \\ 1 & -r \end{bmatrix} \begin{bmatrix} \mathbf{K_a} \\ \mathbf{K_r} \end{bmatrix} = \begin{bmatrix} 1 \\ \tau \end{bmatrix}. \tag{12.26}$$

In this case, the c_m/c_0 ratio is given as

$$c_m/c_0 = \left(\frac{(1+r)}{(1-\tau)}\right)^{2/(r+\tau)}. \tag{12.27}$$

Equation (12.27) makes clear that the underlying reason for variation in the precursor crack extension ratio at indentation flaws in toughened materials is the dissimilar functions of the driving and resistance fracture fields. For example, for $r = 2$ and $\tau = 0.5$, $c_m/c_0 \approx 4.2$. Such a combination of parameters may pertain to very large grain size Al_2O_3 materials in which grain boundary related lateral cracking and chipping at indentations severely disrupt the half-penny crack geometry. The dissimilar nature of the driving and resistance fields, encapsulated in Eq. (12.27), will be used in Section 12.3.1 in consideration of crack propagation in polycrystalline Al_2O_3.

12.2.4 Piecewise Continuous Resistance Field

The final considerations of indentation crack propagation in toughened materials here are the effects of the transition from untoughened to toughened behavior on crack extension. This transition was discussed briefly in Chapter 4 in relation to indentation strength, see Figure 4.38. The crack wake based microstructural toughening analysis of Chapter 4 and Section 12.2.2 is used here to consider the transition. Briefly, for cracks c shorter than a length scale δ the toughness T of the material is given by an invariant base value T_0. For cracks longer than δ, the toughness is given by $T = T_0 + T_\mu(c/\delta)$, where $T_\mu(c/\delta)$ is an increasing function. $T_\mu(c/\delta)$ reaches a plateau value of ΔT for very long cracks, $c \gg \delta$, such that the long crack steady-state toughness is $T_\infty = T_0 + \Delta T$. It is convenient to define an indentation load P^* characterizing the long crack change in toughness by $\chi P^*/\delta^{3/2} = \Delta T$, where χ is a geometrical factor. Under the action of applied stress σ_a, the equilibrium conditions for indentation fracture either side of the transition are then

$$\begin{aligned} \psi\sigma_a c^{1/2} + \chi P/c^{3/2} &= T_0, \quad c < \delta \\ \psi\sigma_a c^{1/2} + \chi (P+P^*)/c^{3/2} &= T_\infty, \quad c \ge \delta, \end{aligned} \tag{12.28}$$

where ψ is another geometrical term. The strength of a Griffith like crack ($P = 0$) of length δ in the toughened region is

$$\sigma_m^* = \frac{\beta T_\infty^{4/3}}{\psi(\chi P^*)^{1/3}}, \tag{12.29}$$

and the crack length at instability is $c_m^* = 4^{2/3}\delta$. To assess the precursor crack extension behavior, it is convenient to use the scaling relations (Chapter 4, P is the effective indentation load)

$$\begin{aligned}
\mathbf{C} &= c/\delta \\
\tilde{\mathbf{P}} &= P/P^* \\
\mathbf{S} &= \sigma/\sigma_m^* \\
\mathbf{T} &= T/T_0.
\end{aligned} \tag{12.30}$$

The scaled equilibrium applied stress as a function of crack length can be obtained by inverting Eq. (12.28) in a piecewise fashion and using Eq. (12.30):

$$\begin{aligned}
\mathbf{S_a} &= \left[\frac{1}{\mathbf{T}_\infty} - \frac{\tilde{\mathbf{P}}(1 - 1/\mathbf{T}_\infty)}{\mathbf{C}^{3/2}}\right]\frac{(1 - 1/\mathbf{T}_\infty)^{1/3}}{\beta\mathbf{C}^{1/2}}, \quad \mathbf{C} < 1 \\
&= \left[1 - \frac{(1 + \tilde{\mathbf{P}})(1 - 1/\mathbf{T}_\infty)}{\mathbf{C}^{3/2}}\right]\frac{(1 - 1/\mathbf{T}_\infty)^{1/3}}{\beta\mathbf{C}^{1/2}}, \quad \mathbf{C} \geq 1.
\end{aligned} \tag{12.31}$$

The untoughened response of Eq. (3.17) is recovered from the second expression for $\mathbf{T}_\infty \gg 1$ and $\tilde{\mathbf{P}} \gg 1$. Similar, unscaled, responses were given in Cook et al. (1987).

Figure 12.3 shows a logarithmic plot of Eq. (12.31) for scaled indentation loads of $\tilde{\mathbf{P}} = 10^{-4}$–$\tilde{\mathbf{P}} = 10$. The vertical and horizontal lines indicate the scaled crack length and applied stress values $\mathbf{C} = 1$ and $\mathbf{S_a} = 1$, respectively. The changes in derivative of the $\mathbf{S_a}(\mathbf{C})$ functions at $\mathbf{C} = 1$ reflect the onset of toughening for $\mathbf{C} \geq 1$. For values of $\tilde{\mathbf{P}} \geq 10^{-1}$, the $\mathbf{S_a}(\mathbf{C})$ functions exhibit single maxima. For values of $\tilde{\mathbf{P}} \leq 10^{-2}$, the $\mathbf{S_a}(\mathbf{C})$ functions exhibit two maxima. It is recognized that in most crack propagation experiments, the applied stress $\mathbf{S_a}$ is the independent variable. The behavior at the extremes of indentation load, small $\tilde{\mathbf{P}} = 10^{-3}$ and $\tilde{\mathbf{P}} = 10^{-4}$ and large $\tilde{\mathbf{P}} = 1$ and $\tilde{\mathbf{P}} = 10$, is shown in a similar logarithmic plot of Figure 12.4.

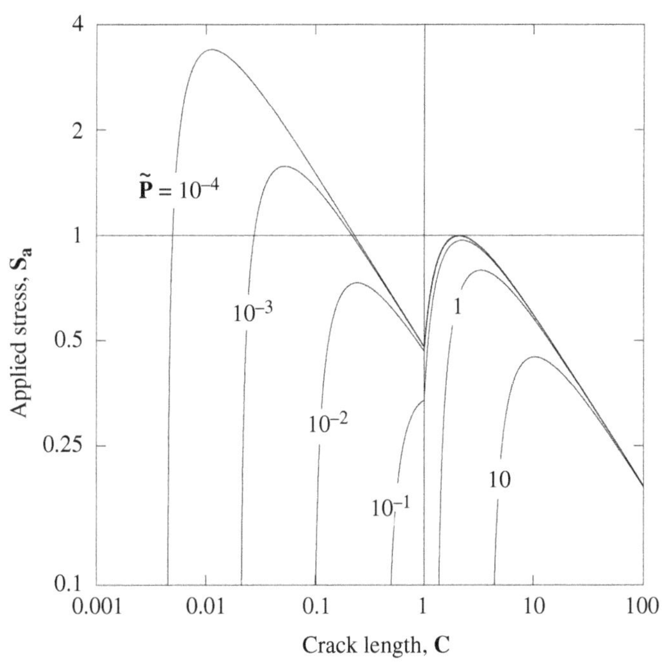

Figure 12.3 Logarithmic plot of equilibrium trajectories of scaled applied stress versus crack length for indentation flaws in a material with piecewise continuous fracture resistance. For cracks shorter than $\mathbf{C} = 1$, an invariant lower-bound toughness pertains. For cracks longer than $\mathbf{C} = 1$, an increasing toughness T curve pertains. Peaks of trajectories are unstable fracture configurations for indentations formed at different contact loads.

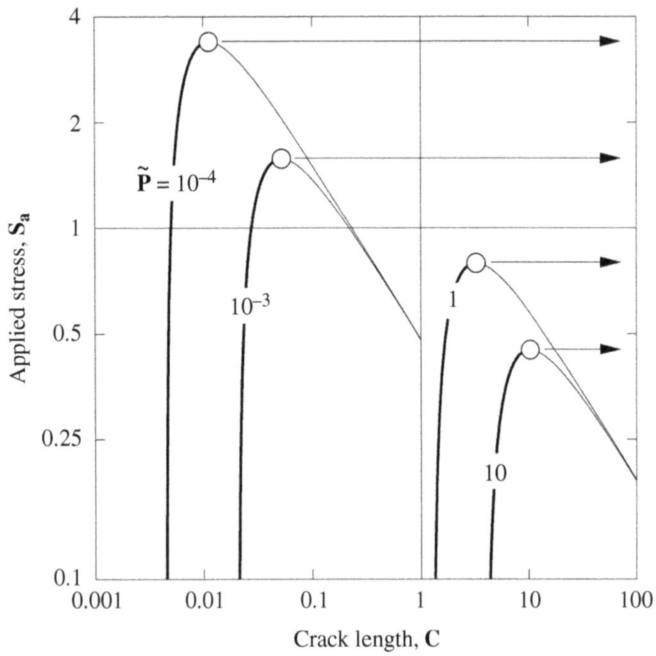

Figure 12.4 Plot of equilibrium trajectories of scaled applied stress versus crack length for indentation flaws exhibiting the extremes of behavior in a material with piecewise continuous fracture resistance. For small indentation loads, failure occurs completely in the lower-bound toughness domain. For large indentation loads, failure occurs completely in the greater toughness *T* curve domain.

The fine lines are as in Figure 12.3, and the maxima are indicated by open symbols. The bold lines indicate the equilibrium crack propagation trajectories, terminating at the maxima of the failure strengths. Conventional indentation crack propagation behavior for polycrystals exhibiting near steady-state toughness is represented by the $\tilde{\mathbf{P}} = 1$ and $\tilde{\mathbf{P}} = 10$ responses. As the applied stress is increased, crack extension is initially small and increases as the failure strength is approached. At failure, crack behavior departs from the equilibrium trajectory and the crack extends rapidly in a dynamic manner (see Chapter 1) at near constant stress, indicated by the horizontal arrows. Similarly, conventional indentation crack propagation behavior for fine-grained polycrystals, single crystals, glasses, and glass-ceramics exhibiting invariant toughness is represented by the $\tilde{\mathbf{P}} = 10^{-4}$ and $\tilde{\mathbf{P}} = 10^{-3}$ responses. The failure process is identical: initially small crack extension that increases as the failure strength is approached followed by rapid crack extension at failure and departure from the equilibrium trajectory indicated by the horizontal arrows. In principle, such failure processes should also occur in large-grained polycrystals that exhibit significant toughening. In practice, such processes are never observed as intrinsic strengths, often determined by flaws originating from microstructural effects, intervene. It is noted that the failure stresses in both the toughened and untoughened regions here that exhibit near constant toughness are well separated for the different indentation loads. Examples of the "conventional" crack propagation processes described are shown in Chapters 11 and 3, respectively.

The behavior at the central indentation loads, $\tilde{\mathbf{P}} = 10^{-2}$ and $\tilde{\mathbf{P}} = 10^{-1}$, is shown in the logarithmic plot of Figure 12.5. These loads lead to crack extension in transitions from untoughened to toughened behavior. The plot of Figure 12.5 is at an expanded scale but otherwise lines and symbols are as before: fine lines indicate the equilibrium behavior of Eq. (12.31) and the maxima are indicated by open symbols. The bold lines indicate equilibrium crack propagation trajectories, terminating at the maxima of the failure strengths. The trajectories are very different from those in Figures 12.3 and 12.4. The crack propagation behavior for $\tilde{\mathbf{P}} = 10^{-1}$ can be regarded as a perturbation from conventional behavior. The initial indentation cracks are formed in the untoughened region, e.g., small cracks restricted to a single grain facet in a polycrystal. On loading by an applied stress, the cracks extend in the untoughened state until a toughening process is activated at a characteristic crack length, here $\mathbf{C} = 1$. The cracks continue to extend but now along a trajectory determined by a transition to the toughened state. The transition is indicated by the shaded symbol in Figure 12.5. Note that there is a discontinuity in only the derivative of the trajectory at this point and that the fracture system remains in equilibrium. Crack extension then proceeds as described previously followed by failure at the lower open symbol and rapid crack extension indicated by the horizontal arrow. The overall trajectory is perturbed

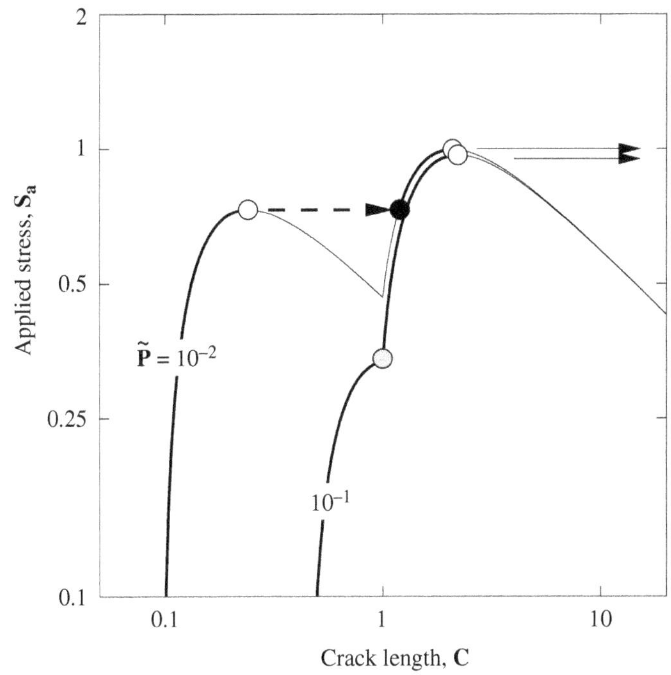

Figure 12.5 Plot of equilibrium trajectories of scaled applied stress versus crack length for indentation flaws exhibiting transitions in toughness behavior in a material with piecewise continuous fracture resistance. At larger loads, the transition is marked by a change in derivative of the trajectory (grayed symbol). At smaller loads, the transition is marked by a nonequilibrium crack extension segment (dashed line between open and filled symbols).

from a conventional form by a midpoint discontinuity in derivative that leads to greater precursor extension prior to failure.

The crack propagation behavior for $\tilde{\mathbf{P}} = 10^{-2}$ is significantly altered from a conventional trajectory. The initial indentation cracks are formed in the untoughened region and on loading by an applied stress extend in the untoughened state until an instability point is reached, indicated by the left most open symbol. Rapid nonequilibrium crack propagation ensues at near constant stress along the horizontal dashed line and arrow (the equilibrium path is indicated by the lower fine line). However, that propagation is halted as the system enters a stable equilibrium configuration associated with the transition to a toughened state. The stability point is indicated by the filled symbol. Crack extension then proceeds as described previously followed by failure at the upper open symbol and rapid crack extension indicated by the horizontal arrow. The overall crack propagation trajectory prior to failure consists of two short stabilized equilibrium branches separated by a longer nonequilibrium segment. In other words, the enforced transition from untoughened to toughened behavior leads to erratic crack advance. In addition, the transition collapses the separation of the failure states such that the failure stresses and crack lengths at failure are similar for the two cases, as indicated by the overlapping open symbols.

12.3 Crack Propagation Observations

12.3.1 Polycrystalline Alumina

Figure 12.6 shows plots of indentation crack propagation in fine-grained polycrystalline Al_2O_3 from the works of (a) Kovar et al. (2000) and (b) Braun et al. (1992), using data derived from the published works cited. The data include observations of (a) P = 5–100 N indentation cracks in dry N_2 and (b) P = 10–200 N indentation cracks in silicone oil and are scaled to the initial crack lengths and failure strengths as in Figure 3.6 and indicated by the vertical and horizontal lines. The curved lines represent the ideal indentation crack propagation responses, given by Eq. (3.26). Both sets of data are well described by the ideal response. There are two minor distinctions from the Si_3N_4 and cordierite responses of Figure 3.6. First, there was no zero stress crack length offset associated with residual stress field relaxation prior to applied stress loading. Second, the experimental scatter was somewhat greater (although there was no systematic trend with indentation load). Overall,

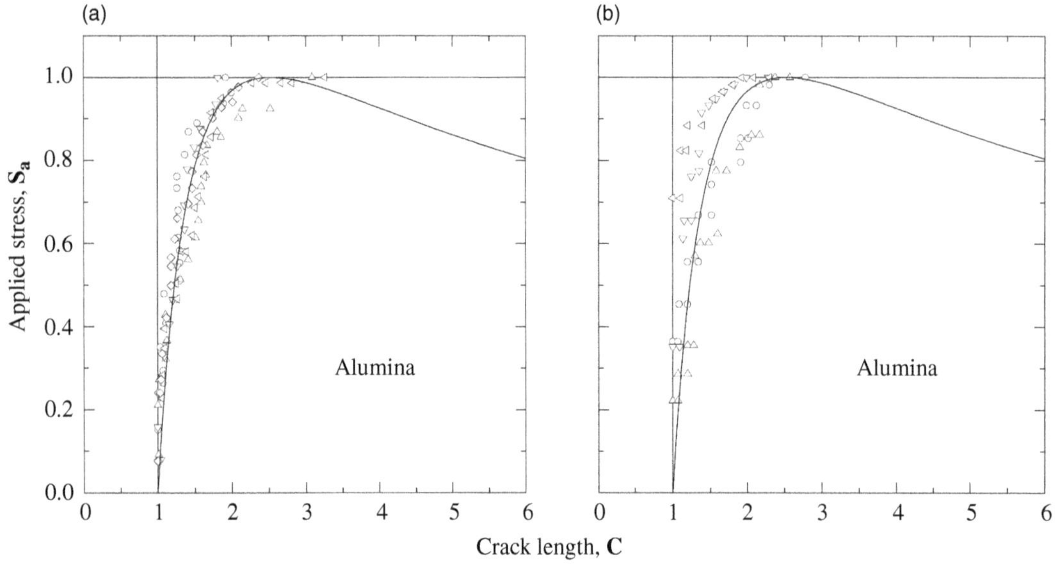

Figure 12.6 Plot of equilibrium trajectories of scaled applied stress versus crack length for indentation flaws in small grained polycrystalline alumina. Responses, although scattered, follow ideal indentation trajectory. Source: Data adapted from Kovar et al. (2000) and Braun et al. (1992).

the polycrystalline crack propagation behavior is well described by Section 12.2.2 in the limit of $P^* = 0$, Section 12.2.3 in the limits of $\tau = 0$ and $r = 3$, and Eq. (12.31) of Section 12.2.4 in the limits of $\tilde{\mathbf{P}} \ll 1$ and $\mathbf{T}_\infty \approx 1$. Figure 12.6 thus serves in the role of a control, illustrating the absence of microstructural effects in indentation crack propagation.

In contrast, Figure 12.7 displays far from ideal crack propagation behavior in a large-grained Al_2O_3 (see Figure 4.10a). The data represent observations of $P = 20$–200 N indentation cracks in air and are shown in logarithmic coordinates to enable comparison between responses at different indentation loads. The responses at increased loads and increased initial crack lengths are to the right. A typical response consists of an initial large increase in applied stress to almost the failure stress with no change in crack length. This is followed by a substantial crack extension with almost no increase

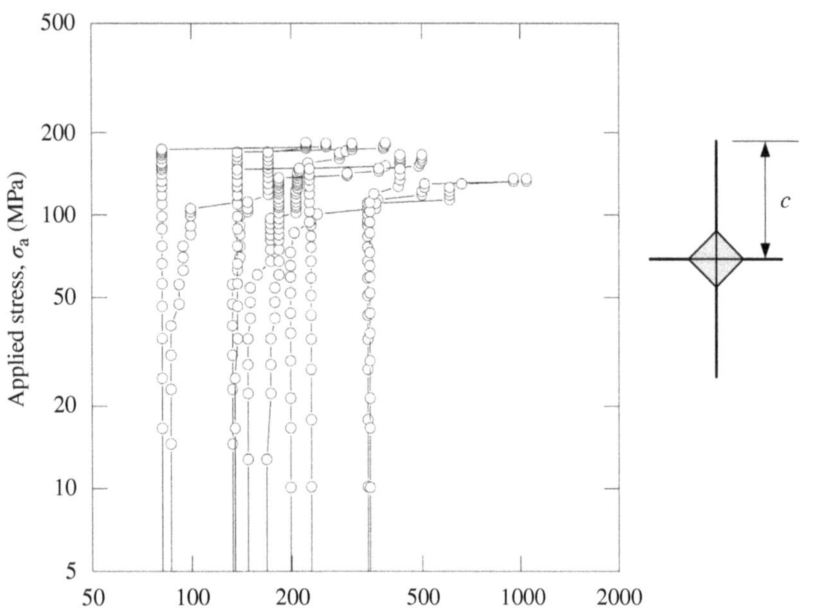

Figure 12.7 Logarithmic plot of equilibrium trajectories of applied stress versus crack length for indentation flaws in large-grained polycrystalline alumina. Responses to the right are for greater indentation loads. Behavior is far from ideal, consisting of initial segments with crack length invariance followed by final segments with applied stress invariance.

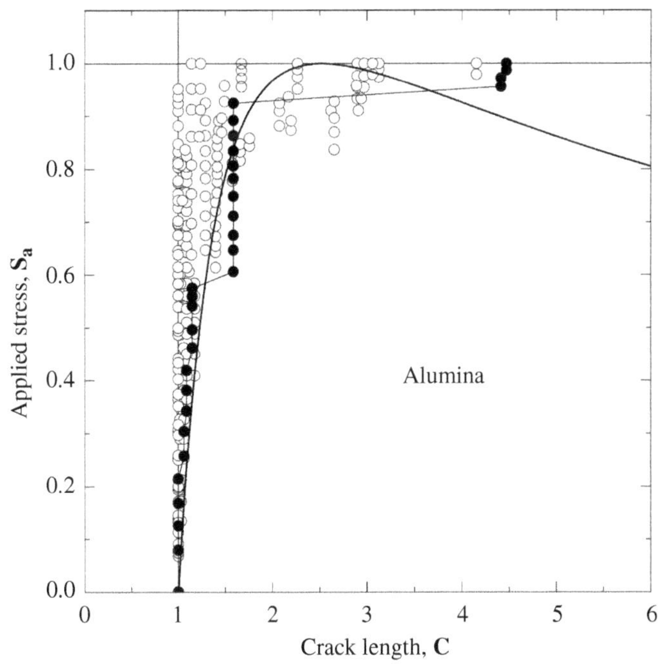

Figure 12.8 Plot of equilibrium trajectories of scaled applied stress versus crack length for indentation flaws in large-grained polycrystalline alumina from Figure 12.7. Note that the ideal response underlies the scatter and erratic behavior of individual indentations (example shown as filled symbols).

in applied stress. A few, smaller, stress increases and crack extensions then follow, also as discrete segments, until failure occurs. The relative length of the predominant crack extension appears to be almost invariant with respect to indentation load, and the failure stresses exhibit a weak decrease with indentation load. Figure 12.8 replots the data of Figure 12.7 in the same scaled form as Figure 12.6. The open symbols represent individual crack length measurements at the applied stresses indicated. The filled symbols are a guide to the eye and are a subset of the data for a single trajectory, highlighting the discrete and erratic nature of the crack propagation. The solid curved line is the ideal indentation fracture response. The mean indentation strength data (for the same material) in Figure 4.10a are well described by the analysis of Eq. (12.18), implying that the mean maximum crack extension response should be described by the same analysis and embodied in Eq. (12.19). This is in fact the case, as shown in Figure 12.8, although the scatter is considerable and the detailed shape of the crack extension responses differs considerably from the predicted mean response. The implication is that the majority of indentation crack extension responses reflected the transitional untoughened–toughened behavior described in Section 12.2.4 superposed on the average microstructural response described in Section 12.2.2.

Figure 12.9 shows plots of indentation crack propagation for (a) large-grained polycrystalline Al_2O_3 from Kovar et al. (2000) and (b) large-grained Al_2O_3–Al_2TiO_5 composite from Braun et al. (1992), using data derived from the published works cited and the same plotting scheme as Figure 12.6. The open symbols represent individual crack length measurements at the applied stresses indicated. The solid curved line is the ideal indentation fracture response. The behaviors for both sets of observations are similar to Figure 12.8 in that there is considerable variability and significant numbers of observations of crack lengths in excess of the ideal response at a given applied stress. In particular, at the maximum stress, the maximum crack length in Figure 12.9 is similar to that in Figure 12.6 and $c_m/c_0 = \mathbf{C} \approx 5$, well in excess of the ideal value of 2.52. In Figure 12.9, the ideal response appears as a lower bound on crack extension behavior. The implication here is that the indentation crack extension responses reflected the transitional untoughened–toughened behavior described in Section 12.2.4. However, distinct from Figure 12.8, the data of Figure 12.9 can also be described as in Section 12.2.3 by an underlying crack resistive field that is dissimilar from the indentation field.

12.3.2 Phase-Transforming Zirconia

Figure 12.10 shows a set of plots of indentation crack propagation in TZP, including data from Braun and Cook (1993). The observations are of $P = 300$ N indentations tested in air. The four materials were all 2.5% by mol fraction Y_2O_3 in ZrO_2, sintered in air to 98% relative density at temperatures of (a) 1300 °C, (b) 1400 °C, (c) 1500 °C, and (d) 1600 °C, giving rise

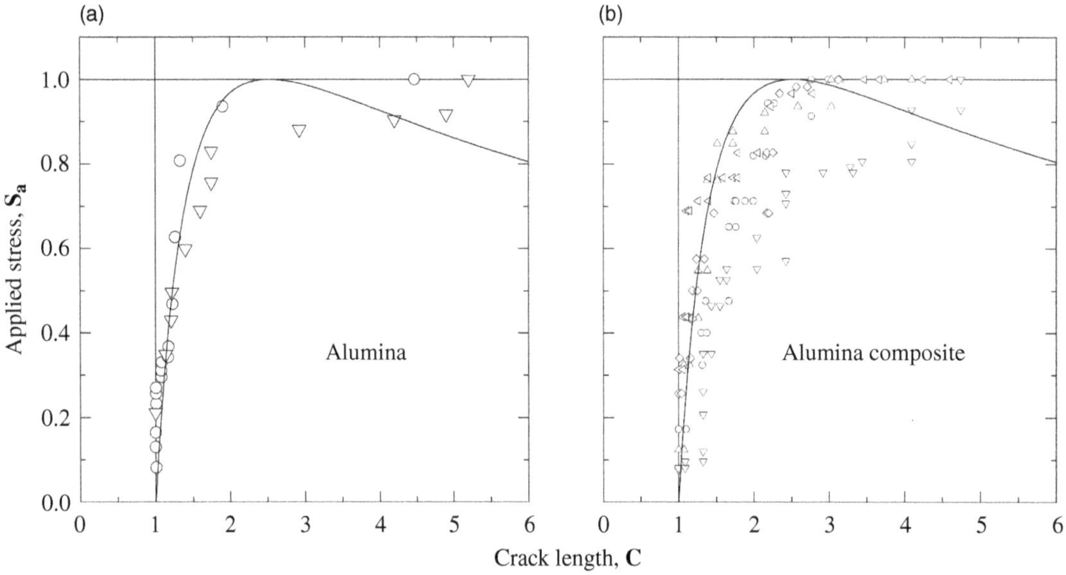

Figure 12.9 Plot of equilibrium trajectories of scaled applied stress versus crack length for indentation flaws in large-grained polycrystalline alumina composites. The ideal response appears as a lower bound to these observations. Source: Data adapted from Kovar et al. (2000) and Braun et al. (1992).

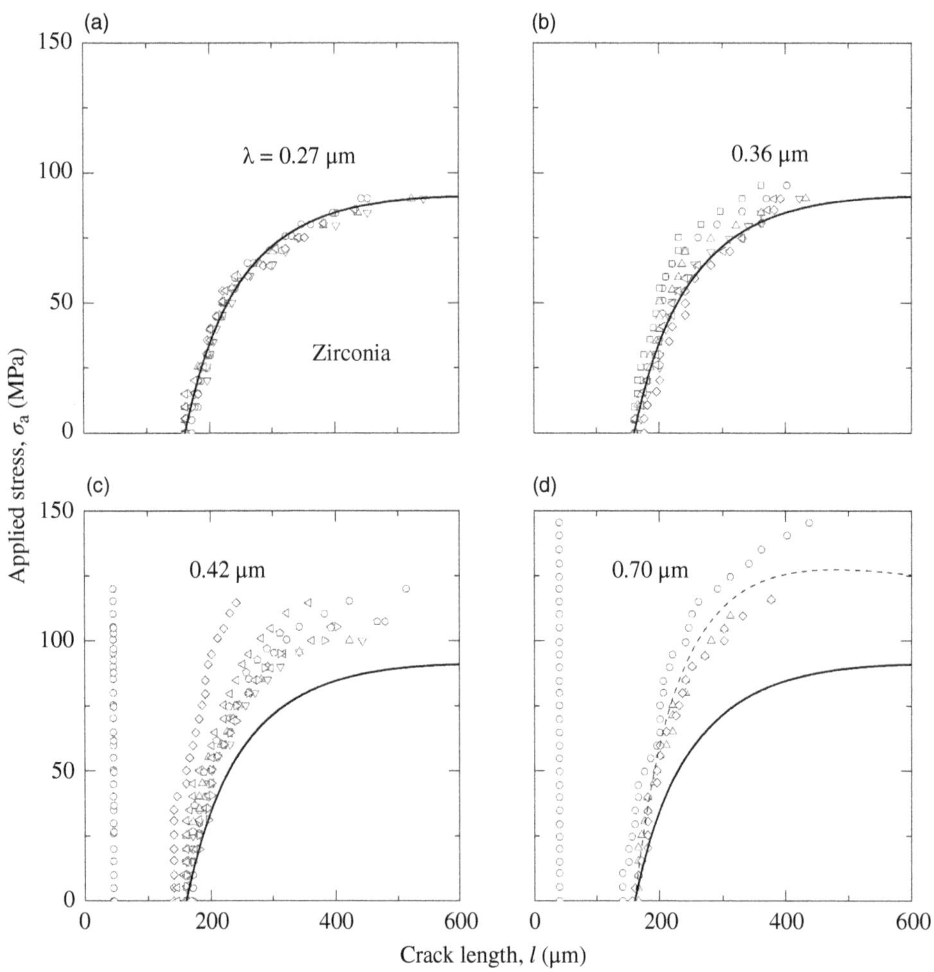

Figure 12.10 Trajectories of applied stress versus crack length for propagation in air of indentation cracks in tetragonal zirconia polycrystals with grain sizes indicted. Bold lines are fit to smallest grain size materials. Dashed line is fit to largest grain size material taking steady-state toughness and crack geometry into account.

to variation in sub-μm grain sizes λ as specified. The symbols represent individual measurements of crack length at the applied stress values indicated. Three to four indentations were tested for each microstructural variant. The solid line is a fit of Eq. (12.8) to the smallest grain size data of Figure 12.10a and is repeated as a guide to the eye in Figure 12.10b–d. Figure 12.10a repeats Figure 12.2. The four-plot format is similar to that in Figure 11.9, in which the increasing volume fraction of whiskers in a glass-ceramic matrix was the microstructural variable. The increase in grain size within the group of zirconia materials here had the effect of increasing the transformability of the tetragonal phase to the monoclinic phase in the presence of an initiating stress field (see Chapter 4). Hence, in crack tip stress fields, there was a greater degree of transformation for larger grain size materials and thus a greater degree of toughening. The steady-state toughness of the materials thus increased in order (a) (4.2 ± 0.2) MPa m $^{1/2}$, (b) (4.9 ± 0.1) MPa m $^{1/2}$, (c) (5.4 ± 0.1) MPa m $^{1/2}$, and (d) (5.9 ± 0.3) MPa m $^{1/2}$. In indentation stress fields, the greater degree of transformation in larger grain size materials lead, to larger zones of transformed material surrounding the indentation contact impressions (Cook et al. 1994; Cook and Braun 1994). The mean zone diameter increased in size relative to the contact impression diameter as (a) 0.43, (b) 0.53, (c) 0.73, and (d) 0.78. The magnitude of the compressive stress exerted by the transformed material within the zone was approximately 120 MPa. The contact impression semi-diagonal for an indentation load of $P = 300$ N for all materials was approximately 100 μm, such that for the larger grain size materials the compressive transformation zone extended approximately 75 μm about the impression.

The crack propagation data of Figure 12.10 exhibit two phenomena associated with two different sizes of cracks observed at TZP indentations. The first phenomenon is associated with large "well-developed" cracks typical of indentation fracture, as shown in Figure 12.1a. Such cracks were much longer than the indentation transformation zones. On application of stress, the extension of these cracks was dependent on the microstructure: As grain size increased, crack extension was contracted to smaller lengths, required greater applied stress, and exhibited greater variability. The dashed line in Figure 12.10d is a visual best fit of Eq. (12.14) to the largest grain size crack propagation data using the measured toughness ratio of $\mathbf{T} = 5.9/4.2 \approx 1.4$ and a residual SIF exponent of $r = 2$. These values compare with $\mathbf{T} = 1$ and $r = 1$ used for the smallest grain size material. The intermediate grain size materials exhibited intermediate behavior. The implication from Figure 12.10 is that the compressive transformation zones partially mitigated the tensile fields associated with plastic deformation zone effects at indentations. This led to crack driving forces that were smaller and more variable as grain size increased.

The second crack propagation phenomenon is associated with small "trapped" cracks that are often observed at indentations in zirconia. Examples of such cracks are shown in Figures 12.11a and 12.12a. These cracks were the focus of earlier works (Cook et al. 1994; Cook and Braun 1994) and are well explained by an indentation fracture model that includes the compressive transformation zone and yields two (small and large) stable equilibrium configurations for indentation cracks in zirconia. The small cracks are smaller than the indentation transformation zone. On application of stress, there was no extension of these cracks. The applied stress simply increased until the failure strength, as shown in Figures 12.10c and 12.10d. Occasionally, a trapped crack would "escape" to the well-developed configuration immediately

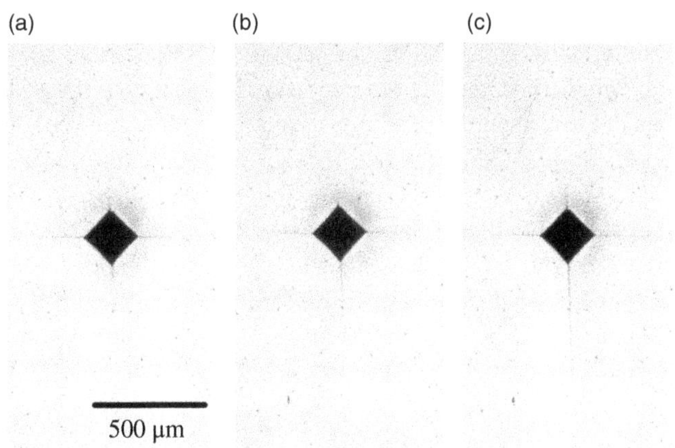

Figure 12.11 In situ indentation crack propagation observations for tetragonal zirconia polycrystal in air. Applied tensile stress direction is horizontal. (a) Initial asymmetric indentation flaw with short trapped crack in upper quadrant. (b) Partially loaded flaw. Lower vertical crack is extended stably; upper crack remains trapped. (c) Flaw loaded immediately prior to failure. Lower vertical crack is extended to the limit of stability; upper crack remains trapped.

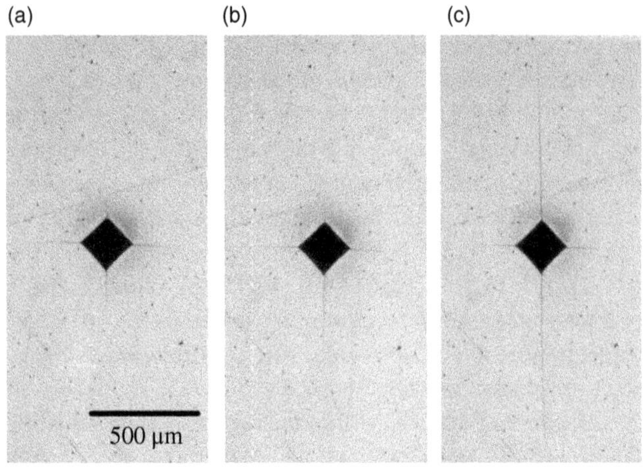

Figure 12.12 In situ indentation crack propagation observations for tetragonal zirconia polycrystal in air. Applied tensile stress direction is horizontal. (a) Initial asymmetric indentation flaw with short trapped crack in upper quadrant. (b) Partially loaded flaw. Lower vertical crack is extended stably; upper crack remains trapped. (c) Flaw loaded immediately prior to failure. Both vertical cracks are extended to the limit of stability; upper crack has "escaped."

prior to failure. Examples of the fully trapped and intermediate escape failure modes are shown in Figures 12.11 and 12.12, respectively.

12.3.3 Ion-Exchanged Glass

Surface-stressed glasses are examples of toughening in which the toughening agent is associated with a component rather than a material. The two most common glass toughening mechanisms are thermal tempering and chemical tempering, the latter often referred to as "ion exchange." In both mechanisms, a limited diffusivity, thermal or chemical, leads to a positive strain imposed on the surface of a glass component relative to the interior bulk. In reaction, a negative stress develops in the surface as a consequence of the component-level constraints of strain compatibility and mechanical equilibrium. The negative stress is compressive and thus opposes fracture and can be analyzed as a toughening effect in a reference frame exterior to the component (e.g., that of the applied stress). Details of the thermal tempering, ion exchange, and related processes can be found elsewhere (Kingery et al. 1976). Briefly, in the ion-exchange process a glass component is immersed in a molten salt bath for a period of time such that large cations from the salt can diffuse into the glass surface and exchange with small cations in the glass, K^+ for Na^+ is common. A positive strain is thus imposed and on removal of the component from the bath a localized, compressive, surface stress results. Ion-exchange strengthened and toughened aluminosilicate glass is a common material in microelectronic display screens. (Thermally tempered soda-lime silicate glass is a common material in automobile windows.) Other component-level toughening processes include surface compression induced by machining and surface lamination. Examples of component-level toughening and the ensuing strengthening are discussed in detail in Chapter 3. A short survey is given here of the effects of ion-exchange based toughening on indentation crack propagation. The framework for the survey is the sequential classification system of indent, exchange, and fracture shown in Figure 3.48.

Figure 12.13 shows plots of indentation crack propagation in (a) soda-lime silicate glass using indentation loads of $P = 10$, 20, 50, and 100 N and (b) aluminosilicate glass using a load of $P = 60$ N. The data are scaled to the initial crack length $\mathbf{C} = 1$ and the failure strength $\mathbf{S_a} = 1$, indicated by the vertical and horizontal lines, respectively. The curved lines are the response of Eq. (3.35) fit to the observations for each glass, treating the indentations stress relaxation parameter $\mathbf{X}$ as a fitting variable. As in Chapter 3, the responses describe the mid domain experimental data and values of $\mathbf{X} = 0.94$ and $\mathbf{X} = 0.97$ were used for soda-lime and aluminosilicate glass, respectively (ideal behavior is given by $\mathbf{X} = 1$). In these cases, however, the relaxed indentation half-penny crack responses do not describe the full crack propagation behavior. Significant contraction in crack extension was observed at large applied stress levels, and there was a clear tendency to nonzero derivative in the stress–crack length response at failure. These deviations from ideality indicate that fracture instability must have been generated by *sub*surface crack extension effects not apparent in the surface trace measurements. As in the case of zirconia in Section 12.3.2, the data can be fit empirically using Eq. (12.14) and here a residual SIF exponent of $r \geq 3$.

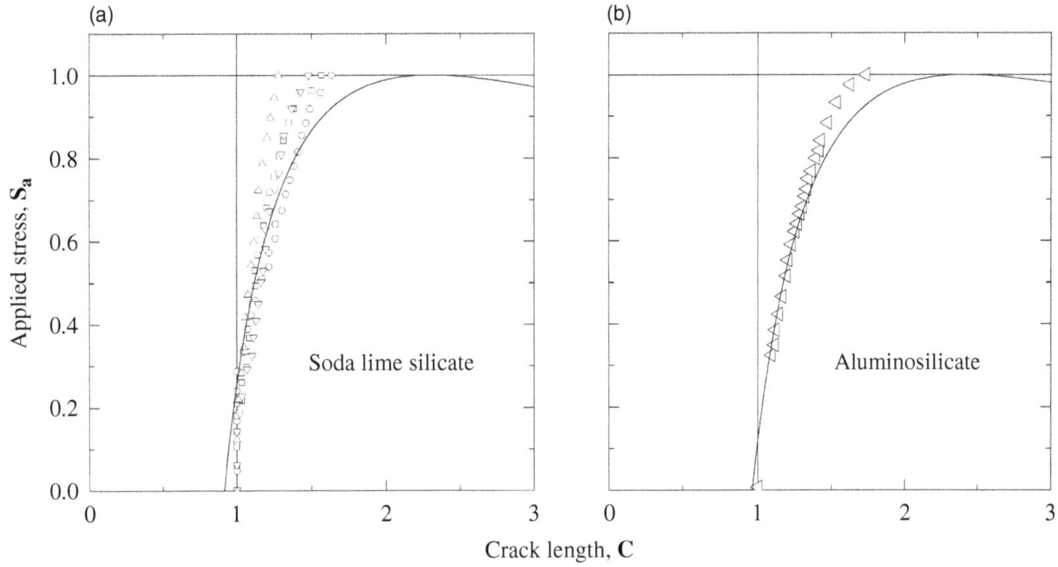

Figure 12.13 Plots of equilibrium trajectories of scaled applied stress versus crack length for indentation flaws in glasses. The ideal response appears as an upper bound to these observations. (a) Soda-lime silicate glass. (b) Aluminosilicate glass.

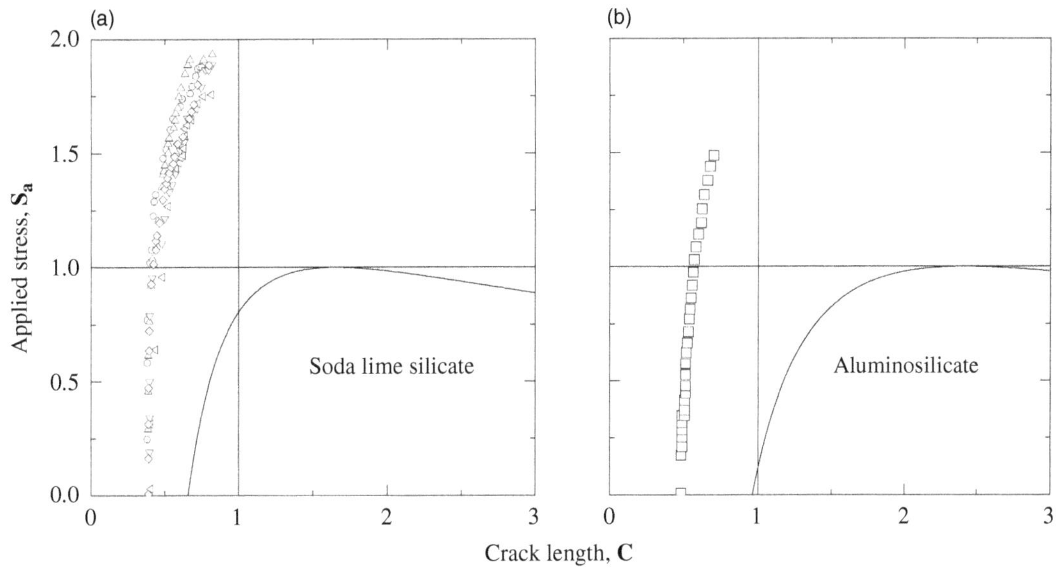

Figure 12.14 Plots of equilibrium trajectories of scaled applied stress versus crack length for indentation flaws in surface compressed glasses corresponding to Case III in Chapter 3 (posttreatment indentation). The base glass responses are shown as curved lines. (a) Thermally tempered soda-lime silicate glass. (b) Ion-exchanged aluminosilicate glass. Source: Includes data adapted from Marshall et al. (1979).

Figure 12.14 shows plots of indentation crack propagation for the most common forms of glass toughening: (a) thermally tempered soda-lime silicate glass using an indentation load of $P = 50$ N (Marshall et al. 1979), using data derived from the published work; and (b) ion-exchanged aluminosilicate glass using a load of $P = 60$ N. These responses correspond to Case III in Figure 3.48: Indentation and fracture *after* implementation of a toughening process. The format of Figure 12.14 is identical to Figure 12.13 with the exception that the stress axes are extended. The data are scaled to the responses of the base materials: (a) Figure 3.22a and (b) Figure 12.13b. The curved lines are the responses of the base materials. In both cases, the initial crack lengths are significantly shortened, the failure stresses are significantly increased, and the crack extension on failure is significantly less. Tempered and exchanged glasses do appear tougher in the frame of reference of the applied loading.

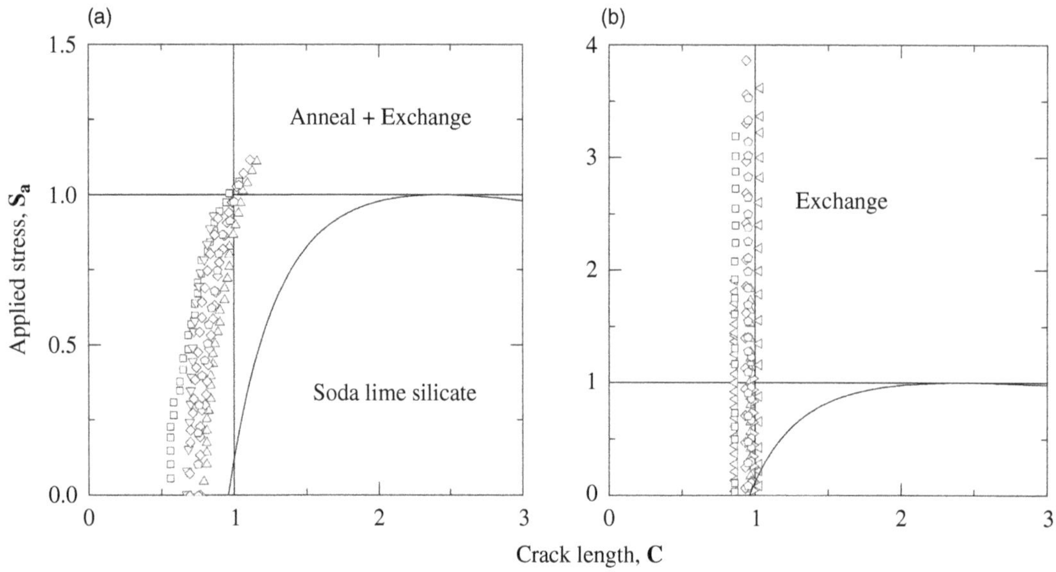

Figure 12.15 Plots of equilibrium trajectories of scaled applied stress versus crack length for indentation flaws in surface-compressed soda-lime silicate glass corresponding to Cases II and I in Chapter 3 (postindentation treatments). The base glass responses are shown as curved lines. (a) Case II. (b) Case I.

Figure 12.15 shows plots of indentation crack propagation for less common forms of ion-exchange toughening of soda-lime silicate glass, using indentation loads of $P = 10$, 20, 50, and 100 N: (a) Case II and (b) Case I, corresponding to flaw formation *prior* to the toughening process. The format of the plots and scaling of the data are identical to those in Figure 12.14. For indentation flaws annealed prior to exchange (a), the cracks contracted, the strengths increased marginally, and the stress-crack length derivative was clearly nonzero at failure. For flaws exchanged after indentation (b), the cracks were not significantly different from the base flaws, the strengths increased substantially, and there was no crack extension during failure. In both Cases I and II, it is clear that sub-surface crack propagation activity during applied stressing is leading to fracture instability and controlling component strength. A similar conclusion was drawn in consideration of fracture of laminated cordierite ceramic in Chapter 3. In cases in which crack propagation under applied stress leads to a transition from a compressive zone to tensile zone, measurement of a single crack dimension, such as the surface trace, is a weak indicator of strength degradation.

12.4 Summary

This chapter has surveyed the experimental observations and analytical descriptions of indentation crack propagation in materials that exhibit toughening. A feature of failure from indentation flaws is that there is a region of stable precursor crack extension prior to failure—from the initial indentation crack length to the final length at system instability. The observation and measurement of crack propagation trajectories as a function of applied stress in materials with invariant toughness were critical elements in the development of indentation fracture mechanics (Marshall et al. 1979, 1983). Subsequent observations of trajectories have verified the importance of the residual field in indentation fracture and pointed to the pervasive effects of indentation lateral cracking in relaxing residual fields and influencing indentation fracture (see Chapter 3). Experimental techniques targeting the end points of the trajectories, the initial and final crack lengths, have verified the application of indentations in measuring the toughness of brittle materials at small, strength relevant, length scales (see Chapter 5). The use of the entire trajectory has enabled the measurement of toughness and toughness variation in materials in which microstructural effects influence crack propagation (Braun et al. 1992; Chapter 11). This chapter has examined this last point in some detail, considering the effects of polycrystalline microstructures, phase-transforming microstructures, and surface compression on indentation crack propagation trajectories. Considerable differences in behavior are observed here in comparison to observations in the early experiments, which were heavily focused on soda-lime glass.

Crack propagation in polycrystalline alumina was marked by erratic crack advance associated with the local transitions of cracks from untoughened to toughened behavior. In fine-grained materials, this behavior lead to mild scatter about the ideal indentation crack propagation trajectory, although it was clear that the effect of microstructure was simply to perturb behavior from the underlying ideal response. The ratio of final crack length to initial length was approximately 2.5, close to the ideal value. In coarse-grained materials, the crack propagation trajectories were altered significantly from the ideal form and exhibited very erratic crack advance distinguished by alternating segments of crack halting and propagation. The ratios of final crack length to initial length varied from near 0 to nearly 5. Transitions from untoughened to untoughened behavior clearly dominated the responses. In contrast, crack propagation in phase-transforming zirconia exhibited little scatter. However, the forms of the crack propagation trajectories were altered from the ideal form based on the half-penny crack geometry. Instead, the indentation-induced phase-transformation zone altered the crack configuration to a more radial like form with attendant change in the crack propagation behavior. In fine-grained material, the crack propagation trajectory was lengthened such that the ratio of final crack length to initial length was approximately 3.1. In large-grained material, this ratio was decreased to about 2.2. In some cases, extremely short indentation cracks were formed that were trapped in the transformation zone. These cracks exhibited no advance on application of an applied stress. Finally, crack propagation behavior in ion-exchanged, surface-compressed, glass, although smooth, exhibited behavior indicative of significant changes of crack shape on exchange and on propagation. For exchanged materials, the surface traces of cracks were shorter and exhibited less extension on application of stress. This contraction reached a limit of no observable surface trace crack extension for components that underwent ion exchange after indentation. In all cases, ion-exchanged components exhibited greater indentation strengths than the unexchanged base components: about a factor of 2 greater for post-exchange indentation and factor of 4 for pre-exchange indentation. Although an effective empirical strengthening mechanism, the indentation fracture mechanics of ion-exchanged surfaces requires significant development.

References

Anderson, R.M. and Braun, L.M. (1990). Technique for the R-curve determination of Y-TZP using indentation-produced flaws. *Journal of the American Ceramic Society* 73: 3059–3062.

Braun, L.M., Bennison, S.J., and Lawn, B.R. (1992). Objective evaluation of short-crack toughness curves using indentation flaws: case study on alumina-based ceramics. *Journal of the American Ceramic Society* 75: 3049–3057.

Braun, L.M. and Cook, R.F. (1993). Effect of stress on trapped cracks in Y-TZP. In *Science and Technology of Zirconia V* pp. 386–400. CRC Press.

Cook, R.F. and Braun, L.M. (1994). Trapped cracks at indentations: Part II Fracture mechanics model. *Journal of Materials Science* 29: 2192–2204.

Cook, R.F., Braun, L.M., and Cannon, W.R. (1994). Trapped cracks at indentations: Part I Experiments on yttria-tetragonal zirconia polycrystals. *Journal of Materials Science* 29: 2133–2142.

Cook, R.F., Fairbanks, C.J., Lawn, B.R., and Mai, Y.W. (1987). Crack resistance by interfacial bridging: its role in determining strength characteristics. *Journal of Materials Research* 2: 345–356.

Kingery, W.D., Bowen, H.K., and Uhlmann, D.R. (1976). *Introduction to Ceramics*. Wiley.

Kovar, D., Bennison, S.J., and Readey, M.J. (2000). Crack stability and strength variability in alumina ceramics with rising toughness-curve behavior. *Acta Materialia* 48: 565–578.

Lin, G.Y., Lakshminarayanan, R., and Shetty, D.K. (2007). Contact damage initiation in silicon nitride in Hertzian indentation: role of microstructure. *Journal of Materials Science* 42: 3508–3519.

Marshall, D.B., Evans, A.G., Khuri-Yakub, B.T., Tien, J.W., and King, G.S. (1983). The nature of machining damage in brittle materials. *Proceedings of the Royal Society London A* 385: 461–475.

Marshall, D.B., Lawn, B.R., and Chantikul, P. (1979). Residual stress effects in sharp contact cracking: Part 2 strength degradation. *Journal of Materials Science* 14: 2225–2235.

Shetty, D.K., Wright, I.G., Mincer, P.N., and Clauer, A.H. (1985). Indentation fracture of WC-Co cermets. *Journal of Materials Science* 20: 1873–1882.

13

Summary and Future

This chapter summarizes the indentation strength results of the previous chapters. The key parameter is the contact strength. Contact strength results include those from materials that exhibited ideal indentation behavior (the minority), those that exhibited lateral crack effects (the majority), and those that exhibited microstructural effects (many). A ranked plot of contact strength is presented, providing a powerful tool for brittle materials designers, developers, manufacturers, and metrologists. The applications of contact strength in predicting minimum and maximum design strengths are demonstrated. A Master Map plotting scheme is presented that allows comparison of failure times of materials in reactive fracture conditions. Future research topics of thermal shock and concurrent flaw distributions that take advantage of indentation fracture techniques are outlined. A "modern day" indentation fracture version of the Griffith strength plot is presented, illustrating the progress in fracture in the century since Griffith.

13.1 Introduction

This chapter summarizes significant points of the book with a focus on indentation strength. The summary draws principally from Chapters 3, 4, and 6 and focuses on practical materials comparisons using indentation fracture techniques. The summary is followed by brief descriptions of some possible areas of future fracture research in which indentation techniques can be applied effectively.

Materials comparisons have been presented at many stages throughout the book, including

- Chapter 2, Figure 2.22. Ranked plot of indentation cracking susceptibility B values. See the database of indentation deformation and fracture properties (Cook 2024) for more details.
- Chapter 3, Tables 3.1 and 3.2. Lists of ideal indentation contact strengths $\hat{S}$ and lateral crack affected contact strengths S.
- Chapter 5, Figures 5.2 and 5.3. Ranked plots of indentation load adjusted initial and instability crack lengths and toughness parameters. Table 5.2. Ranked list of indentation crack length, strength, and toughness parameters.
- Chapter 6, Figure 6.27. Ranked plot of reactive/inert surface energy ratio $2\gamma/2\gamma_0$.
- Chapter 7, Figure 7.14. Ranked plot of crack branching constant S_B.
- Chapter 9, Figure 9.16. Plot of predicted Mohs number.
- Chapter 11, Tables 11.1 and 11.2. Lists of cordierite whisker composite properties.

Lacking from this list is a comprehensive comparison of contact strength parameters S for all materials, including those that exhibit ideal indentation behavior, lateral crack affected behavior, and microstructurally affected behavior. S is arguably the most important and useful parameter evaluated in indentation testing. This chapter provides a comparison of all S values presented in the book, drawing from the extensive range of cited sources. The comparison is used to generate other parameters useful in brittle materials design for performance and reliability, including lower-bound strength values, intrinsic flaw and strength values, and failure times.

Indentation Fracture: Strength and Toughness for Brittle Materials Design, First Edition. Robert F. Cook.

13.2 Materials Comparisons

13.2.1 Fracture in Inert Environments

A fundamental relation in indentation strength testing in inert environments relates strength σ_m to indentation load P by

$$\sigma_m = SP^{-1/3}. \tag{13.1}$$

The linking coefficient is the contact strength S. For materials that exhibit ideal indentation strength behavior, S takes the value $\hat{S}$ and describes indentation strengths over the complete domain of indentation loads. An important feature of $\hat{S}$ is that it is proportional to the invariant material toughness T_0 as $\hat{S} \sim T_0^{4/3}$. Most materials exhibit lateral crack effects at large indentation loads such that indentation strengths deviate significantly from the ideal response to greater values at large loads. For this pervasive phenomenon, S describes the asymptotic indentation strength response at small indentation loads. Many materials exhibit microstructural toughening effects at small indentation loads such that indentation strengths deviate significantly from the ideal response to lesser values at small loads. For this phenomenon, S describes asymptotic indentation strength behavior at moderate indentation loads. In these cases, S is proportional to the upper-bound steady-state toughness T_∞ of the material as $S \sim T_\infty^{4/3}$. For most materials influenced by both lateral crack and microstructural effects, the experimental indentation strength response passes through the intermediate asymptote. In all cases, S provides a measure of the characteristic indentation strength of a material.

Figure 13.1 shows a semilogarithmic plot of ranked S values from the complete set of indentation strength measurements considered in this book, drawing values from Chapters 3, 4, and 11. Symbols with bars represent means and standard deviation limits of one- and two-parameter evaluations of $\hat{S}$ or S from Chapters 3 and 11. Individual symbols represent estimates from three-parameter evaluations of S from Chapter 4. In all cases, evaluations represent the fitted indentation strength response over an indentation load domain, typically 1–100 N, and consequent strength range, typically including 50–100 strength tests. Hence, Figure 13.1 represents an analysis of approximately 7500 strength tests. The ranked contact strength S plot of Figure 13.1 is directly analogous to the ranked cracking susceptibility B plot of Figure 2.22 (that is based on an even greater number of crack length measurements, Cook 2024). The S values in Figure 13.1 extend over a range greater than a factor of 20, comparable to the B range in Figure 2.22. Characteristic materials and values are labeled: SLG, soda-lime silicate glass; cordierite glass-ceramic; AD96, glass-bonded polycrystalline alumina; polycrystalline SiC; Vistal, glass-free translucent polycrystalline alumina; SiN, polycrystalline Si_3N_4. The relative rankings are in accordance with the expectations of Chapters 1 and 2 for the strength properties of functional and structural ceramics.

Figure 13.1 is an extremely powerful tool for designers of components using brittle materials. Materials can be selected for mechanical applications with accuracy (ranking) and precision (S value) and overall component performance optimized, particularly if multiple factors need to be optimized (Ashby 1999; Cook 2015). Figure 13.1 is also a powerful tool for materials developers, manufacturers, and metrologists. For materials developers, Figure 13.1 provides a ready means of assessing the properties of a new material, enabling quantitative comparisons with existing materials and therefore providing guidance for development directions. For materials manufacturers, Figure 13.1 provides a means of assessing whether a production sequence remains within specification. For materials metrologists, Figure 13.1 provides a means of assessing whether a strength measurement method is performing correctly. The design, development, manufacturing, and measurement of brittle components are largely materials engineering applications of Figure 13.1. Figure 13.1 also provides a powerful tool for materials scientists: Chapters 2–5 make clear that S values reflect the combined effects and interactions of material elastic modulus, hardness, and toughness values, and their variations with porosity, grain bridging, phase transformation, and surface stress. For materials scientists, Figure 13.1 thus provides a ready means of assessing the effectiveness of new models and interpretations of contact fracture behavior.

A conservative design approach for structural components uses the minimum strength of materials. For brittle materials with strengths limited by contact flaws, this minimum strength is the lateral crack limited strength σ_L applicable at large sharp contact loads. Figure 13.2 shows a logarithmic plot of σ_L as a function of S value from indentation strength measurements in Chapters 3 and 4. Symbols with bars represent means and standard deviation limits of two-parameter evaluations from Chapter 3, and individual symbols represent estimates from three-parameter evaluations from Chapter 4. In all cases, evaluations represent the result from a fitted strength response to an indentation load domain. Characteristic materials and values are labeled as in Figure 13.1. As might be anticipated, there is a clear correlation between σ_L and S. The line in Figure 13.2 represents an empirical fit to the data consistent with Eq. (13.1) of $\sigma_L \approx SP_L^{-1/3}$ with $P_L = 125$ N, a typical value for brittle materials (see Chapter 3). A consequence of this correlation is that a conservative estimate of brittle material

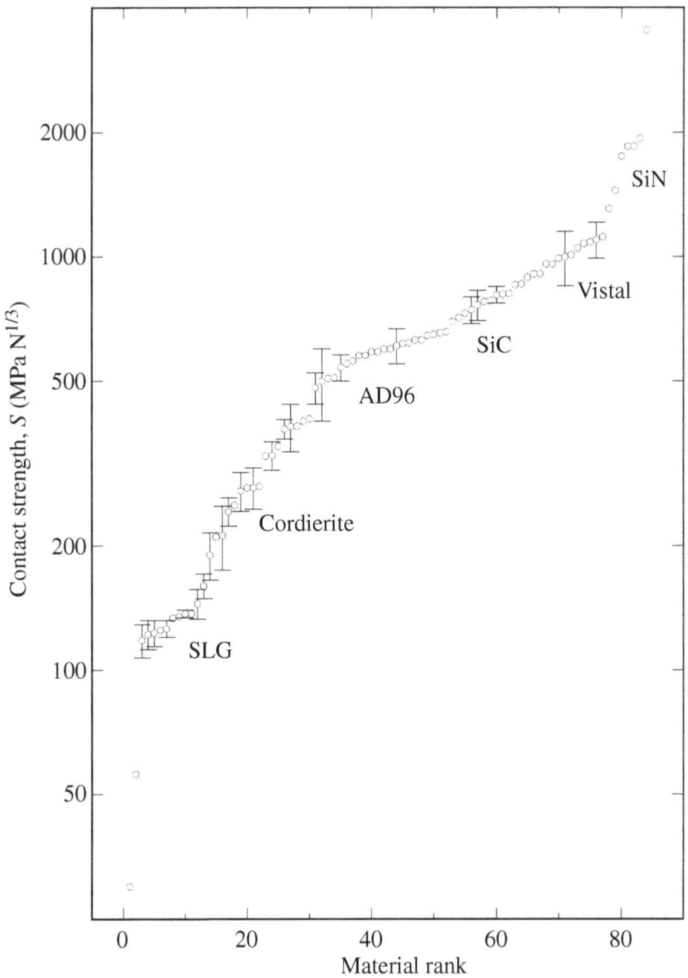

Figure 13.1 Semilogarithmic plot of ranked Vickers indentation contact strength values S for a large set of brittle materials. Ranking increases left to right such that the strongest materials are on the extreme right. Individual materials of note are indicated.

strengths from indentation tests is $S/5$ (omitting units) and that a constraint of design is to not allow any applied stress to exceed this strength level.

A less conservative design approach for brittle structural components is to estimate the contact load at which strengths begin to degrade from the as-received or as-manufactured level. As given in Chapter 3, this contact load is P_i, indicating a contact load corresponding to intrinsic strengths. Again, consistent with Eq. (13.1), for materials exhibiting ideal or lateral crack affected strengths $P_i = (S/\sigma_i)^3$, where σ_i is the intrinsic strength. Figure 13.3 shows a semi-logarithmic plot of P_i as a function of S value from indentation strength measurements in Chapters 3 and 4. Symbols with bars represent means and standard deviation limits of one- and two-parameter evaluations from Chapter 3, and individual symbols represent estimates from three-parameter evaluations from Chapter 4 for materials that exhibited weak microstructural interactions (specifically for materials with $P^* < 3$ N). The evaluations in Figure 13.3 represent results from indentation strength responses extrapolated to intrinsic strength levels. Uncertainties were determined by the propagation of variance. There was no significant trend in P_i across the domain of S examined. A characteristic value of $P_i \approx 2$ N was identified and is emphasized by the shaded band in Figure 13.3 extending over the range 0.8–4 N that includes the results from most materials. The implication is that the intrinsic flaws in most brittle materials are equivalent to 2 N indentations. For a typical hardness value of $H = 10$ GPa and a typical cracking susceptibility of $B = 12\,\mu\text{m N}^{-2/3}$, this value of P corresponds to a $10\,\mu$m contact impression (a) with $20\,\mu$m cracks (c) (see Chapter 2). A constraint of design is then to not allow any contact loads greater than this intrinsic level.

An aggressive, but not necessarily nonconservative, approach to design with brittle materials is to *predict* the intrinsic strength from indentation strength measurements. Such prediction is possible for materials that exhibit significant

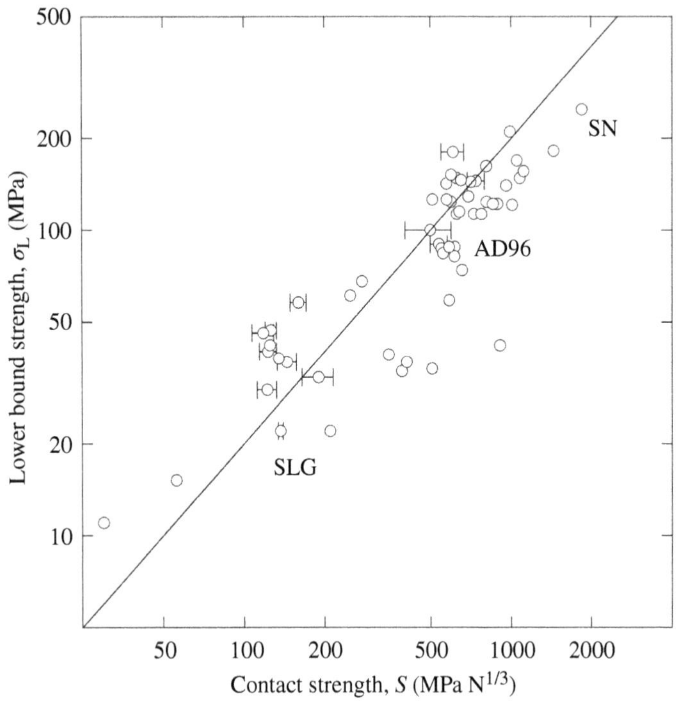

Figure 13.2 Logarithmic plot of lower-bound strength versus contact strength for a large set of brittle materials. Individual materials of note are indicated.

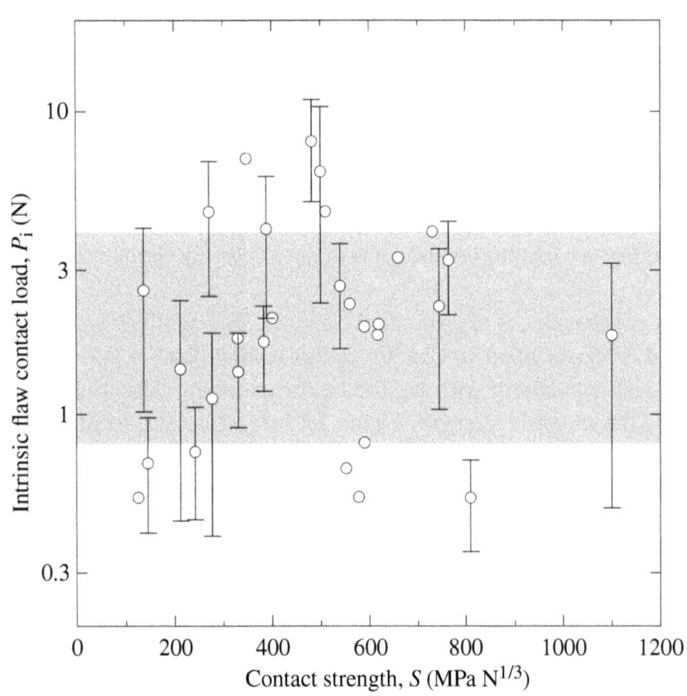

Figure 13.3 Semilogarithmic plot of equivalent intrinsic flaw contact loads versus contact strength for materials with negligible microstructural effects on flaws. Gray band indicates that many materials have intrinsic flaws equivalent to 2 N contacts.

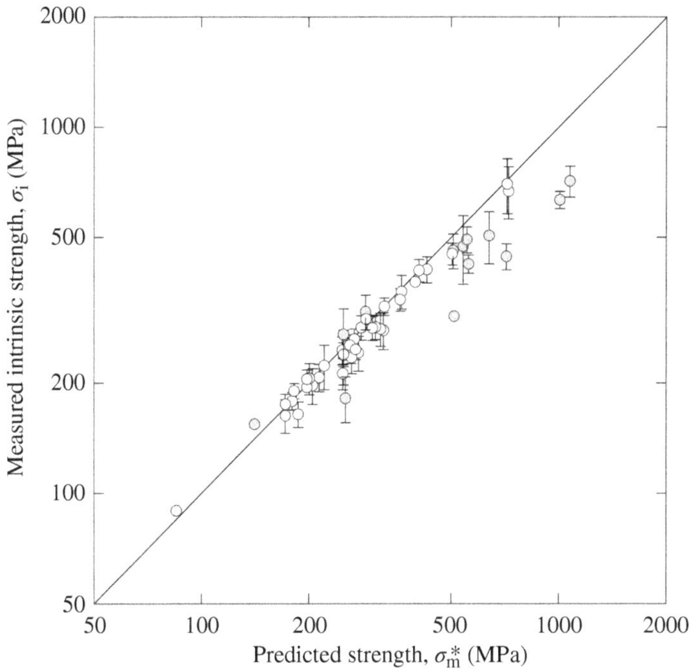

Figure 13.4 Logarithmic plot of independently measured intrinsic strengths versus strengths predicted from indentation tests for a range of materials with clear microstructural effects on fracture.

microstructural effects. The basis of this procedure is considered in Chapter 4. Although strength prediction is a somewhat aggressive design idea, microstructurally controlled strengths are flaw tolerant and hence the prediction is not really that nonconservative. A demonstration of the effectiveness of this approach is given in Figure 13.4 which shows a logarithmic plot of measured intrinsic strength as a function of predicted strength determined from independent indentation tests. Data were obtained from Chapter 4 for materials with pronounced microstructural effects on indentation strength. Predicted strengths σ_m^* were obtained from three-parameter indentation strength fits: open symbols for $P^* \geq 3$ N, grayed symbols for $P^* < 3$ N. Measured strengths σ_i represent the means and standard deviation limits of intrinsic strength measurements. The line indicates ideal correlation. Figure 13.4 greatly extends an earlier similar version of this plot (Cook 2015) in range and the number of observations. An obvious feature of Figure 13.4 is the closeness of the correlation extending over about a factor of 10 in strengths. There is a preponderance of data for strengths of 200–500 MPa, consistent with the strength observations for many brittle materials. For predicted strengths greater than approximately 500 MPa the correlation weakens: measured intrinsic strengths are less than those predicted. Inspection of Figure 13.4 shows that most of the lesser intrinsic strengths were exhibited by materials with weak microstructural interactions and were thus significantly influenced by nonmicrostructural flaws (and were in fact included in Figure 13.3). Overall, Figure 13.4 provides clear support for the design approach of engineering flaw tolerant material microstructures with intrinsic strengths predictable from indentation tests. Figure 13.4 also provides clear support for the materials science aspects of indentation fracture mechanics that incorporate microstructural toughening effects.

13.2.2 Fracture in Reactive Environments: Master Maps

A final consideration in brittle materials comparisons using indentation techniques involves nonequilibrium crack propagation and time-dependent failure. Such crack propagation and failure were examined in detail in Chapter 6: Brittle failure stresses could be reduced by a factor of approximately 2 from the inert strength value if sufficient time was allowed for crack propagation in reactive environments. Time-dependent failure was shown to be studied most easily by replacing time in experiments with applied stressing rate $\dot{\sigma}_a$ as an independent variable and measuring failure stress σ_f as a dependent variable. The decrease of σ_f in reactive environments (usually water) from the inert level with decreasing $\dot{\sigma}_a$ is known as dynamic fatigue. The dynamic time to failure t_D is given simply as $t_D = \sigma_f/\dot{\sigma}_a$. A phenomenon of greater interest in component operation and reliability assessment is static fatigue, in which the time to failure t_f increases with decreasing static applied stress σ_A. As shown in Chapter 6, for fracture systems with similar functional forms of crack driving and resistance fields (see Chapter 12), the analytical relationship between static and dynamic times to failure at a common failure stress, $\sigma_A = \sigma_f$, is a fixed multiple, $t_f \sim t_D$. Experimental verification of this relationship was demonstrated in Chapter 6 for several materials with strengths controlled by intrinsic and indentation flaws.

For many brittle components with strengths controlled by contact flaws such as indentations, it is possible to express the fracture behavior with similar driving and resistance functions using an effective contact load P_{eff} (see Chapters 3, 4, and 12). As a consequence, static and dynamic fatigue behavior can be reduced to contact load-independent responses in load-adjusted coordinates ($t_f P_{eff}^{-2/3}$, $\sigma_A P_{eff}^{1/3}$) and ($\sigma_f P_{eff}^{1/3}$, $\dot{\sigma}_a P_{eff}$), respectively (see Chapter 6). Experimental verification of the effectiveness of load-adjusted coordinates was also demonstrated in Chapter 6. In particular, a universal dynamic fatigue response of the form

$$\sigma_f P_{eff}^{1/3} = S_{min} + [(S - S_{min})/2]\left[1 + \tanh[m_{med} \ln(\dot{\sigma}_a P_{eff}/\dot{S}_{med})]\right] \tag{13.2}$$

was used to describe experimental observations of many material-indentation load combinations. An adjusted dynamic fatigue failure time was implicit as

$$t_D P_{eff}^{-2/3} = \sigma_f P_{eff}^{1/3} / \dot{\sigma}_a P_{eff}. \tag{13.3}$$

In implementing Eq. (13.2), S was a fixed parameter set by inert strength tests, and S_{min}, m_{med}, and $\dot{S}_{med}$ were parameters set by fitting to dynamic fatigue data. The ratio S_{min}/S is a measure of the equilibrium surface energies of a material in the reactive and inert environments and is considered in Figure 6.27. The parameters m_{med} and $\dot{S}_{med}$ are related to the kinetics of material bond rupture and reaction in the environment. Hence, S, S_{min}, m_{med}, and $\dot{S}_{med}$ are all indentation load-independent parameters that describe fatigue behavior of a material in a reactive environment. Comparison of these parameters thus permits the comparison of material fatigue responses. Extension to static fatigue predictions from dynamic fatigue measurements is accomplished by using the proportionality between static and dynamic fatigue failure times noted above. Thus, rewriting Eqs. (13.2) and (13.3) as

$$\sigma_A P_{eff}^{1/3} = S_{min} + [(S - S_{min})/2]\left[1 + \tanh[m_{med} \ln(x/\dot{S}_{med})]\right] \tag{13.4}$$

and

$$t_f P_{eff}^{-2/3} = \sigma_A P_{eff}^{1/3} / x \tag{13.5}$$

gives the load-adjusted static fatigue failure time as a function of the load-adjusted applied stress, where x is a stress rate like parameter.

Graphical representation of Eqs. (13.2) and (13.3) or Eqs. (13.4) and (13.5) enables simultaneous comparison of equilibrium and kinetic fracture properties of materials in "Master Maps." Master Maps for dynamic fatigue properties, assuming the empirical power-law crack velocity function, were developed for ceramics by Multhopp et al. (1983) and later extended to include glasses (from the work of Dabbs) and other ceramics in static fatigue predictions, also with the power-law assumption, by Cook and Lawn (1984). Figure 6.19 uses a graphical application of Eqs. (13.4) and (13.5) for a single material. Figure 13.5 shows a logarithmic Master Map for static fatigue behavior of many glass and ceramic materials predicted from the dynamic fatigue results of Chapter 6. The predicted failure time responses of various materials are shown as functions of applied stress by lines in load-adjusted coordinates. Black and gray lines are used to distinguish materials that are identified by acronyms: SLG, soda-lime silicate glass; BSG, borosilicate glass; PZT, lead zirconate titanate; FSG, fused silica glass; SYN, Synroc; BT, barium titanate; MGC, Macor glass-ceramic; PGC, Pyroceram glass-ceramic; AD96, glass bonded-polycrystalline alumina; FF, low glass polycrystalline alumina; Vi, glass-free polycrystalline alumina. Noting that the characteristic contact loads for intrinsic flaws are $P_i \approx 1$–2 N inferred from Figures 13.3 and 13.4, the periods on the right of the map give characteristic failure times assuming $P_{eff} \approx P_i$.

There are several major features of Figure 13.5 (apart from the resemblance to op art): Rankings of material equilibrium fracture behavior in the reactive environment (here water) can be read from the top of the map. Rankings of material equilibrium fracture behavior in an inert environment can be read from the bottom of the map. The rankings are not quite the same, although the structural ceramic polycrystalline alumina materials exhibit superior fracture properties to the optical glass materials in both inert and reactive environments. The curvatures of the responses are different for each material, indicating different rates of bond rupture kinetics and fatigue crack propagation. For example, soda-lime silicate glass has been known to exhibit pronounced fatigue characteristic since the earliest studies. The fatigue response on the map is clearly more curved relative to other materials. In contrast, Pyroceram glass-ceramic is often regarded as relatively immune

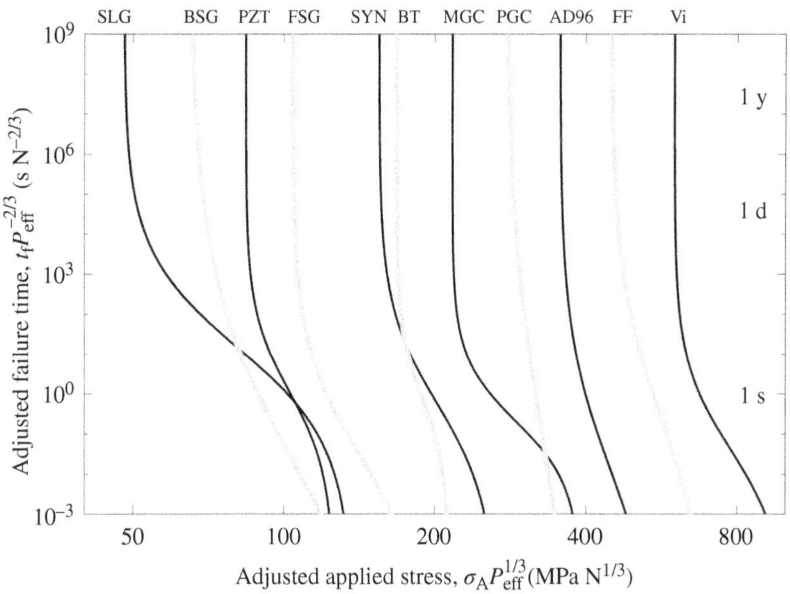

Figure 13.5 Static fatigue Master Map in logarithmic indentation load adjusted coordinates for a range of materials. Fatigue resistance to contact flaws in reactive environments increases from lower left to upper right.

to fatigue effects, and the fatigue response is clearly less curved than that for other materials. The different separations of inert and reactive fracture equilibria and the different curvatures associated with fracture kinetics lead to crossings of fatigue responses for some materials (such that the inert and reactive rankings are not identical). In cases without crossings, materials comparisons are straightforward. For example, Synroc and borosilicate glass were both considered for nuclear waste disposal. If stress corrosion based fatigue cracking is a deciding factor, Synroc is clearly the superior material. In cases with crossings, comparisons are ambiguous. For example, soda-lime silicate glass and borosilicate glass exhibit comparable short-term behavior. For long-term applications, borosilicate glass is the superior material. Useful as the map is, there are several caveats. First, the map will be different for different reactive environments. Second, the map compares material behavior: effective contact loads must be specified to compare component behavior. Third, dynamic fatigue experiments rarely extend to a day and static fatigue experiments rarely extend to a year: extrapolation uncertainty must be considered for long failure times. Nevertheless, the fatigue Master Map of Figure 13.5 is a powerful design tool that is complementary to the contact strength ranking of Figure 13.1.

13.3 Future Topics

As shown throughout this book, considerations of indentation fracture require developments in both analytical and experimental techniques. These developments were required initially to explain indentation deformation and fracture phenomena, e.g., indentation strengths and crack propagation (here Chapters 3, 4, and 12). The techniques developed were used subsequently in the application of indentation fracture to materials and component characterization, e.g., toughness, strength, and failure time estimation (here Chapters 5, 6, 9, and 11). This section describes two areas of future fracture research in which the analytical and experimental techniques developed for indentations could be applied effectively: thermal shock and concurrent flaw distributions. The effectiveness of indentation techniques in advancing other areas of fracture research has been well demonstrated, e.g., fragmentation, chipping, and sawing (here Chapters 7, 8, and 10). The section concludes with a modern-day look at the Griffith strength plot.

13.3.1 Thermal Shock

The phenomenon of thermal shock, shown in schematic form in Figure 13.6, is similar to the two-step indentation strength test sequence of Figure 3.1. In Step 1 of Figure 13.6, a component is equilibrated at an elevated temperature, of order 500 °C;

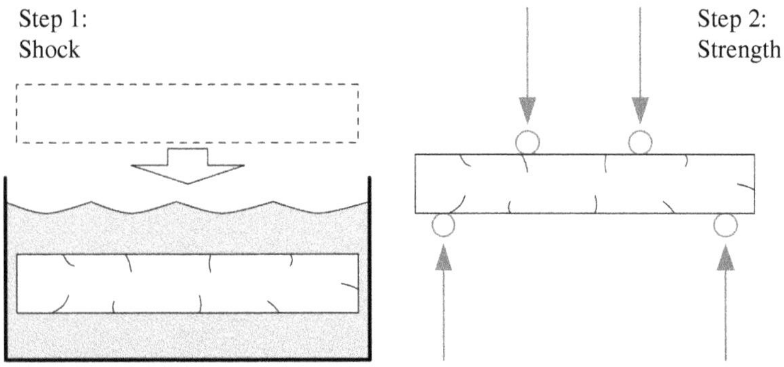

Figure 13.6 Schematic cross-section diagrams of a thermal shock experiment. Step 1: A brittle component is exposed to a rapid change in temperature, usually by immersion in liquid, leading to the formation of surface cracks. Step 2: The fracture strength of the cracked component is measured.

the heated component is indicated by the dashed outline. The component is then thermally "shocked" by rapid exposure or quenching into an environment of much lower temperature, usually by immersion into room temperature water or sometimes silicone oil. If the thermal diffusivity of the material forming the component is small, tensile (positive) stress develops in the surface of the component. The stress is a mechanical reaction to the imposed negative strain arising from the thermal contraction of the surface. Such stresses can lead to surface cracking, often on a length scale comparable to the size of the component. The immersed, shocked, and cracked component is shown in the lower diagram of Step 1 of Figure 13.6. In Step 2 of Figure 13.6, the strength of the shocked component is measured. Thermal shock experiments usually measure the strength of shocked components as a function of the quenching temperature difference. Thermal shock is thus a form of controlled flaw strength testing.

Figure 13.7 shows a semilogarithmic plot of thermal shock strength measurements for polycrystalline alumina rods from the work of Hasselman (1970), using data derived from the published work. Symbols with bars represent the means and standard deviation limits of four-point bending strength measurements at the quenching temperature difference indicated. Open, gray, and filled symbols represent rods of 9.5, 4.7, and 2.0 mm diameter, respectively. The shaded band represents

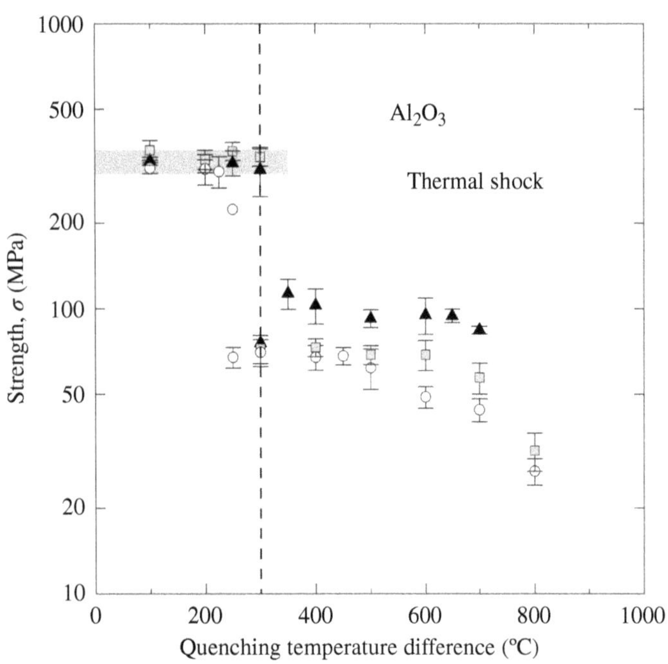

Figure 13.7 Semilogarithmic plot of thermal shock experiments for polycrystalline alumina rods. Different symbols represent different rod diameters. Dashed line is a guide to the eye indicating the shock temperature separating intrinsic strengths from shocked strengths. Source: Data adapted from Hasselman (1970).

the bounds of the intrinsic strengths of the rods (there was no significant effect of diameter). The thermal shock strength degradation plot of Figure 13.7 exhibits similarities and differences relative to the indentation strength degradation plots of Chapter 3. The similarities are that (i) for small quenching temperature differences, the strength is unaltered from the intrinsic level, (ii) for a small domain of moderate quenching temperature differences, a decreased range of strengths occurs, separated from the intrinsic strengths, and finally (iii) for large quenching temperature differences, the strength decreases with increasing temperature difference. Comparison with the subthreshold strength degradation plots of SLG, FSG, and BSG, Figures 3.52–3.54, illustrates the similarities. The differences are that strength degradation on either side of the two-strength domain of moderate quenching temperature difference is small or nonexistent and that there is a clear trend to greater strengths at slightly elevated temperatures with decreasing component size.

Thermal shock is clearly a crack initiation phenomenon driven by a transient, decreasing stress field extending inward from the surface of the shocked component. The stress-intensity factor (SIF) arising from the stress field is characterized by a maximum, leading to crack initiation behavior as discussed in Chapter 1. The lack of strength variation for small quenching temperature differences, the weak strength variation for large quenching temperature differences, and the effects of rod diameter are all simply explained by the transient SIF. The nonmonotonic nature of the thermal shock SIF field with maximum is discussed and briefly reviewed by Swain (1990) in the context of R curve effects in zirconia. The SIF transient is considered in some analytical detail by Wang et al. (2015) in the context of thermal transport. However, there appears to be no framework for consideration of strength degradation by thermal shock along the SIF lines developed in this book. This lack hinders the interpretation of experimental results and design of shock resistant brittle components. In addition, there appears to have been no use of controlled flaws as the initiating agents in Step 1 of shock experiments; experiments overwhelmingly use intrinsic flaws with their inherent scatter. This lack of control hinders definitive comparisons of thermal shock performance of different components and thermal shock properties of different materials. Future work on thermal shock would benefit greatly from indentation fracture ideas and methods: analytically through familiarity with crack initiation and experimentally through familiarity with controlled flaw strength testing.

13.3.2 Concurrent Flaw Distributions

Figure 13.8 illustrates a phenomenon often regarded as annoying in indentation fracture: strength tests in which indented specimens do not fail at the indentation. Such specimens fail from more potent intrinsic flaws, and the specimens are often regarded as wasted. Figure 13.8a shows a semilogarithmic plot of indentation failure probability as a function of scaled indentation load. The symbols indicate probability values from the indentation strength measurements resulting in the ideal indentation responses of Chapter 3. For very small indentation loads, there is 0% probability of indentation failure and only intrinsic failures are observed. This domain is shown shaded, and the upper bound is designated $P_{0\sigma}$. Indentation loads in Figure 13.8 are scaled to this value as $P/P_{0\sigma}$. For $P/P_{0\sigma} > 1$, indentation failures are observed and the probability

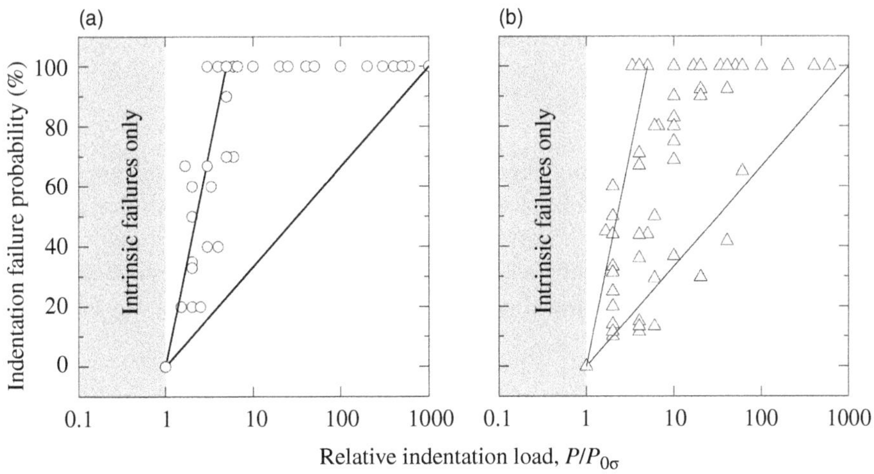

Figure 13.8 Semilogarithmic plots of indentation failure probability versus indentation load. (a) Invariant toughness materials. (b) Toughened materials. The widths of the distributions are an indication of the widths of the distributions of intrinsic and indentation flaws.

of indentation failure increases with increasing $P/P_{0\sigma}$. The left diagonal line in Figure 13.8a is a guide to the eye of this trend, which occupies about a factor of 5 in relative indentation load. For $P/P_{0\sigma} \gg 1$, the probability of indentation failure is 100%. Figure 13.8b shows a similar plot of indentation failure probability as a function of scaled indentation load from the indentation strength measurements of the toughened material responses in Chapter 4. The increasing probability behavior is much more scattered and distributed over broader domain. The right diagonal line in Figure 13.8b is a guide to the eye of this broader domain, which occupies about a factor of 1000 in relative indentation load. The diagonals are repeated in each plot.

The increasing indentation failure probabilities in Figure 13.8 represent competition between concurrent flaw distributions: intrinsic flaws and indentation flaws. The competition is clearly not the same in invariant toughness materials (Figure 13.8a) and toughened materials (Figure 13.8b). Such competitions were the focus in early considerations of optical fiber failure by concurrent subthreshold and postthreshold flaws. As illustrated by the shaded bands in Figure 3.31, concurrent flaw distributions are of greatest concern when the distributions overlap: Strength distributions of components are broadened and manufacturing is impeded by the need to optimize (eradicate) multiple flaw types. Analyses of overall strength distributions resulting from concurrent flaw distributions are well developed (e.g., Johnson 1983; Stoyan et al. 2013) but the field is hampered by a lack of experimental data. In particular, there is an absence of experiments in which the individual concurrent flaw distributions were known in advance. This absence renders analyses of overall composite strength distributions ambiguous. For example, is the broader probability distribution in Figure 13.8b the result of the broadening of the intrinsic flaw distributions by microstructural influences, or is it the result of broadening of both intrinsic and indentation flaw distributions (the latter by lateral crack effects)? Future work on concurrent flaw distributions would benefit greatly from indentation fracture methods, particularly through the ability to control flaw, and thus strength, distributions.

13.3.3 Indentation Measurements

Figure 1.10b is a logarithmic plot from the work of Griffith (1921) of the strengths σ of scratched and annealed glass bulbs and tubes as a function of the crack lengths c resulting from the scratches. The ten symbols represent the behavior of individual glass specimens and are shown to obey the Griffith equation

$$\sigma = Dc^{-1/2}, \tag{13.6}$$

where D is a material- and flaw shape-dependent constant of proportionality (see Chapter 1). A similar plot does not appear in the original publication.

It is interesting to compare the Griffith plot with a similar plot resulting from the indentation fracture experiments presented in this book. Indentation crack lengths were shown to obey the Roesler (1956) equation

$$c = BP^{2/3}, \tag{13.7}$$

where B is a material- and indenter-dependent constant of proportionality (see Chapter 2). Indentation strengths were shown to obey the equation

$$\sigma = SP^{-1/3}, \tag{13.8}$$

where S is a material- and indenter-dependent constant of proportionality (see Chapter 3). Equation (13.8) follows directly from the combination of Eqs. (13.6) and (13.7). Equation (13.7) follows directly from the insight of Griffith that fracture was a manifestation of the First Law of Thermodynamics in which work was converted into elastic energy and then into surface energy, γ. As a consequence, the constants of proportionality in Eqs. (13.7) and (13.8) both include surface energy factors, $B \sim \gamma^{-1/3}$ and $S \sim \gamma^{2/3}$. As a result, the constants of proportionality for indentation fracture are related by Eq. (3.19)

$$S \sim B^{-2}. \tag{13.9}$$

Figure 13.9 shows a logarithmic plot of contact strength S as a function of cracking susceptibility B using results from Chapters 2–5, 10, and 11. The symbols and bars represent the means and standard deviation limits of parameters determined from experimental observations and propagation of variance. The line is of slope −2 (in logarithmic coordinates), an empirical fit of Eq. (13.9), and well describes the data. Figure 13.9 is a modern-day equivalent of the Griffith plot. There are several

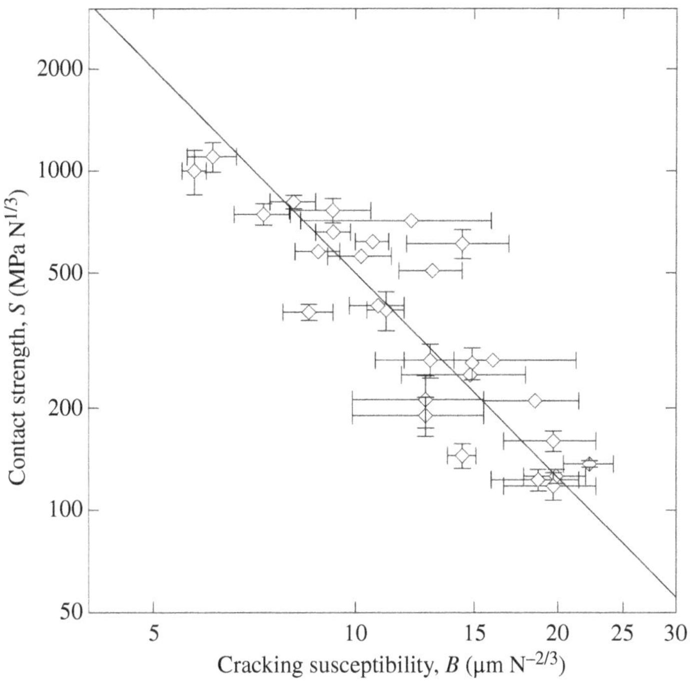

Figure 13.9 Logarithmic plot of contact strength versus cracking susceptibility for indentation flaws in a range of brittle materials. The plot is a modern-day equivalent of the strength versus crack length measurements made by Griffith (1921).

similarities: A strength parameter *S* is plotted against a crack length parameter *B*. The two are related by a simple power law. The crack lengths reflect a controlled flaw approach to strength measurement. There is however a distinct difference leading to several advances: Indentation flaws formed by load-controlled diamond pyramids are more reproducible than scratches. This reproducibility has enabled useful comparison of many observations by many researchers on many more materials. There are about 50 symbols in Figure 13.9. Each symbol represents measurements over an indentation load domain of approximately 100 crack lengths and 100 strengths. Thus, Figure 13.9 comprises about 10,000 observations, about a factor of 1000 more than the 10 published by Griffith. Figure 13.9 should thus be seen as both an indication of how far fracture has advanced in the century since Griffith and a reinforcement of the energy principle developed by Griffith. Future work can further this advance and strengthen this reinforcement by adding to Figure 13.9.

References

Ashby, M.F. (1999). *Materials Selection in Mechanical Design, 2nd ed.* Butterworth-Heinemann.

Cook, R.F. (2015). Multi-scale effects in the strength of ceramics. *Journal of the American Ceramic Society* 98: 2933–2947.

Cook, R.F. (2024). Indentation fracture and deformation v.2. *Zenodo* DOI: 10.5281/zenodo.1149515.

Cook, R.F. and Lawn, B.R. (1984). Controlled indentation flaws for the construction of toughness and fatigue master maps. *Journal of Research of the National Bureau of Standards* 89: 453–465.

Griffith, A.A. (1921). The phenomena of rupture and flow in solids. *Philosophical Transactions of the Royal Society London A* 221: 163–198.

Hasselman, D.P.H. (1970). Strength behavior of polycrystalline alumina subjected to thermal shock. *Journal of the American Ceramic Society*, 53: 490–495.

Johnson, C.A. (1983) Fracture statistics of multiple flaw distributions. In *Fracture Mechanics of Ceramics*, Vol 5. (Eds. Bradt, R.C., Evans, A.G., Hasselman, D.P.H., and Lange, F.F.) pp. 365–386. Plenum.

Multhopp, H., Cook, R.F., and Lawn, B.R. (1983). Universal fatigue curves for ceramics using indentation flaws. *Journal of Materials Science Letters* 2: 683–684.

Roesler, F.C. (1956). Brittle fractures near equilibrium. *Proceedings of the Physical Society. Section B* 69: 981–992.

Stoyan, D., Funke, C., and Rasche, S. (2013). Back to and beyond Weibull–The hazard function approach. *Computational Materials Science* 68: 181–188.

Swain, M.V. (1990). R-curve behavior and thermal shock resistance of ceramics. *Journal of the American Ceramic Society* 73: 621–628.

Wang, B., Li, J.E., and Yang, C. (2015). Thermal shock fracture mechanics analysis of a semi-infinite medium based on the dual-phase-lag heat conduction model. *Proceedings of the Royal Society A: Mathematical, Physical and Engineering Sciences*, 471: 20140595.

Index

Note: Page numbers referring to figures are indicated in *italics*. Page numbers referring to tables are indicated in **bold**.

a
adiabatic 6, 20
adsorption 25
agglomerate *11*
Al_2O_3-TiC *41*, *263*, *267*
alumina, Al_2O_3 *5*, *16*, *42*, *69*, *77*, *111*, *120*, *153*, *177*, *256*, *267*, *359*
aluminosilicate glass, ASG *38*, *91*
aluminum nitride *113*
aluminum titanate *168*

b
backgrinding *304*
bar *2*
barium titanate *112*, *165*
beam *4*
birefringence *43*, *281*
borosilicate glass, BSG *43*, 82, 129
brittle
- components *2*
- failure *2*, *3*
- fracture *5*
- materials *2*

c
calcite *294*
cantilever *283*
cement *198*
ceramic *16*
chemical potential 25
chipping *264*, *267*, *271*
chromia *347*
cladding *122*
compaction 39
compliance 2, 285
composite *69*, *170*, *332*, **333**
cone crack *15*
contact
- angle, ϕ **48**, 79, 108
- blunt 15
- conical *289*
- dragged *291*, *312*
- friction 80
- Hertzian 15
- linear *283*
- radius 80, *110*
- rolled *291*
- sharp 47, 263
- spherical *15*
- strength, *S* 71, *371*
- subthreshold 104
- wedge *280*

controlled flaw 67
cordierite *11*, *41*, *69*, *76*, *96*, *105*, *242*, *245*, *332*
corundum *see* alumina, Al_2O_3
crack 6
- branching *251*
- cone *82*
- equilibrium 12
- extension 74, *76*, *342*, 351
- front 6
- half-penny *15*, *40*
- initiation 28, *60*
- lateral *40*, *280*
- median *40*
- nonequilibrium 20
- nucleus 28
- propagation 20, 210, *352*, 353
- radial *40*
- ring *81*
- stable 14, 288
- susceptibility 55
- tip 6
- trace *38*

Indentation Fracture: Strength and Toughness for Brittle Materials Design, First Edition. Robert F. Cook.

crack (*contd.*)
unstable 9
velocity 20
wake *145*
cracking susceptibility, *B* 14, 38, 56
crack velocity
empirical 210, 212
observations *230*
thermally activated 216
threshold 24, 216

d
density 93
diamond 37, 293
dicing, die *5*, *255*, *264*, *274*, 303
disk *5*, *245*
distribution
complementary 250
concurrent flaw 377
cumulative 250, 325, 344
empirical *248*
fragment *248*
dummy method *190*
dynamic fracture 20, 241

e
edge chipping *264*
elastic
behavior *2*, 3
energy 4, 7
energy density 4
modulus 3
energy
balance 6, 20
density 4
elastic 4, 8
kinetic 20
mechanical 7, 8
surface 6, 8
entropy 22
entropy production rate 24
environment 25, *204*
equilibrium
stable *14*
unstable *9*
extensive 3

f
fatigue
dynamic 206
lifetime 206
observations *221*
predictions *225*
static 205
ferroelastic *165*
fibers 105
field
blister 15
Boussinseq 15
grain 31
Hertz 15
inclusion 28
indentation 44
flaw 6
edge 264
grinding *319*
linear 279, 305, 310
probability density 345
tolerance, 1/*D* 10, *176*
flexure *4*, 242
force
configurational 12, 286
failure 2
line *27*, 285
peak 2
fracture mechanics 7, 282
fracture resistance 12, 23
fragments, fragmentation *5*, *230*, *245*, 246, *248*, *255*
function
complementary cumulative distribution 250
cumulative distribution *248*
empirical distribution *248*
fitting *220*, *250*, 341
probability density 345
state 23
fused silica glass, FSG 43, *69*, *128*

g
glass *11*, *16*, *77*, *98*, *115*, *289*
ion-exchanged *92*, *115*, *365*
tempered *98*, *116*
glass ceramic *11*, *16*, 41, *46*, *76*, *123*, *190*, *198*, 331
grain 31, *145*
Griffith equation 10, *379*

h
half penny crack *15*, *40*, *264*
hardness 38
heat 22
Hertzian contact 15, *82*

i
impact *79*, 94, *115*, *177*
included angle *47*, *283*, *289*

inclusion *27*
indentation
  blunt *15*, 78, 80
  conical 15, 47
  ideal 70
  linear 282
  load, *P* 15
  pyramidal *15*, 47
  sharp 15
  spherical *15*, *78*
  strength, S *69*
  threshold 17, 88
  track 280, 283, *291*
indenter
  angle *47*
  Berkovich *47*
  biconical roller *280*
  conical *289*
  conospherical *78*, 85
  cube corner *16*, 47
  flat punch *16*
  Knoop *11*
  spherical *16*, *110*
  Vickers *16*, *38*, 47
initiation 28, *60*, 287
intensive 3
internal energy 4
intrinsic
  flaws *69*, *372*
  strength *69*, *154*, *343*
irreversible 24
Irwin relation 18
isothermal 22, *61*, 217

k
kinetic fracture 22, 218

l
lateral crack *40*, *87*, *280*, 288
  effects on strength 87, 112
layered components 96
linear flaw *281*
line force 283
loading
  applied 3
  bending *4*
  biaxial *5*, *11*, 241
  combined 70
  cycle 3
  localized 13
  spatially varying 26
  tensile *2*
  uniform *7*

m
machining damage *121*
master maps *375*
mechanical energy release rate 12, 23, 286
median crack, median vent *280*, *310*
MgO *41*, *172*
microcracking 30
microstructure *145*
minerals 293
modulus
  plane strain 4
  Young's 3
Mohs scale 293
mullite *169*

n
nonequilibrium fracture 20, *60*, 216

o
observable 22

p
parameter
  contact strength, *S* **70**, **114**, **193**, *371*
  crack branching S_B **253**
  cracking susceptibility, *B* **55**, **193**
  hardness, *H* **55**
  Mohs number, *M* *298*
  surface energy ratio, $2\gamma/2\gamma_0$ *234*
  toughness, *T* **193**
phase transformation 28, 172
phonons 20
plastic deformation zone, pdz 15, 37
plate *5*
p-norm 276
Poisson's ratio 4
porcelain *112*, *270*
pore *334*
porosity *256*
pressure
  inclusion 27
  partial 25
probability
  complementary 250
  cumulative 248
  density function 345
  empirical 248
  lifetime *232*

q

quartz *294*

R

R-curve 142
residual stress field *43*, 44, *281*
reversible 23
Roesler equation 14, 83
roller flaw *292*

S

sapphire *16*, *42*, *113*, *127*, *130*
scaling 71, 72, 75, 83, 90, 95, 101, 152, 211, 212, 354
scallop *264*, *271*, *275*
scratch *11*, 280, 290, *294*
scratch resistance *297*
shear faults *44*
sigmoidal *153*, *248*, *344*
silica, fused, SiO_2 15, *43*, *69*, *128*
silicon, Si *5*, *11*, *42*, *44*, *88*, *92*, *112*, *255*, *264*, *267*, *317*
silicon carbide, SiC *54*, *113*, *252*
silicon nitride, Si_3N_4 *76*, *161 252*
slow crack growth *see* stress, corrosion
soda lime silicate glass, SLG *43*, *44*, *112*, *116*, *268*, *281*
state function 23
steel *294*
stiffness 2
strain, engineering 3
strength
  brittle fracture 3
  degradation, *D* 10
  edge chip *274*
  empirical distribution function 66
  fatigue *206*, *221*
  grinding controlled 322
  inert 32, 69
  intrinsic *69*
  lateral crack affected 87, *112*, *372*
  microstructure affected *150*, *372*
  prediction *373*
  reactive *206*, *221*, *253*, *375*
  subthreshold *128*
  tensile 2
  threshold 85, 380
stress
  corrosion 203
  engineering 3
  failure 2
  gradient 103, 178
  surface 97
stress-intensity factor, *K* 17
striations *305*
structural component 3
surface energy 8, 25, *234*
SYNROC *69*, *224*

T

T-curve 143, *196*
thermal expansion 3, 27
thermal shock *376*
Thermodynamics
  First Law 4, 6, 286
  Second Law 22
toughening
  crack wake *147*
  frontal zone *173*
toughness, *T* 19, 187, *194*, *342*

V

variable
  external 23
  internal 23
Vickers indentation *38*

W

wafer *304*
whisker *332*
work, *W* 6, 7

Y

Young's modulus, *E* 3
yttria *46*, *113*, *172*

Z

zinc sulphide *86*, *119*
zirconia *41*, *77*, *105*, *130*, *174*, *252*, *270*, *362*

www.ingramcontent.com/pod-product-compliance
Ingram Content Group UK Ltd.
Pitfield, Milton Keynes, MK11 3LW, UK
UKHW050952290926
15470UKWH00005B/1772

9 781394 207206